A Course In
Probability

A Course In
Probability

Neil A. Weiss

Arizona State University

With Contributions From

Paul T. Holmes

Clemson University

Michael Hardy

Massachusetts Institute of Technology

PEARSON

Addison
Wesley

Boston   San Francisco   New York
London   Toronto   Sydney   Tokyo   Singapore   Madrid
Mexico City   Munich   Paris   Cape Town   Hong Kong   Montreal

Publisher: Greg Tobin
Executive Editor: Deirdre Lynch
Managing Editor: Ron Hampton
Editorial Assistant: Sara Oliver
Senior Production Supervisor: Jeffrey Holcomb
Marketing Manager: Phyllis Hubbard
Marketing Assistant: Heather Peck
Senior Manufacturing Buyer: Evelyn Beaton
Senior Prepress Supervisor: Caroline Fell
Associate Producer: Sara Anderson
Photo Research: Beth Anderson
Senior Author Support/Technology Specialist: Joe Vetere
Supplements Production Supervisor: Sheila Spinney
Compositor: Neil A. Weiss
Illustrations: Techsetters, Inc.
Text Designer: Sandra Rigney
Cover Designer: Dardani Gasc Design
Design Supervisor: Barbara T. Atkinson
Clover Close-up © Photodisc Green / Getty Images
Four Leaf Clover © Photodisc Green / Getty Images

Photos for pages 2, 24, 84, 124, 174, 258, 324, 400, 484, 562, 628, and 684 courtesy of
St. Andrew's University Mac Tutor Archive

Many of the designations used by manufacturers and sellers to distinguish their products are
claimed as trademarks. Where those designations appear in this book, and Addison-Wesley
was aware of a trademark claim, the designations have been printed in initial caps or all caps.

Library of Congress Cataloging-in-Publication Data

Weiss, N. A. (Neil A.)
  A course in probability / Neil A. Weiss, with contributions from Paul T. Holmes,
  Michael Hardy.
      p. cm.
  Includes index.
  ISBN 0-201-77471-2
    1. Probabilities. I. Holmes, Paul T. II. Hardy, Michael. III. Title.

QA273.W425 2005
519.2--dc22

                                                                    2004051068

Printed in the United States of America.

To Carol

About the Author

Neil A. Weiss received his Ph.D. from UCLA in 1970 and subsequently accepted an assistant-professor position at Arizona State University (ASU), where he was ultimately promoted to the rank of full professor.

Dr. Weiss has taught probability, statistics, operations research, and mathematics from the freshman level to the advanced graduate level for more than 30 years. In recognition of his excellence in teaching, he received the *Dean's Quality Teaching Award* from the ASU College of Liberal Arts and Sciences and has been twice runner-up for the *Charles Wexler Teaching Award* in the ASU Department of Mathematics and Statistics.

Dr. Weiss has published research papers in both theoretical and applied probability, including applications to engineering, operations research, numerical analysis, and psychology. He has also published several teaching-related papers.

In addition to his numerous research publications, Dr. Weiss has authored or coauthored books in probability, statistics, finite mathematics, and real analysis, and is currently working on a new book in applied regression analysis and the analysis of variance. His texts—well known for their precision, readability, and pedagogical excellence—are used worldwide.

In his spare time, Dr. Weiss enjoys walking, studying and practicing meditation, and playing hold 'em poker. He is married and has two sons.

Preface

This is a book about probability, but it's not an ordinary probability book. Written with the student in mind, this text incorporates pedagogical techniques rarely found in books at this level. The author brings more than 30 years of teaching, research, and writing experience to this project.

A Course in Probability is intended primarily for use in a first course in mathematical probability for students in mathematics, statistics, actuarial science, operations research, engineering, and computer science. However, the book is also appropriate for mathematically oriented students in the physical and social sciences.

The prerequisite material consists of basic set theory (i.e., sets, subsets, unions, intersections, and complements) and a firm foundation in elementary calculus, including infinite series, partial differentiation, and multiple integration. Some exposure to rudimentary linear algebra (e.g., matrices and determinants) is also required for those covering the multivariate normal distribution in Section 12.3.

Key Features

Motivation of concepts. All concepts are motivated, either by example or through appropriate expository discussion. The importance of and rationale behind ideas such as the axioms of probability, conditional probability, independence, random variables, joint and conditional distributions, and expected value are made transparent. Additionally, formulas for probability mass functions and probability density functions are developed and explained instead of only stated. In this way, students will understand how such formulas arise naturally.

Structured pedagogy. We provide many more displays, including definitions, figures, tables, etc., than other probability books at this level. Furthermore, we present material in a step-by-step fashion and avoid cognitive jumps. These strategies make the text effective and the learning process smooth, easy, and enjoyable.

Careful organization. As an aid to learning, we cover one main concept at a time. For instance, we separate the concepts of discrete and continuous random variables and the topic of expectation from the distribution theory.

Illustrative examples. Definitions and results (e.g., theorems and propositions) are followed by one or more examples that illustrate the concept or result in order to solidify it in the mind of the student and to provide a concrete frame of reference.

Abundant and varied exercises. An abundance of exercises (1726, not including parts)—far more than in most other probability books—are presented. Furthermore, the exercises provide an opportunity to vary course coverage and level. These issues are discussed in greater detail later in the Preface.

Applications. A diverse collection of applications appear throughout the text, some as examples or exercises and others as entire sections. The last chapter of the text, Chapter 12, provides introductory materials—independent of one another—for three main follow-up courses: mathematical statistics, stochastic processes, and operations research. See the table of contents for the specific topics covered in this chapter.

Careful referencing. As an aid to effective use of the book, we consistently provide references—including page numbers—to definitions, examples, exercises, and results.

Biographies. Each chapter begins with a one-page biography of a famous probabilist, mathematician, statistician, or scientist who has made substantial contributions to probability theory or its applications. Besides being of general interest, these biographies help the student obtain a perspective on the development of probability and its applications.

Chapter introductions. Each chapter begins with an introduction that provides a preview of the topics to be covered in the chapter.

Chapter reviews. Each chapter ends with a comprehensive chapter review that includes a detailed summary of the chapter, chapter objectives, a list of key terms with page references, and an extensive collection of chapter review problems. These chapter reviews provide the student with an organized strategy for reviewing the material in each chapter.

Extensive Exercises

A Course in Probability contains an unparalleled quantity of high-quality exercises. The exercises in this text play an essential role in the learning process and are also central to varying the level and focus of the course.

Many probability texts include exercises only at the end of each chapter. In contrast, we present exercise sets at the end of each section and provide a comprehensive collection of review problems at the end of each chapter. Furthermore, each exercise set and each collection of review problems contain three groups of problems, specifically designated as follows.

- *Basic Exercises:* These exercises include a diverse set of problems that should be solved by all students as a way to master the concepts presented in the text.

- *Theory Exercises:* These (optional) exercises are available for those courses in which students are expected to write proofs and explore additional aspects of the theory of probability.

- *Advanced Exercises:* These (optional) exercises, which include both applied and theory problems, provide more challenging problems than those found in the basic exercises and, consequently, present an opportunity to increase the level of the course. The advanced exercises also contain problems that examine concepts not specifically covered in the text, at least up to the point where such problems are presented.

The fundamental concepts in a first course in mathematical probability are essentially the same, irrespective of the level at which the course is taught. Thus the instructor can freely vary the desired level and focus of the course through the choice of exercises and be confident that, regardless of his or her choice, the student will be exposed to the probability concepts required for courses having mathematical probability as a prerequisite.

Flexibility of Presentation, Level, and Coverage

Considerable variations exist in instructional presentation method, level, and topic coverage for a first course in mathematical probability. With that in mind, we have organized *A Course in Probability* for maximum flexibility.

- *Flexibility of Presentation:* The instructor can use this book in either a traditional lecture format or an alternative format in which the student takes a more active role in the learning process. Indeed, the text has been written so that the student can read it on his or her own. By having the student do so prior to the class discussion of a topic, the instructor is freed to emphasize that which he or she thinks is most important and to answer any questions from students. Then exercises and other problems can be worked.

- *Flexibility of Level:* Because of the organization of the text, the instructor has several options for the level of presentation. For a basic applied course, the instructor could concentrate on the examples and basic exercises and omit proofs of propositions and theorems as well as the theory and advanced exercises. For an intermediate course, the instructor could cover the proofs and (optionally) assign theory exercises in addition to the basic exercises. For an advanced course, the instructor could include the proofs and also assign the theory and/or advanced exercises in addition to the basic exercises.

- *Flexibility of Coverage:* The organization of the text offers several options for topic coverage. For example, a course in discrete probability (fundamentals of probability, discrete random variables, and expectation of discrete random variables) could be taught by concentrating on the first seven chapters of the book. For those courses in which both discrete and continuous probability are examined, an option could be to cover univariate continuous distributions (Chapter 8) immediately after univariate discrete distributions (Chapter 5).

Solutions Manuals for Instructors and Students

We have prepared detailed and comprehensive solutions manuals—both an Instructor's Solutions Manual (ISM) and a Student's Solutions Manual (SSM). The ISM contains solutions to all exercises in the book. The SSM contains solutions to every fourth end-of-section exercise and to all review exercises. In both manuals, the solutions to the exercises employ precisely the same notation, format, and style as the solutions to the examples in the main text.

Electronic versions (PDF) of the ISM and SSM are available for download from the Instructor Resource Center (www.pearsonhighered.com/irc). Please note that the Instructor Resource Center (IRC) is a password-protected site for instructors only. The instructor may download or print the SSM and share it with students, but IRC log-in information should not be made available to students.

Acknowledgments

Writing a textbook is an extraordinary project in which the contributions made by the reviewers are essential. We are pleased to acknowledge the following reviewers whose comments resulted in significant improvements to the text.

Anna Amirdjanova
University of Michigan, Ann Arbor

John Beckwith
Michigan Technological University

Linda Cave
Western Washington University

Douglas Frank
Indiana University of Pennsylvania

Michael Hardy
Massachusetts Institute of Technology

Paul Holmes
Clemson University

Michael Nussbaum
Cornell University

William Schildknecht
University of Maryland

Kyle Siegrist
University of Alabama, Huntsville

Daniel Weiner
Boston University

Jerome Wolbert
University of Michigan, Ann Arbor

Shunpu Zhang
University of Alaska

To Professors Paul Holmes and Michael Hardy, we convey our sincere appreciation for their contribution of an extensive number of high-quality exercises; their efforts have resulted in a book with a tremendous variety of exercise types, both conceptually and topically. We are also pleased to thank the Society of Actuaries for permission to use problems from the actuarial exams as examples and exercises in the book. Additionally, we are grateful to Dr. Jeffrey Farr for electronically compiling the exercises submitted by Professor Holmes.

Many thanks go to Professor Dan Weiner for accuracy checking the text and to Professor John McDonald for accuracy checking the answers to the exercises that appear in the appendix. Our thanks as well are extended to Professors Douglas Blount, Matthew Hassett, Kyle Siegrist, and Dennis Young with whom we have had many illuminating discussions relevant to the writing of *A Course in Probability*.

Special thanks are due to our executive editor Deirdre Lynch. As always, she provided unflagging encouragement and support from the very beginning of the project to its culmination in the form of a printed book. Deirdre's dedication to quality makes working with her a pleasure.

We are pleased to thank our excellent production supervisors Jeffrey Holcomb and Julie LaChance. To Barbara Atkinson of Addison-Wesley, Sandra Rigney (interior design), and Dardani Gasc Design (cover design), we express our appreciation for an awesome design and cover. And our sincere thanks go as well to Techsetters, Inc., for a terrific job of illustration. We also thank Beth Anderson for her photo research. Thanks as well are extended to our copyeditor Jerrold Moore and to our proofreaders Cindy Bowles and Carol Weiss.

Without the help of many people at Addison-Wesley, this book would not have been possible; to all of them go our heartfelt thanks. We would, however, like to give special thanks to Greg Tobin, Deirdre Lynch, and to the following other people at Addison-Wesley: Ron Hampton, Sara Oliver, Jeffrey Holcomb, Julie LaChance, Phyllis Hubbard, Celena Carr, Heather Peck, Evelyn Beaton, Caroline Fell, Sara Anderson, Joe Vetere, Sheila Spinney, and Barbara Atkinson.

Finally, we convey our appreciation to Carol A. Weiss. Apart from writing the text, she was involved in every aspect of development and production. In addition, Carol's tenacity, patience, and moral support were invaluable in helping to ensure that the text is of the highest quality.

Prescott, Arizona   *N.A.W.*

Contents

PART TWO   Discrete Random Variables

PART FOUR   Limit Theorems and Applications

A Course In
Probability

PART ONE

Fundamentals of Probability

Girolamo Cardano 1501–1576

Girolamo Cardano was born on September 24, 1501, in Pavia, Italy. He was the illegitimate son of Chiara Micheria and Fazio Cardano, a lawyer and mathematician. As a boy, Girolamo worked in his father's law offices and was taught mathematics by him.

Cardano studied medicine, both at Pavia University and the University of Padua. He was a brilliant student, but his outspokenness, tactlessness, and addiction to gambling caused him problems throughout his life, including time in jail.

After Cardano was awarded his doctorate in medicine in 1525, he went to a small village near Padua. There he set up a medical practice and, in 1531, married Lucia Bandarini. Unable to support a wife on the income from his practice, Cardano again turned to gambling. This particular gambling binge landed Cardano and his wife in the Milan poorhouse.

Then Cardano's luck changed. He obtained a position as mathematics lecturer at the Piatti Foundation in Milan, again began to treat patients in his spare time, and, in 1539, was finally admitted to the College of Physicians in Milan, where he had been previously refused membership. He became rector of the College of Physicians and taught medicine.

In 1545, Cardano published his greatest mathematical work, *Ars Magna*, which gave solutions to cubic and quartic equations and presented the first published calculation of complex numbers. He also made important discoveries in probability, hydrodynamics, mechanics, and geology. Many regard his *Liber de Ludo Aleae* ("The Book of Games of Chance") as the beginning of probability theory.

While in Bologna in 1570, Cardano was jailed for heresy. Although he was incarcerated for only a few months, he was prohibited from holding a university position or publishing his works. Upon his release from jail, he went to Rome where, quite unexpectedly, the Pope gave him a pension. Cardano committed suicide on September 21, 1576, in Rome, Italy.

CHAPTER ONE

Probability Basics

INTRODUCTION

In this chapter, we discuss the basic nature of probability and present the prerequisite set theory for constructing a mathematical model for probability. In Section 1.1, we demonstrate that probability can be viewed as a generalization of the simple concept of percentage, thus providing a concrete basis for an intuitive notion of probability. We also discuss the general meaning of probability and indicate the necessity of taking an axiomatic approach to the subject.

In Section 1.2, we provide a review of basic set theory. As we demonstrate in Chapter 2, set theory is of fundamental importance in the axiomatic development and subsequent application of probability theory. Thus a firm grasp of the essentials of set theory—including the primary set operations of complementation, intersection, and union—is required for a proper study of probability.

1.1   From Percentages to Probabilities

Probability is a measure of likelihood. In its simplest form, it is a generalization of the concept of percentage, as illustrated in Example 1.1.

EXAMPLE 1.1 *Probability and Percentage*

Regions of the States The U.S. Census Bureau divides the states in the United States into four regions—Northeast, Midwest, South, and West—as shown in Table 1.1. (The asterisks are for use in Exercise 1.46 of the Chapter Review, so you can ignore them now.)

Table 1.1 Regions of the states in the United States

State	Region	State	Region	State	Region
Alabama*	South	Louisiana*	South	Ohio	Midwest
Alaska	West	Maine	Northeast	Oklahoma	South
Arizona	West	Maryland	South	Oregon	West
Arkansas*	South	Massachusetts	Northeast	Pennsylvania	Northeast
California	West	Michigan	Midwest	Rhode Island	Northeast
Colorado	West	Minnesota	Midwest	South Carolina*	South
Connecticut	Northeast	Mississippi*	South	South Dakota	Midwest
Delaware	South	Missouri	Midwest	Tennessee*	South
Florida*	South	Montana	West	Texas*	South
Georgia*	South	Nebraska	Midwest	Utah	West
Hawaii	West	Nevada	West	Vermont	Northeast
Idaho	West	New Hampshire	Northeast	Virginia*	South
Illinois	Midwest	New Jersey	Northeast	Washington	West
Indiana	Midwest	New Mexico	West	West Virginia	South
Iowa	Midwest	New York	Northeast	Wisconsin	Midwest
Kansas	Midwest	North Carolina*	South	Wyoming	West
Kentucky	South	North Dakota	Midwest		

a) Find the percentage of states that are in the South.

b) If a state is selected **at random**—meaning that each state is equally likely to be the one obtained—find the probability that the state selected is in the South.

Solution a) Referring to Table 1.1, we note that 16 of the 50 states are in the South. So the percentage of states that are in the South is $16/50 = 0.320$, or 32.0%.

b) Because 16 of the 50 states are in the South and each of the states is equally likely to be the one selected, chances are 16 in 50 that the state selected is in the South. In other words, the probability is $16/50 = 0.320$, or 32.0%, which is the same result found in part (a). ∎

In this book, we use the term **population** in a general sense—namely, as a collection of all individuals or items under consideration. Using that terminology, we state the following important fact, illustrated by Example 1.1.

Probability and Percentage

Suppose that a member is selected at random from a finite population. Then the probability that the member obtained has a specified attribute equals the percentage of the population that has that attribute.

In Example 1.1, the population consists of the 50 states in the United States and the specified attribute is "in the South."

The Meaning of Probability

When a member is selected at random from a finite population, as in Example 1.1, probability is nothing more than percentage. But, in general, how do we interpret probability? For instance, what do we mean by saying that

- the probability is 0.314 that the gestation period of a woman will exceed 9 months or
- the probability is 2/3 that the favorite in a horse race finishes in the money (first, second, or third place) or
- the probability is 0.40 that a traffic fatality involves an intoxicated or alcohol-impaired driver or nonoccupant?

In answering the question of how to interpret probability, we use the terms "random experiment" and "event" in their intuitive sense. Basically, by a **random experiment,** we mean an action whose outcome cannot be predicted with certainty. And, by an **event,** we mean some specified result that may or may not occur when the random experiment is performed. For instance, in Example 1.1, the random experiment consists of selecting one state at random from the 50 states of the United States; and the event in question is that the state obtained is in the South.

Some probabilities are easy to interpret: A probability near 0 indicates that the event under consideration is very unlikely to occur when the random experiment is performed, whereas a probability near 1 (100%) suggests that the event is quite likely to occur. To gain further insight into the meaning of probability, we present the **frequentist interpretation of probability.**

Frequentist Interpretation of Probability

The probability of an event is the long-run proportion of times that the event occurs in independent repetitions of the random experiment.

Alternatively, we can express the frequentist interpretation of probability as follows. Let E be an event and let $P(E)$ denote its probability. For n independent repetitions of the random experiment, let $n(E)$ denote the number of times that event E occurs. Then,

$$\frac{n(E)}{n} \approx P(E), \qquad \text{for large } n. \tag{1.1}$$

In words, for large n, the proportion of times that event E occurs in n independent repetitions of the random experiment will approximately equal the probability of event E.

To illustrate, let's consider the random experiment of a single toss of a balanced coin. Because the coin is balanced, we reason that there is a 50–50 chance of a head (i.e., that the coin will land with the head facing up). Thus we attribute probability 0.5 to that event. The frequentist interpretation is: If we toss the coin repeatedly, then, in the long run, the result will be a head half the time; or, in a large number of tosses of the coin, the result will be a head about half the time. We used a computer to perform two simulations of tossing a balanced coin 100 times. The results are displayed in Figure 1.1 and seem to support the frequentist interpretation.

Figure 1.1 Two computer simulations of tossing a balanced coin 100 times

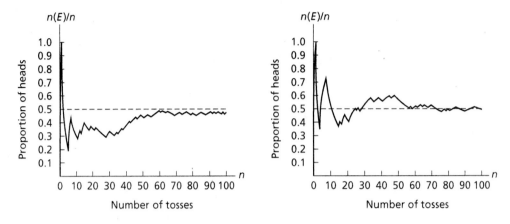

All attempts to use the frequentist interpretation as a definition of probability [by defining the probability of an event E to be $P(E) = \lim_{n \to \infty} n(E)/n$] have failed. Nonetheless, the frequentist interpretation is invaluable for motivational purposes in the axiomatic development of probability theory. Furthermore, as we demonstrate in Chapter 11, once the axioms of probability are in place, a mathematically precise version of Relation (1.1) can be proved as a theorem. That theorem is a special case of one of the most important results in probability theory—the *law of large numbers*.

EXERCISES 1.1   Basic Exercises

1.1 A committee in the U.S. Senate consists of Democrats Graham, Baucus, and Conrad and Republicans Murkowski and Kyl.
a) Identify the population under consideration.
b) If one senator is selected at random to be chair of the committee, what is the probability that the chosen senator is a Democrat?

1.2 Refer to Exercise 1.1. Suppose that a subcommittee is to consist of two senators with equal status, selected at random from the committee of five senators.
a) List the possible outcomes. *Hint:* There are 10 possibilities.
Determine the probability that the subcommittee will consist of
b) two Democrats.   **c)** two Republicans.   **d)** one Democrat and one Republican.

1.3 Refer to Exercise 1.1. Suppose that two members of the five-member committee are selected at random, the first to be chair of the committee and the second to be vice-chair.
a) List the possible outcomes. *Hint:* There are 20 possibilities.
b) Is it more probable, less probable, or equally probable that Senator Graham will be chosen as chair in this scenario than in the scenario described in Exercise 1.1? Explain.
c) What is the probability that the senator chosen to be chair is a Democrat?

1.4 The U.S. Bureau of the Census publishes data on housing units in *American Housing Survey in the United States.* The following table provides a frequency distribution for the number of rooms in U.S. housing units. The frequencies are in thousands.

Rooms	1	2	3	4	5	6	7	8+
No. Units	471	1,470	11,715	23,468	24,476	21,327	13,782	15,647

If a U.S. housing unit is selected at random, find the probability that it has
a) four rooms.   **b)** more than four rooms.   **c)** one or two rooms.
d) fewer than one room.   **e)** one or more rooms.
f) Identify the population under consideration.

1.5 As reported by the Federal Bureau of Investigation in *Crime in the United States,* during one year, the age distribution of murder victims between 20 and 59 years old was as follows.

Age	20–24	25–29	30–34	35–39	40–44	45–49	50–54	55–59
Frequency	2916	2175	1842	1581	1213	888	540	372

A murder case in which the person murdered was between 20 and 59 years old is selected at random. Find the probability that the murder victim was
a) between 40 and 44 years old, inclusive.   **b)** 25 years old or older.
c) between 45 and 59 years old, inclusive.   **d)** under 30 or over 54.
e) Identify the population under consideration.

1.6 This problem requires that you first obtain the gender and handedness of each student in your class. Determine the probability that a randomly selected student in your class is
a) female.   **b)** left-handed.
c) female and left-handed.   **d)** neither female nor left-handed.

1.7 At a certain university in the United States, 62% of the students are bilingual—speaking English and at least one other language. Of the bilingual students, 80% speak Spanish and, of these, 10% also speak French. Determine the probability that a randomly selected student at this university
a) doesn't speak Spanish.   **b)** speaks Spanish and French.

1.8 Consider the random experiment of tossing a balanced coin once. There are two possible outcomes for this experiment—namely, a head (H) or a tail (T).
a) Repeat the random experiment five times—that is, toss a coin five times—and record the information required in the following table. (The third and fourth columns are for running totals and proportions, respectively.)

Toss n	Outcome (H or T)	Number of heads $n(E)$	Proportion of heads $n(E)/n$
1			
2			
3			
4			
5			

b) Based on your five tosses, what estimate would you give for the probability of a head when this coin is tossed once? Explain your answer.

c) Now toss the coin five more times and continue recording in the table so that you now have entries for tosses 1–10. Based on your 10 tosses, what estimate would you give for the probability of a head when this coin is tossed once? Explain your answer.

d) Now toss the coin 10 more times and continue recording in the table so that you now have entries for tosses 1–20. Based on your 20 tosses, what estimate would you give for the probability of a head when this coin is tossed once? Explain your answer.

e) In view of your results in parts (b)–(d), explain why the frequentist interpretation can't be used as the definition of probability—that is, why the probability of an event E can't be defined as $P(E) = \lim_{n\to\infty} n(E)/n$.

1.9 In 2004, according to the National Governors Association, 28 of the state governors were Republicans. Suppose that on each day of 2004, one U.S. state governor was randomly selected to read the invocation on a popular radio program. On approximately how many of those days should we expect that a Republican was chosen?

1.10 Use the frequentist interpretation of probability to interpret each statement.

a) The probability is 0.314 that the gestation period of a woman will exceed 9 months.

b) The probability is 2/3 that the favorite in a horse race finishes in the money (first, second, or third place).

c) The probability is 0.40 that a traffic fatality involves an intoxicated or alcohol-impaired driver or nonoccupant.

1.11 Refer to Exercise 1.10.

a) In 4000 human gestation periods, roughly how many will exceed 9 months?

b) In 500 horse races, roughly how many times will the favorite finish in the money?

c) In 389 traffic fatalities, roughly how many will involve an intoxicated or alcohol-impaired driver or nonoccupant?

Advanced Exercises

Odds. Closely related to probabilities are *odds*. Newspapers, magazines, and other popular publications often express likelihood in terms of odds instead of probabilities, and odds are used much more than probabilities in gambling contexts. If the probability that an event occurs is p, the odds that the event occurs are p to $1 - p$. This fact is also expressed by saying that the odds are p to $1 - p$ *in favor of the event* or that the odds are $1 - p$ to p *against the event*. Conversely, if the odds in favor of an event are a to b (or, equivalently, the odds against it are b to a), the probability that the event occurs is $a/(a + b)$. For example, if an event has probability 0.75 of occurring, the odds that the event occurs are 0.75 to 0.25, or 3 to 1; if the odds against an event are 3 to 2, the probability that the event occurs is $2/(2 + 3)$, or 0.4.

1.12 An American roulette wheel contains 38 numbers, of which 18 are red, 18 are black, and 2 are green. When the roulette wheel is spun, the ball is equally likely to land on any of the 38 numbers. For a bet on red, the house pays even odds (i.e., 1 to 1). What should the odds actually be to make the bet fair?

1.13 Fusaichi Pegasus, the winner of the 2000 Kentucky Derby, was the heavy favorite to win the Preakness on May 20, 2000, with odds at 3 to 5 (against). The second favorite and actual winner, Red Bullet, posted odds at 9 to 2 (against) to win the race. Based on the posted odds, determine the probability that

a) Fusaichi Pegasus would win the race.

b) Red Bullet would win the race.

1.2   Set Theory

A **set** is a collection of elements (members, objects, items, points). If A is a set and x is an element of A, we write $x \in A$. To indicate that x is not an element of A, we write $x \notin A$ and, more generally, we use "/" to signify negation. The symbol $\emptyset$ denotes the **empty set**, a set containing no elements.

Let A and B be sets. If every element of A is an element of B, then A is said to be a **subset** of B, denoted $A \subset B$ or $B \supset A$. Two sets, A and B, are **equal sets** if they contain the same elements—in other words, if $A \subset B$ and $B \subset A$. If $A \subset B$ but $B \not\subset A$, we say that A is a **proper subset** of B.

EXAMPLE 1.2   *Sets and Subsets*

Real Numbers   Some important sets that you will need in the study of probability are

$\mathcal{R}$ = collection of real numbers, $\mathcal{Q}$ = collection of rational numbers,

$\mathcal{Z}$ = collection of integers, $\mathcal{N}$ = collection of positive integers.

Note that $\mathcal{N} \subset \mathcal{Z} \subset \mathcal{Q} \subset \mathcal{R}$ or, equivalently, that $\mathcal{R} \supset \mathcal{Q} \supset \mathcal{Z} \supset \mathcal{N}$. Also, note that $\mathcal{R} \not\subset \mathcal{Q} \not\subset \mathcal{Z} \not\subset \mathcal{N}$ or, equivalently, that $\mathcal{N} \not\supset \mathcal{Z} \not\supset \mathcal{Q} \not\supset \mathcal{R}$. Thus $\mathcal{N}$ is a proper subset of $\mathcal{Z}$, which is a proper subset of $\mathcal{Q}$, which is a proper subset of $\mathcal{R}$.   ■

We use the notation $\{a\}$ to denote the set consisting only of the element a; $\{a, b\}$ to denote the set consisting of the elements a and b; $\{a, b, c\}$ to denote the set consisting of the elements a, b, and c; and so on.

Let U be a set. Subsets of U are often defined in terms of properties that characterize its elements. If $P(x)$ is a statement concerning x, then $\{x \in U : P(x)\}$ is the collection of elements $x \in U$ such that $P(x)$ is true. For instance, $\{x \in \mathcal{N} : x^2 < 5\} = \{1, 2\}$. When no confusion is possible, we frequently abbreviate the set $\{x \in U : P(x)\}$ to $\{x : P(x)\}$.

Of particular importance in probability are **intervals** of real numbers.

> **DEFINITION 1.1   Intervals of Real Numbers**
>
> Let a and b be real numbers. Then we use the following notation.
>
> $(a, b) = \{x \in \mathcal{R} : a < x < b\}$, $[a, b] = \{x \in \mathcal{R} : a \leq x \leq b\}$,
>
> $[a, b) = \{x \in \mathcal{R} : a \leq x < b\}$, $(a, b] = \{x \in \mathcal{R} : a < x \leq b\}$,
>
> $(a, \infty) = \{x \in \mathcal{R} : x > a\}$, $(-\infty, b) = \{x \in \mathcal{R} : x < b\}$,
>
> $[a, \infty) = \{x \in \mathcal{R} : x \geq a\}$, $(-\infty, b] = \{x \in \mathcal{R} : x \leq b\}$.

Note: We call (a, b) a *bounded open interval*, $[a, b]$ a *bounded closed interval*, $[a, b)$ and $(a, b]$ *bounded half-open intervals*, (a, ∞) and $(-\infty, b)$ *unbounded open intervals*, and $[a, \infty)$ and $(-\infty, b]$ *unbounded closed intervals*.

In the remainder of this section, we assume that all sets under consideration are subsets of some fixed set—say, U—often referred to as the **universal set.**

Venn Diagrams

Graphical displays of sets are useful for explaining and understanding set theory and probability. **Venn diagrams,** named after English logician John Venn (1834–1923), are one of the best ways to visually portray sets and relationships among sets. The universal set is depicted as a rectangle, and the various subsets are drawn as disks or other geometric shapes inside the rectangle. In the simplest case, only one subset is displayed, as in Figure 1.2.

Figure 1.2 Venn diagram for the set E

Complement, Intersection, and Union

We now discuss three fundamental operations on sets—*complement, intersection,* and *union.* In Section 2.1, we present the relevance of these operations to probability theory.

Let E, A, and B be subsets of U. The **complement** of E, denoted E^c, is the set of all elements of U that do not belong to E, as shown in Figure 1.3(a); thus $E^c = \{ x : x \notin E \}$. The **intersection** of A and B, denoted $A \cap B$, is the set of all elements of U that belong to both A and B, as shown in Figure 1.3(b); thus $A \cap B = \{ x : x \in A \text{ and } x \in B \}$. The **union** of A and B, denoted $A \cup B$, is the set of all elements of U that belong to either A or B or to both A and B; in other words, those elements that belong to at least one of A and B, as shown in Figure 1.3(c); thus $A \cup B = \{ x : x \in A \text{ or } x \in B \}$. Here and throughout the book, we use "or" in the inclusive sense unless specifically stated otherwise. For instance, "$x \in A$ or $x \in B$" means that x belongs to at least one of A and B. Definition 1.2 summarizes the discussion in this paragraph.

DEFINITION 1.2   Complement, Intersection, and Union

Let U be a set and let A, B, and E be subsets of U.
a) The **complement** of E is $E^c = \{ x : x \notin E \}$.
b) The **intersection** of A and B is $A \cap B = \{ x : x \in A \text{ and } x \in B \}$.
c) The **union** of A and B is $A \cup B = \{ x : x \in A \text{ or } x \in B \}$.

Figure 1.3 Venn diagrams for complement, intersection, and union

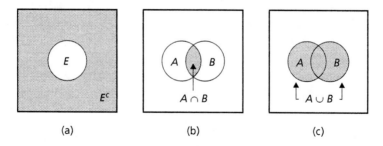

(a)	(b)	(c)

EXAMPLE 1.3 *Complement, Intersection, and Union*

a) Let $U = \{1, 2, 3, 4, 5, 6\}$, $E = \{1, 3\}$, $A = \{1, 2, 3, 4\}$, and $B = \{2, 4, 6\}$. Obtain E^c, $A \cap B$, $A \cup B$, and $E \cap B$.

b) Let $U = \mathcal{R}$, $E = [0, 1)$, $A = (-\infty, 5)$, and $B = (4, \infty)$. Determine E^c, A^c, $A \cap B$, and $A \cup B$.

Solution   a) $E^c = \{2, 4, 5, 6\}$, $A \cap B = \{2, 4\}$, $A \cup B = \{1, 2, 3, 4, 6\}$, and $E \cap B = \emptyset$.

b) $E^c = (-\infty, 0) \cup [1, \infty)$, $A^c = [5, \infty)$, $A \cap B = (4, 5)$, and $A \cup B = \mathcal{R}$.   ■

Two important relationships among the three set operations of union, intersection, and complement are given in Proposition 1.1. These two relationships are known collectively as **De Morgan's laws.**

◆◆◆ **Proposition 1.1   De Morgan's Laws**

Let A and B be subsets of U. Then
a) $(A \cup B)^c = A^c \cap B^c$.
b) $(A \cap B)^c = A^c \cup B^c$.

Proof   We prove part (a) and leave the proof of part (b) to you as Exercise 1.26. Suppose that $x \in (A \cup B)^c$. Then $x \notin A \cup B$ so that $x \notin A$ and $x \notin B$. But then $x \in A^c$ and $x \in B^c$, which means $x \in A^c \cap B^c$. Thus $(A \cup B)^c \subset A^c \cap B^c$. Conversely, suppose that $x \in A^c \cap B^c$. Then $x \in A^c$ and $x \in B^c$ so that $x \notin A$ and $x \notin B$. But then $x \notin A \cup B$, which means $x \in (A \cup B)^c$. Thus $A^c \cap B^c \subset (A \cup B)^c$. We have shown that $(A \cup B)^c \subset A^c \cap B^c$ and $A^c \cap B^c \subset (A \cup B)^c$. Therefore, $(A \cup B)^c = A^c \cap B^c$, as required.

More intuitively, although less rigorously, part (a) of De Morgan's laws can be verified by using Venn diagrams, as shown in Figure 1.4. Venn diagrams (a)–(d) are for $(A \cup B)^c$, and Venn diagrams (a')–(c') are for $A^c \cap B^c$. Note that Venn diagrams (d) and (c') are identical, thus showing that $(A \cup B)^c = A^c \cap B^c$.   ◆

Figure 1.4 Verification of $(A \cup B)^c = A^c \cap B^c$ using Venn diagrams

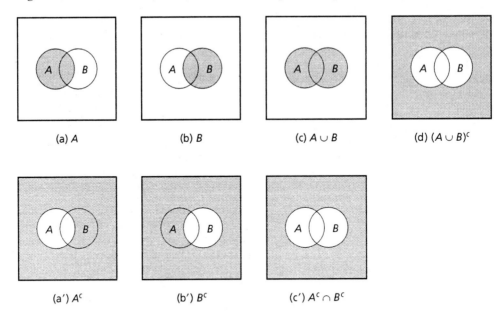

    (a) A (b) B (c) $A \cup B$ (d) $(A \cup B)^c$

    (a') A^c (b') B^c (c') $A^c \cap B^c$

Proposition 1.2 shows that intersection and union obey the **distributive laws.** We leave its proof to you as Exercise 1.27.

◆◆◆ **Proposition 1.2   Distributive Laws**

Let A, B, and C be subsets of U. Then
a) $A \cap (B \cup C) = (A \cap B) \cup (A \cap C)$.
b) $A \cup (B \cap C) = (A \cup B) \cap (A \cup C)$.

Proposition 1.3 states that intersection and union obey the **associative laws** and the **commutative laws.** These laws are intuitively clear. For instance, the commutative law for intersection says that the elements common to both A and B are the same as those common to both B and A. We leave formal proofs to you as Exercise 1.28.

◆◆◆ **Proposition 1.3   Associative and Commutative Laws**

Let A, B, and C be subsets of U. Then

a) $A \cap B = B \cap A$.   *b) $A \cup B = B \cup A$.*
c) $A \cap (B \cap C) = (A \cap B) \cap C$.   *d) $A \cup (B \cup C) = (A \cup B) \cup C$.*

Because of Proposition 1.3(c), $A \cap B \cap C$ can be defined unambiguously as the set of all elements that belong to all three of A, B, and C. Thus

$$A \cap B \cap C = \{ x : x \in A \text{ and } x \in B \text{ and } x \in C \}.$$

Similarly, because of Proposition 1.3(d), $A \cup B \cup C$ can be defined unambiguously as the set of all elements that belong to at least one of A, B, and C. Thus

$$A \cup B \cup C = \{ x : x \in A \text{ or } x \in B \text{ or } x \in C \}.$$

More on Set Operations

We can generalize the notions of intersection and union to arbitrary collections of sets. For our purposes, we need only do so for countable collections of sets—that is, finite or countably infinite collections.[†] For a finite collection of N sets, we usually assume that the sets are indexed by the first N positive integers, $1, 2, \ldots, N$; for a countably infinite collection of sets, by the positive integers, $\mathcal{N}$.

DEFINITION 1.3 Intersection and Union

Let U be a set.

a) For a finite collection, $A_1, A_2, \ldots, A_N$, of subsets of U, we define the **intersection** of the sets to be the set of all elements common to the sets. We denote that intersection $\bigcap_{n=1}^{N} A_n$. Thus

$$\bigcap_{n=1}^{N} A_n = \{ x : x \in A_n \text{ for all } n = 1, 2, \ldots, N \}.$$

We define the **union** of the sets to be the set of all elements belonging to at least one of the sets. We denote that union $\bigcup_{n=1}^{N} A_n$. Thus

$$\bigcup_{n=1}^{N} A_n = \{ x : x \in A_n \text{ for some } n = 1, 2, \ldots, N \}.$$

b) For a countably infinite collection, $A_1, A_2, \ldots$, of subsets of U, we define the **intersection** of the sets to be the set of all elements common to the sets. We denote that intersection $\bigcap_{n=1}^{\infty} A_n$. Thus

$$\bigcap_{n=1}^{\infty} A_n = \{ x : x \in A_n \text{ for all } n \in \mathcal{N} \}.$$

We define the **union** of the sets to be the set of all elements belonging to at least one of the sets. We denote that union $\bigcup_{n=1}^{\infty} A_n$. Thus

$$\bigcup_{n=1}^{\infty} A_n = \{ x : x \in A_n \text{ for some } n \in \mathcal{N} \}.$$

[†]A set is a **finite set** if either it is empty or its elements can be listed in a finite sequence $a_1, a_2, \ldots, a_N$, for some positive integer N. A set is a **countably infinite set** if its elements can be listed in an infinite sequence $a_1, a_2, \ldots$. A set is a **countable set** if it is either finite or countably infinite; otherwise, it is an **uncountable set.** This same terminology applies as well to collections of sets.

The sets $\mathcal{N}$, $\mathcal{Z}$, and $\mathcal{Q}$ are countably infinite sets and hence countable sets. Every nondegenerate interval of real numbers is an uncountable set. The collection of all subsets of a finite set is a finite set and hence a countable set, whereas the collection of all subsets of an infinite (i.e., nonfinite) set is an uncountable set. The advanced exercises for this section investigate some of the most important properties of countability.

Note: It is convenient to use the notations $\bigcap_n A_n$ and $\bigcup_n A_n$ to denote, respectively, an intersection and union of a countable collection of sets, regardless of whether that collection is finite or countably infinite.

EXAMPLE 1.4   *Intersection and Union*

Let $U = \mathcal{R}$. For each $n \in \mathcal{N}$, define $A_n = [1/n, 2 + 1/n)$. Determine

a) $\bigcap_{n=1}^{3} A_n$ and $\bigcup_{n=1}^{3} A_n$.   b) $\bigcap_{n=1}^{\infty} A_n$ and $\bigcup_{n=1}^{\infty} A_n$.

Solution   a) We have

$$\bigcap_{n=1}^{3} A_n = [1, 3) \cap [1/2, 5/2) \cap [1/3, 7/3) = [1, 7/3)$$

and

$$\bigcup_{n=1}^{3} A_n = [1, 3) \cup [1/2, 5/2) \cup [1/3, 7/3) = [1/3, 3).$$

b) We claim that

$$\bigcap_{n=1}^{\infty} A_n = [1, 2].$$

Indeed, $x \in \bigcap_{n=1}^{\infty} A_n$ if and only if $1/n \leq x < 2 + 1/n$ for all $n \in \mathcal{N}$, which is true if and only if $x \geq 1$ and $x \leq 2$—that is, $x \in [1, 2]$. Next we claim that

$$\bigcup_{n=1}^{\infty} A_n = (0, 3).$$

To prove this equality, we first observe that $A_n \subset (0, 3)$ for all $n \in \mathcal{N}$ and therefore $\bigcup_{n=1}^{\infty} A_n \subset (0, 3)$. Conversely, suppose that $x \in (0, 3)$. Let m be the smallest positive integer such that $x \geq 1/m$. If $m = 1$, then $x \in [1, 3) = A_1 = A_m$; if $m \geq 2$, then $1/m \leq x < 1/(m-1) \leq 1 < 2 + 1/m$ and hence $x \in A_m$. Thus, in either case, $x \in A_m \subset \bigcup_{n=1}^{\infty} A_n$. Therefore $(0, 3) \subset \bigcup_{n=1}^{\infty} A_n$.   ∎

De Morgan's laws and the distributive laws hold for any collection of subsets and, so, in particular, for countable collections. We state these laws in Propositions 1.4 and 1.5 and leave their proofs to you as Exercises 1.32 and 1.33, respectively.

◆◆◆ **Proposition 1.4   De Morgan's Laws**

Let $A_1, A_2, \ldots$ be subsets of U. Then

a) $\left(\bigcap_n A_n \right)^c = \bigcup_n A_n^c.$ b) $\left(\bigcup_n A_n \right)^c = \bigcap_n A_n^c.$

◆◆◆ **Proposition 1.5   Distributive Laws**

Let B and $A_1, A_2, \ldots$ be subsets of U. Then

a) $B \cap \left(\bigcup_n A_n \right) = \bigcup_n (B \cap A_n).$ b) $B \cup \left(\bigcap_n A_n \right) = \bigcap_n (B \cup A_n).$

Disjoint Sets

Next we discuss **disjoint sets** and **pairwise disjoint sets.** These concepts, presented in Definition 1.4, are essential in mathematics and, in particular, in probability theory.

DEFINITION 1.4   Disjoint and Pairwise Disjoint Sets

Two sets, A and B, are said to be **disjoint sets** if $A \cap B = \emptyset$—that is, if A and B have no elements in common. More generally, sets $A_1, A_2, \ldots$ are said to be **pairwise disjoint sets** if $A_m \cap A_n = \emptyset$ whenever $m \neq n$.

The Venn diagrams shown in Figure 1.5 graphically portray the difference between two sets that are disjoint and two sets that are not. Figure 1.6 shows three pairwise disjoint sets and two cases where three sets are not pairwise disjoint.

Figure 1.5 (a) Two disjoint sets; (b) two sets that are not disjoint

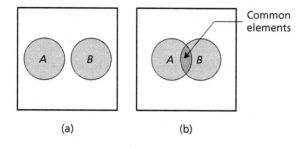

Figure 1.6 (a) Three pairwise disjoint sets; (b) and (c) three sets that are not pairwise disjoint

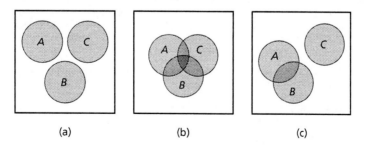

EXAMPLE 1.5   *Disjoint and Pairwise Disjoint Sets*

Let $U = \mathcal{R}$.
a) The sets $\mathcal{Z}$ and $(0, 1)$ are disjoint because $\mathcal{Z} \cap (0, 1) = \emptyset$.

b) The sets $\mathcal{Z}$ and $[0, 1]$ are not disjoint because $\mathcal{Z} \cap [0, 1] = \{0, 1\} \neq \emptyset$.

c) For each $n \in \mathcal{N}$, let $A_n = [n - 1, n)$. These sets are pairwise disjoint because $[m - 1, m) \cap [n - 1, n) = \emptyset$ whenever $m \neq n$.

d) For each $n \in \mathcal{N}$, let $B_n = [n - 1, n]$. These sets are not pairwise disjoint because, for instance, $[0, 1] \cap [1, 2] = \{1\} \neq \emptyset$. Note, however, that the intersection of all B_n is empty—that is, $\bigcap_{n=1}^{\infty} [n - 1, n] = \emptyset$. This result shows that the sets in a collection may not be pairwise disjoint even though the intersection of all the sets is empty. ∎

Cartesian Product

Another set concept important in the study of probability theory is that of **Cartesian product,** named in honor of René Descartes (1596–1650). For our purposes, we need only define the Cartesian product of a finite number of sets, as presented in Definition 1.5.

DEFINITION 1.5   Cartesian Product

Let A and B be two sets. Then the **Cartesian product** of A and B (in that order), denoted $A \times B$, is the set of all ordered pairs (a, b), where $a \in A$ and $b \in B$. Thus

$$A \times B = \{(a, b) : a \in A, \ b \in B\}.$$

More generally, if $A_1, A_2, \ldots, A_N$ are sets, the Cartesian product of those N sets, denoted $A_1 \times A_2 \times \cdots \times A_N$ or $\times_{n=1}^{N} A_n$, is the set of all ordered N-tuples $(a_1, a_2, \ldots, a_N)$, where $a_n \in A_n$ for $n = 1, 2, \ldots, N$. Thus

$$\underset{n=1}{\overset{N}{\times}} A_n = \{(a_1, a_2, \ldots, a_N) : a_n \in A_n, \ 1 \leq n \leq N\}.$$

An important special case of Cartesian product occurs when all sets in the product are identical. If $A_n = A$, for $1 \leq n \leq N$, where A is some set, we write A^N for the Cartesian product. In other words, $A^N = \underbrace{A \times A \times \cdots \times A}_{N \text{ times}}$.

EXAMPLE 1.6   *Cartesian Product*

a) If at least one of A and B is empty, so is $A \times B$.

b) $\{1, 2, 3\} \times \{3, 4\} = \{(1, 3), (1, 4), (2, 3), (2, 4), (3, 3), (3, 4)\}$.

c) Let a, b, c, and d be real numbers with $a < b$ and $c < d$. The Cartesian product of the closed intervals $[a, b]$ and $[c, d]$ is

$$[a, b] \times [c, d] = \{(x, y) : a \leq x \leq b, \ c \leq y \leq d\}.$$

Therefore $[a, b] \times [c, d]$ is the closed rectangle with vertices (a, c), (b, c), (a, d), and (b, d), as shown in Figure 1.7.

Figure 1.7 Cartesian product of two bounded closed intervals

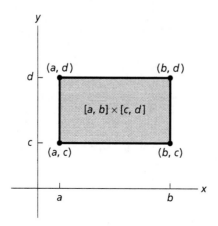

d) Consider the two sets consisting of the first m positive integers and the first n positive integers—namely, $\{1, 2, \ldots, m\}$ and $\{1, 2, \ldots, n\}$. Then

$$\{1, 2, \ldots, m\} \times \{1, 2, \ldots, n\} = \{\, (j, k) \in \mathcal{N} \times \mathcal{N} : 1 \leq j \leq m,\ 1 \leq k \leq n \,\}$$

Thus $\{1, 2, \ldots, m\} \times \{1, 2, \ldots, n\}$ consists of the mn integer lattice points shown in Figure 1.8.

Figure 1.8 Cartesian product of $\{1, 2, \ldots, m\}$ and $\{1, 2, \ldots, n\}$

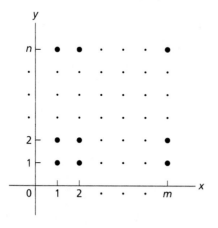

e) $\mathcal{R}^n = \{\, (x_1, x_2, \ldots, x_n) : x_j \in \mathcal{R},\ 1 \leq j \leq n \,\}$ is called *Euclidean n-space*. ■

For future reference, we note that the Cartesian product of a finite number of countable sets is countable. See Exercise 1.41.

EXERCISES 1.2   Basic Exercises

1.14 In Example 1.5(d) on page 16, we showed that it is possible for the sets in a collection to not be pairwise disjoint even though the intersection of all the sets is empty. If the sets in a collection are pairwise disjoint, must the intersection of all the sets be empty?

1.15 Draw a Venn diagram showing three subsets, A, B, and C, such that no two are disjoint but that $A \cap B \cap C = \emptyset$.

1.16 Make a Venn diagram with four subsets, A, B, C, and D, such that $A \cap B \neq \emptyset$, $B \cap C \neq \emptyset$, $C \cap D \neq \emptyset$, and $A \cap D \neq \emptyset$ but that $A \cap B \cap C \cap D = \emptyset$.

1.17 Give an example of a collection of sets satisfying the following properties: The collection contains at least four sets, the sets are not pairwise disjoint, and every three sets have an empty intersection.

1.18 Let $U = \mathcal{R}$ and, for each $n \in \mathcal{N}$, define $A_n = [0, 1/n]$.
a) Determine $\bigcap_{n=1}^{4} A_n$ and $\bigcup_{n=1}^{4} A_n$.   **b)** Determine $\bigcap_{n=1}^{\infty} A_n$ and $\bigcup_{n=1}^{\infty} A_n$.

1.19 Express each of the following sets in a simple form in which "$\bigcup$" and "$\bigcap$" don't occur.
a) $\bigcup_{n=1}^{\infty} [1 + 1/n, 2 - 1/n]$ **b)** $\bigcup_{n=1}^{\infty} [1, 2 - 1/n]$ **c)** $\bigcap_{n=1}^{\infty} (1 - 1/n, 2 + 1/n)$
d) $\bigcap_{n=1}^{\infty} (3 - 1/n, 3 + 1/n)$ **e)** $\bigcap_{n=1}^{\infty} (n, \infty)$ **f)** $\bigcap_{n=1}^{\infty} (5 - 1/n, 5)$
g) $\bigcap_{n=1}^{\infty} (5 - 1/n, 6)$

1.20 Let the universal set U be $\{1, 2, 3, 4, 5\} \times \{1, 2, 3, 4, 5\}$.
a) List the members of $\{1, 2, 3\} \times \{3, 4, 5\}$.
b) List the members of $(\{1, 2, 3\} \times \{3, 4, 5\}) \cup (\{3, 4, 5\} \times \{1, 2, 3\})$.
c) List the members of $A = \big((\{1, 2, 3\} \times \{3, 4, 5\}) \cup (\{3, 4, 5\} \times \{1, 2, 3\})\big)^c$.
d) Write the set A in part (c) as the union of two Cartesian products—that is, in the form (some set × some set) $\cup$ (some set × some set).

1.21 List the members of each of the following sets.
a) $\{0, 1\}^3 = \{0, 1\} \times \{0, 1\} \times \{0, 1\}$ **b)** $\{0, 1\} \times \{0, 1\} \times \{1, 2\}$
c) $(\{a, b\} \cup \{c, d, e\}) \times \{f, g, h\}$ **d)** $(\{a, b\} \times \{f, g, h\}) \cup (\{c, d, e\} \times \{f, g, h\})$

1.22 Simplify the expression $([0, 2] \times [0, 2]) \cap ([1, 3] \times [1, 3])$, writing your answer in the form (some set × some set).

1.23 Find a countable collection of subintervals I_1, I_2, ... of the interval $[0, 1]$ such that $I_j \cap I_k \neq \emptyset$ for all j and k, but $\bigcap_{i=1}^{\infty} I_i = \emptyset$.

1.24 Find a countable collection of intervals I_1, I_2, ... of $\mathcal{R}$ such that $\bigcap_{i=1}^{\infty} I_i$ consists of a single point (i.e., is a singleton set).

1.25 Express $\mathcal{R}$ as a countably infinite union of pairwise disjoint intervals, each of length 1.

Theory Exercises

1.26 Refer to De Morgan's laws, as given in Proposition 1.1 on page 11.
a) Verify part (b) of De Morgan's laws by using Venn diagrams.
b) Prove part (b) of De Morgan's laws mathematically in a manner similar to that done for part (a) of De Morgan's laws on page 11.
c) Prove part (b) of De Morgan's laws by using part (a) of De Morgan's laws.

1.27 Refer to the distributive laws, as given in Proposition 1.2 on page 12.
a) Verify parts (a) and (b) by using Venn diagrams.
b) Prove parts (a) and (b) mathematically.

1.28 Refer to the associative and commutative laws, as given in Proposition 1.3 on page 12.
a) Verify parts (a)–(d) by using Venn diagrams.
b) Prove parts (a)–(d) mathematically.

1.29 Let A, B, and C be subsets of U. Prove each of the following statements.
a) $A \cup \emptyset = A$ **b)** $A \subset A \cup B$ **c)** $A = A \cup B$ if and only if $B \subset A$

1.30 Let A, B, and C be subsets of U. Prove each of the following statements.
a) $A \cap \emptyset = \emptyset$ **b)** $A \supset A \cap B$ **c)** $A = A \cap B$ if and only if $B \supset A$

1.31 Let A and B be subsets of U. Verify each of the following statements.
a) $A = (A \cap B) \cup (A \cap B^c)$ **b)** $A \cap B = \emptyset \Rightarrow A \subset B^c$ **c)** $A \subset B \Rightarrow B^c \subset A^c$

1.32 Prove De Morgan's laws for countable collections of sets, Proposition 1.4 on page 14.

1.33 Prove the distributive laws for countable collections of sets, Proposition 1.5 on page 14.

1.34 Let A_1, A_2, ... be a countable collection of subsets of U. Prove the following.
a) If $B \subset \bigcup_n A_n$, then $B = \bigcup_n (A_n \cap B)$.
b) If $\bigcup_n A_n = U$, then $E = \bigcup_n (A_n \cap E)$ for each subset E of U.
c) If A_1, A_2, ... are pairwise disjoint, so are $A_1 \cap E$, $A_2 \cap E$, ... for each subset E of U.
d) We say that A_1, A_2, ... form a *partition* of U if they are pairwise disjoint and their union is U. Conclude from parts (b) and (c) that, if A_1, A_2, ... form a partition of U, each subset E of U can be expressed as a disjoint union of the sets $A_1 \cap E$, $A_2 \cap E$, ...

Advanced Exercises

1.35 Let A_1, A_2, ... be a countably infinite collection of subsets of U.
a) Prove that

$$\bigcup_{n=1}^{\infty} \left(\bigcap_{k=n}^{\infty} A_k \right) \subset \bigcap_{n=1}^{\infty} \left(\bigcup_{k=n}^{\infty} A_k \right).$$

The set on the left is called the *limit inferior* of the A_ns, denoted $\liminf_{n \to \infty} A_n$; the set on the right is called the *limit superior* of the A_ns, denoted $\limsup_{n \to \infty} A_n$.
b) Describe in words the limit inferior and limit superior of the A_ns, and use that description to interpret the relation in part (a). *Hint:* To how many A_ns must a point belong?
c) Let $U = \mathcal{R}$ and define

$$A_n = \begin{cases} [0, 1 + 1/n], & \text{if } n \text{ is an even positive integer;} \\ [-1 - 1/n, 0], & \text{if } n \text{ is an odd positive integer.} \end{cases}$$

Determine $\liminf_{n \to \infty} A_n$ and $\limsup_{n \to \infty} A_n$.

Finite and infinite sets; countability: Two sets are said to be *equivalent* if there is a one-to-one function from one set onto the other. A set E is *finite* if it is either empty or equivalent to the first N positive integers for some $N \in \mathcal{N}$; E is *infinite* if it is not finite; E is *countably infinite* if it is equivalent to $\mathcal{N}$; E is *countable* if it is either finite or countably infinite; E is *uncountable* if it is not countable.

1.36 Prove that $\mathcal{Z}$ is countable. *Hint:* Define a function that maps the even positive integers onto the positive integers and the odd positive integers onto the nonpositive integers.

1.37 Prove that $\mathcal{N}^2$ is countable. *Hint:* Define $f \colon \mathcal{N}^2 \to \mathcal{N}$ by $f(m, n) = 2^{m-1}(2n - 1)$.

1.38 Prove that a nonempty set is countable if and only if it is the range of an infinite sequence.

1.39 Use Exercise 1.38 to prove that the image (under a function) of a countable set is countable.

1.40 Use Exercise 1.38 to prove that a subset of a countable set is countable.

1.41 Use Exercises 1.37–1.39 to prove that the Cartesian product of two countable sets is countable. Then extend this result to a finite number of countable sets.

1.42 Prove that Q is countable. *Hint:* Use Exercises 1.36 and 1.41.

1.43 Prove that any (nondegenerate) interval of R is uncountable by using the following steps.
a) Show that the interval $[0, 1)$ is uncountable. *Hint:* Suppose to the contrary that $[0, 1)$ is countable and let $\{x_n\}_{n=1}^{\infty}$ be an enumeration of its elements. Construct a number in $[0, 1)$ whose decimal expansion contains only 0s and 1s and, for each $n \in \mathcal{N}$, differs from x_n at the nth decimal digit.
b) Use part (a) to conclude that $(0, 1)$ is uncountable.
c) Use part (b) to show that any bounded interval of the form (a, b) is uncountable. *Hint:* Construct a one-to-one and onto function from $(0, 1)$ to (a, b).
d) Use part (c) to conclude that any interval is uncountable. In particular, R is uncountable.

1.44 Prove that a countable union of countable sets is countable.

1.45 The *Cartesian product* of an indexed collection $\{A_\iota\}_{\iota \in I}$ of sets, denoted $\times_{\iota \in I} A_\iota$, is the set of all functions x on I such that $x(\iota)$ is an element of A_ι for each $\iota \in I$.
a) For finite I, show that this definition of Cartesian product is consistent with Definition 1.5.
b) True or false: The Cartesian product of a countable number of countable sets is countable. *Hint:* Consider decimal expansions of numbers in $[0, 1]$.

CHAPTER REVIEW

Summary

In this chapter, we discussed the nature of probability. First we examined how probability can be considered a generalization of the concept of percentage; in particular, we showed that if one member of a finite population is selected at random, the probability that the member obtained has a specified attribute equals the percentage of the population that has that attribute.

Next we investigated the meaning of probability, which led us to the frequentist interpretation of probability: The probability of an event is the long-run proportion of times that the event occurs in independent repetitions of the random experiment. Although this interpretation can't be used successfully as a definition of probability, it provides an intuitive basis for motivating and developing probability axioms and concepts.

In anticipation of developing an axiomatic framework for probability, we reviewed the basics of set theory. Set theory is fundamental to probability theory because, in the axiomatic framework, an event—some specified result that may or may not occur when the random experiment is performed—corresponds to a unique subset of the set of all possible outcomes of the random experiment; and relationships among events are interpreted mathematically as set operations. Armed with the required set theory background, we can provide a detailed correspondence between events and sets, which we pursue in Chapter 2.

You Should Be Able To ...

1. determine probabilities for an experiment that consists of selecting a member at random from a finite population.

2. state and understand the frequentist interpretation of probability.

3. state and apply the basics of set theory, including set-theory notation, set-theory terminology, set operations, and properties of set operations.

4. define and apply the basics of Cartesian products.

Key Terms

associative laws, *12*	empty set, *9*	proper subset, *9*
at random, *4*	equal sets, *9*	random experiment, *5*
Cartesian product, *16*	event, *5*	set, *9*
commutative laws, *12*	finite set, *13*	subset, *9*
complement, *10*	frequentist interpretation of	uncountable set, *13*
countable set, *13*	probability, *5*	union, *10, 13*
countably infinite set, *13*	intersection, *10, 13*	universal set, *10*
De Morgan's laws, *11, 14*	intervals, *9*	Venn diagrams, *10*
disjoint sets, *15*	pairwise disjoint sets, *15*	
distributive laws, *12, 14*	population, *4*	

Chapter Review Exercises

Basic Exercises

1.46 Refer to Example 1.1 on page 4.
a) Suppose that a state in the South is selected at random. Determine the probability that the state obtained is one that seceded from the Union in 1860 and 1861. The states that seceded are marked with asterisks in Table 1.1.
b) Identify the population in part (a).

1.47 The National Science Foundation collects data on Nobel Prize Laureates in the field of science and the date and location of their award-winning research. A frequency distribution for the number of winners, by country, for the years 1901–1997 are as follows.

Country	USA	UK	Germany	France	Russia	Japan	Other
Winners	190	71	61	25	10	4	87

Suppose that a recipient of a Nobel Prize in science between 1901–1997 is selected at random. Find the probability that the Nobel Laureate is from
a) Japan.   **b)** either France or Germany.
c) any country other than the United States.

1.48 A die is a cube on which a different number of dots, from one to six, is painted on each of the six faces. Suppose that a balanced die is rolled and that we observe the number of dots facing up. Use the frequentist interpretation of probability to interpret each statement.
a) The probability of getting a 3 (i.e., three dots facing up) is 1/6.
b) The probability of getting a 3 or more (i.e., three or more dots facing up) is 2/3.

If the die is rolled 10,000 times, roughly how many times will it come up
c) 3?   **d)** 3 or more?

1.49 Suppose that 60% of the voters in a population will vote *yes* on a particular proposition. A political opinion poll chooses voters one at a time from the population with replacement. Let n be the number of such randomly chosen voters who have been polled so far and let $n(Y)$ be the number of those sampled voters who will vote *yes*. For large n, express $n(Y)$ approximately as a function of n.

1.50 Solve each of the following problems.
a) List explicitly the elements of the set $\{x^2 : x \in \{-2, -1, 0, 1, 2\}\}$.
b) Express $\{x^2 : -2 < x < 2\}$ as a bounded half-open interval.

1.51 Explain how the eight members of $\{0, 1\}^3$ can be regarded as the outcomes of the experiment of tossing a coin three times. Consider those eight members a finite population in the sense of Section 1.1. By examining the list of all eight members of the population (and not by some other method), find the probability that you get two heads and one tail when you toss a balanced coin three times.

1.52 Simplify $(A \cap B) \cup (A \cap B^c)$.

1.53 Solve each of the following problems.
a) Express $\bigcup_{n=1}^{\infty} (1/(n+1), 1/n]$ as a bounded half-open interval.
b) Express $\bigcup_{n=1}^{\infty} [1/(n+1), 1/n]$ as a bounded half-open interval.
c) One of parts (a) and (b) involves a union of pairwise disjoint sets; the other involves a union of sets that are not pairwise disjoint but whose intersection is empty. Which is which? Explain your answer.

1.54 Classify each of the following sets as finite, countably infinite, or uncountable. *Hint:* Refer to the discussion of finite and infinite sets on pages 19 and 20.
a) $\{1, 2, 3\} \times \{2, 3, 4\}$ **b)** $\{1, 2, 3\} \times [2, 4]$ **c)** $[2, 4] \times \{1, 2, 3\}$
d) $\{1, 2, 3\} \times \{1, 2, 3, \ldots\}$ **e)** $[1, 3] \times [2, 4]$ **f)** $\bigcup_{n=1}^{\infty} \{n, n+1, n+2\}$
g) $\bigcap_{n=1}^{\infty} \{n, n+1, n+2\}$

1.55 Suppose that the universal set is $\mathcal{Z}$, the collection of all integers.
a) Use De Morgan's laws to simplify $(\{3, 4\}^c \cap \{4, 5\}^c)^c$ without explicitly mentioning any infinite sets.
b) Simplify the set in part (a) without De Morgan's laws by first finding $\{3, 4\}^c$ and $\{4, 5\}^c$, next finding their intersection, and then finding its complement.

Theory Exercises

1.56 Let $A_1, A_2, \ldots$ be a countably infinite collection of sets.
a) For each positive integer n, determine pairwise disjoint sets, $B_1, B_2, \ldots, B_n$, such that $\bigcup_{j=1}^{n} B_j = \bigcup_{j=1}^{n} A_j$.
b) Find pairwise disjoint sets, $B_1, B_2, \ldots$, such that $\bigcup_{n=1}^{\infty} B_n = \bigcup_{n=1}^{\infty} A_n$.

1.57 Let A, B, and C be subsets of a set U.
a) Show that $(A \cup B) \cap C \subset A \cup (B \cap C)$.
b) Give an example of three sets, A, B, and C, such that $(A \cup B) \cap C \neq A \cup (B \cap C)$.
c) Prove the *modular law:* If $A \subset C$, then $(A \cup B) \cap C = A \cup (B \cap C)$.
d) Show that, if $(A \cup B) \cap C = A \cup (B \cap C)$, then $A \subset C$.

1.58 Let A and B be subsets of a set U.
a) Show that $A \cup B$ is the intersection of all sets $C \subset U$ such that $A \subset C$ and $B \subset C$.

b) If $A \subset D$ and $B \subset D$, determine and interpret the relationship between D and $A \cup B$.

c) Show that $A \cap B$ is the union of all sets $C \subset U$ such that $A \supset C$ and $B \supset C$.

d) If $A \supset D$ and $B \supset D$, determine and interpret the relationship between D and $A \cap B$.

e) Generalize parts (a)–(d) to countable unions and intersections.

1.59 Let A, B, and C be sets. For each of the following equalities, either prove that it's (always) true or find a counterexample.

a) $A \times (B \cup C) = (A \times B) \cup (A \times C)$ **b)** $A \times (B \cap C) = (A \times B) \cap (A \times C)$

c) $A \times B = B \times A$

Advanced Exercises

1.60 The following hypothetical data cross classifies 525 individuals with respect to color-blind status (CBS) and handedness.

Handedness

		Left	Right	Ambidextrous
CBS	Colorblind	71	61	37
	Normal	92	212	52

One person is selected at random from the 525 individuals. Determine the probability that the person obtained is

a) left-handed.   **b)** colorblind but not ambidextrous.

c) If the person selected is left-handed, what is the probability that he or she is colorblind?

d) If the person selected is colorblind, what is the probability that he or she is left-handed?

e) Suppose that the left-handed, colorblind individuals consist of 47 men and 24 women. What is the probability that the person obtained is a left-handed, colorblind woman?

1.61 As found in *USA TODAY*, results of a survey by International Communications Research revealed that roughly 75% of adult females believe that having a "cyber affair"—a romantic relationship over the Internet while in an exclusive relationship in the real world—is cheating. What are the odds against an adult female believing that having a "cyber affair" is cheating?

1.62 Relative complement: Let A and B be subsets of U. The *relative complement* of B with respect to A, denoted $A \setminus B$, is the set of all elements belonging to A that do not belong to B. Thus $A \setminus B = \{ x : x \in A \text{ and } x \notin B \}$.

a) Express B^c as a relative complement.   **b)** Show that $A \setminus B = A \cap B^c$.

c) Show that $(A \setminus B)^c = A^c \cup B$.

1.63 Symmetric difference: Let A and B be subsets of U. The *symmetric difference* of A and B, denoted $A \triangle B$, is the set of all elements belonging to either A or B but not both A and B. Express $A \triangle B$ in terms of union and relative complement.

1.64 Refer to Exercise 1.63. Prove that

a) $A \triangle (B \triangle C) = (A \triangle B) \triangle C$.   **b)** $A \triangle U = A^c$.

c) $A \triangle \emptyset = A$.   **d)** $A \triangle A = \emptyset$.

1.65 Refer to Exercise 1.63.

a) Prove that $A \cap (B \triangle C) = (A \cap B) \triangle (A \cap C)$.

b) What is the relationship between $A \cup (B \triangle C)$ and $(A \cup B) \triangle (A \cup C)$?

c) Precisely when does $A \cup (B \triangle C) = (A \cup B) \triangle (A \cup C)$?

1.66 Refer to Exercise 1.63. Show that $A = A \triangle B$ if and only if $B = \emptyset$.

Andrei Kolmogorov *1903–1987*

Andrei Nikolaevich Kolmogorov was born on April 25, 1903, in Tambov, Russia. At the age of 17, he entered Moscow State University and graduated from there in 1925. His contributions to mathematics, many of which appear in his numerous articles and books, encompass a formidable range of subjects.

Kolmogorov revolutionized the theory of probability by introducing the modern axiomatic approach to probability theory and by proving many of the fundamental theorems that are consequences of that approach. He also developed two systems of partial differential equations, which bear his name. Those systems extended the development of probability theory and allowed its broader application to the fields of physics, chemistry, biology, and civil engineering.

In 1938, Kolmogorov published an extensive article entitled "Mathematics," which appeared in the first edition of the *Bolshaya Sovyetskaya Entsiklopediya* (Great Soviet Encyclopedia). In this article, he discussed the development of mathematics from ancient to modern times. He also clarified the concept of the algorithm, a basic category of mathematics that cannot be defined in any simpler concepts.

Kolmogorov became a member of the faculty at Moscow State University in 1925, at the age of 22. In 1931, he was promoted to professor; in 1933, he was appointed a director of the Institute of Mathematics of the university; and, in 1937, he became Head of the University.

In addition to his work in higher mathematics, Kolmogorov was interested in the mathematical education of schoolchildren. He was chair of the Commission for Mathematical Education under the Presidium of the Academy of Sciences of the U.S.S.R. During his tenure as chair, he was instrumental in the development of a new mathematics training program that was introduced into Soviet schools.

Kolmogorov remained on the faculty at Moscow State University until his death in Moscow on October 20, 1987.

CHAPTER TWO

Mathematical Probability

INTRODUCTION

Now that we have provided an intuitive introduction to probability (Section 1.1) and reviewed the basics of set theory (Section 1.2), we can construct a mathematical model for random experiments and probability. In Section 2.1, we explain how events naturally correspond to subsets of the collection of all possible outcomes of a random experiment; this correspondence permits us to use the power of set theory as an aid to mathematical modeling and in the application of probability.

In Section 2.2, we present the axioms of probability, sometimes referred to as the Kolmogorov axioms in honor of the mathematician who suggested their use. (See the facing page for a biographical sketch of Andrei Kolmogorov.) In Section 2.3, we investigate common methods for specifying and determining probabilities, including equal-likelihood models, empirical probability, and subjective probability. The chapter concludes with Section 2.4 where we derive from the Kolmogorov axioms some of the most useful properties of probability.

2.1 Sample Space and Events

We are now in a position to construct a mathematical model for random experiments and probability. Essentially, there are two interrelated aspects of the model:

- the representation of events by sets and
- the axioms specifying the basic properties that probabilities must satisfy.

We examine the first aspect in this section and the second aspect in Section 2.2.

Sample Space

We begin by recalling that a *random experiment* is an action whose outcome can't be predicted with certainty beforehand. The set of all possible outcomes for a random experiment is called the **sample space** and is represented by the Greek letter Ω (uppercase omega).[†] The possible outcomes themselves—that is, the elements of the sample space—are represented generically by the Greek letter ω (lowercase omega).[‡] Thus the sample space Ω is the universal set for a random experiment and the outcomes ω are the elements of that universal set.

> **DEFINITION 2.1   Sample Space**
> The set of all possible outcomes for a random experiment is called the **sample space** and is represented by the Greek letter Ω.

Before illustrating sample spaces, we need to point out that many of the examples and exercises used in a probability course are purposely chosen for their simplicity or concreteness. For instance, we often present examples and exercises that involve cards, dice, or coin tossing. But, keep in mind that, in practice, sample spaces can range from the extremely simple to the extraordinarily complex.

EXAMPLE 2.1   *Sample Spaces*

Coin Tossing   Suppose that we toss a coin three times and observe the result (a head or a tail) for each toss. Determine the sample space for this random experiment.

Solution   The sample space for the experiment is $\Omega = \{$HHH, HHT, HTH, HTT, THH, THT, TTH, TTT$\}$ where, for instance, HTT denotes the outcome of a head on the first toss and tails on the second and third tosses. ∎

[†]The term *outcome space* is more descriptive than *sample space*. However, we stay with the traditional terminology of *sample space*, which reflects the fact that the possible outcomes of a random experiment are sometimes the possible samples in a statistical study.

[‡]The Greek letters Ω and ω correspond to the English letters O and o, respectively.

EXAMPLE 2.2 *Sample Spaces*

Emergency Room Traffic Suppose that, starting at 6:00 P.M., we observe the elapsed time, in hours, until the first patient arrives at a particular emergency room. Determine the sample space for this random experiment.

Solution We can reasonably assume that eventually someone will arrive at the emergency room. The sample space can be taken to be the nonnegative real numbers—that is, $\Omega = [0, \infty)$. (If it is possible that no one will ever arrive at the emergency room, we can represent that outcome by ∞ and take the sample space to be $[0, \infty] = [0, \infty) \cup \{\infty\}$.) ∎

EXAMPLE 2.3 *Sample Spaces*

Rolling Dice A die is a cube on which a different number of dots, from one to six, is painted on each of the six faces. Suppose that two dice are rolled—one black and the other gray—and that we observe on each die the number of dots facing up. Determine the sample space for this random experiment.

Solution The possible outcomes for rolling two dice are shown in Figure 2.1.

Figure 2.1 Possible outcomes for rolling a pair of dice

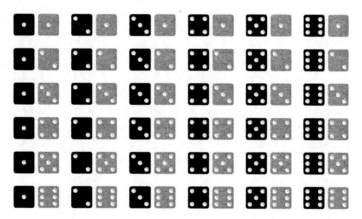

The set of these 36 outcomes is the sample space. Mathematically, we can express the sample space as $\Omega = \{ (x, y) : x, y \in \{1, 2, 3, 4, 5, 6\} \}$—that is, as the set of all ordered pairs of integers between 1 and 6, inclusive. ∎

EXAMPLE 2.4 *Sample Spaces*

Component Analysis Mechanical or electrical units often consist of several components, each of which is subject to failure. Suppose that, at a specified time, we observe a unit that consists of five components and determine which are working and which have failed. Obtain the sample space for this random experiment.

Solution Let s denote success (a working component) and f denote failure. We can represent each outcome as an ordered five-tuple where each entry of the five-tuple is either s or f.

For instance, (s, s, s, f, s) represents the outcome that components 1–3 are working, component 4 has failed, and component 5 is working. Therefore a sample space for this random experiment is $\Omega = \{ (x_1, x_2, x_3, x_4, x_5) : x_j \in \{s, f\}, \ 1 \le j \le 5 \}$. ■

EXAMPLE 2.5   *Sample Spaces*

Blood Type   Blood types of human beings are determined by a pair of alleles, one received from each parent. We denote the three different blood-type alleles a, b, and o. The possible pairs are referred to as *genotypes*. Suppose that a person is selected at random and that his or her genotype is observed. Obtain the sample space for this random experiment.

Solution   The sample space for this random experiment consists of the possible genotypes and can be expressed as $\Omega = \{aa, ab, ao, bb, bo, oo\}$. Note that the order of an allele pair is irrelevant for genotypes so that, for instance, ab and ba yield the same genotype. ■

EXAMPLE 2.6   *Sample Spaces*

Bacteria on a Petri Dish   A petri dish is a small, shallow dish of thin glass or plastic used especially for cultures in bacteriology. Suppose that a petri dish of unit radius, containing nutrients upon which bacteria can multiply, is smeared with a uniform suspension of bacteria. Subsequently, spots indicating colonies of bacteria will appear. Suppose also that we observe the location of the center of the first spot to appear. Obtain the sample space for this random experiment.

Solution   The location of the center of the first spot to appear can be anywhere on the petri dish. If we take the center of the petri dish to be at the origin, the sample space can be represented as the unit disk: $\Omega = \{ (x, y) : x^2 + y^2 < 1 \}$. Thus, for instance, if the horizontal and vertical coordinates of the center of the first spot are at $1/4$ and $5/8$, respectively, the outcome of the random experiment is $(1/4, 5/8)$. ■

EXAMPLE 2.7   *Sample Spaces*

Oklahoma State Officials   As reported by the *World Almanac*, the top five state officials of Oklahoma are as shown in Table 2.1.

Table 2.1 The top five Oklahoma state officials

Governor (G)
Lieutenant Governor (L)
Secretary of State (S)
Attorney General (A)
Treasurer (T)

Suppose that we randomly sample, without replacement, two officials from the five officials. Find the sample space for this random experiment.

Solution For convenience, we use the letters in parentheses after the officials in Table 2.1 to represent the officials. Each possible outcome for this random experiment consists of a sample of two officials from the five. There are 10 such samples, and these 10 samples constitute the sample space:

$$\Omega = \{ \{G, L\}, \{G, S\}, \{G, A\}, \{G, T\}, \{L, S\}, \{L, A\}, \{L, T\}, \{S, A\}, \{S, T\}, \{A, T\} \},$$

where, for instance, $\{G, L\}$ represents the outcome that the two officials selected are the governor and lieutenant governor. ∎

In general, we want the sample space to consist precisely of all possible outcomes of the random experiment. However, in practice, knowing exactly the possible outcomes is often difficult, as illustrated by Example 2.2. Thus we permit as a sample space any set containing at least all possible outcomes of the random experiment.

It is best if the sample space is taken to be the set of all possible outcomes or, if that is not feasible, to be the smallest set that we are certain contains all possible outcomes. In any case, it is essential that all possible outcomes be included as elements of the sample space; not all elements of the sample space have to be possible outcomes.

Events

In Section 1.1, we defined an *event* intuitively as some specified result that may or may not occur when a random experiment is performed. For mathematical purposes, we need to be more precise.

The assumption is that any specified "event" will either occur or not occur when the random experiment is performed. So each event E can be considered a subset of the sample space—namely, the collection of outcomes that satisfy the conditions for the occurrence of E. Using this identification between events and sets, an event E *occurs* if and only if the outcome, ω, of the random experiment is an element of E—that is, $\omega \in E$.

DEFINITION 2.2 Event; Occurrence of an Event

An **event** is a subset of the sample space.[†] We say that an event **occurs** if and only if the outcome of the random experiment is an element of the event.

Note: Generally, we use uppercase English letters near the beginning of the alphabet to represent events.

[†]Technically, it is not always possible to construct a mathematical model for a random experiment in which every subset of the sample space is an event. However, for our purposes, it suffices to assume that every subset of the sample space is an event (and vice versa). More generally, see the advanced exercises in this section for a discussion of the conditions that the collection of all events must satisfy.

EXAMPLE 2.8   *Events*

Coin Tossing   Refer to Example 2.1 on page 26 where we toss a coin three times and observe the result (a head or a tail) for each toss.
a) Determine, as a subset of the sample space, the event that the total number of heads is two.
b) Identify, in words, the event {HHH, HTH}.

Solution a) The event is represented as the set {HHT, HTH, THH}.
b) This event is that the first and third tosses are heads.   ■

EXAMPLE 2.9   *Events*

Emergency Room Traffic   Refer to Example 2.2 on page 27 where, starting at 6:00 P.M., we observe the elapsed time, in hours, until the first patient arrives at a particular emergency room.
a) Determine, as a subset of the sample space, the event that the first patient arrives before 6:15 P.M.
b) Identify, in words, the event [1/12, 1/6].

Solution a) The event is represented as the set [0, 1/4).
b) This event is that the first patient arrives at the emergency room between 6:05 P.M. and 6:10 P.M., inclusive.   ■

EXAMPLE 2.10   *Events*

Rolling Dice   Refer to Example 2.3 on page 27 where two dice are rolled.
a) Determine the event that the sum of the dice is 5.
b) Identify the event $\{(x, x) : x \in \{1, 2, 3, 4, 5, 6\}\}$.

Solution a) The event is represented as the set $\{(1, 4), (2, 3), (3, 2), (4, 1)\}$ or, more compactly, as $\{(x, y) \in \Omega : x + y = 5\}$.
b) This event is that doubles are rolled; that is, both dice come up the same number.   ■

EXAMPLE 2.11   *Events*

Component Analysis   Refer to Example 2.4 on page 27 where we observe a unit that consists of five components and determine, at a specified time, which components are working (s) and which have failed (f).
a) Determine the event that at most one of the components is not working.
b) Identify the event $\{(f, f, f, f, f)\}$.

Solution a) That at most one of the components is not working means that all five components are working or exactly one of the five is not working. Thus the event is given by
$\{(s, s, s, s, s), (s, s, s, s, f), (s, s, s, f, s), (s, s, f, s, s), (s, f, s, s, s), (f, s, s, s, s)\}$.
b) This event is that none of the components are working.   ■

EXAMPLE 2.12   *Events*

Blood Type   Refer to Example 2.5 on page 28 regarding blood types of human beings. The actual blood type of a human being is categorized as A, B, AB, or O, depending on his or her genotype. Table 2.2 shows the relationship between genotype and blood type.

Table 2.2 Genotype and blood type

Genotype	*aa*	*ab*	*ao*	*bb*	*bo*	*oo*
Blood type	A	AB	A	B	B	O

Suppose that a person is selected at random and that his or her genotype is observed.
a) Determine the event that the person has type B blood.
b) Identify, in words, the event $\{aa, ao, bb, bo\}$.

Solution   a) The event that the person has type B blood is $\{bb, bo\}$.
b) This event is that the person has either type A or type B blood.   ■

EXAMPLE 2.13   *Events*

Bacteria on a Petri Dish   Refer to Example 2.6 on page 28 where a petri dish of unit radius, containing nutrients upon which bacteria can multiply, is smeared with a uniform suspension of bacteria. Suppose that we observe the location of the center of the first spot (visible bacteria colony) to appear.
a) Find the event that the center of the first spot is more than 1/4 unit from the center of the petri dish.
b) Identify the event $\{(x, y) \in \Omega : x > 0\}$.

Solution   a) The event that the center of the first spot is more than 1/4 unit from the center of the petri dish is $\left\{(x, y) : 1/4 < \sqrt{x^2 + y^2} < 1\right\}$.
b) This event is that the center of the first spot is on the right half of the petri dish.   ■

EXAMPLE 2.14   *Events*

Oklahoma State Officials   Refer to Example 2.7 on page 28 where we randomly sample, without replacement, two officials from the top five Oklahoma state officials.
a) Determine the event that the attorney general is included in the sample.
b) Identify the event $\{\{L, S\}, \{L, A\}, \{S, A\}\}$.

Solution   a) The event is represented as the set $\{\{G, A\}, \{L, A\}, \{S, A\}, \{A, T\}\}$.
b) This event is that neither the governor nor the treasurer is included in the sample.   ■

Events and Set Operations

Now that we have shown how events naturally correspond to subsets of the sample space, we need to associate relationships among events with the set operations of complementation, intersection, and union. As preliminary to that association, we examine in Proposition 2.1 the meaning of the subset relation with respect to the occurrence of events.

◆◆◆   **Proposition 2.1**   **Subsets and Events**

$A \subset B$ *if and only if event B occurs whenever event A occurs.*

Proof   Suppose that $A \subset B$. If event A occurs, the outcome, ω, of the random experiment is an element of A—that is, $\omega \in A$. Because $A \subset B$, we then have $\omega \in B$—that is, event B occurs. Thus event B occurs whenever event A occurs.

Conversely, suppose that event B occurs whenever event A occurs. Let $\omega \in A$. If the outcome of the random experiment is ω, event A occurs and hence, by assumption, so does event B. This, in turn, implies that $\omega \in B$. Thus $A \subset B$.   ◆

Now let's consider complementation. Suppose that E is an event—that is, a subset of the sample space, Ω. Recall that event E occurs if and only if the outcome, ω, of the random experiment is an element of E—that is, $\omega \in E$. Equivalently, E does not occur if and only if the outcome, ω, of the random experiment isn't an element of E—that is, $\omega \in E^c$. Thus event E^c corresponds to the nonoccurrence of event E or, more simply, E^c is the event that E does not occur.

Next we consider intersection. Suppose that A and B are events. Both events occur when the random experiment is performed if and only if the outcome, ω, is an element of both A and B—that is, $\omega \in A \cap B$. Thus event $A \cap B$ corresponds to the occurrence of both event A and event B or, more simply, $A \cap B$ is the event that both A and B occur.

Now we go to unions. Again, suppose that A and B are events. At least one of these events occurs when the random experiment is performed if and only if the outcome, ω, is an element of either A or B or both—that is, $\omega \in A \cup B$. Thus event $A \cup B$ corresponds to the occurrence of at least one of event A and event B or, more simply, $A \cup B$ is the event that either A or B occurs.

We summarize the preceding discussion of set operations and relationships among events in Table 2.3.

Table 2.3 Correspondence between set operations and relationships among events

Operation	Event	Description
Complement	E^c	E does not occur.
Intersection	$A \cap B$	Both A and B occur.
Union	$A \cup B$	Either A or B or both occur.

EXAMPLE 2.15   *Events and Set Operations*

Coin Tossing   Refer to Example 2.1 on page 26 where we toss a coin three times and observe the result (a head or a tail) for each toss. Recall that the sample space for this random experiment is {HHH, HHT, HTH, HTT, THH, THT, TTH, TTT}. Let

$$A = \text{event exactly one head is tossed,}$$
$$B = \text{event the first two tosses are tails, and}$$
$$C = \text{event all three tosses come up tails.}$$

Determine and describe in words each of the following events.
a) B^c b) $A \cap B$ c) $A \cup C$ d) $A \cap C$

Solution   a) As $B = \{\text{TTH, TTT}\}$, we have $B^c = \{\text{HHH, HHT, HTH, HTT, THH, THT}\}$. Hence B^c is the event that not both the first two tosses are tails; that is, at least one of the first two tosses is heads.

b) We have $A = \{\text{HTT, THT, TTH}\}$ and $B = \{\text{TTH, TTT}\}$, so $A \cap B = \{\text{TTH}\}$. Hence $A \cap B$ is the event that the first two tosses are tails and the third is a head.

c) We have $A = \{\text{HTT, THT, TTH}\}$ and $C = \{\text{TTT}\}$, so $A \cup C = \{\text{HTT, THT, TTH, TTT}\}$. Thus $A \cup C$ is the event that two or more tails are tossed.

d) Referring to part (c), we find that $A \cap C = \emptyset$. This result reflects the fact that it isn't possible to toss a coin three times and get exactly one head and all three tails.   ■

EXAMPLE 2.16   *Events and Set Operations*

Emergency Room Traffic   Refer to Example 2.2 on page 27 where, starting at 6:00 P.M., we observe the elapsed time, in hours, until the first patient arrives at a particular emergency room. Recall that the sample space is $[0, \infty)$. Let

$$A = \text{event the first patient arrives before 6:15 P.M. and}$$
$$B = \text{event the first patient arrives before 6:30 P.M.}$$

Determine and describe in words each of the following events.
a) A^c b) $A \cap B$ c) $A \cup B$ d) $A^c \cap B$ e) $A \cap B^c$

Solution   a) Here $A = [0, 1/4)$, so $A^c = [1/4, \infty)$. In words, A^c is the event that the first patient arrives at or after 6:15 P.M.

b) We first note that $B = [0, 1/2)$, Now referring to the result in part (a), we conclude that $A \cap B = [0, 1/4)$, which is the same as event A. From a set theory point of view, we get this result because $A \subset B$. It also makes sense from an event perspective because the event that the first patient arrives before 6:15 P.M. and before 6:30 P.M. is the same as the event that the first patient arrives before 6:15 P.M.

c) Referring to parts (a) and (b), we observe that $A \cup B = [0, 1/2)$, which is the same as event B. From a set theory point of view, we get this result because $A \subset B$. It also makes sense from an event perspective because the event that the first patient arrives either before 6:15 P.M. or before 6:30 P.M. is the same as the event that the first patient arrives before 6:30 P.M.

d) From parts (a) and (b), we find that $A^c = [1/4, \infty)$ and $B = [0, 1/2)$. Consequently, $A^c \cap B = [1/4, 1/2)$. In words, $A^c \cap B$ is the event that the first patient arrives at or after 6:15 P.M. but before 6:30 P.M.

e) Here $A = [0, 1/4)$ and $B^c = [1/2, \infty)$, so $A \cap B^c = \emptyset$. This result reflects the fact that it is impossible for the first patient to arrive before 6:15 P.M. and at or after 6:30 P.M.   ■

Mutually Exclusive Events

In Section 1.2, we discussed the concepts of disjoint sets and pairwise disjoint sets. These concepts are also important in probability theory, but the terminology used is different—namely, the term "mutually exclusive" is used in place of "disjoint."

DEFINITION 2.3   Mutually Exclusive Events

Events A and B are said to be **mutually exclusive** if they cannot both occur when the random experiment is performed. In terms of sets this means that A and B are disjoint—that is, $A \cap B = \emptyset$. More generally, events $A_1, A_2, \ldots$ are said to be **pairwise mutually exclusive** if no two of them can occur when the random experiment is performed. In terms of sets this means that $A_1, A_2, \ldots$ are pairwise disjoint—that is, $A_m \cap A_n = \emptyset$ whenever $m \neq n$.

Note: The word "pairwise" is commonly omitted from "pairwise mutually exclusive." We follow that convention. In other words, when we say "mutually exclusive," we mean "pairwise mutually exclusive" unless specifically stated otherwise.

The events A and C in Example 2.15 are mutually exclusive, as are the events A and B^c in Example 2.16. From those examples, we see that the empty set, $\emptyset$, corresponds to an event that cannot occur, which is called an **impossible event.** Keeping that and Definition 2.3 in mind, we can think of *mutually exclusive events* in various equivalent ways.

- Event A and event B cannot both occur.
- Sets A and B are disjoint.
- Event A and event B have no common outcomes.
- At most one of event A and event B can occur.
- Event $A \cap B$ is an impossible event.

Similar comments hold for a collection of more than two mutually exclusive events.

EXAMPLE 2.17   *Mutually Exclusive Events*

Playing Cards   A deck of playing cards contains 52 cards, as illustrated in Figure 2.2. When we perform the experiment of randomly selecting one card from the deck, exactly one of the 52 cards will be obtained. The collection of all 52 cards—the possible outcomes—is the sample space for this random experiment. Let

A = event the card selected is a heart,

B = event the card selected is a face card,

C = event the card selected is an ace,

D = event the card selected is an 8, and

E = event the card selected is a 10 or a jack.

Decide which of the following collections of events are mutually exclusive.
a) A and B b) A and C c) B and C
d) B, C, and D e) B, C, D, and E

Figure 2.2 Possible outcomes for selecting a card at random

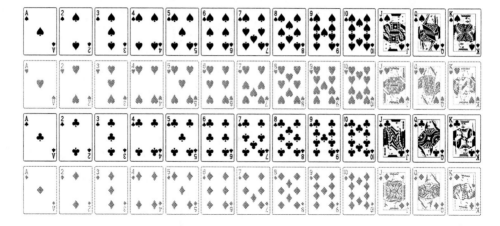

Solution a) Event A and event B are not mutually exclusive because they have the common outcomes "king of hearts," "queen of hearts," and "jack of hearts." Both events occur if the card selected is the king, queen, or jack of hearts.

b) Event A and event C are not mutually exclusive; they have the common outcome "ace of hearts." Both events occur if the card selected is the ace of hearts.

c) Event B and event C are mutually exclusive because they have no common outcomes. They cannot both occur when the experiment is performed because selecting a card that is both a face card and an ace is impossible.

d) Events B, C, and D are mutually exclusive because no two of them can occur simultaneously.

e) Events B, C, D, and E are not mutually exclusive because event B and event E both occur if the card selected is a jack. ■

Basic Exercises

2.1 Construct a Venn diagram representing each event and interpret the event in words.
a) E^c **b)** $A \cup B$ **c)** $A \cap B$ **d)** $A \cap B \cap C$ **e)** $A \cup B \cup C$ **f)** $A^c \cap B$

2.2 Answer true or false to each statement and give reasons for your answers.
a) If event A and event B are mutually exclusive, so are events A, B, and C for every event C.
b) If event A and event B are not mutually exclusive, neither are events A, B, and C for every event C.

2.3 Suppose that, in the petri-dish illustration of Example 2.6 on page 28, you can't observe the location of the spot but you can observe its distance from the center of the petri dish.
a) Determine the sample space for this random experiment.
b) Find, as a subset of the sample space, the event that the spot is between 1/4 and 1/2 unit, inclusive, from the center of the petri dish.
c) Describe, in words, the event $[0, 1/3]$.

2.4 Suppose that one die is rolled and that you observe the number of dots facing up.
a) Obtain a sample space for this random experiment whose elements are integers.
b) Determine as a subset of the sample space each of the events $A =$ die comes up even, $B =$ die comes up at least 4, $C =$ die comes up at most 2, and $D =$ die comes up 3.
c) Determine as a subset of the sample space and describe in words each of the events A^c, $A \cap B$, and $B \cup C$.
d) Determine which of the following collections of events are mutually exclusive and explain your answers: A and B; B and C; A, C, and D.
e) Are there three mutually exclusive events among A, B, C, and D? four?
f) Describe in words each of the events $\{5\}$, $\{1, 3, 5\}$, and $\{1, 2, 3, 4\}$.

2.5 Recent research by Schidt et al. (*African Entomology*, 1999, 7, pp. 107–112) describes the effectiveness of a seed-eating weevil on the population control of a nonnative, invasive species of tree in South Africa, called *Paraserianates lophantha*. A frequency distribution of percent seed damage caused by the weevil for 39 trees is provided in the following table. (The notation $a \leftarrow b$ is shorthand for "a up to, but not including b.")

Percent seed damage	Number of trees	Percent seed damage	Number of trees
0 ← 10	19	40 ← 50	6
10 ← 20	2	50 ← 60	2
20 ← 30	5	60 ← 70	2
30 ← 40	3		

Suppose that one of these 39 trees is selected at random. Let $A =$ event that the tree has less than 40% seed damage, $B =$ event that the tree has at least 20% seed damage, $C =$ event that the tree has at least 30% but less than 60% seed damage, and $D =$ event that the tree has at least 50% seed damage.
a) What is the sample space for this random experiment?
Describe each of the following events in words and determine the number of outcomes (trees) that comprise each event.
b) B^c **c)** $C \cap D$ **d)** $A \cup D$ **e)** C^c **f)** $A \cap D$
g) Among the events A, B, C, and D, identify the collections that are mutually exclusive.

2.6 An urn contains 10 balls, numbered 0, 1, 2, ..., 9. Three balls are removed, one at a time, without replacement.
a) Obtain the sample space for this random experiment.
b) Determine, as a subset of the sample space, the event that an even number of odd-numbered balls are removed from the urn.

2.7 Refer to Example 2.3 on page 27 where two dice are rolled, one black and one gray. For $i = 2, 3, \ldots, 12$, determine explicitly as a subset of the sample space the event A_i that the sum of the faces is i.

2.8 Consider the following random experiment: First a die is rolled and you observe the number of dots facing up; then a coin is tossed the number of times that the die shows and you observe the total number of heads.
a) Determine the sample space for this random experiment.
b) Determine the event that the total number of heads is even.

2.9 George and Laura take turns tossing a coin. The first person to get a tail wins. George goes first. *Note:* You may assume that eventually a tail will be tossed.
a) Describe the sample space for this random experiment.
b) Determine, as a subset of the sample space, the event that Laura wins.

2.10 This exercise considers two random experiments involving the repeated tossing of a coin. *Note:* You may assume that eventually a head will be tossed.
a) If the coin is tossed until the first time a head appears, find the sample space.
b) If the coin is tossed until the second time a head appears, find the sample space.
c) For the experiment in part (a), express the event that the coin is tossed exactly six times in the form $\{\ldots\}$, where in place of "..." you list all of the outcomes in that event.
d) Repeat part (c) for the experiment described in part (b).

2.11 From 10 men and 8 women in a pool of potential jurors, 12 are chosen at random to constitute a jury. Suppose that you observe the number of men who are chosen for the jury. Let A be the event that at least half of the 12 jurors are men and let B be the event that at least half of the 8 women are on the jury.
a) Determine the sample space for this random experiment.
b) Find $A \cup B$, $A \cap B$, and $A \cap B^c$, listing all the outcomes for each of those three events.
c) Are A and B mutually exclusive? A and B^c? A^c and B^c? Explain your answers.

2.12 Let A and B be events of a sample space.
a) Show that, if A and B^c are mutually exclusive, then B occurs whenever A occurs.
b) Show that, if B occurs whenever A occurs, then A and B^c are mutually exclusive.

2.13 Let A, B, and C be events of a sample space. Write a mathematical expression for each of the following events.
a) A occurs, but B doesn't occur.   b) Exactly one of A and B occurs.
c) Exactly one of A, B, and C occurs.   d) At most two of A, B, and C occur.

2.14 Refer to Example 2.17 on page 34, but now suppose that two cards are selected at random, one after the other, without replacement.
a) What is Ω for this random experiment?
b) Let A be the event that at least one of the cards is a face card and let B be the event that at least one of the cards is an ace. Are A and B mutually exclusive? Why or why not?

2.15 Refer to the emergency-room-traffic illustration in Example 2.2 on page 27.

a) Suppose that you observe the elapsed time in hours from 6:00 P.M. until the arrival of the fourth patient but that you don't know when the first three patients arrived (beyond the fact that they arrived before the fourth). Determine the sample space.

b) Suppose that you observe the number of patients who arrive during the first half hour but that you can't tell precisely when they arrived. Determine the sample space.

c) Suppose that you observe the times of arrival of all patients who arrive after 6:00 P.M. Let A be the event that the number of patients who arrive before 6:30 P.M. is five or more, and let B be the event that the fourth patient arrives before 6:30 P.M. Express a simple relationship between A and B by using notation introduced in this section.

d) Determine the sample space in part (c) and write it in the form $\{(\cdots) : \cdots \}$, using mathematical notation and no words.

A researcher named Smith is investigating the quality of treatment given to patients with sprained ankles. Smith waits in the emergency room until the admission of the first patient with a sprained ankle; an inspector named Jones has the job of observing the efficiency with which the first 10 patients are admitted; Jones remains in the emergency room until the admission of the tenth patient. Let A be the event that the first 9 patients don't have sprained ankles; let B be the event that Smith leaves no later than Jones leaves; let C be the event that Jones leaves no later than Smith leaves. Classify each of the following statements as either true or false and explain how you know.

e) A and B are mutually exclusive.   f) A and C are mutually exclusive.

g) B and C are mutually exclusive.   h) A occurs whenever B occurs.

i) B occurs whenever A occurs.   j) A occurs whenever C occurs.

k) C occurs whenever A occurs.   l) B occurs whenever C occurs.

m) C occurs whenever B occurs.

2.16 Suppose that $A_1, A_2, \ldots$ are events of a sample space such that $A_n \subseteq A_{n+1}$ for all $n \in \mathcal{N}$. Show that $A_2 \cap A_1^c$, $A_3 \cap A_2^c$, $A_4 \cap A_3^c$, $\ldots$ are pairwise mutually exclusive.

Advanced Exercises

2.17 Let Ω be the sample space for a random experiment.

a) List all events if Ω has just two elements—say, $\Omega = \{a, b\}$. Include $\emptyset$ and Ω in your list.

b) List all events if $\Omega = \{a, b, c\}$.

c) List all events if $\Omega = \{a, b, c, d\}$.

d) In general, how many events are there if Ω contains exactly n elements?

2.18 Let $A_1, A_2, \ldots$ be events of a sample space and set $A^* = \bigcap_{n=1}^{\infty} \left(\bigcup_{k=n}^{\infty} A_k \right)$. Verify each of the following statements.

a) If $A_2, A_4, A_6, A_8, \ldots$ occur and $A_1, A_3, A_5, A_7, \ldots$ fail to occur, then A^* occurs.

b) If $A_1, A_3, A_5, A_7, \ldots$ occur and $A_2, A_4, A_6, A_8, \ldots$ fail to occur, then A^* occurs.

c) If $A_{10}, A_{100}, A_{1000}, \ldots, A_{10^n}, \ldots$ occur and A_n fails to occur for any other value of n (i.e., for any n not a power of 10), then A^* occurs.

d) If A_{86}, A_{2049}, and $A_{30498541}$ occur and A_n fails to occur for each positive integer n such that $n \notin \{86, 2049, 30498541\}$, then A^* fails to occur.

e) Event A^* occurs if and only if $\{n \in \mathcal{N} : A_n \text{ occurs}\}$ is infinite. Interpret this result.

2.19 σ-Algebra of events: As we mentioned in the footnote on page 29, constructing a mathematical model for a random experiment in which every subset of the sample space is an event isn't always possible. Nonetheless, to work with events mathematically requires that they be closed under the basic set operations of complementation, intersection, and union.

More precisely, the collection $\mathcal{A}$ of events is required to be a σ-*algebra* (or σ-*field*) of subsets of the sample space Ω—that is, $\mathcal{A}$ must satisfy the following conditions: (1) if $E \in \mathcal{A}$, then $E^c \in \mathcal{A}$; (2) if $A_n \in \mathcal{A}$, $n = 1, 2, \ldots$, then $\bigcap_n A_n \in \mathcal{A}$; and (3) if $A_n \in \mathcal{A}$, $n = 1, 2, \ldots$, then $\bigcup_n A_n \in \mathcal{A}$.

a) Prove that conditions (1) and (2) of a σ-algebra imply condition (3).

b) Explain why the collection of all subsets of Ω is a σ-algebra.

c) Show that $\{\Omega, \emptyset\}$ is a σ-algebra.

d) Show that $\{\Omega, E, E^c, \emptyset\}$, where $E \neq \emptyset$ and $E \neq \Omega$, is a σ-algebra.

e) Suppose that $E_1, E_2, \ldots$ is a countable collection of mutually exclusive events and that $\bigcup_n E_n = \Omega$. Let $\mathcal{D}$ be the collection of all countable (including empty) unions of $E_1, E_2, \ldots$. Prove that $\mathcal{D}$ is a σ-algebra.

f) Assume that Ω is an infinite set and let $\mathcal{D} = \{ E \subset \Omega : \text{either } E \text{ or } E^c \text{ is finite} \}$. Is $\mathcal{D}$ a σ-algebra? Explain your answer.

2.2   Axioms of Probability

As we mentioned earlier, there are two interrelated aspects of a mathematical model for random experiments and probability:

- the representation of events by sets and
- the axioms specifying the basic properties that probabilities must satisfy.

We discussed the first aspect in Section 2.1. Now, in this section, we examine the second aspect, thus completing the mathematical model for random experiments and probability.

In Section 1.1, we introduced the frequentist interpretation of probability, which construes the probability of an event to be the long-run proportion of times that the event occurs in independent repetitions of the random experiment. More formally, let E be an event and $P(E)$ its probability. For n independent repetitions of the random experiment, let $n(E)$ denote the number of times that event E occurs. The frequentist interpretation is that, for large n, the proportion of times that event E occurs will approximately equal the probability of event E:

$$\frac{n(E)}{n} \approx P(E), \qquad \text{for large } n. \tag{2.1}$$

As we noted in Section 1.1, although all attempts to use the frequentist interpretation as a definition of probability have failed, it is invaluable in the axiomatic development of probability theory. Specifically, we use the frequentist interpretation of probability as an intuitive basis for identifying the basic properties that probabilities must satisfy.

The Kolmogorov Axioms

In geometry, certain concepts, such as that of a point, are not defined but, rather, are taken as primary, undefined, and underived. Likewise, in the axiomatic treatment of probability, we consider probabilities primitive and focus on the conditions they must meet. How these probabilities are determined is another matter, one that we take up in Section 2.3.

In other words, for the axiomatic development of probability theory, we are concerned with the properties that probabilities must satisfy, not with how probabilities are specified or obtained. Consequently, we assume that, corresponding to each event E, is a number $P(E)$ representing the probability that event E occurs. Mathematically, P is a real-valued function defined on the collection of all events. We now develop the basic properties that probabilities must meet—that is, the conditions that P must satisfy.

Suppose that the random experiment under consideration is repeated a large number of times—say, n. As $n(E)$ denotes the number of times that event E occurs, it is a nonnegative integer. Therefore $n(E)/n \geq 0$ and, consequently, in view of Relation (2.1), we require that

$$P(E) \geq 0, \qquad \text{for each event } E. \tag{2.2}$$

In other words, probabilities should be nonnegative, an obvious restriction for numbers measuring likelihood.

Next we note that, because the sample space, Ω, contains all possible outcomes of the random experiment, it must occur every time the random experiment is performed. Hence $n(\Omega)/n = 1$, which means, in view of Relation (2.1), that we should have

$$P(\Omega) = 1. \tag{2.3}$$

This obvious condition reflects the fact that the probability is 1 (100%) that something happens when the random experiment is performed.

Now suppose that A and B are mutually exclusive events. Then the number of times either event A or event B occurs in the n trials equals the sum of the number of times event A occurs and the number of times event B occurs—that is, $n(A \cup B) = n(A) + n(B)$. Therefore, by Relation (2.1),

$$P(A \cup B) \approx \frac{n(A \cup B)}{n} = \frac{n(A) + n(B)}{n} = \frac{n(A)}{n} + \frac{n(B)}{n} \approx P(A) + P(B).$$

Consequently, for mutually exclusive events, A and B, we should have

$$P(A \cup B) = P(A) + P(B). \tag{2.4}$$

In words, the probability that either event A or event B occurs must equal the sum of their probabilities. We can also demonstrate the reasonableness of Equation (2.4) by referring to the Venn diagram shown in Figure 2.3.

Figure 2.3 Two mutually exclusive events

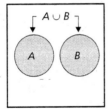

If we think of the shaded regions as probabilities, the shaded disk on the left is $P(A)$, the shaded disk on the right is $P(B)$, and the total shaded region is $P(A \cup B)$. Because event A and event B are mutually exclusive, the total shaded region equals the sum of the two shaded disks—that is, $P(A \cup B) = P(A) + P(B)$.

For mathematical reasons, we impose a stronger condition on probabilities than that given by Equation (2.4). We require that probabilities add not only for any two mutually exclusive events, but also for any countable collection of mutually exclusive events. In other words, if $A_1, A_2, \ldots$ are mutually exclusive, we require that

$$P\left(\bigcup_n A_n\right) = \sum_n P(A_n).^\dagger \qquad (2.5)$$

In summary, probabilities must satisfy the conditions given in Relations (2.2), (2.3), and (2.5). Remarkably, these three simple and intuitively obvious conditions for probabilities are all that we need in order to develop the entire theory of probability. We refer to these three conditions individually as the **nonnegativity axiom,** the **certainty axiom,** and the **additivity axiom,** respectively. Collectively they are known as the **Kolmogorov axioms** (in honor of the mathematician who suggested their use). A function P, defined on the events of a sample space, that satisfies the Kolmogorov axioms is called a **probability measure.**

DEFINITION 2.4   Kolmogorov Axioms for a Probability Measure

Let Ω be the sample space for a random experiment. A function P defined on the events of Ω is called a **probability measure** if it satisfies the following three conditions.

a) *Nonnegativity:* $P(E) \geq 0$ for each event E.

b) *Certainty:* $P(\Omega) = 1$.

c) *Additivity:* If $A_1, A_2, \ldots$ is a countable collection of mutually exclusive events, then

$$P\left(\bigcup_n A_n\right) = \sum_n P(A_n).$$

From a mathematical or probabilistic point of view, a function P defined on the events of a sample space is acceptable for specifying probabilities provided only that it is a probability measure—that is, it satisfies the three Kolmogorov axioms. From a practical point of view, however, we want the specification of probabilities to reflect reality. Otherwise, even though the probabilities may be mathematically correct, the resulting probability model will be of little use in applications.

---

†Using mathematical induction, we can show that Equation (2.4) implies that Equation (2.5) holds for all finite collections of mutually exclusive events. However, it doesn't imply that for countably infinite collections.

In Section 2.3, we discuss methods for specifying and determining probabilities in ways that reflect the underlying realities of the random experiment. But, for now, we concentrate on understanding the mathematical requirements for probabilities.

Countable Sample Spaces

As a first application of the Kolmogorov axioms, we consider **countable sample spaces,** sample spaces arising from random experiments that have a countable (finite or countably infinite) number of possible outcomes. Such sample spaces are important in probability theory and its applications because they occur so frequently in practice. Of particular significance is the special case of **finite sample spaces,** sample spaces arising from random experiments that have only a finite number of possible outcomes.

To begin, we adopt a convention when referring to probabilities of events that consist of a single outcome. Recall that each event is a subset of the sample space and, as such, is a collection of outcomes. An event consisting of only one outcome is called a **simple event.**

For instance, consider the random experiment of rolling a die once and observing the number of dots facing up. A sample space for this experiment is $\{1, 2, 3, 4, 5, 6\}$. Here there are six simple events, one corresponding to each of the six possible outcomes. Thus, $\{2\}$ is the simple event corresponding to the outcome 2—it is the event that the die comes up 2. Note the difference between $\{2\}$ and 2: The first is the event that the die comes up 2, whereas the second is the outcome 2 itself.

Although a simple event and the outcome of which it consists are technically different, it is natural to identify the two. In particular, we often use the phrase "the probability of the outcome" when we really mean the probability of the simple event corresponding to that outcome. This slight abuse of terminology is generally considered acceptable as it provides a less awkward way of referring to probabilities of simple events.

Probability of an Outcome

When we refer to the probability of an outcome ω, we really mean the probability of the simple event $\{\omega\}$ corresponding to that outcome.

We now show that, for a countable sample space, the probabilities of the individual outcomes determine the probabilities of all events. Specifically, we have Proposition 2.2.

◆◆◆   **Proposition 2.2   Probabilities for Countable Sample Spaces**

Let Ω be a countable sample space. Then, for each event E, we have

$$P(E) = \sum_{\omega \in E} P(\{\omega\}).$$

In words, the probability of an event equals the sum of the probabilities of the outcomes that constitute the event.

Proof   Let E be an event—say, $E = \{\omega_1, \omega_2, \ldots\}$, where $\omega_1, \omega_2, \ldots$ may be a finite or infinite sequence. Noting that distinct simple events are mutually exclusive and applying the additivity axiom (third Kolmogorov axiom), we get

$$P(E) = P\left(\bigcup_n \{\omega_n\}\right) = \sum_n P(\{\omega_n\}) = \sum_{\omega \in E} P(\{\omega\}),$$

as required.   ◆

Consequently, when considering countable sample spaces, you need only know the probabilities of the individual outcomes or, more precisely, of the simple events. In view of Proposition 2.2, those probabilities determine the probability of every event.

Applying Proposition 2.2 with $E = \Omega$, we obtain Corollary 2.1, which provides a necessary condition on the probabilities of the individual outcomes of a countable sample space.

◆◆◆   Corollary 2.1

Let Ω be a countable sample space. Then

$$\sum_{\omega \in \Omega} P(\{\omega\}) = 1.$$

In words, the probabilities of the individual outcomes must sum to 1.

Proof   The certainty axiom (second Kolmogorov axiom) requires that $P(\Omega) = 1$. Using that equality and Proposition 2.2 with $E = \Omega$, we find that

$$1 = P(\Omega) = \sum_{\omega \in \Omega} P(\{\omega\}),$$

as required.   ◆

The nonnegativity axiom (first Kolmogorov axiom) implies that the probability of each individual outcome must be a nonnegative number: $P(\{\omega\}) \geq 0$ for each $\omega \in \Omega$. And, for a countable sample space, Corollary 2.1 reveals that the sum of those probabilities must equal 1. Now the question is whether, for a countable sample space, those two conditions determine a legitimate probability assignment—that is, a probability measure. The answer is *yes,* as Proposition 2.3 demonstrates.

◆◆◆   Proposition 2.3

Let Ω be a countable sample space—say, $\Omega = \{\omega_1, \omega_2, \ldots\}$. Suppose that $p_1, p_2, \ldots$ is a sequence of real numbers that satisfies
a) $p_n \geq 0$ for each n.
b) $\sum_n p_n = 1$.
Then there is a unique probability measure P on the events of Ω such that $P(\{\omega_n\}) = p_n$ for all n.

Proof   In view of Proposition 2.2, if the required probability measure exists, it must be given as

$$P(E) = \sum_{\omega_n \in E} p_n. \tag{2.6}$$

Thus we need only show that Equation (2.6) actually defines a probability measure—in other words, that it satisfies the three Kolmogorov axioms.

Because $p_n \geq 0$ for each n, it follows from Equation (2.6) that $P(E) \geq 0$ for each event E, satisfying the nonnegativity axiom (first Kolmogorov axiom). And, because $\sum_n p_n = 1$, it follows from Equation (2.6) that $P(\Omega) = 1$, satisfying the certainty axiom (second Kolmogorov axiom).

It remains to verify the additivity axiom (third Kolmogorov axiom). Let $A_1, A_2, \ldots$ be mutually exclusive events. We can write $A_1 = \{\omega_{11}, \omega_{12}, \ldots\}$, $A_2 = \{\omega_{21}, \omega_{22}, \ldots\}$, and so on. Then, by Equation (2.6), we have

$$P(A_n) = \sum_k p_{nk}, \qquad \text{for each } n. \tag{2.7}$$

However, as $A_1, A_2, \ldots$ are mutually exclusive, we also know that

$$\bigcup_n A_n = \{\omega_{11}, \omega_{12}, \ldots, \omega_{21}, \omega_{22}, \ldots\}$$

and, consequently, by Equation (2.6),

$$P\left(\bigcup_n A_n\right) = p_{11} + p_{12} + \cdots + p_{21} + p_{22} + \cdots$$

$$= \underbrace{p_{11} + p_{12} + \cdots}_{P(A_1)} + \underbrace{p_{21} + p_{22} + \cdots}_{P(A_2)} + \cdots = \sum_n P(A_n),$$

as required.   ◆

EXAMPLE 2.18   *Probabilities for Countable Sample Spaces*

Coin Tossing   Consider again the random experiment of tossing a coin three times and observing the result (a head or a tail) for each toss. Recall that the sample space for this random experiment is $\Omega = \{$HHH, HHT, HTH, HTT, THH, THT, TTH, TTT$\}$, where, for instance, HTT denotes the outcome of a head on the first toss and tails on the second and third tosses. Table 2.4 provides six different potential probability assignments to the outcomes. (Don't be concerned right now about how we arrived at these assignments.)

Table 2.4 Six potential probability assignments

Outcome	#1	#2	#3	#4	#5	#6
HHH	0.125	1	0.008	0.135	0.375	0.008
HHT	0.125	0	0.032	0.220	0.375	0.096
HTH	0.125	0	0.032	0.050	0.375	−0.032
HTT	0.125	0	0.128	0.110	0.375	0.128
THH	0.125	0	0.032	0.110	0.375	0.032
THT	0.125	0	0.128	0.235	0.375	0.128
TTH	0.125	0	0.128	0.000	0.375	0.128
TTT	0.125	0	0.512	0.140	0.375	0.512

a) Which of assignments #1–#6 are legitimate probability assignments?

b) Determine the probability of obtaining exactly two heads by using each of the legitimate probability assignments.

Solution a) In view of Proposition 2.3, we need only check each assignment for nonnegativity and summing to 1. Doing so, we find that assignments #1–#4 are legitimate probability assignments. Assignment #5 consists of nonnegative numbers, but those numbers don't sum to 1; therefore this assignment is not legitimate. Regarding assignment #6, although the numbers sum to 1, not all of them are nonnegative, so this assignment is also not legitimate. (Again, just because assignments #1–#4 are each legitimate from a probabilistic point of view, we can't conclude that each assignment is reasonable from a practical point of view. At most one—and perhaps none—of those assignments reflects the true nature of the coin and the random experiment.)

b) The event of obtaining exactly two heads is $E = \{\text{HHT, HTH, THH}\}$. Applying Proposition 2.2 on page 42, we have

$$P(E) = P(\{\text{HHT, HTH, THH}\}) = P(\{\text{HHT}\}) + P(\{\text{HTH}\}) + P(\{\text{THH}\}).$$

Referring to Table 2.4, we can now determine the probability of event E by using each of the legitimate probability assignments—namely, assignments #1–#4. We have for those four assignments, respectively,

$$P(E) = 0.125 + 0.125 + 0.125 = 0.375,$$
$$P(E) = 0 + 0 + 0 = 0,$$
$$P(E) = 0.032 + 0.032 + 0.032 = 0.096, \text{ and}$$
$$P(E) = 0.220 + 0.050 + 0.110 = 0.380.$$

As expected, the probability of event E depends on the probability measure being employed. ∎

EXERCISES 2.2 **Basic Exercises**

2.20 The U.S. Coast Guard maintains a database of the number, source, and location of oil spills in U.S. navigable and territorial waters. According to the *Statistical Abstract of the United States,* a probability distribution for location of oil spill events is as follows.

Location	Probability	Location	Probability	Location	Probability
Atlantic Ocean	0.011	Great Lakes	0.018	Bays and sounds	0.094
Pacific Ocean	0.059	Other lakes	0.003	Harbors	0.099
Gulf of Mexico	0.271	Rivers and canals	0.211	Other	0.234

a) Explain why this probability assignment is legitimate from a probabilistic point of view.

Determine the probability that an oil spill in U.S. navigable and territorial waters

b) occurs in an ocean.   c) occurs in a lake or harbor.

d) doesn't occur in a lake, ocean, river, or canal.

2.21 Consider the experiment of tossing a coin once and observing whether it comes up a head (H) or a tail (T). The sample space for this random experiment is $\Omega = \{\text{H,T}\}$. Let p be a real number with $0 \leq p \leq 1$.

a) Explain why the probability assignment $P(\{H\}) = p$ and $P(\{T\}) = 1 - p$ determines a unique probability measure on Ω.

b) Based on the probability assignment in part (a), find the probability of each of the four events of this random experiment.

2.22 Suppose that one die is rolled and that you observe the number of dots facing up. A sample space for this random experiment is $\Omega = \{1, 2, 3, 4, 5, 6\}$. The following table provides five different potential probability assignments to the possible outcomes.

Outcome	#1	#2	#3	#4	#5
1	1/6	0.10	0.2	1/2	1/16
2	1/6	0.15	0.2	1/4	1/8
3	1/6	0.40	0.2	−1/4	1/4
4	1/6	0.05	0.2	1/2	0
5	1/6	0.10	0.2	−1/8	7/16
6	1/6	0.20	0.2	1/8	1/8

a) Which of assignments #1–#5 are legitimate probability assignments? Explain.

b) Let A = die comes up even, B = die comes up at least 4, C = die comes up at most 2, and D = die comes up 3. Determine the probability of each of these four events by using each of the legitimate probability assignments.

c) Determine the probability that the die comes up odd by using each of the legitimate probability assignments in the table.

d) If the die is balanced, which probability assignment should be used? Explain.

e) Suppose that experience indicates that the percentages of occurrence of the numbers 1, 2, 3, 4, 5, and 6 are 10%, 15%, 40%, 5%, 10%, and 20%, respectively. Which probability assignment should be used? Explain.

2.23 A sequence of attempts to accomplish some goal continues until it succeeds. Let s and f denote success and failure, respectively, and let $0 < p \le 1$. The following table provides five different potential probability assignments to the possible outcomes.

Outcome	#1	#2	#3	#4	#5
s	1/2	1/3	1/2	0	p
f, s	1/4	1/3	1/3	0	$p(1 - p)$
f, f, s	1/8	1/3	1/4	1	$p(1 - p)^2$
f, f, f, s	1/16	0	1/5	0	$p(1 - p)^3$
⋮	⋮	⋮	⋮	⋮	⋮
$f, f, \ldots, f, s$ $n-1$ times	$1/2^n$	0	$1/n$	0	$p(1 - p)^{n-1}$
⋮	⋮	⋮	⋮	⋮	⋮

a) Which of these proposed probability assignments are legitimate? Explain your answers.

b) For each of the legitimate probability assignments in the table, determine the probability that a success occurs by the second attempt; by the fourth attempt; by the sixth attempt.

c) Now suppose that $p = 0$. Is assignment #5 still legitimate? What does $p = 0$ mean in terms of this random experiment? How can the sample space and assignment #5 be extended so as to obtain a legitimate probability assignment?

2.24 Let Ω be the sample space for a random experiment and let P be a probability measure on Ω. Use the Kolmogorov axioms to verify the following for events A and B.
a) $P(B) = P(B \cap A) + P(B \cap A^c)$
b) $P(A \cup B) = P(A) + P(B \cap A^c)$
c) Suppose that at least one of events A and B must occur—that is, $A \cup B = \Omega$. Show that the probability that both events occur is $P(A) + P(B) - 1$.

2.25 Consider the experiment of rolling two dice. The possible outcomes are shown in Figure 2.1 on page 27.
a) Assign each outcome a probability of 1/36. Show that this probability assignment is legitimate.
b) Based on the probability assignment in part (a), determine the probability of the event A_i that the sum of the faces is i, for each $i = 2, 3, \ldots, 12$.
c) Provide another probabilistically legitimate assignment to the 36 possible outcomes, and then repeat part (b) for that assignment.
d) Assuming that the die is balanced, is your probability assignment in part (c) reasonable? What about the one in part (a)? Explain your answers.

2.26 A number is chosen at random from the integers $1, 2, \ldots, 100$. The sample space is the set $\Omega = \{1, 2, \ldots, 100\}$, and each outcome is assigned probability 0.01.
a) Show that this probability assignment is legitimate.
b) Let A be the event that the number chosen is even, B be the event that the number chosen is at most 10π, and C be the event that the number chosen is prime. Determine the probabilities of events A, B, and C.

2.27 An urn contains four balls numbered 1, 2, 3, and 4. A ball is chosen at random, its number noted, and the ball is replaced in the urn. This process is repeated one more time.
a) Determine the sample space Ω.
b) If each outcome is assigned the same probability, what is that common probability?
c) Using the probability assignment in part (b), find the probability that the two numbers chosen are different.

2.28 Suppose that Ω is a finite sample space—say, with N possible outcomes. Further suppose that those N possible outcomes are equally likely.
a) What common probability should be assigned to each possible outcome?
b) Determine the probability of an event that consists of m outcomes.

Theory Exercises

2.29 A special case of a relation called *Boole's inequality* is that, for each positive integer n,

$$P(A_1 \cup \cdots \cup A_n) \le P(A_1) + \cdots + P(A_n), \qquad (*)$$

for all events $A_1, \ldots, A_n$ of a sample space.
a) If $P(A_i) = 1/6$ for each i and $n = 10$, Relation $(*)$ is trivial. However, if $P(A_i) = 1/60$ for each i and $n = 10$, or if $P(A_i) = 1/6$ for each i and $n = 4$, Relation $(*)$ conveys more substantial information than it does in the trivial case. Explain the difference between the trivial case and the two cases in which more substantial information is conveyed. Why is one case "trivial" while the other two are "more substantial"?
b) Prove Relation $(*)$. *Hint:* Use Exercise 2.24, the nonnegativity axiom, and mathematical induction.

Advanced Exercises

2.30 A point is chosen from the lattice points in the first quadrant. The sample space for this random experiment is $\Omega = \{(i, j) : i, j \in \{0, 1, 2, \ldots\}\}$. Assign the outcome (i, j) probability $e^{-5} 2^i 3^j / i! j!$.

a) Show that this probability assignment is legitimate.

Find the probability that the point selected

b) is on the x-axis.

c) is in the closed square of side length 3 with lower left vertex at the origin.

d) has x coordinate i, where i is a nonnegative integer.

e) has y coordinate j, where j is a nonnegative integer.

2.31 Suppose that $\Omega_1 = \{\omega_{11}, \omega_{12}, \omega_{13}, \ldots\}$ and $\Omega_2 = \{\omega_{21}, \omega_{22}, \omega_{23}, \ldots\}$ are countably infinite sample spaces. Suppose further that $p_{11}, p_{12}, p_{13}, \ldots$ is a legitimate probability assignment for Ω_1 and that $p_{21}, p_{22}, p_{23}, \ldots$ is a legitimate probability assignment for Ω_2. Now let $\Omega = \Omega_1 \times \Omega_2$.

a) For each $m, n \in \mathcal{N}$, define $P(\{(\omega_{1m}, \omega_{2n})\}) = p_{1m} p_{2n}$. Show that this probability assignment for Ω is legitimate.

Based on the probability assignment in part (a), determine the probability of each of the following events of Ω.

b) $\{(\omega_{11}, \omega_{21}), (\omega_{12}, \omega_{24})\}$ **c)** $\{(\omega_{1m}, \omega_{2n}) : n \in \mathcal{N}\}$ **d)** $\{(\omega_{1m}, \omega_{2n}) : m \in \mathcal{N}\}$

2.32 Specify a legitimate probability assignment on the rational numbers for which no outcome has zero probability.

2.33 An infinitely thin dart is thrown at the unit square $[0, 1] \times [0, 1]$. (More prosaically, one point is chosen at random from the unit square.) The probability that the dart falls in any particular region within the square equals the area of the region. The area of any region consisting of just one point is 0. Use these facts and Corollary 2.1 on page 43 to prove that the set of points in the unit square is uncountable.

2.34 Bruno de Finetti's thought experiment: Consider the price you would be willing to pay to be promised that, if the Red Sox win next year's World Series, you will receive \$1. Suppose that, once you have announced that price, your opponent has a choice: Either your opponent sells you that promise for the price you have chosen or your opponent buys that promise from you at that same price. The price you choose while knowing that your opponent will then make that choice is your "operational subjective probability" that the Red Sox will win.

a) Show that setting a price higher than \$1 is bad for you.

b) Show that setting a price lower than \$0 is bad for you.

c) Suppose that you are absolutely certain that the Red Sox will win. What decision by your opponent would you fear if you set a price lower than \$1 but not if the price you set is exactly \$1? Similarly, if you are absolutely certain that the Red Sox will not win, what decision by your opponent would you fear if you set a price higher than \$0 but not if the price you set is \$0?

Now complicate the scenario. You must choose the price of a promise to pay \$1 if the Red Sox win the World Series, the price of a promise to pay \$1 if the Diamondbacks win the World Series, and the price of a promise to pay \$1 if either the Red Sox or the Diamondbacks win the World Series. (You are setting these prices before the season, so it's possible that some other team will win, but don't consider any bets on them.) For each of these three promises separately, your opponent will be able to choose to buy the promise from you at the price you have set or to sell the promise to you at that same price or not to buy or sell the promise.

d) Show that setting the three prices in the following way is bad for you:

Price(Red Sox) + Price(Diamondbacks) > Price(Red Sox or Diamondbacks).

e) Show that setting the three prices in the following way is bad for you:

Price(Red Sox) + Price(Diamondbacks) < Price(Red Sox or Diamondbacks).

Note: The moral of all this scenario is that, although you are free to set the prices—which are analogous to probabilities—in such a way that they violate Kolmogorov's axioms, nonetheless to do so is demonstrably bad strategy.

2.3   Specifying Probabilities

In Section 2.2, we concentrated on the conditions that probabilities must satisfy. We showed that, mathematically, only three requirements are necessary for a probability measure—namely, the three Kolmogorov axioms of nonnegativity, certainty, and additivity. However, in reality, probabilities must not only be mathematically valid, but they must also reflect the underlying nature of the random experiment under consideration. Otherwise, the results obtained by using those probabilities will not be useful in practice.

Three common ways of specifying or obtaining probabilities so that they reflect the underlying nature of the random experiment are

- probability models,
- empirical probability, and
- subjective probability.

In practice, these three methods often are combined in some manner to yield an appropriate probability assignment.

We begin by discussing and providing some important examples of probability models. A **probability model** is a mathematical description of the random experiment based on certain primary aspects and assumptions. In this chapter, we concentrate on one of the most widely used and vintage probability models, the *equal-likelihood model.* Later in the book we discuss several other probability models.

The Equal-Likelihood Model: Classical Probability

The primary aspect and assumption of the **equal-likelihood model** is that the possible outcomes of the random experiment are equally likely to occur. For finite sample spaces (i.e., random experiments with a finite number of possible outcomes), the equal-likelihood model corresponds to selecting a member at random from a finite population. In other words, it is the model for which probabilities are simply percentages.

The equal-likelihood model for finite sample spaces was the first probability model and, consequently, is often referred to as **classical probability.** We now examine classical probability in detail. From Proposition 2.2 on page 42, we know that the probabilities of the individual outcomes determine the probabilities of all events. Thus we can concentrate on obtaining the probabilities of the individual outcomes.

Before deriving a general result, let's look at a simple example. Consider the random experiment of tossing a *balanced* coin once and observing the side landing face up (head or tail). The sample space for this experiment is $\{H, T\}$. A key word in the description of the experiment is "balanced." It means that when the coin is tossed, the side landing face up is equally likely to be either one of the two possibilities. Thus a classical probability model is appropriate here. Because the two possible outcomes are equally likely, they have the same probability. As you might guess, that common probability is $1/2$, the reciprocal of the number of possible outcomes.

In general, suppose that Ω is a finite sample space with, say, N equally likely possible outcomes. We write $\Omega = \{\omega_1, \omega_2, \ldots, \omega_N\}$ and use p to denote the common probability of the individual outcomes. Referring to Corollary 2.1 on page 43, we get

$$1 = \sum_{n=1}^{N} P(\{\omega_n\}) = \sum_{n=1}^{N} p = Np.$$

Consequently, $p = 1/N$, the reciprocal of the number of possible outcomes, as expected.

Knowing the probabilities of the individual outcomes, we can obtain the probabilities of all events. In doing so, we use the notation $N(E)$ to denote the number of outcomes comprising an event E—that is, the number of ways that E can occur.

◆◆◆ Proposition 2.4   Classical Probability

Let Ω be a finite sample space with equally likely outcomes. Then, for each event E, we have

$$P(E) = \frac{N(E)}{N(\Omega)}. \tag{2.8}$$

In words, the probability of an event equals the number of ways the event can occur divided by the total number of possible outcomes of the random experiment.

Proof   Recalling that the common probability of the individual outcomes is the reciprocal of the total number of possible outcomes and applying Proposition 2.2 on page 42, we get

$$P(E) = \sum_{\omega \in E} P(\{\omega\}) = \sum_{\omega \in E} \frac{1}{N(\Omega)} = \frac{1}{N(\Omega)} \sum_{\omega \in E} 1 = \frac{N(E)}{N(\Omega)},$$

as required.   ◆

EXAMPLE 2.19   *Classical Probability*

Regions of the States   Refer to Example 1.1, where the random experiment consists of selecting one of the 50 states at random. Table 1.1 on page 4 shows the region for each of the 50 states. Using that table, determine the probability that the state selected is in the South.

Solution The sample space for this random experiment consists of the 50 states. Because a state is being selected at random, each is equally likely to be the one obtained. In other words, a classical probability model is appropriate here. So, for each event E, Equation (2.8) gives

$$P(E) = \frac{N(E)}{N(\Omega)} = \frac{N(E)}{50}.$$

Let S denote the event that the state selected is in the South. Referring to Table 1.1, we find that S can occur in 16 ways—it occurs if and only if the state selected is one of the 16 states in the South. Consequently,

$$P(S) = \frac{N(S)}{50} = \frac{16}{50} = 0.320.$$

In words, the probability is 0.320 that the state selected is in the South. Of course, this is the same answer that we obtained in Example 1.1 by proceeding intuitively. ■

EXAMPLE 2.20 *Classical Probability*

Rolling Dice Refer to Example 2.3 on page 27, where the random experiment consists of rolling two dice and observing on each die the number of dots facing up. Let's suppose, in addition, that the dice are balanced. Determine the probability of the event that
a) the sum of the dice is 5.
b) doubles are rolled—that is, both dice come up the same number.

Solution A sample space for this experiment is $\Omega = \{\, (x, y) : x, y \in \{1, 2, 3, 4, 5, 6\} \,\}$, the set of all ordered pairs of integers between 1 and 6. This sample space is portrayed graphically in Figure 2.1 on page 27. Because the dice are balanced, each of the 36 possible outcomes is equally likely to occur. In other words, a classical probability model is appropriate here. So, for each event E, Equation (2.8) gives

$$P(E) = \frac{N(E)}{N(\Omega)} = \frac{N(E)}{36}.$$

We can now obtain the required probabilities.
a) Let A denote the event that the sum of the dice is 5. Then we can express the event A as $A = \{(1, 4), (2, 3), (3, 2), (4, 1)\}$. Thus

$$P(A) = \frac{N(A)}{36} = \frac{4}{36} = \frac{1}{9} = 0.111.$$

There is an 11.1% chance that the sum of the dice will be 5.
b) Let's use B to denote the event that doubles are rolled. Then we can express the event B as $B = \{(1, 1), (2, 2), (3, 3), (4, 4), (5, 5), (6, 6)\}$. Consequently,

$$P(B) = \frac{N(B)}{36} = \frac{6}{36} = \frac{1}{6} = 0.167.$$

There is a 16.7% chance that doubles will be rolled. ■

EXAMPLE 2.21   *Classical Probability*

Oklahoma State Officials   Refer to Example 2.7 on page 28, where the experiment consists of randomly sampling, without replacement, two officials from the top five Oklahoma state officials: governor (G), lieutenant governor (L), secretary of state (S), attorney general (A), and treasurer (T). Determine the probability of the event that
a) the attorney general is included in the sample.
b) neither the governor nor the treasurer are included in the sample.

Solution   We recall that the sample space for this experiment is

$$\Omega = \{\, \{G, L\}, \{G, S\}, \{G, A\}, \{G, T\}, \{L, S\}, \{L, A\}, \{L, T\}, \{S, A\}, \{S, T\}, \{A, T\}\,\},$$

where, for instance, $\{G, L\}$ represents the outcome that the two officials selected are the governor and lieutenant governor. Because the sampling is done randomly, each of the 10 possible samples is equally likely to be the one obtained. In other words, a classical probability model is appropriate here. So, for each event E, Equation (2.8) gives

$$P(E) = \frac{N(E)}{N(\Omega)} = \frac{N(E)}{10}.$$

We can now obtain the required probabilities.
a) Let A denote the event that the attorney general is included in the sample. Then we have $A = \{\{G, A\}, \{L, A\}, \{S, A\}, \{A, T\}\}$. Therefore

$$P(A) = \frac{N(A)}{10} = \frac{4}{10} = 0.4.$$

There is a 40% chance that the attorney general will be included in the sample.
b) Let B denote the event that neither the governor nor the treasurer are included in the sample. Then $B = \{\{L, S\}, \{L, A\}, \{S, A\}\}$. Consequently,

$$P(B) = \frac{N(B)}{10} = \frac{3}{10} = 0.3.$$

There is a 30% chance that neither the governor nor the treasurer will be included in the sample.   ■

Our final example of classical probability involves the analysis of data obtained from two variables. In it, we also introduce some important probability terminology that we use throughout the text.

EXAMPLE 2.22   *Classical Probability*

Age and Rank of Faculty   Table 2.5 provides information on age and rank for the faculty members at a university. Because that table contains cross-classified data on two variables (age and rank), it is called a **two-way table** or **contingency table.**

Table 2.5 Age and rank of faculty members

Rank

	Full professor R_1	Associate professor R_2	Assistant professor R_3	Instructor R_4	Total
Under 30 A_1	2	3	57	6	68
30–39 A_2	52	170	163	17	402
40–49 A_3	156	125	61	6	348
50–59 A_4	145	68	36	4	253
60 & over A_5	75	15	3	0	93
Total	430	381	320	33	1164

Age

Suppose that a faculty member is selected at random. Determine the probability that the faculty member obtained is

a) an associate professor.

b) under the age of 30.

c) an associate professor under the age of 30.

Solution   To begin, we interpret the entries in Table 2.5 and introduce some associated terminology. The small boxes inside the rectangle formed by the heavy lines are called **cells.** The number 2 in the upper left cell indicates that two faculty members are full professors under the age of 30. The number 170, diagonally below and to the right of the 2, shows that 170 faculty members are associate professors in their 30s.

The row total in the first row reveals that 68 ($2 + 3 + 57 + 6$) of the faculty members are under the age of 30. Similarly, the column total in the second column shows that 381 of the faculty members are associate professors. The number 1164 in the lower right corner gives the total number of faculty. That total can be found by summing either the row totals or the column totals; it can also be found by summing the frequencies in the 20 cells of the table.

Note that the rows and columns of Table 2.5 are labeled with subscripted letters. The subscripted letter A_1, labeling the first row, represents the event that the faculty member obtained is under the age of 30; similarly, the subscripted letter R_2, labeling the second column, represents the event that the faculty member obtained is an associate professor;

and so on. Clearly, the events A_1, A_2, A_3, A_4, and A_5 are mutually exclusive, as are the events R_1, R_2, R_3, and R_4.

In addition to considering events A_1 through A_5 and R_1 through R_4 separately, we can also consider them jointly. For example, the event that the faculty member obtained is under the age of 30 (event A_1) *and* is also an associate professor (event R_2) can be expressed as $A_1 \cap R_2$. This event is represented by the cell in the first row and second column. Here there are 20 different joint events, one for each cell of the contingency table.

We can now obtain the required probabilities. First we note that, because a faculty member is being selected at random, each of the 1164 faculty members is equally likely to be the one obtained. Consequently, a classical probability model is appropriate here. So, for each event E, Equation (2.8) gives

$$P(E) = \frac{N(E)}{N(\Omega)} = \frac{N(E)}{1164}.$$

a) Referring to the preceding equation and Table 2.5, we can determine the probability that the faculty member obtained is an associate professor:

$$P(R_2) = \frac{N(R_2)}{1164} = \frac{381}{1164} = 0.327.$$

In other words, 32.7% of the faculty are associate professors.

b) Likewise, the probability that the faculty member obtained is under the age of 30 is

$$P(A_1) = \frac{N(A_1)}{1164} = \frac{68}{1164} = 0.058.$$

Thus 5.8% of the faculty are under the age of 30.

c) Similarly, the probability that the faculty member obtained is an associate professor under the age of 30 is

$$P(A_1 \cap R_2) = \frac{N(A_1 \cap R_2)}{1164} = \frac{3}{1164} = 0.003.$$

Thus 0.3% of the faculty are associate professors under the age of 30.

In Table 2.6, we have replaced the joint frequency distribution in Table 2.5 with a **joint probability distribution.** Each probability in Table 2.6 is determined in the same way as the three probabilities we just computed—namely, by dividing the corresponding entry in Table 2.5 by the grand total of 1164.

Each probability of the form $P(A_i \cap R_j)$, where $1 \le i \le 5$ and $1 \le j \le 4$, is called a **joint probability** because it is a probability involving the events A_i and R_j jointly. The joint probabilities are displayed in the cells of Table 2.6.

Note that we changed the row and column labels "Total" in Table 2.5 to $P(R_j)$ and $P(A_i)$, respectively, in Table 2.6. The reason is that the last row of Table 2.6 gives the probabilities of events R_1 through R_4 and that the last column of Table 2.6 gives the probabilities of events A_1 through A_5. Each of those probabilities is called a **marginal probability** because it is in the margin of the joint probability distribution.

The sum of the joint probabilities in a row or column of a joint probability distribution equals the marginal probability in that row or column, with any discrepancy being due

Table 2.6 Joint probability distribution corresponding to Table 2.5

Rank

	Full professor R_1	Associate professor R_2	Assistant professor R_3	Instructor R_4	$P(A_i)$
Under 30 A_1	0.002	0.003	0.049	0.005	0.058
30–39 A_2	0.045	0.146	0.140	0.015	0.345
40–49 A_3	0.134	0.107	0.052	0.005	0.299
50–59 A_4	0.125	0.058	0.031	0.003	0.217
60 & over A_5	0.064	0.013	0.003	0.000	0.080
$P(R_j)$	0.369	0.327	0.275	0.028	1.000

Age

to roundoff error.[†] For instance, consider the A_4 row of Table 2.6. The sum of the joint probabilities in that row is

$$0.125 + 0.058 + 0.031 + 0.003 = 0.217,$$

which is the marginal probability at the end of the A_4 row.   ■

Although Equation (2.8) on page 50 provides a genuine probability measure for any random experiment with finitely many possible outcomes, the resulting probabilities will not correctly reflect reality unless the equal-likelihood model is appropriate. In other words, Equation (2.8) applies only to random experiments in which the possible outcomes are equally likely to occur.

The Equal-Likelihood Model: Geometric Probability

We can develop an equal-likelihood model for certain types of infinite sample spaces. In this book, we do that for sample spaces that are subsets of n-dimensional Euclidean space, $\mathcal{R}^n$. Because of the geometric nature of such sample spaces, the equal-likelihood model in this context is often referred to as **geometric probability.**

[†]We show in Section 2.4 that this fact is a consequence of a general property of probability, called the *law of partitions.*

To illustrate, let Ω be a subset of $\mathcal{R}^2$ with finite nonzero area, such as a disk or a rectangle. We consider the experiment of selecting a point at random from Ω. The possible outcomes for this random experiment are the elements of Ω itself, so Ω is the sample space. As always, an event E is a subset of the sample space. In this experiment, E is the event that the point selected is an element of the subset E.

Because the point is selected at random, the probability of event E (i.e., the probability that the point selected is an element of the subset E) is proportional to the area of E, which we denote $|E|$. In other words, there is a constant k such that $P(E) = k \cdot |E|$ for each event E. Applying the certainty axiom, we can determine k:

$$1 = P(\Omega) = k \cdot |\Omega|.$$

Thus $k = 1/|\Omega|$, the reciprocal of the area of Ω. And from this fact it follows that

$$P(E) = \frac{|E|}{|\Omega|}, \qquad \text{for each event } E.$$

This result stands to reason: If we select a point at random from the set Ω, the probability that the point obtained is an element of a subset E of Ω should equal the proportion of the area of Ω occupied by E, that is, the ratio of the area of E to the area of Ω.

In general, we use the notation $|E|$ to denote the n-dimensional volume of a subset E of $\mathcal{R}^n$. Of course, one-dimensional volume is length and two-dimensional volume is area. Consider, for the moment, one-dimensional Euclidean space, $\mathcal{R}$. Although strictly speaking, length is defined only for intervals, we can use advanced mathematics to extend the concept of length to all subsets of $\mathcal{R}$ that arise in practice.[†] For instance, the "length" of the union of two disjoint intervals is the sum of the lengths of the individual intervals. When we use the term *length*, we do so in this extended sense; likewise we do so for the term *area* in two dimensions, the term *volume* in three dimensions, and so on.

By using the same argument in n-dimensional Euclidean space as we did above for two-dimensional Euclidean space, we obtain Proposition 2.5.

◆◆◆ **Proposition 2.5   Geometric Probability**

Let Ω be a subset of $\mathcal{R}^n$ with finite nonzero n-dimensional volume. Suppose that a point is selected at random from Ω. Then the probability that the point obtained is an element of the subset E of Ω is

$$P(E) = \frac{|E|}{|\Omega|}. \tag{2.9}$$

In words, the probability of an event equals the n-dimensional volume of the event divided by the n-dimensional volume of the sample space.

Note the similarity between Equation (2.9)—used for geometric probability—and Equation (2.8) on page 50—used for classical probability. In classical probability we measure size by number of elements, whereas in geometric probability we measure size by n-dimensional volume.

[†]A detailed discussion of how this extension is obtained can be found on pages 104–125 of the book *A Course in Real Analysis* by John N. McDonald and Neil A. Weiss (San Diego: Academic Press, 1999).

EXAMPLE 2.23   *Geometric Probability*

Random Number Generators   Suppose that a number is selected at random from the interval (0, 1), a task that a basic random number generator aims to accomplish. Determine the probability that the number obtained is
a) 0.25 or greater.
b) between 0.1 and 0.4, inclusive.
c) either less than 0.1 or greater than 0.4.

Solution   We first note that the sample space for this experiment consists of all real numbers between 0 and 1, so $\Omega = (0, 1)$. Because we are selecting a number at random, a geometric probability model is appropriate here. Hence, for each event E, Equation (2.9) yields

$$P(E) = \frac{|E|}{|\Omega|} = \frac{|E|}{1} = |E|,$$

where $|E|$ denotes the length (in the extended sense) of the set E. We can now obtain the required probabilities.

a) Let A denote the event that the number selected is 0.25 or greater. Then $A = [0.25, 1)$, and so

$$P(A) = |A| = |[0.25, 1)| = 1 - 0.25 = 0.75.$$

Chances are 75% that the number selected will be 0.25 or greater.

b) Let B denote the event that the number selected is between 0.1 and 0.4, inclusive. Then $B = [0.1, 0.4]$ and, consequently,

$$P(B) = |B| = |[0.1, 0.4]| = 0.4 - 0.1 = 0.3.$$

Chances are 30% that the number selected will be between 0.1 and 0.4, inclusive.

c) Let C denote the event that the number selected is either less than 0.1 or greater than 0.4. Then $C = (0, 0.1) \cup (0.4, 1)$, and so

$$P(C) = |C| = |(0, 0.1)| + |(0.4, 1)| = 0.1 + 0.6 = 0.7.$$

Chances are 70% that the number selected will be either less than 0.1 or greater than 0.4.   ∎

EXAMPLE 2.24   *Geometric Probability*

Bacteria on a Petri Dish   Suppose that a petri dish of unit radius, containing nutrients upon which bacteria can multiply, is smeared with a uniform suspension of bacteria. Subsequently, spots indicating colonies of bacteria will appear. Determine the probability that the center of the first spot to appear is
a) more than 1/4 unit from the center of the petri dish.
b) on the right half of the petri dish.

Solution   As we showed in Example 2.6, we can take the sample space for this random experiment to be the unit disk: $\Omega = \{(x, y) : x^2 + y^2 < 1\}$. Because the dish is smeared with a uniform suspension of bacteria, it is reasonable to use a geometric probability model here. Specifically, we can think of the location of the center of the first spot (visible bacteria colony) as a point selected at random from the unit disk. Consequently, for each event E, Equation (2.9) yields

$$P(E) = \frac{|E|}{|\Omega|} = \frac{|E|}{\pi} = \frac{1}{\pi}|E|,$$

where $|E|$ denotes the area (in the extended sense) of the set E. We can now obtain the required probabilities.

a) Let A denote the event that the center of the first spot to appear is more than 1/4 unit from the center of the petri dish. Then $A = \{(x, y) \in \Omega : 1/4 < \sqrt{x^2 + y^2} < 1\}$. Referring to Figure 2.4(a), we conclude that

$$P(A) = \frac{1}{\pi}|A| = \frac{1}{\pi}\left(\pi - \frac{\pi}{16}\right) = \frac{15}{16} = 0.938.$$

In other words, 93.8% of the time the center of the first spot to appear will be more than 1/4 unit from the center of the petri dish.

Figure 2.4 Event that the center of the first spot to appear is (a) more than 1/4 unit from the center of the petri dish; (b) on the right half of the petri dish

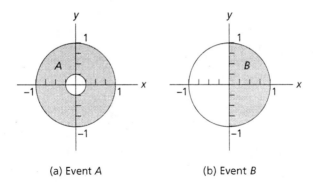

(a) Event A   (b) Event B

b) Let B denote the event that the center of the first spot to appear is on the right half of the petri dish. Then $B = \{(x, y) \in \Omega : x > 0\}$. Referring to Figure 2.4(b), we conclude that

$$P(B) = \frac{1}{\pi}|B| = \frac{1}{\pi} \cdot \frac{\pi}{2} = 0.5.$$

In other words, half the time the center of the first spot to appear will be on the right half of the petri dish, a result that we could have easily guessed.   ∎

Empirical Probability

As we have demonstrated, probability models provide a method for specifying and obtaining probabilities based on certain primary aspects and assumptions about the random experiment. Often, however, the primary aspects are not obvious and reasonable assumptions are not evident.

For example, consider the experiment of selecting a person at random and observing the person's birthday. The possible outcomes of the experiment are the 365 days of the year (ignoring leap years). It might seem reasonable to use an equal-likelihood model here and assign probability 1/365 to each of the 365 possible outcomes. If we use this model, the probability that the person selected was born in, say, September would be $30/365 = 0.082$.

However, we can argue that the equal-likelihood model is inappropriate because, for instance, conception rates might differ during different parts of the year—they might be higher in the winter months. Instead, it might be more reasonable to consult birth statistics to assign (approximately correct) probabilities. Doing so reveals, for example, that the probability a randomly selected person was born in September is roughly 0.089, rather than 0.082. Consulting birth statistics to assign probabilities for birthdays is an example of *empirical probability*.

Empirical probability is based on the frequentist interpretation of probability or its mathematical counterpart, the law of large numbers. Thus the method of empirical probability relies on repeating the random experiment a large number of times and using the proportions of occurrence of the various outcomes or events as probabilities, keeping in mind that these probabilities are generally only approximately correct.

EXAMPLE 2.25   *Empirical Probability*

Number of Children Per Birth   Consider the random experiment of observing the number of children born at a given U.S. birth where, for simplicity, we take the possible outcomes to be one child, two children, and three or more children. A sample space for this experiment is $\Omega = \{1, 2, 3+\}$. Obviously, an equal-likelihood model—assigning probability 1/3 to each of the three possible outcomes—is inappropriate here. Use empirical probability to assign (approximately correct) probabilities.

Solution   From the publication "Births: Final Data for 2002" (*National Vital Statistics Report*, Vol. 52, No. 10, 2003) by J. Martin et al., we obtained data on single and multiple live births for 2002, as presented in Table 2.7. The table shows, for instance, that 62,567 women gave birth to (live) twins in 2002.

Table 2.7 Data and empirical probabilities for number of children per birth

Number of children	Number of births	Empirical probability
1	3,889,191	0.9836
2	62,567	0.0158
3+	2,467	0.0006
	3,954,225	1.0000

For these data, we are repeating the experiment of observing the number of children born at a given U.S. birth 3,954,225 times; thus $n = 3,954,225$. Let's obtain the empirical probability for, say, the outcome 2 (twins). Let T denote the event of twins. In view of the frequentist interpretation of probability and the data in Table 2.7, we have

$$P(T) \approx \frac{n(T)}{n} = \frac{62,567}{3,954,225} = 0.0158.$$

So the empirical probability for twins is 0.0158, as recorded in the third column of Table 2.7. The empirical probabilities for the other two outcomes are obtained similarly.

We can use the empirical probabilities in the third column of Table 2.7 as approximations to the true probabilities. The accuracy of these approximations is a problem addressed in inferential statistics. ∎

Whenever possible, it is best to use a probability model. However, as we have just shown, that method for specifying probabilities is frequently not feasible. In such cases, the method of empirical probability often provides a useful alternative for the specification of probabilities. Computers sometimes make possible the use of simulation to obtain empirical probabilities, even for extremely complex random experiments.

Subjective Probability

Another method that is often used to specify probabilities is based on the experience of the person assigning the probabilities. It may take the form of an educated guess, indirect evidence, or intuition. Because such probabilities are subjective, this method of probability assignment is often referred to as **subjective probability.**

For instance, consider a quote by Lynn Reaser, chief economist at Banc of America Capital Management, in a March 2001 *Associated Press* article regarding the chances of a Federal Reserve interest rate cut: "I'd say there is a 60% probability of a three-quarter point reduction on Tuesday primarily because of the low rate of underlying inflation, the weakness in the factory sector and the negative feedback that could come from a continued slide in the stock market."

Subjective probability is particularly appropriate when a probability model is infeasible and there is little or no empirical information. Generally, you should resort to the method of subjective probability only when other methods for probability assignment don't apply.

EXERCISES 2.3   **Basic Exercises**

2.35 An American roulette wheel contains 38 numbers, of which 18 are red, 18 are black, and 2 are green. When the roulette wheel is spun, the ball is equally likely to land on any of the 38 numbers. Determine the probability that the number on which the ball lands is
a) red.   **b)** black.   **c)** green.
d) black or green.   **e)** not black.   **f)** not green.

2.36 Refer to Example 2.4 on page 27, where we considered the random experiment of observing a mechanical or electrical unit that consists of five components and determining which components are working and which have failed. Assuming that the components work or fail independently of each other, answer these questions.

a) Under what conditions would an equal-likelihood model be appropriate?

b) Presuming that an equal-likelihood model is appropriate, write the formula for obtaining the probability of each event *E*. *Hint:* The number of possible outcomes of the random experiment is 32.

Presuming that an equal-likelihood model is appropriate, determine the probability that the number of nonworking components is

c) at most one. **d)** at least one. **e)** zero.

2.37 Refer to Example 2.5 on page 28, where we considered the random experiment of selecting a person at random and observing his or her blood genotype.

a) Under what condition would it be reasonable to use an equal-likelihood model for this random experiment?

b) Assuming that the condition in part (a) actually holds, determine the probability that a randomly selected person has type B blood. *Hint:* See Table 2.2 on page 31.

c) Do you think that the condition you specified in part (a) holds? Why or why not?

The actual percentages for blood genotypes in the general population are presented in the following table.

Genotype	aa	ab	ao	bb	bo	oo
Percent	5.0	3.8	31.0	0.7	11.6	47.9

d) Using the actual percentages for genotypes, determine the probability that a randomly selected person has type B blood. Compare your answer to the one obtained in part (b).

2.38 Explain why the joint events in a contingency table are mutually exclusive.

2.39 The National Football League updates team rosters and posts them on its Web site at www.nfl.com. The following contingency table provides a cross-classification of players on the New England Patriots roster as of May 24, 2000, by weight and years of experience.

Years of experience

	Rookie Y_1	1–5 Y_2	6–10 Y_3	10+ Y_4	Total
Under 200 W_1	9	9	2	1	21
200–300 W_2	22	25	5	2	54
Over 300 W_3	11	7	1	0	19
Total	42	41	8	3	94

Weight (lb)

a) How many cells are in this contingency table?

b) How many players are on the New England Patriots roster as of May 24, 2000?

c) How many players are rookies?

d) How many players weigh between 200 and 300 lb?

e) How many players are rookies who weigh between 200 and 300 lb?

Suppose that a player on the New England Patriots is selected at random.

f) Describe the events Y_3, W_2, and $W_1 \cap Y_2$ in words.

g) Compute the probability of each event in part (f). Interpret your answers in terms of percentages.

h) Construct a joint probability distribution similar to Table 2.6 on page 55.

i) Verify that the sum of each row and column of joint probabilities equals the marginal probability in that row or column. *Note:* Rounding may cause slight deviations.

2.40 Refer to Example 2.3 on page 27, where two dice are rolled, and assume that the dice are balanced.

a) Is a classical probability model appropriate here? Explain your answer.

Find the probability of each of the following events.

b) The sum of the two numbers is 4.   **c)** The sum of the two numbers is 5.

d) The minimum of the two numbers is 4.   **e)** The maximum of the two numbers is 4.

2.41 Two Republicans and two Democrats sit on a committee in the Senate. A subcommittee of two senators is chosen at random from among these four. Find the probability that the number of Republicans on the subcommittee is

a) two.   **b)** one.   **c)** zero.

2.42 A commuter train arrives punctually at a station every half hour. Each morning, a commuter named John leaves his house and casually strolls to the train station. Find the probability that John waits for the train

a) between 10 and 15 minutes.   **b)** at least 10 minutes.

2.43 Refer to Example 2.24 on page 57, where a petri dish of unit radius, containing nutrients upon which bacteria can multiply, is smeared with a uniform suspension of bacteria. Determine the probability of each of the following events.

a) The distance from the center to the first spot is less than half the distance from the center to the rim.

b) The distance from the center to the first spot is more than half the distance from the center to the rim.

c) The center of the first spot to appear is not in the inscribed square centered at the origin with vertices on the circle $x^2 + y^2 = 1$ and sides parallel to the coordinate axes.

d) The center of the first spot to appear is on the strip $\{ (x, y) : -1/4 < x < 1/4 \}$.

2.44 A point is chosen at random in the unit square $\Omega = \{ (x, y) : 0 \le x \le 1, 0 \le y \le 1 \}$. Describe each of the following events in words and determine the probability of each.

a) $A = \{ (x, y) \in \Omega : x > 1/3 \}$ **b)** $B = \{ (x, y) \in \Omega : y \le 0.7 \}$

c) $C = \{ (x, y) \in \Omega : x + y > 1.2 \}$ **d)** $D = \{ (x, y) \in \Omega : |y - x| < 1/10 \}$

e) $E = \{ (x, y) \in \Omega : x = y \}$

2.45 Refer to Exercise 2.44. Determine the probability that, for the point obtained,

a) the x coordinate is less than the y coordinate.

b) the smaller of the two coordinates is less than 1/2.

c) the smaller of the two coordinates exceeds 1/2.

d) the larger of the two coordinates is less than 1/2.

e) the larger of the two coordinates exceeds 1/2.

f) the sum of the two coordinates is between 1 and 1.5.

g) the sum of the two coordinates is between 1.5 and 2.

2.46 If a number is chosen at random from the interval $(0, 1)$, find the probability that

a) the first digit of its decimal expansion is 7.

b) the second digit of its decimal expansion is 7.

c) the second digit of the decimal expansion of its square root is 7.

2.47 A point is chosen at random from the unit sphere $\{ (x, y, z) \in \mathcal{R}^3 : x^2 + y^2 + z^2 \leq 1 \}$. Determine the probability that the point obtained is

a) more than 1/4 unit from the origin.

b) in a cube of side length 1 centered at the origin.

c) on the surface of the sphere.

2.48 If a point is selected at random from the interior of a triangle with base b and height h, find the probability that its distance from the base of the triangle is

a) at most x.   **b)** between x and y.

2.49 If a point is selected at random from one side of an equilateral triangle with common side length ℓ, find the probability that its distance from the opposite vertex is

a) at most x.   **b)** between x and y.

2.50 According to the *National Vital Statistics Report,* there were 4,058,814 live births in the United States in 2000. Of those births, 3,194,005 were to white mothers and 622,598 were to black mothers.

a) Use the data to obtain the approximate probability that a live birth in the United States is to a white mother; black mother; other-race mother.

b) What type of probabilities did you obtain in part (a)? Explain.

2.51 Suppose that an engineer has a job interview. Based on her previous limited interviewing experience, knowledge of the job market, and partial information about the company, she estimates her probability of being offered a job at 0.70, or 70%. What type of probability is she specifying here? Explain.

2.52 In 2001, the author of this book and his wife put their home in Sun Lakes, Arizona, up for sale. On Sunday, November 4, their realtor called with an update about a possible sale. She said that there was a 90% chance that an acceptable offer would be made later in the day. What type of probability was she specifying here? Explain your answer.

Advanced Exercises

2.53 If a number is selected at random from the interval $(0, 1)$, what is the probability that it is rational?

2.54 Show that having an equal-likelihood model for a countably infinite sample space isn't possible.

2.55 Suppose that three numbers are selected at random from the interval $(0, \ell)$. Determine the probability that a triangle can be formed from the three line segments whose lengths are the three numbers obtained.

2.56 Suppose that two numbers are selected at random from the interval $(0, \ell)$. Find the probability that a triangle can be formed from the resulting three line segments.

2.57 Buffon's needle problem: A floor is ruled with a series of parallel lines, d units apart. Suppose that a needle of length ℓ, where $\ell < d$, is dropped randomly onto the floor. Determine the probability that the needle crosses one of the parallel lines. *Hint:* Formulate the problem in terms of selecting a point (x, y) at random from the rectangle $[0, d/2] \times [0, \pi]$, where x denotes the distance from the center of the needle to the closest line and y denotes the angle that the needle forms with that line.

2.4 Basic Properties of Probability

Earlier we stated that the Kolmogorov axioms are all that we need in order to develop the entire theory of probability. In this section, we apply those axioms to derive some of the most important and useful properties of probability measures.

As a first application, we prove that

$$P(\emptyset) = 0. \tag{2.10}$$

Recall that the empty set, $\emptyset$, represents an impossible event—that is, an event that cannot occur—so it makes sense that its probability is 0. To prove that, we note that $\Omega = \Omega \cup \emptyset$ and that the two events on the right of this equation are mutually exclusive. Applying the certainty and additivity axioms, respectively, we conclude that

$$1 = P(\Omega) = P(\Omega \cup \emptyset) = P(\Omega) + P(\emptyset) = 1 + P(\emptyset),$$

or $P(\emptyset) = 0$.

Domination Principle

Let A and B be events of a sample space Ω. Recall from Proposition 2.1 that $A \subset B$ if and only if event B occurs whenever event A occurs. It stands to reason then that $A \subset B$ implies $P(A) \leq P(B)$. We call this fact the **domination principle** and prove it in Proposition 2.6.

◆◆◆ **Proposition 2.6   Domination Principle**

Let A and B be events. Then

$$A \subset B \quad \Rightarrow \quad P(A) \leq P(B). \tag{2.11}$$

In words, if one event always occurs whenever another event occurs, then the probability of the latter event is at most equal to the probability of the former event.

Proof   A Venn diagram for this situation is shown in Figure 2.5. From it, we can obtain an informal proof of $P(A) \leq P(B)$. Indeed, if we think of the shaded regions as probabilities, the darkly shaded disk is $P(A)$ and the total shaded region is $P(B)$. Because the darkly shaded region is smaller than the total shaded region, we conclude that $P(A) \leq P(B)$.

Figure 2.5 Venn diagram for $A \subset B$

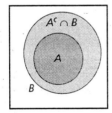

For a formal proof, we first note that A and $A^c \cap B$ are mutually exclusive and, because $A \subset B$, we have $B = A \cup (A^c \cap B)$. Applying the additivity and nonnegativity axioms, respectively, we get $P(B) = P(A) + P(A^c \cap B) \geq P(A)$, as required. ◆

EXAMPLE 2.26 *The Domination Principle*

Playing Cards A deck of playing cards contains 52 cards, as shown in Figure 2.2 on page 35. When we perform the experiment of randomly selecting one card from the deck, the sample space consists of the collection of all 52 cards. Let

$$A = \text{event the card selected is a jack and}$$

$$B = \text{event the card selected is a face card.}$$

a) Without doing any calculations, explain why $P(A) \leq P(B)$.
b) Verify numerically that $P(A) \leq P(B)$.

Solution a) Because a jack is a face card, event B must occur whenever event A occurs and, consequently, $A \subset B$. Therefore the domination principle implies that $P(A) \leq P(B)$.
b) Because we are selecting a card at random, a classical probability model is appropriate here. Thus, for each event E,

$$P(E) = \frac{N(E)}{N(\Omega)} = \frac{N(E)}{52}.$$

Referring to Figure 2.2, we see that event A consists of the four jacks and event B consists of the 12 face cards. So, $P(A) = 4/52 = 0.077$ and $P(B) = 12/52 = 0.231$. Therefore we see numerically that $P(A) \leq P(B)$. ■

Let E be any event of a sample space Ω. Because $E \subset \Omega$, the domination principle and the certainty axiom imply that $P(E) \leq P(\Omega) = 1$. Thus

$$P(E) \leq 1, \qquad \text{for each event } E. \tag{2.12}$$

Relation (2.12) makes sense because the chance that an event occurs can't exceed 100%.

Complementation Rule

Another rule of probability is the **complementation rule,** which we establish in Proposition 2.7.

◆◆◆ **Proposition 2.7 Complementation Rule**

For each event E,

$$P(E) = 1 - P(E^c). \tag{2.13}$$

In words, the probability that an event occurs equals 1 minus the probability that it doesn't occur.

Proof A Venn diagram for this situation is shown in Figure 2.6. From it, we can obtain an informal proof of the required result. Indeed, if we think of the regions in Figure 2.6 as probabilities, the entire region enclosed by the rectangle is the probability of the sample space, or 1. Furthermore, the shaded region is $P(E)$ and the unshaded region is $P(E^c)$. Hence $P(E) + P(E^c) = 1$ or, equivalently, $P(E) = 1 - P(E^c)$.

Figure 2.6 An event E and its complement

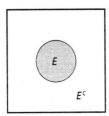

For a formal proof, we first note that E and E^c are mutually exclusive. Moreover, because either E occurs or it doesn't, $E \cup E^c = \Omega$. Applying the certainty and additivity axioms, respectively, we get

$$1 = P(\Omega) = P(E \cup E^c) = P(E) + P(E^c),$$

as required.   ◆

The complementation rule is useful because computing the probability that a specified event doesn't occur is sometimes easier than computing the probability that the event does occur. In such cases, we can, in view of the complementation rule, obtain the probability of the specified event by first computing the probability that it doesn't occur and then subtracting the result from 1. Example 2.27 illustrates this fact.

EXAMPLE 2.27   *The Complementation Rule*

Coin Tossing Suppose that we toss a coin three times and observe the result (a head or a tail) for each toss. Let's suppose also that the coin is balanced. Determine the probability of the event that at least one head is tossed.

Solution Because the coin is balanced, the eight possible outcomes are equally likely. Thus a classical probability model is appropriate here: $P(E) = N(E)/8$ for each event E.

Let E denote the event that at least one head is tossed. We note that E^c is the event that no heads are tossed—that is, all three tosses are tails. Thus E^c is the simple event $\{TTT\}$, whose probability is 1/8, or 0.125. Applying the complementation rule, we conclude that

$$P(E) = 1 - P(E^c) = 1 - 0.125 = 0.875.$$

The probability is 0.875 that at least one head is tossed.   ■

Law of Partitions

In preparation for presenting our next important property of probabilities, we first consider the concept of exhaustive events. Events A_1, A_2, ... are said to be **exhaustive events** if at least one of them must occur when the random experiment is performed. Mathematically, this condition means that $\bigcup_n A_n = \Omega$, where Ω is the sample space of the random experiment.

For instance, the National Governors' Association classifies governors as Democrat, Republican, or Independent. Suppose that a governor is selected at random; let A_1, A_2, and A_3 denote the events that the governor selected is a Democrat, a Republican, and an Independent, respectively. Events A_1, A_2, and A_3 are exhaustive because at least one of them must occur when a governor is selected—the governor obtained must be a Democrat, a Republican, or an Independent. Events A_1, A_2, and A_3 are not only exhaustive, but they are also mutually exclusive because a governor cannot have more than one political party classification at a time.

Generally speaking, if events of a sample space are both exhaustive and mutually exclusive, then exactly one of them must occur when the random experiment is performed. Indeed, at least one of the events must occur (because the events are exhaustive) and at most one of the events can occur (because the events are mutually exclusive).

Figure 2.7(a) portrays three events, A_1, A_2, and A_3, that are both mutually exclusive and exhaustive. The three events don't overlap, indicating that they are mutually exclusive; furthermore, they fill the entire region enclosed by the heavy rectangle (i.e., the sample space), indicating that they are exhaustive.

Figure 2.7 (a) Three mutually exclusive and exhaustive events; (b) an event B and three mutually exclusive and exhaustive events

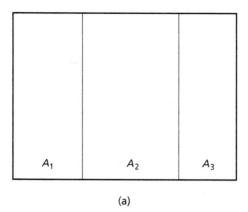
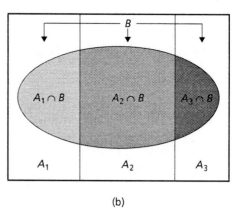

(a)   (b)

As described in Definition 2.5, we use a special terminology—**partition of a sample space**—for events that are both mutually exclusive and exhaustive.

DEFINITION 2.5 Partition of a Sample Space

Events $A_1, A_2, \ldots$ are said to form a **partition of a sample space** Ω if they are both mutually exclusive and exhaustive. That is,

a) $A_m \cap A_n = \emptyset$ for $m \neq n$.

b) $\bigcup_n A_n = \Omega$.

Note: Any event and its complement form a partition. That is, if E is an event, the two events E and E^c form a partition—they are mutually exclusive (E cannot both occur and not occur) and exhaustive (either E occurs or it doesn't).

Now consider a partition consisting of, say, three mutually exclusive and exhaustive events, A_1, A_2, and A_3, and any event B, as portrayed in Figure 2.7(b). Event B can be decomposed into the three mutually exclusive events $A_1 \cap B$, $A_2 \cap B$, and $A_3 \cap B$, shown shaded. This decomposition reflects the fact that event B must occur in conjunction with exactly one of the events A_1, A_2, and A_3. Applying the additivity axiom, we conclude that

$$P(B) = P(A_1 \cap B) + P(A_2 \cap B) + P(A_3 \cap B).$$

More generally, we have Proposition 2.8, known as the **law of partitions.**

◆◆◆ **Proposition 2.8 Law of Partitions**

Suppose that $A_1, A_2, \ldots$ form a partition of the sample space Ω. Then, for each event B,

$$P(B) = \sum_n P(A_n \cap B). \tag{2.14}$$

In particular, if E is an event, then

$$P(B) = P(E \cap B) + P(E^c \cap B), \tag{2.15}$$

for each event B.

Proof Because events $A_1, A_2, \ldots$, are exhaustive,

$$B = \Omega \cap B = \left(\bigcup_n A_n \right) \cap B = \bigcup_n (A_n \cap B).$$

Furthermore, because events $A_1, A_2, \ldots$ are mutually exclusive, so are events $A_1 \cap B$, $A_2 \cap B$, ... Applying the additivity axiom to the preceding display, we get Equation (2.14). Applying Equation (2.14) to the partition consisting of an event E and its complement, we get Equation (2.15). ◆

EXAMPLE 2.28 *The Law of Partitions*

Institutions of Higher Education The U.S. National Center for Education Statistics compiles information on institutions of higher education and publishes its findings in the *Digest of Education Statistics.* According to that document, 8.1% of institutions of higher education are public schools in the Northeast, 11.0% are public schools in the Midwest, 16.3% are public schools in the South, and 9.6% are public schools in the West. If a U.S. institution of higher education is selected at random, determine the probability that it is public.

Solution First we rephrase the problem in terms of probability. Let

$$A_1 = \text{event the school selected is in the Northeast,}$$
$$A_2 = \text{event the school selected is in the Midwest,}$$
$$A_3 = \text{event the school selected is in the South,}$$
$$A_4 = \text{event the school selected is in the West,}$$

and let B denote the event that the school selected is public. According to the data provided, we have $P(A_1 \cap B) = 0.081$, $P(A_2 \cap B) = 0.110$, $P(A_3 \cap B) = 0.163$, and $P(A_4 \cap B) = 0.096$.

Because a U.S. school must be in exactly one of the four U.S. regions, events A_1, A_2, A_3, and A_4 form a partition. Applying the law of partitions, we conclude that

$$P(B) = P(A_1 \cap B) + P(A_2 \cap B) + P(A_3 \cap B) + P(A_4 \cap B)$$
$$= 0.081 + 0.110 + 0.163 + 0.096 = 0.450.$$

Thus 45.0% of institutions of higher education in the United States are public. ∎

General Addition Rule

The additivity axiom implies that the probability of the union of two mutually exclusive events equals the sum of their probabilities—that is, $P(A \cup B) = P(A) + P(B)$, whenever A and B are mutually exclusive. For events that are not mutually exclusive, we must take a different approach—namely, we must apply the **general addition rule,** presented in Proposition 2.9.

◆◆◆ **Proposition 2.9** **General Addition Rule**

For any two events A and B, we have

$$P(A \cup B) = P(A) + P(B) - P(A \cap B).$$

In words, for any two events, the probability that one or the other (or both) occurs equals the sum of the individual probabilities less the probability that both occur.

Proof A Venn diagram for this situation is shown in Figure 2.8. From it, we can obtain an informal proof of the required result. Indeed, if we think of the shaded regions as probabilities, the shaded disk on the left is $P(A)$, the shaded disk on the right is $P(B)$, and the total shaded region is $P(A \cup B)$. To obtain the total shaded region, $P(A \cup B)$, we first sum the two shaded disks, $P(A)$ and $P(B)$. In doing so, the common shaded region, $P(A \cap B)$, is counted twice. To account for that fact, we must subtract $P(A \cap B)$ from the sum. Hence $P(A \cup B) = P(A) + P(B) - P(A \cap B)$.

Figure 2.8 Two non–mutually exclusive events

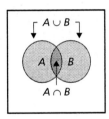

For a formal proof, we first note that $A \cup B = A \cup (A^c \cap B)$, as can be seen easily from a Venn diagram. The advantage of expressing $A \cup B$ in this way is that events A and $A^c \cap B$ are mutually exclusive. So, by the additivity axiom, we have

$$P(A \cup B) = P(A) + P(A^c \cap B). \tag{2.16}$$

Next we apply the law of partitions, specifically, Equation (2.15) with $E = A$, to conclude that $P(B) = P(A \cap B) + P(A^c \cap B)$. Equivalently,

$$P(A^c \cap B) = P(B) - P(A \cap B).$$

Substituting the right side of the preceding equation for $P(A^c \cap B)$ in Equation (2.16), we obtain the required result.   ◆

EXAMPLE 2.29   *The General Addition Rule*

Playing Cards Consider again the experiment of selecting a card at random from a deck of 52 playing cards. Find the probability that the card selected is either a spade or a face card
a) without using the general addition rule.   b) by using the general addition rule.

Solution   a) Let

$$E = \text{event the card selected is either a spade or a face card.}$$

Event E consists of 22 cards: the 13 spades plus the other 9 face cards that are not spades. Consequently,

$$P(E) = \frac{N(E)}{52} = \frac{22}{52} = 0.423.$$

b) To determine $P(E)$ by using the general addition rule, we write $E = C \cup D$, where

$$C = \text{event the card selected is a spade and}$$
$$D = \text{event the card selected is a face card.}$$

Event C consists of the 13 spades and event D consists of the 12 face cards. Also, event $C \cap D$ consists of the three spades that are face cards—the jack, queen, and king of spades. Applying the general addition rule, we get

$$P(E) = P(C \cup D) = P(C) + P(D) - P(C \cap D)$$

$$= \frac{13}{52} + \frac{12}{52} - \frac{3}{52} = 0.250 + 0.231 - 0.058 = 0.423,$$

which agrees with the answer obtained in part (a).   ■

In Example 2.29, we computed the probability of selecting either a spade or a face card in two ways: first without using the general addition rule and then using it. There, computing the probability was somewhat simpler without using the general addition rule. Frequently, however, the general addition rule is the easier or even the only way to compute a probability. We illustrate this point in Example 2.30.

EXAMPLE 2.30   *The General Addition Rule*

Characteristics of People Arrested   Data on people arrested are published by the U.S. Federal Bureau of Investigation in *Crime in the United States*. One year, 79.6% of people arrested were male, 18.3% were under the age of 18, and 13.5% were males under the age of 18. Suppose that a person arrested that year is selected at random. Find the probability that the person obtained is either male or under the age of 18.

Solution   Let

$$M = \text{event the person obtained is male and}$$

$$E = \text{event the person obtained is under the age of 18.}$$

The event that the person obtained is either male or under the age of 18 can be expressed as $M \cup E$. We want to determine $P(M \cup E)$. From the given data, we know that $P(M) = 0.796$, $P(E) = 0.183$, and $P(M \cap E) = 0.135$. Applying the general addition rule, we conclude that

$$P(M \cup E) = P(M) + P(E) - P(M \cap E)$$

$$= 0.796 + 0.183 - 0.135 = 0.844.$$

In terms of percentages, 84.4% of those arrested in the year in question were either male or under the age of 18.   ■

The general addition rule is consistent with the additivity axiom: If two events are mutually exclusive, both formulas yield the same result. See Exercise 2.70.

Inclusion–Exclusion Principle

We can extend the general addition rule to any finite number of events. To show how this method works, we consider three events, A, B, and C. The correct result is that

$$P(A \cup B \cup C) = P(A) + P(B) + P(C)$$
$$- P(A \cap B) - P(A \cap C) - P(B \cap C) \qquad \textbf{(2.17)}$$
$$+ P(A \cap B \cap C).$$

A Venn diagram for this situation is shown in Figure 2.9. From it, we can obtain an informal proof of Equation (2.17). Indeed, if we think of the shaded regions as probabilities, the total shaded region is $P(A \cup B \cup C)$. To obtain its total, we first sum the three shaded disks, $P(A)$, $P(B)$, and $P(C)$. In doing so, the region corresponding to $P(A \cap B)$ is counted twice, once with $P(A)$ and once with $P(B)$. To account for that fact, we must subtract $P(A \cap B)$ from the sum. Likewise, we must subtract both $P(A \cap C)$ and $P(B \cap C)$ from the sum. We have added $P(A \cap B \cap C)$ three times—once each with $P(A)$, $P(B)$, and $P(C)$—and subtracted it three times—once each with $P(A \cap B)$, $P(A \cap C)$, and $P(B \cap C)$. So, to include $P(A \cap B \cap C)$, we must again add it. In summary, we get Equation (2.17).

Figure 2.9 Three non–mutually exclusive events

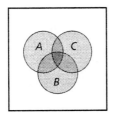

For a formal proof of Equation (2.17), we first apply the general addition rule to the events $A \cup B$ and C to get

$$
\begin{aligned}
P(A \cup B \cup C) &= P((A \cup B) \cup C) \\
&= P(A \cup B) + P(C) - P((A \cup B) \cap C) \\
&= P(A) + P(B) - P(A \cap B) + P(C) \\
&\quad - P((A \cup B) \cap C).
\end{aligned}
\tag{2.18}
$$

To expand the last probability in Equation (2.18), we first apply the distributive law for sets and then apply the general addition rule to $A \cap C$ and $B \cap C$:

$$
\begin{aligned}
P((A \cup B) \cap C) &= P((A \cap C) \cup (B \cap C)) \\
&= P(A \cap C) + P(B \cap C) - P((A \cap C) \cap (B \cap C)) \\
&= P(A \cap C) + P(B \cap C) - P(A \cap B \cap C).
\end{aligned}
$$

Substituting the right side of the preceding equation for $P((A \cup B) \cap C)$ in Equation (2.18), we get Equation (2.17), the inclusion–exclusion principle for three events.

We can use mathematical induction and the techniques just used to obtain the general form of the **inclusion–exclusion principle,** presented in Proposition 2.10. We leave the details to you as Exercise 2.72.

◆◆◆ **Proposition 2.10**   **Inclusion–Exclusion Principle**

Let $A_1, A_2, \ldots, A_N$ be events. Then

$$P\left(\bigcup_{n=1}^{N} A_n\right) = \sum_{k=1}^{N} P(A_k) - \sum_{k_1 < k_2} P(A_{k_1} \cap A_{k_2}) + \cdots$$

$$+ (-1)^{n+1} \sum_{k_1 < k_2 < \cdots < k_n} P(A_{k_1} \cap A_{k_2} \cap \cdots \cap A_{k_n})$$

$$+ \cdots + (-1)^{N+1} P(A_1 \cap A_2 \cap \cdots \cap A_N).$$

In words, the probability that at least one of a finite number of events occurs equals the sum of the probabilities of these events taken one at a time, minus the sum of the probabilities of these events taken two at a time, plus the sum of the probabilities of these events taken three at a time, and so on.

EXAMPLE 2.31   *The Inclusion–Exclusion Principle*

Newspaper Subscription   Suppose that a certain city has three major newspapers, the *Times*, *Herald*, and *Examiner*. Circulation information indicates that 47.0% of households get the *Times*, 33.4% get the *Herald*, 34.6% get the *Examiner*, 11.9% get the *Times* and *Herald*, 15.1% get the *Times* and *Examiner*, 10.4% get the *Herald* and *Examiner*, and 4.8% get all three. If a household in this city is selected at random, determine the probability that it gets at least one of the three major newspapers.

Solution   Let

$$T = \text{event the household selected gets the } Times,$$
$$H = \text{event the household selected gets the } Herald, \text{ and}$$
$$E = \text{event the household selected gets the } Examiner.$$

We want to obtain the probability that the household selected gets at least one of the three major newspapers—that is, $P(T \cup H \cup E)$. Applying the inclusion–exclusion principle with $N = 3$ or, equivalently, Equation (2.17) on page 71, yields

$$P(T \cup H \cup E) = P(T) + P(H) + P(E)$$
$$- P(T \cap H) - P(T \cap E) - P(H \cap E)$$
$$+ P(T \cap H \cap E)$$
$$= 0.470 + 0.334 + 0.346 - 0.119 - 0.151 - 0.104 + 0.048$$
$$= 0.824.$$

The probability is 0.824 that a household selected at random gets at least one of the three major newspapers. In other words, 82.4% of households in the city get at least one of the three major newspapers.   ∎

Continuity Properties

Probability measures have important "continuity properties," as presented in Proposition 2.11. These properties are especially useful in developing the theory of probability. We leave the proof of the proposition to you as Exercise 2.73.

◆◆◆ **Proposition 2.11   Continuity Properties of Probability Measures**

Let $A_1, A_2, \ldots$ be events of a sample space Ω.
a) If $A_1 \subset A_2 \subset \cdots$, then

$$P\left(\bigcup_{n=1}^{\infty} A_n \right) = \lim_{n \to \infty} P(A_n).$$

b) If $A_1 \supset A_2 \supset \cdots$, then

$$P\left(\bigcap_{n=1}^{\infty} A_n \right) = \lim_{n \to \infty} P(A_n).$$

EXAMPLE 2.32   *Continuity Properties of Probability Measures*

Coin Tossing   Suppose that a balanced coin is tossed repeatedly. Determine the probability that, eventually, a head will be tossed—that is, the probability that, eventually, a toss of the coin will land with the head facing up.

Solution   Let A denote the event that a head will eventually be tossed and, for each $n \in \mathcal{N}$, let A_n denote the event of at least one head in the first n tosses. Then $A_1 \subset A_2 \subset \cdots$ and $A = \bigcup_{n=1}^{\infty} A_n$. Hence, by Proposition 2.11(a), $P(A) = \lim_{n \to \infty} P(A_n)$.

We now obtain $P(A_n)$ for each $n \in \mathcal{N}$. As it is simpler to compute $P(A_n^c)$—the probability that the first n tosses all result in tails—we use the complementation rule to determine $P(A_n)$.

- $n = 1$: When a balanced coin is tossed once, two equally likely outcomes are possible—namely, H and T. Therefore $P(A_1^c) = P(\{T\}) = 1/2$ and, consequently, $P(A_1) = 1 - P(A_1^c) = 1 - 1/2$.
- $n = 2$: When a balanced coin is tossed twice, four equally likely outcomes are possible—namely, HH, HT, TH, TT. Therefore $P(A_2^c) = P(\{TT\}) = 1/4$ and, consequently, $P(A_2) = 1 - P(A_2^c) = 1 - 1/4$.
- $n = 3$: When a balanced coin is tossed three times, eight equally likely outcomes are possible—namely, HHH, HHT, HTH, HTT, THH, THT, TTH, TTT. Therefore $P(A_3^c) = P(\{TTT\}) = 1/8$ and, consequently, $P(A_3) = 1 - P(A_3^c) = 1 - 1/8$.

We can now see a pattern (which can be established by induction or by using a simple counting technique discussed in Chapter 3): When a balanced coin is tossed n times, 2^n equally likely outcomes are possible and hence $P(A_n) = 1 - P(A_n^c) = 1 - 1/2^n$. Therefore

$$P(A) = \lim_{n \to \infty} P(A_n) = \lim_{n \to \infty} \left(1 - \frac{1}{2^n} \right) = 1.$$

The probability that a head will eventually be tossed is 1. In other words, we will eventually get a head when we repeatedly toss a balanced coin. As we demonstrate in Example 4.17, this result holds regardless of whether the coin is balanced, provided only that the probability is not 0 of getting a head when the coin is tossed once.   ▪

EXERCISES 2.4   Basic Exercises

2.58 Let A and B be events of a sample space. Provide an example where, as sets, A is a proper subset of B, but $P(A) = P(B)$.

2.59 Give an example to show that the converse of the domination principle fails.

2.60 A person is selected at random from among the inhabitants of a state. Which is more probable: that the person so chosen is a lawyer, or that the person so chosen is a Republican lawyer? Explain your answer.

2.61 Refer to Exercise 2.41 on page 62. Use the complementation rule to find the probability that at least one Republican will be on the subcommittee. Why would use of the complementation rule for this problem make things easier than if that rule weren't used?

2.62 Refer to Exercise 2.20 on page 45. Determine the probability that an oil spill in U.S. navigable and territorial waters doesn't occur in the Gulf of Mexico
a) without use of the complementation rule   **b)** by using the complementation rule.
c) Compare the work done in your solutions in parts (a) and (b).

2.63 Refer to Exercise 2.39 on page 61. Suppose that a player on the New England Patriots is selected at random. Determine the probability that the player obtained
a) has at least 1 year of experience.   **b)** weighs at most 300 lb.
c) is either a rookie or weighs more than 300 lb. Solve this problem both with and without use of the general addition rule and compare your work.

2.64 If a point is selected at random from the unit square $\{ (x, y) : 0 \le x \le 1, 0 \le y \le 1 \}$, find the probability that the magnitude of the difference between the x and y coordinates of the point obtained is at most 1/4. Solve this problem both with and without use of the complementation rule and compare your work.

2.65 According to *Current Population Reports*, published by the U.S. Bureau of the Census, 51.0% of U.S. adults are female, 7.1% are divorced, and 4.1% are divorced females. Determine the probability that a U.S. adult selected at random is
a) either female or divorced.   **b)** a male.
c) a female but not divorced.   **d)** a divorced male.

2.66 Let A and B be events such that $P(A) = 1/4$, $P(B) = 1/3$, and $P(A \cup B) = 1/2$.
a) Are events A and B mutually exclusive? Explain your answer.
b) Determine $P(A \cap B)$.

2.67 Let A and B be events such that $P(A) = 1/3$, $P(A \cup B) = 5/8$, and $P(A \cap B) = 1/10$. Determine
a) $P(B)$.   **b)** $P(A \cap B^c)$.   **c)** $P(A \cup B^c)$.   **d)** $P(A^c \cup B^c)$.

2.68 Gerald Kushel, Ed.D., was interviewed by *Bottom Line/Personal* on the secrets of successful people. To study success, Kushel questioned 1200 people, among whom were lawyers, artists, teachers, and students. He found that 15% enjoy neither their jobs nor their personal lives, 80% enjoy their jobs but not their personal lives, and 4% enjoy both their jobs and their personal lives. Obtain the percentage of the 1200 people interviewed who enjoy

a) either their jobs or their personal lives.   **b)** their jobs.

c) their personal lives but not their jobs.   **d)** their personal lives.

2.69 Refer to Example 2.31 on page 73. Determine the probability that a household selected at random gets

a) either the *Times* or the *Herald,* but not both.

b) exactly one of the three newspapers.

c) none of the three newspapers.

d) the *Times* and the *Herald,* but not the *Examiner.*

e) exactly two of the three newspapers.

2.70 This exercise deals with the relationships among the general addition rule, the inclusion–exclusion principle, and the additivity axiom.

a) Show that, for two events, the inclusion–exclusion principle reduces to the general addition rule.

b) Show that the general addition rule is consistent with the additivity axiom—that is, for two mutually exclusive events, the general addition rule reduces to the additivity axiom when applied to two events.

c) More generally than in part (b), show that the inclusion–exclusion principle is consistent with the additivity axiom—that is, for N mutually exclusive events, the inclusion–exclusion principle reduces to the additivity axiom when applied to N events.

2.71 A quiz was administered to four students. Somehow the quizzes got shuffled, and the one at the top of the stack was returned to the first student, the one below it was returned to the second student, and so on. For $i \in \{1, 2, 3, 4\}$, let A_i be the event that the ith student got his own quiz back. Without trying to evaluate the result, use the inclusion–exclusion principle to write an expression for the probability that at least one student got the right quiz back. *Hint:* The expression you get should have 15 terms.

Theory Exercises

2.72 Use mathematical induction to prove the inclusion–exclusion principle.

2.73 In this exercise, you are to prove the continuity properties of a probability measure, Proposition 2.11 on page 74. Let $A_1, A_2, \ldots$ be events of a sample space Ω. Suppose first that $A_1 \subset A_2 \subset \cdots$. Let $B_1 = A_1$ and, for $n \geq 2$, let $B_n = A_n \cap A_{n-1}^c$.

a) Interpret B_n in words.

b) Prove that $B_1, B_2, \ldots$ are mutually exclusive and that $\bigcup_{n=1}^{\infty} B_n = \bigcup_{n=1}^{\infty} A_n$.

c) Prove that, for $n \geq 2$, $P(B_n) = P(A_n) - P(A_{n-1})$.

d) Show that $P\left(\bigcup_{n=1}^{\infty} B_n\right) = \lim_{n \to \infty} P(A_n)$.

e) Deduce that Proposition 2.11(a) holds.

f) Use Proposition 2.11(a) to deduce Proposition 2.11(b). *Hint:* Consider complements.

2.74 Bonferroni's inequality: A useful relation in probability theory is *Bonferroni's inequality,* named for its discoverer Carlo Emilio Bonferroni (1892–1960). This inequality provides a lower bound on the probability of the simultaneous occurrence of a finite number of events in terms of the individual probabilities of the events. Specifically, it states that, for each positive integer N and events $A_1, A_2, \ldots, A_N$,

$$P\left(\bigcap_{n=1}^{N} A_n\right) \geq \sum_{n=1}^{N} P(A_n) - (N - 1).$$

a) Prove Bonferroni's inequality for $N = 2$.

b) Use mathematical induction to prove that Bonferroni's inequality holds for all positive integers N.

2.75 Boole's inequality: Another useful relation in probability theory is *Boole's inequality*, named for its discoverer George Boole (1815–1864). This inequality provides an upper bound on the probability of the occurrence of at least one of a countable number of events in terms of the individual probabilities of the events. Specifically, it states that, if $A_1, A_2, \ldots$ are events, then

$$P\left(\bigcup_n A_n\right) \leq \sum_n P(A_n).$$

a) Prove Boole's inequality for two events.

b) Use mathematical induction to prove that Boole's inequality holds for a finite number of events.

c) Use Proposition 2.11 on page 74 to prove that Boole's inequality holds for a countably infinite number of events.

d) Derive Bonferroni's inequality (Exercise 2.74) from Boole's inequality.

Advanced Exercises

2.76 Refer to Example 2.32 on page 74, where a balanced coin is tossed repeatedly. Let A denote the event that a head will eventually be tossed.

a) Express A as a countably infinite union of mutually exclusive events and use that representation to show that $P(A) = 1$. *Hint:* Let A_n be the event that the first head occurs on the nth toss.

b) Determine the probability that the first head occurs on an odd-numbered toss.

c) Determine the probability that the first head occurs on an even-numbered toss.

2.77 First Borel–Cantelli lemma: Let $A_1, A_2, \ldots$ be a countably infinite sequence of events and let A^* be the event that infinitely many of the A_ns occur. Prove that, if $\sum_{n=1}^{\infty} P(A_n) < \infty$, then $P(A^*) = 0$. This result is often referred to as the *first Borel–Cantelli lemma. Hint:* Refer to Exercise 2.18(e) on page 38 and to Exercise 2.75.

2.78 Suppose that instead of Kolmogorov's axiom of countable additivity [Definition 2.4(c) on page 41], we assume only the axiom of *finite additivity:* If A and B are two mutually exclusive events, then $P(A \cup B) = P(A) + P(B)$.

a) Use mathematical induction to prove that, if $A_1, A_2, \ldots, A_N$ are mutually exclusive events, then

$$P\left(\bigcup_{n=1}^{N} A_n\right) = \sum_{n=1}^{N} P(A_n).$$

Note that this equation says something only about finite sequences of mutually exclusive events, whereas the statement of countable additivity also includes infinite sequences of mutually exclusive events.

b) Now suppose that, in addition to assuming the axiom of finite additivity, we take the continuity property (Proposition 2.11 on page 74) as an axiom as well. Deduce countable additivity from these two axioms.

c) Conclude now that finite additivity and continuity together are equivalent to countable additivity.

CHAPTER REVIEW

Summary

We began this chapter by establishing a representation of events by sets. In particular, we showed that each event corresponds to a unique subset of the sample space—the set of all possible outcomes of the random experiment. We also demonstrated that relationships among events are interpreted mathematically as set operations.

We then turned our attention to the axioms of probability—the basic properties that probabilities must satisfy. Using the frequentist interpretation of probability as an intuitive basis, we developed the required axioms—the three Kolmogorov axioms of nonnegativity, certainty, and additivity.

Mathematically, probabilities need only satisfy the three Kolmogorov axioms. However, to be useful practically, they must also reflect the underlying nature of the random experiment under consideration. Otherwise, results derived by employing those probabilities will be useless.

Next we presented three methods for specifying or obtaining probabilities so that they reflect the underlying nature of the random experiment. These three methods are probability models, empirical probability, and subjective probability. Our discussion of probability models concentrated on two of the most important ones: classical probability and geometric probability. The classical probability model is used for random experiments in which there are a finite number of equally likely outcomes; geometric probability can be used to model random experiments in which a point is selected at random from a subset of n-dimensional Euclidean space that has finite, nonzero n-dimensional volume.

Finally, we applied the Kolmogorov axioms to derive some of the most important and widely used properties of probabilities. In doing so, we also introduced the concept of a partition of the sample space—a collection of countably many events that are both mutually exclusive and exhaustive.

You Should Be Able To ...

1. define and apply the concepts of sample space and events.

2. identify whether events are mutually exclusive.

3. state, understand, and apply the three Kolmogorov axioms.

4. compute probabilities for events of a finite sample space, knowing only the probabilities of the individual outcomes.

5. decide whether a probability assignment to the outcomes of a finite sample space is mathematically valid.

6. identify methods for specifying and obtaining probabilities.

7. construct, use, and apply a classical probability model.

8. construct, use, and apply a geometric probability model.

9. define and understand empirical probability and subjective probability.

10. state, derive, and apply basic properties of probability.

11. explain and use the concept of partition of a sample space.

Key Terms

additivity axiom, *41*

cells, *53*

certainty axiom, *41*

classical probability, *49*

complementation rule, *65*

contingency table, *52*

countable sample space, *42*

domination principle, *64*

empirical probability, *59*

equal-likelihood model, *49*

event, *29*

exhaustive events, *67*

finite sample space, *42*

general addition rule, *69*

geometric probability, *55*

impossible event, *34*

inclusion–exclusion principle, *73*

joint probability, *54*

joint probability distribution, *54*

Kolmogorov axioms, *41*

law of partitions, *68*

marginal probability, *54*

mutually exclusive events, *34*

nonnegativity axiom, *41*

occurrence of an event, *29*

pairwise mutually exclusive
   events, *34*

partition of a sample space, *68*

probability measure, *41*

probability model, *49*

sample space, *26*

simple event, *42*

subjective probability, *60*

two-way table, *52*

Chapter Review Exercises

Basic Exercises

2.79 Refer to Example 2.3 on page 27. Using the ordered-pair representation of the sample space given there, determine the event that the black die comes up 2.

2.80 A small club has six members—three men (say, a, b, and c) and three women (say, d, e, and f). Two members of the six are chosen at random to be co-chairs.
a) Determine the sample space for this random experiment.
b) Determine, as a subset of the sample space, the event that at least one of a and f is chosen as a co-chair.
c) Repeat parts (a) and (b) if one member is chosen at random from among the three men and one member is chosen at random from among the three women.
d) Repeat part (a) if two members of the six are chosen at random to be chair and vice-chair. Then determine, as a subset of the sample space, the event that at least one of a and f is chosen as one of the two officers.

2.81 You intend to observe 100 cars passing a busy intersection and to classify each car as being foreign-made or U.S.-made, and also according to size—compact, midsize, or fullsize. Obtain a sample space for which each outcome represents the number of the 100 cars observed in each of the possible cross-classification categories.

2.82 A jury of 12 will be chosen randomly from a panel of 10 men and 10 women. Which of the following are mutually exclusive collections of events? Explain your answers.
a) At least 5 men are chosen; at least 5 women are chosen.
b) At least 5 men are chosen; at least 8 women are chosen.
c) Five men and 7 women are chosen; 4 men and 8 women are chosen; 3 men and 9 women are chosen.
d) The first person on the list of 20 is among those chosen; the second person on the list of 20 is among those chosen; the third person on the list of 20 is among those chosen.

2.83 Twelve jurors and three alternates have been chosen. The three alternates are, say, A, B, and C. One of these three will be designated the first alternate, who will become a member of the jury if at some time during the trial one of the 12 jurors fails to attend; another will be designated the second alternate, who will become a member of the jury if at some time during the trial two of the 12 jurors fail to attend; and similarly the last will be the third alternate. Who among A, B, and C will be the first alternate, who will be the second, and who will be the third will be decided by chance.

a) Obtain the sample space for the random experiment of assigning alternates.

b) Determine, as a subset of the sample space, the event that A is the first alternate.

Let A = event that A is the first alternate, let B = event that B is the second alternate, and let C = event that C is the third alternate. Describe in words and obtain as a subset of the sample space each of the following events.

c) $A^c \cap B^c \cap C^c$ **d)** $A \cap B \cap C$ **e)** $A \cap C$

f) What probability assignment is appropriate for this random experiment? Explain.

g) Using the probability assignment in part (f), determine the probability of each of the events in parts (c)–(e).

2.84 This problem involves the use of simulation to provide an intuitive justification for the additivity axiom (third Kolmogorov axiom) of a probability measure and requires a computer or calculator with a random-number generator. Consider the experiment of randomly selecting a decimal digit—that is, an integer between 0 and 9, inclusive. Let A = event that the digit is 3 or less and let B = event that the digit is 7 or 8.

a) Use a random-number generator to obtain 1000 random decimal digits and then find the proportion of times that event A occurred. What is this proportion estimating?

b) Repeat part (a) for event B.

c) Repeat part (a) for event $A \cup B$.

d) Are your results in parts (a)–(c) consistent with the additivity axiom? Explain.

e) Repeat parts (a)–(d) with 10,000 random digits instead of 1000.

2.85 Complete the following table.

Event	Description
$A \cap B^c$	A occurs but B doesn't.
$A \cup B^c$	——
——	Neither A nor B occurs.
$(A \cap B^c) \cup (A^c \cap B)$	——
——	Events A, B, and C occur.
$A \cap (B \cup C)$	——
——	Either A occurs or both B and C occur.
$A \cup B \cup C$	——
——	At least one of A, B, and C does not occur.
$\bigcap_n A_n$	——
——	At least one of $A_1, A_2, \ldots$ occur.

2.86 As you know, $A \cup B$ is the event that at least one of event A and event B occurs. Using set notation, express the event that exactly one of event A and event B occurs.

2.87 Let A, B, and C be events. Answer true or false to each statement and explain your reasoning.

a) If events A, B, and C are mutually exclusive, so are events A and B.

b) If events A and B are mutually exclusive, so are events A, B, and C.

2.88 Consider the random experiment of tossing a coin twice and observing on each toss whether the coin comes up a head (H) or a tail (T). A sample space for this random experiment is $\Omega = \{\text{HH, HT, TH, TT}\}$. Let p be a real number with $0 \le p \le 1$.

a) Does the assignment $P(\{HH\}) = p^2$, $P(\{HT\}) = p(1 - p)$, $P(\{TH\}) = p(1 - p)$, and $P(\{TT\}) = (1 - p)^2$ determine a unique probability measure on Ω? Explain.

b) Based on the probability assignment in part (a), find the probability of each of the 16 events of this random experiment.

2.89 A device contains three components, each of which is equally likely to be working or broken at any given time.

a) Construct a probability model for this situation.

b) Determine the probability that all three components are broken.

c) At least two of the three components must be working for the device to function. What is the probability of that happening?

2.90 A point is chosen at random from the unit disk $\{ (x, y) : x^2 + y^2 < 1 \}$. Find the probability that

a) its distance from the center of the disk exceeds 1/2.

b) the sum of the magnitudes of its two coordinates exceeds 1/2.

c) the maximum of the magnitudes of its two coordinates exceeds 1/2.

2.91 Let $\Omega = \mathcal{N}$ and assign the outcome n probability $1/2^n$.

a) Verify that this probability assignment is legitimate.

b) Determine the probability that the outcome is an odd number.

2.92 If the outcome of the toss of a balanced coin is a head, you take one step to the right; if it's a tail, you take two steps to the left. The coin is tossed once every second. Your goal is to reach a position three steps to the right of your starting point. For $n = 3, 4, 5, \ldots$, let B_n be the event that, after the nth toss, you have still not reached your goal.

a) Explain why $P(B_5) \geq P(B_6)$.

b) Express the probability that you will never reach your goal by writing an expression in which $P(B_n)$ appears.

c) Explain why the probability that you will never reach your goal is strictly less than 7/8.

2.93 Two women agree to meet at a specified place between 3:00 P.M. and 5:00 P.M. They further agree that the first one to arrive will wait a maximum of 40 minutes, after which she will leave. Assuming that each arrival time is equally likely between 3:00 P.M. and 5:00 P.M. and that the time of arrival of one woman doesn't affect that of the other, determine the probability that the two women meet.

2.94 The following statement was heard on the radio one morning: "The probability of rain in upstate South Carolina today is 40%." What type of probability is this? Explain.

2.95 In each case, decide which method of specifying probabilities would most likely be used— probability model, empirical probability, or subjective probability.

a) The probability of a defective bolt in a manufacturing process

b) The probability that, in a particular horse race, a specified horse will finish first

c) The probability that, in general, the favorite in a horse race will finish first

d) The chance that it rains tomorrow in your local area

2.96 In a certain town, 80% of the households own an automobile, 45% own a home, and 35% own both an automobile and a home. Determine the probability that a household in this town selected at random

a) owns an automobile or a home but not both.

b) owns neither an automobile nor a home.

2.97 The U.S. Federal Highway Administration compiles information on motor vehicle use around the world and publishes its findings in *Highway Statistics*. The following table gives the number of motor vehicles in use in North America by country and type of vehicle. Frequencies are in thousands.

Vehicle type	Country: U.S. C_1	Canada C_2	Mexico C_3
Automobiles V_1	129,728	13,138	8,607
Motorcycles V_2	3,871	320	270
Trucks V_3	75,940	6,933	4,287

Obtain the row and column totals and then determine how many vehicles are

a) not automobiles. **b)** Canadian. **c)** motorcycles.

d) Canadian motorcycles. **e)** either Canadian or motorcycles.

Suppose that a North American vehicle is selected at random.

f) Describe the events C_1, V_3, and $C_1 \cap V_3$ in words.

g) Compute the probability of each event in part (f).

h) Compute $P(C_1 \cup V_3)$ directly from the table.

i) Compute $P(C_1 \cup V_3)$, using the general addition rule and your answers from part (g).

j) Construct a joint probability distribution.

2.98 An urn contains three balls, numbered 1, 2, and 3. Three balls are drawn at random without replacement. For $i = 1, 2,$ and 3, let A_i be the event that ball i occurs on draw i.

a) Find $P(A_i)$ for $i = 1, 2,$ and 3.

b) Are A_1, A_2 and A_3 mutually exclusive? Explain.

2.99 A balanced coin is tossed four times. What is the probability that

a) the first tail is followed by two consecutive heads?

b) a run of three or more heads occurs?

2.100 Three couples are paired at random on the dance floor, with each pair consisting of one man and one woman. Determine the probability of each of the following events.

a) Each wife dances with her own husband.

b) No wife dances with her own husband.

c) At least one wife dances with her own husband.

2.101 A customer in an appliance store will purchase a washer with probability 0.4, a dryer with probability 0.3, and an iron with probability 0.23. He will purchase both a washer and dryer with probability 0.15, both a washer and an iron with probability 0.13, both a dryer and an iron with probability 0.09, and all three items with probability 0.05. Determine the probability that the customer will purchase

a) none of the items. **b)** two or more of the items. **c)** exactly one of the items.

2.102 Suppose that you classify each e-mail message as nonlegitimate (e.g., unwanted or unsolicited) or legitimate. Further suppose that, during any given hour, the probability of getting a nonlegitimate e-mail message is 0.5, the probability of getting a legitimate e-mail message is 0.7, and the probability of getting some of each is 0.4. What is the probability of receiving no e-mail message during a given hour?

Theory Exercises

2.103 Use mathematical induction to prove that Equation (2.4) on page 40 implies finite additivity; that is, it implies that Equation (2.5) on page 41 holds for all finite collections of mutually exclusive events. Why then is it necessary to assume Equation (2.5) instead of simply Equation (2.4)?

2.104 Let A and B be events of a sample space.
a) Prove that, if A and B both have probability 0, so does $A \cup B$.
b) Prove that, if A and B both have probability 1, so does $A \cap B$.
c) Prove or give a counterexample: Parts (a) and (b) hold for countably many events.
d) Prove or give a counterexample: Parts (a) and (b) hold for uncountably many events.

Advanced Exercises

2.105 Two cards are drawn at random from an ordinary deck of 52 playing cards. Determine the probability that the second card drawn is an ace
a) if the first card is replaced in the deck before the second card is drawn.
b) if the first card is not replaced in the deck before the second card is drawn.

2.106 Relative complement of events: Suppose that A and B are events of a sample space.
a) Interpret the event $A \setminus B$. *Note:* Refer to Exercise 1.62 on page 23.
b) Prove that $P(A \setminus B) = P(A) - P(A \cap B)$.
c) Prove that, if $B \subset A$, then $P(A \setminus B) = P(A) - P(B)$.

2.107 Symmetric difference of events: Suppose that A and B are events of a sample space.
a) Interpret the event $A \bigtriangleup B$. *Note:* Refer to Exercise 1.63 on page 23.
b) Prove that $P(A \bigtriangleup B) = P(A) + P(B) - 2P(A \cap B)$.

2.108 Interpret, in words, each of the following events.

a) $\displaystyle\bigcup_{n=1}^{\infty}\left(\bigcap_{k=n}^{\infty} A_k\right)$ **b)** $\displaystyle\bigcap_{n=1}^{\infty}\left(\bigcup_{k=n}^{\infty} A_k\right)$

c) What is the relationship between the events in parts (a) and (b)? Explain your answer.

2.109 A balanced die is thrown eight times. Your naive colleague wrote

$P(\text{“5” on first throw OR “5” on second throw OR } \ldots \text{ OR “5” on eighth throw})$

$= P(\text{“5” on first throw}) + P(\text{“5” on second throw}) + \cdots + P(\text{“5” on eighth throw})$

$= 1/6 + 1/6 + \cdots + 1/6 = 8/6 = 1.333\ldots$

Explain why this result is impossible and what's wrong with your colleague's reasoning.

2.110 A balanced die is thrown three times. Using the inclusion–exclusion principle, find

$P(\text{“5” on first throw OR “5” on second throw OR “5” on third throw}).$

2.111 Tom, Dick, and Harry alternate tossing a balanced die—first Tom, next Dick, then Harry, then Tom, then Dick, then Harry, and so on. The game stops when the first six appears, the winner being the person who tossed the six.
a) Obtain a sample space for this random experiment.
b) Assign probabilities to the individual outcomes. *Hint:* When a die is tossed n times, there are 6^n outcomes possible. Of those 6^n possible outcomes, 5^{n-1} have the property that the first $n - 1$ tosses are not six and the nth toss is six.
c) Determine the probability that, eventually, the game stops.
d) Determine the probability that the winner is Tom; Dick; Harry.

Jacob Bernoulli *1654–1705*

Jacob (aka Jacques, Jakob, James) Bernoulli was born on December 27, 1654, in Basle, Switzerland. He was the first of the Bernoulli family of mathematicians. Jacob's younger brother, Johann, and various nephews and grandnephews also became renowned mathematicians. His father, Nicolaus Bernoulli (1623–1708), planned the ministry as Jacob's career. However, Jacob rebelled because, to him, mathematics was much more interesting.

Although Bernoulli was schooled in theology, he studied mathematics on his own. He was especially fascinated with calculus. In a 1690 issue of the journal *Acta eruditorum*, Bernoulli used the word *integral* to describe the inverse of differential. The results of his studies of calculus and the catenary (the curve formed by a cord freely suspended between two fixed points) were soon applied to the building of suspension bridges.

Some of Bernoulli's most important work was published posthumously in *Ars Conjectandi* (The Art of Conjecturing) in 1713. This book contains his theory of permutations and combinations, the Bernoulli numbers, and his writings on probability, which include the weak law of large numbers for Bernoulli trials. *Ars Conjectandi* has been regarded as the beginning of the theory of probability.

Both Jacob and his brother Johann were highly accomplished mathematicians. But rather than collaborate in their work, they most often competed. Jacob would publish a question inviting solutions in a professional journal. Johann would reply in the same journal with a solution, only to find that an ensuing issue would contain another article by Jacob, telling him that he was wrong. In their later years, they communicated only in this manner.

Bernoulli began lecturing in natural philosophy and mechanics at the University of Basle in 1682 and became a Professor of Mathematics there in 1687. He remained at the university until his death of a "slow fever" on August 10, 1705.

CHAPTER THREE

Combinatorial Probability

INTRODUCTION

In Chapter 2, we discussed methods for specifying and obtaining probabilities. As you learned, one common method for specifying probabilities is through the use of a *probability model*, a mathematical description of the random experiment based on certain primary aspects and assumptions.

The *classical probability model*—which applies to random experiments with a finite number of equally likely outcomes—is one of the most widely used probability models. For such random experiments, as we demonstrated in Proposition 2.4, probabilities are obtained from the formula

$$P(E) = \frac{N(E)}{N(\Omega)}, \qquad \text{for each event } E, \qquad (3.1)$$

where $N(E)$ denotes the number of ways that E can occur. In words, Equation (3.1) states that the probability of an event equals the number of ways the event can occur divided by the total number of possible outcomes of the random experiment.

Thus, in classical probability, the computation of probabilities reduces to counting the number of ways events can occur. Sometimes this counting process is easy; for instance, we can often obtain the number of ways an event can occur by simply listing the outcomes of which it is comprised, as we did in Examples 2.20 and 2.21 on pages 51 and 52.

But frequently the counting process is much more challenging. That happens, in particular, when the number of ways an event can occur is so large that a listing of outcomes is prohibitive. For such cases, we must instead employ *counting rules* to determine the number of ways an event can occur.

In the first two sections of this chapter, we examine some of the most important counting rules. Then, in the following section, we apply those counting rules to compute probabilities. And, throughout the text, we often rely on counting rules in our study of probability.

3.1   The Basic Counting Rule

We often need to determine the number of ways something can happen: the number of possible outcomes for an experiment, the number of ways an event can occur, the number of ways a certain task can be performed, and the like. Sometimes we can list the possibilities and then count them; but, in most cases, the number of possibilities is so large that a direct listing is impractical.

Consequently, we need to develop techniques that do not rely on a direct listing for determining the number of ways something can happen. Such techniques are usually referred to as **counting rules** or, more formally, as **combinatorial analysis.** In this section, we examine the *basic counting rule,* a counting rule that provides the foundation for all the counting techniques we discuss.

EXAMPLE 3.1   *Introducing the Basic Counting Rule*

Home Options   Robson Communities, Inc., builds new-home communities in several parts of Arizona. In Sun Lakes, Arizona, four models are offered—the Shalimar, Palacia, Valencia, and Monterey—each in three different elevations, designated A, B, and C. How many choices are there for the selection of a home, including both model and elevation?

Solution   We first use the **tree diagram,** shown in Figure 3.1, to systematically obtain a direct listing of the possibilities. In the tree diagram, we have used S for Shalimar, P for Palacia, V for Valencia, and M for Monterey.

Each branch of the tree corresponds to one possibility for model and elevation. For instance, the first branch of the tree, ending in SA, corresponds to the Shalimar model with the A elevation. The total number of possibilities can be obtained by counting the number of branches at the end of the tree. Thus there are 12 choices for the selection of a home, including both model and elevation.

Although the tree-diagram approach for determining the number of possibilities is a direct listing, it provides us with a clue for obtaining the number of possibilities without resorting to a direct listing. Specifically, there are four possibilities for model, indicated by the four subbranches emanating from the starting point of the tree; for each model possibility, there are three elevation possibilities, indicated by the three subbranches

Figure 3.1 Tree diagram for model and elevation possibilities

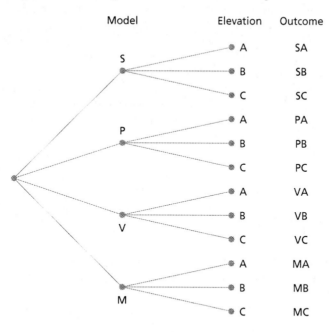

emanating from the end of each model subbranch. Consequently, there are

$$\underbrace{3 + 3 + 3 + 3}_{4 \text{ times}} = 4 \cdot 3 = 12$$

possibilities altogether. Hence the total number of possibilities can be obtained by multiplying the number of possibilities for the model by the number of possibilities for the elevation. ∎

In Example 3.1, there are two actions—choice of model and choice of elevation—and we multiplied the number of possibilities for each action to obtain the total number of possibilities. This same multiplication principle applies regardless of the number of actions. We call that principle the **basic counting rule** and develop it formally in Proposition 3.1.

◆◆◆ **Proposition 3.1** **The Basic Counting Rule (BCR)**[†]

Suppose that r actions (choices, experiments) are to be performed in a definite order. Further suppose that there are m_1 possibilities for the first action, and that corresponding to each of these possibilities there are m_2 possibilities for the second action, and so on. Then there are $m_1 \cdot m_2 \cdots m_r$ possibilities altogether for the r actions.

[†]The basic counting rule is also known as the *basic principle of counting*, the *fundamental counting rule*, and the *multiplication rule*.

Proof To prove the basic counting rule, we can first establish it for two actions and then use mathematical induction on r to obtain it in the general case. We leave the induction proof to you as Exercise 3.20.

So, we consider two actions ($r = 2$) to be performed in a definite order. Suppose that there are m_1 possibilities for the first action and that corresponding to each of these possibilities there are m_2 possibilities for the second action. Let $f_1, f_2, \ldots, f_{m_1}$ denote the m_1 possibilities for the first action and let $s_1, s_2, \ldots, s_{m_2}$ denote the m_2 possibilities for the second action. We enumerate the possibilities for the two actions as follows:

$$
\begin{array}{cccc}
f_1 s_1 & f_1 s_2 & \cdots & f_1 s_{m_2} \\
f_2 s_1 & f_2 s_2 & \cdots & f_2 s_{m_2} \\
\vdots & \vdots & \ddots & \vdots \\
f_{m_1} s_1 & f_{m_1} s_2 & \cdots & f_{m_1} s_{m_2}
\end{array}
$$

In the display, $f_1 s_1$ represents selecting possibility f_1 for the first action and possibility s_1 for the second action, $f_1 s_2$ represents selecting possibility f_1 for the first action and possibility s_2 for the second action, and so on.

Each row in the display gives m_2 possibilities for the two actions. Because there are m_1 rows in the display, we conclude that the total number of possibilities for the two actions is

$$
\underbrace{m_2 + m_2 + \cdots + m_2}_{m_1 \text{ times}} = m_1 \cdot m_2,
$$

as required.   ◆

As we noted in Example 3.1, there are two actions ($r = 2$): selecting a model and selecting an elevation. Because there are four possibilities for the model, $m_1 = 4$; and because there are three possibilities for the elevation of each model, $m_2 = 3$. Therefore, by the BCR, the total number of possibilities, including both model and elevation, is $m_1 \cdot m_2 = 4 \cdot 3 = 12$, as we demonstrated in Example 3.1.

Because the number of possibilities in the model–elevation problem is quite small, determining that number by a direct listing, as we did in the tree diagram in Figure 3.1, is relatively simple. Nonetheless, obtaining the number of possibilities by applying the BCR is even easier. Moreover, in problems where the number of possibilities is large, a direct listing isn't feasible and the BCR is the only practical way to proceed.

EXAMPLE 3.2   ***The BCR***

License Plates   Arizona license plates consist of three digits followed by three letters.
a) How many different license plates are possible?
b) How many different license plates are possible in which no digit or letter is repeated?

Solution For both parts (a) and (b), we apply the BCR with six actions ($r = 6$).
a) There are 10 possibilities for the first digit, 10 for the second digit, and 10 for the third digit; and there are 26 possibilities for the first letter, 26 for the second letter,

and 26 for the third letter. Hence, by the BCR, there are

$$m_1 \cdot m_2 \cdot m_3 \cdot m_4 \cdot m_5 \cdot m_6 = 10 \cdot 10 \cdot 10 \cdot 26 \cdot 26 \cdot 26 = 17{,}576{,}000$$

possibilities altogether for different license plates. Obviously, it wouldn't be practical to obtain the number of possibilities by a direct listing—a tree diagram, for instance, would have 17,576,000 branches!

Drawing a simple diagram can be helpful in organizing the application of the BCR. One way of doing so is to draw one box for each action and think of the actions as filling in the boxes. In this problem, we would draw six boxes, each of the first three to be filled in with a digit and each of the last three to be filled in with a letter. Application of the BCR can then be represented as shown in Figure 3.2.

Figure 3.2 Diagram for obtaining the number of possible license plates

Action	Digit #1	Digit #2	Digit #3	Letter #1	Letter #2	Letter #3
	□	□	□	□	□	□
	↑	↑	↑	↑	↑	↑
Possibilities	10	10	10	26	26	26

b) For this part, there are again 10 possibilities for the first digit. For each possibility for the first digit, there are 9 possibilities for the second digit because the second digit cannot be the same as the first. And for each possibility for the first two digits, there are 8 possibilities for the third digit because the third digit cannot be the same as either the first or the second. Similarly, there are 26 possibilities for the first letter, 25 for the second letter, and 24 for the third letter. So, by the BCR, there are

$$m_1 \cdot m_2 \cdot m_3 \cdot m_4 \cdot m_5 \cdot m_6 = 10 \cdot 9 \cdot 8 \cdot 26 \cdot 25 \cdot 24 = 11{,}232{,}000$$

possibilities for license plates in which no digit or letter is repeated. This application of the BCR can be represented schematically as shown in Figure 3.3.

Figure 3.3 Diagram for obtaining the number of possible license plates in which no digit or letter is repeated

Action	Digit #1	Digit #2	Digit #3	Letter #1	Letter #2	Letter #3
	□	□	□	□	□	□
	↑	↑	↑	↑	↑	↑
Possibilities	10	9	8	26	25	24

To obtain the total number of license plates in which no digit or letter is repeated, we simply multiply the six numbers at the bottom of the diagram.   ■

EXAMPLE 3.3 *The BCR*

Coin Tossing Consider the random experiment of tossing a coin repeatedly and observing the result (a head or a tail) for each toss.

a) If the coin is tossed three times, determine the number of possible outcomes by a direct listing.

b) Solve part (a) by using the BCR.

c) If the coin is tossed n times, determine the number of possible outcomes.

Solution

a) In Example 2.1 on page 26, we listed the possible outcomes of tossing a coin three times: HHH, HHT, HTH, HTT, THH, THT, TTH, TTT. Thus eight outcomes are possible .

b) To apply the BCR, we first note that there are three actions—namely, the three tosses of the coin. There are 2 possibilities for the first toss (H or T), 2 for the second toss, and 2 for the third toss. Applying the BCR yields $2 \cdot 2 \cdot 2 = 2^3 = 8$ possibilities altogether. Thus eight outcomes are possible when a coin is tossed three times. This result agrees with the answer obtained in part (a), but here no listing is required.

c) For this part, we proceed as in part (b), except that here the coin is tossed n times instead of three times. There are 2 possibilities for the first toss, 2 for the second toss, ..., and 2 for the nth toss. Hence, by the BCR, there are

$$\underbrace{2 \cdot 2 \cdots 2}_{n \text{ times}} = 2^n$$

possible outcomes when a coin is tossed n times. For instance, $2^{25} = 33,554,432$ outcomes are possible when a coin is tossed 25 times. ∎

Applying the BCR to Sampling With Replacement

Consider a population with N members. Suppose that n members of the population are selected one at a time with identity noted and that each member selected is returned to the population for possible reselection. This selection procedure is called **sampling with replacement.** The collection of members obtained, including the order of selection, is called an **ordered sample of size n with replacement.** In statistics, we refer to N as the **population size** and n as the **sample size.**

EXAMPLE 3.4 *The BCR*

Sampling With Replacement Suppose that a sample is drawn with replacement from a population.

a) Use a direct listing to determine the number of possible ordered samples of size 3 with replacement from a population of size 4.

b) Solve part (a) by using the BCR.

c) In general, determine the number of possible ordered samples of size n with replacement from a population of size N.

Solution

a) We denote the four members of the population a, b, c, and d. The possible ordered samples of size 3 with replacement are displayed in Table 3.1.

Table 3.1 Possible ordered samples of size 3 with replacement

aaa	*aba*	*aca*	*ada*	*baa*	*bba*	*bca*	*bda*	*caa*	*cba*	*cca*	*cda*	*daa*	*dba*	*dca*	*dda*
aab	*abb*	*acb*	*adb*	*bab*	*bbb*	*bcb*	*bdb*	*cab*	*cbb*	*ccb*	*cdb*	*dab*	*dbb*	*dcb*	*ddb*
aac	*abc*	*acc*	*adc*	*bac*	*bbc*	*bcc*	*bdc*	*cac*	*cbc*	*ccc*	*cdc*	*dac*	*dbc*	*dcc*	*ddc*
aad	*abd*	*acd*	*add*	*bad*	*bbd*	*bcd*	*bdd*	*cad*	*cbd*	*ccd*	*cdd*	*dad*	*dbd*	*dcd*	*ddd*

Thus there are 64 possible ordered samples of size 3 with replacement from a population of size 4.

b) To apply the BCR, we note that there are three actions: selection of the first, second, and third members of the sample. There are 4 possibilities for the first member of the sample, 4 possibilities for the second member, and 4 possibilities for the third member. Thus, by the BCR, there are $4 \cdot 4 \cdot 4 = 4^3 = 64$ possibilities altogether. This result agrees with the answer obtained in part (a), but here no listing is required.

c) We proceed as in part (b). There are N possibilities for the first member of the sample, N possibilities for the second member, ..., and N possibilities for the nth member. Thus, by the BCR, there are

$$\underbrace{N \cdot N \cdots N}_{n \text{ times}} = N^n$$

possibilities altogether. Hence there are N^n possible ordered samples of size n with replacement from a population of size N.   ■

The coin-tossing problem in Example 3.3 can be solved in the context of sampling with replacement. Think of the population as consisting of the two letters H and T. Tossing the coin n times is equivalent to selecting an ordered sample of size n with replacement from that population of size 2. Applying Example 3.4(c) with $N = 2$ gives 2^n possible outcomes, the same result as that obtained in Example 3.3(c).

EXERCISES 3.1   **Basic Exercises**

3.1 Refer to Example 3.1 on page 86. Suppose that the developer discontinues the Shalimar model but provides an additional elevation choice, D, for each of the remaining three models.
a) Draw a tree diagram similar to Figure 3.1 on page 87 showing the possible choices for the selection of a home, including both model and elevation.
b) Use the tree diagram in part (a) to determine the total number of choices for the selection of a home, including both model and elevation.
c) Use the BCR to determine the total number of choices for the selection of a home, including both model and elevation.

3.2 Refer to Example 3.1 on page 86. Suppose that the developer retains all four models but provides an additional elevation choice, D, for the Shalimar and Palacia. Determine, without using a direct listing, the total number of choices for the selection of a home, including both model and elevation.

3.3 The menu at a restaurant has five choices of beverage, three salads, six entrees and four desserts. How many complete meals are available with beverage, salad, entree, and dessert?

3.4 An advertisement for *i*Dolls^{TM} states: "Choose from 69 billion combinations to create a one-of-a-kind doll." The ad goes on to say that there are 39 choices for hairstyle, 19 for eye color, 8 for hair color, 6 for face shape, 24 for lip color, 5 for freckle pattern, 5 for line of clothing, 6 for blush color, and 5 for skin tone. Exactly how many possibilities are there for these options?

3.5 An identification number consists of an ordered arrangement of eight decimal digits. How many identification numbers can be formed if
a) there are no restrictions?   **b)** no digit can occur twice?
c) no digit can agree with its predecessor?
d) no digit can agree with either of its two immediate predecessors?

3.6 How many batting orders are possible for the nine starting players on a baseball team?

3.7 The author of this book spoke with a representative of the United States Postal Service and obtained the following information about zip codes. A five-digit zip code consists of five digits of which the first three give the sectional center and the last two the post office or delivery area. In addition to the five-digit zip code, there is a trailing *plus four zip code*. The first two digits of the plus four zip code give the sector or several blocks and the last two the segment or side of the street. For the five-digit zip code, the first four digits can be any of the digits 0–9 and the fifth any of the digits 1–8. For the plus four zip code, the first three digits can be any of the digits 0–9 and the fourth any of the digits 1–9.
a) How many possible five-digit zip codes are there?
b) How many possible plus four zip codes are there?
c) How many possibilities are there including both the five-digit zip code and the plus four zip code?

3.8 Telephone numbers in the United States consist of a three-digit area code followed by a seven-digit local number. Suppose that neither the first digit of an area code nor the first digit of a local number can be a zero but that all other choices are acceptable.
a) How many different area codes are possible?
b) For a given area code, how many local telephone numbers are possible?
c) How many telephone numbers are possible?

3.9 In Example 2.3 on page 27, we considered the random experiment of rolling two dice. Use the BCR to determine the number of possible outcomes
a) for this random experiment.   **b)** in which the sum of the dice is 5.
c) in which doubles are rolled.   **d)** in which the sum of the dice is even.

3.10 In Example 2.4 on page 27, we considered the random experiment of observing a mechanical or electrical unit consisting of five components and determining which components are working and which have failed. Use the BCR to find the number of possible outcomes
a) for this random experiment.
b) in which exactly one of the five components is not working.
c) in which at least one of the five components is not working.
d) in which at most one of the five components is not working.

3.11 An alphabet has six letters a, b, c, d, e, and f. How many four-letter words (i.e., ordered arrangements of four of the six letters) can be formed if
a) there are no restrictions?   **b)** no letter can occur twice?
c) the first letter must be a or b?   **d)** the letter c must occur at least once?

3.12 Let A and B be two finite sets with the same number of elements—say, n.
a) How many functions are there from A to B?
b) How many of these functions are one-to-one?

3.13 Five people—say, a, b, c, d, and e—are arranged in a line. How many arrangements are there in which
a) c is before d? **b)** d is not first?

3.14 Suppose that you and your best friend are among n people to be arranged in a line. How many arrangements are possible in which exactly k people are between you and your friend?

3.15 Four married couples attend a banquet.
a) How many ways can they be seated on one side of a straight (i.e., noncircular) table in such a way that each husband sits next to his wife?
b) Repeat part (a) for a circular table.

3.16 Consider a domino that consists of two subrectangles, each marked with a number from 1 to n. How many such dominos are possible? *Note:* A number pair is not ordered.

3.17 If n people attend a party and each pair of people shake hands, how many handshakes will there be?

3.18 A poker hand consists of 5 cards dealt from an ordinary deck of 52 playing cards.
a) In five-card stud, the order in which the cards are dealt matters. How many five-card stud hands are possible?
b) In five-card draw, the order in which the cards are dealt doesn't matter. How many five-card draw hands are possible?

3.19 In how many ways can n distinguishable balls be arranged in n distinguishable boxes so that
a) no box is empty? **b)** exactly one box is empty? **c)** at least one box is empty?

Theory Exercises

3.20 Use mathematical induction to complete the proof of the BCR, as given in Proposition 3.1 on page 87.

Advanced Exercises

3.21 In this exercise, you are to obtain the number of subsets of a finite set, Ω.
a) Suppose that Ω consists of three elements—say, $\{a, b, c\}$. List the possible subsets of Ω and, from your list, determine the number of subsets.
b) In part (a), determine the number of subsets by using the BCR. *Hint:* For a given subset, each element of Ω either is a member of that subset or it isn't.
c) Suppose that Ω consists of n elements. Determine the number of subsets of Ω.

3.22 This exercise provides an alternative derivation of the result in Exercise 3.21(c). The eight subsets of $\{a, b, c\}$ are $\emptyset$, $\{a\}$, $\{b\}$, $\{c\}$, $\{a, b\}$, $\{a, c\}$, $\{b, c\}$, and $\{a, b, c\}$. Although listing all 16 subsets of $\{a, b, c, d\}$ is relatively easy, you are to answer the following questions by other methods, which can be applied in more complicated problems where you may not be able to write a complete list easily.
a) To create a subset of the set $\{a, b, c, d\}$, first choose a subset of $\{a, b, c\}$, and then decide whether to add d to that subset. Use that idea and the BCR to find the number of subsets of $\{a, b, c, d\}$.

b) Suppose that you have written the 8 subsets of $\{a, b, c\}$ in a column on the left side of a page and that to the right of each of those 8 subsets you write the union of that subset with the set $\{d\}$. The two columns together then amount to a list of all 16 subsets of $\{a, b, c, d\}$. Does the same thing work with sets of other sizes?

c) Use the result of part (b) and mathematical induction to prove that the number of subsets of a set with n elements is 2^n.

3.2 Permutations and Combinations

To introduce some additional counting rules, we first recall the definition of **factorial.**

DEFINITION 3.1   Factorial

The product of the first k positive integers is called k **factorial** and is denoted $k!$. In symbols,

$$k! = k(k - 1) \cdots 2 \cdot 1.$$

We also define $0! = 1$.

EXAMPLE 3.5   *Factorials*

Determine $3!$, $4!$, and $5!$.

Solution   Referring to Definition 3.1, we obtain that $3! = 3 \cdot 2 \cdot 1 = 6$, $4! = 4 \cdot 3 \cdot 2 \cdot 1 = 24$, and $5! = 5 \cdot 4 \cdot 3 \cdot 2 \cdot 1 = 120$. ∎

Note, for instance, that $6! = 6 \cdot 5!$, $6! = 6 \cdot 5 \cdot 4!$, $6! = 6 \cdot 5 \cdot 4 \cdot 3!$, and so on. In general, if $j \leq k$, then $k! = k(k - 1) \cdots (k - j + 1)(k - j)!$.

Permutations

We now discuss permutations. In Definition 3.2, we define **permutation** and present the symbol used for the number of possible permutations.

DEFINITION 3.2   Permutation

A **permutation** of r objects from a collection of m objects is any *ordered* arrangement of r distinct objects from the m objects. The number of possible permutations of r objects that can be formed from a collection of m objects is denoted $(m)_r$ or $_mP_r$.

EXAMPLE 3.6   *Introducing Permutations*

Consider a collection consisting of the five letters *a, b, c, d,* and *e*.
a) List all possible permutations of three letters from this collection of five letters.
b) Use part (a) to determine the number of possible permutations of three letters that can be formed from the collection of five letters; that is, find $(5)_3$.
c) Use the BCR to determine the number of possible permutations of three letters that can be formed from the collection of five letters.

Solution   a) For this part, we need to list all ordered arrangements of three letters from the first five letters of the English alphabet. We do so in Table 3.2.

Table 3.2 Possible permutations of three letters from the collection of five letters

abc	abd	abe	acd	ace	ade	bcd	bce	bde	cde
acb	adb	aeb	adc	aec	aed	bdc	bec	bed	ced
bac	bad	bae	cad	cae	dae	cbd	cbe	dbe	dce
bca	bda	bea	cda	cea	dea	cdb	ceb	deb	dec
cab	dab	eab	dac	eac	ead	dbc	ebc	ebd	ecd
cba	dba	eba	dca	eca	eda	dcb	ecb	edb	edc

b) Table 3.2 reveals that there are 60 possible permutations of three letters from the collection of five letters; in other words, $(5)_3 = 60$.
c) Here we want to use the BCR to determine the number of possible permutations of three letters from the collection of five letters. There are five possibilities for the first letter, four possibilities for the second, and three possibilities for the third. Hence, by the BCR, there are $5 \cdot 4 \cdot 3 = 60$ possibilities altogether. So, again, $(5)_3 = 60$. ■

We can make two relevant observations from Example 3.6. First, listing all possible permutations under consideration is generally tedious or impractical. Second, listing the possible permutations to determine how many there are isn't necessary—we can use the BCR to count them. In that regard, we now use the BCR to establish a general formula, called the **permutations rule,** for determining the number of possible permutations.

◆◆◆   **Proposition 3.2   Permutations Rule**

The number of possible permutations of r objects from a collection of m objects is

$$(m)_r = \frac{m!}{(m-r)!}. \tag{3.2}$$

Proof   There are m possibilities for the first object, $m - 1$ for the second object, $\ldots$, and $m - r + 1$ for the rth object. Applying the BCR, we find that the number of possible permutations of r objects from a collection of m objects is $m(m - 1) \cdots (m - r + 1)$. If we now multiply the preceding quantity by $(m - r)!/(m - r)!$, we obtain the right side of Equation (3.2). ◆

Definition 3.2 assumes tacitly that r is an integer between 1 and m. However, the right side of Equation (3.2) makes sense as well for $r = 0$, and its value is 1; thus we extend Definition 3.2 by defining $(m)_0 = 1$.

EXAMPLE 3.7   *The Permutations Rule*

Exacta Wagering   In an exacta wager at the race track, the bettor picks the two horses that she thinks will finish first and second, in a specified order. For a race with 12 entrants, determine the number of possible exacta wagers.

Solution   Selecting 2 horses from the 12 horses for an exacta wager is equivalent to specifying a permutation of 2 objects from a collection of 12 objects; the first object is the horse selected to finish first and the second object is the horse selected to finish second. Thus the number of possible exacta wagers is $(12)_2$—the number of possible permutations of 2 objects from a collection of 12 objects. Applying the permutations rule, with $m = 12$ and $r = 2$, we obtain

$$(12)_2 = \frac{12!}{(12 - 2)!} = \frac{12!}{10!} = \frac{12 \cdot 11 \cdot \cancel{10!}}{\cancel{10!}} = 12 \cdot 11 = 132.$$

In a 12-horse race, 132 exacta wagers are possible.   ■

Applying the Permutations Rule to Sampling Without Replacement

Consider a population with N members. Suppose that n members of the population are selected one at a time with identity noted and that a member selected is not returned to the population for possible reselection. This selection procedure is called **sampling without replacement.** The collection of members obtained, including the order of selection, is called an **ordered sample of size n without replacement.**

EXAMPLE 3.8   *The Permutations Rule*

Sampling Without Replacement   Suppose that a sample is drawn without replacement from a population.
a) Determine the number of possible ordered samples of size n without replacement from a population of size N.
b) Apply part (a) to obtain the number of possible ordered samples of size 5 without replacement from a population of size 52.

Solution   a) An ordered sample of size n without replacement from a population of size N is a permutation of n objects from a collection of N objects. Hence, by the permutations rule, there are

$$(N)_n = \frac{N!}{(N - n)!}$$

possible ordered samples of size n without replacement from a population of size N.

b) Substituting $N = 52$ and $n = 5$ into the formula in part (a), we find that

$$(52)_5 = \frac{52!}{(52 - 5)!} = \frac{52!}{47!} = 311,875,200.$$

There are 311,875,200 possible ordered samples of size 5 without replacement from a population of size 52. ∎

Further Applications of the Permutations Rule

We now present some additional applications of the permutations rule, including an important special case of that rule.

EXAMPLE 3.9   *The Permutations Rule*

Arranging Books in a Bookcase   A student has 10 books to arrange on a shelf of a bookcase. In how many ways can the 10 books be arranged?

Solution   Any particular arrangement of the 10 books on the shelf is a permutation of 10 objects from a collection of 10 objects. So, for this problem, we need to determine $(10)_{10}$, the number of possible permutations of 10 objects from a collection of 10 objects, more commonly expressed as the number of possible permutations of 10 objects among themselves. Applying the permutations rule, we get

$$(10)_{10} = \frac{10!}{(10 - 10)!} = \frac{10!}{0!} = \frac{10!}{1} = 10! = 3,628,800.$$

There are 3,628,800 ways to arrange the 10 books on the shelf. It doesn't seem possible that there could be this many ways, but there are! ∎

We can generalize Example 3.9 to find the number of possible permutations of m objects among themselves. Specifically, we have Proposition 3.3, known as the **special permutations rule.**

◆◆◆   **Proposition 3.3   Special Permutations Rule**

The number of possible permutations of m objects among themselves is $m!$.

Proof   By definition, the number of possible permutations of m objects among themselves is the number of possible permutations of m objects from a collection of m objects. Applying the permutations rule, we find that number to be

$$(m)_m = \frac{m!}{(m - m)!} = \frac{m!}{0!} = \frac{m!}{1} = m!,$$

as required.   ◆

EXAMPLE 3.10   *The Special Permutations Rule*

Arranging Books in a Bookcase   Refer to Example 3.9. Suppose that among the 10 books, 5 are mathematics books, 2 are English books, and 3 are history books.

Suppose further that the student wants to keep like-subject books together. In how many ways can the 10 books be arranged?

Solution   We use both the special permutations rule and the BCR to solve this problem. First we note that four actions determine the final arrangement of the 10 books on the shelf: arranging the 3 subject groups among themselves, the 5 mathematics books among themselves, the 2 English books among themselves, and the 3 history books among themselves.

Applying the special permutations rule gives 3! ways to arrange the 3 subject groups among themselves, 5! ways to arrange the 5 mathematics books among themselves, 2! ways to arrange the 2 English books among themselves, and 3! ways to arrange the 3 history books among themselves. Now applying the BCR, we conclude that there are $3! \cdot 5! \cdot 2! \cdot 3! = 8640$ possible arrangements of the 10 books in which like-subject books are kept together.   ∎

Combinations

Now we discuss combinations. In Definition 3.3, we define **combination** and present the symbols used for the number of possible combinations.

DEFINITION 3.3   Combination

A **combination** of r objects from a collection of m objects is any *unordered* arrangement of r distinct objects from the m objects. The number of possible combinations of r objects that can be formed from a collection of m objects is denoted $\binom{m}{r}$ or $_mC_r$.

Note the following.

- Order matters in permutations but not in combinations.
- The symbol $\binom{m}{r}$ or $_mC_r$ is often read as "m choose r" because each symbol represents the number of ways that r distinct objects can be chosen from m objects without regard to order.
- A combination of r objects from a collection of m objects is a subset of r objects from the collection of m objects, and vice versa.

EXAMPLE 3.11   *Introducing Combinations*

Let's return to Example 3.6 on page 95 where we considered the collection consisting of the five letters a, b, c, d, and e.
a) List all possible combinations of three letters from this collection of five letters.
b) Use part (a) to determine the number of possible combinations of three letters that can be formed from the collection of five letters; that is, find $\binom{5}{3}$.

Solution a) For this part, we need to list all unordered arrangements of three letters from the first five letters in the English alphabet. We do so in Table 3.3.

Table 3.3 Possible combinations of three letters from the collection of five letters

$\{a,b,c\}$	$\{a,b,d\}$	$\{a,b,e\}$	$\{a,c,d\}$	$\{a,c,e\}$	$\{a,d,e\}$	$\{b,c,d\}$	$\{b,c,e\}$	$\{b,d,e\}$	$\{c,d,e\}$

b) Table 3.3 reveals that there are 10 possible combinations of three letters from the collection of five letters; in other words, $\binom{5}{3} = 10$. ■

In Example 3.11, we obtained the number of possible combinations by a direct listing. We can avoid a direct listing by deriving a formula for the number of possible combinations. To show how, we return to Example 3.11.

Look at the first combination in Table 3.3, $\{a, b, c\}$. By the special permutations rule, there are $3! = 6$ permutations of these three letters among themselves; they are *abc, acb, bac, bca, cab,* and *cba.* These six permutations are displayed in the first column of Table 3.2 on page 95. Similarly, there are $3! = 6$ permutations of the three letters in the second combination in Table 3.3, $\{a, b, d\}$. These six permutations are displayed in the second column of Table 3.2. The same comments apply to the other eight combinations in Table 3.3.

Hence, to each combination of three letters from the collection of five letters, there correspond 3! permutations of three letters from the collection of five letters. Moreover, any such permutation is accounted for in this way. Consequently, there must be 3! times as many permutations as combinations. Equivalently, the number of possible combinations of three letters from the collection of five letters must equal the number of possible permutations of three letters from the collection of five letters divided by 3!, or

$$\binom{5}{3} = \frac{(5)_3}{3!} = \frac{5!/(5-3)!}{3!} = \frac{5!}{3!\,(5-3)!} = \frac{5 \cdot 4 \cdot \cancel{3!}}{\cancel{3!}\,2!} = \frac{5 \cdot 4}{2} = 10,$$

which is the number we obtained in Example 3.11 by a direct listing.

The same argument holds in general. In Proposition 3.4, we present a general formula, called the **combinations rule,** for determining the number of possible combinations.

◆◆◆ **Proposition 3.4   Combinations Rule**

The number of possible combinations of r objects from a collection of m objects is

$$\binom{m}{r} = \frac{m!}{r!\,(m-r)!}. \tag{3.3}$$

Numbers of this kind are referred to as **binomial coefficients.**

Proof   To each combination of r objects from the collection of m objects, there correspond $r!$ permutations—namely, the $r!$ permutations of the r objects among themselves. Hence there must be $r!$ times as many permutations of r objects from the m objects as there are combinations of r objects from the m objects. Therefore

$$\binom{m}{r} = \frac{(m)_r}{r!} = \frac{m!/(m-r)!}{r!} = \frac{m!}{r!\,(m-r)!},$$

as required.   ◆

Definition 3.3 assumes tacitly that r is an integer between 1 and m. However, the right side of Equation (3.3) makes sense as well for $r = 0$, and its value is 1; thus we extend Definition 3.3 by defining $\binom{m}{0} = 1$.

EXAMPLE 3.12   *The Combinations Rule*

Compact-Disc Club   In an attempt to recruit new members, a compact-disc club advertises a special introductory offer: A new member agrees to buy 1 compact disc at regular club prices and receives free any 4 compact discs of his choice from a collection of 69 compact discs. How many possibilities does a new member have for the selection of the 4 free compact discs?

Solution   Any particular selection of 4 compact discs from 69 compact discs is a combination of 4 objects from a collection of 69 objects. So, by the combinations rule, the number of possible selections is

$$\binom{69}{4} = \frac{69!}{4!\,(69-4)!} = \frac{69!}{4!\,65!} = \frac{69 \cdot 68 \cdot 67 \cdot 66 \cdot \cancel{65!}}{4!\,\cancel{65!}} = 864,501.$$

There are 864,501 possibilities for the selection of 4 compact discs from the collection of 69 compact discs.   ■

Applying the Combinations Rule to Sampling Without Replacement

Consider a population with N members. Suppose that n members of the population are selected one at a time with identity noted and that a member selected is not returned to the population for possible reselection. As we mentioned earlier, this selection procedure is called *sampling without replacement*. The collection of members obtained, irrespective of the order of selection, is called an **unordered sample of size n without replacement.**

EXAMPLE 3.13   *The Combinations Rule*

Sampling Without Replacement   Suppose that a sample is drawn without replacement from a population.
a) Determine the number of possible unordered samples of size n without replacement from a population of size N.
b) Apply part (a) to obtain the number of possible unordered samples of size 5 without replacement from a population of size 52.

Solution a) An unordered sample of size n without replacement from a population of size N is a combination of n objects from a collection of N objects. Hence, by the combinations rule, there are

$$\binom{N}{n} = \frac{N!}{n!\,(N-n)!}$$

possible unordered samples of size n without replacement from a population of size N.

b) Substituting $N = 52$ and $n = 5$ into the formula in part (a), we find that

$$\binom{52}{5} = \frac{52!}{5!\,(52-5)!} = 2{,}598{,}960.$$

There are 2,598,960 possible unordered samples of size 5 without replacement from a population of size 52. ∎

When we refer simply to "a sample" from a finite population, we mean an unordered sample without replacement unless specifically stated otherwise. As we showed in Example 3.13(a), there are $\binom{N}{n}$ possible samples of size n from a population of size N.

Combining Counting Rules

In Example 3.14, we illustrate how counting rules are sometimes combined to obtain a required result.

EXAMPLE 3.14 *The Combinations Rule*

Five-Card Draw Poker A five-card draw poker hand consists of 5 cards dealt from an ordinary deck of 52 playing cards. The order in which the cards are dealt is unimportant.
a) How many possible five-card draw poker hands are there?
b) How many different hands are there consisting of three kings and two queens?
c) The hand in part (b) is an example of a full house: three cards of one denomination (face value) and two of another. How many different full houses are there?

Solution a) A five-card draw poker hand is a combination of 5 objects (the hand) from a collection of 52 objects (the deck). Hence, by the combinations rule,

$$\binom{52}{5} = \frac{52!}{5!\,(52-5)!} = 2{,}598{,}960$$

hands are possible.

b) For this part, we use both the combinations rule and the BCR. There are three actions: the choice of which three of the four kings are to appear in the hand, the choice of which two of the four queens are to appear in the hand, and the choice of which zero of the other (nonking and nonqueen) 44 cards are to appear in the hand. By the combinations rule, we have $\binom{4}{3}$ possibilities for the first action, $\binom{4}{2}$ possibilities for the second action, and $\binom{44}{0}$ possibilities for the third action. Hence, by the BCR, there are

$$\binom{4}{3} \cdot \binom{4}{2} \cdot \binom{44}{0} = 4 \cdot 6 \cdot 1 = 24$$

possibilities altogether for the three actions. In other words, there are 24 different hands consisting of three kings and two queens. We illustrate this discussion schematically in Figure 3.4. That diagram is helpful for organizing and summarizing the required steps for solving the problem.

Figure 3.4 Diagram for obtaining the number of possible hands containing three kings and two queens

	Cards		
Groups	Kings 4	Queens 4	Others 44
	↓	↓	↓
Number to be chosen	Kings 3	Queens 2	Others 0
Number of ways	$\binom{4}{3}$	$\binom{4}{2}$	$\binom{44}{0}$
Total possibilities		$\binom{4}{3} \cdot \binom{4}{2} \cdot \binom{44}{0}$	

c) This problem is similar to part (b) except that here we have two additional actions—namely, the choice of denomination for the three of a kind and the choice of denomination for the pair. By the combinations rule, we have $\binom{13}{1} = 13$ possibilities for the first additional action and then $\binom{12}{1} = 12$ possibilities for the second additional action. And, as in part (a), once we have decided on the denominations of the three of a kind and pair, there are 24 possibilities for the choice of the cards in those denominations. Hence, by the BCR, there are $13 \cdot 12 \cdot 24 = 3744$ possible full houses. ∎

Ordered Partition of a Collection of Objects

Frequently we want to count the number of ways that some collection of objects can be partitioned (divided) into distinct subgroups. In Definition 3.4, we define **ordered partition of a collection of objects** and present the symbol used for the number of possible ordered partitions.

> **DEFINITION 3.4   Ordered Partition of a Collection of Objects**
>
> An **ordered partition** of m objects into k distinct groups of sizes m_1, m_2, $\ldots$, m_k is any division of the m objects into a combination of m_1 objects constituting the first group, m_2 objects constituting the second group, $\ldots$, and m_k objects constituting the kth group. The number of such partitions possible is denoted $\binom{m}{m_1, m_2, \ldots, m_k}$.

EXAMPLE 3.15   *Introducing Ordered Partitions of a Collection of Objects*

Consider the collection of objects consisting of the six letters a, b, c, d, e, and f.

a) List all possible ordered partitions of these six letters into three distinct groups of sizes 2, 1, and 3, respectively.

b) Use part (a) to determine the number of possible ordered partitions of the six letters into three groups of sizes 2, 1, and 3, respectively.

c) Use the combinations rule and the BCR to determine the number of possible ordered partitions of the six letters into three groups of sizes 2, 1, and 3, respectively.

Solution   In the notation of Definition 3.4, we have $m = 6$, $k = 3$, $m_1 = 2$, $m_2 = 1$, and $m_3 = 3$.

a) A direct listing of the possibilities is provided in Table 3.4. Note, for instance, that the first entry is an ordered partition of the six letters into the three groups $\{a, b\}$, $\{c\}$, and $\{d, e, f\}$.

Table 3.4 Possible ordered partitions of six letters into three groups of sizes 2, 1, and 3

$\{a, b\}, \{c\}, \{d, e, f\}$	$\{b, c\}, \{a\}, \{d, e, f\}$	$\{c, e\}, \{a\}, \{b, d, f\}$
$\{a, b\}, \{d\}, \{c, e, f\}$	$\{b, c\}, \{d\}, \{a, e, f\}$	$\{c, e\}, \{b\}, \{a, d, f\}$
$\{a, b\}, \{e\}, \{c, d, f\}$	$\{b, c\}, \{e\}, \{a, d, f\}$	$\{c, e\}, \{d\}, \{a, b, f\}$
$\{a, b\}, \{f\}, \{c, d, e\}$	$\{b, c\}, \{f\}, \{a, d, e\}$	$\{c, e\}, \{f\}, \{a, b, d\}$
$\{a, c\}, \{b\}, \{d, e, f\}$	$\{b, d\}, \{a\}, \{c, e, f\}$	$\{c, f\}, \{a\}, \{b, d, e\}$
$\{a, c\}, \{d\}, \{b, e, f\}$	$\{b, d\}, \{c\}, \{a, e, f\}$	$\{c, f\}, \{b\}, \{a, d, e\}$
$\{a, c\}, \{e\}, \{b, d, f\}$	$\{b, d\}, \{e\}, \{a, c, f\}$	$\{c, f\}, \{d\}, \{a, b, e\}$
$\{a, c\}, \{f\}, \{b, d, e\}$	$\{b, d\}, \{f\}, \{a, c, e\}$	$\{c, f\}, \{e\}, \{a, b, d\}$
$\{a, d\}, \{b\}, \{c, e, f\}$	$\{b, e\}, \{a\}, \{c, d, f\}$	$\{d, e\}, \{a\}, \{b, c, f\}$
$\{a, d\}, \{c\}, \{b, e, f\}$	$\{b, e\}, \{c\}, \{a, d, f\}$	$\{d, e\}, \{b\}, \{a, c, f\}$
$\{a, d\}, \{e\}, \{b, c, f\}$	$\{b, e\}, \{d\}, \{a, c, f\}$	$\{d, e\}, \{c\}, \{a, b, f\}$
$\{a, d\}, \{f\}, \{b, c, e\}$	$\{b, e\}, \{f\}, \{a, c, d\}$	$\{d, e\}, \{f\}, \{a, b, c\}$
$\{a, e\}, \{b\}, \{c, d, f\}$	$\{b, f\}, \{a\}, \{c, d, e\}$	$\{d, f\}, \{a\}, \{b, d, f\}$
$\{a, e\}, \{c\}, \{b, d, f\}$	$\{b, f\}, \{c\}, \{a, d, e\}$	$\{d, f\}, \{b\}, \{a, c, e\}$
$\{a, e\}, \{d\}, \{b, c, f\}$	$\{b, f\}, \{d\}, \{a, c, e\}$	$\{d, f\}, \{c\}, \{a, b, e\}$
$\{a, e\}, \{f\}, \{b, c, d\}$	$\{b, f\}, \{e\}, \{a, c, d\}$	$\{d, f\}, \{e\}, \{a, b, c\}$
$\{a, f\}, \{b\}, \{c, d, e\}$	$\{c, d\}, \{a\}, \{b, e, f\}$	$\{e, f\}, \{a\}, \{b, c, d\}$
$\{a, f\}, \{c\}, \{b, d, e\}$	$\{c, d\}, \{b\}, \{a, e, f\}$	$\{e, f\}, \{b\}, \{a, c, d\}$
$\{a, f\}, \{d\}, \{b, c, e\}$	$\{c, d\}, \{e\}, \{a, b, f\}$	$\{e, f\}, \{c\}, \{a, b, d\}$
$\{a, f\}, \{e\}, \{b, c, d\}$	$\{c, d\}, \{f\}, \{a, b, e\}$	$\{e, f\}, \{d\}, \{a, b, c\}$

b) Table 3.4 reveals that there are 60 possible ordered partitions of the six letters into three groups of sizes 2, 1, and 3, respectively; in other words, $\binom{6}{2,1,3} = 60$.

c) There are three actions: the choice of two letters from the six to form the first group, the choice of one letter from the remaining four to form the second group, and the choice of three letters from the remaining three to form the third group. By the combinations rule, there are $\binom{6}{2}$ possibilities for the first action, $\binom{4}{1}$ possibilities for the second action, and $\binom{3}{3}$ possibilities for the third action. Hence, by the BCR, there are

$$\binom{6}{2} \cdot \binom{4}{1} \cdot \binom{3}{3} = 15 \cdot 4 \cdot 1 = 60$$

possibilities altogether. So, again, $\binom{6}{2,1,3} = 60$. ∎

Note that the term on the left side of the preceding displayed equation can be expressed as

$$\binom{6}{2} \cdot \binom{4}{1} \cdot \binom{3}{3} = \frac{6!}{2!\,4!} \cdot \frac{4!}{1!\,3!} \cdot \frac{3!}{3!\,0!} = \frac{6!}{2!\,1!\,3!}.$$

Thus, in terms of the notation introduced in Definition 3.4,

$$\binom{6}{2,\ 1,\ 3} = \frac{6!}{2!\,1!\,3!}.$$

Keep this equation in mind as we now present a general formula, called the **partitions rule,** for determining the number of possible ordered partitions.

◆◆◆ **Proposition 3.5   Partitions Rule**

Let $m_1, m_2, \ldots, m_k$ be nonnegative integers whose sum is m. The number of possible ordered partitions of m objects into k distinct groups of sizes $m_1, m_2, \ldots, m_k$ is

$$\binom{m}{m_1, m_2, \ldots, m_k} = \frac{m!}{m_1!\,m_2! \cdots m_k!}.$$

Numbers of this kind are referred to as **multinomial coefficients.**

Proof   We proceed as in Example 3.15(c), first noting that there are k actions here: the choice of m_1 objects from the m objects to form the first group, the choice of m_2 objects from the remaining $m - m_1$ objects to form the second group, …, and the choice of m_k objects from the remaining $m - m_1 - m_2 - \cdots - m_{k-1}$ objects to form the kth group.

By the combinations rule, we have $\binom{m}{m_1}$ possibilities for the first action, $\binom{m-m_1}{m_2}$ possibilities for the second action, …, and $\binom{m-m_1-m_2-\cdots-m_{k-1}}{m_k}$ possibilities for the kth action.

Hence, by the BCR, there are

$$\binom{m}{m_1} \cdot \binom{m - m_1}{m_2} \cdots \binom{m - m_1 - m_2 - \cdots - m_{k-1}}{m_k}$$

$$= \frac{m!}{m_1!\,(m - m_1)!} \cdot \frac{(m - m_1)!}{m_2!\,(m - m_1 - m_2)!} \cdots \frac{(m - m_1 - m_2 - \cdots - m_{k-1})!}{m_k!\,(m - m_1 - m_2 - \cdots - m_k)!}$$

$$= \frac{m!}{m_1!\,m_2! \cdots m_k!}$$

possibilities altogether. Note that, in the last equation, we used the fact that

$$(m - m_1 - m_2 - \cdots - m_k)! = (m - m)! = 0! = 1.$$

The proof of Proposition 3.5 is now complete. ◆

EXAMPLE 3.16 *The Partitions Rule*

Cooperative Learning In a cooperative-learning environment, the students in a class break into groups and the students in each group work together. A class has 20 students. The professor decides to partition the students into five groups, with each group assigned to a different project. The group sizes are 3, 4, 5, 4, and 4 students. How many possibilities are there for the formation of the five groups?

Solution We apply the partitions rule with $m = 20$, $k = 5$, $m_1 = 3$, $m_2 = 4$, $m_3 = 5$, $m_4 = 4$, and $m_5 = 4$. The number of possibilities is

$$\binom{20}{3,\ 4,\ 5,\ 4,\ 4} = \frac{20!}{3!\,4!\,5!\,4!\,4!} = 244{,}432{,}188{,}000.$$

There are roughly 250 billion ways to partition 20 students into five distinct groups of sizes 3, 4, 5, 4, and 4. ■

In our discussion of partitions, we assumed that the groups are distinguishable. If they aren't, we must adjust accordingly. For instance, suppose that in Example 3.16 the three groups of four students are each assigned an identical project. Then the number of possibilities is one-sixth ($1/3!$) of that obtained in Example 3.16 because each of the 3! permutations of any particular three groups of four students yields the same assignments for those groups.

EXAMPLE 3.17 *The Partitions Rule*

Finishing Places at a Track Meet At a track meet, three countries are represented: the United States, Canada, and Mexico. Eight athletes are competing in the 100-yard dash, three from the United States, two from Canada, and three from Mexico. In terms of total score for each of the three countries, and assuming no ties, how many possible results are there for the finishing places?

Solution   We can think of the possible results as partitioning the eight finishing places into three groups, one of size 3 for the U.S. athletes, one of size 2 for the Canadian athletes, and one of size 3 for the Mexican athletes. For instance, the three groups of finishing places $\{2, 5, 7\}$, $\{3, 4\}$, and $\{1, 6, 8\}$ would correspond to the three U.S. athletes finishing in second, fifth, and seventh places; the two Canadian athletes finishing in third and fourth places; and the three Mexican athletes finishing in first, sixth, and eighth places. Hence, by the partitions rule,

$$\binom{8}{3,\ 2,\ 3} = \frac{8!}{3!\,2!\,3!} = 560$$

results are possible.   ∎

Note that multinomial coefficients reduce to binomial coefficients in the case of two groups ($k = 2$). Indeed, if m_1 and m_2 are nonnegative integers whose sum is m, then

$$\binom{m}{m_1,\ m_2} = \frac{m!}{m_1!\,m_2!} = \frac{m!}{m_1!\,(m - m_1)!} = \binom{m}{m_1}.$$

This result makes sense from a conceptual point of view: The number of possible ordered partitions of m objects into two distinct groups of sizes m_1 and m_2 equals the number of ways that we can choose m_1 objects from the m objects to constitute the first group, which is $\binom{m}{m_1}$. The remaining $m - m_1 = m_2$ objects must constitute the second group.

EXERCISES 3.2   **Basic Exercises**

3.23 Show for $k, j \in \mathcal{N}$, with $j \leq k$, that $k! = k(k-1) \cdots (k - j + 1)(k - j)!$.

3.24 Determine the value of
a) $(7)_3$.   **b)** $(5)_2$.   **c)** $(8)_4$.   **d)** $(6)_0$.   **e)** $(9)_9$.

3.25 At a movie festival, a team of judges is to pick the first, second, and third place winners from the 18 films entered. Use permutation notation to express the number of possibilities and then evaluate that expression.

3.26 Investment firms usually have a large selection of mutual funds from which an investor can choose. One such firm has 30 mutual funds. Suppose that you plan to invest in 4 of these mutual funds, 1 during each quarter of next year. Use permutation notation to express the number of possibilities and then evaluate that expression.

3.27 The sales manager of a clothing company needs to assign seven salespeople to seven different territories. How many possibilities are there for the assignments?

3.28 Determine the value of
a) $\binom{7}{3}$.   **b)** $\binom{5}{2}$.   **c)** $\binom{8}{4}$.   **d)** $\binom{6}{0}$.   **e)** $\binom{9}{9}$.

3.29 The Internal Revenue Service (IRS) decides that it will audit the returns of 3 people from a group of 18. Use combination notation to express the number of possibilities and then evaluate that expression.

3.30 At a lottery, 100 tickets were sold and three prizes are to be given. How many possible outcomes are there if
a) the prizes are equivalent?   **b)** there is a first, second, and third prize?

3.31 An economics professor is using a new method to teach a junior-level course with an enrollment of 42 students. The professor wants to conduct in-depth interviews with the students to get feedback on the new teaching method but doesn't want to interview all 42 of them. She decides to interview a sample of 5 students from the class. How many different samples are possible?

3.32 The Powerball® is a multistate lottery that was introduced in April 1992 and is now sold in 24 states, the District of Columbia, and the Virgin Islands. To play the game, a player first selects five numbers from the numbers 1–53 and then chooses a Powerball number, which can be any number between 1 and 42, inclusive. How many possibilities are there?

3.33 In the game of *keno,* there are 80 balls, numbered 1–80. From these 80 balls, 10 are selected at random.
a) How many different outcomes are possible?
b) If a player specifies 20 numbers, in how many ways can he get all 10 numbers selected?

3.34 A club has 14 members.
a) How many ways can a governing committee of size 3 be chosen?
b) How many ways can a president, vice president, and treasurer be chosen?
c) How many ways can a president, vice president, and treasurer be chosen if two specified club members refuse to serve together?

3.35 How many license plates are there consisting of three digits and three letters if there is no restriction on where the digits and letters are placed?

3.36 A five-card draw poker hand consists of 5 cards dealt from an ordinary deck of 52 playing cards. The order in which the cards are received is unimportant. Note that, in sequence, an ace can play as either the lowest or highest card. In other words, the hierarchy of card denominations, from lowest to highest, is ace, 2, 3, . . . , 10, jack, queen, king, ace. Determine the number of possible hands of the specified type.
a) Straight flush: five cards of the same suit in sequence
b) Four of a kind: $\{w, w, w, w, x\}$, where w and x are distinct denominations
c) Full house: $\{w, w, w, x, x\}$, where w and x are distinct denominations
d) Flush: five cards of the same suit, not all in sequence
e) Straight: five cards in sequence, not all of the same suit
f) Three of a kind: $\{w, w, w, x, y\}$, where $w, x,$ and y are distinct denominations
g) Two pair: $\{w, w, x, x, y\}$, where $w, x,$ and y are distinct denominations
h) One pair: $\{w, w, x, y, z\}$, where $w, x, y,$ and z are distinct denominations

3.37 Repeat Exercise 3.36 for the game of five-card stud, where the order in which the cards are received matters.

3.38 Refer to the inclusion–exclusion principle, Proposition 2.10 on page 73. Determine the number of summands in each sum.

3.39 The U.S. Senate consists of 100 senators, 2 from each state. A committee consisting of 5 senators is to be formed.
a) How many different committees are possible?
b) How many are possible if no state can have more than 1 senator on the committee?

3.40 Refer to Example 3.17 on page 105. How many possible results are there in which the United States has exactly two finishers in the top three and one in the bottom three?

3.41 Without doing any calculations, explain why

a) $\dbinom{n}{k} = \dbinom{n}{n-k}.$ **b)** $\dbinom{n}{k} = \dbinom{n}{k,\, n-k}.$

3.42 A teacher plans to construct a 25-question exam from previous exams, in which there are 50 true–false questions and 80 multiple-choice questions. Ignoring the order of the questions,
a) how many exams can be constructed?
b) how many exams can be constructed that consist of exactly 13 true–false questions?
c) how many exams can be constructed that consist of all true–false questions?

3.43 How many distinct arrangements are there of the letters in the word PERSEVERE?

3.44 Consider a group of 30 football players from which 8 are to be selected as quarterbacks, 7 as fullbacks, and 4 as centers.
a) In how many ways can this selection be made?
b) Now suppose that the group of 30 football players consists of 15 quarterbacks, 9 fullbacks, and 6 centers. Further suppose that the 8 quarterbacks to be selected must be chosen from among the 15 quarterbacks and likewise for the fullbacks and centers. How many different selections are possible?

3.45 From a list of 20 candidates, you must choose a committee of 8 and, from among those 8, you must choose a president, a secretary, and a treasurer. In parts (a)–(c), imagine (but don't draw) a tree diagram representing the process whose first and second steps are as specified. Then use the BCR in a way suggested by the structure of the tree diagram, together with permutation and combination rules, to write an expression for the number of ways to make the necessary choices. Finally, evaluate that expression.
a) The first step is the choice of the 8 committee members; the second step is the choice of the 3 officers from among those 8.
b) The first step is the choice of the 5 nonofficer committee members; the second step is the choice of the 3 officers from among the 15 remaining candidates.
c) The first step is the choice of the 3 officers; the second step is the choice of the 5 nonofficer committee members from among the 17 remaining candidates.
d) Rather than 20 candidates, suppose that there are n candidates; rather than 3 officers, j officers; and rather than 8 committee members, including the 3 officers, k committee members, including the j officers. Without mentioning factorials, explain how parts (a), (b), and (c) suggest the truth of the identities

$$\binom{n}{k}(k)_j = \binom{n}{k-j}(n-(k-j))_j = (n)_j \binom{n-j}{k-j}.$$

e) Now establish the identities in part (d) by a different method—namely, by using the permutations and combinations rules and some algebra.

Theory Exercises

3.46 Binomial theorem: Use combinatorial analysis to prove the *binomial theorem;* that is,

$$(a+b)^n = \sum_{k=0}^{n} \binom{n}{k} a^k b^{n-k}.$$

for all numbers a and b and positive integers n.

3.47 Pascal's triangle: In this exercise, you are asked to examine an interesting pattern that occurs when the coefficients of the expansions of $(a+b)^n$, for $n = 0, 1, 2, \ldots$, are written in the form of a triangle, called *Pascal's triangle.* Subsequently, you are to prove an identity involving binomial coefficients that establishes the pattern mathematically.
a) Consider the expressions $(a+b)^n$, for $n = 0, 1, 2, \ldots$. Use the binomial theorem (Exercise 3.46) to expand each expression up to and including $n = 6$.

b) Arrange the coefficients in the expansions in part (a) in an isosceles triangle by placing the coefficients for each n in a row, with $n = 0$ as the top row, $n = 1$ as the second row, and so on.

c) Observe that, in the triangle found in part (b), each entry can be obtained by adding the two nearest entries in the row above. Prove this result mathematically by establishing the identity

$$\binom{n}{k-1} + \binom{n}{k} = \binom{n+1}{k}$$

for all $n, k \in \mathcal{N}$, with $k \leq n$.

3.48 Vandermonde's identity: In this exercise, you are to establish *Vandermonde's identity*, which states that

$$\sum_{j=0}^{k} \binom{n}{j}\binom{m}{k-j} = \binom{n+m}{k}.$$

This formula uses the convention that $\binom{b}{a} = 0$ if either $a > b$ or $a < 0$.

a) Prove Vandermonde's identity by using mathematical induction on n. *Hint:* Refer to Exercise 3.47(c).

b) Let $\Omega = \{a_1, \dots, a_n, b_1, \dots, b_m\}$, $\Omega_1 = \{a_1, \dots, a_n\}$, and $\Omega_2 = \{b_1, \dots, b_m\}$. Prove Vandermonde's identity by using the fact that each subset of Ω consisting of k elements is, for some j, the union of a subset of Ω_1 of size j and a subset of Ω_2 of size $k - j$.

3.49 Multinomial theorem: Use combinatorial analysis to prove the *multinomial theorem*; that is,

$$(a_1 + a_2 + \cdots + a_m)^n = \sum \binom{n}{n_1, n_2, \dots, n_m} a_1^{n_1} a_2^{n_2} \cdots a_m^{n_m},$$

where the sum is taken over all nonnegative integers, $n_1, n_2, \dots, n_m$, whose sum equals n.

Advanced Exercises

3.50 In Exercises 3.21 and 3.22, you were asked to prove that the number of subsets of a set with n elements is 2^n. Prove that result by determining, for each k, the number of subsets of size k and then adding the results.

3.51 Let m and r be positive integers with $r \leq m$. Prove that there are $\binom{m-1}{r-1}$ positive-integer solutions to the equation $x_1 + \cdots + x_r = m$.

3.52 Let m and r be positive integers. Prove that there are $\binom{m+r-1}{r-1}$ nonnegative-integer solutions to the equation $x_1 + \cdots + x_r = m$.

For Exercises 3.53–3.55, refer to Exercises 3.51 and 3.52.

3.53 Determine the number of possible solutions to the equation $x_1 + x_2 + x_3 + x_4 = 10$, where the x_js are
a) nonnegative integers.   **b)** positive integers.

3.54 Refer to Exercise 2.81 on page 79. Determine the number of possible outcomes for the random experiment.

3.55 Suppose that n balls are distributed randomly into N distinguishable urns, where an urn can contain multiple balls. How many possible outcomes are there if the balls are
a) distinguishable?   **b)** indistinguishable?

3.3   Applications of Counting Rules to Probability

Suppose that an experiment has a finite number of equally likely possible outcomes. Then, according to Proposition 2.4 on page 50, the probability that a specified event E occurs equals the number of ways, $N(E)$, that the event can occur divided by the total number of possible outcomes, $N(\Omega)$. Mathematically,

$$P(E) = \frac{N(E)}{N(\Omega)}, \qquad \text{for each event } E. \tag{3.4}$$

For the probability problems considered in Chapter 2, it was easy to determine $N(\Omega)$ and $N(E)$ by a direct listing. However, that usually isn't the case. You must often use counting rules to obtain the number of possible outcomes for the random experiment and the number of ways that a specified event can occur. Examples 3.18–3.20 illustrate the application of counting rules to solve probability problems for random experiments in which there are a finite number of equally likely possible outcomes.

EXAMPLE 3.18   *Applying Counting Rules to Obtain Probabilities*

Drug Effectiveness   A drug known to have a 50% effectiveness rate in curing a certain disease is administered to 20 people who have the disease. Determine the probability that
a) exactly 10 of the 20 people will be cured.
b) at least three-fourths of the 20 people will be cured.

Solution   Because the drug has a 50% effectiveness rate, the possible outcomes of the experiment (cure–noncure results) are equally likely. Hence we can apply Equation (3.4) to obtain the required probabilities.

It is convenient to think of the outcome of this random experiment as the result of filling each of 20 boxes (representing the 20 people) with either a c (cure) or an n (noncure). So, for instance, the outcome that the first 10 people are cured and that the last 10 people aren't cured corresponds to the result that the first 10 boxes are filled with cs and the last 10 with ns.

First we determine the number of possible outcomes, $N(\Omega)$. It is just the total number of ways that we can fill each of the 20 boxes with either a c or an n. There are two possibilities for the first box (c or n), two for the second box (c or n), and so on. Applying the BCR, we conclude that there are

$$\underbrace{2 \cdot 2 \cdots 2}_{20 \text{ times}} = 2^{20} = 1{,}048{,}576$$

possibilities altogether. Thus $N(\Omega) = 1{,}048{,}576$.

a) Let E denote the event that exactly 10 of the 20 people are cured. Then $N(E)$ equals the number of ways that we can fill the 20 boxes with exactly 10 cs (and 10 ns) or, equivalently, the number of ways that we can choose exactly 10 boxes from the 20 boxes to contain cs. Thus $N(E)$ equals the number of possible combinations of 10 objects from 20 objects, which by the combinations rule is $\binom{20}{10} = 184{,}756$.

Applying Equation (3.4) yields

$$P(E) = \frac{N(E)}{N(\Omega)} = \frac{184{,}756}{1{,}048{,}576} = 0.176.$$

There is a 17.6% chance that exactly 10 of the 20 people will be cured.

b) We now want to obtain the probability that at least three-fourths of the 20 people are cured, that is, 15 or more. We can use either of two methods to solve this problem. The first method is to determine the number of ways it can happen that 15 or more of the 20 people are cured and then divide that number by $N(\Omega)$. The second method is to proceed as in part (a) to find the individual probabilities of exactly 15 cures, exactly 16 cures, ..., exactly 20 cures and then, applying the additivity property of a probability measure, sum those six individual probabilities.

However, regardless of which of these two methods we use, we must determine, for $k = 15, 16, \ldots, 20$, the number of ways it can happen that exactly k of the 20 people are cured or, equivalently, the number of ways that we can choose exactly k boxes from the 20 boxes to contain cs. Proceeding as in part (a), we find that the number of ways is $\binom{20}{k}$. We leave it to you to verify that the required probability is 0.021. There is only a 2.1% chance that at least three-fourths of the 20 people will be cured.   ∎

EXAMPLE 3.19   *Applying Counting Rules to Obtain Probabilities*

Statistical Quality Control   Statistical quality control (SQC) often involves sampling items from a *lot* of items and making inferences about the entire lot based on information obtained from the sample. Understanding this type of procedure requires knowledge of how probabilities are calculated in such situations.

To illustrate, suppose that the quality assurance engineer of a company that manufactures TV sets inspects finished products in lots of 100. He selects 5 of the 100 TVs at random and inspects them thoroughly. If, in fact, 6 of the 100 TVs in the current lot are actually defective, find the probability that exactly 2 of the 5 TVs selected by the engineer are defective.

Solution   Because the engineer makes his selection at random, each possible collection of 5 TVs from the 100 TVs is equally likely to be the one obtained. Thus we should use a classical probability model and compute probabilities by applying Equation (3.4).

First we determine the number of possible outcomes for the random experiment. It is the number of ways that 5 TVs can be selected from the 100 TVs—the number of possible combinations of 5 objects from a collection of 100 objects. Applying the combinations rule yields

$$\binom{100}{5} = \frac{100!}{5!\,(100 - 5)!} = \frac{100!}{5!\,95!} = 75{,}287{,}520. \tag{3.5}$$

Thus $N(\Omega) = 75{,}287{,}520$.

Next we determine the number of ways the specified event—which we denote E—can occur. In other words, we must find the number of outcomes in which exactly 2 of the 5 TVs selected are defective. To do so, we think of the 100 TVs as divided into two groups—namely, the defective TVs and the nondefective TVs—as shown in the top portion of Figure 3.5.

Figure 3.5 Diagram for obtaining the number of outcomes in which exactly 2 of the 5 TVs selected are defective

TVs	
Defective 6	Nondefective 94

Groups

$\downarrow$ $\downarrow$

Defective 2	Nondefective 3

Number to be chosen

Number of ways $\binom{6}{2}$ $\binom{94}{3}$

Total possibilities $\binom{6}{2} \cdot \binom{94}{3}$

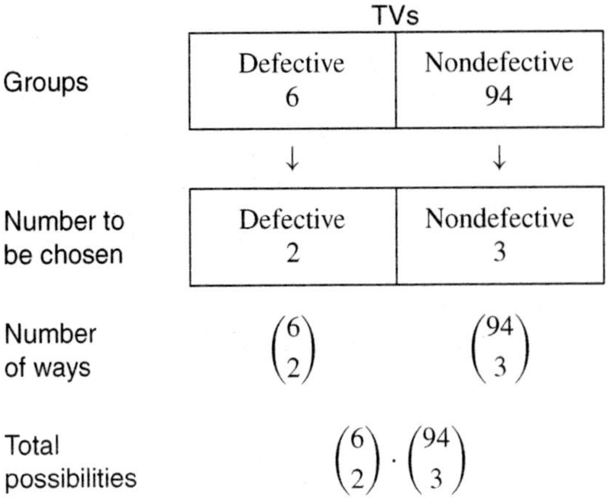

There are 6 TVs in the first group, of which 2 are to be selected; we can do so in $\binom{6}{2} = 15$ ways. There are 94 TVs in the second group, of which 3 are to be selected; we can do so in $\binom{94}{3} = 134,044$ ways. Consequently, by the BCR, there are

$$\binom{6}{2} \cdot \binom{94}{3} = 15 \cdot 134,044 = 2,010,660 \qquad \textbf{(3.6)}$$

outcomes in which exactly 2 of the 5 TVs selected are defective, or $N(E) = 2,010,660$. Figure 3.5 summarizes the calculations performed in this paragraph.

Applying Equation (3.4), we conclude that

$$P(E) = \frac{N(E)}{N(\Omega)} = \frac{2,010,660}{75,287,520} = 0.027.$$

Chances are 2.7% that exactly 2 of the 5 TVs selected will be defective.   ∎

We note a useful check for problems of the type encountered in Example 3.19. Specifically, we observe that the top and bottom numbers in the two binomial coefficients on the left side of Equation (3.6) sum to the top and bottom numbers, respectively, in the binomial coefficient on the left side of Equation (3.5). Can you explain why?

Random Sampling

Suppose that n members of a population of N members are selected one at a time with identity noted. Previously, we discussed three types of samples.

- *Ordered sample with replacement:* The collection of members obtained, including the order of selection, when each member selected is returned to the population for possible reselection.

- *Ordered sample without replacement:* The collection of members obtained, including the order of selection, when each member selected is not returned to the population for possible reselection.

- *Unordered sample without replacement:* The collection of members obtained, not including the order of selection, when each member selected is not returned to the population for possible reselection.

For any of these three sampling schemes, we say that we have conducted **random sampling** if each of the possible samples is equally likely to be the one obtained.[†] The sample obtained in random sampling is called a **random sample.**

EXAMPLE 3.20   *Applying Counting Rules to Obtain Probabilities*

Random Sampling   For a random sample of size n, determine the probability that a specified member will be included in the sample in the case of an
a) ordered sample with replacement.
b) ordered sample without replacement.
c) unordered sample without replacement.

Solution   Let E denote the event that a specified member will be included in the sample. Because the sampling is random, a classical probability model is appropriate in each case. Thus we can use Equation (3.4) to determine the required probabilities.

a) Example 3.4(c) on page 91 indicates that the number of possible samples is N^n, that is, $N(\Omega) = N^n$. To obtain $P(E)$, we use the complementation rule. Event E doesn't occur if and only if the specified member isn't included in the sample. For this to happen, there are $N - 1$ possibilities for the first member selected (any member other than the specified one), $N - 1$ possibilities for the second member, ..., and $N - 1$ possibilities for the nth member. Applying the BCR, we conclude that $N(E^c) = (N - 1)^n$. Consequently, by the complementation rule and Equation (3.4),

$$P(E) = 1 - P(E^c) = 1 - \frac{(N-1)^n}{N^n} = 1 - \left(1 - \frac{1}{N}\right)^n.$$

Note that, for fixed N, this probability approaches 1 as n gets large; and for fixed n, this probability approaches 0 as N gets large. You should be able to explain why these two results are reasonable.

b) Example 3.8(a) on page 96 indicates that the number of possible samples is $(N)_n$, that is, $N(\Omega) = (N)_n$. To obtain $P(E)$, we again use the complementation rule. Event E doesn't occur if and only if the specified member isn't included in the sample. For this to happen, there are $N - 1$ possibilities for the first member selected (any member other than the specified one), $N - 2$ possibilities for the second member (any member other than the specified one and the first one selected), ..., and $N - n$ possibilities for the nth member. Applying the BCR, we find that $N(E^c) = (N - 1)_n$. Consequently,

---

[†] To be perfectly correct, this type of sampling is called *simple random sampling.*

by the complementation rule, Equation (3.4), and the permutations rule,

$$P(E) = 1 - P(E^c) = 1 - \frac{(N-1)_n}{(N)_n}$$

$$= 1 - \left(\frac{(N-1)!/(N-1-n)!}{N!/(N-n)!}\right) = 1 - \frac{N-n}{N} = \frac{n}{N}.$$

We can also compute $P(E)$ without using the complementation rule. Exercise 3.67 discusses two ways to do that.

c) Example 3.13(a) on page 101 indicates that the number of possible samples is $\binom{N}{n}$, that is, $N(\Omega) = \binom{N}{n}$. To obtain $N(E)$, we think of the population as divided into two groups, the specified member and the remaining $N - 1$ members. For event E to occur, we must select one member from the first group and $n - 1$ members from the second group. Applying the combinations rule twice and then the BCR gives $N(E) = \binom{1}{1}\binom{N-1}{n-1}$. Therefore, by Equation (3.4),

$$P(E) = \frac{\binom{1}{1}\binom{N-1}{n-1}}{\binom{N}{n}} = \frac{\dfrac{1!}{1!\,0!} \cdot \dfrac{(N-1)!}{(n-1)!\,(N-n)!}}{\dfrac{N!}{n!\,(N-n)!}} = \frac{n}{N}.$$

Note that this probability is the same as that obtained in part (b). For random samples without replacement, you can compute probabilities based on either ordered or unordered samples provided that you are consistent throughout the solution. ∎

Matching Problems

Counting techniques are frequently applied to solve *matching problems*. We illustrate matching problems and their solution in Example 3.21.

EXAMPLE 3.21   *Applying Counting Rules to Obtain Probabilities*

Matching Problems Suppose that each of N women throws her house key in a box and then each woman randomly selects one key from the box. Find the probability that
a) a specified woman gets her own key.
b) n specified women get their own keys.
c) at least one woman gets her own key.

Solution   Because the selection is random, a classical probability model is appropriate. Consequently, we can apply Equation (3.4) to determine probabilities.

For convenience, we think of the N women as being numbered 1, 2, ..., N and their keys as being numbered correspondingly; that is, woman 1's key is numbered 1, woman 2's key is numbered 2, and so on. We can then consider the random selection of keys by the women as randomly filling each of N boxes (corresponding to the N women) with a different number between 1 and N (corresponding to the N keys). Using this scheme, we see, for instance, that woman 1 gets her own key if and only if box 1 is filled with the number 1.

First we obtain the number of possible outcomes. It is the number of ways that the N boxes can be filled with the numbers 1 through N, with a different number in each box. But that number of ways is just the number of possible permutations of the numbers 1 through N among themselves, which by the special permutations rule is $N!$.

a) The number of ways that a specified woman gets her own key, or that a specified box gets filled with its own number, equals the number of ways that the remaining $N - 1$ boxes can be filled with the remaining $N - 1$ numbers, which is $(N - 1)!$. Hence the required probability is $(N - 1)!/N! = 1/N$.[†]

b) The number of ways that n specified women get their own keys, or that n specified boxes get filled with their own numbers, equals the number of ways that the remaining $N - n$ boxes can be filled with the remaining $N - n$ numbers, which is $(N - n)!$. Hence the required probability is $(N - n)!/N!$.

c) Let E denote the event that at least one woman gets her own key. Also, for $n = 1, 2,$ $\ldots, N$, let A_n denote the event that woman n gets her own key. At least one woman gets her own key if and only if woman 1 gets her own key or woman 2 gets her own key ... or woman N gets her own key. In other words, $E = \bigcup_{n=1}^{N} A_n$.

Events $A_1, A_2, \ldots, A_N$ aren't mutually exclusive, so we apply the inclusion–exclusion principle (Proposition 2.10 on page 73) to determine $P\left(\bigcup_{n=1}^{N} A_n\right)$. In doing so, we must obtain $P(A_{k_1} \cap A_{k_2} \cap \cdots \cap A_{k_n})$ for each collection of integers $1 \le k_1 < k_2 < \cdots < k_n \le N$, where $1 \le n \le N$.

We first note that $A_{k_1} \cap A_{k_2} \cap \cdots \cap A_{k_n}$ is the event that women $k_1, k_2, \ldots, k_n$ get their own keys. Part (b) indicates that the probability of that event is $(N - n)!/N!$, which does not depend on the particular n specified women. Next we note that the number of terms appearing in the corresponding sum in the inclusion–exclusion principle is $\binom{N}{n}$, the number of possible combinations of n integers from N integers. Therefore

$$\sum_{k_1 < k_2 < \cdots < k_n} P(A_{k_1} \cap A_{k_2} \cap \cdots \cap A_{k_n}) = \binom{N}{n} \cdot \frac{(N - n)!}{N!}$$

$$= \frac{N!}{n!\,(N - n)!} \cdot \frac{(N - n)!}{N!} = \frac{1}{n!}.$$

Applying the inclusion–exclusion principle, we conclude that

$$P(E) = P\left(\bigcup_{n=1}^{N} A_n\right) = \sum_{n=1}^{N} \frac{(-1)^{n+1}}{n!}.$$

Interestingly, the right side of this equation gives the first $N + 1$ terms in the Taylor series expansion of $1 - e^{-1}$. Hence, for a large number of women (i.e., large N), the probability that at least one woman gets her own key is approximately $1 - e^{-1}$, or roughly 0.632. ■

[†] We can also obtain this probability by noting the following. Because the numbers are randomly distributed into the boxes, the number that goes into the specified box is equally likely to be any of the N numbers. So the probability that the specified box gets its own number is $1/N$ (one chance in N).

EXERCISES 3.3   **Basic Exercises**

3.56 Provide the details for the solution of Example 3.18(b) on page 111.

3.57 Four cards are dealt from an ordinary deck of 52 playing cards. What is the probability that the denominations (face values) of the cards are
a) all the same?   **b)** all different?

3.58 In a small lottery, 10 tickets—numbered $1, 2, \ldots, 10$—are sold. Two numbers are drawn at random for prizes. You hold tickets numbered 1 and 2. What is the probability that you win at least one prize?

3.59 An ordinary deck of 52 playing cards is shuffled and dealt. What is the probability that
a) the seventh card dealt is an ace?   **b)** the first ace occurs on the seventh card dealt?

3.60 From an urn containing M red balls and $N - M$ black balls, a random sample of size n is taken without replacement. Find the probability that exactly j black balls are in the sample.

3.61 The birthday problem: A probability class has 38 students.
a) Find the probability that at least 2 students in the class have the same birthday. For simplicity, assume that there are always 365 days in a year and that birth rates are constant throughout the year. *Hint:* Use the complementation rule.
b) Repeat part (a) if the class has N students.
c) Evaluate the probability obtained in part (b) for $N = 1, 2, \ldots, 70$. Use a computer or calculator to do the number crunching.
d) What is the smallest class size for which the probability that at least 2 students in the class have the same birthday exceeds 0.5?

3.62 An urn contains four red balls and six black balls. Balls are drawn one at a time at random until three red balls have been drawn. Determine the probability that a total of seven balls is drawn if the sampling is
a) without replacement.   **b)** with replacement.

3.63 Four mathematicians, three chemists, and five physicists are seated randomly in a row. Find the probability that all the members of each discipline sit together.

3.64 Suppose that a random sample of size n without replacement is taken from a population of size N. For $k = 1, 2, \ldots, n$, determine the probability that k specified members of the population will be included in the sample.

3.65 Suppose that a random sample of size n with replacement is taken from a population of size N.
a) Determine the probability that no member of the population is selected more than once.
b) Show that the probability in part (a) approaches 1 as $N \to \infty$. Interpret this result.

3.66 Refer to Example 3.19 on page 111. Determine the probability that the number of defective TVs selected is
a) exactly one.   **b)** at most one.   **c)** at least one.

3.67 Refer to Example 3.20(b) on page 113. In this exercise, you are to obtain $P(E)$—the probability that a specified member of the population will be included in the sample—in two additional ways.
a) Compute $N(E)$ directly and then apply Equation (3.4) on page 110 to determine $P(E)$.
b) For $k = 1, 2, \ldots, n$, let A_k denote the event that the kth member selected is the specified member. Without doing any computations, explain why $P(A_k) = 1/N$, for $k = 1, 2, \ldots, n$. Conclude that $P(E) = n/N$. Explain your reasoning.

3.68 Refer to Example 3.21 on page 114.

a) For $N = 1, 2, \ldots, 6$, determine the probability that at least one woman gets her own key.

b) Compare your answers in part (a) to the approximate probability of $1 - e^{-1}$ given at the end of Example 3.21.

c) Comment on the accuracy of using $1 - e^{-1}$ to approximate the probability that at least one woman gets her own key.

3.69 In an extrasensory-perception (ESP) experiment, a psychologist takes 10 cards, numbered 1–10, and shuffles them. Then, as she looks at each card, the subject writes the number he thinks is on the card.

a) How many possibilities exist for the order in which the subject writes the numbers?

b) If the subject doesn't have ESP, what is the probability that he writes the numbers in the correct order—that is, in the order that the cards are actually arranged?

3.70 Suppose that you have a key ring with N keys, exactly one of which is your house key. Further suppose that you get home after dark and can't see the keys on the key ring. You randomly try one key at a time until you get the correct key, being careful not to mix the keys you have already tried with the ones you haven't.

a) Use counting techniques to determine the probability that you get the correct key on the nth try, where n is an integer between 1 and N, inclusive.

b) Solve part (a) without doing any computations.

c) Determine the probability that you get the correct key on or before the nth try by using (i) a direct counting and (ii) your result from part (a) and the additivity property of a probability measure.

3.71 Refer to Exercise 3.70, but now suppose that you mix the keys you have already tried with the ones you haven't.

a) Use counting techniques to determine the probability that you get the correct key for the first time on the nth try, where n is any positive integer.

b) Determine the probability that you get the correct key on or before the nth try by using (i) a direct counting and (ii) your result from part (a) and the additivity property of a probability measure. Here n is any positive integer.

3.72 A student takes a true–false test consisting of 15 questions. Assuming that the student guesses at each question, find the probability that the student gets

a) at least 1 question correct.   **b)** a 60% or better on the exam.

3.73 Refer to Exercise 3.36 on page 107.

a) Determine the probability of each of the five-card draw poker hands considered there.

b) For the hands considered in part (a), are the five-card stud poker probabilities (where the order in which the cards are received matters) different from the five-card draw poker probabilities? Explain your answer.

3.74 A gene consists of 10 subunits, each of which is normal or mutant. For a particular cell, that gene consists of 3 mutant subunits and 7 normal subunits. Before the cell divides into two daughter cells—say, cell 1 and cell 2—the gene duplicates. The corresponding gene of cell 1 consists of 10 subunits chosen at random from the 6 mutant subunits and 14 normal subunits; cell 2 gets the remaining subunits. Determine the probability that one of the daughter cells consists of all normal subunits.

3.75 If n balls are distributed randomly into n boxes, what is the probability that exactly one box is empty?

3.76 Suppose that you and your best friend are among n people to be arranged randomly in a line. Determine the probability that exactly k people are between you and your friend.

Theory Exercises

3.77 Consider a population of size N in which $100p\%$ of the population have a specified attribute. Suppose that a random sample of size n ($1 \le n \le N$) is taken from the population. Let E_k denote the event that exactly k members of the sample have the specified attribute.
a) Determine $P(E_k)$ if the sampling is without replacement.
b) Determine $P(E_k)$ if the sampling is with replacement.
c) Prove that, for fixed n, the probability in part (a) converges to that in part (b) as $N \to \infty$.
d) Interpret and explain the result in part (c).

Advanced Exercises

3.78 Refer to Example 3.21 on page 114. Determine the probability that exactly n of the N women get their own keys, where $n = 1, 2, \ldots, N$.

3.79 According to the Center for Political Studies at the University of Michigan, Ann Arbor, roughly 50% of U.S. adults are Democrats. Suppose that 10 U.S. adults are selected at random without replacement. Without referring to the number of U.S. adults, determine the approximate probability that
a) exactly 5 are Democrats.   b) 8 or more are Democrats.
c) Why are the probabilities that you obtained in parts (a) and (b) only approximately correct?
d) Obtain expressions for the exact probabilities in parts (a) and (b) in terms of the number of U.S. adults, N.

3.80 Suppose that you take an ordinary deck of 52 playing cards and turn 1 card over at a time until all 52 cards are exposed. Find the probability that the card following the first ace is
a) the ace of spades.   b) the jack of hearts.   c) an ace.   d) a jack.

3.81 Suppose that you have an infinitely large box and countably infinitely many basketballs, numbered $1, 2, \ldots$. For each scenario, determine how many basketballs are in the box at noon.
a) At 1 minute to noon, balls 1–10 are placed in the box and ball 10 is removed; at $\frac{1}{2}$ minute to noon, balls 11–20 are placed in the box and ball 20 is removed; at $\frac{1}{4}$ minute to noon, balls 21–30 are placed in the box and ball 30 is removed; and so on.
b) At 1 minute to noon, balls 1–10 are placed in the box and ball 1 is removed; at $\frac{1}{2}$ minute to noon, balls 11–20 are placed in the box and ball 2 is removed; at $\frac{1}{4}$ minute to noon, balls 21–30 are placed in the box and ball 3 is removed; and so on.
c) At 1 minute to noon, balls 1–10 are placed in the box and then a ball is selected at random from the box and removed; at $\frac{1}{2}$ minute to noon, balls 11–20 are placed in the box and then a ball is selected at random from the box and removed; at $\frac{1}{4}$ minute to noon, balls 21–30 are placed in the box and then a ball is selected at random from the box and removed; and so on.

CHAPTER REVIEW

Summary

In this chapter, we identified and described techniques for counting the number of ways something can happen: the number of possible outcomes for an experiment, the number of ways an event can occur, the number of ways a certain task can be performed, and the like. Such techniques are called counting rules or combinatorial analysis.

The basic counting rule (BCR) is fundamental to all counting techniques. It states that, if two or more actions are performed in a definite order, the total number of possibilities for all the actions can be obtained by multiplying the number of possibilities for the individual actions.

A permutation of r objects from a collection of m objects is any ordered arrangement of r distinct objects from the m objects; a combination is any unordered arrangement. We applied the BCR to derive the permutations rule, which gives the number of possible permutations of r objects from a collection of m objects, denoted $(m)_r$. Using that result, we also derived the combinations rule, which gives the number of possible combinations of r objects from a collection of m objects, denoted $\binom{m}{r}$.

We developed further counting techniques by combining the BCR with either the permutations rule or the combinations rule. In particular, we considered ordered partitions. An ordered partition of m objects into k distinct groups of sizes $m_1, m_2, \ldots, m_k$ is any division of the m objects into an unordered arrangement of m_1 objects constituting the first group, m_2 objects constituting the second group, $\ldots$, and m_k objects constituting the kth group. We derived the partitions rule, which gives the number of possible ordered partitions, denoted $\binom{m}{m_1, m_2, \ldots, m_k}$.

We applied counting techniques to solve probability problems. Specifically, we used counting rules and Equation (3.1) on page 85 to determine probabilities for various types of random experiments in which there are a finite number of equally likely outcomes.

You Should Be Able To ...

1. state and apply the basic counting rule, BCR.

2. define permutation, combination, and ordered partition.

3. find the number of possible permutations, combinations, and ordered partitions.

4. solve counting problems that require the application of several of the fundamental counting techniques.

5. apply counting techniques to obtain probabilities for a random experiment with a finite number of equally likely outcomes.

Key Terms

basic counting rule (BCR), *87*

binomial coefficients, *99*

combination, *98*

combinations rule, *99*

combinatorial analysis, *86*

counting rules, *86*

factorial, *94*

multinomial coefficients, *104*

ordered partition of a collection
    of objects, *103*

ordered sample with
    replacement, *90*

ordered sample without
    replacement, *96*

partitions rule, *104*

permutation, *94*

permutations rule, *95*

population size, *90*

random sample, *113*

random sampling, *113*

sample size, *90*

sampling with replacement, *90*

sampling without replacement, *96*

special permutations rule, *97*

tree diagram, *86*

unordered sample without
    replacement, *100*

Chapter Review Exercises

Basic Exercises

3.82 Certain offices must be filled by appointing members from a list of five candidates—say, a, b, c, d, and e.

a) Draw a tree diagram like that shown in Figure 3.1 on page 87, representing the process whose first step is the choice of a president and whose second step is the choice of a secretary, assuming that the president and the secretary must be two different candidates.

b) Draw a tree diagram like that shown in Figure 3.1, representing the process whose first step is the choice of a secretary and whose second step is the choice of a president, assuming that the president and the secretary must be two different candidates.

c) The set of all size-2 subsets of the set $\{a, b, c, d, e\}$ of all five candidates is

$$\big\{\{a, b\}, \{a, c\}, \{a, d\}, \{a, e\}, \{b, c\}, \{b, d\}, \{b, e\}, \{c, d\}, \{c, e\}, \{d, e\}\big\}.$$

Draw a tree diagram representing the process whose first step is the choice of two of the five candidates and whose second step is the choice of which of the two will be president and which will be secretary.

d) Assume that the offices of president, secretary, and treasurer must be filled and that a, b, and c are to fill these offices, but which of these three will fill which office hasn't yet been decided. Draw a tree diagram representing the process of choosing first the president, next the secretary, and then the treasurer, from among these three.

e) Suppose that a, b, and c are Democrats and that d and e are Republicans. Further suppose that two Democrats and one Republican must be chosen as officers. The list of all ways to choose two Democrats is $\{a, b\}, \{a, c\}, \{b, c\}$. Draw a tree diagram representing the process whose first step is the choice of the two Democrats and whose second step is the choice of the one Republican. Draw another tree diagram representing the process whose first step is the choice of the one Republican and whose second step is the choice of the two Democrats.

3.83 Consider the first 10 letters of the English alphabet, $a, b, c, \ldots, i, j$.

a) Find the number of length-6 sequences if no letter can be used more than once.

b) Repeat part (a) if letters can be used more than once.

c) If a length-6 sequence of letters is randomly chosen from among the letters a through j, with repetitions allowed, find the probability that there will be no repetitions.

3.84 *Scientific Computing & Automation* magazine offers free subscriptions to the scientific community. The magazine does ask, however, that a person answer six questions: primary title, type of facility, area of work, brand of computer used, type of operating system in use, and type of instruments in use. There are 6 options for the first question, 8 for the second, 5 for the third, 19 for the fourth, 16 for the fifth, and 14 for the sixth. How many possibilities are there for answering all six questions?

3.85 A football coach must choose a first-string quarterback, a second-string quarterback, and a third-string quarterback from a group of 15 applicants. Use permutation notation to express the number of possibilities and then evaluate that expression.

3.86 How many samples of size 6 are possible from a population with 45 members?

3.87 A baseball team has 8 pitchers, 3 catchers, and 10 outfielders. The manager must select the starting pitcher, the starting catcher, and the 3 starting outfielders. How many possibilities are there if at the present time the fields for the 3 outfielders

a) are specified?   **b)** aren't specified?

3.88 From a group of 50 basketball players, 5 are to be selected as centers, 10 as forwards, and 10 as guards. In how many ways can this selection be made?

3.89 From the 100 members of the U.S. Senate, four committees are to be formed consisting of 10, 7, 15, and 8 members. If no senator is permitted to serve on more than one committee, in how many ways can these committees be formed?

3.90 In Example 3.7 on page 96, we considered exacta wagering in horse racing. Two similar wagers are the quinella and the trifecta. In a quinella wager, the bettor picks the two horses that she believes will finish first and second, but not in a specified order. In a trifecta wager, the bettor picks the three horses she thinks will finish first, second, and third in a specified order. For a 12-horse race, determine the number of possible
a) quinella wagers.   **b)** trifecta wagers.

3.91 If a die is rolled eight times, in how many ways can it come up 4 twice, 5 three times, 6 once, and 1 twice?

3.92 A company employs 10 stenographers.
a) If 3 are to be assigned to the executive suite, 3 to the marketing department, and 4 to a general stenographic pool, in how many ways can the assignments be made?
b) If the 3 stenographers in the executive suite are to be assigned to the president, executive vice president, and financial vice president, with the remaining 7 stenographers assigned as in part (a), in how many different ways can they now be assigned?
c) If the 3 stenographers in the executive suite are to be assigned as in part (b), the 3 in the marketing department are to be assigned to the general manager, manager for domestic marketing, and manager for foreign marketing, and the remaining 4 as in part (a), in how many different ways can they now be assigned?

3.93 Determine the number of distinguishable arrangements of the letters in the word
a) MISSISSIPPI.   **b)** MASSACHUSETTS.

3.94 In five-card draw poker, what is the probability that you get at least one ace on the deal?

3.95 A bridge hand consists of 13 cards dealt from an ordinary deck of 52 playing cards. Determine the probability that a bridge hand
a) contains exactly two of the four aces.
b) has an 8-4-1 distribution—eight cards of one suit, four of another, and one of another.
c) has a 5-5-2-1 distribution.
d) is void in a specified suit.   **e)** is void in at least one suit.

3.96 Refer to Exercise 3.95. In the actual game of bridge, there are four players—designated North, South, East, and West. The 52 cards are dealt one by one, alternating from one player to the next, until all 52 cards are distributed. Determine the probability that
a) each player gets 1 ace.
b) North and South together have exactly 3 aces.

3.97 Balls are drawn at random without replacement from an urn containing 10 red balls and 16 black balls. Find the probability that it takes at least three draws to get the first red ball.

3.98 All dice considered in this problem are balanced.
a) If two dice are rolled—one red and one green—find the probability that the green die shows a larger number than the red die.
b) If three dice are rolled, find the probability that at least two show the same number.

3.99 A bowl contains 10 chips, numbered 1–10. Two chips are randomly selected from the bowl. Determine the probability that the sum of the numbers on the chips obtained is 10 if the sampling is
a) without replacement.   **b)** with replacement.

3.100 An urn contains 10 red, 20 white, and 30 blue balls. If 6 balls are selected at random from the urn, determine the probability that 1 red, 2 white, and 3 blue balls are obtained if the sampling is
a) without replacement.   **b)** with replacement.

3.101 In a lot containing N items, exactly M are defective. Items are inspected one at a time without replacement. Determine the probability that the kth item inspected ($k \geq M$) will be the last defective one in the lot.

3.102 An Arizona state lottery, called *Lotto,* is played as follows: The player selects six numbers from the numbers 1–42 and buys a ticket for $1. There are six winning numbers, which are selected at random from the numbers 1–42. To win a prize, a *Lotto* ticket must contain three or more of the winning numbers. A ticket with exactly three winning numbers is paid $2. The prize for a ticket with exactly four, five, or six winning numbers depends on sales and on how many other tickets were sold that have exactly four, five, or six winning numbers, respectively. If you buy one *Lotto* ticket, determine the probability that
a) you win the jackpot; that is, your six numbers are the same as the six winning numbers.
b) your ticket contains exactly four winning numbers.
c) you don't win a prize.
d) Repeat part (a) if you buy N tickets.

3.103 Four married couples are randomly paired at a dinner table—the four women on one side of the table and the four men on the other side. What is the probability that no man sits across from his wife?

3.104 An elevator starts with six passengers and stops at eight floors. Find the probability that no two passengers get off on the same floor.

3.105 A random sample of size n is taken without replacement from a population of size N. Find the probability that k specified members of the population are included in the sample.

3.106 A random sample of size n is taken without replacement from a population of size N. The n members obtained are noted and then returned to the population. Subsequently, a random sample of size m is taken without replacement from the population, and the m members obtained are noted.
a) Determine the probability that the two samples have exactly k members in common by considering the first sample as "marking" the members of the population.
b) Repeat part (a) by considering the second sample as "marking" the members of the population.
c) Use parts (a) and (b) to establish the following combinatorial identity without doing any calculations.

$$\frac{\binom{n}{k}\binom{N-n}{m-k}}{\binom{N}{m}} = \frac{\binom{m}{k}\binom{N-m}{n-k}}{\binom{N}{n}}$$

d) Establish the combinatorial identity in part (c) by using the combinations rule and elementary algebra.

3.107 Suppose that a die is tossed n times.
a) How many possible outcomes are there?
b) How many outcomes are there for which none of the first $n-1$ tosses are 6 and the nth toss is a 6?
c) If the die is balanced, determine the probability that none of the first $n-1$ tosses are 6 and the nth toss is a 6.

Advanced Exercises

3.108 Suppose that a balanced die is tossed repeatedly. Determine the probability that
a) a 6 is eventually tossed. *Hint:* Refer to Exercise 3.107.
b) the first 6 occurs on an odd-numbered toss; an even-numbered toss.

3.109 Capture–recapture method: There are N deer in the woods. You capture M of them at random and then tag and release them. In a few days you capture n deer at random.
a) What is the probability of getting k tagged deer in the second sample?
b) Suppose that N is unknown, $M = 10$, $n = 8$, and $k = 3$ (i.e., exactly three tagged deer are in the second sample). How could you use this information to estimate N? (This procedure is called the *capture–recapture method.*)

3.110 According to Definition 3.1 on page 94, $0! = 1$. To some people, this definition is mysterious.
a) Show that, if $0!$ is defined to be any number other than 1, an exception exists to the equation $(n + 1)! = (n + 1) \cdot n!$.
b) Imagine a calculator that does nothing but multiplication. Its advertising brochure says that, if you press the CLEAR key and subsequently "14, ENTER, 3, ENTER, 6, ENTER," the display reads 252 because $14 \times 3 \times 6 = 252$. If any number appears in the display and you enter another number, the product of those two numbers will appear in the display. If the calculator is to function as advertised, what number must appear in the display when you press the CLEAR key? In other words, what number do you get if you multiply no numbers at all? Explain your reasoning and how the result justifies the conclusions that $0! = 1$ and $9^0 = 1$.
c) Consider a group of five men and two women. To choose three from among them is to choose [3 men and 0 women] OR [2 men and 1 woman] OR [1 man and 2 women]. However, that process is the same as choosing [3 men] OR [2 men and 1 woman] OR [1 man and 2 women]. In particular, the number of ways to choose three men and no women must be the same as the number of ways to choose three men. Conjoin this fact with the BCR and combinations rule to provide yet another explanation of why $0!$ must be 1.

3.111 Suppose that $2N$ people are waiting in line to see a movie for which a ticket costs $10. Further suppose that the box office opens with no cash, that half the people have only $10 bills, and that the other half have only $20 bills. Let E denote the event that no one will be required to wait for change. Proceed as follows to determine $P(E)$. If the first person in line has only $10 bills, draw a line segment from the origin in the x-y plane to the point $(1, -1)$; otherwise, draw a line segment from the origin to the point $(1, 1)$. In general, if a person in line has only $10 bills, draw a line segment from the current point to the point 1 unit to the right and 1 unit down; otherwise, draw a line segment from the current point to the point 1 unit to the right and 1 unit up.
a) How many polygonal lines are possible, beginning with the first person in line and ending with the last person in line?
b) At what point must each polygonal line terminate?
c) Explain why event E occurs if and only if the polygonal line doesn't go above the x-axis.
d) Determine the number of polygonal lines that go above the x-axis. *Hint:* To each polygonal line L that goes above the x-axis, consider the polygonal line that coincides with L up to the first point at which L goes above the x-axis and then is the mirror image of L relative to the line $y = 1$ from then on.
e) Obtain $P(E^c)$ and, subsequently, $P(E)$.

Thomas Bayes 1702–1761

Thomas Bayes was born in London in 1702, the exact date seems to be unknown. His father was among the first Nonconformist ministers to be ordained in England; his church was located in Holborn.

Bayes's education was private, and nothing definite is known of his tutors. However, G. A. Barnard in "Thomas Bayes—A Biographical Note" (*Biometrika*, 45, pp. 293–315, 1958) speculates that Abraham De Moivre may have been among Bayes's teachers as De Moivre was tutoring in London at the time.

Like his father, Thomas Bayes was an ordained Nonconformist minister. He first assisted at his father's church at Holborn and then, in the late 1720s, was assigned to the Presbyterian Chapel in Tunbridge Wells, where he remained until his retirement in 1752.

Bayes is unique in that he was elected a Fellow of the Royal Society (in 1742) despite the fact that, while living, nothing in mathematics was published under his name. He did publish, anonymously in 1736, an article entitled *An Introduction to the Doctrine of Fluxions, and a Defence of the Mathematicians Against the Objections of the Author of The Analyst*.

Bayes's contributions to probability theory are contained in "Essay Towards Solving a Problem in the Doctrine of Chances," published in 1764 in the *Philosophical Transactions of the Royal Society of London*. This paper was found after Bayes's death by his friend Richard Price who subsequently submitted it to the Royal Society.

Laplace, in a 1781 memoir, approved Bayes's conclusions on probability theory. However, Boole, in his *Laws of Thought*, challenged these conclusions, and they have been considered controversial ever since.

Thomas Bayes died on April 17, 1761, in Tunbridge Wells, England.

CHAPTER FOUR

Conditional Probability and Independence

INTRODUCTION

In Chapter 1, you learned that the probability of an event is a measure of the likelihood of its occurrence. Frequently, additional information concerning a random experiment is available in the form of knowing that a specific event occurs. This additional information often affects the likelihood of occurrence of other events; that is, it frequently alters their probabilities. These new probabilities are called *conditional probabilities* because they measure likelihood under the condition that a specific event occurs.

For instance, consider the random experiment of selecting an adult at random. We might be interested in the probability that the person selected is colorblind—that is, in the chance that an adult is colorblind. But we also might want to know the probability that the person selected is colorblind, given that the person selected is female—that is, the chance that a woman is colorblind. This latter probability is an example of a conditional probability, conditional on the event that the adult selected is a woman.

In this chapter, we introduce conditional probability and some of its most important properties and applications. In Section 4.1, we define conditional probability, give examples of its basic uses, and identify several of its characteristics. In Section 4.2, we discuss the general multiplication rule, which gives a method for obtaining joint probabilities (probabilities of the intersection of two or more events) in terms of conditional probabilities and marginal probabilities.

In Section 4.3, we examine one of the most important concepts in probability theory—namely, that of *independent events*. We show how this concept arises naturally from that of conditional probability and then develop the basic techniques for its use. In Section 4.4 on Bayes's rule, we discuss a widely used application of conditional probability, one that occurs in probability, statistics, and many other fields of study.

4.1   Conditional Probability

The **conditional probability** of an event is the probability that the event occurs under the condition that another event occurs. In Definition 4.1, we present the notation and terminology used for conditional probability.

DEFINITION 4.1   Conditional Probability

Let A and B be events of a sample space. The probability that event B occurs, given that event A occurs, is called a **conditional probability.** It is denoted $P(B \mid A)$, which is read "the probability of B given A." The event on the right of the "$\mid$" is called the **given event** or the **event being conditioned on.**

Thus conditional probabilities are no different than ordinary (unconditional) probabilities except that the sample space changes from the original one to the event being conditioned on. Examples 4.1–4.3 illustrate this point.

EXAMPLE 4.1   *Conditional Probability*

Coin Tossing   Consider the random experiment of tossing a balanced coin three times and observing the result (a head or a tail) for each toss. Let

A = event that exactly two heads are tossed,

B = event the first toss is a head, and

C = event the second toss is a head.

Determine the following probabilities:
a) $P(B)$, the probability that the first toss is a head.
b) $P(B \mid A)$, the conditional probability that the first toss is a head, given that exactly two heads are tossed.
c) $P(B \mid C)$, the conditional probability that the first toss is a head, given that the second toss is a head.

Solution   a) As we showed previously, a sample space for this random experiment is given by $\Omega = \{$HHH, HHT, HTH, HTT, THH, THT, TTH, TTT$\}$, where, for instance, HTT denotes the outcome of a head on the first toss and tails on the second and third tosses. Because the coin is balanced, the eight possible outcomes are equally likely to occur; hence a classical probability model is appropriate here. Event B can occur in four ways: HHH, HHT, HTH, or HTT. Consequently,

$$P(B) = \frac{4}{8} = 0.5.$$

There is a 50% chance that the first toss is a head.

b) We have $A = \{$HHT, HTH, THH$\}$. So, given that event A occurs, there are no longer eight possible outcomes; rather there are only three—namely, the three outcomes comprising event A. In other words, given that event A occurs, the sample space is now A instead of Ω. Under these circumstances, event B (first toss is a head) can occur in two ways: HHT or HTH. Consequently,

$$P(B \mid A) = \frac{2}{3} = 0.667.$$

Given that exactly two heads are tossed, there is a 66.7% chance that the first toss is a head. Comparing this probability with the one obtained in part (a) reveals that $P(B \mid A) \neq P(B)$; the conditional probability that the first toss is a head, given that exactly two heads are tossed, isn't the same as the (unconditional) probability that the first toss is a head. Knowing that event A occurs affects the probability of event B.

c) We have $C = \{$HHH, HHT, THH, THT$\}$. So, given that event C occurs, there are no longer eight possible outcomes; rather there are only four—namely, the four outcomes comprising event C. In other words, given that event C occurs, the sample space is now C instead of Ω. Under these circumstances, event B (first toss is a head) can occur in two ways: HHH or HHT. Consequently,

$$P(B \mid C) = \frac{2}{4} = 0.5.$$

Given that the second toss is a head, there is a 50% chance that the first toss is a head. Comparing this probability with the one obtained in part (a) reveals that $P(B \mid C) = P(B)$; the conditional probability that the first toss is a head, given that the second toss is a head, is the same as the (unconditional) probability that the first toss is a head. Knowing that event C occurs doesn't affect the probability of event B.   ∎

Referring to Example 4.1, we observe that the probability of an event may or may not be affected by additional information in the form of knowing that a specific event occurs. In other words, the conditional probability of an event may or may not be the same as its unconditional probability.

EXAMPLE 4.2   *Conditional Probability*

Bacteria on a Petri Dish   Suppose that a petri dish of unit radius, containing nutrients upon which bacteria can multiply, is smeared with a uniform suspension of bacteria.

Subsequently, spots indicating colonies of bacteria will appear. Suppose that we observe the location of the center of the first spot to appear.

a) Find the probability that the center of the first spot is more than 1/4 unit from the center of the petri dish.

b) Find the conditional probability that the center of the first spot is more than 1/4 unit from the center of the petri dish, given that it is within 1/2 unit.

c) Find the conditional probability that the center of the first spot is more than 1/4 unit from the center of the petri dish, given that it is on the right half of the petri dish.

Solution   a) A sample space for this random experiment is $\Omega = \{ (x, y) \in \mathcal{R}^2 : x^2 + y^2 < 1 \}$, the unit disk. Because the dish is smeared with a uniform suspension of bacteria, use of a geometric probability model is reasonable. Specifically, we can think of the location of the center of the first spot as a point selected at random from the unit disk. Let A denote the event that the center of the first spot to appear is more than 1/4 unit from the center of the petri dish: $A = \{ (x, y) : 1/4 < \sqrt{x^2 + y^2} < 1 \}$. Then

$$P(A) = \frac{\pi - \pi/16}{\pi} = \frac{15}{16} = 0.938.$$

b) Let B denote the event that the center of the first spot to appear is within 1/2 unit of the center of the petri dish: $B = \{ (x, y) : \sqrt{x^2 + y^2} \le 1/2 \}$. Given that event B occurs, it is the new sample space for the random experiment. Under these circumstances, event A occurs if and only if the center of the first spot to appear is at a distance of between 1/4 unit and 1/2 unit from the center of the petri dish. Consequently,

$$P(A \mid B) = \frac{\pi/4 - \pi/16}{\pi/4} = \frac{3}{4} = 0.75.$$

Comparing this probability with the one in part (a) reveals that $P(A \mid B) \neq P(A)$; the conditional probability that the center of the first spot is more than 1/4 unit from the center of the petri dish, given that it is within 1/2 unit, isn't the same as the (unconditional) probability that the center of the first spot is more than 1/4 unit from the center of the petri dish. Knowing that event B occurs affects the probability of event A.

c) Let C denote the event that the center of the first spot to appear is on the right half of the petri dish: $C = \{ (x, y) \in \Omega : x > 0 \}$. Given that event C occurs, it is the new sample space for the random experiment. Under these circumstances, event A occurs if and only if the center of the first spot to appear is more than 1/4 unit from the center and on the right side of the petri dish. Therefore

$$P(A \mid C) = \frac{(\pi - \pi/16)/2}{\pi/2} = \frac{15}{16} = 0.938.$$

Comparing this probability with the one in part (a) reveals that $P(A \mid C) = P(A)$; the conditional probability that the center of the first spot is more than 1/4 unit from the center of the petri dish, given that it is on the right half of the petri dish, is the same as the (unconditional) probability that the center of the first spot is more than 1/4 unit from the center of the petri dish. Knowing that event C occurs doesn't affect the probability of event A.   ∎

EXAMPLE 4.3   *Conditional Probability*

> **Age and Rank of Faculty**   Data on age and rank of the faculty members at a university are provided in Table 4.1. Suppose that a faculty member is selected at random.
> a) Determine the (unconditional) probability that the faculty member obtained is in his or her 50s.
> b) Determine the (conditional) probability that the faculty member obtained is in his or her 50s, given that an assistant professor is selected.
> c) Interpret the probabilities found in parts (a) and (b) in terms of percentages.

Table 4.1 Table for age and rank of faculty members

Rank

	Full professor R_1	Associate professor R_2	Assistant professor R_3	Instructor R_4	Total
Under 30 A_1	2	3	57	6	68
30–39 A_2	52	170	163	17	402
40–49 A_3	156	125	61	6	348
50–59 A_4	145	68	36	4	253
60 & over A_5	75	15	3	0	93
Total	430	381	320	33	1164

(Age)

Solution   The sample space for this random experiment is the set whose elements are the 1164 faculty members at the university. Because a faculty member is being selected at random, each of the 1164 faculty members is equally likely to be the one obtained. Consequently, a classical probability model is appropriate here.

a) We observe from Table 4.1 that A_4 denotes the event that the faculty member obtained is in his or her 50s and that this event can occur in 253 ways—namely, if any one of the 253 faculty members in their 50s is selected. Consequently,

$$P(A_4) = \frac{253}{1164} = 0.217.$$

b) We are to find the probability that the faculty member selected is in his or her 50s (event A_4), given that an assistant professor is selected (event R_3). In other words, we want to determine $P(A_4 \mid R_3)$. Given that event R_3 occurs, the new sample space for the random experiment is the set whose elements are the 320 assistant professors. Under these circumstances, event A_4 occurs if and only if the faculty member obtained is an assistant professor in his or her 50s. Hence, restricting our attention to the assistant professor column of Table 4.1, we find that

$$P(A_4 \mid R_3) = \frac{36}{320} = 0.113.$$

c) In terms of percentages, $P(A_4) = 0.217$ means that 21.7% of the faculty are in their 50s, whereas $P(A_4 \mid R_3) = 0.113$ means that 11.3% of the assistant professors are in their 50s. ∎

We now elaborate on Example 4.3 for the purpose of introducing some useful probability terminology. Proceeding as in part (b), we can determine all conditional probabilities of age group given rank, as displayed in Table 4.2. Each column is obtained by dividing the corresponding column in Table 4.1 by its column total.

Table 4.2 Conditional distributions of age by rank

Rank

Age	Full professor R_1	Associate professor R_2	Assistant professor R_3	Instructor R_4	$P(A_i)$
Under 30 A_1	0.005	0.008	0.178	0.182	0.058
30–39 A_2	0.121	0.446	0.509	0.515	0.345
40–49 A_3	0.363	0.328	0.191	0.182	0.299
50–59 A_4	0.337	0.178	0.113	0.121	0.217
60 & over A_5	0.174	0.039	0.009	0.000	0.080
Total	1.000	1.000	1.000	1.000	1.000

The first column of Table 4.2 gives the probability distribution of age for full professors. Because this probability distribution is obtained by conditioning on a faculty member being a full professor, it is called a **conditional probability distribution.** Likewise, the second, third, and fourth columns give the conditional probability distributions of age for associate professors, assistant professors, and instructors, respectively.

The $P(A_i)$-column of Table 4.2 provides the (unconditional) probability distribution of age for all faculty members at the university. In this context, that age distribution is called a **marginal probability distribution** because it occurs in the margin of the table. This age distribution is the same as that given in the last column of Table 2.6 on page 55.

Similarly, we can obtain the conditional probability distributions of rank by age and the marginal probability distribution of rank. We leave the details of determining those probability distributions to you in Exercise 4.5.

The Conditional Probability Rule

Examples 4.1–4.3 used the **direct method for obtaining conditional probabilities**— that is, we first obtained the new sample space determined by the event being conditioned on (the given event) and then, using this new sample space, we calculated probabilities in the usual manner. For instance, in part (b) of Example 4.1 on page 127, where a balanced coin is tossed three times, we computed the conditional probability that the first toss is a head, given that exactly two heads are tossed. To do that, we first obtained the new sample space—in this case, {HHT, HTH, THH}—and then went on from there to determine the required probability in the usual manner.

Sometimes we can't determine conditional probabilities directly but must instead compute them in terms of unconditional probabilities. We can develop the formula for doing so by again referring to the frequentist interpretation of probability. Let's assume that the random experiment under consideration is repeated a large number of times— say, n times. Let A be an event with nonzero probability. Given that event A occurs, an event B will occur if and only if event $A \cap B$ occurs. Consequently, in the n repetitions of the random experiment, the proportion of times that event B occurs among those times in which event A occurs is $n(A \cap B)/n(A)$. Therefore, in view of the frequentist interpretation of probability,

$$P(B \mid A) \approx \frac{n(A \cap B)}{n(A)} = \frac{n(A \cap B)/n}{n(A)/n} \approx \frac{P(A \cap B)}{P(A)}.$$

Thus we have an alternative definition of conditional probability, the **conditional probability rule,** presented in Definition 4.2.

DEFINITION 4.2   The Conditional Probability Rule

Let A and B be events of a sample space with $P(A) > 0$. The conditional probability that event B occurs given that event A occurs can be obtained by using the formula

$$P(B \mid A) = \frac{P(A \cap B)}{P(A)}.$$

Note the following.

- When applying the conditional probability rule, divide by the probability of the event being conditioned on—that is, the event on the right of the vertical bar.
- From now on, whenever you see an expression of the form $P(B \mid A)$, assume tacitly that $P(A) > 0$, unless it is explicitly stated otherwise.

EXAMPLE 4.4   *The Conditional Probability Rule*

Coin Tossing   Refer to Example 4.1 on page 126 where a balanced coin is tossed three times. Let A denote the event that exactly two heads are tossed and let B denote the event that the first toss is a head.

a) Use the conditional probability rule to obtain $P(B \mid A)$.

b) Compare the answer in part (a) to the one obtained in Example 4.1(b).

Solution   a) We have $B = \{\text{HHH, HHT, HTH, HTT}\}$ and $A = \{\text{HHT, HTH, THH}\}$. Applying the conditional probability rule, we obtain

$$P(B \mid A) = \frac{P(A \cap B)}{P(A)} = \frac{P(\{\text{HHT, HTH}\})}{P(\{\text{HHT, HTH, THH}\})} = \frac{2/8}{3/8} = \frac{2}{3} = 0.667.$$

b) The conditional probability obtained in part (a) by using the conditional probability rule agrees with that found in Example 4.1(b) where we computed the conditional probability directly.   ■

For the coin-tossing illustration considered in Examples 4.1 and 4.4, conditional probabilities can be obtained either directly or by applying the conditional probability rule. However, as Examples 4.5 and 4.6 show, the conditional probability rule is sometimes the only way in which conditional probabilities can be determined.

EXAMPLE 4.5   *The Conditional Probability Rule*

Educational Institutions   The National Center for Education Statistics publishes data about educational institutions in the *Digest of Education Statistics.* According to that publication, 13.4% of all students attend private schools, and 4.5% of all students attend private colleges. What percentage of students attending private schools are in college?

Solution   To solve this problem, we first translate it into the language of probability. Assume that a student is selected at random and let

$$C = \text{event the student selected is in college and}$$

$$D = \text{event the student selected attends a private school.}$$

We need to determine $P(C \mid D)$. From the percentage data supplied in the statement of the problem, we know that $P(D) = 0.134$ and $P(D \cap C) = 0.045$. Hence, by the conditional probability rule,

$$P(C \mid D) = \frac{P(D \cap C)}{P(D)} = \frac{0.045}{0.134} = 0.336.$$

Thus 33.6% of students attending private schools are in college.   ■

EXAMPLE 4.6   *The Conditional Probability Rule*

Marital Status by Sex   Data on the marital status of U.S. adults can be found in *Current Population Reports,* a publication of the U.S. Bureau of the Census. Table 4.3 provides a joint probability distribution for the marital status of U.S. adults by sex. We have used "Single" as an abbreviation for "Never married." If a U.S. adult is selected at random, determine the probability that
a) the adult selected is divorced, given that the adult selected is a male.
b) the adult selected is a male, given that the adult selected is divorced.

Table 4.3 Joint probability distribution of marital status and sex for U.S. adults

Marital status

		Single M_1	Married M_2	Widowed M_3	Divorced M_4	$P(S_i)$
Sex	Male S_1	0.129	0.298	0.013	0.040	0.480
	Female S_2	0.104	0.305	0.057	0.054	0.520
	$P(M_j)$	0.233	0.603	0.070	0.095	1.000

Solution   Unlike our previous illustrations with contingency tables, we don't have frequency data here—only probability (percentage) data. Because of that, we can't compute conditional probabilities directly; we must use the conditional probability rule.
a) We want $P(M_4 \mid S_1)$. Using the conditional probability rule and Table 4.3, we get

$$P(M_4 \mid S_1) = \frac{P(S_1 \cap M_4)}{P(S_1)} = \frac{0.040}{0.480} = 0.083.$$

In other words, 8.3% of adult males are divorced.
b) We want $P(S_1 \mid M_4)$. Using the conditional probability rule and Table 4.3, we get

$$P(S_1 \mid M_4) = \frac{P(M_4 \cap S_1)}{P(M_4)} = \frac{0.040}{0.095} = 0.421.$$

In other words, 42.1% of divorced adults are males.   ∎

Conditional Probability Is a Probability Measure

In Chapter 2, we discussed the Kolmogorov axioms for a probability measure—that is, the conditions that probabilities must satisfy. Proposition 4.1, whose proof we leave to you in Exercise 4.18, shows that, for an event A with positive probability, the set function $P(\cdot \mid A)$ is a probability measure. It provides the likelihood of events under the condition that event A occurs.

◆◆◆ **Proposition 4.1   Conditional Probability Is a Probability Measure**

Let A be an event with positive probability. Define P_A by

$$P_A(E) = P(E \mid A), \qquad \text{for each event } E.$$

Then P_A is a probability measure.

Consequently, all properties of (unconditional) probabilities also hold for conditional probabilities when the event being conditioned on is held fixed. For instance, we have the conditional version of the complementation rule:

$$P(E \mid A) = 1 - P(E^c \mid A),$$

for all events E. Likewise, the law of partitions, general addition rule, and inclusion–exclusion principle also hold for conditional probabilities.

EXERCISES 4.1   Basic Exercises

4.1 Give an example, other than one presented in the book, where the conditional probability of an event is the same as the unconditional probability of the event.

4.2 Let Ω be the sample space of a random experiment.
a) Without doing any mathematics or computations, explain why conditional probabilities, given the original sample space Ω, are identical to (unconditional) probabilities—that is, indicate why $P(E \mid \Omega) = P(E)$ for all events E.
b) Use the conditional probability rule to show that $P(E \mid \Omega) = P(E)$ for all events E.
c) Discuss the statement, "All probabilities are conditional."

4.3 One card is selected at random from an ordinary deck of 52 playing cards. Let A be the event that a face card is selected, B be the event that a king is selected, and C be the event that a heart is selected. Find the following probabilities and express your results in words. Compute the conditional probabilities directly; do not use the conditional probability rule.
a) $P(B)$ **b)** $P(B \mid A)$ **c)** $P(B \mid C)$ **d)** $P(B \mid A^c)$
e) $P(A)$ **f)** $P(A \mid B)$ **g)** $P(A \mid C)$ **h)** $P(A \mid B^c)$

4.4 The U.S. Bureau of the Census publishes data on housing units in *American Housing Survey in the United States*. The following table provides a frequency distribution, with frequencies in thousands, for the number of rooms in U.S. housing units.

Rooms	1	2	3	4	5	6	7	8+
No. Units	471	1,470	11,715	23,468	24,476	21,327	13,782	15,647

Compute the following conditional probabilities directly; do not use the conditional probability rule. For a U.S. housing unit selected at random, find the probability that the unit has
a) exactly four rooms.   **b)** exactly four rooms, given that it has at least two rooms.
c) at most four rooms, given that it has at least two rooms.
d) Interpret your answers in parts (a)–(c) in terms of percentages.

4.5 Refer to the illustration on age and rank of faculty members in Example 4.3 on page 129.
a) Determine the probability that the faculty member obtained is an assistant professor.

b) Determine the probability that the faculty member obtained is an assistant professor, given that the faculty member is in his or her 50s.

c) Interpret the probabilities found in parts (a) and (b) in terms of percentages.

d) Obtain a table similar to Table 4.2 on page 130 for the conditional probability distributions of rank by age and the marginal probability distribution of rank.

4.6 The National Football League updates team rosters and posts them on its Web site at www.nfl.com. The contingency table in Exercise 2.39 on page 61 provides a cross-classification of players on the New England Patriots roster as of May 24, 2000, by weight and years of experience. A player on the New England Patriots is selected at random. Compute the following conditional probabilities directly; that is, do not use the conditional probability rule. Find the probability that the player selected

a) is a rookie.   **b)** weighs more than 300 pounds.

c) is a rookie, given that he weighs more than 300 pounds.

d) weighs more than 300 pounds, given that he is a rookie.

e) Interpret your answers in parts (a)–(d) in terms of percentages.

4.7 Refer to Exercise 4.6.

a) Determine the probability distribution of weight for rookies; that is, construct a table showing the conditional probabilities that a rookie weighs less than 200 pounds, between 200 and 300 pounds, and more than 300 pounds.

b) Determine the probability distribution of years of experience for players who weigh more than 300 pounds.

c) Construct a table similar to Table 4.2 on page 130 for the conditional probability distributions of weight by years of experience and the marginal probability distribution of weight.

d) Construct a table for the conditional probability distributions of years of experience by weight and the marginal probability distribution of years of experience.

4.8 As reported in the U.S. Census Bureau's *Current Population Reports,* a joint probability distribution for living arrangement and age of U.S. citizens 15 years of age and older is as follows.

Age (yrs)

Living arrangement		15–24 A_1	25–44 A_2	45–64 A_3	Over 64 A_4	$P(L_j)$
	Alone L_1	0.006	0.038	0.035	0.047	0.126
	With spouse L_2	0.017	0.241	0.187	0.084	0.529
	With others L_3	0.154	0.122	0.047	0.022	0.345
	$P(A_i)$	0.177	0.401	0.269	0.153	1.000

A U.S. citizen 15 years of age or older is selected at random. Determine the probability that the person selected

a) lives with spouse.   **b)** is older than 64.   **c)** lives with spouse and is older than 64.

d) lives with spouse, given that the person is older than 64.

e) is older than 64, given that the person lives with spouse.

f) Interpret your answers in parts (a)–(e) in terms of percentages.
g) Construct a table for the conditional probability distributions of age by living arrangement and the marginal probability distribution of age.
h) Construct a table for the conditional probability distributions of living arrangement by age and the marginal probability distribution of living arrangement.

4.9 Four cards are dealt at random without replacement from an ordinary deck of 52 cards.
a) If it is known that the four cards have different face values, what is the probability that the first card is a spade?
b) If it is known that the four cards come from different suits, what is the probability that the first card is a face card?

4.10 A balanced die is tossed 12 times. Given that a 3 occurs at least once, what is the probability that it occurs four times or more?

4.11 A king has two children. What is the probability that both children are boys, given that
a) the first child born is a boy?   **b)** at least one child is a boy?

4.12 A balanced die is rolled until the first 6 occurs. If that happens on the eighth toss, what is the probability that there are exactly two 4s among the first seven tosses?

4.13 As reported by the U.S. Federal Bureau of Investigation (FBI) in *Crime in the United States*, 4.9% of property crimes are committed in rural areas and 1.9% of property crimes are burglaries committed in rural areas. What percentage of property crimes committed in rural areas are burglaries?

4.14 A balanced coin is tossed eight times. Find the probability that
a) no heads appear on the first four tosses.
b) no heads appear on the first four tosses, given that exactly three heads appear in all eight tosses.
c) at least one head appears in the first four tosses, given that exactly three heads appear in all eight tosses.

4.15 A dart is thrown randomly at the unit square $\{ (x, y) : 0 \le x \le 1, 0 \le y \le 1 \}$, and the point at which it hits is observed.
a) Determine the probability that the x coordinate of the point exceeds 1/2.
b) Determine the probability that the x coordinate of the point exceeds its y coordinate.
c) Determine the conditional probability that the x coordinate of the point exceeds its y coordinate, given that the x coordinate of the point exceeds 1/2.

4.16 Repeat Exercise 4.15 if the dart is thrown at the unit disk, $\{ (x, y) : x^2 + y^2 < 1 \}$, instead of the unit square.

Theory Exercises

4.17 Let A and B be events of a sample space with positive probability. Prove that
a) $P(B \mid A) > P(B)$ if and only $P(A \mid B) > P(A)$. (In this case, events A and B are said to be *positively correlated.*)
b) $P(B \mid A) < P(B)$ if and only $P(A \mid B) < P(A)$. (In this case, events A and B are said to be *negatively correlated.*)
c) $P(B \mid A) = P(B)$ if and only $P(A \mid B) = P(A)$. (In this case, events A and B are said to be *independent.*)
d) Interpret the results in parts (a)–(c).

4.18 Prove Proposition 4.1 on page 134 by showing that $P_A(\cdot) = P(\cdot \mid A)$ is a probability measure—that is, prove that $P_A(\cdot)$ satisfies the three Kolmogorov axioms.

4.19 Let A, B, and C be events of a sample space and assume that $A \cap B$ has positive probability. Prove that

$$P(C \mid A \cap B) = \frac{P(B \cap C \mid A)}{P(B \mid A)}.$$

4.20 Let A and B be events with $P(A \cap B) > 0$. Prove that $P_A(C \mid B) = P(C \mid A \cap B)$ for each event C. Interpret this result. *Hint:* Use Exercise 4.19.

Advanced Exercises

4.21 Suppose that the inhabitants of a certain community are classified according to their religion as Religion 1 (R_1), Religion 2 (R_2), or Religion 3 (R_3), and according to their occupation as white-collar (W), blue-collar (B), or other (O). Assume that the conditional probability distributions of occupation by religion are as follows.

$$P(W \mid R_1) = 0.12 \qquad P(W \mid R_2) = 0.18 \qquad P(W \mid R_3) = 0.13$$
$$P(B \mid R_1) = 0.81 \qquad P(B \mid R_2) = 0.72 \qquad P(B \mid R_3) = 0.75$$
$$P(O \mid R_1) = 0.07 \qquad P(O \mid R_2) = 0.10 \qquad P(O \mid R_3) = 0.12$$

Assume in addition that the marginal probability distribution of religion is $P(R_1) = 0.35$, $P(R_2) = 0.60$, and $P(R_3) = 0.05$.
a) Use the information provided to determine the nine joint probabilities for religion and occupation. *Hint:* Algebraically manipulate the conditional probability rule.
b) Find the marginal probability distribution of occupation.
c) Obtain the conditional probability distributions of religion by occupation and construct a table similar to Table 4.2 on page 130.
d) If you are told that a randomly chosen member of this community is a white-collar worker, is it more likely or less likely that the person is of Religion 1 than it would be if you didn't know that person's occupational classification? Explain.

4.22 Each student is allowed two attempts to pass an exam. Experience shows that 60% of the students pass on the first try and that, for those who don't, 80% pass on the second try.
a) What is the probability that a student passes the exam?
b) If a student passed, what is the probability that he or she passed on the first attempt?

4.2   The General Multiplication Rule

As we demonstrated in Section 4.1, the conditional probability rule is used to obtain conditional probabilities in terms of unconditional probabilities:

$$P(B \mid A) = \frac{P(A \cap B)}{P(A)}. \qquad (4.1)$$

More precisely, the conditional probability rule provides a formula for computing conditional probabilities in terms of joint probabilities and marginal probabilities.

Multiplying both sides of Equation (4.1) by $P(A)$, we obtain a formula, called the **general multiplication rule,** for computing joint probabilities in terms of marginal and conditional probabilities:

$$P(A \cap B) = P(A)P(B \mid A). \tag{4.2}$$

The conditional probability rule and the general multiplication rule are simply variations of each other. When joint and marginal probabilities are known or easily determined directly, we can use the conditional probability rule to obtain conditional probabilities. When marginal and conditional probabilities are known or easily determined directly, we can use the general multiplication rule to obtain joint probabilities.

We can extend the general multiplication rule to any finite number of events. To illustrate, let's consider the case of three events—say, A, B, and C. The correct result is

$$P(A \cap B \cap C) = P(A)P(B \mid A)P(C \mid A \cap B). \tag{4.3}$$

To prove Equation (4.3), we first apply Equation (4.2) to $A \cap B$ and C and then to A and B:

$$P(A \cap B \cap C) = P(A \cap B)P(C \mid A \cap B) = P(A)P(B \mid A)P(C \mid A \cap B).$$

In general, we have Proposition 4.2.

◆◆◆   **Proposition 4.2   The General Multiplication Rule**

Let A and B be events with $P(A) > 0$. Then

$$P(A \cap B) = P(A)P(B \mid A). \tag{4.4}$$

More generally, if $A_1, A_2, \ldots, A_N$ are events with $P(A_1 \cap A_2 \cap \cdots \cap A_{N-1}) > 0$, then

$$P\left(\bigcap_{n=1}^{N} A_n\right) = P(A_1)P(A_2 \mid A_1) \cdots P(A_N \mid A_1 \cap A_2 \cap \cdots \cap A_{N-1}). \tag{4.5}$$

Proof   We can prove Equation (4.5) by using mathematical induction. Alternatively, we can expand the right side of Equation (4.5) by using the conditional probability rule to obtain

$$P(A_1)P(A_2 \mid A_1)P(A_3 \mid A_1 \cap A_2) \cdots P(A_n \mid A_1 \cap A_2 \cap \cdots \cap A_{N-1})$$

$$= P(A_1) \cdot \frac{P(A_1 \cap A_2)}{P(A_1)} \cdot \frac{P(A_1 \cap A_2 \cap A_3)}{P(A_1 \cap A_2)} \cdots \frac{P(\bigcap_{n=1}^{N} A_n)}{P(\bigcap_{n=1}^{N-1} A_n)} = P\left(\bigcap_{n=1}^{N} A_n\right),$$

as required.   ◆

EXAMPLE 4.7   *The General Multiplication Rule*

Composition of Congress   The U.S. Congress, Joint Committee on Printing, provides information on the composition of Congress in the *Congressional Directory*. For the 107th Congress, 18.7% of members are senators and 50% of the senators are Democrats. What is the probability that a randomly selected member of the 107th Congress is a Democratic senator?

Solution   Let

$$D = \text{event the member selected is a Democrat and}$$

$$S = \text{event the member selected is a senator.}$$

The event that the member selected is a Democratic senator can be expressed as $S \cap D$. We want to determine the probability of that event.

Because 18.7% of members are senators, $P(S) = 0.187$; and because 50% of the senators are Democrats, $P(D \mid S) = 0.500$. Applying the general multiplication rule gives

$$P(S \cap D) = P(S) \cdot P(D \mid S) = 0.187 \cdot 0.500 = 0.094.$$

The probability is 0.094 that a randomly selected member of the 107th Congress is a Democratic senator. Expressed in terms of percentages, 9.4% of members of the 107th Congress are Democratic senators. ■

The general multiplication rule is also applied to problems that involve sampling two or more members from a population. Example 4.8 illustrates such an application.

EXAMPLE 4.8   *The General Multiplication Rule*

Male and Female   In one of Professor Weiss's classes, there are a total of 40 students. The numbers of males and females are as shown in the frequency distribution in Table 4.4.

Table 4.4 Frequency distribution of males and females in one of Professor Weiss's classes

Sex	Frequency
Male	17
Female	23

Two students are selected at random without replacement. Find the probability that
a) the first student obtained is male.
b) the first student obtained is female and the second is male.
c) the first student obtained is male and the second is male.
d) the second student obtained is male.

Solution   Let F_j denote the event that the jth student obtained is female and let M_j denote the event that the jth student obtained is male.
a) Because 17 of the 40 students are male, the probability that the first student obtained is male is

$$P(M_1) = \frac{17}{40} = 0.425.$$

b) We want to determine $P(F_1 \cap M_2)$. By the general multiplication rule,

$$P(F_1 \cap M_2) = P(F_1)P(M_2 \mid F_1).$$

Determining the two probabilities on the right side of this equation is easy. To find $P(F_1)$—the probability that the first student obtained is female—we note from

Table 4.4 that 23 of the 40 students are female, so $P(F_1) = 23/40$. Next we find $P(M_2 \mid F_1)$—the conditional probability that the second student obtained is male given that the first one obtained is female. Given that the first student obtained is female, there are 39 students remaining in the class, of which 17 are males, so $P(M_2 \mid F_1) = 17/39$. Consequently,

$$P(F_1 \cap M_2) = P(F_1)P(M_2 \mid F_1) = \frac{23}{40} \cdot \frac{17}{39} = 0.251.$$

When two students are randomly selected from the class, the probability is 0.251 that the first student obtained is female and the second is male.[†]

c) The probability that both the first and second students obtained are male is

$$P(M_1 \cap M_2) = P(M_1)P(M_2 \mid M_1) = \frac{17}{40} \cdot \frac{16}{39} = 0.174.$$

d) To determine the probability that the second student obtained is male, we can use the law of partitions and the results from parts (b) and (c). The first student selected is either male or female (but not both) so that events F_1 and M_1 form a partition of the sample space. Applying first the law of partitions and then the results from parts (b) and (c) gives

$$P(M_2) = P(F_1 \cap M_2) + P(M_1 \cap M_2) = 0.251 + 0.174 = 0.425.$$

The probability that the second student obtained is male—which we just found—is identical to the probability that the first student obtained is male—which we found in part (a). This result is no accident. In fact, we can more easily find the probability that the second student obtained is male by reasoning as follows.

1. Because the sampling is random, the second student obtained is equally likely to be any of the 40 students in the class.

2. Because 17 of the students in the class are male, the probability that the second student obtained is male is $17/40 = 0.425$.

The technique just described—referred to as a **symmetry argument**—can frequently be applied to solve seemingly difficult problems quickly and easily. You will often encounter problems for which this technique is suitable.   ■

Drawing a tree diagram when applying the general multiplication rule is often helpful. An appropriate tree diagram for Example 4.8 is shown in Figure 4.1. Each branch of the tree corresponds to one possibility in terms of gender for selecting two students at random from the class. For instance, the second branch of the tree corresponds to event $F_1 \cap M_2$—the event that the first student obtained is female (event F_1) and that the second student obtained is male (event M_2).

[†]This problem can also be solved by using counting techniques. The number of possible outcomes is $40 \cdot 39 = 1560$, and the number of ways that the event in question can occur is $23 \cdot 17 = 391$. Hence the required probability is $391/1560 = 0.251$.

Figure 4.1 Tree diagram for the student-selection problem in Example 4.8

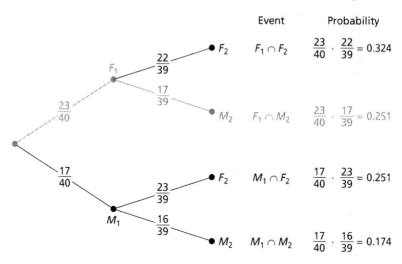

	Event	Probability
	$F_1 \cap F_2$	$\frac{23}{40} \cdot \frac{22}{39} = 0.324$
	$F_1 \cap M_2$	$\frac{23}{40} \cdot \frac{17}{39} = 0.251$
	$M_1 \cap F_2$	$\frac{17}{40} \cdot \frac{23}{39} = 0.251$
	$M_1 \cap M_2$	$\frac{17}{40} \cdot \frac{16}{39} = 0.174$

Starting from the left on that branch, the number $\frac{23}{40}$ is the probability that the first student obtained is female, or $P(F_1)$; and the number $\frac{17}{39}$ is the conditional probability that the second student obtained is male given that the first student obtained is female, or $P(M_2 \mid F_1)$. The product of those two probabilities is, by the general multiplication rule, the probability that the first student obtained is female and that the second is male, or $P(F_1 \cap M_2)$. The second entry in the Probability column of Figure 4.1 shows that this probability is 0.251, as we discovered in Example 4.8(b).

Using this tree diagram, we can also find $P(M_2)$—the probability that the second student obtained is male. We just add the numbers in the Probability column that correspond to the two branches that end with M_2. The result is $P(M_2) = 0.251 + 0.174 = 0.425$, as we found in Example 4.8(d).

Pólya's Urn Scheme

Urn models are used to describe the aftereffects of conditions such as contagious diseases or accidents in an industrial plant. One such model is called *Pólya's urn scheme*, named in honor of its developer, George Pólya (1887–1985), a Hungarian mathematician who later came to the United States and became a professor at Stanford University.

We can describe Pólya's urn scheme as follows. An urn contains b black balls and r red balls. One ball is selected at random, its color noted, and then it and c balls of the same color are added to the urn; the process is then repeated. Here b, r, and c are positive integers.[†]

[†]More general urn schemes are also used. For instance, some permit c to be any integer. Note that if $c = 0$, we have sampling with replacement, whereas if $c = -1$, we have sampling without replacement.

EXAMPLE 4.9   *The General Multiplication Rule*

Pólya's Urn Scheme   Determine the probability that, in three draws, the first two balls are black and the third is red.

Solution   For each $j \in \mathcal{N}$, let B_j denote the event that the jth ball selected is black and let R_j denote the event that the jth ball selected is red. We want to compute $P(B_1 \cap B_2 \cap R_3)$. Applying the general multiplication rule for the case of three events, we get

$$P(B_1 \cap B_2 \cap R_3) = P(B_1)P(B_2 \mid B_1)P(R_3 \mid B_1 \cap B_2). \tag{4.6}$$

Recalling that the urn initially contains b black balls and r red balls, we have that $P(B_1) = b/(b+r)$. Given that the first ball selected is black, the urn then contains $b + c$ black balls and r red balls; hence $P(B_2 \mid B_1) = (b+c)/(b+r+c)$. And, given that the first two balls selected are black, the urn then contains $b + 2c$ black balls and r red balls; hence $P(R_3 \mid B_1 \cap B_2) = r/(b+r+2c)$. Consequently, from Equation (4.6),

$$P(B_1 \cap B_2 \cap R_3) = \frac{b}{b+r} \cdot \frac{b+c}{b+r+c} \cdot \frac{r}{b+r+2c},$$

as required.   ∎

The Law of Total Probability

From the law of partitions and the general multiplication rule, we can derive Proposition 4.3, often referred to as the **law of total probability.**[†]

◆◆◆   **Proposition 4.3**   **Law of Total Probability**

Suppose that $A_1, A_2, \ldots$ form a partition of the sample space. Then, for each event B,

$$P(B) = \sum_n P(A_n)P(B \mid A_n). \tag{4.7}$$

In particular, if E is an event, then

$$P(B) = P(E)P(B \mid E) + P(E^c)P(B \mid E^c), \tag{4.8}$$

for each event B.

Proof   If we first apply the law of partitions (Proposition 2.8 on page 68) and then the general multiplication rule, we obtain

$$P(B) = \sum_n P(A_n \cap B) = \sum_n P(A_n)P(B \mid A_n),$$

as required.   ◆

Note: Computing probabilities with the law of total probability is often referred to as "conditioning on which member of the partition occurs." When the partition consists of an event E and its complement, we are "conditioning on whether event E occurs."

[†]Some mathematicians and statisticians refer to the law of total probability as the *stratified sampling theorem* or the *law of alternatives*.

Referring to the details of the first solution of Example 4.8(d) on page 140, we see that the solution is actually an application of the law of total probability. Examples 4.10 and 4.11 further illustrate use of the law of total probability.

EXAMPLE 4.10   *The Law of Total Probability*

Belief in Aliens   According to an Opinion Dynamics Poll published in *USA TODAY,* roughly 54% of U.S. men and 33% of U.S. women believe in aliens. And the U.S. Bureau of the Census reveals in *Current Population Reports* that 48% of U.S. adults are men. What percentage of U.S. adults believe in aliens?

Solution   We begin by translating the data into the language of probability. Assume that a U.S. adult is selected at random and let

$$A = \text{event the adult selected believes in aliens and}$$
$$M = \text{event the adult selected is a man.}$$

We want to determine $P(A)$. According to the data, $P(M) = 0.48$ which, in turn, implies that $P(M^c) = 0.52$. We also know from the data that $P(A \mid M) = 0.54$ and that $P(A \mid M^c) = 0.33$. Applying the law of total probability yields

$$P(A) = P(M)P(A \mid M) + P(M^c)P(A \mid M^c) = 0.48 \cdot 0.54 + 0.52 \cdot 0.33 = 0.431.$$

In other words, 43.1% of U.S. adults believe in aliens.   ■

EXAMPLE 4.11   *The Law of Total Probability*

Revising Textbooks   Textbook publishers must estimate the sales of new (first-edition) books. The records of one major publishing company indicate that, for a new book, there is a 9.8% chance it will sell more than projected, a 31.6% chance that it will sell close to projected, and a 58.6% chance that it will sell less than projected. Of the new books that sell more than projected, 70% are revised for a second edition, as are 50% of those that sell close to projected, and 20% of those that sell less than projected. Find the probability that a new book published by this company will go to a second edition.

Solution   Let S_1, S_2, and S_3 denote the events that a new book sells more than, close to, and less than projected, respectively. Also, let R denote the event that a new book goes to a second edition. From the information provided,

$$\begin{array}{ll} P(S_1) = 0.098; & P(R \mid S_1) = 0.70; \\ P(S_2) = 0.316; & P(R \mid S_2) = 0.50; \\ P(S_3) = 0.586; & P(R \mid S_3) = 0.20. \end{array}$$

Applying the law of total probability, we conclude that

$$P(R) = P(S_1)P(R \mid S_1) + P(S_2)P(R \mid S_2) + P(S_3)P(R \mid S_3)$$
$$= 0.098 \cdot 0.70 + 0.316 \cdot 0.50 + 0.586 \cdot 0.20 = 0.344.$$

Hence the probability is 0.344—a 34.4% chance—that a new book published by this company will go to a second edition.   ■

EXERCISES 4.2   Basic Exercises

4.23 An article in *Science News* (2000, Vol. 157, p. 135), reported on research by InterSurvey and the Quantitative Study of Society on the effects of regular Internet usage. According to the article, 36% of Americans with Internet access are regular Internet users, meaning that they log on for at least 5 hours per week. Among regular Internet users, 25% say that the Web has reduced their social contacts.
a) Determine the probability that a randomly selected American with Internet access is a regular Internet user who feels that the Web has reduced his or her social contact.
b) Interpret your answer in part (a) in terms of percentages.

4.24 A woman has agreed to participate in an ESP experiment. She's asked to pick, randomly, two numbers between 1 and 6. The second number must be different from the first. Determine the probability that
a) the first number picked is 3 and the second number picked exceeds 4.
b) both numbers picked are less than 3; greater than 3.

4.25 Refer to Table 1.1 on page 4, which gives the regions of the states in the United States. Suppose that two U.S. states are selected at random without replacement.
a) Find the probability that the first is in the Northeast and the second is in the West.
b) Find the probability that both are in the South.
c) Draw a tree diagram for this problem similar to the one shown in Figure 4.1 on page 141.
d) What is the probability that the two states selected are in the same region?
e) What is the probability that one of the states selected is in the Midwest and the other is in the West?

4.26 Suppose that, in Exercise 4.25, three states (instead of two) are selected at random without replacement.
a) Find the probability that the first is in the Northeast and the second two are in the South.
b) Find the probability that all three are in the West.
c) Without drawing a tree diagram for this problem, determine how many branches it would have.
d) What is the probability that the three states selected are all from different regions?
e) What is the probability that two of the states selected are in the South and one is in the Midwest?

4.27 Students are given three chances to pass a basic skills exam for permission to enroll in Calculus I. Sixty percent of the students pass on the first try; of those that fail on the first try, 54% pass on the second try; and, of those remaining, 48% pass on the third try.
a) What is the probability that a student passes on the second try?
b) What is the probability that a student passes on the third try?
c) What percentage of students pass?

4.28 Suppose that you have a key ring with N keys, exactly one of which is your house key. Further suppose that you get home after dark and can't see the keys on the key ring. You randomly try one key at a time until you get the correct key, being careful not to mix the keys you have already tried with the ones you haven't. Let n be an integer between 1 and N, inclusive. Determine the probability that you get the correct key on the nth try by using
a) combinatorial analysis.
b) the general multiplication rule.
c) a symmetry argument.

4.29 The National Sporting Goods Association collects and publishes data on participation in selected sports activities. For Americans 7 years old or older, 17.4% of males and 4.5% of females play golf. And, according to the U.S. Census Bureau's *Current Population Reports,* of Americans 7 years old or older, 48.6% are male and 51.4% are female. From among those Americans who are 7 years old or older, one is selected at random. Find the probability that the person selected plays golf. Interpret your answer in terms of percentages.

4.30 A survey conducted by TELENATION/Market Facts, Inc., combined with information from the U.S. Census Bureau's *Current Population Reports,* yielded the following table. The first two columns provide a percentage distribution of adults by age group; the third column gives the percentage of people in each age group who go to the movies at least once a month—people referred to here as *moviegoers*. Find the percentage of adult moviegoers.

Age	Percentage of adults	Percentage moviegoers
18–24	12.7	83
25–34	20.7	54
35–44	22.0	43
45–54	16.5	37
55–64	10.9	27
65 & over	17.2	20

4.31 A balanced die is rolled and then a balanced coin is tossed the number of times that the die shows. Find the probability of obtaining exactly two heads.

4.32 Refer to Pólya's urn scheme on page 141. Determine the probability that
a) the second ball drawn is red.
b) the third ball drawn is red.
c) the first ball drawn is red, given that the second ball drawn is red.
d) the second ball drawn is black and the third ball drawn is red.

4.33 Urn I contains two white balls and one black ball; Urn II contains one white ball and five black balls. A ball is randomly chosen from Urn I and placed in Urn II. If a ball is then randomly chosen from Urn II, what is the probability it is black?

4.34 Choose a number at random from $\{0, 1, 2, \ldots, 9\}$, call it j. Next, choose a number at random from $\{j, j + 1, \ldots, 9\}$. Find the probability that the second number chosen is k, where $k = 0, 1, \ldots, 9$.

4.35 In the student-selection process in Example 4.8 on page 139, find the conditional probability that the first student is female, given that the second is male.

Theory Exercises

4.36 Prove Equation (4.5) on page 138 by using mathematical induction.

Advanced Exercises

4.37 Research by B. Hatchwell et al. on divorce rates among the long-tailed tit (*Aegithalos caudatus*) appeared in *Science News* (2000, Vol. 157, No. 20, p. 317). Tracking birds in Yorkshire from one breeding season to the next, the researchers noted that 63% of pairs divorced and that "... compared with moms whose offspring had died, nearly twice the percentage of females that raised their youngsters to the fledgling stage moved out of the family flock and took mates elsewhere the next season—81% versus 43%." For the females in this study, find the percentage

a) whose offspring died.

b) that divorced and whose offspring died.

c) whose offspring died among those that divorced.

4.38 Urn I contains two red balls and three green balls, Urn II contains four red balls and one green ball, and Urn III contains two red balls and two green balls. A ball is chosen at random from Urn I and placed in Urn II. Next, a ball is chosen at random from Urn II and placed in Urn III. Then a ball is chosen at random from Urn III. What is the probability that it is red?

4.39 In Pólya's urn scheme (page 141), use mathematical induction to prove that the probability is $r/(b + r)$ that the nth ball drawn is red.

4.40 Suppose that, for each $k \in \{0, 1, 2, \ldots, n\}$, there is a bowl containing k red marbles and $n - k$ green marbles. One bowl is chosen at random from these $n + 1$ bowls and then two marbles are selected at random without replacement from the chosen bowl. Determine the probability that both marbles are red.

4.41 Tom, Dick, and Harry compete in a game that consists of a succession of rounds. In each round, two of the players compete, the winner being decided by the toss of a balanced coin; the third player sits on a bench and waits. In each round after the first, the loser of the previous round sits on the bench while the other two compete. The first player to win two consecutive rounds wins the game. The problem is to determine the probability that Tom wins the game, the probability that Dick wins the game, and the probability that Harry wins the game. Proceed as follows.

a) In each round after the first, call the winner of the previous round the incumbent, the person against whom the incumbent competes in that round the challenger, and the player who sits out that round the benchwarmer. Let p, q, and r denote the probabilities that the current incumbent, challenger, and benchwarmer, respectively, win the game. Determine p, q, and r. *Hint:* Use the law of total probability, conditioning on whether the current incumbent does or does not win the current round.

b) Suppose that, in the first round, Tom and Dick play. Determine the probability that Tom wins the game, the probability that Dick wins the game, and the probability that Harry wins the game. *Hint:* Use the law of total probability, conditioning on whether Tom does or does not win the first round.

c) Now suppose that the initial benchwarmer is obtained by choosing one of Tom, Dick, and Harry at random. Use two different methods to determine the probability that Tom wins the game, the probability that Dick wins the game, and the probability that Harry wins the game: (i) a symmetry argument; (ii) conditioning on whether the specified contestant is chosen as the benchwarmer.

4.42 An ordinary deck of 52 playing cards is randomly divided into two equal stacks of 26 cards each. A card randomly drawn from the first stack turns out to be the queen of spades. That queen is placed in the second stack and then a card is randomly drawn from that stack. What is the probability that the card obtained is a queen? *Note:* There is an efficient way to do this problem by conditioning suitably and using symmetry.

4.3 Independent Events

One of the most important concepts in probability theory is that of *independent events*. For two events, we can define independence as in Definition 4.3.

DEFINITION 4.3   Independence of One Event Relative to Another

Let A and B be events of a sample space with $P(A) > 0$. We say that event B is **independent** of event A if the occurrence of event A doesn't affect the probability that event B occurs. In symbols,

$$P(B \mid A) = P(B).$$

That is, knowing that event A occurs provides no probabilistic information about the occurrence of event B.

EXAMPLE 4.12   *Independence*

Playing Cards   Consider the experiment of randomly selecting one card from a deck of 52 playing cards. Let

$$F = \text{event a face card is selected,}$$
$$K = \text{event a king is selected, and}$$
$$H = \text{event a heart is selected.}$$

a) Determine whether event K is independent of event F.
b) Determine whether event K is independent of event H.

Solution   Event K can occur in four ways (the four kings). Thus the unconditional probability that event K occurs is

$$P(K) = \frac{4}{52} = 0.077.$$

a) To determine whether event K is independent of event F, we must compute $P(K \mid F)$ and compare it to $P(K)$. If those two probabilities are equal, event K is independent of event F; otherwise, event K is not independent of event F. Now, given that event F occurs, 12 outcomes are possible (four jacks, four queens, and four kings), and event K can occur in four of those 12 ways. Hence

$$P(K \mid F) = \frac{4}{12} = \frac{1}{3} = 0.333,$$

which isn't equal to $P(K)$; so event K isn't independent of event F. This lack of independence stems from the fact that the percentage of kings among the face cards (33.3%) is not the same as the percentage of kings among all the cards (7.7%).

b) We need to compute $P(K \mid H)$ and compare it to $P(K)$. Given that event H occurs, 13 outcomes are possible (the 13 hearts), and event K can occur in one of those 13 ways. Therefore

$$P(K \mid H) = \frac{1}{13} = 0.077,$$

which equals $P(K)$; so event K is independent of event H. This independence stems from the fact that the percentage of kings among the hearts is the same as the percentage of kings among all the cards; namely, 7.7%. ∎

The Special Multiplication Rule

Recall that the general multiplication rule states that $P(A \cap B) = P(A)P(B \mid A)$ for any two events A and B where $P(A) > 0$. If event B is independent of event A, then $P(B \mid A) = P(B)$. Consequently, for the special case of independent events, we can replace the term $P(B \mid A)$ in the general multiplication rule with the term $P(B)$. This substitution yields the equation $P(A \cap B) = P(A)P(B)$, which provides an alternative way to define independence for two events, known as the **special multiplication rule.**

DEFINITION 4.4   The Special Multiplication Rule

Two events, A and B, are said to be **independent events** if

$$P(A \cap B) = P(A)P(B).$$

In words, two events are independent if their joint probability equals the product of their marginal probabilities.

Note the following.

- Other terms used for *independent events* are **stochastically independent events** and **statistically independent events.**
- If events are not independent, they are said to be **dependent events.**

On the one hand, if event B is independent of event A in the sense of Definition 4.3, then the two events are independent in the sense of Definition 4.4. On the other hand, if $P(A) > 0$ and events A and B are independent in the sense of Definition 4.4, then event B is independent of event A in the sense of Definition 4.3. (See Exercise 4.43 for details). Using Definition 4.4 rather than Definition 4.3 as the definition of independence has several advantages, including the following.

- Definition 4.4 does not require either of the events to have positive probability, whereas Definition 4.3 does.
- Definition 4.4 is symmetric in the two events, whereas Definition 4.3 is not.

If A and B are independent events, knowing that event A occurs doesn't affect the probability of occurrence of event B. Therefore, knowing that event A doesn't occur (i.e., that event A^c occurs) shouldn't affect the probability of occurrence of event B. In other words, events A^c and B should also be independent. We state this and related facts formally in Proposition 4.4.

◆◆◆ **Proposition 4.4**

If A and B are independent events, then so are each of the three following pairs of events: A^c and B; A and B^c; and A^c and B^c.

Proof Applying first the law of partitions, next the assumed independence of events A and B, and then the complementation rule, we obtain

$$P(A^c \cap B) = P(B) - P(A \cap B) = P(B) - P(A)P(B)$$
$$= (1 - P(A))P(B) = P(A^c)P(B).$$

Hence A^c and B are independent events. That the other two pairs of events are independent follows easily and is left to you as Exercise 4.58.   ◆

Independence for Three or More Events

To extend the definition of independent events to include more than two events, let's consider three events, A, B, and C. One approach to defining independence for these three events would be to require that any two of the three events are independent; that is, events A and B are independent, events A and C are independent, and events B and C are independent. In this case, we say that A, B, and C are **pairwise independent events.**

However, pairwise independence doesn't conform to our intuitive notion of three or more events being independent. For instance, we would like independence of three events, A, B, and C, to imply that event $A \cap B$ is independent of event C. Pairwise independence doesn't imply this condition, as Example 4.13 shows.

EXAMPLE 4.13 *Pairwise Independent Events*

Rolling Dice Suppose that two balanced dice are rolled, one black and one gray. Let

$$A = \text{event the black die comes up even,}$$
$$B = \text{event the gray die comes up even, and}$$
$$C = \text{event the sum of the dice is even.}$$

a) Verify that A, B, and C are pairwise independent events.
b) Show that $A \cap B$ and C aren't independent events.

Solution As shown in Figure 2.1 on page 27, there are 36 equally likely outcomes. Referring to that figure, we find that

$$P(A) = \frac{18}{36} = 0.5, \quad P(B) = \frac{18}{36} = 0.5, \quad P(C) = \frac{18}{36} = 0.5,$$

$$P(A \cap B) = \frac{9}{36} = 0.25, \quad P(A \cap C) = \frac{9}{36} = 0.25, \quad P(B \cap C) = \frac{9}{36} = 0.25,$$

$$P(A \cap B \cap C) = \frac{9}{36} = 0.25.$$

a) From the display, $P(A \cap B) = 0.25 = 0.5 \cdot 0.5 = P(A)P(B)$, so events A and B are independent. Similarly, events A and C are independent, as are events B and C. Hence A, B, and C are pairwise independent events.

b) Again, from the display,

$$P((A \cap B) \cap C) = P(A \cap B \cap C) = 0.25 \neq 0.25 \cdot 0.5 = P(A \cap B)P(C).$$

Therefore $A \cap B$ and C aren't independent events.   ∎

The definition of independence for three or more events that conforms to our intuition requires that each subcollection of events satisfy the multiplication property. For three events, A, B, and C, this requirement means that the events must be pairwise independent and that $P(A \cap B \cap C) = P(A)P(B)P(C)$. In other words,

$$P(A \cap B) = P(A)P(B), \quad P(A \cap C) = P(A)P(C), \quad P(B \cap C) = P(B)P(C),$$

and

$$P(A \cap B \cap C) = P(A)P(B)P(C).$$

More generally, we have Definition 4.5 for **mutually independent events.**

DEFINITION 4.5   Mutually Independent Events

Events $A_1, A_2, \ldots, A_N$ are said to be **mutually independent events** (or, simply, **independent events**) if each subcollection of the N events satisfies the multiplication property. That is, for each integer n, where $2 \leq n \leq N$,

$$P(A_{k_1} \cap A_{k_2} \cap \cdots \cap A_{k_n}) = P(A_{k_1})P(A_{k_2}) \cdots P(A_{k_n}),$$

where $k_1, k_2, \ldots, k_n$ are distinct integers between 1 and N. The events of an infinite collection are said to be *independent events* if the events of every finite subcollection of those events are independent.

For three or more events, note the following.

- Mutually independent events are necessarily pairwise independent.

- Pairwise independent events aren't necessarily mutually independent, as Example 4.13 shows.

- We can't conclude that events are mutually independent just because the multiplication property holds for all of the events in question, as Example 4.14 shows.

EXAMPLE 4.14 *Nonindependence*

Rolling Dice Refer to Example 4.13 where two balanced dice are rolled, one black and one gray. Let

$$D = \text{event the black die comes up 1, 2, or 3,}$$
$$E = \text{event the black die comes up 3, 4, or 5, and}$$
$$F = \text{event the sum of the dice is 5.}$$

Show that $P(D \cap E \cap F) = P(D)P(E)P(F)$ but that D, E, and F aren't (mutually) independent events.

Solution To begin, we note that $P(D) = P(E) = 1/2$, $P(F) = 1/9$, and $P(D \cap E \cap F) = 1/36$. Therefore $P(D \cap E \cap F) = P(D)P(E)P(F)$. However, we have, for instance,

$$P(D \cap F) = \frac{1}{12} \neq \frac{1}{18} = \frac{1}{2} \cdot \frac{1}{9} = P(D)P(F)$$

so that $P(D \cap F) \neq P(D)P(F)$. This result implies that D, E, and F aren't (mutually) independent events. ∎

Suppose that A, B, and C are independent events. As a first consequence of Definition 4.5, we show that event $A \cap B$ is independent of event C. Indeed,

$$P((A \cap B) \cap C) = P(A \cap B \cap C) = P(A)P(B)P(C) = P(A \cap B)P(C).$$

As a second consequence of Definition 4.5, we show that event $A \cup B$ is independent of event C. Indeed, by the distributive law for sets, the general addition rule, and Definition 4.5, we have

$$\begin{aligned}
P((A \cup B) \cap C) &= P((A \cap C) \cup (B \cap C)) \\
&= P(A \cap C) + P(B \cap C) - P(A \cap B \cap C) \\
&= P(A)P(C) + P(B)P(C) - P(A)P(B)P(C) \\
&= (P(A) + P(B) - P(A)P(B))P(C) \\
&= (P(A) + P(B) - P(A \cap B))P(C) = P(A \cup B)P(C).
\end{aligned}$$

As a third consequence of Definition 4.5, we show that A^c, B, and C^c are independent events. First we apply Definition 4.5 and Proposition 4.4 on page 149 to conclude that A^c, B, and C^c are pairwise independent events. Next we use the result of the preceding paragraph and Proposition 4.4 to conclude that $(A \cup C)^c$ and B are independent events. Therefore, by De Morgan's laws and Proposition 4.4,

$$\begin{aligned}
P(A^c \cap B \cap C^c) &= P((A \cup C)^c \cap B) = P((A \cup C)^c)P(B) = P(A^c \cap C^c)P(B) \\
&= P(A^c)P(C^c)P(B) = P(A^c)P(B)P(C^c).
\end{aligned}$$

We have demonstrated that A^c, B, and C^c are pairwise independent events and that $P(A^c \cap B \cap C^c) = P(A^c)P(B)P(C^c)$. So, events A^c, B, and C^c satisfy the conditions of Definition 4.5 and hence are independent.

Proposition 4.5, which we state without proof, includes as special cases the results obtained in the preceding three paragraphs.

◆◆◆   **Proposition 4.5**

*Events formed from disjoint subcollections of independent events are themselves inde-
pendent events. That is, if we have disjoint subcollections of independent events and
within each subcollection we form an event from the events of the subcollection by using
the set operations of complementation, intersection, and union, then the resulting events
are independent.*

Using Contextual Independence to Obtain Probabilities

Frequently, we know or can reasonably assume from the context of a problem that
some particular events are independent. From that knowledge—which we refer to as
contextual independence—we can then obtain probabilities involving those events. Examples 4.15–4.18 illustrate this idea.

EXAMPLE 4.15   *Contextual Independence*

Coin Tossing   Consider again the random experiment of tossing a coin three times and
observing the result (a head or a tail) for each toss. A sample space for this random
experiment is $\Omega = \{$HHH, HHT, HTH, HTT, THH, THT, TTH, TTT$\}$. From the context, we can
reasonably assume that successive tosses of the coin are independent of one another.

a) If the coin is balanced, determine the probabilities of the individual outcomes, using
   the fact that the eight possible outcomes are equally likely.
b) If the coin is balanced, determine the probabilities of the individual outcomes, using
   the fact that successive tosses are independent of one another.
c) If the coin has probability 0.2 of landing heads (face up) on any given toss, determine
   the probabilities of the individual outcomes.
d) If the coin has probability p ($0 \leq p \leq 1$) of landing heads on any given toss, determine
   the probabilities of the individual outcomes.
e) Repeat part (d) if the coin is tossed n times.

Solution   a) Because the coin is balanced, each of the eight possible outcomes is equally likely to be
   the one that occurs. In particular, each possible outcome has probability $1/8 = 0.125$
   of occurring. Note that this situation corresponds to assignment #1 in Table 2.4
   on page 44.

b) For $1 \leq j \leq 3$, we let A_j denote the event that the jth toss is a head; then A_1, A_2,
   and A_3 are independent events. Moreover, because the coin is balanced, we have
   $P(A_j) = 1/2 = 0.5$ for each j. From these facts, we can obtain the probabilities
   of the individual outcomes. Consider, for instance, the possible outcome HTH. Because $\{$HTH$\} = A_1 \cap A_2^c \cap A_3$, we find that

$$P(\{\text{HTH}\}) = P(A_1 \cap A_2^c \cap A_3) = P(A_1)P(A_2^c)P(A_3)$$
$$= P(A_1)\big(1 - P(A_2)\big)P(A_3) = 0.5 \cdot (1 - 0.5) \cdot 0.5 = 0.125.$$

Using similar arguments, we conclude that each possible outcome has probability 0.125 of occurring. Note that this result agrees with the result of part (a), although
the approach is different.

c) We proceed as in part (b), but now $P(A_j) = 0.2$ for each j. So, for instance,

$$P(\{\text{HTH}\}) = P(A_1 \cap A_2^c \cap A_3) = P(A_1)P(A_2^c)P(A_3)$$
$$= P(A_1)\big(1 - P(A_2)\big)P(A_3) = 0.2 \cdot (1 - 0.2) \cdot 0.2 = 0.032.$$

Using similar arguments, we can obtain the probabilities of the other seven individual outcomes. Doing so, we find that the probabilities are as given in assignment #3 in Table 2.4.

d) We now assume that the coin has probability p of landing heads, which implies that $P(A_j) = p$ for each j. So, for instance,

$$P(\{\text{HTH}\}) = P(A_1 \cap A_2^c \cap A_3) = P(A_1)P(A_2^c)P(A_3)$$
$$= P(A_1)\big(1 - P(A_2)\big)P(A_3) = p \cdot (1 - p) \cdot p = p^2(1 - p).$$

In general, the probability of any individual outcome for which the coin lands heads exactly k times in the three tosses is $p^k(1 - p)^{3-k}$.

e) We now assume that the coin has probability p of landing heads and that the coin is tossed n times. As in part (d), the probability of any individual outcome for which the coin lands heads exactly k times in the n tosses is $p^k(1 - p)^{n-k}$.   ∎

EXAMPLE 4.16   *Contextual Independence*

Component Analysis   Mechanical or electrical units often consist of several components, each of which is subject to failure. We consider a unit whose components act independently of one another—that is, whether one component is working is independent of the status of the other components. Let p_j denote the probability that, at a specified time, component j is working.

a) A unit is said to be a *series system* if it functions only when all the components are working. Determine the probability that, at the specified time, a series system consisting of five components is functioning.

b) A unit is said to be a *parallel system* if it functions when at least one of the components is working. Determine the probability that, at the specified time, a parallel system consisting of five components is functioning.

c) Repeat parts (a) and (b) for a unit consisting of n components.

Solution   Let A_j denote the event that component j is working. By assumption, $A_1, A_2, \ldots$ are independent events with probabilities $p_1, p_2, \ldots$, respectively.

a) Let S denote the event that the series system is functioning; then $S = \bigcap_{j=1}^{5} A_j$. By independence,

$$P(S) = P\left(\bigcap_{j=1}^{5} A_j\right) = P(A_1)P(A_2) \cdots P(A_5) = p_1 p_2 \cdots p_5 = \prod_{j=1}^{5} p_j.$$

b) Let L denote the event that the parallel system is functioning; then $L = \bigcup_{j=1}^{5} A_j$. We can determine $P(L)$ directly by using the inclusion–exclusion principle, but using the complementation rule is easier. Applying the complementation rule, De Morgan's laws, and Proposition 4.5 on page 152, we conclude that

$$P(L) = 1 - P(L^c) = 1 - P\left(\bigcap_{j=1}^{5} A_j^c\right) = 1 - P(A_1^c)P(A_2^c)\cdots P(A_5^c)$$

$$= 1 - (1 - p_1)(1 - p_2)\cdots(1 - p_5) = 1 - \prod_{j=1}^{5}(1 - p_j).$$

c) We proceed in the same way as in parts (a) and (b) except that here we have n components instead of five. We find that

$$P(S) = \prod_{j=1}^{n} p_j \quad \text{and} \quad P(L) = 1 - \prod_{j=1}^{n}(1 - p_j).$$

We leave the details to you as Exercise 4.53.   ■

EXAMPLE 4.17   *Contextual Independence*

Coin Tossing In Example 2.32 on page 74, we discussed the random experiment of repeatedly tossing a balanced coin. We used a continuity property of probabilities to show that the probability is 1 that eventually a head will be tossed. We mentioned that this result holds regardless of whether the coin is balanced, provided only that the probability is not 0 of getting a head when the coin is tossed once. Verify this fact.

Solution   Let A denote the event that a head will eventually be tossed and, for each $n \in \mathcal{N}$, let A_n denote the event of at least one head in the first n tosses. Then $A_1 \subset A_2 \subset \cdots$ and $A = \bigcup_{n=1}^{\infty} A_n$. Hence, by Proposition 2.11(a) on page 74, $P(A) = \lim_{n\to\infty} P(A_n)$.

We now obtain $P(A_n)$ for each $n \in \mathcal{N}$. As computing $P(A_n^c)$—the probability that the first n tosses all result in tails—is simpler, we use the complementation rule to determine $P(A_n)$. We let B_j denote the event that the jth toss is a tail. Note that $B_1, B_2, \ldots$ are independent events and that $A_n^c = \bigcap_{j=1}^{n} B_j$. Also, if we let p denote the probability of a head on any particular toss, then $P(B_j) = 1 - p$ for all j. Therefore

$$P(A_n^c) = P\left(\bigcap_{j=1}^{n} B_j\right) = \prod_{j=1}^{n} P(B_j) = (1 - p)^n$$

and hence $P(A_n) = 1 - (1 - p)^n$. Consequently,

$$P(A) = \lim_{n\to\infty} P(A_n) = \lim_{n\to\infty} 1 - (1 - p)^n = \begin{cases} 1, & \text{if } 0 < p \le 1; \\ 0, & \text{if } p = 0. \end{cases}$$

This result shows that the probability is 1 that a head will eventually be tossed unless the probability is 0 of getting a head when the coin is tossed once; in this latter case, the probability is 0 that a head will eventually be tossed.   ■

We can easily generalize the results of Example 4.17 to obtain Proposition 4.6, a useful and interesting fact. We leave the details of verification to you as Exercise 4.59.

◆◆◆ **Proposition 4.6**

Let E be an event with positive probability. If the random experiment is repeated independently and indefinitely, the probability is 1 that event E will eventually occur.

To further illustrate Proposition 4.6, let's consider the Powerball® multi-state lottery. To play the game, a player first selects five distinct numbers from the integers 1–53 and then chooses a Powerball number, which can be any integer between 1 and 42, inclusive. To win the jackpot, the player's five numbers and Powerball number must match those obtained in the drawing.

Using counting techniques, we can easily show that the probability of winning the jackpot is approximately 0.00000000830, or a little less than one in 100 million. Although this probability is quite small, it is positive. Thus Proposition 4.6 implies that, if a player could continually play the Powerball, he would eventually win the jackpot.[†]

EXAMPLE 4.18 *Contextual Independence*

Craps The game of *craps* is played by rolling two balanced dice. A first roll of a sum of 7 or 11 wins; a first roll of a sum of 2, 3, or 12 loses. To win with any other first sum, that sum must be repeated before a sum of 7 is thrown. Suppose that a player rolls a sum of 8 on the first roll. Determine the probability that the player wins.

Solution We need to find the probability that, upon successive rolls of a pair of balanced dice, a sum of 8 is rolled before a sum of 7, an event that we denote W. To determine $P(W)$, we can use two different methods. The first is straightforward, although lengthy; the second uses conditioning and is trickier, although quicker and less computationally intensive.

Method 1: For each $n \in \mathcal{N}$, we let W_n denote the event that a sum of 8 occurs on the nth roll of the dice and that neither a sum of 8 nor a sum of 7 occurs before that. Events $W_1, W_2, \ldots$ are mutually exclusive and $W = \bigcup_{n=1}^{\infty} W_n$. Therefore

$$P(W) = \sum_{n=1}^{\infty} P(W_n). \tag{4.9}$$

To determine $P(W_n)$, we let E_j and S_j denote the events of a sum of 8 and a sum of 7 on the jth roll, respectively; we also let $N_j = (E_j \cup S_j)^c$. From Figure 2.1 on page 27, $P(E_j) = 5/36$ and $P(S_j) = 6/36$ for each j and, moreover, because E_j and S_j are mutually exclusive,

$$P(N_j) = 1 - P(E_j \cup S_j) = 1 - \big(P(E_j) + P(S_j)\big) = 1 - \left(\frac{5}{36} + \frac{6}{36}\right) = \frac{25}{36}.$$

[†]If a player could indefinitely play one ticket once a week, then, on average, it would take roughly 2.3 million years before the player would win the jackpot. This result can be obtained by applying techniques discussed in Chapter 7.

Now, $W_n = \left(\bigcap_{j=1}^{n-1} N_j\right) \cap E_n$ and, because the rolls of the dice are independent of one another, we get

$$P(W_n) = \left(\prod_{j=1}^{n-1} P(N_j)\right) P(E_n) = \left(\frac{25}{36}\right)^{n-1} \cdot \frac{5}{36} = \frac{5}{36} \left(\frac{25}{36}\right)^{n-1}.$$

In view of this equation and Equation (4.9),

$$P(W) = \sum_{n=1}^{\infty} P(W_n) = \frac{5}{36} \sum_{n=1}^{\infty} \left(\frac{25}{36}\right)^{n-1} = \frac{5}{36} \cdot \frac{1}{1 - \frac{25}{36}} = \frac{5}{11}.$$

Method 2:   Here we condition on the result of the first roll (i.e., the second roll by the player). Let E and S denote the events of a sum of 8 and a sum of 7 on the first roll, respectively. Events E, S, and $N = (E \cup S)^c$ form a partition of the sample space, and $P(E) = 5/36$, $P(S) = 6/36$, and $P(N) = 25/36$. Applying the law of total probability yields

$$P(W) = P(E)P(W \mid E) + P(S)P(W \mid S) + P(N)P(W \mid N).$$

Now, clearly, $P(W \mid E) = 1$ and $P(W \mid S) = 0$. Moreover, because the rolls of the dice are independent, we have $P(W \mid N) = P(W)$. Therefore

$$P(W) = \frac{5}{36} \cdot 1 + \frac{6}{36} \cdot 0 + \frac{25}{36} \cdot P(W).$$

Solving for $P(W)$ in this equation gives $P(W) = 5/11$. This answer is the same as that obtained by using Method 1, but Method 2 is quicker and computationally easier.   ■

We can apply the same type of reasoning as that in Example 4.18 to obtain Proposition 4.7. We leave the verification to you as Exercise 4.60.

◆◆◆   **Proposition 4.7**

Suppose that A and B are mutually exclusive events, at least one of which has positive probability. If the random experiment is repeated independently, the probability is

$$\frac{P(A)}{P(A) + P(B)}$$

that event A occurs before event B occurs.

In Example 4.18, the random experiment consists of rolling two balanced dice, and the problem is to compute the probability that a sum of 8 is rolled before a sum of 7. To apply Proposition 4.7, we let E and S denote the events that a sum of 8 is rolled and a sum of 7 is rolled, respectively. We have $P(E) = 5/36$ and $P(S) = 6/36$. Therefore, by Proposition 4.7, the probability that a sum of 8 is rolled before a sum of 7 is

$$\frac{P(E)}{P(E) + P(S)} = \frac{\frac{5}{36}}{\frac{5}{36} + \frac{6}{36}} = \frac{5}{11},$$

which agrees with the result obtained in Example 4.18.

Mutually Exclusive Versus Independent Events

The terms *mutually exclusive* and *independent* refer to different concepts. Mutually exclusive events are those that can't occur simultaneously. Independent events are those for which the occurrence of some doesn't affect the probabilities of the others occurring. If two or more events are mutually exclusive, the occurrence of one precludes the occurrence of the others. Hence, two or more events with positive probabilities can't be both mutually exclusive and independent. See Exercise 4.55 for more on this issue.

EXERCISES 4.3   **Basic Exercises**

4.43 Verify the following statements made on page 148.
a) If event B is independent of event A in the sense of Definition 4.3, then the two events are independent in the sense of Definition 4.4.
b) If $P(A) > 0$ and events A and B are independent in the sense of Definition 4.4, then event B is independent of event A in the sense of Definition 4.3.

4.44 The U.S. National Center for Health Statistics compiles data on injuries and publishes the information in *Vital and Health Statistics*. A contingency table for injuries in the United States, by circumstance and sex, is as follows. Frequencies are in millions.

Circumstance

		Work C_1	Home C_2	Other C_3	Total
Sex	Male S_1	8.0	9.8	17.8	35.6
	Female S_2	1.3	11.6	12.9	25.8
	Total	9.3	21.4	30.7	61.4

a) Are events C_1 and S_2 independent? Explain.
b) Is the event that an injured person is male independent of the event that an injured person was hurt at home? Explain.

4.45 Refer to the joint probability distribution in Exercise 4.8 on page 135 for living arrangement and age of U.S. citizens 15 years of age and older. Are events A_2 and L_1 independent? Interpret your answer.

4.46 In the game of Yahtzee, five balanced dice are rolled. What is the probability
a) of rolling all 2s?
b) that all the dice come up the same number?
c) of getting a full house—three of one number and two of another?

4.47 Let A be an event of a sample space. Verify the following statements.
a) If $P(A) = 0$ or $P(A) = 1$, then, for each event B of the sample space, A and B are independent events.
b) If A and A are independent events, then $P(A) = 0$ or $P(A) = 1$.

4.48 Suppose that a number is chosen at random from the interval (0, 1). For the number obtained, let A be the event that the first decimal digit is 6 and let B be the event that the second decimal digit is 4. Determine whether events A and B are independent.

4.49 Respond *true* or *false* and explain your answer: If events A and B are independent and events B and C are independent, then events A and C are independent.

4.50 According to *Accident Facts*, published by the National Safety Council, a probability distribution of age group for drivers at fault in fatal crashes is as follows.

Age (yrs)	Probability
16–24	0.255
25–34	0.238
35–64	0.393
65 & over	0.114

Of three fatal automobile crashes, find the probability that
a) the drivers at fault in the first, second, and third crashes are in the age groups 16–24, 25–34, and 35–64, respectively.
b) two of the drivers at fault are between 16 and 24 years old and one of the drivers at fault is 65 years old or older.

4.51 Determine the number of equations for checking mutual independence of n events.

4.52 Refer to Example 4.15 on page 152 and also to Example 2.18 on page 44, where we provided six potential probability assignments (Table 2.4) and discovered that only assignments #1–#4 are legitimate.
a) Of the assignments #1–#4, which correspond to independent tosses of the coin? Explain your reasoning.
b) For those assignments that correspond to independent tosses of the coin, identify the probability of a head on any given toss.

4.53 Refer to Example 4.16 on page 153 regarding component analysis.
a) Solve part (b) by using the inclusion–exclusion principle.
b) Provide the details for part (c).

4.54 Referring to Example 4.18 on page 155, determine the probability of winning a game of craps.

4.55 In this exercise, you are to further examine the concepts of independent events and mutually exclusive events.
a) If two events are mutually exclusive, determine their joint probability.
b) If two events with positive probability are independent, explain why their joint probability is not 0. Conclude that the two events can't be mutually exclusive.
c) Give an example of two events that are neither mutually exclusive nor independent.

4.56 Suppose that you have two coins, one balanced and the other with probability p of a head. You select one of the coins at random and toss it twice. Determine the probability that
a) the first toss is a head.   b) the second toss is a head.   c) both tosses are heads.
d) Show that the events "first toss a head" and "second toss a head" are independent if and only if the second coin is also balanced.

4.57 Suppose that, for children born to a certain couple, the events "first is a boy," "second is a boy," "third is a boy," and "fourth is a boy" are mutually independent. Further suppose that the probability of a boy on any specified birth is p. For $1 \leq j \leq 4$, let B_j and G_j denote the events that the jth child born is a boy and girl, respectively. Also, for $0 \leq k \leq 4$, let S_k denote the event that exactly k of the four children born are boys.

a) Explain and interpret the equation

$$S_2 = (B_1 \cap B_2 \cap G_3 \cap G_4) \cup (B_1 \cap G_2 \cap B_3 \cap G_4) \cup (B_1 \cap G_2 \cap G_3 \cap B_4)$$
$$\cup (G_1 \cap B_2 \cap B_3 \cap G_4) \cup (G_1 \cap B_2 \cap G_3 \cap B_4) \cup (G_1 \cap G_2 \cap B_3 \cap B_4).$$

b) Use part (a) to determine $P(S_2)$. Justify all steps.

c) Obtain a formula for $P(S_k)$ that holds for $k = 0, 1, 2, 3, 4$.

Theory Exercises

4.58 Complete the proof of Proposition 4.4 on page 149.

4.59 Prove Proposition 4.6 on page 155: Let E be an event with positive probability. If the random experiment is repeated independently and indefinitely, the probability is 1 that event E will eventually occur.

4.60 Prove Proposition 4.7 on page 156: Suppose that A and B are mutually exclusive events, at least one of which has positive probability. If the random experiment is repeated independently, the probability that event A occurs before event B occurs is $P(A)/\big(P(A) + P(B)\big)$.

Advanced Exercises

4.61 Conditional independence: Let A, B, and C be events of a sample space with $P(A) > 0$. We say that events B and C are *conditionally independent given event A* if $P(B \cap C \mid A) = P(B \mid A)P(C \mid A)$. Answer *true* or *false* to each of the following statements. If true, prove it; if false, provide a counterexample.

a) If B and C are conditionally independent given A, then $P(C \mid A \cap B) = P(C \mid A)$.

b) If B and C are independent, then they are conditionally independent given A.

c) If B and C are conditionally independent given A, then they are independent.

4.62 Second Borel–Cantelli lemma: Let $A_1, A_2, \ldots$ be a countably infinite sequence of events and let A^* be the event that infinitely many of the A_ns occur. In Exercise 2.77 on page 77, you were asked to prove the first Borel–Cantelli lemma: If $\sum_{n=1}^{\infty} P(A_n) < \infty$, then $P(A^*) = 0$.

a) Now prove the *second Borel–Cantelli lemma:* If $A_1, A_2, \ldots$ are independent events and $\sum_{n=1}^{\infty} P(A_n) = \infty$, then $P(A^*) = 1$. *Hint:* Use the fact that $1 - x \leq e^{-x}$.

b) Provide an example to show that the second Borel–Cantelli lemma fails without the independence assumption.

4.63 Suppose that each second an amoeba either splits into two amoebas or dies, with probabilities r and $1 - r$, respectively.

a) Starting with one amoeba, determine the probability of eventual extinction. *Hint:* Condition on whether the initial amoeba splits. You'll get a quadratic equation in the probability of extinction. Analyze the results obtained to decide which value is appropriate.

b) Use part (a) to determine the probability of eventual extinction if the probability that an amoeba splits rather than dies is 1/3; 1/2; 3/4; 9/10.

c) How is contextual independence used in this problem?

4.64 Binomial distribution: Let E be an event of a random experiment and set $p = P(E)$. For n independent repetitions of the random experiment, let $n(E)$ denote the number of times that event E occurs. Prove that

$$P\big(n(E) = k\big) = \binom{n}{k} p^k (1 - p)^{n-k}, \qquad k = 0, 1, 2, \ldots, n.$$

Hint: Refer to Example 4.15(e) on page 152.

4.65 Jan and Jean each independently toss a balanced coin n times. What is the probability that they get the same number of heads? *Hint:* Refer to Exercise 4.64 and to Exercise 3.48 on page 109.

4.4 Bayes's Rule

We now introduce the rule of probability developed by Thomas Bayes, an eighteenth-century clergyman. This rule is aptly called **Bayes's rule.** A primary use of Bayes's rule is to revise probabilities in accordance with newly acquired information. Such revised probabilities are actually conditional probabilities and, so, in some sense, we have already examined much of the material to be covered in this section. However, as we show, Bayes's rule involves some new concepts, techniques, and terminology.

In anticipation of stating and proving Bayes's rule, we present Example 4.19 to preview the pertinent ideas.

EXAMPLE 4.19   *Introducing Bayes's Rule*

U.S. Resident Population   The U.S. Bureau of the Census collects data on the resident population, by age and region of residence, and presents its findings in *Current Population Reports*. The first two columns of Table 4.5 display a percentage distribution for region of residence; the third column displays the percentage of seniors (aged 65 or over) in each region.

Table 4.5 Percentage distribution for region of residence and percentage of seniors in each region

Region	Percentage of U.S. population	Percentage seniors
Northeast	19.0	13.8
Midwest	23.1	13.0
South	35.5	12.8
West	22.4	11.1

For instance, Table 4.5 shows that 13.8% of Northeast residents are seniors. Now we ask: What percentage of seniors are Northeast residents?

Solution   We first translate the information displayed in Table 4.5 into the language of probability. Suppose that a U.S. resident is selected at random. Let

$$S = \text{event the resident selected is a senior}$$

and

$$R_1 = \text{event the resident selected lives in the Northeast,}$$
$$R_2 = \text{event the resident selected lives in the Midwest,}$$
$$R_3 = \text{event the resident selected lives in the South, and}$$
$$R_4 = \text{event the resident selected lives in the West.}$$

The percentages shown in the second and third columns of Table 4.5 translate into the probabilities displayed in Table 4.6.

Table 4.6 Probabilities derived from Table 4.5

$P(R_1) = 0.190$	$P(S \mid R_1) = 0.138$
$P(R_2) = 0.231$	$P(S \mid R_2) = 0.130$
$P(R_3) = 0.355$	$P(S \mid R_3) = 0.128$
$P(R_4) = 0.224$	$P(S \mid R_4) = 0.111$

The problem is to determine the percentage of seniors that live in the Northeast region or, in terms of probability, $P(R_1 \mid S)$. Note that we already know the "reverse" conditional probability $P(S \mid R_1)$. The idea therefore is to "switch" the events in the former conditional probability, which we do by first applying the conditional probability rule and then invoking the general multiplication rule:

$$P(R_1 \mid S) = \frac{P(S \cap R_1)}{P(S)} = \frac{P(R_1 \cap S)}{P(S)} = \frac{P(R_1)P(S \mid R_1)}{P(S)}. \qquad (4.10)$$

From Table 4.6, we already know the two probabilities in the numerator of the fraction on the right side of Equation (4.10). Thus we need only determine $P(S)$.

We first observe that R_1, R_2, R_3, and R_4 form a partition of the sample space. Therefore, by the law of total probability,

$$P(S) = \sum_{n=1}^{4} P(R_n)P(S \mid R_n)$$
$$= 0.190 \cdot 0.138 + 0.231 \cdot 0.130 + 0.355 \cdot 0.128 + 0.224 \cdot 0.111$$
$$= 0.127.$$

Using this result (which, by the way, shows that 12.7% of U.S. residents are seniors), we obtain from Equation (4.10) and Table 4.6 that

$$P(R_1 \mid S) = \frac{P(R_1)P(S \mid R_1)}{P(S)} = \frac{0.190 \cdot 0.138}{0.127} = 0.206.$$

Thus 20.6% of seniors are Northeast residents.   ■

With Example 4.19 in mind, we now state and prove Bayes's rule as Proposition 4.8.

◆◆◆ **Proposition 4.8 Bayes's Rule**

Suppose that $A_1, A_2, \ldots$ form a partition of the sample space. Then, for each event B,

$$P(A_j \mid B) = \frac{P(A_j)P(B \mid A_j)}{\sum_n P(A_n)P(B \mid A_n)},$$

where A_j can be any one of the events $A_1, A_2, \ldots$.

Proof The proof follows the argument presented in Example 4.19. Applying, in turn, the conditional probability rule, the general multiplication rule, and the law of total probability, we obtain

$$P(A_j \mid B) = \frac{P(B \cap A_j)}{P(B)} = \frac{P(A_j)P(B \mid A_j)}{P(B)} = \frac{P(A_j)P(B \mid A_j)}{\sum_n P(A_n)P(B \mid A_n)},$$

as required. ◆

Bayes's rule applies when the probabilities $P(A_1), P(A_2), \ldots$ and the conditional probabilities $P(B \mid A_1), P(B \mid A_2), \ldots$ are known. The denominator of the fraction on the right side of Bayes's rule is simply $P(B)$ expressed in terms of the known unconditional and conditional probabilities by applying the law of total probability.

EXAMPLE 4.20 *Bayes's Rule*

The Monty Hall Problem Several years ago, in a column published by Marilyn vos Savant in *Parade* magazine, an interesting probability problem was posed. That problem is now referred to as the *Monty Hall Problem* because of its origins from the television show *Let's Make a Deal*. Following is a version of the Monty Hall Problem.

On a game show, there are three doors, behind each of which is one prize. Two of the prizes are worthless and one is valuable. A contestant selects one of the doors, following which, the game-show host, who knows where the valuable prize lies, opens one of the remaining two doors to reveal a worthless prize. The host then offers the contestant the opportunity to change his selection. Should the contestant switch?

Solution By symmetry, we can assume that the contestant selects Door 1. We first determine the probability that the valuable prize is behind Door 3, given that the game-show host opens Door 2. Let T denote the event that the game-show host opens Door 2 and, for $1 \leq n \leq 3$, let A_n denote the event that the valuable prize is behind Door n. We want to find $P(A_3 \mid T)$.

Assuming, as we are, that initially the valuable prize is equally likely to be behind each of the three doors, we have $P(A_n) = 1/3$ for each n. Also, assuming, as we do, that the game-show host is equally likely to open either Door 2 or Door 3 if the valuable prize is behind Door 1, we have $P(T \mid A_1) = 1/2$. Moreover, it is clear that $P(T \mid A_2) = 0$ and that $P(T \mid A_3) = 1$. Applying Bayes's rule gives

$$P(A_3 \mid T) = \frac{P(A_3)P(T \mid A_3)}{\sum_{n=1}^{3} P(A_n)P(T \mid A_n)} = \frac{\frac{1}{3} \cdot 1}{\frac{1}{3} \cdot \frac{1}{2} + \frac{1}{3} \cdot 0 + \frac{1}{3} \cdot 1} = \frac{2}{3}.$$

Likewise, we find that $P(A_1 \mid T) = 1/3$ and that $P(A_2 \mid T) = 0$. Thus the contestant should switch from Door 1 to Door 3 because, given that the game-show host opens Door 2, the probability of the valuable prize being behind Door 3 is twice that of being behind Door 1. A similar argument yields analogous results under the condition that the game-show host opens Door 3. ■

EXAMPLE 4.21 *Bayes's Rule*

Smoking and Lung Disease According to the Arizona Chapter of the American Lung Association, 7.0% of the population has lung disease. Of those people having lung disease, 90.0% are smokers; and of those not having lung disease, 74.7% are non-smokers. What are the chances that a smoker has lung disease?

Solution Suppose that a person is selected at random. Let S denote the event that the person selected is a smoker and let L denote the event that the person selected has lung disease. We want to determine the probability that a randomly selected smoker has lung disease—that is, $P(L \mid S)$.

From the information provided, we know that $P(L) = 0.070$, $P(S \mid L) = 0.900$, and $P(S^c \mid L^c) = 0.747$. We apply Bayes's rule—with L and its complement as the partition—to obtain $P(L \mid S)$. By the complementation rule and its conditional form, $P(L^c) = 1 - P(L) = 0.930$ and $P(S \mid L^c) = 1 - P(S^c \mid L^c) = 0.253$. Hence

$$P(L \mid S) = \frac{P(L)P(S \mid L)}{P(L)P(S \mid L) + P(L^c)P(S \mid L^c)}$$

$$= \frac{0.070 \cdot 0.900}{0.070 \cdot 0.900 + 0.930 \cdot 0.253} = 0.211.$$

Consequently, the probability is 0.211 that a randomly selected smoker has lung disease. In terms of percentages, 21.1% of smokers have lung disease. ■

Example 4.21 indicates that the rate of lung disease among smokers (21.1%) is more than three times the rate of lung disease among the general population (7.0%). Using arguments similar to those in Example 4.21, we can show that the probability is 0.010 that a randomly selected nonsmoker has lung disease; in other words, 1.0% of nonsmokers have lung disease. Therefore the rate of lung disease among smokers (21.1%) is more than 20 times that among nonsmokers (1.0%).

Prior and Posterior Probabilities

Two important terms associated with Bayes's rule are *prior probability* and *posterior probability*. We discuss these terms by referring to Example 4.21.

From the information provided, we know that the probability is 0.070 that a randomly selected person has lung disease: $P(L) = 0.070$. This probability doesn't take into consideration whether the person is a smoker. It is therefore called a **prior probability** because it represents the probability that the person selected has lung disease *before* we know whether the person is a smoker.

Now suppose that the person selected is a smoker. On the basis of this additional information, we can revise the probability that the person has lung disease. We do so by determining the conditional probability that a randomly selected person has lung disease, given that the person selected is a smoker: $P(L \mid S) = 0.211$ (from Example 4.21). This revised probability is called a **posterior probability** because it represents the probability that the person selected has lung disease *after* we know that the person is a smoker.

EXERCISES 4.4   Basic Exercises

4.66 The National Sporting Goods Association collects and publishes data on participation in selected sports activities. For Americans 7 years old or older, 17.4% of males and 4.5% of females play golf. And, according to the U.S. Census Bureau's *Current Population Reports,* of Americans 7 years old or older, 48.6% are male and 51.4% are female. From among those Americans who are 7 years old or older, one is selected at random. Find the probability that the person selected
a) plays golf, given that the person is a female.
b) is a female, given that the person plays golf.
c) Interpret your answers in parts (a) and (b) in terms of percentages.

4.67 A survey conducted by TELENATION/Market Facts, Inc., combined with information from the U.S. Census Bureau's *Current Population Reports,* yielded the table given in Exercise 4.30 on page 145. What percentage of adult moviegoers are between 25 and 34 years old?

4.68 In a certain population of registered voters, 40% are Democrats, 32% are Republicans, and 28% are Independents. Sixty percent of the Democrats, 80% of the Republicans, and 30% of the Independents favor increased spending to combat terrorism. If a person chosen at random from this population favors increased spending to combat terrorism, what is the probability that the person is a Democrat?

4.69 An insurance company classifies people as *normal* or *accident prone.* Suppose that the probability that a normal person has an accident in a specified year is 0.2 and that for an accident prone person this probability is 0.6. Further suppose that 18% of the policyholders are accident prone. A policyholder had no accidents in a specified year. What is the probability that he or she is accident prone?

4.70 Refer to Example 4.18 on page 155. If you win a game of craps, what is the probability that your first toss resulted in a sum of 4?

4.71 An urn contains one red marble and nine green marbles; a second urn contains two red marbles and eight green marbles; and a third urn contains three red marbles and seven green marbles. An urn is chosen at random, and then one marble is randomly selected from the chosen urn.
a) Given that the marble obtained is red, what is the probability that the chosen urn was the first one; the second one; the third one?
b) Modify the problem by adding one million additional green marbles to each urn. Given the unlikely result that the marble selected is red, what are the posterior probabilities of the three urns? How is the result affected by the addition of the extra green marbles?
c) In part (a), find the posterior probabilities of the three urns, given that the marble selected is green. Then do the same after the additional green marbles of part (b) have been introduced. How is this result affected by the addition of the extra green marbles?

4.72 Suppose that you have four chests, each with two drawers, and that each drawer contains either a gold coin or a silver coin. Chests A and B each have one gold and one silver coin, chest C has two silver coins, and chest D has two gold coins. You select a chest at random, open one of the drawers, and find a silver coin within. What is the probability that the other drawer also contains a silver coin?

4.73 At a grocery store, eggs come in cartons that hold a dozen eggs. Experience indicates that 78.5% of the cartons have no broken eggs, 19.2% have one broken egg, 2.2% have two broken eggs, 0.1% have three broken eggs, and that the percentage of cartons with four or more broken eggs is negligible. An egg selected at random from a carton is found to be broken. What is the probability that this egg is the only broken one in the carton?

4.74 If you toss a balanced coin, in the long run, you get a head half the time. If you toss a certain unbalanced coin, in the long run, you get a head two-thirds of the time. First you choose one of these two coins at random, then you toss the chosen coin twice. Find the conditional probability that the balanced coin is chosen, given that
a) the first toss is a head and the second toss is a tail.
b) the second toss is a head and the first toss is a tail.
c) exactly one of the two tosses is a head.

4.75 Refer to Exercise 4.74. Find the conditional probability that the second toss is a head, given that the first toss is a head.

Advanced Exercises

4.76 Medical tests are frequently used to decide whether a person has a particular disease. The *sensitivity* of a test is the probability that a person having the disease will test positive; the *specificity* of a test is the probability that a person not having the disease will test negative. A test for a certain disease has been used for many years. Experience with the test indicates that its sensitivity is 0.96 and that its specificity is 0.98. Furthermore, it is known that roughly 1 in 1000 people has the disease.
a) Interpret the sensitivity and specificity of this test in terms of percentages.
b) Determine the probability that a person testing positive actually has the disease.
c) Interpret your answer from part (b) in terms of percentages.
d) Your naive colleague objects to the result of part (b), saying, "A positive test should be evidence of illness, not evidence of wellness. But, if there is only a 4.6% chance of illness—and therefore a 95.4% chance of wellness—after the patient tests positive, then a positive test would be evidence against illness." Set your colleague straight by comparing and contrasting the prior and posterior probabilities of illness, given a positive test.
e) The application of Bayes's rule in this exercise can be abbreviated as

$$(0.001, \ 0.999) \cdot (0.96, \ 0.02) \mapsto (1, \ 999) \cdot (96, \ 2) \mapsto (96, \ 1998)$$

$$\mapsto \left(\frac{96}{96 + 1998}, \ \frac{1998}{96 + 1998} \right) = (0.046, \ 0.954).$$

Explain why this formulation is really the same thing as Bayes's rule, as stated in Proposition 4.8 on page 162.

4.77 Stratification: For purposes of statistically analyzing a specified attribute (e.g., female), statisticians often divide the population under consideration into subpopulations called *strata*. A finite population is divided into m strata such that, for $1 \le k \le m$, stratum k has N_k members of which $100 p_k\%$ have the specified attribute. What percentage of members having the specified attribute belong to stratum j ($1 \le j \le m$)?

4.78 Bayesian analysis: This exercise involves the use of *Bayesian analysis*. Machines 1 and 2 produce coins that have probabilities 0.40 and 0.55 of a head, respectively. You have a coin but don't know which machine produced it, though it's as likely to have been produced by Machine 1 as Machine 2. Take the position that the probability of a head, p, is equally likely to be 0.40 or 0.55—that is, $P(p = 0.40) = P(p = 0.55) = 0.5$. These are the prior probabilities on p.

a) Now suppose that you toss the coin 10 times and observe 6 heads. How does this information change the probability distribution of p—that is, what are the posterior probabilities on p, given that you observed 6 heads in 10 tosses?

b) Suppose that you make another 10 tosses and this time observe 8 heads. What are the probabilities on p now?

CHAPTER REVIEW

Summary

In this chapter, we introduced the concepts of conditional probability and independence and presented some of the most important properties associated with those concepts.

The conditional probability of an event is the probability that the event occurs under the assumption that another event occurs. Conditional probabilities can be computed in two ways. One way is the direct method, which uses the event being conditioned on as the new sample space. The other way relies on the conditional probability rule, which provides a formula for computing conditional probabilities in terms of unconditional probabilities.

The general multiplication rule gives a formula for expressing the joint probability of two or more events in terms of marginal and conditional probabilities. It also leads to an important result called the law of total probability. The law of total probability provides a method for obtaining the probability of an event by conditioning on which event among a partition occurs.

Two or more events are independent events if the knowledge that some of these events have occurred doesn't affect the probabilities of occurrence of the other events. More precisely, the events of a collection are independent if the multiplication property holds for each finite subcollection of these events—that is, each joint probability equals the product of the marginal probabilities. It is often clear from the context of a problem that some particular events are independent, in which case the multiplication property can be used to obtain appropriate joint probabilities.

Bayes's rule is a formula used to compute unknown conditional probabilities in terms of known marginal and conditional probabilities. In this context, the known marginal probabilities are often referred to as prior probabilities, and the conditional probabilities obtained from Bayes's rule are often called posterior probabilities.

You Should Be Able To ...

1. determine conditional probabilities by using the direct method.

2. determine conditional probabilities by using the conditional probability rule.

3. state and apply the general multiplication rule for two or more events.

4. state and apply the law of total probability.

5. define and use the concept of independent events.

6. distinguish between pairwise independence and (mutual) independence.

7. identify situations in which the context of a problem implies independence of particular events (contextual independence).

8. state and apply Bayes's rule.

Key Terms

Bayes's rule, *162*

conditional probability, *126*

conditional probability
distribution, *130*

conditional probability rule, *131*

dependent events, *148*

direct method for obtaining
conditional probabilities, *131*

event being conditioned on, *126*

general multiplication rule, *138*

given event, *126*

independent events, *148, 150*

law of total probability, *142*

marginal probability
distribution, *131*

mutually independent events, *150*

pairwise independent events, *149*

posterior probability, *164*

prior probability, *163*

special multiplication rule, *148*

statistically independent
events, *148*

stochastically independent
events, *148*

symmetry argument, *140*

Chapter Review Exercises

Basic Exercises

4.79 A bowl contains 10 red marbles and 5 green marbles. The red marbles are numbered 1 through 10, and the green marbles are numbered 11 through 15. One of the 15 marbles is drawn at random and not replaced, and then a second marble is drawn.
a) Find the conditional probability that the first marble drawn is red, given that it is one of those numbered 8 through 15.
b) Find the conditional probability that the first marble drawn is one of those numbered 8 through 15, given that it is red.
c) Find the conditional probability that the second marble drawn is red, given that the first marble drawn is red; given that the first marble drawn is green.
d) Use the general multiplication rule to find the probability that both marbles drawn are red.
e) By conditioning on the result of the first draw, find the probability that the second marble drawn is red.
f) Use the results of parts (d) and (e) and the conditional probability rule to find the conditional probability that the first marble drawn is red, given that the second marble drawn is red.
g) Use the result of part (f) to find the conditional probability that the first marble drawn is green, given that the second marble drawn is red.

4.80 In a bridge hand, North has two aces. What is the probability that South has exactly one of the other aces?

4.81 The U.S. National Center for Education Statistics publishes information about school enrollment in the *Digest of Education Statistics.* Following is a contingency table for enrollment in public and private schools by level. The cell frequencies are given in thousands of students.

Level

Type		Elementary L_1	High school L_2	College L_3	Total
	Public T_1	33,903	13,537	11,626	59,066
	Private T_2	4,640	1,366	3,263	9,269
	Total	38,543	14,903	14,889	68,335

A student is selected at random.
a) Determine directly from the table the conditional probability that the student obtained is in college, given that he or she attends a private school.
b) Are the events "is in college" and "attends a private school" independent? Explain.
c) Determine the conditional distributions of type by level and the marginal distribution of type by constructing a table similar to Table 4.2 on page 130.
d) Determine the conditional distributions of level by type and the marginal distribution of level.

4.82 During one year, the College of Public Programs at Arizona State University awarded 3 master of arts degrees, 28 master of public administration degrees, and 19 master of science degrees. Three students who received such master's degrees are selected at random without replacement. Determine the probability that
a) the first student received a master of arts and the second and third a master of science.
b) one student received a master of arts and two a master of science.
c) all three students received a master of public administration.
d) all three students received the same type of degree.
e) Construct a tree diagram for this problem similar to that shown in Figure 4.1 on page 141.

4.83 Complete and explain this statement: Sampling from a finite population without replacement is to the general multiplication rule as sampling with replacement is to the _____.

4.84 Gerald Kushel, Ed.D., was interviewed by *Bottom Line/Personal* on the secrets of successful people. To study success, Kushel questioned 1200 people, among whom were lawyers, artists, teachers, and students. He found that 15% enjoy neither their jobs nor their personal lives, 80% enjoy their jobs but not their personal lives, and 4% enjoy both their jobs and their personal lives. What percentage of the people interviewed who enjoy their personal lives also enjoy their jobs?

4.85 Two balanced dice are thrown. What is the probability that the first die comes up 1, given that the sum of the dice is 9; given that the sum of the dice is 4?

4.86 Draw a Venn diagram illustrating and give an example of three events that are
a) mutually exclusive but not exhaustive.
b) exhaustive but not mutually exclusive.

4.87 In a lot containing N items, exactly M are defective. Items are inspected one at a time without replacement. Determine the probability that the kth item inspected $(k \geq M)$ will be the last defective one in the lot. In Exercise 3.101 on page 122, you were asked to solve this problem by using only basic counting techniques. Now solve it by using conditional probability.

4.88 A graduate probability class has 10 students, 4 of whom are women. Three students are selected at random without replacement from the class, their gender unnoted. Determine the probability that a fourth student selected from the class is a woman by using
a) the law of total probability and the general multiplication rule.
b) a symmetry argument.
c) mathematical induction on the number of students in the class.

4.89 A balanced die is thrown four times. Determine the probability that at least one 5 is obtained by using the
a) inclusion–exclusion principle.
b) complementation rule.
c) Discuss the relevance of contextual independence to this problem.

4.90 Suppose that you have probability 1/4 of hitting a target. If you fire eight shots, what is the probability that you hit the target
a) at least once?
b) at least twice?
c) at least twice, given that you hit it at least once?
d) What is the probability that your first shot was a miss, given that you hit the target at least twice?

4.91 According to Maureen and Jay Neitz of the Medical College of Wisconsin Eye Institute, men have a 9% chance of being colorblind. For four randomly selected men, determine the probability that
a) none are colorblind.
b) the first three aren't colorblind and the fourth is.
c) exactly one of the four is colorblind.
d) exactly two of the four are colorblind.

4.92 The National Center for Health Statistics compiles information on activity limitations. Results are published in *Vital and Health Statistics*. The data show that 13.6% of males and 14.4% of females have an activity limitation. Are gender and activity limitation statistically independent? Explain your answer.

4.93 A point is selected at random from the unit square, $\{(x, y) : 0 \leq x \leq 1, 0 \leq y \leq 1\}$. Let a and b be real numbers between 0 and 1. In each part, determine whether the specified events are independent.
a) The x coordinate is less than a; the y coordinate is less than b.
b) The x coordinate is less than a; the y coordinate is greater than b.

4.94 A point is selected at random from the triangle $\{(x, y) : 0 \leq y \leq x \leq 1\}$. Let a and b be real numbers strictly between 0 and 1. In each part, determine whether the specified events are independent.
a) The x coordinate is less than a; the y coordinate is less than b.
b) The x coordinate is less than a; the y coordinate is less than b times the x coordinate.

4.95 In a letter to the editor that appeared in the February 23, 1987, issue of *U.S. News and World Report,* a reader discussed the issue of space shuttle safety. Each "criticality 1" item must have 99.99% reliability, according to NASA standards, meaning that the probability of failure for such an item is 0.0001. Mission 25, the mission in which the Challenger exploded, had 748 "criticality 1" items. Determine and interpret the probability that at least one "criticality 1" item would fail.

4.96 The National Center for Education Statistics publishes information on U.S. engineers and scientists in *Digest of Education Statistics*. Of U.S. engineers and scientists, 47.1% are engineers and 52.9% are scientists. The following table presents relative-frequency distributions for highest degree obtained.

Highest degree	Proportion for engineers	Proportion for scientists
Bachelors	0.728	0.546
Masters	0.208	0.276
Doctorate	0.036	0.172
Other	0.028	0.006

Suppose that a person is selected at random from among engineers and scientists in the United States.
a) Determine the probability that the highest degree of the person obtained is a masters.
b) Determine the prior probability that the person obtained is an engineer.
c) Determine the posterior probability that the person obtained is an engineer, given that his or her highest degree is a masters.
d) Interpret the probabilities obtained in parts (a)–(c) in terms of percentages.

4.97 You have two cards—one is red on both sides, the other is red on one side and black on the other. After shuffling the cards behind your back, you select one of them at random and place it on your desk with your hand covering it. Upon lifting your hand, you observe that the face showing is red.
a) What is the probability that the other side is red?
b) Provide an intuitive explanation for the result in part (a).

4.98 Factory I produces 3% defective items, Factory II produces 2% defective items, and Factory III produces 1% defective items. In a large shipment of items, 50% came from Factory I, 30% from Factory II, and 20% from Factory III.
a) What percentage of items in this shipment are defective?
b) What percentage of defective items in this shipment came from Factory II?

4.99 One coin has two heads, another coin has two tails, and a third coin is an ordinary balanced coin. One of these three coins is randomly chosen and tossed; the result is a head. Determine the probability that the other side of the coin, which is now facing downward, is a head.

4.100 A multiple-choice question has five options for the answer. Experience indicates that $100p\%$ of all students know the answer. Students who don't know the answer guess at random among the five choices.
a) What percentage of students who answer correctly know the answer?
b) Suppose that only one student in a million knows the answer. Show that the probability is about five in one million that a student who answers correctly knows the answer.
c) Suppose that only one student in a million is ignorant of the answer. Show that the probability is about one in five million that a student who answers correctly is ignorant of the answer.

Theory Exercises

4.101 Suppose that $A_1, A_2, \ldots$ are independent events. Prove that

a) $P\left(\bigcap_{n=1}^{\infty} A_n\right) = \prod_{n=1}^{\infty} P(A_n).$ **b)** $P\left(\bigcup_{n=1}^{\infty} A_n\right) = 1 - \prod_{n=1}^{\infty}(1 - P(A_n)).$

4.102 Suppose that A and B are events of a sample space, both with positive probability and with $P(A) < 1$. Prove that

$$\frac{P(A \mid B)}{P(A^c \mid B)} = \frac{P(A)}{P(A^c)} \cdot \frac{P(B \mid A)}{P(B \mid A^c)}.$$

This result can be stated as follows: The posterior odds favoring A equals the product of the prior odds favoring A and the likelihood ratio.

Advanced Exercises

4.103 A shipment of N items contains d defectives. A random sample of size n is taken without replacement from the shipment. Then a random sample of size k is taken without replacement from the first sample. Let p_i denote the probability of exactly i defective items in the second sample.
a) Obtain a formula for p_i.
b) Use a computer or calculator to evaluate p_i explicitly for $N = 500$, $n = 50$, $k = 10$, and $d = 5$.
c) If there are no defective items in the second sample, what is the probability that there were no defective items in the first sample?

4.104 Repeat Exercise 4.103 if the first sample is taken with replacement. (This assumption is reasonable if n is small relative to N.)

4.105 A balanced die is tossed until the first 6 appears. If the number of tosses required is even, what is the probability that the first 6 occurred on the fourth toss?

4.106 Prisoner's paradox: Three men—say, a, b, and c—are in jail. One prisoner has been chosen at random for execution; the other two will go free. The jailer knows which prisoner has been selected for execution. Prisoner c asks the jailer to tell him the name of one of his comrades who will be freed. The jailer declines to divulge this information, reasoning that, if c knew the name of one of his comrades who will be freed, the probability of c being executed would increase from 1/3 to 1/2. Is the jailer correct in his reasoning? Explain.

4.107 In a population of size N, $100p\%$ of the members have a specified attribute. Members of the population are randomly selected one at a time without replacement. Determine the probability that the kth member selected has the specified attribute.

4.108 Each of n urns contains r red balls and g green balls. A ball is chosen at random from Urn 1 and placed in Urn 2, next a ball is chosen at random from Urn 2 and placed in Urn 3, and so on, until a ball is chosen at random from Urn $n - 1$ and placed in Urn n. Then a ball is chosen at random from Urn n. What is the probability that it is red?

4.109 Laplace's law of succession: Suppose that, for each $k \in \{0, 1, 2, \ldots, N\}$, we have a family with k boys and $N - k$ girls. A family is chosen at random and, from the family obtained, n children are selected at random with replacement.
a) Given that all n children obtained are boys, what is the probability that another randomly selected child from the same family is a boy?
b) Show that, for large N, the probability in part (a) is approximately $(n + 1)/(n + 2)$. *Hint:* Use Riemann integral approximations.

4.110 Multinomial distribution: Let $E_1, \ldots, E_m$ be mutually exclusive and exhaustive events of a random experiment with probabilities $p_1, \ldots, p_m$, respectively. In n independent repetitions of this random experiment, let $n(E_j)$ denote the number of times that event E_j occurs $(1 \leq j \leq m)$. Prove that

$$P\big(n(E_1) = k_1, \ldots, n(E_m) = k_m\big) = \binom{n}{k_1, \ldots, k_m} p_1^{k_1} \cdots p_m^{k_m},$$

where $k_1, k_2, \ldots, k_m$ are nonnegative integers whose sum is n.

4.111 This problem is a special case of the so-called *gambler's ruin problem.* A balanced coin is tossed. If it comes up heads, you win \$1 from your opponent; otherwise, you lose \$1 to your opponent. Initially, you have \$i and your opponent has \$(10 − i)$. The game continues until one of you goes broke. Determine the probability that you win. *Hint:* Condition on the result of the first toss. You will obtain a second-order difference equation that is relatively easy to solve.

4.112 Randomized response: The law of total probability is often used by statisticians in surveys that ask sensitive *yes/no*-questions such as "Do you use drugs?" They apply a technique called *randomized response.* Basically, the idea involves using two questions, the sensitive one and an innocuous one whose probability of a *yes* answer is known. A random device, such as tossing a coin, is used by the respondent to determine which question he or she will answer. The statistician implementing the procedure can observe only the total number of people in the sample and the total number of positive responses. Thus, because only the respondent knows which question is being answered, truthful answers to the sensitive question are more likely to occur. Explain how the technique of randomized response can be applied to estimate the probability that a person uses drugs.

4.113 An experiment requires a balanced coin. You have a coin but don't know whether it is balanced. How can you use your coin to simulate a balanced coin? *Hint:* If you toss the coin twice, what is the relationship between the probability of getting first a head and then a tail and the probability of getting first a tail and then a head?

PART TWO

Discrete Random Variables

Siméon-Denis Poisson 1781–1840

Siméon Poisson was born in Pithviers, France, on June 21, 1781. He was home-schooled by his father, an ex-private in the army who had become an administrator in the local government. Although his father encouraged him to become a physician, Poisson found after only a little experience in medicine that he had no strong interest in the profession.

When Poisson was 17, he entered École Polytechnique as a student of Laplace and Lagrange, both of whom became his lifelong friends. Mathematics was both work and play for Poisson. In fact, according to Boyer and Merzbach (*A History of Mathematics*, 2nd ed., New York: Wiley, 1991, p. 530), Poisson "is said to have once remarked that life is only good for two things: to do mathematics and to teach it." Poisson also said that, in his infancy, his nurse would suspend him by a cord from a nail in the wall to protect him from the animals that roamed the floors of the house; his movement swung him from side to side and thus began his study of the pendulum.

Poisson finished his studies at the Polytechnique in 1802 and became an instructor there until 1808. After a 1-year position at the Bureau des Longitudes as an astronomer, he was given the chair of pure mathematics at the Faculté des Sciences. He held various positions during his life, mostly either government scientific jobs or professorships.

Poisson made significant contributions to many areas of mathematics and its applications—in particular, to probability theory. In 1837, he published his *Recherches sur la probabilité des jugements*, where he introduced and developed what has become known as the Poisson distribution, one of the most important probability distributions. Poisson also derived a generalization of Bernoulli's version of the weak law of large numbers. The Poisson process, which is based on the Poisson distribution, is considered one of the principal stochastic processes.

Poisson died near Paris on April 25, 1840.

CHAPTER FIVE

Discrete Random Variables and Their Distributions

INTRODUCTION

When analyzing a random experiment, we may not be interested in the outcome itself but, rather, in some numerical characteristic, or function, of the outcome. For instance, in the basic game of blackjack, a player is dealt two cards at random from an ordinary deck of 52 playing cards. The outcome of this random experiment consists of the two cards received by the player. However, in general, we are interested not in the two cards themselves, but in the sum of their face values. This sum is a numerical function of the outcome of the random experiment and is an example of a *random variable*. It is also an example of a *discrete random variable* because its possible values form a countable set.

In Section 5.1, we first reveal how, in the context of randomness, variables give rise naturally to random variables; we then provide precise definitions of the terms *random variable* and *discrete random variable*. In Section 5.2, we introduce the probability mass function of a discrete random variable, a function that gives the likelihood of each possible value of the random variable.

In Sections 5.3–5.6, we present four of the most important discrete random variables: the *binomial random variable*, the *hypergeometric random variable*, the *Poisson random variable*, and the *geometric random variable*. In Section 5.7, we investigate some other widely used discrete random variables. In the final section of this chapter, Section 5.8, we discuss the process of taking a function of a discrete random variable to form another discrete random variable.

5.1   From Variables to Random Variables

In statistics and related fields, the term **variable** denotes a characteristic that varies from one person or thing to another. Examples of variables for human beings are height, weight, number of siblings, gender, marital status, and eye color. The first three of these variables yield numerical information and are examples of **quantitative variables;** the last three yield nonnumerical information and are examples of **qualitative variables** (or **categorical variables**). In this book, we concentrate mostly on quantitative variables.

EXAMPLE 5.1   *Introducing Random Variables*

Number of Siblings   Professor Weiss asked the 40 students in one of his classes to state how many siblings they have. The following table displays their responses.

1	1	0	1	3	1	1	0	2	4
1	1	1	2	2	1	2	1	2	1
0	1	2	0	1	2	1	3	2	0
0	0	2	2	1	3	1	1	0	2

From this table, we obtain Table 5.1, which presents frequency, relative-frequency, and percent information for the number-of-siblings data; for instance, 11 of the 40 students, or 27.5%, have two siblings.[†] Because the "number of siblings" varies from student to student, it is a variable.

Table 5.1 Frequency, relative-frequency, and percentage distributions for number of siblings

Siblings x	Frequency of students	Relative frequency	Percentage of students
0	8	0.200	20.0
1	17	0.425	42.5
2	11	0.275	27.5
3	3	0.075	7.5
4	1	0.025	2.5
	40	1.000	100.0

a) Identify the variable "number of siblings" as either a quantitative variable or a qualitative variable.

b) Discuss this variable in the context of selecting a student at random from the class.

[†]Each *frequency* (or *count*) gives the number of students in the class that have a specified number of siblings. Each *relative frequency* is obtained by dividing the corresponding frequency by the total number of students in the class. A relative frequency is simply a percentage expressed as a decimal.

Solution   a) Because the "number of siblings" provides numerical information, it is a quantitative variable.

b) If a student is selected at random, the "number of siblings" of the student obtained is called a *random variable* because its value depends on chance—namely, on which student is selected.   ■

Example 5.1(b) indicates that a random variable is a quantitative variable whose value depends on chance. But to use this concept mathematically, we must be more precise. To do that, let's again refer to Example 5.1. The random experiment is selecting a student at random from the class and the sample space Ω is the set whose members are the 40 students in the class. The random variable "number of siblings" is actually a real-valued function whose domain is that sample space.

Recall that we generally use letters near the end of the alphabet—such as x, y, and z—to denote variables. We also use such letters to represent random variables but, in that context, we usually make the letters uppercase. For instance, we might use x to denote the variable "number of siblings." But, in the context of randomness, we would use X.

We summarize the discussion of a **random variable** in Definition 5.1. The footnote covers some technical mathematical aspects of that term.

DEFINITION 5.1   Random Variable

A **random variable** is a real-valued function whose domain is the sample space of a random experiment. In other words, a random variable is a function $X : \Omega \to \mathcal{R}$, where Ω is the sample space of the random experiment under consideration.[†]

Example 5.1 shows how random variables arise naturally as quantitative variables of finite populations in the context of randomness. But random variables occur in many other ways. Four examples are

- the sum of the dice when a pair of dice is rolled,

- the number of puppies in a litter,

- the amount of an insurance claim, and

- the lifetime of a flashlight battery.

[†] As we mentioned in Chapter 2, we can't always construct a mathematical model for a random experiment in which every subset is an event. Consequently, to be completely precise, the definition of a random variable should also include the condition that $\{ \omega \in \Omega : X(\omega) \le x \}$ is an event for each real number x. This condition ensures that it makes sense to consider probabilities of all subsets of the sample space of the form $\{ \omega \in \Omega : X(\omega) \in A \}$, where A is any set of real numbers of the type encountered in calculus.

Random-Variable Notation

Using **random-variable notation,** we develop useful shorthand for discussing and analyzing random variables and their probability distributions. To illustrate, we refer again to Example 5.1 and let X denote the number of siblings of a randomly selected student.

The event that the student obtained has, say, two siblings is $\{\omega \in \Omega : X(\omega) = 2\}$ or, more concisely, $\{X = 2\}$, read "X equals 2." Table 5.1 indicates that the event $\{X = 2\}$ is the set whose members are the 11 students in the class who have two siblings. We express the probability of that event as $P(X = 2)$, read "the probability that X equals 2."

The event that the student obtained has at least two siblings is $\{\omega \in \Omega : X(\omega) \geq 2\}$ or, more concisely, $\{X \geq 2\}$, read "X is greater than or equal to 2." Table 5.1 indicates that the event $\{X \geq 2\}$ is the set whose members are the 15 students in the class who have two or more siblings. The probability of that event is expressed as $P(X \geq 2)$, read "the probability that X is greater than or equal to 2."

Let A be, say, the set of odd positive integers—that is, $A = \{1, 3, 5, \ldots\}$. The event that the student obtained has an odd number of siblings is $\{\omega \in \Omega : X(\omega) \in A\}$, or, more concisely, $\{X \in A\}$, read "X is in A." Table 5.1 indicates that the event $\{X \in A\}$ is the set whose members are the 20 students in the class who have an odd number of siblings. The probability of that event is expressed as $P(X \in A)$, read "the probability that X is in A."

In general, for any random variable X and any subset A of real numbers, the event that the observed value of X is in A is denoted $\{X \in A\}$ and the probability of that event is expressed as $P(X \in A)$. Observe, for instance, that, if A is the set consisting of only the number 2 (i.e., $A = \{2\}$), then $\{X \in A\} = \{X = 2\}$; if A is the set consisting of all real numbers greater than or equal to 2 (i.e., $A = [2, \infty)$), then $\{X \in A\} = \{X \geq 2\}$.

Discrete Random Variables

There are two main types of random variables, namely, *discrete random variables* and *continuous random variables.* In this chapter and Chapters 6 and 7, we consider only discrete random variables.

A discrete random variable often involves a count of something, such as the number of cars owned by a randomly selected family, the number of people waiting for a haircut in a barbershop, or the number of households in a sample that own a computer. More generally, we define **discrete random variable** as follows.

DEFINITION 5.2   Discrete Random Variable

A random variable X is called a **discrete random variable** if there is a countable set K of real numbers such that $P(X \in K) = 1$.

Consider the random variable "number of siblings" in Example 5.1, which we denote X. The possible values of X are 0, 1, 2, 3, and 4; that is, the range of X is the set $\{0, 1, 2, 3, 4\}$. Let K be that set. Clearly, then, K is a countable set of real numbers and we have $P(X \in K) = P(\Omega) = 1$. Hence X is a discrete random variable. Essentially the same argument shows that:

> Any random variable with a countable range—that is, whose possible values form a finite or countably infinite set—is a discrete random variable.

Note, however, that discrete random variables with uncountable ranges do exist. Exercise 5.15 provides an example.

In the advanced exercises for this section, we ask you to prove the following.

- X is a discrete random variable if and only if $\sum_{x \in \mathcal{R}} P(X = x) = 1$, that is, 100% of the probability is accounted for by "point masses."

- X is a discrete random variable if and only if there is a random variable X_0 with a countable range such that $P(X_0 = x) = P(X = x)$ for all $x \in \mathcal{R}$.

As a consequence of the second bulleted item, we can say the following from a probability distribution point of view:

> We can assume without loss of generality that a discrete random variable has a countable range.

We make that assumption throughout this book!

Examples 5.2 and 5.3 further illustrate the concept of a discrete random variable and the use of random-variable notation.

EXAMPLE 5.2 *Discrete Random Variables*

Coin Tossing Consider the random experiment of tossing a coin three times and observing the result (a head or a tail) for each toss. Let X denote the total number of heads obtained in the three tosses of the coin.

a) Construct a table that shows the value of the random variable X for each possible outcome of the random experiment.
b) Explain why X is a discrete random variable.
c) Identify the event $\{X = 1\}$ in words and as a subset of the sample space.
d) Identify the event $\{X \leq 1\}$ in words and as a subset of the sample space.

Solution   We first recall that a sample space for this random experiment is

$$\Omega = \{\text{HHH, HHT, HTH, HTT, THH, THT, TTH, TTT}\},$$

where, for instance, HTT denotes the outcome of a head on the first toss and tails on the second and third tosses.

a) We have $X(\text{HHH}) = 3$, $X(\text{HHT}) = 2$, and so on. Table 5.2 shows the value of X for each possible outcome of the random experiment.
b) Because X is a real-valued function on the sample space, it is a random variable. Furthermore, as revealed in Table 5.2, the possible values of X are 0, 1, 2, and 3; so, the range of X is the set $\{0, 1, 2, 3\}$, which is a finite and hence countable subset of $\mathcal{R}$. This result shows that X is discrete.

Table 5.2 Values of the random variable X, the total number of heads in three tosses of a coin

Outcome ω	HHH	HHT	HTH	HTT	THH	THT	TTH	TTT
Value of X $X(\omega)$	3	2	2	1	2	1	1	0

c) $\{X = 1\}$ is the event that exactly one of the three tosses of the coin results in a head. Table 5.2 shows that, as a subset of the sample space,

$$\{X = 1\} = \{\omega \in \Omega : X(\omega) = 1\} = \{\text{HTT, THT, TTH}\}.$$

d) $\{X \leq 1\}$ is the event that at most one of the three tosses of the coin results in a head. Table 5.2 indicates that, as a subset of the sample space,

$$\{X \leq 1\} = \{\omega \in \Omega : X(\omega) \leq 1\} = \{\text{HTT, THT, TTH, TTT}\}. \qquad \blacksquare$$

EXAMPLE 5.3 *Discrete Random Variables*

Oklahoma State Officials According to the *World Almanac*, as of 1998, the five top state officials of Oklahoma and their annual salaries were as shown in Table 5.3.

Table 5.3 Five top Oklahoma state officials and their annual salaries

Official	Salary ($1000s)
Governor (G)	70
Lieutenant Governor (L)	63
Secretary of State (S)	44
Attorney General (A)	75
Treasurer (T)	70

Suppose that we take a random sample without replacement of two of the five officials. Let Y denote the mean (average) salary of the two officials obtained.

a) Construct a table that shows the value of the random variable Y for each possible outcome of the random experiment.

b) Explain why Y is a discrete random variable.

Identify in words and as a subset of the sample space each of the following events.

c) $\{Y = 57.0\}$ d) $\{Y = 64.4\}$ e) $\{Y > 64.4\}$

Solution We first recall that a sample space for this random experiment is

$$\Omega = \{\{G, L\}, \{G, S\}, \{G, A\}, \{G, T\}, \{L, S\}, \{L, A\}, \{L, T\}, \{S, A\}, \{S, T\}, \{A, T\}\},$$

where, for instance, $\{G, L\}$ represents the outcome that the two officials selected are the governor and lieutenant governor.

a) From Table 5.3, $Y(\{G, L\}) = (70 + 63)/2 = 66.5$, $Y(\{G, S\}) = (70 + 44)/2 = 57.0$, and so on. Table 5.4 shows the value of Y for each possible outcome of the random experiment.

Table 5.4 Values of the random variable Y, the mean salary of the random sample of two officials

Outcome ω	{G, L}	{G, S}	{G, A}	{G, T}	{L, S}	{L, A}	{L, T}	{S, A}	{S, T}	{A, T}
Value of Y $Y(\omega)$	66.5	57.0	72.5	70.0	53.5	69.0	66.5	59.5	57.0	72.5

b) Because Y is a real-valued function on the sample space, it is a random variable. Furthermore, as indicated in Table 5.4, the possible values of Y are 53.5, 57.0, 59.5, 66.5, 69.0, 70.0, and 72.5. Therefore the range of Y is a finite, and hence countable, subset of $\mathcal{R}$. This result shows that Y is discrete.

c) $\{Y = 57.0\}$ is the event that the mean salary of the two officials selected is 57.0 (i.e., \$57,000). Table 5.4 reveals that $\{Y = 57.0\} = \{\{G, S\}, \{S, T\}\}$.

d) $\{Y = 64.4\}$ is the event that the mean salary of the two officials selected is 64.4 (i.e., \$64,400). According to Table 5.4 or part (b), the mean salary of the two officials selected can't be 64.4. Thus $\{Y = 64.4\} = \emptyset$.

e) $\{Y > 64.4\}$ is the event that the mean salary of the two officials selected exceeds 64.4. From Table 5.4, $\{Y > 64.4\} = \{\{G, L\}, \{G, A\}, \{G, T\}, \{L, A\}, \{L, T\}, \{A, T\}\}$. ■

In Examples 5.2 and 5.3, we described events specified by value(s) of a random variable explicitly as subsets of the sample space. We were able to do so because, in each example, the number of possible outcomes (i.e., size of the sample space) is small. More commonly, the number of possible outcomes is quite large. In such cases, we usually describe events in words.

EXERCISES 5.1   Basic Exercises

5.1 Provide an example, other than one discussed in the text, of a random variable that doesn't arise from a quantitative variable of a finite population in the context of randomness.

5.2 Suppose that two balanced dice are rolled. Let X be the sum of the two faces showing.
a) Is X a random variable? Is it discrete? Explain your answers.

Identify in words and as a subset of the sample space each of the following events. *Note:* Use the ordered-pair representation of the sample space given in Example 2.3 on page 27.
b) $\{X = 7\}$ **c)** $\{X > 10\}$ **d)** $\{X = 2 \text{ or } 12\}$ **e)** $\{4 \le X \le 6\}$
f) $\{X \in A\}$, where A is the set of even integers.

5.3 Refer to the component-analysis illustration in Example 2.4 on page 27. Let X denote the number of components that are working at the specified time.
a) Is X a random variable? Is it discrete? Explain your answers.

Identify in words and as a subset of the sample space each of the following events.

b) $\{X \geq 4\}$ **c)** $\{X = 0\}$ **d)** $\{X = 5\}$ **e)** $\{X \geq 1\}$

Use random variable terminology to represent each of the following events. The number of working components is

f) two.   **g)** at least two.   **h)** at most two.   **i)** between two and four, inclusive.

5.4 Suppose that a family is selected at random from among those in a population that have three children.

a) Obtain a sample space for this random experiment that takes into account gender and order of birth.

b) Let Y denote the number of female children in the family obtained. Construct a table showing the value of Y for each outcome in the sample space.

c) Is Y a discrete random variable? Explain your answer.

5.5 An archer shoots an arrow into a square target 6 feet on a side whose center we call the origin. The outcome of this random experiment is the point in the target hit by the arrow. The archer scores 10 points if she hits the bull's eye—a disk of radius 1 foot centered at the origin; she scores 5 points if she hits the ring with inner radius 1 foot and outer radius 2 feet centered at the origin; and she scores 0 points otherwise. For one arrow shot, let S be the score.

a) Determine the value of S for each possible outcome of the random experiment.

b) Is S a discrete random variable? Explain your answer.

Identify in words and as a subset of the sample space each of the following events.

c) $\{S = 5\}$ **d)** $\{S > 0\}$ **e)** $\{S \leq 7\}$
f) $\{5 < S \leq 15\}$ **g)** $\{S < 15\}$ **h)** $\{S < 0\}$

5.6 A factory received a shipment of 100 parts that are vital to the construction of its product. The shipment contains an unknown number (possibly 0) of defective parts. The quality-control inspector decides to take a random sample of 5 parts without replacement and to accept the shipment if and only if at most 1 of the parts in the sample is defective. Let X denote the number of defective parts in the sample and set $Y = 0$ if the shipment is rejected and $Y = 1$ if the shipment is accepted.

a) Describe the sample space of this random experiment.

b) Express the random variable Y in terms of the random variable X.

5.7 Urn I contains four red balls and one white ball; Urn II contains one red ball and four white balls. An urn is chosen at random and then two balls are drawn without replacement from the chosen urn. Let X be the number of red balls drawn.

a) Construct a sample space for this random experiment whose outcomes specify which urn is chosen, the color of the first ball drawn, and the color of the second ball drawn.

b) Construct a table showing the value of X for each possible outcome.

5.8 Six men and five women apply for a job at Alpha, Inc. Three of the applicants are selected for interviews. Let X denote the number of women in the interview pool.

a) What are the possible values of the random variable X?

b) Obtain the number of interview pools corresponding to each possible value of X; that is, for each possible value x of X, find the number of ways that event $\{X = x\}$ can occur.

c) Describe the event $\{X \leq 1\}$ in words and find the number of ways in which it can occur.

5.9 Refer to the coin tossing illustration in Example 5.2 on page 179. Let Y denote the difference between the number of heads obtained and the number of tails obtained.

a) Construct a table showing the value of Y for each possible outcome.

b) Identify in words and as a subset of the sample space the event $\{Y = 0\}$.

5.10 The cafeteria at a school has two salads that cost $1.00 and $1.50; three entrees that cost $2.50, $3.00, and $3.50; and two desserts that cost $1.50 each. A person chooses a meal consisting of at most one selection from each of the three categories. Let X be the cost of the meal, in dollars.
a) Determine the value of X for each possible meal.
b) Identify in words and as a subset of the sample space the event $\{X \leq 3\}$.

5.11 Suppose that you play a certain lottery by buying one ticket per week. Let W be the number of weeks until you win a prize.
a) Is W a random variable? Explain.
b) Is W a discrete random variable? Explain.

Identify in words each of the following events.
c) $\{W > 1\}$ **d)** $\{W \leq 10\}$ **e)** $\{15 \leq W < 20\}$

5.12 Suppose that a number is selected at random from the interval $(0, 1)$, a task that a basic random number generator aims to accomplish. Let X denote the number obtained. For integers m and n with $m < n$, set $Y = \lfloor m + (n - m + 1)X \rfloor$, where $\lfloor x \rfloor$ denotes the *floor function*—the greatest integer smaller than or equal to x. (The floor function is identical to the greatest-integer function.)
a) What are the possible values of Y?
b) Is Y a discrete random variable? Explain.
c) Let y be a possible value of Y. Express the event $\{Y = y\}$ in terms of X.

Theory Exercises

5.13 Prove that a random variable with countable range is a discrete random variable in the sense of Definition 5.2 on page 178. That is, if X is a random variable whose range is countable, there is a countable set K of real numbers such that $P(X \in K) = 1$.

5.14 Prove that, for any random variable X, the set $\{x \in \mathcal{R} : P(X = x) \neq 0\}$ is countable. Use the following steps.
a) Show that, for each $n \in \mathcal{N}$, no more than $n - 1$ values of $x \in \mathcal{R}$ can satisfy the inequality $P(X = x) > 1/n$.
b) Use the result of part (a) to deduce that $\{x \in \mathcal{R} : P(X = x) \neq 0\}$ is countable. *Hint:* A countable union of countable sets is countable.
c) Interpret the result in part (b).

Advanced Exercises

Note: In the remaining exercises for this section, adhere strictly to Definition 5.2 on page 178 of a discrete random variable.

5.15 Construct a discrete random variable with an uncountable range. *Hint:* Use the fact that there is an uncountable subset C of $[0, 1]$ that has length 0 (in the extended sense).

5.16 Prove, for any random variable X, that $0 \leq \sum_x P(X = x) \leq 1$, where the notation $\sum_x$ means the sum over all $x \in \mathcal{R}$. *Hint:* Refer to Exercise 5.14.

5.17 Prove that a random variable X is discrete if and only if $\sum_x P(X = x) = 1$.

5.18 Prove that X is a discrete random variable if and only if there is a random variable X_0 with a countable range such that $P(X_0 = x) = P(X = x)$ for all $x \in \mathcal{R}$.

In Exercises 5.19–5.22, you may want to refer to Exercise 5.17.

5.19 Suppose that a number is selected at random from the interval $(0, 1)$, a task that a basic random number generator aims to accomplish. Let X denote the number obtained. Answer the following questions and explain your answers.
a) Is X a random variable?   **b)** Is X a discrete random variable?

5.20 Suppose that a petri dish of unit radius, containing nutrients upon which bacteria can multiply, is smeared with a uniform suspension of bacteria. Let Z denote the distance from the center of the petri dish to the center of the first spot (bacteria colony) to appear. Answer the following questions and explain your answers.
a) Is Z a random variable?   **b)** Is Z a discrete random variable?

5.21 The *World Almanac* provides information on past and projected total solar eclipses.
a) Let X denote the duration, in minutes, of a total solar eclipse. Is X a discrete random variable? *Hint:* Use the frequentist interpretation of probability (page 5).
b) Let Y denote the duration, to the nearest minute, of a total solar eclipse. Is Y a discrete random variable? Explain.

5.22 Desert Samaritan Hospital in Mesa, Arizona, keeps records of its emergency-room traffic. Suppose that you begin observing the emergency-room traffic at a specified time and measure time in hours. Identify each of the following random variables as discrete or not discrete and explain your answers.
a) The time T_6 at which the sixth patient arrives. *Hint:* Use the frequentist interpretation of probability (page 5).
b) The number X_3 of patients who arrive during the first 3 hours.
c) The number S of "full hours" up to and including the time of arrival of the sixth patient. For instance, if the sixth patient arrives 4 hours and 23 minutes from the time you begin observing, then $S = 4$.
d) The elapsed time between the arrivals of the sixth and eighth patients.
e) The elapsed time between the arrival of the last patient to arrive before 3:00 A.M. and the arrival of the first patient to arrive after 3:00 A.M.
f) Explain why $\{X_3 \geq 6\} = \{T_6 \leq 3\}$.

5.23 Suppose that you toss a coin with probability p of a head $(0 < p < 1)$ until the first time a head appears. Let N denote the number of times that you toss the coin. Once you know the value of N, choose a number M at random from $\{1, \ldots, N\}$. Set $X = M/N$.
a) Explain why X is a discrete random variable.
b) Suppose that x is a possible value of X. Explain why there can be no next possible value of X larger than x—that is, if y is a possible value of X and $y > x$, then there exists a possible value z of X such that $x < z < y$.

5.2   Probability Mass Functions

In Section 5.1, we introduced the concept of a discrete random variable—a real-valued function on the sample space of a random experiment whose possible values (range) can be assumed to be a countable set. In probability theory, we not only want to know the possible values of a discrete random variable but also the probabilities associated with the possible values. This information is provided by the **probability mass function,** which we define as follows.

> **DEFINITION 5.3   Probability Mass Function**
>
> Let X be a discrete random variable. Then the **probability mass function** of X, denoted p_X, is the real-valued function defined on $\mathcal{R}$ by
>
> $$p_X(x) = P(X = x).$$
>
> We use **PMF** as an abbreviation for "probability mass function."

Note the following.

- The PMF of a discrete random variable is represented by a lowercase p subscripted with the letter representing the discrete random variable. Therefore p_X denotes the PMF of the discrete random variable X, p_Y denotes the PMF of the discrete random variable Y, and so on.

- If x is not a possible value (i.e., isn't in the range) of a discrete random variable X, then $p_X(x) = 0$. Indeed, for such a value of x, we have

$$p_X(x) = P(X = x) = P(\{\,\omega \in \Omega : X(\omega) = x\,\}) = P(\emptyset) = 0.$$

- Other terms used for *probability mass function* are **discrete density function, probability function,** and **density function.**

We can visually portray the PMF of a discrete random variable by constructing a **probability histogram**—a graph that displays the possible values of the discrete random variable on the horizontal axis and the probabilities of those values on the vertical axis. The probability of each value is represented by a vertical bar whose height is equal (or proportional) to the probability.

EXAMPLE 5.4   *Probability Mass Functions and Histograms*

Number of Siblings   Refer to Example 5.1 on page 176. Let X denote the number of siblings of a randomly selected student. Then X is a discrete random variable whose possible values, as shown in Table 5.1, are 0, 1, 2, 3, and 4.

a) Determine $p_X(2)$.

b) Obtain the PMF of X.

c) Construct a probability histogram for X.

Solution   Because a student is being selected at random from the class, each of the 40 students is equally likely to be the one obtained. In other words, a classical probability model is appropriate here. So, for each event E,

$$P(E) = \frac{N(E)}{N(\Omega)} = \frac{N(E)}{40}.$$

a) Here we want to determine $p_X(2)$—that is, $P(X = 2)$—the probability that the student selected has two siblings. From Table 5.1, the event $\{X = 2\}$ that the student selected has two siblings can occur in 11 ways—namely, if any one of the 11 students with two siblings is the student selected. Therefore

$$p_X(2) = P(X = 2) = \frac{N(\{X = 2\})}{40} = \frac{11}{40} = 0.275.$$

b) To obtain the PMF of X, we need to compute $p_X(x)$ for each of the possible values of X. In part (a), we did so for the possible value 2. Proceeding as we did in part (a), we find that the PMF of X is as given in Table 5.5.

Table 5.5 PMF of the discrete random variable X, the number of siblings of a randomly selected student

Siblings x	Probability $p_X(x)$
0	0.200
1	0.425
2	0.275
3	0.075
4	0.025
	1.000

c) To construct a probability histogram for X, we draw a graph that displays the possible values of X on the horizontal axis and the probabilities of those values on the vertical axis. Referring to Table 5.5, we obtain the probability histogram for X shown in Figure 5.1.

Figure 5.1 Probability histogram of the discrete random variable X, the number of siblings of a randomly selected student

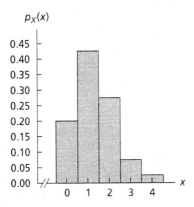

This probability histogram provides a quick and easy way to visualize how the probabilities of the random variable X are distributed.

Because the variable "number of siblings" is a variable of a finite population, probabilities for the corresponding random variable are identical to relative frequencies. Consequently, the probability distribution in the two columns of Table 5.5 is identical to the relative-frequency distribution shown in the first and third columns of Table 5.1 on page 176. This statement holds in general—that is, the probability and relative-frequency distributions are identical for any variable of a finite population.

Note also that the probabilities in the second column of Table 5.5 sum to 1. This property always holds for discrete random variables, being a consequence of the certainty and additivity properties of a probability measure. In fact, we have an important result concerning probability mass functions, presented as Proposition 5.1.

◆◆◆ **Proposition 5.1   Basic Properties of a PMF**

The probability mass function p_X of a discrete random variable X satisfies the following three properties.

a) *$p_X(x) \geq 0$ for all $x \in \mathcal{R}$; that is, a PMF is a nonnegative function.*

b) *$\{x \in \mathcal{R} : p_X(x) \neq 0\}$ is countable; that is, the set of real numbers for which a PMF is nonzero is countable.*

c) *$\sum_x p_X(x) = 1$; that is, the sum of the values of a PMF equals 1.[†]*

Proof   a) For each $x \in \mathcal{R}$, we have $p_X(x) = P(X = x) \geq 0$, where the inequality follows from the nonnegativity property of a probability measure.

b) As we know, $p_X(x) = 0$ if x isn't a possible value (i.e., isn't in the range) of X. Because X is a discrete random variable, its possible values form a countable set. Thus $\{x \in \mathcal{R} : p_X(x) \neq 0\}$ is countable.

c) We first note that

$$\Omega = \{X \in \mathcal{R}\} = \bigcup_x \{X = x\} \tag{5.1}$$

and that the events in the union on the right side of Equation (5.1) are mutually exclusive. Now, because $\{X = x\} \neq \emptyset$ if and only if x is a possible value of X, the union on the right side of Equation (5.1) reduces to a countable union over the possible values of X. And, in view of part (b), the sum $\sum_x p_X(x)$ reduces to a countable sum over the possible values of X. Keeping these facts and Equation (5.1) in mind, we apply the certainty and additivity properties of a probability measure to obtain

$$1 = P(\Omega) = P\left(\bigcup_x \{X = x\}\right) = \sum_x P(X = x) = \sum_x p_X(x). \qquad ◆$$

In Exercise 5.36, we ask you to show that a function $p: \mathcal{R} \to \mathcal{R}$ that satisfies properties (a), (b), and (c) of Proposition 5.1 is the PMF of some discrete random variable. Therefore, for any such function, we can say, "Let X be a discrete random variable with PMF p." This statement make sense regardless of whether we explicitly give X and the sample space on which it is defined.

[†]The expression $\sum_x$ means that we are summing over all real numbers x. Likewise, $\bigcup_x$ indicates a union over all real numbers x.

EXAMPLE 5.5   *Probability Mass Functions*

Coin Tossing   Refer to Example 5.2 on page 179. Let X denote the total number of heads obtained in the three tosses of the coin. Determine the PMF of X when
a) the coin is balanced.
b) the coin has probability 0.2 of a head on any given toss.
c) the coin has probability p $(0 \le p \le 1)$ of a head on any given toss.

Solution   A sample space for this experiment is $\Omega = \{$HHH, HHT, HTH, HTT, THH, THT, TTH, TTT$\}$. In Example 4.15 on page 152, we showed how to determine the probabilities of the individual outcomes for this random experiment with different values of p. Table 5.6 provides these probabilities for $p = 0.5$ (balanced case), $p = 0.2$, and general p.

Table 5.6 Probabilities of individual outcomes for coin tossing experiment

Outcome	Probability		
	$p = 0.5$	$p = 0.2$	General p
HHH	0.125	0.008	$p^3(1 - p)^0$
HHT	0.125	0.032	$p^2(1 - p)^1$
HTH	0.125	0.032	$p^2(1 - p)^1$
HTT	0.125	0.128	$p^1(1 - p)^2$
THH	0.125	0.032	$p^2(1 - p)^1$
THT	0.125	0.128	$p^1(1 - p)^2$
TTH	0.125	0.128	$p^1(1 - p)^2$
TTT	0.125	0.512	$p^0(1 - p)^3$

The possible values of the discrete random variable X—the total number of heads obtained in the three tosses of the coin—are 0, 1, 2, and 3. We have, for instance, that

$$p_X(1) = P(X = 1) = P(\{\omega \in \Omega : X(\omega) = 1\}) = P(\{\text{HTT, THT, TTH}\})$$
$$= P(\{\text{HTT}\}) + P(\{\text{THT}\}) + P(\{\text{TTH}\}).$$

From Table 5.6, we obtain $p_X(1)$ for each of the three cases of interest:

$$p_X(1) = P(\{\text{HTT}\}) + P(\{\text{THT}\}) + P(\{\text{TTH}\})$$

$$= \begin{cases} 0.125 + 0.125 + 0.125 = 3 \cdot 0.125 = 0.375, & \text{if } p = 0.5; \\ 0.128 + 0.128 + 0.128 = 3 \cdot 0.128 = 0.384, & \text{if } p = 0.2; \\ p^1(1 - p)^2 + p^1(1 - p)^2 + p^1(1 - p)^2 = 3p^1(1 - p)^2, & \text{general } p. \end{cases}$$

Similarly, we can determine $p_X(x)$ for the other three possible values x of the random variable X for each of the three cases of interest. We provide the results separately.

a) For a balanced coin, $p = 0.5$. The PMF of X is as shown in Table 5.7.

Table 5.7 PMF of X (number of heads in three tosses) for a balanced coin ($p = 0.5$)

x	0	1	2	3
$p_X(x)$	0.125	0.375	0.375	0.125

b) For a coin with probability 0.2 of a head on any given toss, $p = 0.2$. The PMF of X is as shown in Table 5.8.

Table 5.8 PMF of X (number of heads in three tosses) in case $p = 0.2$

x	0	1	2	3
$p_X(x)$	0.512	0.384	0.096	0.008

c) For a coin with probability p of a head on any given toss, the PMF of X is as shown in Table 5.9.

Table 5.9 PMF of X (number of heads in three tosses) for general p

x	0	1	2	3
$p_X(x)$	$p^0(1-p)^3$	$3p^1(1-p)^2$	$3p^2(1-p)^1$	$p^3(1-p)^0$

Note that we can obtain Tables 5.7 and 5.8 from Table 5.9 by substituting $p = 0.5$ and $p = 0.2$, respectively.   ■

Fundamental Probability Formula for a Discrete Random Variable

Once we have the PMF of a discrete random variable, we can easily determine any probability involving that random variable. We do so by using the additivity property of a probability measure, as shown in Example 5.6.

EXAMPLE 5.6   *Motivates the Fundamental Probability Formula*

Number of Siblings   Consider again the random experiment of selecting a student at random from Professor Weiss's class. Let X denote the number of siblings of the student obtained. Use the PMF of X, given in Table 5.5 on page 186, to determine $P(X \leq 2)$—the probability that the student obtained has at most two siblings.

Solution   We first recall that the possible values of X are 0, 1, 2, 3, and 4 and that $\{X = x\} = \emptyset$ if x isn't a possible value of X. Hence

$$\{X \le 2\} = \bigcup_{x \le 2}\{X = x\} = \{X = 0\} \cup \{X = 1\} \cup \{X = 2\}. \tag{5.2}$$

Clearly, the events in the union on the right side of Equation (5.2) are mutually exclusive. In view of Equation (5.2), the additivity property of a probability measure, and Table 5.5, we conclude that

$$\begin{aligned}
P(X \le 2) &= P\big(\{X = 0\} \cup \{X = 1\} \cup \{X = 2\}\big) \\
&= P(X = 0) + P(X = 1) + P(X = 2) \\
&= p_X(0) + p_X(1) + p_X(2) \\
&= 0.200 + 0.425 + 0.275 = 0.900,
\end{aligned} \tag{5.3}$$

as required.   ■

Example 5.6 suggests how to prove a general result for obtaining any probability for a discrete random variable in terms of the PMF of that random variable. We refer to this result as the **fundamental probability formula,** which we abbreviate as **FPF** and present formally as Proposition 5.2.

◆◆◆   **Proposition 5.2   Fundamental Probability Formula: Univariate Discrete Case**

Suppose that X is a discrete random variable. Then, for any subset A of real numbers,

$$P(X \in A) = \sum_{x \in A} p_X(x). \tag{5.4}$$

In words, the probability that a discrete random variable takes a value in a specified subset of real numbers can be obtained by summing the PMF of the random variable over that subset of real numbers.

Proof   We first observe that

$$\{X \in A\} = \bigcup_{x \in A}\{X = x\} \tag{5.5}$$

and that the events in the union on the right side of Equation (5.5) are mutually exclusive. Now, because $\{X = x\} \ne \emptyset$ if and only if x is a possible value of X, the union on the right side of Equation (5.5) reduces to a countable union over the possible values of X that lie in A. Keeping these facts and Equation (5.5) in mind, we apply the additivity property of a probability measure to obtain

$$P(X \in A) = P\Big(\bigcup_{x \in A}\{X = x\}\Big) = \sum_{x \in A} P(X = x) = \sum_{x \in A} p_X(x),$$

as required.   ◆

In Example 5.6, the problem was to obtain $P(X \le 2)$, where X is the number of siblings of a randomly selected student from Professor Weiss's class. In this case, with reference to the FPF, we have $A = \{x \in \mathcal{R} : x \le 2\}$. Applying the FPF and recalling that the possible values of X are 0, 1, 2, 3, and 4, we get

$$P(X \le 2) = \sum_{x \le 2} p_X(x) = p_X(0) + p_X(1) + p_X(2)$$

$$= 0.200 + 0.425 + 0.275 = 0.900.$$

Compare the previous display to that in Equation (5.3).

EXAMPLE 5.7 *The Fundamental Probability Formula*

Coin Tossing Consider again the random experiment of tossing a coin three times and observing the result (a head or a tail) for each toss. Let X denote the total number of heads obtained in the three tosses of the coin. Determine $P(1 \le X < 3)$ when
a) the coin is balanced.
b) the coin has probability 0.2 of a head on any given toss.
c) the coin has probability p ($0 \le p \le 1$) of a head on any given toss.

Solution   In Example 5.5 on page 188, we found the PMF of the random variable X for each of the three cases under consideration. To obtain $P(1 \le X < 3)$ in each case, we apply the FPF to the corresponding PMF. We have

$$P(1 \le X < 3) = \sum_{1 \le x < 3} p_X(x) = p_X(1) + p_X(2). \tag{5.6}$$

a) For a balanced coin, the PMF of X is as shown in Table 5.7 on page 189. Hence, by Equation (5.6),

$$P(1 \le X < 3) = p_X(1) + p_X(2) = 0.375 + 0.375 = 0.750.$$

b) For a coin with probability 0.2 of a head on any given toss, the PMF of X is as shown in Table 5.8 on page 189. Hence, by Equation (5.6),

$$P(1 \le X < 3) = p_X(1) + p_X(2) = 0.384 + 0.096 = 0.480.$$

c) For a coin with probability p of a head on any given toss, the PMF of X is as shown in Table 5.9 on page 189. Hence, by Equation (5.6),

$$P(1 \le X < 3) = p_X(1) + p_X(2) = 3p^1(1-p)^2 + 3p^2(1-p)^1 = 3p(1-p).$$

Note that we can obtain the probabilities required in parts (a) and (b) by substituting $p = 0.5$ and $p = 0.2$, respectively, in the answer to part (c). ∎

Interpretation of PMFs

Recall that the frequentist interpretation of probability construes the probability of an event to be the long-run proportion of times that the event occurs in independent repetitions of the random experiment. Using that interpretation, we clarify the meaning of probability mass functions in Example 5.8.

EXAMPLE 5.8   *Interpreting a Probability Mass Function*

Coin Tossing   Consider again the discrete random variable X discussed in Example 5.7, the number of heads obtained in three tosses of a coin. For this example, we presume that the coin is balanced, but the same ideas hold regardless of whether that is the case.

Suppose that we repeat the experiment of observing the number of heads obtained in three tosses of a balanced coin a large number of times. Then the proportion of those times in which, say, no heads are obtained (i.e., $X = 0$) should be approximately equal to the probability of that event [i.e., $P(X = 0)$]. Analogous statements hold for the other three possible values of the random variable X.

We used a computer to simulate 1000 observations of the number of heads obtained in three tosses of a balanced coin. (Simulating a random variable means using a computer or calculator to generate observations of the random variable.) The second and third columns of Table 5.10 show the frequencies and proportions, respectively, for the numbers of heads obtained in the 1000 observations. The fourth column provides the actual probabilities, repeated from Table 5.7. For instance, 136 of the 1000 observations resulted in no heads in three tosses, giving a proportion of 0.136; the actual probability of getting no heads is 0.125.

Table 5.10 Frequencies, proportions, and probabilities for the number of heads in three tosses of a balanced coin for 1000 observations

Heads	Frequency	Proportion	Probability
0	136	0.136	0.125
1	377	0.377	0.375
2	368	0.368	0.375
3	119	0.119	0.125
	1000	1.000	1.000

As expected, the proportions in the third column of Table 5.10 are fairly close to the actual probabilities in the fourth column. You can see these results more easily by comparing a histogram of the proportions to the probability histogram of the random variable X, as shown in Figure 5.2(a) and (b), respectively.

If we simulated, say, 10,000 observations instead of 1000, the proportions that would appear in the third column of Table 5.10 would most likely be even closer to the actual probabilities listed in the fourth column.   ■

In view of Example 5.8, we can make the following statement.

Interpretation of a PMF

In a large number of independent observations of a discrete random variable X, the proportion of times that each possible value occurs will approximate the PMF at that value. Equivalently, a histogram of the proportions will approximate the probability histogram for X.

Figure 5.2 (a) Histogram of proportions for the number of heads obtained in three tosses of a balanced coin for 1000 observations; (b) probability histogram for the number of heads obtained in three tosses of a balanced coin

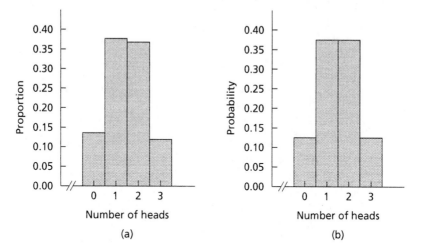

EXERCISES 5.2   Basic Exercises

Note: Many of the exercises in this section continue those presented in Section 5.1.

5.24 Suppose that two balanced dice are rolled. Let X be the sum of the two faces showing.
a) Obtain the PMF of the random variable X.
b) Construct a probability histogram for X.

In the game of craps, a first roll of a sum of 7 or 11 wins, whereas a first roll of a sum of 2, 3, or 12 loses. To win with any other first sum, that sum must be repeated before a sum of 7 is rolled. Use part (b) and the FPF to determine the probability of
c) a win on the first roll.   **d)** a loss on the first roll.
e) For each of parts (c) and (d), identify the set A appearing in Equation (5.4) on page 190.

5.25 Refer to the component-analysis illustration in Example 2.4 on page 27. Assume that the components act independently and that, for $1 \le j \le 5$, the probability is 0.8 that, at the specified time, component j is working. Let X denote the number of components that are working at the specified time.
a) Obtain the PMF of the random variable X.
b) Use part (a) and the FPF to obtain the probability that between 1 and 3 inclusive of the components are working at the specified time. In this case, identify the set A appearing in Equation (5.4) on page 190.
c) Suppose that the unit is a series system—that is, it functions only when all five components are working. Express the event that the unit is functioning at the specified time in terms of the random variable X. Then use part (a) to find the probability of that event.
d) Suppose that the unit is a parallel system—that is, it functions when at least one of the five components is working. Express the event that the unit is functioning at the specified time in terms of the random variable X. Then use part (a) to find the probability of that event.
e) Repeat parts (a)–(d) if the probability is p that, at the specified time, any particular component is working.

5.26 Suppose that, in a large population, a family is selected at random from among those that have three children. Further suppose that genders in successive births are independent and that the probability of a female child is p. Let Y denote the number of female children in the family obtained.

a) Obtain the PMF of the random variable Y.

Use the FPF and part (a) to determine the probability that the number of female children in the family obtained is

b) exactly one.   **c)** at most one.   **d)** at least one.

5.27 An archer shoots an arrow into a square target 6 feet on a side whose center we call the origin. The outcome of this random experiment is the point in the target hit by the arrow. The archer scores 10 points if she hits the bull's eye—a disk of radius 1 foot centered at the origin; she scores 5 points if she hits the ring with inner radius 1 foot and outer radius 2 feet centered at the origin; and she scores 0 points otherwise. For one arrow shot, let S be the score. Assume that the archer will actually hit the target and is equally likely to hit any portion of the target.

a) Obtain and interpret the PMF of the random variable S.

Use the FPF to determine and interpret the probability of each of the following events.

b) $\{S = 5\}$ **c)** $\{S > 0\}$ **d)** $\{S \leq 7\}$
e) $\{5 < S \leq 15\}$ **f)** $\{S < 15\}$ **g)** $\{S < 0\}$

5.28 Urn I contains four red balls and one white ball; Urn II contains one red ball and four white balls. A coin with probability p of a head is tossed. If the result is a head, Urn I is chosen; otherwise, Urn II is chosen. Subsequently, two balls are drawn at random without replacement from the chosen urn. Let X be the number of red balls drawn. Obtain the PMF of X when

a) $p = 0.5$ (balanced coin).
b) $p = 0.4$ (biased coin).

5.29 Six men and five women apply for a job at Alpha, Inc. Three of the applicants are selected for interviews. Let X denote the number of women in the interview pool.

a) Assuming that the selection is done randomly, find the PMF of X.
b) Use part (a) and the FPF to find the probability that either one or two women are included in the interview pool.

5.30 Refer to the coin tossing illustration of Example 5.2 on page 179. Let Y denote the difference between the number of heads obtained and the number of tails obtained. Determine the PMF of the random variable Y when the probability of a head on any given toss is

a) 0.5 (balanced coin).   **b)** 0.2.   **c)** p $(0 \leq p \leq 1)$.
d) Obtain the results of parts (a) and (b) from that of part (c).

5.31 Suppose that you play a certain lottery by buying one ticket per week. Let W be the number of weeks until you win a prize and let p $(0 < p < 1)$ be the probability of winning a prize during any particular week.

a) Obtain the PMF of the random variable W.
b) Let $n \in \mathcal{N}$. Use part (a), the FPF, and a geometric formula to obtain $P(W > n)$.
c) Let $n \in \mathcal{N}$. What must happen in the first n weeks for the event $\{W > n\}$ to occur?
d) Use your answer from part (c) to obtain $P(W > n)$. Compare your work and result with those in part (b).

Use part (b) to determine the probability of each of the following events.

e) $\{W > 1\}$ **f)** $\{W \leq 10\}$ **g)** $\{15 < W \leq 20\}$ **h)** $\{15 \leq W < 20\}$

5.32 Random number generation: Suppose that a number is selected at random from the interval $(0, 1)$, a task that a basic random number generator aims to accomplish. Let X denote the number obtained. For integers m and n with $m < n$, set $Y = \lfloor m + (n - m + 1)X \rfloor$, where $\lfloor x \rfloor$ denotes the *floor function*—the greatest integer smaller than or equal to x.
a) Determine the PMF of the random variable Y.
b) Explain the significance of the result in part (a) with respect to random number generation.

5.33 Simulation: This exercise requires access to a computer.
a) Repeat the simulation performed in Example 5.8 on page 192 of 1000 observations of the number of heads obtained in three tosses of a balanced coin. Construct a table and graph similar to Table 5.10 and Figure 5.2, respectively.
b) Repeat part (a) with 10,000 observations.
c) Repeat part (a) for a coin with probability 0.4 of a head.
d) Repeat part (a) for a coin with probability 0.7 of a head.

5.34 Suppose that you have two balanced coins and that you toss each coin twice. Let X and Y be the number of times a head appears for the first and second coins, respectively.
a) Find $p_X(x)$ for $x \in \{0, 1, 2\}$.
b) Find $p_Y(y)$ for $y \in \{0, 1, 2\}$.
c) Find $p_X(y)$ for $y \in \{0, 1, 2\}$. (This is an exercise in understanding the notation.)
d) Find $p_{X+Y}(z)$ for all values of z for which that number is positive.
e) Find $p_X(z) + p_Y(z)$ for all values of z for which that sum is positive.
f) Find $p_{XY}(z)$ for all values of z for which that number is positive.
g) Find $p_{X^2}(z)$ for all values of z for which that number is positive.

5.35 Suppose that three balanced dice are rolled. Let Y be the sum of the three faces showing.
a) Determine the PMF of Y.   b) Construct a probability histogram for Y.
c) Compare the probability histogram here with that in Exercise 5.24.

Theory Exercises

5.36 Show that a function $p: \mathcal{R} \to \mathcal{R}$ that satisfies properties (a), (b), and (c) of Proposition 5.1 on page 187 is the PMF of some discrete random variable. *Hint:* Consider the sample space $\Omega = \{x \in \mathcal{R} : p(x) \neq 0\}$ and refer to Proposition 2.3 on page 43.

Advanced Exercises

5.37 A sequence of attempts to accomplish some goal continues until it succeeds (e.g., tossing a coin until a head appears or playing a lottery until winning a prize). Assume that the results of the attempts are independent and that the probability is p $(0 < p \leq 1)$ that the goal is realized on any particular attempt. Let X denote the number of attempts required to realize the goal.
a) Obtain the PMF of the random variable X.
b) Specialize your answer in part (a) to $p = 0.25$, $p = 0.5$, and $p = 1$. Interpret your results.
c) Show that $P(X < \infty) = 1$ and interpret this result.
d) Use your answer from part (a) and the FPF to determine the probability that X is even.
e) Solve part (d) by conditioning on the result of the first attempt.
f) Suppose now that $p = 0$. Is X a random variable? Explain.

5.38 Suppose that you toss a coin with probability p of a head $(0 < p < 1)$ until the first time a head appears. Let N denote the number of times that you toss the coin. Once you know the value of N, choose a number M at random from $\{1, \ldots, N\}$. Determine the PMF of the random variable $X = M/N$.

5.3   Binomial Random Variables

In Sections 5.1 and 5.2, we examined discrete random variables and their probability mass functions. In doing so, we considered examples of these concepts in a specific context (e.g., number of siblings of a randomly selected student, total number of heads in three tosses of a coin, and mean income of a sample of two Oklahoma state officials).

Different specific contexts often result in probability mass functions that have the same mathematical form. In such cases, analyzing the general context that gives rise to the *family* of probability mass functions is more efficient than repeatedly "reinventing the wheel" by analyzing each specific context separately.

In this section, we introduce the most important family of probability mass functions—namely, the family of probability mass functions corresponding to binomial random variables. In preparation, we discuss some terminology associated with binomial random variables, terminology that is important in its own right.

Bernoulli Trials

Many problems in probability theory and its applications involve independently repeating a random experiment and observing at each repetition whether a specified event occurs. In this context, occurrence of the specified event is called a **success** and nonoccurrence of the specified event (i.e., occurrence of its complement) is called a **failure.** Note that the words *success* and *failure* are used simply as labels and don't necessarily signify the ordinarily positive and negative connotations, respectively, of those words.

Each repetition of the random experiment is called a **trial** and, collectively, the trials are called **Bernoulli trials,** named in honor of the Swiss mathematician Jacob Bernoulli. (See the biography on page 84 for more information about Jacob Bernoulli.)

It is customary to denote the probability that the specified event occurs (i.e., the probability of a success on any particular trial) by the letter p and refer to it as the **success probability.** Likewise, we denote the probability that the complement of the specified event occurs (i.e., the probability of a failure on any particular trial) by the letter q and refer to it as the **failure probability.** Of course, we have $p + q = 1$, so we often write $1 - p$ instead of q. We summarize the discussion of Bernoulli trials in Definition 5.4.

DEFINITION 5.4   Bernoulli Trials

Repeated trials of a random experiment are called **Bernoulli trials** if the following conditions hold.
a) The trials are independent of one another.
b) The result of each trial is classified as a success or failure, depending on whether or not a specified event occurs, respectively.
c) The success probability—and hence the failure probability—remains the same from trial to trial.

Note: If $p = 0$, the specified event never occurs, and if $p = 1$, the specified event always occurs. Thus, to avoid trivialities, we assume, unless explicitly stated otherwise, that the success probability is strictly between 0 and 1, that is, $0 < p < 1$.

EXAMPLE 5.9   *Bernoulli trials*

In each case, identify the random experiment, the trials, a success, a failure, the success probability, and the failure probability. Explain why the trials constitute Bernoulli trials.
a) *Tossing a coin:* A coin is tossed repeatedly and on each toss it is observed whether the result is a head or not.
b) *Betting on red in roulette:* A player makes repeated bets on red in roulette and, on each bet, observes whether he wins or not.
c) *Testing the effectiveness of a drug:* Several patients with identical infirmities take a drug and, for each patient, it is observed whether the drug is effective or not effective.

Solution   a) Here the random experiment consists of tossing a coin once; each toss constitutes one trial. The specified event is a head; thus a success is a head and a failure is not a head (a tail). The success probability, p, is the probability that, on any particular toss, the result is a head; the failure probability, $q = 1 - p$, is the probability that, on any particular toss, the result is a tail. If the coin is balanced, $p = q = 0.5$; otherwise, p and q are different. Finally, it is reasonable to assume that successive tosses of the coin are independent of one another. Hence the conditions for Bernoulli trials (Definition 5.4) are met.

b) A U.S. roulette wheel contains 38 numbers of which 18 are red, 18 are black, and two are green. When the roulette wheel is spun, the ball is equally likely to land on any of the 38 numbers. Here the random experiment consists of one play at a roulette wheel; each play constitutes one trial. The specified event is that the ball lands on a red number; thus a success is red and a failure is not red (black or green). The success probability, p, is the probability that, on any particular play, the ball lands on a red number; the failure probability, $q = 1 - p$, is the probability that, on any particular play, the ball lands on a nonred (black or green) number. If we assume that the wheel is balanced, we have $p = 18/38$ and $q = 20/38$. Finally, we can reasonably assume that successive plays at the wheel are independent of one another. Hence the conditions for Bernoulli trials (Definition 5.4) are met.

c) Here the random experiment consists of administering a drug to a patient with a particular infirmity; each administration constitutes one trial. The specified event is that the patient is cured; thus a success is a cure and a failure is a noncure. The success probability, p, is the effectiveness of the drug—that is, the probability that the drug cures a patient; the failure probability, $q = 1 - p$, is the probability that the drug doesn't cure a patient. Finally, we can reasonably assume that the results of administering the drug are independent from one patient to another. Hence the conditions for Bernoulli trials (Definition 5.4) are met.   ∎

Sometimes you won't be able to tell whether trials constitute Bernoulli trials. Example 5.10 illustrates two such situations.

EXAMPLE 5.10   *Bernoulli Trials*

In each case, discuss whether presuming Bernoulli trials is reasonable.

a) *Hits by a baseball player:* Suppose that we observe the at-bats of a particular baseball player, noting for each at-bat whether the player gets a hit (single, double, etc.).

b) *Quality control:* In quality control, we are interested in whether a particular product satisfies certain specifications. Suppose, for instance, that we observe bolts coming off an assembly line for conformity to diameter tolerance specifications.

Solution   a) Here the random experiment consists of observing an at-bat of the player, and the specified event (success) is that the player gets a hit. We might reasonably regard the player's batting average as the success probability, but whether successive trials (at-bats) are independent isn't at all clear.

b) Here the random experiment consists of observing a bolt coming off the assembly line, and the specified event (success) is that the bolt is in conformance. Depending on the actual production process, presuming independence from one bolt to the next may or may not be reasonable. And, owing to such factors as machine wear, whether the success probability (probability of conformance) remains the same from trial to trial is also not clear.   ▪

To explain Bernoulli trials further, we consider random sampling, both with and without replacement, in Example 5.11.

EXAMPLE 5.11   *Bernoulli Trials*

Random Sampling and Bernoulli Trials   Consider a population with N members in which each member of the population can be classified as either having or not having a specified attribute. For a human population, the specified attribute might be "has Internet access," "is a college graduate," or "is married." For the population of U.S. businesses, the specified attribute might be "is incorporated," "is technology based," or "is minority owned." In each part, decide whether the sampling process constitutes Bernoulli trials.

a) *Random sampling with replacement:* Suppose that n members of the population are randomly selected one at a time with identity noted and that each member selected is returned to the population for possible reselection. Further suppose that we are interested in whether each member selected has or doesn't have the specified attribute.

b) *Random sampling without replacement:* Suppose that n members of the population are randomly selected one at a time with identity noted and that each member selected isn't returned to the population for possible reselection. Further suppose that we are interested in whether each member selected has or doesn't have the specified attribute.

Solution   In both sampling processes—random sampling with replacement and random sampling without replacement—the specified event is that the member selected has the specified attribute. Thus a success is that the member selected has the specified attribute, and a failure is that the member selected doesn't have the specified attribute.

Let M denote the number of members of the population that have the specified attribute. We claim that, for random sampling both with and without replacement, the probability is M/N that the kth member sampled has the specified attribute. The easiest way to verify this claim is to use a symmetry argument as follows. Because the sampling is random, the kth member sampled is equally likely to be any of the N members of the population. And, because M members of the population have the specified attribute, the probability is M/N that the kth member sampled has the specified attribute. See Exercise 5.40 for another way to calculate this probability.

Thus, in both cases, the success probability is $p = M/N$, which is the proportion of members of the population that have the specified attribute. The failure probability is $q = 1 - p = 1 - M/N = (N - M)/N$, which is the proportion of members of the population that don't have the specified attribute. Hence conditions (b) and (c) for Bernoulli trials are satisfied for both sampling with and without replacement. Consequently, it remains only to consider condition (a), the independent-trials condition.

a) For random sampling with replacement, the trials are independent—so the sampling process constitutes Bernoulli trials.
b) For random sampling without replacement, the trials aren't independent—so the sampling process doesn't constitute Bernoulli trials.   ∎

Binomial Random Variables

Now that we have covered Bernoulli trials, we can introduce binomial random variables. Frequently, we are interested in the number of successes in a finite sequence of Bernoulli trials. The random variable "number of successes" is called a *binomial random variable*.

We already encountered a binomial random variable when we considered the number of heads obtained in three tosses of a coin. The PMF of this random variable is given in Tables 5.7, 5.8, and 5.9, respectively, for a balanced coin, a coin with probability 0.2 of a head on any given toss, and a coin with probability p of a head on any given toss.

Before deriving a general formula for the PMF of a binomial random variable, let's consider, in Example 5.12, the case of a sequence of four Bernoulli trials.

EXAMPLE 5.12   *Illustrates the PMF of a Binomial Random Variable*

Consider a sequence of four Bernoulli trials in which the success probability is p and let X denote the number of successes in the four trials. Find the PMF of the random variable X.

Solution   Let s and f denote success and failure, respectively. Each possible outcome of the four Bernoulli trials can be regarded as a four-tuple in which each entry is either an s or an f. For instance, (s, s, f, s) represents the outcome that the first two trials result in success, the third in failure, and the fourth in success. With this representation, a sample space for the possible outcomes of the four Bernoulli trials is

$$\Omega = \{ (x_1, x_2, x_3, x_4) : x_j \in \{s, f\}, \ 1 \le j \le 4 \}.$$

Because there are two possibilities for the result of each trial (s or f), the number of possible outcomes in the four Bernoulli trials is $2^4 = 16$. The 16 possible outcomes are listed in the Outcome columns of Table 5.11.

Table 5.11 Probabilities of individual outcomes in four Bernoulli trials

Outcome	Probability	Outcome	Probability	Outcome	Probability	Outcome	Probability
(s, s, s, s)	$p^4(1-p)^0$	(s, f, s, s)	$p^3(1-p)^1$	(f, s, s, s)	$p^3(1-p)^1$	(f, f, s, s)	$p^2(1-p)^2$
(s, s, s, f)	$p^3(1-p)^1$	(s, f, s, f)	$p^2(1-p)^2$	(f, s, s, f)	$p^2(1-p)^2$	(f, f, s, f)	$p^1(1-p)^3$
(s, s, f, s)	$p^3(1-p)^1$	(s, f, f, s)	$p^2(1-p)^2$	(f, s, f, s)	$p^2(1-p)^2$	(f, f, f, s)	$p^1(1-p)^3$
(s, s, f, f)	$p^2(1-p)^2$	(s, f, f, f)	$p^1(1-p)^3$	(f, s, f, f)	$p^1(1-p)^3$	(f, f, f, f)	$p^0(1-p)^4$

For $1 \le j \le 4$, let A_j denote the event that the jth trial results in a success. Then, by the independence assumption for Bernoulli trials, A_1, A_2, A_3, and A_4 are independent events. Moreover, because the success probability is p, we have $P(A_j) = p$ for each j. From these facts, we can obtain the probabilities of the individual outcomes in the four Bernoulli trials.

Consider, for instance, the possible outcome (s, s, f, s). We can express that outcome in terms of the A_js as $\{(s, s, f, s)\} = A_1 \cap A_2 \cap A_3^c \cap A_4$. Consequently,

$$P(\{(s, s, f, s)\}) = P(A_1 \cap A_2 \cap A_3^c \cap A_4) = P(A_1)P(A_2)P(A_3^c)P(A_4)$$
$$= p \cdot p \cdot (1-p) \cdot p = p^3(1-p)^1.$$

Proceeding in the same way, we find that any outcome consisting of exactly 3 successes (and exactly $4 - 3 = 1$ failure) has this same probability—namely, $p^3(1-p)^1$. More generally, any outcome consisting of exactly k successes (and exactly $4 - k$ failures) has the same probability—namely, $p^k(1-p)^{4-k}$. Table 5.11 provides the probabilities of all 16 possible outcomes.

From Table 5.11, we can easily obtain the PMF of X. For instance,

$$p_X(3) = P(X = 3) = P(\{(s, s, s, f), (s, s, f, s), (s, f, s, s), (f, s, s, s)\})$$
$$= P(\{(s, s, s, f)\}) + P(\{(s, s, f, s)\}) + P(\{(s, f, s, s)\}) + P(\{(f, s, s, s)\})$$
$$= p^3(1-p)^1 + p^3(1-p)^1 + p^3(1-p)^1 + p^3(1-p)^1 = 4p^3(1-p)^1.$$

The coefficient 4 appearing in the last term signifies that there are four possible outcomes in which exactly three of the four trials result in success. We can obtain that coefficient without resorting to a direct listing by arguing as follows. The number of outcomes in which there are exactly three successes equals the number of ways that we can choose exactly three of the four entries in a four-tuple to contain an s, which is $\binom{4}{3} = 4$.

Similarly, we obtain $p_X(x)$ for the other four possible values of X. Table 5.12 displays the results and, thereby, provides the PMF of the random variable X. ∎

Table 5.12 PMF of the random variable X, the number of successes in four Bernoulli trials

x	0	1	2	3	4
$p_X(x)$	$p^0(1-p)^4$	$4p^1(1-p)^3$	$6p^2(1-p)^2$	$4p^3(1-p)^1$	$p^4(1-p)^0$

With Example 5.12 in mind, we now derive a general formula for the PMF of a binomial random variable, which we present in Proposition 5.3.

◆◆◆   **Proposition 5.3   PMF of a Binomial Random Variable**

Let X denote the number of successes in a sequence of n Bernoulli trials with success probability p. Then the PMF of the random variable X is

$$p_X(x) = \binom{n}{x} p^x (1-p)^{n-x}, \qquad x = 0, 1, \ldots, n, \tag{5.7}$$

and $p_X(x) = 0$ otherwise. The random variable X is called a **binomial random variable** *and is said to have the* **binomial distribution with parameters n and p.** *For convenience, we often write $X \sim \mathcal{B}(n, p)$ to indicate that X has the binomial distribution with parameters n and p.*

Proof   We can take the sample space for the n Bernoulli trials to be

$$\Omega = \{ (x_1, x_2, \ldots, x_n) : x_j \in \{s, f\}, \ 1 \le j \le n \};$$

that is, the possible n-tuples in which each entry is either an s (success) or an f (failure).

Let k be an integer between 0 and n, inclusive. The event $\{X = k\}$—exactly k successes in the n trials—consists of all n-tuples in which exactly k entries are s and exactly $n - k$ entries are f. Because the trials are independent, we conclude that any particular outcome in which exactly k entries are s and exactly $n - k$ entries are f has probability $p^k(1 - p)^{n-k}$. Moreover, the number of such outcomes equals the number of ways that we can choose k entries from n as the entries to contain an s, which is $\binom{n}{k}$. Hence

$$p_X(k) = P(X = k) = \binom{n}{k} p^k (1-p)^{n-k},$$

as required.   ◆

Note: A binomial random variable with parameters $n = 1$ (i.e., one Bernoulli trial) and p is usually called a **Bernoulli random variable** and is said to have the **Bernoulli distribution with parameter p.**

To apply Proposition 5.3, it is useful to have a well-organized strategy, such as that presented in Procedure 5.1.

PROCEDURE 5.1   To Find a Binomial PMF

Step 1 *Identify a success.*

Step 2 *Determine p, the success probability.*

Step 3 *Determine n, the number of trials.*

Step 4 *Apply Proposition 5.3 to obtain the binomial PMF.*

EXAMPLE 5.13   *Binomial Random Variables*

Mortality   Mortality tables enable actuaries to obtain the probabilities that people at various ages will live specified numbers of years. Life-insurance premiums, retirement pensions, annuity payments, and related items are based on these probabilities.

According to tables provided by the U.S. National Center for Health Statistics in *Vital Statistics of the United States,* there is about an 80% chance that a person aged 20 will be alive at age 65. Suppose that three people aged 20 are selected at random. Find the probability that the number still alive at age 65 (i.e., 45 years later) will be

a) exactly two.   b) at most one.   c) at least one.

Solution   Each trial consists of observing, for a person currently aged 20, his or her life–death status at age 65 (i.e., 45 years later). We can reasonably presume that the three trials are independent. Let X denote the number of people of the three that are still alive at age 65. We apply Procedure 5.1.

Step 1 *Identify a success.*

Here a success is that a person currently aged 20 is still alive at age 65.

Step 2 *Determine p, the success probability.*

The probability that a person currently aged 20 will still be alive at age 65 is 80%, so $p = 0.8$.

Step 3 *Determine n, the number of trials.*

The number of trials equals the number of people being observed, or three, so $n = 3$.

Step 4 *Apply Proposition 5.3 to obtain the binomial PMF.*

Referring to Steps 1–3, we conclude that $X \sim \mathcal{B}(3, 0.8)$; that is, the random variable X has the binomial distribution with parameters $n = 3$ and $p = 0.8$. Proposition 5.3 gives the PMF of X as

$$p_X(x) = \binom{3}{x}(0.8)^x (0.2)^{3-x}, \qquad x = 0, 1, 2, 3, \tag{5.8}$$

and $p_X(x) = 0$ otherwise.

We can now easily determine the required binomial probabilities.

a) From Equation (5.8),

$$P(X = 2) = p_X(2) = \binom{3}{2}(0.8)^2(0.2)^1 = 0.384.$$

Chances are 38.4% that exactly two of the three people will still be alive at age 65.

b) Applying the FPF and Equation (5.8), we get

$$P(X \le 1) = \sum_{x \le 1} p_X(x) = p_X(0) + p_X(1)$$

$$= \binom{3}{0}(0.8)^0(0.2)^3 + \binom{3}{1}(0.8)^1(0.2)^2 = 0.008 + 0.096 = 0.104.$$

Chances are 10.4% that at most one of the three people will still be alive at age 65.

c) Here we want $P(X \geq 1)$. We can obtain this probability by first using the FPF to write

$$P(X \geq 1) = \sum_{x \geq 1} p_X(x) = p_X(1) + p_X(2) + p_X(3)$$

and then applying Equation (5.8) to calculate each of the three probabilities on the right of the previous display. However, using the complementation rule is easier:

$$P(X \geq 1) = 1 - P(X < 1) = 1 - P(X = 0) = 1 - p_X(0)$$

$$= 1 - \binom{3}{0}(0.8)^0(0.2)^{3-0} = 1 - 0.008 = 0.992.$$

Chances are 99.2% that at least one of the three people will still be alive at age 65.   ∎

The Shape of a Binomial Distribution

It is useful to understand the behavior of the PMF of a binomial random variable—in particular, the shape of the corresponding probability histogram. In Example 5.13, we obtained, for three people currently aged 20, the probability distribution of the number, X, that will still be alive at age 65. We found that $X \sim \mathcal{B}(3, 0.8)$, whose PMF is given by Equation (5.8). Using that equation, we get Table 5.13 and Figure 5.3.

Table 5.13 PMF of the random variable X, the number of three 20-year-olds that are still alive at age 65

x	0	1	2	3
$p_X(x)$	0.008	0.096	0.384	0.512

Figure 5.3 Probability histogram of the random variable X, the number of three 20-year-olds that are still alive at age 65

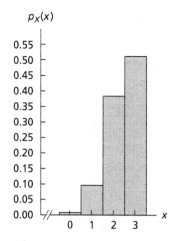

Figure 5.3 shows that, for three people currently aged 20, the probability distribution of the number still alive at age 65 is left-skewed; the reason is that the success probability, $p = 0.8$, exceeds 0.5. More generally, a binomial distribution is right-skewed if $p < 0.5$, is symmetric if $p = 0.5$, and is left-skewed if $p > 0.5$. Figure 5.4 illustrates these facts for binomial distributions with $n = 6$ and $p = 0.25$, $p = 0.5$, and $p = 0.75$.

Figure 5.4 Probability histograms for binomial distributions with parameters $n = 6$ and (a) $p = 0.25$, (b) $p = 0.5$, and (c) $p = 0.75$

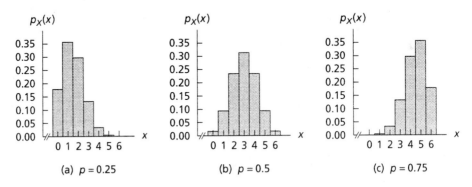

What we have just presented graphically is made more precise in Proposition 5.4.

◆◆◆   **Proposition 5.4   Shape of a Binomial Distribution**

Suppose that X has a binomial distribution with parameters n and p. Let k represent an integer between 0 and n and set $m = \lfloor (n + 1)p \rfloor$, where $\lfloor x \rfloor$ denotes the floor function—the greatest integer smaller than or equal to x. (The floor function is identical to the greatest-integer function.)

a) If $(n + 1)p$ isn't an integer, then $p_X(k)$ is strictly increasing as k goes from 0 to m and is strictly decreasing as k goes from m to n.

b) If $(n + 1)p$ is an integer, then $p_X(k)$ is strictly increasing as k goes from 0 to $m - 1$, $p_X(m - 1) = p_X(m)$, and $p_X(k)$ is strictly decreasing as k goes from m to n.

In either case, p_X attains its maximum value at $x = m = \lfloor (n + 1)p \rfloor$.

Proof   Recalling that $q = 1 - p$, we have, for $k = 1, 2, \ldots, n$,

$$\frac{p_X(k)}{p_X(k - 1)} = \frac{\binom{n}{k}p^k q^{n-k}}{\binom{n}{k-1}p^{k-1}q^{n-(k-1)}} = \frac{n - k + 1}{k} \cdot \frac{p}{q}$$

$$= \frac{(n - k + 1)p}{kq} = \frac{kq + (n - k + 1)p - kq}{kq} = 1 + \frac{(n + 1)p - k}{kq}.$$

It follows that

$$p_X(k) > p_X(k - 1), \quad \text{if } k < (n + 1)p;$$
$$p_X(k) = p_X(k - 1), \quad \text{if } k = (n + 1)p; \qquad \textbf{(5.9)}$$
$$p_X(k) < p_X(k - 1), \quad \text{if } k > (n + 1)p.$$

Let's now consider two cases.

Case 1: $(n + 1)p$ isn't an integer.

We have $k < (n + 1)p$ if and only if $k \leq m$ and we have $k > (n + 1)p$ if and only if $k \geq m + 1$. Thus, from Relations (5.9), we see that $p_X(k)$ is strictly increasing as k goes from 0 to m and is strictly decreasing as k goes from m to n.

Case 2: $(n + 1)p$ is an integer.

We have $k < (n + 1)p$ if and only if $k \leq m - 1$ and we have $k > (n + 1)p$ if and only if $k \geq m + 1$. Thus, from Relations (5.9), we see that $p_X(k)$ is strictly increasing as k goes from 0 to $m - 1$, that $p_X(m - 1) = p_X(m)$, and that $p_X(k)$ is strictly decreasing as k goes from m to n.   ♦

Let's apply Proposition 5.4 to the binomial distributions shown in Figure 5.4. In all three graphs, $n = 6$. For the probability histogram in Figure 5.4(a), we have $p = 0.25$ and so $(n + 1)p = 1.75$. Because $(n + 1)p$ isn't an integer, Proposition 5.4(a) implies that the PMF has a unique maximum occurring at $\lfloor (n + 1)p \rfloor = \lfloor 1.75 \rfloor = 1$, as you can also see from Figure 5.4(a). Similar remarks apply to the other two graphs.

EXERCISES 5.3   Basic Exercises

5.39 Pinworm infestation, commonly found in children, can be treated with the drug pyrantel pamoate. According to the *Merck Manual,* the treatment is effective in 90% of cases. Suppose that three children with pinworm infestation are given pyrantel pamoate.
a) Considering a success in any particular case to be "a cure," formulate the process of observing which children are cured and which aren't as a sequence of three Bernoulli trials.
b) Construct a table similar to Table 5.11 on page 200 for the three cases. Display the probabilities to three decimal places.
c) List the outcomes in which exactly two of the three children are cured.
d) Find the probability of each outcome in part (c). Why are those probabilities all the same?
e) Use parts (c) and (d) to determine the probability that exactly two of the three children will be cured.
f) Without using the binomial probability formula, obtain the probability distribution of the random variable X, the number of children out of three that are cured.
g) Use Procedure 5.1 on page 201 to solve part (f).

5.40 Refer to Example 5.11 on page 198. Using counting arguments, determine the probability that the kth member sampled has the specified attribute when the sampling is
a) with replacement.   b) without replacement.

5.41 Refer to Example 5.11 on page 198. Explain why the trials are independent for random sampling with replacement but not for random sampling without replacement.

5.42 According to the *Daily Racing Form,* the probability is about 0.67 that the favorite in a horse race will finish in the money (first, second, or third place). In the next five races, what is the probability that the favorite finishes in the money
a) exactly twice?   b) exactly four times?   c) at least four times?
d) between two and four times, inclusive?
e) Determine the probability distribution of the random variable X, the number of times the favorite finishes in the money in the next five races.

f) Identify the probability distribution of X as right-skewed, symmetric, or left-skewed without consulting its probability distribution or drawing its probability histogram.

g) Draw a probability histogram for X.

5.43 Sickle cell anemia is an inherited blood disease that occurs primarily in blacks. In the United States, about 15 of every 10,000 black children have sickle cell anemia. The red blood cells of an affected person are abnormal; the result is severe chronic anemia (inability to carry the required amount of oxygen), which causes headaches, shortness of breath, jaundice, increased risk of pneumococcal pneumonia and gallstones, and other severe problems. Sickle cell anemia occurs in children who inherit an abnormal type of hemoglobin, called hemoglobin S, from both parents. If hemoglobin S is inherited from only one parent, the person is said to have sickle cell trait and is generally free from symptoms. There is a 50% chance that a person who has sickle cell trait will pass hemoglobin S to an offspring.

a) Obtain the probability that a child of two people who have sickle cell trait will have sickle cell anemia.

b) If two people who have sickle cell trait have five children, determine the probability that at least one of the children will have sickle cell anemia.

c) If two people who have sickle cell trait have five children, find the probability distribution of the number of those children who will have sickle cell anemia.

d) Construct a probability histogram for the probability distribution in part (c).

5.44 If all sex distributions are equally likely, what proportion of families with five children have three girls and two boys?

5.45 A baseball player has a batting average of .260. Suppose that you observe successive at-bats of the player and note for each at-bat whether the player gets a hit.

a) Under what conditions is the assumption of Bernoulli trials appropriate?

b) Assuming your conditions in part (a), what is the probability that the player will get two or more hits in his next four times at-bat?

5.46 Sixty percent of all voters in a state intend to vote "yes" in a referendum. An opinion poll took a sample of 20 voters. No precaution was taken against the unlikely event of choosing the same voter more than once. Find the probability that in the sample of 20 there are more voters who intend to vote "yes" than voters who don't intend to vote "yes."

5.47 Consider 15 Bernoulli trials with success probability p.

a) Assuming that $p = 0.5$, what is the probability that the number of successes is between six and nine, inclusive? What is the most likely number of successes?

b) Repeat part (a) for $p = 0.4$.

5.48 Availability of statistical software or binomial tables is useful for this problem. Suppose that you throw a balanced die 100 times.

a) What is the probability that you get a 1 at most 4 times? What would you think if that happened?

b) What is the probability that you get a 1 more than 32 times? What would you think if that happened?

5.49 Simultaneously and independently, each of n people make a single toss of a coin with probability p of a head. What is the probability of an "odd man"—that is, one person gets a different result from all the other people?

5.50 How long must a sequence of random decimal digits be so that the probability of getting a 6 or a 7 will exceed 0.95?

5.51 Three numbers are selected independently and randomly from the interval $(0, 1)$.
a) Explain why the probability that the median of the three numbers exceeds 0.7 is the same as the probability of two or more successes in three Bernoulli trials, with a certain success probability p.
b) In part (a), what is the value of p?

5.52 Six marbles are randomly selected with replacement from an urn. The urn is equally likely to be Urn I (which contains 1 red marble and 1 green marble) or Urn II (which contains 11 red marbles and 9 green marbles). Find the conditional probability that the urn from which the marbles are selected is Urn II, given that the number of red marbles selected is x.

Theory Exercises

5.53 Let n be a positive integer and let $0 < p < 1$.
a) Use the binomial theorem to prove that $\sum_{k=0}^{n} \binom{n}{k} p^k (1-p)^{n-k} = 1$.
b) Establish the identity in part (a) without using the binomial theorem or doing any computations, but rather by using a probabilistic argument.

5.54 Let $X \sim \mathcal{B}(n, p)$. We say that X is *symmetric* if $p_X(x) = p_X(n-x)$ for all x. Prove that X is symmetric if and only if $p = 1/2$.

5.55 Let $n \in \mathcal{N}$, let $0 \le k \le n$ be a nonnegative integer, and let $0 < p < 1$. Set $q = 1 - p$.
a) Prove the combinatorial identities

$$(n-k)\binom{n}{k} = n\binom{n-1}{k} \quad \text{and} \quad k\binom{n}{k} = n\binom{n-1}{k-1}.$$

b) Use integration by parts to verify that, for $0 \le k \le n-1$,

$$\int_0^q t^{n-k-1}(1-t)^k \, dt = \frac{1}{n-k} p^k q^{n-k} + \frac{k}{n-k} \int_0^q t^{n-k}(1-t)^{k-1} \, dt.$$

c) Let $X \sim \mathcal{B}(n, p)$. Prove that

$$P(X \le k) = n\binom{n-1}{k} \int_0^q t^{n-k-1}(1-t)^k \, dt = n\binom{n-1}{k} B_q(n-k, k+1),$$

where $B_x(\alpha, \beta) = \int_0^x t^{\alpha-1}(1-t)^{\beta-1} \, dt$, the so-called *incomplete beta function*.
d) Let $X \sim \mathcal{B}(n, p)$. Prove that

$$P(X > k) = n\binom{n-1}{k} \int_0^p t^k(1-t)^{n-k-1} \, dt = n\binom{n-1}{k} B_p(k+1, n-k).$$

Hint: Consider the random variable $Y = n - X$ and refer to part (c).

Advanced Exercises

5.56 An urn contains k red marbles and $12 - k$ green marbles. Five times, a marble is chosen randomly and then replaced.
a) What value of k maximizes the probability that the number of red marbles among the 5 obtained is exactly 2?
b) Referring to part (a), what is that maximum probability?

5.57 One rocket has two engines and another has four. All engines are identical. Each rocket will achieve its mission if and only if at least half its engines work. Both rockets have the same nonzero probability of achieving its mission. What is the probability than any particular engine works?

5.4   Hypergeometric Random Variables

In this section, we discuss the family of *hypergeometric random variables,* a family closely related to the family of binomial random variables.[†] Hypergeometric random variables are particularly important in quality control and in the statistical estimation of population proportions.

In Example 5.11 on page 198, we considered random sampling from a finite population in which each member of the population can be classified as either having or not having a specified attribute. We demonstrated that, if sampling is with replacement, the process of observing whether or not each member selected has the specified attribute constitutes Bernoulli trials. Hence, for random sampling with replacement, the number of members obtained that have the specified attribute is a binomial random variable. Specifically, the number of members obtained that have the specified attribute has a binomial distribution with parameters n and p, where n is the sample size and p is the proportion of the population that has the specified attribute.

However, if sampling is without replacement, the trials aren't independent. Hence the process of observing whether or not each member selected has the specified attribute doesn't constitute Bernoulli trials. In particular, then, for random sampling without replacement, the number of members obtained that have the specified attribute isn't a binomial random variable; rather, it is a *hypergeometric random variable.*

In Example 5.14, we introduce hypergeometric random variables with an application from Chapter 3 (Example 3.19 on page 111).

EXAMPLE 5.14   *Introduces Hypergeometric Random Variables*

Statistical Quality Control   A quality assurance engineer of a company that manufactures TV sets inspects finished products in lots of 100. He selects 5 of the 100 TVs at random and inspects them thoroughly. Let X denote the number of defective TVs obtained. If, in fact, 6 of the 100 TVs in the current lot are actually defective, find the PMF of the random variable X.

Solution   The possible values of the random variable X are 0, 1, ..., 5. Because the engineer makes his selection at random, each possible collection of 5 TVs from the 100 TVs is equally likely to be the one obtained. Hence we should use a classical probability model:

$$P(E) = \frac{N(E)}{N(\Omega)}, \qquad \text{for each event } E. \tag{5.10}$$

To begin, the number of possible outcomes—that is, the number of possible samples of 5 TVs from the 100 TVs—equals $\binom{100}{5}$. Thus $N(\Omega) = \binom{100}{5}$.

We want to determine $P(X = k)$ for $k = 0, 1, \ldots, 5$. Equation (5.10) requires obtaining $N(\{X = k\})$ for $k = 0, 1, \ldots, 5$. In other words, we need to find the number

---

[†]The term *hypergeometric* is used for these random variables because of a relationship they have with special mathematical functions called *hypergeometric functions.*

of outcomes in which exactly k of the 5 TVs selected are defective. To do so, we think of the 100 TVs as divided into two groups—namely, the defective TVs and the nondefective TVs, as shown in the top portion of Figure 5.5.

Figure 5.5 Diagram for obtaining the number of outcomes in which exactly k of the 5 TVs selected are defective

	TVs	
Groups	Defective 6	Nondefective 94
	↓	↓
Number to be chosen	Defective k	Nondefective $5-k$
Number of ways	$\binom{6}{k}$	$\binom{94}{5-k}$
Total possibilities	$\binom{6}{k} \cdot \binom{94}{5-k}$	

There are 6 TVs in the first group, of which k are to be selected; we can do so in $\binom{6}{k}$ ways. There are 94 TVs in the second group, of which $5-k$ are to be selected; we can do so in $\binom{94}{5-k}$ ways. Thus, by the BCR, there are a total of $\binom{6}{k}\binom{94}{5-k}$ outcomes in which exactly k of the 5 TVs selected are defective; in other words, $N(\{X = k\}) = \binom{6}{k}\binom{94}{5-k}$. Figure 5.5 summarizes the discussion in this paragraph.

Applying Equation (5.10) gives

$$P(X = k) = \frac{N(\{X = k\})}{N(\Omega)} = \frac{\binom{6}{k}\binom{94}{5-k}}{\binom{100}{5}}, \qquad k = 0, 1, \ldots, 5. \tag{5.11}$$

In other words, the PMF of the random variable X is given by

$$p_X(x) = \frac{\binom{6}{x}\binom{94}{5-x}}{\binom{100}{5}}, \qquad x = 0, 1, \ldots, 5,$$

and $p_X(x) = 0$ otherwise. ■

In Proposition 5.5, we define a hypergeometric random variable and obtain its PMF. Because the derivation of the hypergeometric PMF follows the same basic steps as those used in Example 5.14, we leave that derivation to you as Exercise 5.69.

◆◆◆   **Proposition 5.5   PMF of a Hypergeometric Random Variable**

Consider a finite population of size N in which each member is classified as either having or not having a specified attribute; represent by p the proportion of the population having the specified attribute. Suppose that a random sample of size n (where n ≤ N) is taken without replacement from the population and let X denote the number of members sampled that have the specified attribute. Then the PMF of the random variable X is

$$p_X(x) = \frac{\binom{Np}{x}\binom{N(1-p)}{n-x}}{\binom{N}{n}}, \qquad x = 0, 1, \ldots, n, \qquad (5.12)$$

and $p_X(x) = 0$ otherwise. The random variable X is called a **hypergeometric random variable** *and is said to have the* **hypergeometric distribution with parameters N, n, and p.** *For convenience, we sometimes write $X \sim \mathcal{H}(N, n, p)$ to indicate that X has the hypergeometric distribution with parameters N, n, and p.*

In Example 5.14, 5 TVs are randomly selected without replacement from a lot of 100 TVs in which 6 are defective, and X denotes the number of defective TVs obtained. Here the population consists of the lot of 100 TVs, the specified attribute is "defective," the proportion of the population that has the specified attribute is 6/100, and the sample size is 5. Thus $N = 100$, $n = 5$, and $p = 6/100 = 0.06$ and, hence, $X \sim \mathcal{H}(100, 5, 0.06)$—that is, X has the hypergeometric distribution with parameters $N = 100$, $n = 5$, and $p = 0.06$. From Proposition 5.5, the PMF of X is

$$p_X(x) = \frac{\binom{100 \cdot 0.06}{x}\binom{100 \cdot (1 - 0.06)}{5 - x}}{\binom{100}{5}} = \frac{\binom{6}{x}\binom{94}{5 - x}}{\binom{100}{5}},$$

for $x = 0, 1, \ldots, 5$, and $p_X(x) = 0$ otherwise, as we demonstrated in Example 5.14.

In general, Np is the number of members of the population that have the specified attribute. Sometimes, as in Example 5.14, we are explicitly given that number instead of the population proportion, p. In such cases, we can use an alternative form of a hypergeometric PMF. Specifically, if M denotes the number of members of the population that have the specified attribute, we can write the hypergeometric PMF as

$$p_X(x) = \frac{\binom{M}{x}\binom{N - M}{n - x}}{\binom{N}{n}}, \qquad x = 0, 1, \ldots, n, \qquad (5.13)$$

and $p_X(x) = 0$ otherwise.

Extending the Definition of Binomial Coefficients

In Equation (5.13), we listed the possible values of X—the number of members sampled that have the specified attribute—as $x = 0, 1, \ldots, n$, because n is the sample size.

However, because we must necessarily have $0 \le x \le M$ and $0 \le n - x \le N - M$, the possible values of X are actually the integers x that satisfy

$$\max\{0, n - (N - M)\} \le x \le \min\{n, M\}.$$

Fortunately, we can keep the simpler form ($x = 0, 1, \ldots, n$) for the possible values of X by extending the definition of binomial coefficients.

Up to now, we have defined the binomial coefficient $\binom{m}{r}$ only when m is a positive integer and r is a nonnegative integer less than or equal to m. Thus, at this point, $\binom{m}{r}$ doesn't make sense when, for instance, $m = 3/2$ or $m = -4$.

To extend the definition of binomial coefficients, we first recall that, for positive integers m and r with $1 \le r \le m$,

$$\binom{m}{r} = \frac{m!}{r!\,(m-r)!} = \frac{m(m-1)\cdots(m-r+1)}{r!}. \tag{5.14}$$

We observe, however, that the right side of Equation (5.14) makes sense for any real number m and any positive integer r. In particular, if m is a positive integer less than r, the right side of Equation (5.14) equals 0; thus, in that case, we define $\binom{m}{r} = 0$. More generally, we extend the definition of the **binomial coefficient** as follows.

DEFINITION 5.5   Binomial Coefficient

Let a denote a real number and let r denote an integer. Then we define the **binomial coefficient** $\binom{a}{r}$ by

$$\binom{a}{0} = 1, \qquad \binom{a}{r} = 0 \quad \text{if } r < 0,$$

and

$$\binom{a}{r} = \frac{a(a-1)\cdots(a-r+1)}{r!} \qquad \text{if } r > 0.$$

Binomial coefficients are left undefined for noninteger values of r.

For a positive integer m and an integer r, Definition 5.5 yields

$$\binom{m}{r} = \begin{cases} \dfrac{m!}{r!\,(m-r)!}, & \text{if } 1 \le r \le m; \\[2mm] 1, & \text{if } r = 0; \\[2mm] 0, & \text{if } r > m \quad \text{or} \quad r < 0. \end{cases}$$

From the first two lines on the right of the previous display, we see that Definition 5.5 really is an extension of our earlier definition of binomial coefficients.

Applications of the Hypergeometric Distribution

In Examples 5.15 and 5.16, we provide two of the countless applications of the hypergeometric distribution.

EXAMPLE 5.15   *Hypergeometric Random Variables*

The Lotto   An Arizona state lottery, called *Lotto*, is played as follows: A player specifies six numbers of her choice from the numbers 1–42; these six numbers constitute the player's "ticket" for which the player pays $1. In the lottery drawing, six winning numbers are chosen at random without replacement from the numbers 1–42. To win a prize, a *Lotto* ticket must contain three or more of the winning numbers.

a) Determine the PMF of the random variable X, the number of winning numbers on the player's ticket.

b) If the player buys one *Lotto* ticket, determine the probability that she wins a prize.

c) If the player buys one *Lotto* ticket per week for a year, determine the probability that she wins a prize at least once in the 52 tries.

Solution   The problem can be formulated in a couple of ways. One way is to regard the 42 numbers as the population of interest and the specified attribute as "is one of the player's six numbers." The random experiment then consists of the lottery drawing of the six winning numbers from the population of 42 numbers. Let X denote the number of those six numbers that have the specified attribute (i.e., that are among the player's six numbers). Note that X also equals the number of winning numbers on the player's ticket.

a) From the formulation of the problem, X has a hypergeometric distribution with parameters $N = 42$, $n = 6$, and $p = 6/42$. The "6" in "$n = 6$" represents the fact that, in the lottery drawing, six numbers are sampled from the population of 42 numbers, whereas the "6" in the "$p = 6/42$" represents the fact that the six numbers selected by the player constitute the members of the population that have the specified attribute. In any case, the PMF of X is

$$p_X(x) = \frac{\binom{Np}{x}\binom{N(1-p)}{n-x}}{\binom{N}{n}} = \frac{\binom{6}{x}\binom{36}{6-x}}{\binom{42}{6}}, \qquad x = 0, 1, \ldots, 6,$$

and $p_X(x) = 0$ otherwise. Using this formula, we obtain Table 5.14.

b) The probability that the player's ticket contains three or more winning numbers, and hence that she wins a prize, is

$$P(X \geq 3) = \sum_{x \geq 3} p_X(x) = p_X(3) + p_X(4) + p_X(5) + p_X(6) = 0.0290647,$$

or roughly a 2.9% chance.

c) Let Y denote the total number of times that the player wins a prize if she buys one ticket per week for a year. Noting that the results from one week to the next are independent, we conclude, in view of part (b), that the random variable Y has

Table 5.14 Probability mass function of the random variable X, the number of winning numbers on the player's ticket

Winning numbers x	Probability $p_X(x)$
0	0.3713060
1	0.4311941
2	0.1684352
3	0.0272219
4	0.0018014
5	0.0000412
6	0.0000002

the binomial distribution with parameters $n = 52$ and $p = 0.029$ (to three decimal places). Consequently,

$$p_Y(y) = \binom{52}{y}(0.029)^y(0.971)^{52-y}, \qquad y = 0, 1, \ldots, 52,$$

and $p_Y(y) = 0$ otherwise. To find $P(Y \geq 1)$, we apply the complementation rule:

$$P(Y \geq 1) = 1 - P(Y < 1) = 1 - P(Y = 0) = 1 - p_Y(0)$$

$$= 1 - \binom{52}{0}(0.029)^0(0.971)^{52} = 0.784.$$

There is roughly a 78.4% chance that the player will win a prize at least once if she buys one ticket per week for a year. ∎

EXAMPLE 5.16 **_Hypergeometric Random Variables_**

Estimating the Size of a Population Suppose that an unknown number, N, of animals inhabit a region and that we want to estimate that number. One procedure for doing so, often referred to as the *capture–recapture method*, is to proceed as follows.

1. Capture M of the animals, mark them in some (nonharmful) way, and then release the animals back into the region and give them time to disperse.

2. Capture n of the animals and note the number X that are marked—that is, the number that are recaptures.

3. Use the data to estimate N.

Explain how this procedure can be used to estimate N, the size of the animal population. Apply the capture–recapture method to estimate N when $M = 50, n = 100$, and $X = 3$.

Solution We begin by obtaining a heuristic estimate of N. After we apply step 1, we have both marked and unmarked animals in the animal population. We take the specified attribute to be "is marked." The *population proportion*—that is, the proportion of marked animals in the entire population—is M/N. After we apply step 2, we can determine the *sample proportion*—the proportion of marked animals in the sample—namely, k/n, where k is the observed value of X. If n isn't too small, the sample proportion should estimate the population proportion—that is, $k/n \approx M/N$. Then we can, as required in step 3, estimate N to be Mn/k:

$$N \approx Mn/k. \tag{5.15}$$

To be more precise, we use the *method of maximum likelihood* to obtain an estimate of N, the size of the animal population. The idea is to choose that value of N for which the observed value of X has the highest probability. Here X has the hypergeometric distribution with parameters N (which is unknown), n, and $p = M/N$. Consequently, referring to the form of the hypergeometric PMF given in Equation (5.13) on page 210, we want to choose N so as to maximize the function

$$L_k(N) = \frac{\binom{M}{k}\binom{N-M}{n-k}}{\binom{N}{n}}, \qquad N \in \mathcal{N},$$

where k is the observed value of X. Using this equation and some straightforward calculations, we get

$$\frac{L_k(N)}{L_k(N-1)} = \frac{(N-M)(N-n)}{N(N-M-n+k)}. \tag{5.16}$$

Adding and subtracting $N(N-M-n+k)$ in the numerator of the fraction on the right side of Equation (5.16), we obtain, after some additional computation,

$$\frac{L_k(N)}{L_k(N-1)} = 1 + \frac{Mn-kN}{N(N-M-n+k)}. \tag{5.17}$$

It follows that

$$L_k(N) > L_k(N-1), \quad \text{if } N < Mn/k;$$
$$L_k(N) < L_k(N-1), \quad \text{if } N > Mn/k.$$

Arguing as in Proposition 5.4 on page 204, we find that $L_k(N)$ attains its maximum value when $N = \lfloor Mn/k \rfloor$, where $\lfloor \ \rfloor$ denotes the floor function. Compare this estimate to the rough one given by Relation (5.15). In any case, the *maximum likelihood estimate* of N, denoted $\widehat{N}$, is the random variable $\widehat{N} = \lfloor Mn/X \rfloor$.

Finally, we apply our result to estimate N when $M = 50$, $n = 100$, and $X = 3$; that is, in step 1, 50 animals are captured, marked, and released; and, in step 2, 100 animals are captured of which 3 are found to be marked. Our maximum likelihood estimate for N is

$$\widehat{N} = \lfloor Mn/X \rfloor = \lfloor 50 \cdot 100/3 \rfloor = \lfloor 1666.67 \rfloor = 1666.$$

Consequently, we estimate that there are 1666 animals in the population.   ∎

The Binomial Approximation to the Hypergeometric Distribution

As Proposition 5.5 shows, the hypergeometric distribution gives the exact probability distribution for the number of successes (members that have a specified attribute) when random sampling is done without replacement from a finite population. If the sample size is small relative to the population size, there is little difference between sampling without replacement and sampling with replacement. In this latter case, the binomial distribution is the correct probability distribution for the number of successes.

Therefore, if the sample size is small relative to the size of the population, a hypergeometric distribution can be approximated by a binomial distribution. More precisely, we have Proposition 5.6 whose proof we leave to you as Exercise 5.70.

◆◆◆ **Proposition 5.6   Binomial Approximation to the Hypergeometric**

Suppose that X has the hypergeometric distribution with parameters N, n, and p. Then

$$\lim_{N \to \infty} p_X(x) = \lim_{N \to \infty} \frac{\binom{Np}{x}\binom{N(1-p)}{n-x}}{\binom{N}{n}} = \binom{n}{x} p^x (1-p)^{n-x},$$

for x = 0, 1, \ldots, n.

Proposition 5.6 implies that, if n is small relative to N, the hypergeometric distribution with parameters N, n, and p can be approximated by the binomial distribution with the same n and p parameters. As a rule of thumb, a hypergeometric distribution can be adequately approximated by a binomial distribution, provided that the sample size doesn't exceed 5% of the population size—that is, $n \le 0.05N$.

EXAMPLE 5.17   *Binomial Approximation to the Hypergeometric*

Construct a table to compare the hypergeometric distribution with parameters $N = 1000$, $n = 8$, and $p = 0.2$ to the binomial distribution with the parameters $n = 8$ and $p = 0.2$.

Solution   Table 5.15 provides a comparison of the two distributions. To obtain a clearer comparison, we rounded the probabilities to four decimal places instead of the usual three.

Table 5.15 Comparison of the hypergeometric and binomial distributions: $N = 1000$, $n = 8$, and $p = 0.2$.

Successes x	Hypergeometric probability	Binomial probability
0	0.1666	0.1678
1	0.3361	0.3355
2	0.2949	0.2936
3	0.1469	0.1468
4	0.0454	0.0459
5	0.0089	0.0092
6	0.0011	0.0011
7	0.0001	0.0001
8	0.0000	0.0000

Table 5.15 reveals that the binomial distribution does a good job of approximating the hypergeometric distribution. In this case, the sample size is 0.8% of the population size, easily satisfying the "at most 5%" rule of thumb. ◼

That a hypergeometric distribution can be approximated by a binomial distribution has important ramifications in statistics. Specifically, in view of Proposition 5.6 and our previous discussions of sampling from a finite population, we can make the following statements.

Sampling and the Binomial Distribution

Suppose that a random sample of size n is taken from a finite population in which the proportion of members having a specified attribute is p. Then the number of members sampled that have the specified attribute has

- *exactly a binomial distribution with parameters n and p if the sampling is done with replacement.*
- *approximately a binomial distribution with parameters n and p if the sampling is done without replacement and the sample size is small relative to the population size.*

For example, according to the Census Bureau publication *Current Population Reports*, 81.7% of U.S. adults have completed high school. Suppose that eight U.S. adults are randomly selected without replacement and let X denote the number of those sampled who have completed high school. The random variable X has the hypergeometric distribution with parameters N, $n = 8$, and $p = 0.817$, where N is the number of U.S. adults (which currently is some 204 million). However, because the sample size doesn't exceed 5% of the population size, we can approximate the probability distribution of X by the binomial distribution with parameters $n = 8$ and $p = 0.817$.

EXERCISES 5.4   Basic Exercises

5.58 Refer to Definition 5.5 on page 211. Show that, for a positive integer m and an integer r,

$$
\binom{m}{r} = \begin{cases} \dfrac{m!}{r!\,(m-r)!}, & \text{if } 1 \le r \le m; \\[2mm] 1, & \text{if } r = 0; \\[2mm] 0, & \text{if } r > m \quad \text{or} \quad r < 0. \end{cases}
$$

5.59 Evaluate these binomial coefficients.

a) $\binom{7}{3}$ b) $\binom{7}{-3}$ c) $\binom{-7}{3}$ d) $\binom{-7}{-3}$ e) $\binom{3}{7}$ f) $\binom{3}{-7}$

g) $\binom{-3}{7}$ h) $\binom{-3}{-7}$ i) $\binom{1/4}{3}$ j) $\binom{1/4}{-3}$ k) $\binom{-1/4}{3}$ l) $\binom{-1/4}{-3}$

m) Can you evaluate $\binom{3}{1/4}$? Explain.

5.60 Let r be a nonnegative integer. Verify these combinatorial identities.

a) $\binom{-1}{r} = (-1)^r$ b) $\binom{-m}{r} = (-1)^r \binom{m+r-1}{r}$, for each $m \in \mathcal{N}$

5.61 Five cards are selected at random without replacement from an ordinary deck of 52 playing cards.
a) What is the probability that exactly 3 face cards are obtained?
b) Identify and provide a formula for the probability distribution of the number of face cards obtained.

5.62 Estimating a population proportion: Many statistical studies are concerned with the proportion of members of a finite population that have a specified attribute, called the *population proportion*. In practice, we mostly rely on sampling and use the sample data to estimate the population proportion. Suppose that a random sample of size n is taken without replacement from a population of size N in which the proportion of members having the specified attribute is p. Intuitively, it makes sense to estimate the population proportion, p, by the *sample proportion*, $\hat{p} = X/n$, where X denotes the number of members sampled that have the specified attribute. Determine the PMF of the random variable $\hat{p}$.

5.63 From 10 pills, of which 5 are placebos, you are to randomly select and take 5 as part of an experiment on the effectiveness of a new treatment. Find the probability that
a) you select at least 2 placebos.
b) the first three pills you select are placebos.

5.64 As reported by Television Bureau of Advertising, Inc., in *Trends in Television*, 84.2% of U.S. households have a VCR. If six U.S. households are randomly selected without replacement, what is the (approximate) probability that the number of households sampled that have a VCR will be
a) exactly four?   b) at least four?   c) at most four?
d) between two and five, inclusive?
e) Determine a formula that approximates the PMF of the random variable Y, the number of households of the six sampled that have a VCR.
f) Strictly speaking, why is the PMF that you obtained in part (e) only approximately correct?
g) Determine a formula that provides the exact PMF of the random variable Y in terms of the number, N, of U.S. households.

5.65 Bin I contains 20 parts, of which 5 are defective. Bin II contains 15 parts, of which 4 are defective. One of these two bins is chosen at random and 3 parts are randomly selected from the bin chosen. If 2 of the 3 parts obtained are defective, what is the probability that Bin I was chosen?

5.66 An upper-level probability class has six undergraduate students and four graduate students. A random sample of three students is taken from the class. Let X denote the number of undergraduate students selected. Identify, obtain a formula for, and tabulate the PMF of the random variable X if the sampling is
a) without replacement.   b) with replacement.
c) Compare your results in parts (a) and (b).

5.67 Refer to Example 5.14 on page 208, where X denotes the number of defective TVs obtained when 5 TVs are randomly selected without replacement from a lot of 100 TVs in which 6 are defective.
a) Construct a table for the PMF of the random variable X similar to Table 5.14 on page 213. Round each probability to eight decimal places.
b) Approximate the PMF by the appropriate binomial distribution. Construct a table similar to Table 5.15 on page 215.
c) Comment on the accuracy here of the binomial approximation to the hypergeometric distribution.

5.68 A dictionary contains 80,000 words, of which a certain student knows 30,000. One hundred of the 80,000 words in the dictionary are chosen at random without replacement.
a) Determine the probability that the student knows exactly 40 of the 100 words obtained.
b) Determine the probability that the student knows at least 35, but no more than 40, of the 100 words obtained.
c) Would a binomial approximation to the hypergeometric distribution be justified in this case? Explain your answer.
d) Another student knows 38 of the 100 words obtained. Estimate the size of that student's vocabulary, assuming that the student knows very few words that aren't in this particular dictionary.

Theory Exercises

5.69 Prove Proposition 5.5 on page 210—that is, derive Equation (5.12) which gives the PMF of a hypergeometric random variable with parameters N, n, and p.

5.70 Prove Proposition 5.6 on page 215, the binomial approximation to the hypergeometric distribution.

5.71 Consider a finite population of size N in which each member is classified as either having or not having a specified attribute; use p to denote the proportion of the population having the specified attribute. Suppose that a random sample of size n is taken from the population and let X denote the number of members sampled that have the specified attribute. Set $q = 1 - p$.
a) If the sampling is with replacement, prove that

$$p_X(x) = \binom{n}{x} \frac{(Np)^x (Nq)^{n-x}}{(N)^n}, \qquad x = 0, 1, \ldots, n,$$

and $p_X(x) = 0$ otherwise.
b) Verify that the PMF in part (a) is that of a binomial distribution with parameters n and p. Thus part (a) gives an alternative form for a binomial PMF in the case of sampling with replacement from a finite population.
c) If the sampling is without replacement, prove that

$$p_X(x) = \binom{n}{x} \frac{(Np)_x (Nq)_{n-x}}{(N)_n}, \qquad x = 0, 1, \ldots, n,$$

and $p_X(x) = 0$ otherwise.
d) Verify that the PMF in part (c) is that of a hypergeometric distribution with parameters N, n, and p. Thus part (c) gives an alternative form for a hypergeometric PMF.

5.72 Let N, M, and n be positive integers with $M \le N$ and $n \le N$.
a) Prove that $\sum_{k=0}^{n} \binom{M}{k}\binom{N-M}{n-k}/\binom{N}{n} = 1$ by applying Vandermonde's identity, presented in Exercise 3.48 on page 109.
b) Establish the identity in part (a) without using Vandermonde's identity or doing any computations, but rather by using a probabilistic argument.
c) Use part (b) to establish Vandermonde's identity.

Advanced Exercises

5.73 An office has 20 employees, 12 men and 8 women. A committee of size 5 that was chosen to organize the office Christmas party has 4 women and 1 man. Do the men have reason to complain that the committee was not randomly selected? Explain your answer.

5.74 Urn I contains eight red balls and four green balls; Urn II contains six red balls and six green balls.

a) Six balls are randomly sampled without replacement from Urn I. Find the probability that the number of red balls in the sample is x.

b) Eight balls are randomly sampled without replacement from Urn II. Find the probability that the number of red balls in the sample is x.

c) For which values of x is the answer to part (a) greater than the answer to part (b), and for which is it smaller? *Hint:* Refer to Exercise 3.106 on page 122.

5.75 One college class has 8 lower-division students and 4 upper-division students; another college class has 80 lower-division students and 40 upper-division students. From each class, 6 students are to be randomly selected without replacement.

a) For each class, identify the set of possible values of the number of lower-division students obtained from among the 6 sampled.

b) For which class is there a greater difference between the probability distributions of the number of lower-division students obtained for sampling with and without replacement? Explain your answer.

5.76 A company receives shipments of parts in lots of 100. A shipment is considered a "good shipment" if and only if it has at most 3% defective parts. To decide whether to reject a shipment, 10 parts are randomly selected from the 100 parts; the shipment is rejected if and only if the number of defectives in the sample is k or larger. Determine the smallest value of k that yields acceptance of good shipments at least 95% of the time.

5.5   Poisson Random Variables

The next family of discrete random variables we consider is the family of *Poisson random variables*. These random variables are named for Siméon-Denis Poisson (1781–1840), who introduced them. (See the biography on page 174 for more about Poisson.)

Poisson distributions comprise one of the most important families of probability distributions. Indeed, as the great twentieth century probabilist William Feller stated in his classic book *An Introduction to Probability Theory and Its Applications* (New York: Wiley, 1968, Vol. I, 3rd ed., p. 156):

> We have here a special case of the remarkable fact that there exist a few distributions of great universality which occur in a surprisingly great variety of problems. The three principal distributions, with ramifications throughout probability theory, are the binomial distribution, the normal distribution, and the Poisson distribution.

The Poisson Approximation to the Binomial Distribution

One way to introduce the Poisson distribution is with regard to the binomial distribution. In doing so, we use a relation from calculus:

$$e^t \approx 1 + t, \qquad \text{for } t \approx 0. \tag{5.18}$$

Let X have the binomial distribution with parameters n and p. For k a fixed nonnegative integer, we examine the behavior of $p_X(k)$ when n (the number of trials) is large and p (the success probability) is small, that is, close to 0.

First we consider the case $k = 0$. In view of Relation (5.18),

$$p_X(0) = \binom{n}{0} p^0 (1 - p)^n = (1 - p)^n \approx (e^{-p})^n = e^{-np}. \tag{5.19}$$

Now we assume that k is positive but relatively small compared to n. From the second line of the proof of Proposition 5.4 on page 204,

$$\frac{p_X(j)}{p_X(j-1)} = \frac{n-j+1}{j} \cdot \frac{p}{1-p},$$

for each $j = 1, 2, \ldots, n$. Therefore, if j is small relative to n,

$$\frac{p_X(j)}{p_X(j-1)} \approx \frac{np}{j},$$

where we have used the fact that, because p is small, $p/(1-p) \approx p$. Taking the product in the preceding display from $j = 1$ to $j = k$ and applying Relation (5.19), we obtain

$$p_X(k) = \frac{p_X(k)}{p_X(0)} p_X(0) = \left(\prod_{j=1}^{k} \frac{p_X(j)}{p_X(j-1)} \right) p_X(0) \approx \left(\prod_{j=1}^{k} \frac{np}{j} \right) e^{-np} = \frac{(np)^k}{k!} e^{-np},$$

or

$$p_X(k) \approx e^{-np} \frac{(np)^k}{k!}. \tag{5.20}$$

Finally, if k isn't small relative to n, both sides of Relation (5.20) are close to 0 and hence to each other. In summary, we have provided a heuristic verification of the **Poisson approximation to the binomial distribution,** presented formally in Proposition 5.7.

◆◆◆   **Proposition 5.7   Poisson Approximation to the Binomial Distribution**

Suppose that X has the binomial distribution with parameters n and p. Then

$$p_X(x) \approx e^{-np} \frac{(np)^x}{x!}, \qquad x = 0, 1, \ldots, n. \tag{5.21}$$

The approximation works well if n is large and p is small. In fact, the approximation is excellent, provided only that np^2 is small.

Note: It can be shown that, if $X \sim \mathcal{B}(n, p)$, then for any subset A of nonnegative integers,

$$\left| P(X \in A) - \sum_{x \in A} e^{-np} \frac{(np)^x}{x!} \right| \le np^2.$$

See, for instance, p. 420 of Sheldon Ross's *A First Course in Probability, fifth edition* (Upper Saddle River, NJ: Prentice-Hall, 1998). In particular, then,

$$\left| p_X(x) - e^{-np} \frac{(np)^x}{x!} \right| \le np^2, \qquad x = 0, 1, \ldots, n, \tag{5.22}$$

which justifies Proposition 5.7.

EXAMPLE 5.18   *Poisson Approximation to the Binomial*

Infant Mortality in Sweden   According to data obtained from the *International Data Base* and published in the *Statistical Abstract of the United States,* the infant mortality rate in Sweden is 4.5 per 1000 live births. Use the Poisson approximation to determine the probability that, of 500 randomly selected live births in Sweden, there are
a) no infant deaths.   b) at most three infant deaths.

Solution   Let X denote the number of infant deaths in 500 live births. Then the random variable X has the binomial distribution with parameters $n = 500$ and $p = 4.5/1000 = 0.0045$. Thus

$$p_X(x) = \binom{500}{x}(0.0045)^x(0.9955)^{500-x}, \qquad x = 0, 1, \ldots, 500. \tag{5.23}$$

We note, however, that $n = 500$ is large, that $p = 0.0045$ is small, and, more important, that $np^2 = 0.010$ is small. Thus, instead of applying Equation (5.23), we can use the Poisson approximation to obtain the required probabilities more easily. Specifically, noting that $np = 2.25$, we can employ the approximation

$$p_X(x) \approx e^{-2.25}\frac{(2.25)^x}{x!}, \qquad x = 0, 1, \ldots, 500. \tag{5.24}$$

a) From Relation (5.24),

$$P(X = 0) = p_X(0) \approx e^{-2.25}\frac{(2.25)^0}{0!} = e^{-2.25} = 0.105.$$

There is roughly a 10.5% chance of no infant deaths in 500 live births.
b) From the FPF and Relation (5.24),

$$P(X \le 3) = \sum_{x \le 3} p_X(x) = p_X(0) + p_X(1) + p_X(2) + p_X(3)$$

$$\approx e^{-2.25}\left(\frac{(2.25)^0}{0!} + \frac{(2.25)^1}{1!} + \frac{(2.25)^2}{2!} + \frac{(2.25)^3}{3!}\right) = 0.809.$$

There is roughly an 81% chance that three or fewer infant deaths will occur in 500 live births.   ■

Referring to Example 5.18, we now illustrate the accuracy of the Poisson approximation to the binomial distribution. We used a computer to obtain both the binomial distribution with parameters $n = 500$ and $p = 0.0045$ and the values of the Poisson approximation $e^{-2.25}(2.25)^x/x!$, as displayed in Table 5.16. Notice that we stopped listing probabilities once they became zero to four decimal places. In any case, Table 5.16 shows how well the Poisson approximates the binomial.

For large n and small p, you can't always use a computer to determine a required binomial distribution—sometimes n is so large or p is so small that even a computer can't handle the computations to obtain a binomial distribution. Nonetheless, the Poisson approximation will still be easy to apply.

Table 5.16 Binomial probabilities and Poisson approximation for $n = 500$ and $p = 0.0045$

Successes x	Binomial probability	Poisson approximation
0	0.1049	0.1054
1	0.2370	0.2371
2	0.2673	0.2668
3	0.2006	0.2001
4	0.1127	0.1126
5	0.0505	0.0506
6	0.0188	0.0190
7	0.0060	0.0061
8	0.0017	0.0017
9	0.0004	0.0004
10	0.0001	0.0001
11	0.0000	0.0000

The Poisson Distribution

From Proposition 5.7, we know that, under certain conditions, binomial probabilities can be well approximated by quantities of the form $e^{-\lambda}\lambda^k/k!$. These quantities are useful in many other contexts.

To begin, we show that, if λ is a positive real number, then the numbers $e^{-\lambda}\lambda^k/k!$, $k = 0, 1, 2, \ldots$, constitute a genuine probability mass function. In other words, the function

$$p(x) = \begin{cases} e^{-\lambda} \dfrac{\lambda^x}{x!}, & \text{if } x = 0, 1, 2, \ldots; \\ 0, & \text{otherwise,} \end{cases} \tag{5.25}$$

satisfies properties (a)–(c) of Proposition 5.1 on page 187. Property (a) obviously holds. For property (b), we note that $\{x \in \mathcal{R} : p(x) \neq 0\}$ consists of the nonnegative integers, which form a countable set. To verify property (c), we first recall from calculus the infinite series for the exponential function:

$$e^t = \sum_{n=0}^{\infty} \frac{t^n}{n!}, \qquad t \in \mathcal{R}. \tag{5.26}$$

From Equations (5.25) and (5.26),

$$\sum_x p(x) = \sum_{x=0}^{\infty} e^{-\lambda} \frac{\lambda^x}{x!} = e^{-\lambda} \sum_{x=0}^{\infty} \frac{\lambda^x}{x!} = e^{-\lambda} e^{\lambda} = 1.$$

Consequently, the function defined by Equation (5.25) is a probability mass function. A random variable with that PMF is called a **Poisson random variable.**

DEFINITION 5.6 Poisson Random Variable

A discrete random variable X is called a **Poisson random variable** if its PMF is of the form

$$p_X(x) = e^{-\lambda} \frac{\lambda^x}{x!}, \qquad x = 0, 1, 2, \ldots, \tag{5.27}$$

and $p_X(x) = 0$ otherwise, where λ is some positive real number. We say that X has the **Poisson distribution with parameter λ.** For convenience, we sometimes write $X \sim \mathcal{P}(\lambda)$ to indicate that X has the Poisson distribution with parameter λ.

Note: As we demonstrate in Chapter 7, the parameter λ of a Poisson random variable represents its average value.

The Poisson distribution is often used to model the frequency with which a specified event occurs during a particular period of time. For instance, we might employ the Poisson distribution when analyzing

- the number of patients that arrive at an emergency room between 6:00 P.M. and 7:00 P.M. Here the specified event is a patient arriving at the emergency room, and the particular period of time is between 6:00 P.M. and 7:00 P.M.

- the number of telephone calls received per day at a switchboard. Here the specified event is a telephone call being received at the switchboard, and the particular period of time is 1 day.

- the number of alpha particles emitted per minute by a radioactive substance. Here the specified event is an alpha particle emission by the radioactive substance, and the particular period of time is 1 minute.

The Poisson distribution is also used to model various spatial phenomena, such as the number of cars located in a particular section of highway at a given time or the number of bacterial colonies appearing in various parts of a petri dish smeared with a bacterial suspension.

EXAMPLE 5.19 *Poisson Random Variables*

Emergency Room Traffic Desert Samaritan Hospital in Mesa, Arizona, keeps records of its emergency-room traffic. From those records, we find that the number of patients arriving between 6:00 P.M. and 7:00 P.M. has a Poisson distribution with parameter $\lambda = 6.9$. Determine the probability that, on a given day, the number of patients arriving at the emergency room between 6:00 P.M. and 7:00 P.M. will be
a) exactly four.
b) at least two.
c) between four and 10, inclusive.

Solution   Let X denote the number of patients arriving between 6:00 P.M. and 7:00 P.M. We know that $X \sim \mathcal{P}(6.9)$; that is, the random variable X has the Poisson distribution with parameter $\lambda = 6.9$. From Definition 5.6, the PMF of X is

$$p_X(x) = e^{-6.9}\frac{(6.9)^x}{x!}, \qquad x = 0, 1, 2, \ldots, \tag{5.28}$$

and $p_X(x) = 0$ otherwise. Using Equation (5.28), we can solve the problems posed in parts (a)–(c).

a) From Equation (5.28),

$$P(X = 4) = p_X(4) = e^{-6.9}\frac{(6.9)^4}{4!} = 0.095.$$

There is a 9.5% chance that exactly four patients will arrive at the emergency room between 6:00 P.M. and 7:00 P.M.

b) The probability of at least two arrivals is most easily obtained by applying the complementation rule and the FPF:

$$P(X \geq 2) = 1 - P(X < 2) = 1 - \sum_{x<2} p_X(x) = 1 - \big(p_X(0) + p_X(1)\big)$$

$$= 1 - \left(e^{-6.9}\frac{(6.9)^0}{0!} + e^{-6.9}\frac{(6.9)^1}{1!}\right) = 0.992.$$

There is a 99.2% chance that at least two patients will arrive at the emergency room between 6:00 P.M. and 7:00 P.M.

c) The probability of between four and 10 arrivals, inclusive, is easily obtained by using the FPF and Equation (5.28):

$$P(4 \leq X \leq 10) = \sum_{4 \leq x \leq 10} p_X(x) = p_X(4) + p_X(5) + \cdots + p_X(10)$$

$$= e^{-6.9}\left(\frac{(6.9)^4}{4!} + \frac{(6.9)^5}{5!} + \cdots + \frac{(6.9)^{10}}{10!}\right) = 0.821,$$

or an 82.1% chance.   ■

The Shape of a Poisson Distribution

It is useful to understand the behavior of the PMF of a Poisson random variable—in particular, the shape of the corresponding probability histogram. In Example 5.19, we discussed the probability distribution of the number of patients arriving at an emergency room between 6:00 P.M. and 7:00 P.M. Specifically, if X denotes that number, X has a Poisson distribution with parameter $\lambda = 6.9$.

The PMF of X is given by Equation (5.28). Using it, we get Table 5.17 and Figure 5.6, which portray the PMF of X in tabular and graphic form, respectively. Note that we stopped presenting probabilities when they become zero to three decimal places.

Table 5.17 Partial PMF of the random variable X, the number of patients arriving at an emergency room between 6:00 P.M. and 7:00 P.M

Number arriving x	Probability $p_X(x)$	Number arriving x	Probability $p_X(x)$	Number arriving x	Probability $p_X(x)$
0	0.001	6	0.151	12	0.025
1	0.007	7	0.149	13	0.013
2	0.024	8	0.128	14	0.006
3	0.055	9	0.098	15	0.003
4	0.095	10	0.068	16	0.001
5	0.131	11	0.043	17	0.001

Figure 5.6 Partial probability histogram for the random variable X, the number of patients arriving at an emergency room between 6:00 P.M. and 7:00 P.M

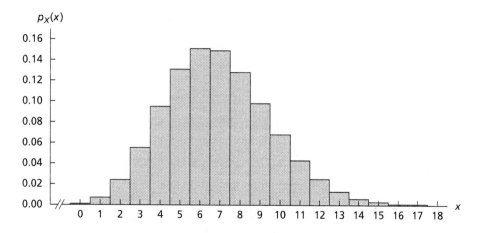

We see from Figure 5.6 that the probability distribution of the number of patients arriving at the emergency room between 6:00 P.M. and 7:00 P.M. is right-skewed. This property holds for all Poisson distributions, as Proposition 5.8 shows. We leave the proof of this proposition to you as Exercise 5.87. Note that Proposition 5.8 also shows that the smaller the value of λ, the greater the skewness.

◆◆◆ Proposition 5.8   Shape of a Poisson Distribution

Suppose that X has the Poisson distribution with parameter λ. Let k represent a non-negative integer and set $m = \lfloor \lambda \rfloor$, where $\lfloor \ \rfloor$ denotes the floor function.

a) If λ is not an integer, then $p_X(k)$ is strictly increasing as k goes from 0 to m and is strictly decreasing as k goes from m to ∞.

b) If λ is an integer, then $p_X(k)$ is strictly increasing as k goes from 0 to $m - 1$, $p_X(m - 1) = p_X(m)$, and $p_X(k)$ is strictly decreasing as k goes from m to ∞.

In either case, p_X attains its maximum value at $x = m = \lfloor \lambda \rfloor$.

Let's apply Proposition 5.8 to the Poisson distribution shown in Figure 5.6. In this case, $\lambda = 6.9$ and, so, according to Proposition 5.8, the probabilities increase until $x = \lfloor 6.9 \rfloor = 6$ and then decrease thereafter. This fact is borne out by both Table 5.17 and Figure 5.6.

EXERCISES 5.5   Basic Exercises

5.77 Suppose that X has the Poisson distribution with parameter $\lambda = 3$. Determine
a) $P(X = 3)$.   **b)** $P(X < 3)$.   **c)** $P(X > 3)$.   **d)** $P(X \leq 3)$.   **e)** $P(X \geq 3)$.

5.78 A paper by L. F. Richardson, published in the *Journal of the Royal Statistical Society,* analyzed the distribution of wars over time. The data indicate that the number of wars that begin during a given calendar year has approximately the Poisson distribution with parameter $\lambda = 0.7$. If a calendar year is selected at random, find the probability that the number of wars that begin during that calendar year will be
a) zero.   **b)** at most two.   **c)** between one and three, inclusive.

5.79 M. F. Driscoll and N. A. Weiss discussed the modeling of motel reservation networks in "An Application of Queuing Theory to Reservation Networks" (*TIMS*, 1976, Vol. 22, pp. 540–546). They defined a Type 1 call to be a call from a motel's computer terminal to the national reservation center. For a certain motel, the number of Type 1 calls per hour has a Poisson distribution with parameter $\lambda = 1.7$. Find the probability that the number of Type 1 calls made from this motel during a period of 1 hour will be
a) exactly one.   **b)** at most two.   **c)** at least two.

Let X denote the number of Type 1 calls made by the motel during a 1-hour period.
d) Construct a table of probabilities for the random variable X. Compute the probabilities until they are zero to three decimal places.
e) Draw a histogram of the probabilities in part (d).

5.80 The second leading genetic cause of mental retardation is Fragile X Syndrome, named for the fragile appearance of the tip of the X chromosome in affected individuals. One in 1500 males are affected worldwide, with no ethnic bias. For a sample of 10,000 males, use the Poisson approximation to the binomial distribution to determine the probability that the number who have Fragile X Syndrome
a) exceeds 7.   **b)** is at most 10.
c) What is the maximum possible error that you made in parts (a) and (b) by using the Poisson approximation to the binomial distribution?

5.81 A most amazing event occurred during the second round of the 1989 U.S. Open at Oak Hill in Pittsford, New York. Four golfers—Doug Weaver, Mark Wiebe, Jerry Pate, and Nick Price—made holes in one on the sixth hole. According to the experts, the odds against a PGA golfer making a hole in one are 3708 to 1—that is, the probability of making a hole in one is 1/3709. Determine the probability to nine decimal places that at least four of the 155 golfers playing the second round would get a hole in one on the sixth hole by using
a) the binomial distribution.   **b)** the Poisson approximation to the binomial.
c) What assumptions are you making in obtaining your answers in parts (a) and (b)? Do you think that those assumptions are reasonable? Explain your reasoning.

5.82 At a service counter, arrivals occur at an average rate of 20 per hour and follow a Poisson distribution. Let X denote the number of arrivals during a particular hour.

a) Using statistical software, we found that $P(X \leq 20) = 0.559$. Based on that result, find $P(X \geq 20)$.

b) If there were at least 20 arrivals during a particular hour, what is the probability that there were exactly 25 arrivals?

5.83 In a certain population, the number of colds a person gets in a year has a $\mathcal{P}(3)$ distribution. A new anti-cold drug lowers λ from 3 to 0.75 and is effective for 8 out of 10 people. At the beginning of last year, the entire population was given the drug. At the end of last year, one person was selected at random from the population and was found to have had only one cold during the year. What is the probability that the drug was effective for this person?

5.84 Characteristic α occurs in about 0.5% of a population. A random sample is to be taken without replacement from the population. How large must the sample size be to ensure that the chances exceed 90% of the sample containing a person with characteristic α? Answer this question by using both the exact distribution and the Poisson approximation. Compare your answers.

5.85 The number of eggs that the female grunge beetle lays in her nest has a $\mathcal{P}(4)$ distribution. Assuming that an empty nest cannot be recognized, determine the PMF of the number of eggs observed in a nest.

Theory Exercises

5.86 Poisson approximation to the binomial: Let λ be a positive real number. Suppose that, for each $n \in \mathcal{N}$, the random variable X_n has the binomial distribution with parameters n and p_n, where $np_n \to \lambda$ as $n \to \infty$.

a) Prove that

$$\lim_{n \to \infty} p_{X_n}(x) = e^{-\lambda} \frac{\lambda^x}{x!}, \qquad x = 0, 1, 2, \ldots .$$

Hint: Use the fact that, if $\{a_n\}_{n=1}^{\infty}$ is a sequence of real numbers that converges to a, then $(1 + a_n/n)^n \to e^a$ as $n \to \infty$.

b) Compare the result in part (a) with Proposition 5.7 on page 220.

5.87 Prove Proposition 5.8 on page 225.

5.88 Let k be a nonnegative integer and let a be a nonnegative real number.

a) Use integration by parts to verify the identity

$$\frac{1}{(k+1)!} \int_a^{\infty} t^{k+1} e^{-t} \, dt = e^{-a} \frac{a^{k+1}}{(k+1)!} + \frac{1}{k!} \int_a^{\infty} t^k e^{-t} \, dt.$$

b) Use part (a) and mathematical induction to prove that $\int_0^{\infty} t^k e^{-t} \, dt = k!$.

c) Let $X \sim \mathcal{P}(\lambda)$. Prove that

$$P(X \leq k) = \frac{1}{k!} \int_{\lambda}^{\infty} t^k e^{-t} \, dt = \frac{1}{k!} \Gamma(k+1, \lambda),$$

where $\Gamma(\alpha, x) = \int_x^{\infty} t^{\alpha-1} e^{-t} \, dt$ is called an *incomplete gamma function*.

d) Let $X \sim \mathcal{P}(\lambda)$. Prove that

$$P(X > k) = \frac{1}{k!} \int_0^{\lambda} t^k e^{-t} \, dt = \frac{1}{k!} \gamma(k+1, \lambda),$$

where $\gamma(\alpha, x) = \int_0^x t^{\alpha-1} e^{-t} \, dt$ is also called an *incomplete gamma function*.

Advanced Exercises

5.89 A hardware manufacturer knows that 1.5% of the bolts produced are defective in some way. In a box of 100 bolts, what is the probability that the number of defectives is
a) 0?   **b)** at most 2?
c) What minimum number of bolts must be placed in a box to ensure that 100 or more of the bolts will be nondefective at least 95% of the time?

5.90 During any particular day, the number of customers that arrive at Grandma's Fudge Shoppe has a $\mathcal{P}(\lambda)$ distribution. Some customers simply watch the fudge makers work and leave without making a purchase. Suppose that an arriving customer makes a purchase with probability p, independent of other arriving customers. Determine and identify the probability distribution of the number of purchasing customers in a day.

5.91 Suppose that X and Y are random variables defined on the same sample space, that $X \sim \mathcal{P}(\lambda)$, and that $Y \sim \mathcal{P}(\mu)$. Suppose further that, for any two nonnegative integers x and y, the events $\{X = x\}$ and $\{Y = y\}$ are independent.
a) Determine and identify the PMF of the random variable $X + Y$. *Hint:* Apply the law of total probability by conditioning on the value of X and also apply the binomial theorem.
b) Find $P(X + Y = 3 \mid X = 1)$.

5.92 Let $X \sim \mathcal{P}(\lambda)$. Determine $P(X \text{ is even})$ in closed form.

5.6 Geometric Random Variables

We next examine the family of *geometric random variables*. To introduce geometric random variables, we present Example 5.20.

EXAMPLE 5.20   *Introduces Geometric Random Variables*

Coin Tossing   Suppose that a coin with probability p of a head is tossed repeatedly. Let X denote the number of tosses until the first head is obtained. Determine the PMF of the random variable X.

Solution   For $j = 1, 2, \ldots$, let A_j denote the event that the jth toss is a head. Then $A_1, A_2, \ldots$ are independent events and $P(A_j) = p$ for each j.

The possible values of the random variable X are the positive integers. For a positive integer k, the event $\{X = k\}$ occurs if and only if the first head is obtained on the kth toss, which means that the first $k - 1$ tosses are tails and the kth toss is a head. We have $\{X = k\} = \left(\bigcap_{j=1}^{k-1} A_j^c \right) \cap A_k$ and, consequently, by independence and the complementation rule,

$$P(X = k) = P\left(\left(\bigcap_{j=1}^{k-1} A_j^c \right) \cap A_k \right) = \left(\prod_{j=1}^{k-1} P(A_j^c) \right) P(A_k)$$

$$= \left(\prod_{j=1}^{k-1} (1 - P(A_j)) \right) P(A_k) = \left(\prod_{j=1}^{k-1} (1 - p) \right) \cdot p = (1 - p)^{k-1} p.$$

Therefore the PMF of the random variable X is

$$p_X(x) = p(1 - p)^{x-1}, \qquad x = 1, 2, \ldots,$$

and $p_X(x) = 0$ otherwise.   ∎

The argument given in Example 5.20 applies to any sequence of Bernoulli trials, not just to coin tossing. Thus we have Proposition 5.9, which provides the PMF of a geometric random variable.

◆◆◆ Proposition 5.9   PMF of a Geometric Random Variable

Consider repeated Bernoulli trials with success probability p. Let X denote the number of trials up to and including the first success. Then the PMF of the random variable X is

$$p_X(x) = p(1 - p)^{x-1}, \qquad x = 1, 2, \ldots, \tag{5.29}$$

and $p_X(x) = 0$ otherwise. The random variable X is called a **geometric random variable** *and is said to have the* **geometric distribution with parameter p.** *We often write $X \sim \mathcal{G}(p)$ to indicate that X has the geometric distribution with parameter p.*

For a geometric random variable with parameter p, the first bar of its probability histogram has height p, the second bar has height $p(1 - p)$, the third bar has height $p(1 - p)^2$, and so on. Thus the bars of the probability histogram decay geometrically by a factor of $1 - p$ from left to right. In this sense, all geometric distributions have basically the same shape. Figure 5.7 shows a (partial) probability histogram of a generic geometric random variable.

Figure 5.7 Partial probability histogram of a geometric random variable

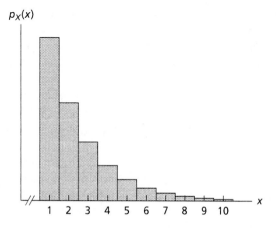

In applying geometric random variables, we often rely on geometric formulas such as

$$\sum_{j=0}^{n} r^j = \frac{1 - r^{n+1}}{1 - r} \quad \text{and} \quad \sum_{j=0}^{\infty} r^j = \frac{1}{1 - r}. \tag{5.30}$$

The first formula holds if $r \neq 1$, whereas the second formula works only for $|r| < 1$. If $|r| < 1$, we obtain the second formula by letting $n \to \infty$ in the first formula. For our applications, we always have $0 < r < 1$ so that both formulas apply.

Other geometric formulas are easily obtained from those in Equations (5.30). For instance, if $r \neq 1$ and a and b are nonnegative integers with $a \leq b$,

$$\sum_{j=a}^{b} r^j = r^a \sum_{j=a}^{b} r^{j-a} = r^a \sum_{j=0}^{b-a} r^j = r^a \cdot \frac{1 - r^{b-a+1}}{1 - r} = \frac{r^a - r^{b+1}}{1 - r}.$$

Also, assuming that $|r| < 1$, we have

$$\sum_{j=a}^{\infty} r^j = r^a \sum_{j=0}^{\infty} r^j = \frac{r^a}{1 - r},$$

for each nonnegative integer a.

Tail Probabilities and Geometric Random Variables

In working with random variables (not just geometric random variables), it is frequently helpful to consider tail probabilities. For a random variable X, its **tail probabilities** are $P(X > x)$, where x is any real number.

A useful identity involving tail probabilities is

$$P(a < X \leq b) = P(X > a) - P(X > b), \tag{5.31}$$

for any real numbers a and b with $a < b$. In words, this equation states that, to find the probability that a random variable takes a value that exceeds a but doesn't exceed b, first find the probability that it exceeds a and then "throw away" the probability that it exceeds b. (Draw a picture!) We leave the rigorous proof to you as Exercise 5.105.

At this point, we concentrate on integer-valued random variables—that is, random variables whose possible values form a subset of the integers. Let X be an integer-valued random variable and n an integer. As X is integer-valued, $\{X = n\} = \{n - 1 < X \leq n\}$ and, consequently, from Equation (5.31),

$$p_X(n) = P(X > n - 1) - P(X > n), \qquad n \in \mathcal{Z}. \tag{5.32}$$

Equation (5.32) shows that we can obtain the PMF of an integer-valued random variable from its tail probabilities.

Now let's find the tail probabilities of a geometric random variable, X. If $n \in \mathcal{N}$, the event $\{X > n\}$ is that the first success occurs after the nth trial. The tail probability $P(X > n)$ is therefore the probability that the first success occurs after the nth trial. Proposition 5.10 provides the tail probabilities for a geometric random variable.

◆◆◆   **Proposition 5.10**   **Tail Probabilities of a Geometric Random Variable**

Suppose that X has the geometric distribution with parameter p. Then

$$P(X > n) = (1 - p)^n, \qquad n \in \mathcal{N}. \tag{5.33}$$

Proof   Equation (5.33) can be proved in several ways. Here we give two proofs.

Method 1: Applying first the FPF and then a geometric formula, we obtain

$$P(X > n) = \sum_{x>n} p_X(x) = \sum_{x=n+1}^{\infty} p(1 - p)^{x-1}$$

$$= p \sum_{x=n}^{\infty} (1 - p)^x = p \cdot \frac{(1 - p)^n}{1 - (1 - p)} = (1 - p)^n.$$

Method 2: Event $\{X > n\}$ occurs if and only if the first success occurs after the nth trial or, equivalently, the first n trials all result in failure. Thus $P(X > n)$ is the probability that the first n trials all result in failure, which is $(1 - p)^n$.   ◆

EXAMPLE 5.21   *Geometric Random Variables*

The Lotto   Refer to Example 5.15 on page 212. As we demonstrated in part (b) of that example, if a player buys one ticket, the probability that she wins a prize is about 0.029. Now suppose that a player repeatedly buys one *Lotto* ticket per week.
a) Find the probability that it takes exactly 1 year for the player to win a prize.
b) Find the probability that it takes no more than 1 year for the player to win a prize.
c) Find the probability that it takes more than 1 year but less than 2 years for the player to win a prize.
d) Determine the number of weeks required for the player to be (at least) 90% sure of winning a prize by that time.

Solution   The lottery results from one week to the next are independent. If we let a success correspond to the player winning a prize on any given week, we have Bernoulli trials with success probability $p = 0.029$. Consequently, the number of weeks, X, until the player wins a prize has the geometric distribution with parameter $p = 0.029$. From Proposition 5.9, the PMF of X is

$$p_X(x) = p(1 - p)^{x-1} = 0.029(0.971)^{x-1}, \qquad x = 1, 2, \ldots, \tag{5.34}$$

and $p_X(x) = 0$ otherwise.
a) From Equation (5.34), the probability that it takes exactly 1 year (52 weeks) for the player to win a prize is

$$P(X = 52) = p_X(52) = 0.029(0.971)^{52-1} = 0.006.$$

b) Here we need to find $P(X \leq 52)$. We can do so by using the FPF, Equation (5.34), and a geometric formula. However, because we have already obtained an expression for the tail probabilities of a geometric random variable, using that expression and the complementation rule is easier. From Proposition 5.10,

$$P(X \leq 52) = 1 - P(X > 52) = 1 - (1 - p)^{52} = 1 - (0.971)^{52} = 0.784.$$

There is roughly a 78.4% chance that the player will win a prize within 1 year.

c) Here we need to find $P(52 < X < 104)$. Again, although we can obtain this probability by using the FPF, Equation (5.34), and a geometric formula, using tail probabilities is easier. From Equation (5.31) on page 230 and Proposition 5.10,

$$P(52 < X < 104) = P(52 < X \leq 103) = P(X > 52) - P(X > 103)$$
$$= (1 - p)^{52} - (1 - p)^{103} = (0.971)^{52} - (0.971)^{103} = 0.168.$$

Chances are about 16.8% that it will take more than 1 year but less than 2 years for the player to win a prize.

d) Here we need to determine the smallest positive integer n that satisfies the inequality $P(X \leq n) \geq 0.90$ or, equivalently, $P(X > n) \leq 0.10$, which, in view of Proposition 5.10, means that $(1 - p)^n \leq 0.10$. Recalling that $p = 0.029$, we must determine the smallest positive integer n such that $(0.971)^n \leq 0.10$. Taking logarithms of both sides of this inequality and solving for n yields $n \geq 78.2$. So the player can be 90% sure that she will win a prize by the 79th week. ∎

Lack of Memory of Geometric Random Variables

Geometric random variables have an interesting and important property called the *lack-of-memory property*. We introduce that property by returning to the *Lotto* example and then develop it formally. Before doing so, however, we need to present some additional probability notation used for random variables.

Let X and Y be random variables defined on the same sample space and let A and B be subsets of real numbers. As you know, $\{X \in A\}$ is the event that X takes a value in A, and $\{Y \in B\}$ is the event that Y takes a value in B. The joint event that X takes a value in A *and* Y takes a value in B is the intersection $\{X \in A\} \cap \{Y \in B\}$. For ease of use, however, it is customary to write the intersection as $\{X \in A, Y \in B\}$, read "X is in A and Y is in B." Just remember to think of the comma in $\{X \in A, Y \in B\}$ as "and."

EXAMPLE 5.22   *Introduces the Lack-of-Memory Property*

The Lotto   Refer to Example 5.21. Suppose that a player buys one ticket per week.

a) Find the probability that it takes exactly 2 weeks for the player to win a prize.
b) Given that it takes more than 3 weeks for the player to win a prize, find the (conditional) probability that it takes exactly 2 additional weeks for the player to win a prize.
c) Compare the answers in parts (a) and (b).

Solution   As we have seen, the probability is 0.029 that, in any particular week, the player wins a prize and, furthermore, the number of weeks, X, until the player wins a prize has the geometric distribution with parameter $p = 0.029$. The PMF of X is presented in Equation (5.34) on page 231.

a) From Equation (5.34),

$$P(X = 2) = p_X(2) = 0.029(0.971)^{2-1} = 0.028.$$

The probability is 0.028 that it takes exactly 2 weeks for the player to win a prize.

b) Given that it takes more than 3 weeks for the player to win a prize, the event that it takes exactly 2 additional weeks for the player to win a prize occurs if and only if it takes exactly 5 weeks for the player to win a prize. Thus we want to determine the conditional probability $P(X = 5 \mid X > 3)$. By the conditional probability rule,

$$P(X = 5 \mid X > 3) = \frac{P(X > 3, \, X = 5)}{P(X > 3)}.$$

The event $\{X > 3, \, X = 5\}$ is identical to the event $\{X = 5\}$ because if $X = 5$, then necessarily $X > 3$. Using Equation (5.34) and Proposition 5.10, we conclude that

$$P(X = 5 \mid X > 3) = \frac{P(X > 3, \, X = 5)}{P(X > 3)} = \frac{P(X = 5)}{P(X > 3)}$$

$$= \frac{0.029(0.971)^{5-1}}{(0.971)^3} = 0.028.$$

The probability is 0.028 that it takes exactly 2 additional weeks for the player to win a prize, given that it takes more than 3 weeks for the player to win a prize.

c) The answers to parts (a) and (b) are identical. This result is no coincidence, as you will soon see. ■

To understand the lack-of-memory property for geometric random variables, recall that, in Bernoulli trials, the trials are independent and the success probability, p, remains the same from trial to trial. Hence, starting at any particular trial, the subsequent process of Bernoulli trials is independent of the prior trials and behaves the same probabilistically as the process of Bernoulli trials starting from the beginning.

Consequently, given that the first success hasn't occurred by the nth trial, the number of trials from the nth trial up to and including the first success has the same probability distribution as the number of trials up to and including the first success starting from the beginning. This property is called the **lack-of-memory property** because, with regard to the number of trials required until the first success, "the process doesn't remember that it has already been running for n trials."

For positive-integer valued random variables, we can express the lack-of-memory property in the form

$$P(X = n + k \mid X > n) = P(X = k), \qquad n, k \in \mathcal{N}. \tag{5.35}$$

In Proposition 5.11, not only do we establish Equation (5.35) for geometric random variables, but we also prove that they are the only positive-integer valued random variables that satisfy that equation. In other words, among positive-integer valued random variables, the lack-of-memory property is unique to geometric random variables.

◆◆◆ **Proposition 5.11** **Lack-of-Memory Property**

A positive-integer valued random variable has the lack-of-memory property if and only if it has a geometric distribution.

Proof   Suppose that X has a geometric distribution with, say, parameter p. Then, by the conditional probability rule and Propositions 5.9 and 5.10 (pages 229 and 231), we have

$$P(X = n + k \mid X > n) = \frac{P(X > n, \, X = n + k)}{P(X > n)} = \frac{P(X = n + k)}{P(X > n)}$$

$$= \frac{p(1 - p)^{n+k-1}}{(1 - p)^n} = p(1 - p)^{k-1} = P(X = k).$$

Hence a geometric random variable has the lack-of-memory property.

Conversely, suppose that X is a positive-integer valued random variable with the lack-of-memory property. Let $p = P(X = 1)$ and let $q_n = P(X > n)$ for each $n \in \mathcal{N}$. We invoke Equation (5.35) with $k = 1$. Applying the conditional probability rule and Equation (5.32) on page 230, we find that, with $k = 1$, the left side of Equation (5.35) is

$$P(X = n + 1 \mid X > n) = \frac{P(X > n, \, X = n + 1)}{P(X > n)}$$

$$= \frac{P(X = n + 1)}{P(X > n)} = \frac{p_X(n + 1)}{q_n} = \frac{q_n - q_{n+1}}{q_n}.$$

Substituting $k = 1$ in the right side of Equation (5.35) gives $P(X = 1)$, or p. Hence $(q_n - q_{n+1})/q_n = p$ or, equivalently,

$$\frac{q_{n+1}}{q_n} = 1 - p, \qquad n \in \mathcal{N}. \tag{5.36}$$

We observe that Equation (5.36) is also true for $n = 0$ because $q_0 = P(X > 0) = 1$ and $q_1 = P(X > 1) = 1 - P(X = 1) = 1 - p$. Taking the product in Equation (5.36) from $n = 0$ to $n = j - 1$, we obtain

$$q_j = \frac{q_j}{q_0} = \prod_{n=0}^{j-1} \frac{q_{n+1}}{q_n} = (1 - p)^j.$$

Therefore, by Equation (5.32) and the previous displayed equation

$$p_X(j) = P(X > j - 1) - P(X > j) = q_{j-1} - q_j$$

$$= (1 - p)^{j-1} - (1 - p)^j = p(1 - p)^{j-1},$$

for each positive integer j. Thus X has a geometric distribution.   ◆

An Alternative Definition of Geometric Random Variables

We have introduced the geometric distribution as the probability distribution of the number of trials, X, up to and including the first success in a sequence of Bernoulli trials. The associated PMF is

$$p_X(x) = p(1 - p)^{x-1}, \qquad x = 1, 2, \ldots, \tag{5.37}$$

and $p_X(x) = 0$ otherwise. Here, of course, p denotes the success probability.

Instead of considering the number of trials, X, up to and including the first success, we can consider the number of trials, Y, before the first success or, equivalently, the number of failures before the first success. We can determine the PMF of Y directly, or we can use Equation (5.37) and the fact that $Y = X - 1$. In any case, we find that the PMF of the random variable Y is

$$p_Y(y) = p(1 - p)^y, \qquad y = 0, 1, \ldots, \tag{5.38}$$

and $p_Y(y) = 0$ otherwise.

Some researchers and textbook authors allude to the geometric distribution as that given by Equation (5.38) instead of Equation (5.37). That is, they define a geometric random variable as the number of failures before the first success rather than the number of trials until the first success. Be sure that you know which form is being considered when you see a reference to a geometric distribution or a geometric random variable.

EXERCISES 5.6   **Basic Exercises**

5.93 According to the *Daily Racing Form,* the probability is about 0.67 that the favorite in a horse race will finish in the money (first, second, or third place). Suppose that you always bet the favorite "across the board," which means that you win something if the favorite finishes in the money. Let X denote the number of races that you bet until you win something.
a) Determine and identify the PMF of the random variable X.
b) Find the probability that the number of races that you bet until you win something is exactly three; at least three; at most three.
c) How many races must you bet to be at least 99% sure of winning something?

5.94 A baseball player has a batting average of .260. Suppose that you observe successive at-bats of the player and note for each at-bat whether the player gets a hit. Presuming that the assumption of Bernoulli trials is appropriate, what is the probability that the first hit by the player occurs
a) on his fifth at-bat?
b) after his fifth at-bat?
c) between his third and tenth at-bats, inclusive?

5.95 Let $X \sim \mathcal{G}(p)$. Determine
a) $P(X \text{ is even})$.   **b)** $P(X \text{ is odd})$.   **c)** $P(2 \leq X \leq 9 \mid X \geq 4)$.
d) $P(X = k \mid X \leq n)$ for $k = 1, 2, \ldots, n$.
e) $P(X = n - k \mid X < n)$ for $k = 1, 2, \ldots, n - 1$.

5.96 Let X be a positive-integer valued random variable. If X has the lack-of-memory property, which of these two equations must hold:

$$P(X = 10 \mid X > 6) = P(X = 4) \qquad \text{or} \qquad P(X = 10 \mid X > 6) = P(X = 10)?$$

Explain.

5.97 Let X have the geometric distribution with parameter p. For k and n positive integers, with $k > n$, determine $P(X = k \mid X > n)$
a) without using the lack-of-memory property.
b) by using the lack-of-memory property.

5.98 Coins I and II have probabilities p_1 and p_2 of a head, respectively. One of these two coins is selected at random and tossed until the first tail occurs. Let X denote the number of tosses required.
a) Determine the PMF of the random variable X.
b) Determine a sufficient condition on p_1 and p_2 that makes X a geometric random variable. In that case, what is the parameter of the geometric distribution for X?

5.99 Consider a finite population of size N in which the proportion of the population that has a specified attribute is p. Suppose that members of the population are randomly sampled one at a time until a member with the specified attribute is obtained. Let X denote the number of members sampled. Determine the PMF of the random variable X if the sampling is
a) with replacement.
b) without replacement.
c) Use the result from part (b) to solve Exercise 4.28 on page 144.

5.100 Tom has a coin with probability p_1 of a head and Dick has a coin with probability p_2 of a head. Tom and Dick alternate tossing their coins, with Tom going first. The first person to get a head wins.
a) Find the probability that Tom wins. *Hint:* First obtain the probability that Tom wins on toss $2n + 1$.
b) Evaluate the probability in part (a) if both coins are balanced.
c) What must be the relationship between p_1 and p_2 to make the game fair?
d) If Dick's coin is balanced, what must be the probability of a head for Tom's coin to make the game fair?

5.101 Refer to Exercise 5.100, but assume the person who goes first is chosen at random.
a) Find the probability that Tom wins.
b) What must be the relationship between p_1 and p_2 to make the game fair?

5.102 Refer to Exercise 5.85 on page 227. Let Y denote the number of nests examined until a nest containing six eggs is observed. Identify the PMF of the random variable Y.

Theory Exercises

5.103 Verify that Equation (5.29) on page 229 really does define a probability mass function.

5.104 Proposition 4.6 on page 155 states that, if E is an event with positive probability and the random experiment is repeated independently and indefinitely, the probability is 1 that event E will eventually occur.
a) Use the forementioned result to deduce that, in Bernoulli trials, the probability is 1 that a success will eventually occur.
b) Use the fact that a geometric PMF really does define a probability mass function to obtain the result of part (a).

5.105 Verify Equation (5.31) on page 230—that is, show that, for any random variable X,

$$P(a < X \leq b) = P(X > a) - P(X > b),$$

for all real numbers a and b with $a < b$.

5.106 Show that a positive-integer valued random variable X satisfies Equation (5.35) on page 233 if and only if it satisfies

$$P(X > n + k \mid X > n) = P(X > k), \qquad n, k \in \mathcal{N}.$$

Note: This equation is sometimes used in place of Equation (5.35) as the criterion for a positive-integer valued random variable to have the lack-of-memory property.

Advanced Exercises

5.107 Let X be a positive-integer valued random variable. Consider the following two equations, where m and n denote positive integers with $m > n$.

$$P(X > m \mid X > n) = P(X > m - n) \tag{$*$}$$
$$P(X > m \mid X > n) = P(X > m) \tag{$**$}$$

a) Which equation is assured to hold by the lack-of-memory property?
b) Is it possible for X to satisfy the equation not assured to hold by the lack-of-memory property? Explain your answer.

5.108 Suppose that X and Y are random variables defined on the same sample space, both having the geometric distribution with parameter p. Suppose further that, for any two positive integers x and y, the events $\{X = x\}$ and $\{Y = y\}$ are independent.
a) Determine the PMF of $X + Y$. *Hint:* Apply the law of total probability by conditioning on the value of X.
b) Does $X + Y$ have a geometric distribution? Explain your answer.
c) Determine and identify the PMF of $\min\{X, Y\}$. *Hint:* Work with tail probabilities.

5.109 Consider Bernoulli trials with success probability p. Let X denote the number of trials until the first success and let Y denote the number of trials from the first success until the second success.
a) Prove that, for any two positive integers x and y, the events $\{X = x\}$ and $\{Y = y\}$ are independent.
b) Obtain the PMF of the random variable $X + Y$. *Hint:* Refer to Exercise 5.108.
c) Interpret the random variable $X + Y$.
d) Use your interpretation in part (c) to obtain the PMF of $X + Y$. Compare your answer with that in part (b).

5.7 Other Important Discrete Random Variables

We have examined some widely used families of discrete random variables: the binomial, the hypergeometric, the Poisson, and the geometric. Although these are some of the most important discrete random variables, many others are worthy of consideration. In this section, we discuss a few of them.

Indicator Random Variables

An *indicator random variable* indicates whether a specified event occurs when a random experiment is performed. More precisely, if E is an event of a sample space, we define the **indicator random variable of E,** denoted I_E, by

$$I_E(\omega) = \begin{cases} 1, & \text{if } \omega \in E; \\ 0, & \text{if } \omega \notin E. \end{cases} \qquad (5.39)$$

Other notations are used for the indicator random variable of an event E. Two common notations are 1_E and χ_E.[†]

An indicator random variable of an event E is quite simple—its possible values are 0 and 1, depending on whether event E doesn't or does occur. Nonetheless, indicator random variables are used extensively in probability theory. Proposition 5.12, whose proof we leave to you as Exercise 5.123, gives the PMF of an indicator random variable.

◆◆◆   **Proposition 5.12   PMF of an Indicator Random Variable**

Let E be an event and let I_E denote its indicator random variable. Then the PMF of the random variable I_E is

$$p_{I_E}(x) = \begin{cases} 1 - P(E), & \text{if } x = 0; \\ P(E), & \text{if } x = 1, \end{cases}$$

and $p_{I_E}(x) = 0$ otherwise.

As you now know, the possible values of an indicator random variable are 0 and 1. Conversely, any random variable, X, whose possible values are 0 and 1 is an indicator random variable—namely, $I_{\{X=1\}}$, the indicator random variable of the event $\{X = 1\}$.

In this regard, we note the relationship between an indicator random variable and a binomial random variable. Specifically, $I_E \sim \mathcal{B}(1, P(E))$—that is, I_E has the binomial distribution with parameters $n = 1$ and $p = P(E)$. Equivalently, I_E has the Bernoulli distribution with parameter $p = P(E)$.

Discrete Uniform Random Variables

Another widely used family of discrete random variables is the family of *discrete uniform random variables.* The term *uniform* refers to the fact that the possible values of such a random variable are equally likely. We introduce discrete uniform random variables in Example 5.23.

EXAMPLE 5.23   *Introduces Discrete Uniform Random Variables*

Rolling a Die   Suppose that a balanced die is rolled once and let X denote the observed number of dots on the side facing up. Obtain the PMF of the random variable X.

Solution   We note that X is a discrete random variable with possible values $1, 2, \ldots, 6$. Moreover, because the die is balanced, each of these possible values has equal probability, $1/6$.

[†] In other mathematical disciplines, indicator functions are sometimes referred to as *characteristic functions.* However, in probability theory, we reserve this latter terminology for another type of function, which we examine briefly in Section 11.1.

Thus the PMF of X is given by

$$p_X(x) = \frac{1}{6}, \qquad x = 1, 2, \ldots, 6,$$

and $p_X(x) = 0$ otherwise.   ■

With Example 5.23 in mind, we now define **discrete uniform random variable**, following which we specify its probability mass function.

DEFINITION 5.7   Discrete Uniform Random Variable

A random variable X is called a **discrete uniform random variable** if its possible values form a finite set S of real numbers and those values are equally likely. We say that X has the **discrete uniform distribution on the set S** or that X is **uniformly distributed on the finite set S**.

Suppose that X has the discrete uniform distribution on the set S. To obtain the PMF of X, we first note that, because X is equally likely to take any value in S, there is a constant c such that $p_X(x) = c$ for $x \in S$ and $p_X(x) = 0$ otherwise. Because a PMF must sum to 1, it follows easily that $c = 1/N(S)$, where $N(S)$ denotes the number of elements of S. Therefore the PMF of X is

$$p_X(x) = \frac{1}{N(S)}, \qquad \text{if } x \in S, \tag{5.40}$$

and $p_X(x) = 0$ otherwise.

Using the FPF and Equation (5.40), we can easily obtain the probability that X takes a value in any specified subset A of real numbers. We have

$$P(X \in A) = \sum_{x \in A} p_X(x) = \sum_{x \in A \cap S} \frac{1}{N(S)} = \frac{N(A \cap S)}{N(S)}. \tag{5.41}$$

In particular, if $A \subset S$, then $P(X \in A) = N(A)/N(S)$. Note the similarity between discrete uniform random variables and the classical probability model for finite sample spaces with equally likely outcomes. See Exercise 5.114 for more about this similarity.

EXAMPLE 5.24   *Discrete Uniform Random Variables*

A Raffle   A charitable organization is conducting a raffle in which the grand prize is a new car. Five thousand tickets, numbered $0001, 0002, \ldots, 5000$, are sold at $10 each. At the grand-prize drawing, one ticket stub will be selected at random from the 5000 tickets stubs. Let X denote the number on the ticket stub obtained.
a) Find the PMF of the random variable X.
b) Suppose that you hold tickets numbered 1003–1025. Express the event that you win the grand prize in terms of the random variable X, and then compute the probability of that event.

Solution   a) The possible values of X are $1, 2, \ldots, 5000$. Because the winning ticket is selected at random, those values are equally likely. Hence the random variable X has a discrete uniform distribution on the set $S = \{1, 2, \ldots, 5000\}$. Its PMF is therefore

$$p_X(x) = \frac{1}{N(S)} = \frac{1}{5000}, \qquad \text{if } x = 1, 2, \ldots, 5000,$$

and $p_X(x) = 0$ otherwise.

b) Let $A = \{1003, 1004, \ldots, 1025\}$. Then the event that you win the prize is $\{X \in A\}$. From Equation (5.41), the probability of that event is

$$P(X \in A) = \frac{N(A \cap S)}{N(S)} = \frac{N(A)}{N(S)} = \frac{23}{5000} = 0.0046.$$

Your chances of winning the car are less than one-half of 1 percent.   ■

Negative Binomial Random Variables

Next we consider the family of *negative binomial random variables,* which includes as a special case the family of geometric random variables. Example 5.25 introduces negative binomial random variables.

EXAMPLE 5.25   *Introduces Negative Binomial Random Variables*

Coin Tossing   Suppose that a coin with probability p of a head is tossed repeatedly. Let X denote the number of tosses until the eighth head is obtained. Determine the PMF of the random variable X.

Solution   The possible values of X are $8, 9, \ldots$. For such a positive integer k, the event $\{X = k\}$ occurs if and only if the eighth head is obtained on the kth toss. That happens if and only if (1) among the first $k - 1$ tosses, exactly seven are heads, and (2) the kth toss is a head. We denote the former event by A and the latter event by B.

We have $\{X = k\} = A \cap B$ and, because the tosses are independent, event A and event B are independent. Hence

$$P(X = k) = P(A \cap B) = P(A)P(B). \tag{5.42}$$

The number of heads in the first $k - 1$ tosses has the binomial distribution with parameters $k - 1$ and p. So, by Proposition 5.3 on page 201, $P(A) = \binom{k-1}{7}p^7(1 - p)^{(k-1)-7}$. Also, the probability that the kth toss is a head is p, so $P(B) = p$. Therefore, from Equation (5.42),

$$P(X = k) = \binom{k - 1}{7}p^7(1 - p)^{(k-1)-7} \cdot p = \binom{k - 1}{7}p^8(1 - p)^{k-8}.$$

Thus the PMF of the random variable X is

$$p_X(x) = \binom{x - 1}{7}p^8(1 - p)^{x-8}, \qquad x = 8, 9, \ldots,$$

and $p_X(x) = 0$ otherwise.   ■

The argument given in Example 5.25 applies to any sequence of Bernoulli trials, not just to coin tossing. Furthermore, there is nothing special about the eighth success; that is, a similar argument can be used to obtain the PMF of the number of trials required until the rth success, where r is any positive integer. Thus we have Proposition 5.13.

◆◆◆ **Proposition 5.13   PMF of a Negative Binomial Random Variable**

Consider repeated Bernoulli trials with success probability p. Let X denote the number of trials up to and including the r th success, where r is a positive integer. Then the PMF of the random variable X is

$$p_X(x) = \binom{x-1}{r-1} p^r (1-p)^{x-r}, \qquad x = r, r+1, \ldots, \tag{5.43}$$

and $p_X(x) = 0$ *otherwise. The random variable X is called a* **negative binomial random variable** *and is said to have the* **negative binomial distribution with parameters r and p.** *For convenience, we sometimes write* $X \sim \mathcal{NB}(r, p)$ *to indicate that X has the negative binomial distribution with parameters r and p.*

Note: The terminology "negative binomial" is used because we can express the PMF in Equation (5.43) in terms of a binomial coefficient containing a negative term. Specifically, if $X \sim \mathcal{NB}(r, p)$, the PMF of X can be written in the alternative form

$$p_X(x) = \binom{-r}{x-r} p^r (p-1)^{x-r}, \qquad x = r, r+1, \ldots, \tag{5.44}$$

and $p_X(x) = 0$ otherwise. We leave the verification of this fact to you as Exercise 5.121.

We stated earlier that the family of negative binomial random variables includes as a special case the family of geometric random variables. A geometric random variable gives the number of trials up to and including the first success, and a negative binomial random variable gives the number of trials up to and including the rth success. Therefore a geometric random variable is also a negative binomial random variable with parameter $r = 1$. We can see this explicitly by setting $r = 1$ in Equation (5.43):

$$\binom{x-1}{1-1} p^1 (1-p)^{x-1} = \binom{x-1}{0} p^1 (1-p)^{x-1} = p(1-p)^{x-1}.$$

The expression on the right is the PMF of a geometric random variable.

The function given by Equation (5.43) really does define a probability mass function; that is, it satisfies properties (a)–(c) of Proposition 5.1 on page 187. The only nontrivial aspect of verifying this fact is to show that the sum of the values equals 1. To do so, we use the **binomial series:**

$$(1+t)^a = \sum_{j=0}^{\infty} \binom{a}{j} t^j, \qquad |t| < 1, \tag{5.45}$$

where a is any real number. We leave the details to you as Exercise 5.125.

EXAMPLE 5.26   *Negative Binomial Random Variables*

Telemarketing   From past experience, it is known that a telemarketer makes a sale with probability 0.2. Assuming that results from one call to the next are independent, determine the probability that the telemarketer makes
a) the second sale on the fifth call.
b) the second sale by the fifth call.
c) the second sale on the fifth call and the fifth sale on the fifteenth call.

Solution   If we let a success correspond to a sale on any particular call, we have Bernoulli trials with success probability $p = 0.2$. Let X denote the call on which the telemarketer makes the second sale. Then $X \sim \mathcal{NB}(2, 0.2)$. From Proposition 5.13, the PMF of X is

$$p_X(x) = \binom{x-1}{2-1}(0.2)^2(1-0.2)^{x-2} = (x-1)(0.2)^2(0.8)^{x-2}, \tag{5.46}$$

for $x = 2, 3, \ldots,$ and $p_X(x) = 0$ otherwise.
a) From Equation (5.46),

$$P(X = 5) = p_X(5) = (5-1)(0.2)^2(0.8)^{5-2} = 0.082.$$

The probability is 0.082 that the telemarketer makes the second sale on the fifth call.
b) To determine $P(X \leq 5)$, we apply the FPF and Equation (5.46):

$$P(X \leq 5) = \sum_{x \leq 5} p_X(x) = \sum_{x=2}^{5}(x-1)(0.2)^2(0.8)^{x-2} = 0.263.$$

The probability is 0.263 that the telemarketer makes the second sale by the fifth call.
c) Here we want to find the probability that the telemarketer (1) makes the second sale on the fifth call and (2) makes the fifth sale on the fifteenth call. Letting A denote the first event and B the second, we want to find $P(A \cap B)$. We use the general multiplication rule, $P(A \cap B) = P(A)P(B \mid A)$, to calculate the required probability.

We have $A = \{X = 5\}$ and hence by part (a), $P(A) = P(X = 5) = 0.082$. Given that event A occurs, event B occurs if and only if among calls 6 through 15 the third sale occurs on the fifteenth call. Because the calls constitute Bernoulli trials, the probability of that happening is the same as the probability of making the third sale on the tenth call. Let Y denote the call on which the telemarketer makes the third sale. Then $Y \sim \mathcal{NB}(3, 0.2)$. Therefore

$$P(B \mid A) = P(Y = 10) = \binom{10-1}{3-1}(0.2)^3(1-0.2)^{10-3} = 0.060.$$

So $P(A \cap B) = P(A)P(B \mid A) = 0.082 \cdot 0.060 = 0.005$. The probability is 0.005 that the telemarketer makes the second sale on the fifth call and the fifth sale on the fifteenth call.   ■

An Alternate Definition of Negative Binomial Random Variables

We have introduced the negative binomial distribution as the probability distribution of the number of trials, X, up to and including the rth success in a sequence of Bernoulli trials. The associated PMF is

$$p_X(x) = \binom{x-1}{r-1} p^r (1-p)^{x-r}, \qquad x = r, r+1, \ldots, \tag{5.47}$$

and $p_X(x) = 0$ otherwise. Here, p denotes the success probability.

Instead of considering the number of trials, X, up to and including the rth success, we can consider the number of failures, Y, that precede the rth success. We can determine the PMF of Y directly or we can use Equation (5.47) and the fact that $Y = X - r$. In any case, we find that the PMF of the random variable Y is

$$p_Y(y) = \binom{r+y-1}{r-1} p^r (1-p)^y, \qquad y = 0, 1, \ldots, \tag{5.48}$$

and $p_Y(y) = 0$ otherwise. Equivalently, using Equation (5.44) on page 241, we can write

$$p_Y(y) = \binom{-r}{y} p^r (p-1)^y, \qquad y = 0, 1, \ldots, \tag{5.49}$$

and $p_Y(y) = 0$ otherwise.

Some researchers and textbook authors allude to the negative binomial distribution as that given by Equation (5.48) or (5.49) instead of Equation (5.47). That is, they define a negative binomial random variable as the number of failures before the rth success rather than the number of trials until the rth success. Be sure that you know which form is being considered when you see a reference to a negative binomial distribution or a negative binomial random variable.

Finally, referring to the right side of Equation (5.49), the binomial series implies that the function defined by

$$p(y) = \binom{-\alpha}{y} p^\alpha (p-1)^y, \qquad y = 0, 1, \ldots, \tag{5.50}$$

and $p(y) = 0$ otherwise, is a probability mass function, provided only that α is a positive real number; it need not be a positive integer. Some researchers and textbook authors refer to the distribution in Equation (5.50) as the negative binomial distribution and, in the special case where α is a positive integer, they call it the **Pascal distribution.**

EXERCISES 5.7   **Basic Exercises**

5.110 Let X be a random variable whose possible values are 0 and 1. Show that X is an indicator random variable—specifically, that $X = I_{\{X=1\}}$.

5.111 Let E and F be events of a sample space.
a) Show that $I_{E \cap F} = I_E \cdot I_F$.
b) Under what conditions is it true that $I_{E \cup F} = I_E + I_F$?
c) Obtain a general formula for $I_{E \cup F}$.

5.112 Suppose that $A_1, A_2, \ldots$ is a countable collection of mutually exclusive events of a sample space. Show that $I_{\bigcup_n A_n} = \sum_n I_{A_n}$.

5.113 Suppose that X has the binomial distribution with parameters n and p.
a) Express X as the sum of n indicator random variables. Interpret your answer.
b) Express X as the sum of n Bernoulli random variables.

5.114 Let $\Omega = \{\omega_1, \ldots, \omega_N\}$ be a finite sample space with equally likely outcomes (i.e., a classical probability model). Define the random variable X on Ω by $X(\omega_k) = k$ for $k = 1, 2, \ldots, N$. Determine and identify the PMF of X.

5.115 Let S consist of the 10 decimal digits. Suppose that a number X is chosen according to the discrete uniform distribution on S and then a number Y is chosen according to the discrete uniform distribution on S with X removed. Determine and identify the PMF of the random variable Y
a) by conditioning on the value of X.
b) by using a symmetry argument.

5.116 According to the *Daily Racing Form*, the probability is about 0.67 that the favorite in a horse race will finish in the money (first, second, or third place). Suppose that you always bet the favorite "across the board," which means that you win something if the favorite finishes in the money. Let X denote the number of races that you bet until you win something three times.
a) Determine and identify the PMF of the random variable X.
b) Find the probability that the number of races that you bet until you win something three times is exactly four; at least four; at most four.

5.117 A baseball player has a batting average of .260. Suppose that you observe successive at-bats of the player and note for each at-bat whether the player gets a hit. Presuming that the assumption of Bernoulli trials is appropriate, what is the probability that the second hit by the player occurs
a) on his fifth at-bat?
b) after his fifth at-bat?
c) between his third and tenth at-bats, inclusive?

5.118 Let X have the negative binomial distribution with parameters r and p. For what values of r does X have the lack-of-memory property? Explain your answer.

5.119 Two balanced dice are rolled until the fourth time a sum of 7 or 11 occurs. What is the probability that it will take more than six rolls?

5.120 Suppose that $X \sim \mathcal{NB}(r, p)$ and that $Y \sim \mathcal{B}(n, p)$.
a) Give a probabilistic argument to show that $P(X > n) = P(Y < r)$.
b) Use the FPF to express the equality in part (a) in terms of PMFs.
c) Using the complementation rule, how many terms of the PMF of X must be evaluated to determine $P(X > n)$?
d) How many terms of the PMF of Y must be evaluated to determine $P(Y < r)$?
e) Use your answers from parts (c) and (d) to comment on the computational savings of using the result of part (a) to evaluate $P(X > n)$ when n is large relative to r.

5.121 Verify that the PMF of a negative binomial random variable with parameters r and p can be expressed in the form given in Equation (5.44) on page 241.

5.122 Refer to Example 5.26 on page 242. Find the probability that the telemarketer makes
a) the second sale by the fifth call and the fifth sale on the fifteenth call.
b) the second sale by the fifth call and the fifth sale by the fifteenth call.

Theory Exercises

5.123 Prove Proposition 5.12 on page 238 by deriving the PMF of an indicator random variable.

5.124 Prove Proposition 5.13 on page 241 by deriving the PMF of a negative binomial random variable with parameters r and p.

5.125 Verify that Equation (5.43) on page 241 defines a probability mass function.

5.126 Proposition 4.6 on page 155 states that, if E is an event with positive probability and the random experiment is repeated independently and indefinitely, the probability is 1 that event E will eventually occur.
a) Use the forementioned result to deduce that, in Bernoulli trials, the probability is 1 that an rth success will eventually occur.
b) Use Exercise 5.125 to obtain the result of part (a).

5.127 Verify that the function given by Equation (5.50) on page 243 is a probability mass function for any positive real number α.

Advanced Exercises

5.128 Banach's matchbox problem: Suppose that a smoker has two matchboxes, one in his left pocket and one in his right. Each matchbox initially contains N matches. When the smoker needs a match, he randomly selects a pocket and then takes a match out of the matchbox in the pocket chosen. At the moment when the smoker finds one of the matchboxes to be empty, let Y denote the number of matches in the other matchbox.
a) Determine the PMF of the random variable Y.
b) Tabulate the values of p_Y when $N = 5$ and when $N = 10$.
c) Repeat part (a) if the smoker chooses his right pocket with probability p.
d) Repeat part (b) if the smoker chooses his right pocket 60% of the time.
e) Repeat part (b) if the smoker chooses his right pocket 40% of the time.

5.129 In the game of ping-pong (table tennis), two players compete. Suppose that the first player to get 21 points wins the game. Jane and Joan are playing ping-pong. At each trial, Jane has probability p of winning 1 point, independently of previous trials.
a) Determine the probability that Jane wins the game. *Hint:* Think of the game being continued until Jane gets 21 points, regardless of how many points Joan gets.
b) Without doing any calculations, use the result of part (a) to explain why

$$\sum_{x=21}^{41} \binom{x-1}{20} 2^{-x} = \frac{1}{2}.$$

5.130 Poisson approximation to the negative binomial: Let λ be a positive real number. Suppose that, for each $r \in \mathcal{N}$, the random variable Y_r has the negative binomial distribution with parameters r and p_r in the form of Equation (5.49) on page 243. Further suppose that $r(1 - p_r) \to \lambda$ as $r \to \infty$.
a) Prove that

$$\lim_{r \to \infty} p_{Y_r}(y) = e^{-\lambda} \frac{\lambda^y}{y!}, \qquad y = 0, 1, 2, \ldots.$$

b) What value must p_r approach as $r \to \infty$? Justify your answer.
c) Interpret the result in part (a).

5.8 Functions of a Discrete Random Variable

In probability theory and its applications, we often consider random variables obtained as functions of other random variables. Example 5.27 introduces some pertinent ideas.

EXAMPLE 5.27   *Introduces Functions of a Discrete Random Variable*

Let X be a discrete random variable with PMF as displayed in Table 5.18. Define the random variable Y by $Y = X^2$, and obtain its PMF.

Table 5.18 PMF of the random variable X

x	-2	-1	0	1	2	3
$p_X(x)$	0.10	0.20	0.20	0.15	0.15	0.20

Solution   Because the possible values of X are $-2, -1, 0, 1, 2$, and 3, the possible values of Y are $0, 1, 4$, and 9. From Table 5.18 and the FPF,

$$p_Y(0) = P(Y = 0) = P(X^2 = 0) = P(X = 0) = P(X \in \{0\})$$
$$= \sum_{x \in \{0\}} p_X(x) = p_X(0) = 0.20$$

and

$$p_Y(1) = P(Y = 1) = P(X^2 = 1) = P(X = \pm 1) = P(X \in \{-1, 1\})$$
$$= \sum_{x \in \{-1, 1\}} p_X(x) = p_X(-1) + p_X(1) = 0.20 + 0.15 = 0.35$$

and

$$p_Y(4) = P(Y = 4) = P(X^2 = 4) = P(X = \pm 2) = P(X \in \{-2, 2\})$$
$$= \sum_{x \in \{-2, 2\}} p_X(x) = p_X(-2) + p_X(2) = 0.10 + 0.15 = 0.25$$

and

$$p_Y(9) = P(Y = 9) = P(X^2 = 9) = P(X = \pm 3) = P(X \in \{-3, 3\})$$
$$= \sum_{x \in \{-3, 3\}} p_X(x) = p_X(-3) + p_X(3) = 0.00 + 0.20 = 0.20.$$

Consequently, we get Table 5.19 and have thus obtained the PMF of the random variable Y, that is, of the random variable X^2.

Table 5.19 PMF of the random variable $Y = X^2$

y	0	1	4	9
$p_Y(y)$	0.20	0.35	0.25	0.20

Referring to Example 5.27, by defining $g(x) = x^2$, we can write $Y = g(X)$. Note that $g(X)$ is the composition of the function g with the function (random variable) X. Thus, as in calculus, we could also write $Y = g \circ X$ but, in probability, it is customary to use the notation $Y = g(X)$. With this notation, we can express generically the computations done in Example 5.27 to obtain the PMF of Y:

$$p_Y(y) = P(Y = y) = P(g(X) = y) = P\big(X \in g^{-1}(\{y\})\big) = \sum_{x \in g^{-1}(\{y\})} p_X(x),$$

where $g^{-1}(\{y\}) = \{ x : g(x) = y \}$.

This same argument applies to any discrete random variable X and any real-valued function g for which the composition $g(X)$ makes sense. Thus we have Proposition 5.14.

◆◆◆   **Proposition 5.14   PMF of a Function of a Discrete Random Variable**

Let X be a discrete random variable and let g be a real-valued function defined on the range of X.[†] Then the PMF of the random variable $Y = g(X)$ is

$$p_Y(y) = \sum_{x \in g^{-1}(\{y\})} p_X(x), \tag{5.51}$$

for y in the range of Y, and $p_Y(y) = 0$ otherwise. In words, if y is in the range of Y, we obtain the probability that $Y = y$—that is, the probability that $g(X) = y$—by summing the PMF of X over all real numbers x such that $g(x) = y$.

Note: We can express Equation (5.51) in the alternate form

$$p_{g(X)}(y) = \sum_{g(x)=y} p_X(x), \tag{5.52}$$

where $\displaystyle\sum_{g(x)=y}$ indicates that the sum is taken over all real numbers x such that $g(x) = y$.

In the special case where g is one-to-one on the range of X, we get the following simplification of Proposition 5.14.

◆◆◆   **Corollary 5.1   PMF of a One-to-One Function of a Discrete Random Variable**

Let X be a discrete random variable and let g be a real-valued function defined and one-to-one on the range of X. Then the PMF of the random variable $Y = g(X)$ is

$$p_Y(y) = p_X(x), \tag{5.53}$$

for y in the range of Y, where x is the unique real number in the range of X such that $g(x) = y$; otherwise, $p_Y(y) = 0$.

[†]Technically, the function g must also be such that the composition $g(X)$ is a random variable in the sense given in the footnote on page 177; that is, $\{ \omega \in \Omega : g(X(\omega)) \le x \}$ must be an event for each real number x.

Proof   To avoid notational confusion, we use a subscript 0 here for particular values. Let y_0 be in the range of Y and let x_0 be the unique real number in the range of X such that $g(x_0) = y_0$. Then, by Proposition 5.14,

$$p_Y(y_0) = \sum_{x \in g^{-1}(\{y_0\})} p_X(x) = \sum_{x \in \{x_0\}} p_X(x) = p_X(x_0),$$

as required.   ◆

Note: We can express Equation (5.53) in the alternate form

$$p_{g(X)}(y) = p_X\left(g^{-1}(y)\right), \tag{5.54}$$

where, in this equation, g^{-1} is the inverse function of g.

EXAMPLE 5.28   *Functions of a Discrete Random Variable*

Bernoulli Trials   For Bernoulli trials with success probability p, let X denote the number of trials up to and including the rth success, where r is a fixed positive integer. As we know, X has a negative binomial distribution with parameters r and p:

$$p_X(x) = \binom{x-1}{r-1} p^r(1-p)^{x-r}, \qquad x = r, r+1, \ldots, \tag{5.55}$$

and $p_X(x) = 0$ otherwise. Let Y denote the number of failures preceding the rth success.
a) Express the random variable Y as a function of the random variable X.
b) Use the result in part (a) to obtain the PMF of Y.

Solution   a) The number of failures preceding the rth success plus r must equal the number of trials up to and including the rth success. In other words, $Y + r = X$. Hence we can express the random variable Y as a function of the random variable X—namely, $Y = X - r$. Alternatively, if we let $g(x) = x - r$, we can write $Y = g(X)$.
b) Because the possible values of X are $r, r+1, \ldots$, the possible values of Y are $0, 1, \ldots$. The function $g(x) = x - r$ is one-to-one on all of $\mathcal{R}$ and hence, in particular, on the range of X. For each real number y, we obtain the unique real number x such that $g(x) = y$ by solving for x in that equation. We have $g(x) = y$ if and only if $x - r = y$ if and only if $x = r + y$. Hence, from Corollary 5.1 and Equation (5.55), the PMF of Y is

$$p_Y(y) = p_X(r+y) = \binom{(r+y)-1}{r-1} p^r(1-p)^{(r+y)-r} = \binom{r+y-1}{r-1} p^r(1-p)^y,$$

for $y = 0, 1, \ldots$, and $p_Y(y) = 0$ otherwise. Compare this result with Equation (5.48) on page 243.   ■

Note: When finding the PMF of a function of a discrete random variable, an important first step is to identify the possible values of the function of the random variable, as done in Examples 5.27 and 5.28.

EXERCISES 5.8   Basic Exercises

5.131 Refer to Example 5.27 on page 246.
a) Determine the PMF of the random variable $Y = 3X$.
b) A transformation of the form $Y = bX$, where b is a positive real number, is called a *change of scale*. Explain why that terminology is used.
c) Does Corollary 5.1 on page 247 apply to a change of scale? Explain your answer.

5.132 Refer to Example 5.27 on page 246.
a) Determine the PMF of the random variable $Y = X - 2$.
b) A transformation of the form $Y = a + X$, where a is a real number, is called a *change of location*. Explain why that terminology is used.
c) Does Corollary 5.1 on page 247 apply to a change of location? Explain your answer.

5.133 For the random variable X in Example 5.27 on page 246,
a) determine the PMF of the random variable $Y = 3X - 2$. *Note:* This transformation involves both scale and location changes.
b) determine the PMF of the random variable $Y = a + bX$, where a is a real number and b is a positive real number.
c) Does Corollary 5.1 on page 247 apply to simultaneous location and scale changes? Explain your answer.

5.134 For the random variable X in Example 5.27 on page 246, determine the PMF of
a) $4X^3 - 1$.   b) X^4.   c) $|X|$.   d) $2|X| + 4$.   e) $\sqrt{3X^2 + 5}$.
f) For which of parts (a)–(e) does Corollary 5.1 on page 247 apply? Explain your answers.

5.135 Suppose that X is a random variable whose range is the set of positive integers. Let $Y = X/(X + 1)$.
a) What is the range of Y?
b) Determine the PMF of Y.
c) Repeat parts (a) and (b) for the random variable $Z = (X + 1)/X$.

5.136 Suppose that X has the geometric distribution with parameter p. For a positive integer m, determine the PMF of the random variable
a) $\min\{X, m\}$.   b) $\max\{X, m\}$.

5.137 If X has the Poisson distribution with parameter $\lambda = 3$, what is the PMF of the random variable $Y = |X - 3|$?

5.138 Suppose that X has the discrete uniform distribution on $\{ k\pi/4 : k \in \{0, 1, \ldots, 8\} \}$. Determine the PMF of
a) $\sin X$.   b) $\cos X$.
c) Is $\tan X$ a random variable? Explain.

5.139 Let X denote the number of trials until the third success in a sequence of Bernoulli trials with success probability p.
a) Express the proportion of failures, Y, in terms of the random variable X.
b) Obtain the PDF of the random variable Y in part (a).

5.140 Suppose that you are observing Bernoulli trials, waiting for the fifth success. You decide to stop observing after the seventeenth trial if the fifth success hasn't yet occurred. Determine the PMF of the number of trials that you observe.

Advanced Exercises

5.141 Let $X \sim \mathcal{G}(p)$ and let $Y = X \pmod 3$, that is, Y is the remainder when X is divided by 3. Determine the PMF of Y
a) by obtaining $p_Y(y)$ individually for each possible value y of Y.
b) by expressing the probability of each possible value of Y in terms of $p_Y(0)$. *Hint:* No infinite series calculations are required here.

CHAPTER REVIEW

Summary

In this chapter, we introduced discrete random variables and examined some of the most important concepts associated with them. Intuitively, a random variable is a quantitative variable whose value depends on chance. But, more precisely, a random variable is a real-valued function whose domain is the sample space of the random experiment. A discrete random variable is a random variable with a countable range—that is, whose possible values form a finite or countably infinite set.

In analyzing and applying random variables, it is useful to employ random-variable notation. For instance, $\{X = x\}$ represents the event that the random variable X takes the value x; that is, $\{X = x\}$ comprises the set of all outcomes ω of the sample space for which $X(\omega) = x$. The probability of the event $\{X = x\}$ is denoted $P(X = x)$. More generally, if A is a subset of real numbers, then $\{X \in A\}$ is the event that X takes a value in A and $P(X \in A)$ is the probability of that event.

The primary object used for studying discrete random variables is the probability mass function (PMF), denoted p_X. The PMF of a discrete random variable X gives the probabilities associated with the possible values of X; thus $p_X(x) = P(X = x)$. Using the PMF and the fundamental probability formula (FPF), we can obtain any probability for a discrete random variable.

We examined four of the most widely used families of discrete random variables: the binomial, hypergeometric, Poisson, and geometric families. We also discussed three other important families of discrete random variables: the indicator, discrete uniform, and negative binomial families.

Finally, we introduced the concept of taking functions of a discrete random variable to form other discrete random variables. Of particular importance is how the PMF of a function of a discrete random variable relates to the PMF of the original random variable.

You Should Be Able To ...

1. define a random variable and a discrete random variable.

2. apply random variable notation.

3. define, state the basic properties of, and apply the probability mass function of a discrete random variable.

4. construct and interpret a probability histogram for a discrete random variable.

5. apply the fundamental probability formula for discrete random variables.

6. define Bernoulli trials.

7. explain and apply binomial random variables.

8. explain and apply hypergeometric random variables.

9. define and use general binomial coefficients.

10. state and use the binomial approximation to the hypergeometric distribution.

11. state and use the Poisson approximation to the binomial distribution.

12. apply Poisson random variables.

13. define, state the basic properties of, and apply tail probabilities.

14. explain and apply geometric random variables, including the lack-of-memory property.

15. explain and apply indicator random variables, discrete uniform random variables, and negative binomial random variables.

16. obtain the PMF of a function of a discrete random variable.

Key Terms

Bernoulli distribution, *201*

Bernoulli random variable, *201*

Bernoulli trials, *196*

binomial coefficient, *211*

binomial distribution, *201*

binomial random variable, *201*

binomial series, *241*

categorical variable, *176*

density function, *185*

discrete density function, *185*

discrete random variable, *178*

discrete uniform distribution, *239*

discrete uniform random variable, *239*

failure, *196*

failure probability, *196*

fundamental probability formula, *190*

geometric distribution, *229*

geometric random variable, *229*

hypergeometric distribution, *210*

hypergeometric random variable, *210*

indicator random variable, *238*

lack-of-memory property, *233*

negative binomial distribution, *241*

negative binomial random variable, *241*

Pascal distribution, *243*

Poisson approximation to the binomial distribution, *220*

Poisson distribution, *223*

Poisson random variable, *223*

probability function, *185*

probability histogram, *185*

probability mass function, *185*

qualitative variable, *176*

quantitative variable, *176*

random variable, *177*

random-variable notation, *178*

success, *196*

success probability, *196*

tail probabilities, *230*

trial, *196*

variable, *176*

Chapter Review Exercises

Basic Exercises

5.142 According to the *Arizona State University Main Facts Book*, a frequency distribution for the number of undergraduate students attending the main campus during one semester, by class level, is as shown in the following table. Here, 1 = freshman, 2 = sophomore, 3 = junior, and 4 = senior.

Class level	1	2	3	4
No. of students	6,159	6,790	8,141	11,220

For the semester under consideration, let X denote the class level of a randomly selected ASU undergraduate.
a) What are the possible values of the random variable X?
b) Use random-variable notation to represent the event that the student selected is a junior.
c) Determine $P(X = 3)$ and interpret your answer in terms of percentages.
d) Determine the PMF of the random variable X.
e) Construct a probability histogram for the random variable X.

5.143 An accounting office has six incoming telephone lines. The PMF of the number of busy lines, Y, is provided in the following table.

y	0	1	2	3	4	5	6
$p_Y(y)$	0.052	0.154	0.232	0.240	0.174	0.105	0.043

Use random-variable notation to express the event that the number of busy lines is
a) exactly four.   b) at least four.   c) between two and four, inclusive.
d) Use the FPF to obtain the probability of each event in parts (a)–(c).

5.144 Suppose that a balanced coin is tossed until the first head appears. Let X denote the number of tosses required.
a) Construct enough of the probability histogram of X to provide an impression of the pattern it follows.
b) In part (a), why you can't you draw the entire probability histogram of X?
c) Find $P(X > 4)$ directly by using the FPF.
d) Find $P(X > 4)$ by first applying the FPF to obtain $P(X \leq 4)$ and then applying the complementation rule.
e) Repeat parts (c) and (d) for $P(X > x)$, where x is a positive integer.
f) Show that, to three significant digits, the probability is 0.415 that X is prime. *Note:* Recall that positive prime numbers are integers greater than 1 that can't be expressed as the product of two positive integers both greater than 1.

5.145 According to *Reader's Digest*, there is a 40% chance that a traffic fatality involves an intoxicated or alcohol-impaired driver or nonoccupant. For brevity, we use the word *drinker* to designate an intoxicated or alcohol-impaired driver or nonoccupant.
a) Suppose that, for a given traffic fatality, we regard a success, s, to be that a drinker is involved. Identify the success probability, p.
b) Construct a table similar to Table 5.11 on page 200—but with three decimal-place numbers in the Probability columns—that shows the possible success–failure results and their probabilities for three traffic fatalities.
c) List the outcomes in which exactly two of three traffic fatalities involve a drinker.
d) Determine the probability of each outcome in part (c). Explain why those probabilities are equal.
e) Use parts (c) and (d) to determine the probability that exactly two of three traffic fatalities involve a drinker.

f) Of three traffic fatalities, let X denote the number that involve a drinker. Identify the probability distribution of the random variable X.

g) Without using the formula for a binomial PMF, but rather by proceeding as in part (e), obtain the PMF of X.

h) Use Procedure 5.1 on page 201 to solve part (g).

5.146 Refer to Exercise 5.145. If traffic fatalities are examined one at a time, obtain the probability that the number examined up to and including the first one that involves a drinker is

a) exactly four.   **b)** at most four.   **c)** at least four.

d) How many traffic fatalities can be examined and still have there be better than a 20% chance of not observing one that involves a drinker?

e) Let Y denote the number of traffic fatalities examined up to and including the first one that involves a drinker. Determine and identify the PMF of the random variable Y.

5.147 A sales representative for a tire manufacturer claims that the company's steel-belted radials last at least 35,000 miles. A tire dealer decides to check that claim by testing eight of the tires. If 75% or more of the eight tires he tests last at least 35,000 miles, he will purchase tires from the sales representative. If, in fact, 90% of the steel-belted radials produced by the manufacturer last at least 35,000 miles, what is the probability that the tire dealer will purchase tires from the sales representative?

5.148 From past experience, the owner of a restaurant knows that, on average, 4% of the parties that make reservations never show. How many reservations can the owner accept and still be at least 80% sure that all parties that make a reservation will show?

5.149 According to The Center for Housing Policy, affordable housing is becoming increasingly scarce in many U.S. cities for working-class families. The study revealed that 76% of the roughly 3-million moderate-income families spend more than half their income on housing. As Michael Stegman, a professor at the University of North Carolina at Chapel Hill and lead author of the study, wrote: "Even families who work and play by the rules don't have a decent place to live." If five moderate-income families in the United States are randomly selected without replacement, what is the (approximate) probability that the number of families spending more than half their income on housing will be

a) exactly three?   **b)** at most three?   **c)** at least four?

d) between one and four, inclusive?

e) Let X denote the number of moderate-income families of five sampled that spend more than half their income on housing. Determine explicitly the (approximate) PMF of the random variable X.

f) Why is the PMF that you obtained in part (e) only approximately correct?

g) Determine a formula for the exact PMF of X in terms of the number, N, of moderate-income families in the United States.

5.150 Suppose that n Bernoulli trials are performed and that exactly k successes are obtained. Determine the probability that a success occurred at trial j ($1 \le j \le n$) by using

a) the conditional probability rule.   **b)** a symmetry argument.

5.151 To estimate the unknown probability that an event E occurs, you independently repeat the random experiment n times and observe the number of times, $n(E)$, that event E occurs in the n trials.

a) Intuitively, how should you estimate $P(E)$ based on observing $n(E)$? *Hint:* Recall the frequentist interpretation of probability.

b) Obtain the maximum likelihood estimate of $P(E)$ based on observing $n(E)$.

c) Compare your results in parts (a) and (b).

5.152 Evaluate the following binomial coefficients.

a) $\binom{9}{4}$ b) $\binom{9}{-4}$ c) $\binom{-9}{4}$ d) $\binom{-9}{-4}$ e) $\binom{4}{9}$ f) $\binom{4}{-9}$

g) $\binom{-4}{9}$ h) $\binom{-4}{-9}$ i) $\binom{1/2}{4}$ j) $\binom{1/2}{-4}$ k) $\binom{-1/2}{4}$ l) $\binom{-1/2}{-4}$

5.153 A random sample of n people is taken from a very large population. The n people sampled will undergo blood tests for a certain disease. The blood test is unfailing, meaning that a person with the disease always tests positive and a person without the disease always tests negative. If the n people are tested separately, then n tests are required. Alternatively, all the individual blood samples could be pooled and tested as a whole. If the test is negative, one test suffices for the entire sample. If the test is positive, each person sampled is tested separately and, therefore, $n + 1$ tests are needed. Let p denote the proportion of people in the population who have the disease. Determine the PMF of the number of tests required under the alternative scheme.

5.154 On average, 6.9 patients arrive at the emergency room of a large hospital each hour. The number who arrive during any particular hour has a Poisson distribution. A doctor arrives for work at the emergency room. Find the probability that
a) the time from the doctor's arrival until the arrival of the first patient exceeds 1 hour.
b) the number of patients arriving during the first hour of the doctor's shift differs from the most probable number by more than 1.
c) at least five patients arrive during the first hour of the doctor's shift.

At a small hospital, an average of 2.6 patients arrive at the hospital emergency room each hour. The number who arrive during any particular hour has a Poisson distribution. A state inspector randomly inspects hospitals. One day, the inspector tosses a balanced coin to decide whether to inspect this small hospital or the large hospital.
d) If the inspector remains in the emergency room for 1 hour, what is the probability that she will observe the arrival of exactly three patients?
e) If during her inspection of the emergency room, the inspector observed the arrival of exactly three patients, what is the probability that she inspected the small hospital?

5.155 An average page in a book contains about 2500 characters, and each character has probability 0.001 of being mistyped. Assuming that typographical errors are independent of one another, do the following.
a) Identify and provide a formula for the probability distribution of the number of typographical errors, X, on a page.
b) Identify and provide a formula for the appropriate Poisson distribution to use in approximating the probability distribution in part (a).
c) Determine the probability of five or more typographical errors on a page by using the probability distribution in part (a); in part (b). Express each of your two answers to four decimal places and compare them.
d) Discuss the assumption that typographical errors are independent of one another.

5.156 A classic study by F. Thorndike on the number of calls to a wrong number appeared in the paper "Applications of Poisson's Probability Summation" (*Bell Systems Technical Journal*, 1926, Vol. 5, pp. 604–624). The study examined the number of calls to a wrong number from coin-box telephones in a large transportation terminal. According to the paper, the number of calls to a wrong number, X, in a 1-minute period has the Poisson distribution with parameter $\lambda = 1.75$. Find the probability that during a 1-minute period the number of calls to a wrong number will be
a) exactly two.
b) between four and six, inclusive.

c) at least one.

d) Obtain a table of probabilities for X, stopping when the probabilities become zero to three decimal places.

e) Use part (d) to construct a partial probability histogram for the random variable X.

f) Identify the shape of the probability distribution of X. Is this shape typical of Poisson distributions? Explain your answer.

5.157 A hand of five-card draw poker consists of 5 cards dealt at random from an ordinary deck of 52 playing cards.

a) Determine to five decimal places the probability of being dealt four of a kind.

b) Use the Poisson approximation to find the probability of being dealt four of a kind exactly twice in 10,000 hands of five-card poker; at least twice.

5.158 Proposition 5.7 on page 220 states that you can approximate a binomial distribution by a Poisson distribution when the number of trials, n, is large and the success probability, p, is small (i.e., near 0). Explain how to use a Poisson distribution to approximate a binomial distribution when n is large and p is large (i.e., near 1).

5.159 A gas station's storage tank has a capacity of m gallons. Once each week the storage tank is filled. The weekly demand, to the nearest gallon, has a Poisson distribution with parameter λ.

a) Determine the probability distribution of weekly gas sales, to the nearest gallon.

b) Determine the probability that the gas station runs out of gas before the end of the week.

5.160 On average, 6.9 patients arrive at a hospital emergency room each hour. The number who arrive during any particular hour has a Poisson distribution. Assume that the following events are independent: exactly seven patients arrive between 12:00 P.M. and 1:00 P.M., exactly seven patients arrive between 1:00 P.M. and 2:00 P.M., exactly seven patients arrive between 2:00 P.M. and 3:00 P.M., and so on. Starting at 12:00 P.M., let X be the number of the first full hour during which exactly seven patients arrive; for instance, if fewer than seven patients arrive between 12:00 P.M. and 1:00 P.M. and exactly seven patients arrive between 1:00 P.M. and 2:00 P.M., then $X = 2$. Determine

a) $P(X = 4)$.

b) $P(X > 6)$.

c) $P(X > 15 \mid X > 9)$.

5.161 Suppose that $X \sim \mathcal{G}(p)$ and that $Y \sim \mathcal{B}(6, p)$. Verify that $P(X > 6) = P(Y = 0)$

a) by using formulas for evaluating these two quantities.

b) by showing that events $\{X > 6\}$ and $\{Y = 0\}$ can be regarded as the same event.

5.162 In Bernoulli trials with success probability p, let X denote the number of trials until the third success.

a) Identify and determine the PMF of the random variable X.

b) How many terms of the PMF of X must be added to obtain $P(X \le 40)$? Without evaluating the result, use the FPF to express $P(X \le 40)$ in terms of the PMF of X.

c) Explain how to obtain $P(X \le 40)$ by adding three terms of the PMF of another random variable and then doing one subtraction. Without evaluating the result, write the required expression explicitly.

d) Compare the amount of work required to evaluate $P(X \le 40)$ in parts (b) and (c).

5.163 At a factory, there is a 40% chance that at least one serious accident occurs during any particular year. Assuming independence of serious accidents from one year to the next, determine the probability that there will be at least 3 years in which no serious accidents occur before the third year in which at least one serious accident occurs.

Advanced Exercises

5.164 Members of a population are numbered $1, 2, \ldots, N$. A random sample of size n is taken with replacement. Determine the PMF of the
a) largest numbered member obtained.   **b)** smallest numbered member obtained.

5.165 Repeat Exercise 5.164 if the sampling is without replacement.

5.166 Suppose that X and Y are random variables defined on the same sample space, that $X \sim \mathcal{B}(n, p)$, and that $Y \sim \mathcal{B}(m, p)$. Suppose further that, for any two nonnegative integers x and y, the events $\{X = x\}$ and $\{Y = y\}$ are independent.
a) Determine and identify the PMF of the random variable $X + Y$. *Hint:* Apply the law of total probability by conditioning on the value of X and also apply Vandermonde's identity (Exercise 3.48 on page 109).
b) Let z be a positive integer at most $n + m$. Given that $X + Y = z$, determine and identify the (conditional) PMF of X by finding $P(X = x \mid X + Y = z)$.

5.167 A lot of 12 items contains exactly 8 nondefectives. A quality control engineer who doesn't know the number of nondefective items must estimate that number based on a sample of six drawn from the lot. Contrast the probability distribution of the number of nondefective items if the sample is drawn with replacement with the probability distribution of the number of nondefective items if the sample is drawn without replacement. Use this contrast to suggest whether a more accurate estimate will result from sampling with or without replacement.

5.168 This problem should be done using a computer or statistical calculator. Refer to Example 5.17 on page 215.
a) If $n = 50$ instead of $n = 8$, would you expect the binomial approximation to be better or worse? Explain your answer.
b) Construct a table similar to Table 5.15 on page 215, but with $n = 50$ instead of $n = 8$.
c) Examine and comment on the accuracy of the binomial approximation considered in part (b), where $n = 50$, compared to that in Example 5.17, where $n = 8$.

5.169 Consider a finite population of size N in which the number of members, M, having a specified attribute is unknown. Suppose that a random sample of size n is taken without replacement from the population. Let X denote the number of members sampled that have the specified attribute.
a) Obtain the maximum likelihood estimate $\hat{M}$ of M based on N, n, and X.
b) Apply your result in part (a) to work Exercise 5.68(d) on page 218.

5.170 Consider Bernoulli trials with success probability p. Let X denote the number of failures before the first success. Suppose that, after observing the value of X, you perform X more Bernoulli trials and let Y be the number of successes in those X Bernoulli trials.
a) Determine the PMF of the random variable Y. *Hint:* Condition on the value of X and use the fact that the values of a negative-binomial PMF sum to 1.
b) Determine and identify the PMF of the random variable $Y + 1$.

5.171 Each manufactured item coming off an assembly line has probability p of being defective and probability q of being inspected, independent of all other items coming off the assembly line. An item is rejected if and only if it is defective and inspected.
a) What is the probability that the nth item is the kth defective one?
b) What is the probability that the nth item is the kth rejected one?
c) Let W denote the number of defective items preceding the first rejected item. Determine the PMF of the random variable W.
d) Determine, identify, and interpret the PMF of the random variable $W + 1$.

e) Repeat parts (b)–(d) if the probability is r that an inspected defective item will be detected as defective and hence rejected.

5.172 A gambler noted that there is 1 chance in 6 of getting a 6 when one balanced die is thrown and that 4 is 2/3 of 6. He also noted that there is 1 chance in 36 of getting a pair of 6s when two balanced dice are thrown and that 24 is 2/3 of 36.

a) The gambler concluded that the probability of getting at least one 6 when one balanced die is thrown four times should be the same as the probability of getting at least one pair of 6s when two balanced dice are thrown 24 times. Is he right? Explain your answer.

b) Determine the probability of getting a 6 at least twice when a balanced die is thrown four times, and the probability of getting a pair of 6s at least twice when two balanced dice are thrown 24 times.

5.173 Consider two experiments: (1) a balanced die is tossed until the first time a "1" appears; (2) a balanced coin is tossed until the first time a head appears. One of these experiments is chosen at random, each having probability 1/2 of being chosen. Let X be the number of tosses of the coin or die.

a) Determine $P(X = 8 \mid X > 5$ and the die experiment was chosen).

b) Determine $P(X = 8 \mid X > 5$ and the coin experiment was chosen).

c) Determine P(the die experiment was chosen $\mid X > 5$).

d) Determine P(the coin experiment was chosen $\mid X > 5$).

e) Determine $P(X = 8 \mid X > 5)$ by using the results of parts (a)–(d) and the conditional form of the law of total probability. *Hint:* Refer to Exercise 4.20 on page 137.

f) Determine $P(X = 3)$.

g) Does the random variable X have the lack-of-memory property? Explain your answer.

h) Not knowing which of the two experiments has been chosen, you are told that five failures have occurred. That is, if the die experiment is being performed, no "1" has appeared in five tosses; if the coin experiment is being performed, no head has appeared in five tosses. Give an intuitive explanation, without mathematical formulas, of why you should feel less sure of a success on the sixth trial, given these five failures, than you were of a success on the first trial.

i) Knowing that the die experiment has been chosen, you have been told that five failures have occurred—that is, that no "1" has appeared in five tosses. Give an intuitive explanation, without mathematical formulas, of why you should feel just as sure of a success on the sixth trial, given these five failures, as you were of a success on the first trial.

5.174 In the long run, the result of tossing a balanced coin is a head half the time and the result of tossing a particular biased coin is a head four-fifths of the time. Your (prior) uncertainty about which of these two coins is the one in your hand is expressed as $P(\text{balanced}) = 0.4$ and $P(\text{biased}) = 0.6$. Two experiments are contemplated for the coin in your hand: (1) toss the coin six times; and (2) toss the coin repeatedly until the third time a head appears. The result of either experiment could be three heads and three tails. Let E denote that event.

a) Find the probability of event E, assuming that you have the balanced coin and the first experiment was performed. Do the same for the biased coin.

b) Find the probability of event E, assuming that you have the balanced coin and the second experiment was performed. Do the same for the biased coin.

c) Find the posterior probability $P(\text{balanced} \mid E)$, assuming that the first experiment was performed.

d) Find the posterior probability $P(\text{balanced} \mid E)$, assuming that the second experiment was performed.

e) Comment on the relationship between the answers to parts (c) and (d).

Blaise Pascal *1623–1662*

Blaise Pascal was born in Clermont, France, on June 19, 1623. He lived in Clermont with his family until the age of 8, when they moved to Paris because of the expanded studying opportunities, both for Pascal and for his father, Etienne. Etienne, a judge by profession, was also interested in mathematics. The "limaçon of Pascal" was named by Roberval (a contemporary geometrician) after Etienne Pascal because of the latter's intensive study of the curve.

Pascal was forbidden by his father to study mathematics because his father feared that it was too rigorous a subject for the frail Blaise. So, until the age of 12, Blaise Pascal was tutored only in languages. Naturally, Pascal was fascinated by the mysterious mathematics; at 12, he gave up his play time to study geometry. His father was properly impressed by the genius that Pascal displayed and gave him Euclid's *Elements* to study.

By the time Pascal was 14 years old, his father allowed him to join weekly meetings with Roberval, Mersenne, and other French geometricians—meetings that eventually evolved into the French Academy. Pascal wrote his first mathematical paper at the age of 16; it was a short essay on conic sections that appeared to be built on the work of Desargues and contained what is now known as *Pascal's theorem*.

Two years later, Pascal designed and built a calculating machine of which he sold about 50 units. In 1653, he employed what has come to be known as *Pascal's triangle*, in part, to determine the number of combinations of m objects taken r at a time. Pascal corresponded regularly with Pierre de Fermat; it was in connection with a problem posed by a gambler, the Chevalier de Méré, to Fermat that Pascal, in 1654, delineated the principles of his theory of probability, the theory for which he is most famous.

Late in 1654, Pascal was involved in a horse and carriage accident which convinced him that he should forego worldly activities and devote himself to religious pursuits. He made only one further, brief foray into the world of mathematics when, in 1658, he wrote an essay on the cycloid. Pascal died at Port Royal on August 19, 1662.

CHAPTER SIX

Jointly Discrete Random Variables

INTRODUCTION

In Chapter 5, we introduced discrete random variables. There we considered only one discrete random variable at a time. However, we frequently need to examine two or more discrete random variables simultaneously.

For instance, two important factors for realtors and home buyers are number of bedrooms and number of bathrooms. Although we can study each of those two discrete random variables separately, it may be more useful to study them jointly in order to discover relationships between them.

In this chapter, we discuss the simultaneous analysis of two or more discrete random variables. We begin in Section 6.1 by examining the joint probability mass function of two discrete random variables. In Section 6.2, we study the general discrete case by considering the joint probability mass function of several discrete random variables.

In Section 6.3, we investigate conditional discrete distributions, considering the conditional PMF of one discrete random variable when the value of another discrete random variable is known. In Section 6.4, we introduce the concept of independence of random variables (of any type) and apply that concept to discrete random variables.

In Section 6.5, we discuss the process of forming a discrete random variable as a function of two or more discrete random variables. As an important special case of this process, we consider, in Section 6.6, forming a discrete random variable by taking the sum of discrete random variables.

6.1 Joint and Marginal Probability Mass Functions: Bivariate Case

In Chapter 5, we introduced the concept of the probability mass function (PMF) of a discrete random variable, which gives the likelihood of each possible value of the random variable. Specifically, if X is a discrete random variable, the PMF of X, denoted p_X, is defined as

$$p_X(x) = P(X = x), \qquad x \in \mathcal{R}. \tag{6.1}$$

Of course, if x isn't a possible value (i.e., isn't in the range) of X, then $p_X(x) = 0$.

To analyze two or more discrete random variables simultaneously, we use the **joint probability mass function.** For the most part, we concentrate on the case of two discrete random variables. The generalization to more than two discrete random variables is conceptually straightforward.

DEFINITION 6.1   Joint Probability Mass Function: Bivariate Case

Let X and Y be two discrete random variables defined on the same sample space. Then the **joint probability mass function** of X and Y, denoted $p_{X,Y}$, is the real-valued function defined on $\mathcal{R}^2$ by

$$p_{X,Y}(x, y) = P(X = x, Y = y). \tag{6.2}$$

We often abbreviate "joint probability mass function" as **joint PMF.**

Note the following.

- The condition that X and Y are defined on the same sample space means that the two random variables simultaneously provide numerical information about the outcome of the same random experiment.
- The joint PMF of two discrete random variables is represented by a lowercase p subscripted with the two letters representing the two discrete random variables. For instance, $p_{X,Y}$ denotes the joint PMF of the discrete random variables X and Y; $p_{U,V}$ denotes the joint PMF of the discrete random variables U and V.
- $\{X = x, Y = y\}$ is the intersection of the events $\{X = x\}$ and $\{Y = y\}$. That is, $\{X = x, Y = y\}$ is the event that X takes the value x *and* Y takes the value y.
- If either x isn't a possible value of X or y isn't a possible value of Y, then we have $\{X = x, Y = y\} = \emptyset$, so that $p_{X,Y}(x, y) = P(X = x, Y = y) = P(\emptyset) = 0$.

We can visually portray the joint PMF of two discrete random variables by constructing a **joint probability histogram**—a three-dimensional graph that displays the possible pairs of values of the two discrete random variables in the x-y plane and the probabilities of those values on the z-axis. The probability of each pair of values is represented by a vertical bar whose height is equal (or proportional) to that probability.

EXAMPLE 6.1   *Joint Probability Mass Functions*

Bedrooms and Bathrooms   Data supplied by *Realty 2000/Better Homes and Gardens* of Prescott, Arizona, resulted in the contingency table displayed as Table 6.1 for number of bedrooms and number of bathrooms for 50 homes currently for sale. Suppose that one of these 50 homes is selected at random. Let X and Y denote the number of bedrooms and number of bathrooms, respectively, of the home obtained.
a) Determine $p_{X,Y}(3, 2)$.
b) Obtain the joint PMF of X and Y.

Table 6.1 Contingency table for number of bedrooms and number of bathrooms for 50 homes currently for sale

	Bathrooms, y				
Bedrooms, x	2	3	4	5	Total
2	3	0	0	0	3
3	14	12	2	0	28
4	2	11	5	1	19
Total	19	23	7	1	50

Solution   Because a home is being selected at random, each of the 50 homes is equally likely to be the one obtained. In other words, a classical probability model is appropriate here. So, for each event E,

$$P(E) = \frac{N(E)}{N(\Omega)} = \frac{N(E)}{50}.$$

a) Table 6.1 indicates that $\{X = 3,\ Y = 2\}$—the event that the home selected has three bedrooms and two bathrooms—can occur in 14 ways, namely, if any one of the 14 homes with three bedrooms and two bathrooms is the home selected. Therefore

$$p_{X,Y}(3, 2) = P(X = 3,\ Y = 2) = \frac{N(\{X = 3,\ Y = 2\})}{50} = \frac{14}{50} = 0.28.$$

The probability is 0.28 that the home selected has three bedrooms and two bathrooms.
b) To obtain the joint PMF of X and Y, we need to compute $p_{X,Y}(x, y)$ for each of the possible pairs of values of X and Y. In part (a), we did so for the possible pair of values 3 and 2. Similarly, we determine $p_{X,Y}(x, y)$ for the other possible pairs of x-y values. The results are presented within the heavy lines of Table 6.2.   ∎

Proposition 5.1 on page 187 provides the basic properties of the PMF of one discrete random variable. Proposition 6.1, on the next page, gives them for the joint PMF of two discrete random variables. The proof of Proposition 6.1 is quite similar to that of Proposition 5.1. We leave its verification to you as Exercise 6.16.

Table 6.2 Joint and marginal PMFs of the discrete random variables X and Y, the number of bedrooms and number of bathrooms of a randomly selected home

		Bathrooms, y				
		2	3	4	5	$p_X(x)$
Bedrooms, x	2	0.06	0.00	0.00	0.00	0.06
	3	0.28	0.24	0.04	0.00	0.56
	4	0.04	0.22	0.10	0.02	0.38
	$p_Y(y)$	0.38	0.46	0.14	0.02	1.00

◆◆◆ **Proposition 6.1   Basic Properties of a Joint PMF: Bivariate Case**

The joint probability mass function $p_{X,Y}$ of two discrete random variables X and Y satisfies the following three properties.

a) *$p_{X,Y}(x, y) \geq 0$ for all $(x, y) \in \mathcal{R}^2$; that is, a joint PMF is a nonnegative function.*

b) *$\{(x, y) \in \mathcal{R}^2 : p_{X,Y}(x, y) \neq 0\}$ is countable; that is, the set of pairs of real numbers for which a joint PMF is nonzero is countable.*

c) *$\sum \sum_{(x,y)} p_{X,Y}(x, y) = 1$; that is, the sum of the values of a joint PMF equals 1.[†]*

In Exercise 6.17, we ask you to show that a function $p: \mathcal{R}^2 \to \mathcal{R}$ that satisfies properties (a), (b), and (c) of Proposition 6.1 is the joint PMF of some pair of discrete random variables. Therefore, for any such function, we can say, "Let X and Y be discrete random variables with joint PMF p." This statement makes sense regardless of whether we explicitly give X and Y and the sample space on which they are defined.

Marginal Probability Mass Functions

In Example 6.1, as we already noted, the values of the joint PMF, $p_{X,Y}(x, y)$, fall inside the heavy lines in Table 6.2. For instance, $p_{X,Y}(4, 3) = 0.22$.

Let's now examine the values that fall outside those heavy lines. The row and column heads "Total" in Table 6.1 were changed in Table 6.2 to $p_Y(y)$ and $p_X(x)$, respectively. Thus, in Table 6.2, the last row gives the PMF of Y, and the last column gives the PMF of X. In this context, for purposes of clarity and because these two probability mass functions are in the margins of the table, the PMF of each is called a **marginal probability mass function**. Technically, however, the adjective "marginal" is redundant.

Note that the sum of the values of the joint PMF in a row or column equals the value of the marginal PMF at the end (right or bottom) of that row or column. For instance, the sum of the values of the joint PMF in the column labeled "3" is $0.00 + 0.24 + 0.22 = 0.46$, which is the value of the marginal PMF at the bottom of that column.

[†]The expression $\sum \sum_{(x,y)}$ means that we are summing over all points (x, y) in $\mathcal{R}^2$.

To see why, we first observe that, because the possible values of X are 2, 3, and 4, the events $\{X = 2\}$, $\{X = 3\}$, and $\{X = 4\}$ form a partition of the sample space. Therefore, by the law of partitions (Proposition 2.8 on page 68), we have

$$p_Y(3) = P(Y = 3) = \sum_{x=2}^{4} P(X = x,\ Y = 3) = \sum_{x=2}^{4} p_{X,Y}(x, 3),$$

as required. Thus we can obtain each marginal probability by summing the appropriate row or column of joint probabilities. More generally, we have Proposition 6.2 whose proof, which we leave to you as Exercise 6.18, is essentially the same as the argument just given.

◆◆◆ **Proposition 6.2**   **Obtaining Marginal PMFs From the Joint PMF**

Let X and Y be discrete random variables defined on the same sample space. Then

$$p_X(x) = \sum_{y} p_{X,Y}(x, y), \qquad x \in \mathcal{R}, \tag{6.3}$$

and

$$p_Y(y) = \sum_{x} p_{X,Y}(x, y), \qquad y \in \mathcal{R}. \tag{6.4}$$

In words, we can obtain the (marginal) PMF of X by summing on y the joint PMF of X and Y and, likewise, we can obtain the (marginal) PMF of Y by summing on x the joint PMF of X and Y.

EXAMPLE 6.2   *Obtaining Marginal PMFs From the Joint PMF*

Lifetimes of Electrical Components   Consider two identical electrical components. Let X and Y denote the respective lifetimes of the two components observed at discrete time units (e.g., every hour). Assume that the joint PMF of X and Y is

$$p_{X,Y}(x, y) = p^2(1 - p)^{x+y-2}, \qquad x, y \in \mathcal{N},$$

and $p_{X,Y}(x, y) = 0$ otherwise, where $0 < p < 1$. Find and identify the marginal PMFs of X and Y.

Solution   The possible values of X are the positive integers, so we can concentrate on values of $x \in \mathcal{N}$. From Equation (6.3), the fact that $p_{X,Y}(x, y) = 0$ unless y is a positive integer, and a geometric series formula, we obtain

$$p_X(x) = \sum_{y} p_{X,Y}(x, y) = \sum_{y=1}^{\infty} p^2(1 - p)^{x+y-2} = p^2(1 - p)^{x-2} \sum_{y=1}^{\infty} (1 - p)^y$$

$$= p^2(1 - p)^{x-2} \cdot \frac{1 - p}{1 - (1 - p)} = p(1 - p)^{x-1},$$

for each $x \in \mathcal{N}$. Consequently, $p_X(x) = p(1 - p)^{x-1}$ for $x = 1, 2, \ldots$, and $p_X(x) = 0$ otherwise. Thus X has the geometric distribution with parameter p.

Likewise, applying Equation (6.4) yields $p_Y(y) = p(1 - p)^{y-1}$ for $y = 1, 2, \ldots$, and $p_Y(y) = 0$ otherwise. Thus Y also has the geometric distribution with parameter p.   ■

Marginal PMFs Don't Determine the Joint PMF

Proposition 6.2 implies that the joint PMF of two discrete random variables determines the marginal PMFs—if we know the joint PMF of X and Y, we can use it to obtain both the marginal PMF of X and the marginal PMF of Y. Thus two pairs of discrete random variables that have the same joint PMF must have the same marginal PMFs.

What about the converse? Do the marginal PMFs determine the joint PMF—if we know the marginal PMF of X and the marginal PMF of Y, can we use them to obtain the joint PMF of X and Y? If two pairs of discrete random variables have the same marginal PMFs, must they have the same joint PMF? In general, the answer is *no*, as Example 6.3 shows.

EXAMPLE 6.3   *Marginal PMFs Don't Determine the Joint PMF*

Bedrooms and Bathrooms   Table 6.3 provides hypothetical data for the number of bedrooms and number of bathrooms for 5000 homes.

Table 6.3 Contingency table for number of bedrooms and number of bathrooms for 5000 homes (hypothetical data)

Bathrooms, v

Bedrooms, u	2	3	4	5	Total
2	114	138	42	6	300
3	1064	1288	392	56	2800
4	722	874	266	38	1900
Total	1900	2300	700	100	5000

Suppose that one of these 5000 homes is selected at random. Let U and V denote the number of bedrooms and number of bathrooms, respectively, of the home obtained.
a) Determine the joint PMF and marginal PMFs of the random variables U and V.
b) Show that U and V have the same (marginal) PMFs as X and Y, respectively, where X and Y are the random variables defined in Example 6.1 on page 261.
c) Show that the joint PMF of U and V is different from the joint PMF of X and Y.

Solution   a) Proceeding as in Example 6.1, we obtain Table 6.4, which gives the joint and marginal probability mass functions of the random variables U and V.
b) Comparing Table 6.4 with Table 6.2 on page 262 reveals that U and X have the same (marginal) PMF and that V and Y have the same (marginal) PMF.
c) Again comparing Table 6.4 with Table 6.2 shows that the joint PMF of U and V isn't the same as the joint PMF of X and Y.   ∎

Table 6.4 Joint and marginal probability mass functions of U and V

Bathrooms, v

		2	3	4	5	$p_U(u)$
Bedrooms, u	2	0.0228	0.0276	0.0084	0.0012	0.0600
	3	0.2128	0.2576	0.0784	0.0112	0.5600
	4	0.1444	0.1748	0.0532	0.0076	0.3800
$p_V(v)$		0.3800	0.4600	0.1400	0.0200	1.0000

From Example 6.3, we conclude that, in general, marginal PMFs don't determine the joint PMF—two pairs of discrete random variables with the same marginal PMFs can have different joint PMFs. Consequently, we can't, in general, obtain the joint PMF of two discrete random variables by knowing only their marginal PMFs.

FPF for Two Discrete Random Variables

As for the case of one discrete random variable, once we have the joint PMF of two discrete random variables, we can determine any probability involving those two random variables. Specifically, we again have a **fundamental probability formula** (FPF). The proof of the FPF for two discrete random variables (Proposition 6.3) is quite similar to that of the FPF for one discrete random variable (Proposition 5.2 on page 190). We leave its verification to you as Exercise 6.19.

◆◆◆   **Proposition 6.3   Fundamental Probability Formula: Bivariate Discrete Case**

Suppose that X and Y are two discrete random variables defined on the same sample space. Then, for any subset $A \subset \mathcal{R}^2$, we have

$$P\big((X, Y) \in A\big) = \sum_{(x,y) \in A} \sum p_{X,Y}(x, y). \tag{6.5}$$

In words, the probability that a pair of discrete random variables takes a value in a specified subset of the plane can be obtained by summing the joint PMF of the two random variables over that subset.

EXAMPLE 6.4   *The Fundamental Probability Formula*

Bedrooms and Bathrooms   Refer to Example 6.1 on page 261. Suppose that one of the 50 homes is selected at random. Determine the probability that the home obtained
a) has the same number of bedrooms and bathrooms.
b) has more bedrooms than bathrooms.

Solution   As previously, we let X and Y denote the number of bedrooms and number of bathrooms, respectively, of the home obtained.

a) We want to find $P(X = Y)$. From the FPF and the joint PMF of X and Y given in Table 6.2 on page 262,

$$P(X = Y) = \sum_{x=y}\sum p_{X,Y}(x, y)$$
$$= p_{X,Y}(2, 2) + p_{X,Y}(3, 3) + p_{X,Y}(4, 4)$$
$$= 0.06 + 0.24 + 0.10$$
$$= 0.40.$$

The probability is 0.40 that the home obtained has the same number of bedrooms and bathrooms. In terms of percentages, 40% of the 50 homes have the same number of bedrooms and bathrooms.

b) We want to find $P(X > Y)$. From the FPF and the joint PMF of X and Y given in Table 6.2,

$$P(X > Y) = \sum_{x>y}\sum p_{X,Y}(x, y)$$
$$= p_{X,Y}(3, 2) + p_{X,Y}(4, 2) + p_{X,Y}(4, 3)$$
$$= 0.28 + 0.04 + 0.22$$
$$= 0.54.$$

The probability is 0.54 that the home obtained has more bedrooms than bathrooms. In terms of percentages, 54% of the 50 homes have more bedrooms than bathrooms.   ∎

EXAMPLE 6.5   *The Fundamental Probability Formula*

Lifetimes of Electrical Components   Refer to Example 6.2 on page 263, where we considered the lifetimes X and Y of two identical electrical components. Determine the probability that

a) both electrical components last longer than 4 time units.

b) one of the electrical components lasts at least twice as long as the other.

Solution   We recall that the joint PMF of X and Y is

$$p_{X,Y}(x, y) = p^2(1 - p)^{x+y-2}, \qquad x, y \in \mathcal{N},$$

and $p_{X,Y}(x, y) = 0$ otherwise, where $0 < p < 1$.

a) Here we want $P(X > 4, Y > 4)$. Applying the FPF, we get

$$P(X > 4, Y > 4) = \sum_{x>4,\, y>4}\sum p_{X,Y}(x, y). \tag{6.6}$$

Figure 6.1 shows the set over which the double summation is taken.

Figure 6.1

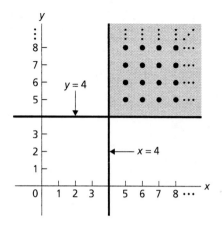

There are several ways to evaluate the double sum in Equation (6.6). Here is the most straightforward way, although not the quickest. Iterating the sums and applying geometric series formulas twice, we get

$$\sum_{x>4,\,y>4}\sum p_{X,Y}(x,y) = \sum_{x=5}^{\infty}\left(\sum_{y=5}^{\infty} p^2(1-p)^{x+y-2}\right)$$

$$= \frac{p^2}{(1-p)^2}\sum_{x=5}^{\infty}(1-p)^x\left(\sum_{y=5}^{\infty}(1-p)^y\right)$$

$$= \frac{p^2}{(1-p)^2}\sum_{x=5}^{\infty}(1-p)^x \cdot \frac{(1-p)^5}{1-(1-p)}$$

$$= p(1-p)^3\sum_{x=5}^{\infty}(1-p)^x$$

$$= p(1-p)^3 \cdot \frac{(1-p)^5}{1-(1-p)}$$

$$= (1-p)^8.$$

The probability is $(1-p)^8$ that both components last longer than 4 time units.

b) The event that one of the electrical components lasts at least twice as long as the other can be expressed as $\{X \geq 2Y\} \cup \{Y \geq 2X\}$. The two events in the union are mutually exclusive and, by symmetry of the PMF, have the same probability. Hence the required probability equals $2P(X \geq 2Y)$. Applying the FPF yields

$$P(X \geq 2Y) = \sum_{x\geq 2y}\sum p_{X,Y}(x,y).$$

Figure 6.2 shows the set over which the double summation is taken.

Figure 6.2

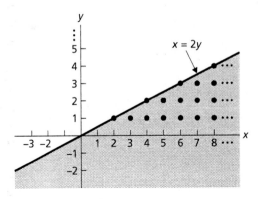

In view of Figure 6.2, we can evaluate the double sum as follows:

$$\sum_{x \geq 2y}\sum p_{X,Y}(x, y) = \sum_{y=1}^{\infty}\left(\sum_{x=2y}^{\infty} p^2(1 - p)^{x+y-2}\right)$$

$$= \frac{p^2}{(1 - p)^2}\sum_{y=1}^{\infty}(1 - p)^y\left(\sum_{x=2y}^{\infty}(1 - p)^x\right)$$

$$= \frac{p^2}{(1 - p)^2}\sum_{y=1}^{\infty}(1 - p)^y \cdot \frac{(1 - p)^{2y}}{1 - (1 - p)}$$

$$= \frac{p}{(1 - p)^2}\sum_{y=1}^{\infty}\left((1 - p)^3\right)^y = \frac{p}{(1 - p)^2} \cdot \frac{(1 - p)^3}{1 - (1 - p)^3}$$

$$= \frac{p(1 - p)}{1 - (1 - p)^3} = \frac{1 - p}{3 - 3p + p^2}.$$

The probability that one of the components lasts at least twice as long as the other is $2(1 - p)/(3 - 3p + p^2)$. ∎

EXERCISES 6.1   Basic Exercises

6.1 Refer to Examples 6.2 and 6.5 on pages 263 and 266, respectively, where X and Y denote the lifetimes of two identical electrical components observed at discrete time units. The joint PMF of X and Y is

$$p_{X,Y}(x, y) = p^2(1 - p)^{x+y-2}, \qquad x, y \in \mathcal{N},$$

and $p_{X,Y}(x, y) = 0$ otherwise, where $0 < p < 1$.
a) Determine and interpret $P(X = Y)$.
b) Use the FPF to determine $P(X > Y)$. Interpret your answer.
c) Without doing any computations, explain why $P(X > Y) = P(X < Y)$.
d) Use the results of parts (a) and (c) to obtain $P(X > Y)$. Compare your answer to that obtained in part (b).

6.2 In Example 5.1 on page 176, we considered the number of siblings for each of 40 students in one of Professor Weiss's classes. The following contingency table cross classifies the number of siblings and number of sisters of the 40 students.

Sisters, y

		0	1	2	3	Total
	0	8	0	0	0	8
	1	8	9	0	0	17
Siblings, x	2	1	8	2	0	11
	3	0	1	2	0	3
	4	0	0	0	1	1
	Total	17	18	4	1	40

Suppose that one of the 40 students is selected at random. Let X and Y denote the number of siblings and number of sisters, respectively, of the student obtained.
a) Determine $p_{X,Y}(2, 1)$ and interpret your answer.
b) Obtain the joint PMF of the random variables X and Y.
c) Construct a table similar to Table 6.2 on page 262.
d) Explain why the marginal PMF of X obtained in part (c) is the same as the PMF in Table 5.5 on page 186.
e) Use random variable notation to express the event that the student obtained has no brothers and then determine the probability of that event.
f) Use random variable notation to express the event that the student obtained has no sisters and then determine the probability of that event.

6.3 Let S consist of the 10 decimal digits. Suppose that a number X is chosen according to the discrete uniform distribution on S and then a number Y is chosen according to the discrete uniform distribution on S with X removed.
a) Obtain the joint PMF of X and Y. *Hint:* Use the general multiplication rule.
b) Determine $P(X = Y)$ first by using the FPF and then without doing any computations.
c) Determine $P(X > Y)$ first by using the FPF and then without doing any computations.
d) Find the marginal PMF of Y by using Proposition 6.2 on page 263.
e) Find the marginal PMF of Y by using a symmetry argument.

6.4 Let E and F be events of a sample space.
a) Determine the joint and marginal PMFs of I_E and I_F. Construct a table similar to Table 6.2.
b) You can apply Proposition 6.2 on page 263 to conclude that the sum of the values of this joint PMF in a row or column equals the value of the marginal PMF at the end (right or bottom) of that row or column. However, in this case, you can apply another result to get that conclusion. What is that result?
c) On page 264, we noted that, in general, the marginal PMFs of two discrete random variables don't determine the joint PMF. What does that mean in the context of indicator random variables?
d) Under what condition is the joint PMF of two indicator random variables determined by the marginal PMFs?

6.5 Proposition 6.2 on page 263 provides formulas for finding the marginal PMFs from the joint PMF. Derive these formulas by using the FPF.

6.6 A balanced die is tossed twice. Let X and Y denote the smaller and larger of the two faces, respectively.
a) Find the joint PMF of X and Y.
b) Construct a table similar to Table 6.2 on page 262 that gives the joint and marginal PMFs of X and Y.
Determine and interpret each of the following probabilities.
c) $P(X = Y)$ d) $P(X < Y)$ e) $P(X > Y)$
f) $P(Y = 2X)$ g) $P(Y \geq 2X)$ h) $P(Y - X \geq 3)$

6.7 Among eight people, two are Greens, two are Democrats, and four are Republicans. Three of these eight people are selected at random with replacement. Let X be the number of Greens in the sample and let Y be the number of Democrats in the sample.
a) Find the joint PMF of X and Y.
b) Construct a table that gives the joint and marginal PMFs of X and Y.
c) Determine and interpret $P(X > Y)$.
d) Express the event that no Republicans are chosen in terms of X and Y.
e) Use the FPF and the result of parts (a) and (d) to determine the probability that no Republicans are chosen.
f) Identify the probability distribution of the number of Republicans in the sample and use it to solve part (e).
g) Without doing any computations, identify and obtain the (marginal) PMF of X.
h) Without doing any computations, identify and obtain the PMF of $X + Y$.
i) Use the FPF and the result of part (a) to find $P(1 \leq X + Y \leq 2)$.
j) Solve part (i) by using the result of part (h).

6.8 Repeat Exercise 6.7 when the sampling is without replacement.

6.9 Consider a sequence of Bernoulli trials with success probability p. Let X denote the number of trials up to and including the first success and let Y denote the number of trials up to and including the second success.
a) Identify the (marginal) PMF of X.   b) Identify the (marginal) PMF of Y.
c) Determine the joint PMF of X and Y.
d) Use Proposition 6.2 on page 263 and the result of part (c) to obtain the marginal PMFs of X and Y. Compare your results with those obtained in parts (a) and (b).
e) Interpret the random variable $Y - X$.
f) Without doing any computations, identify and obtain the PMF of $Y - X$.
g) Solve part (f) by using the FPF to obtain $P(Y - X = z)$ for each positive integer z.

6.10 Jan and Jean play a coin tossing game with a balanced coin. If the coin comes up a head, Jan wins \$1 from Jean; if a tail, Jean wins \$1 from Jan. Let X and Y represent Jan's total winnings after the second and third tosses, respectively, where losses are negative winnings.
a) Obtain a sample space for the head–tail outcomes in three tosses of the coin.
b) Determine the joint PMF of X and Y.
c) Determine $P(X > Y)$ by using the FPF and the result of part (b).
d) Determine $P(X > Y)$ without doing any computations. Explain your reasoning.
e) Obtain the marginal PMF of X by using the result of part (b).
f) Obtain the marginal PMF of X by expressing X as a function of an appropriate binomial random variable and then applying Corollary 5.1 on page 247.
g) Obtain the marginal PMF of Y by using the result of part (b).

h) Obtain the marginal PMF of Y by expressing Y as a function of an appropriate binomial random variable and then applying Corollary 5.1 on page 247.
i) Without doing any computations, find the PMF of $Y - X$.
j) Solve part (i) by using the FPF to obtain $P(Y - X = z)$ for each possible value z of $Y - X$.

6.11 Repeat Exercise 6.10 if the probability of a head is p.

6.12 Repeat Exercise 6.10 if X and Y represent Jan's total winnings after the third and fourth tosses, respectively, where losses are negative winnings.

6.13 Arrivals at the emergency rooms of two hospitals occur at average hourly rates of λ and μ, respectively. The joint PMF of the numbers of arrivals during any particular hour is

$$p_{X,Y}(x, y) = e^{-(\lambda+\mu)} \frac{\lambda^x \mu^y}{x!y!}, \qquad x, y = 0, 1, 2, \ldots,$$

and $p_{X,Y}(x, y) = 0$ otherwise.
a) Determine and identify the marginal PMFs of X and Y.
b) What is the relationship between the joint PMF and the marginal PMFs?

6.14 Let X and Y be discrete random variables. Show that the function $p: \mathcal{R}^2 \to \mathcal{R}$ defined by $p(x, y) = p_X(x)p_Y(y)$ is a joint PMF by verifying that it satisfies properties (a)–(c) of Proposition 6.1 on page 262. *Hint:* A subset of a countable set is countable.

6.15 Let X and Y be discrete random variables. Suppose that the joint PMF of X and Y can be factored into a function of x alone and a function of y alone; that is, there exist two real-valued functions, q and r, of one real variable such that $p_{X,Y}(x, y) = q(x)r(y)$ for all real numbers x and y.
a) Obtain the marginal PMFs of X and Y.
b) Verify that $p_{X,Y}(x, y) = p_X(x)p_Y(y)$ for all real numbers x and y. Interpret this result.
c) Under what conditions are q and r the marginal PMFs of X and Y, respectively?

Theory Exercises

6.16 Prove Proposition 6.1 on page 262 by showing that the joint PMF of two discrete random variables X and Y satisfies properties (a), (b), and (c) of that proposition. *Hint:* For property (b), use the facts that the Cartesian product of two countable sets is countable (Exercise 1.41) and that a subset of a countable set is countable (Exercise 1.40).

6.17 Show that a function $p: \mathcal{R}^2 \to \mathcal{R}$ that satisfies properties (a), (b), and (c) of Proposition 6.1 is the joint PMF of some pair of discrete random variables. *Hint:* Consider the sample space $\Omega = \{(x, y) \in \mathcal{R}^2 : p(x, y) \neq 0\}$ and refer to Proposition 2.3 on page 43.

6.18 Prove Proposition 6.2 on page 263 by verifying Equations (6.3) and (6.4).

6.19 Prove the FPF for two discrete random variables, Proposition 6.3 on page 265.

Advanced Exercises

6.20 A number is chosen at random from the first N positive integers—call it X. Subsequently, a number is chosen at random from the first X positive integers—call it Y.
a) Obtain the joint PMF of X and Y. *Hint:* Use the general multiplication rule.
b) Obtain the marginal PMF of Y and verify that its values sum to 1.
c) Determine $P(Y = X)$.
d) Show that $P(Y = X) \sim (\ln N)/N$ as $N \to \infty$, where "$\sim$" means that the limit of the ratio of the two sides is 1. *Hint:* Use the fact that $\sum_{n=1}^{N} n^{-1} \sim \ln N$ as $N \to \infty$.

6.2 Joint and Marginal Probability Mass Functions: Multivariate Case

So far, we have concentrated on two discrete random variables. Now we examine the case of any finite number of discrete random variables. For the most part, this case is a straightforward generalization of the case of two discrete random variables, so the discussion is brief and we leave the proofs of the relevant propositions to you.

To begin, we introduce some terminology that is commonly employed in probability, statistics, and other related areas. The distribution of one random variable is often referred to as a **univariate distribution,** that for two random variables as a **bivariate distribution,** and that for three random variables as a **trivariate distribution.** More generally, we use **multivariate distribution** to refer to the distribution of more than one random variable. For the multivariate case, we define the **joint probability mass function** as follows.

DEFINITION 6.2   Joint PMF: Multivariate Case

Let $X_1, \ldots, X_m$ be m discrete random variables defined on the same sample space. Then the **joint probability mass function** of $X_1, \ldots, X_m$, denoted $p_{X_1, \ldots, X_m}$, is the real-valued function defined on $\mathcal{R}^m$ by

$$p_{X_1, \ldots, X_m}(x_1, \ldots, x_m) = P(X_1 = x_1, \ldots, X_m = x_m).$$

We use **joint PMF** as an abbreviation for "joint probability mass function."

Proposition 6.1 on page 262 gives the basic properties of the PMF of two discrete random variables. Proposition 6.4, whose proof we leave to you as Exercise 6.35, gives the analogous properties for the multivariate case.

◆◆◆   **Proposition 6.4   Basic Properties of a Joint PMF: Multivariate Case**

The joint probability mass function $p_{X_1, \ldots, X_m}$ of m discrete random variables $X_1, \ldots, X_m$ satisfies the following three properties.

a) $p_{X_1, \ldots, X_m}(x_1, \ldots, x_m) \geq 0$ for all $(x_1, \ldots, x_m) \in \mathcal{R}^m$.

b) $\{(x_1, \ldots, x_m) \in \mathcal{R}^m : p_{X_1, \ldots, X_m}(x_1, \ldots, x_m) \neq 0\}$ is countable.

c) $\sum \cdots \sum_{(x_1, \ldots, x_m)} p_{X_1, \ldots, X_m}(x_1, \ldots, x_m) = 1$.

In Exercise 6.36, we ask you to show that a function $p: \mathcal{R}^m \to \mathcal{R}$ that satisfies properties (a), (b), and (c) of Proposition 6.4 is the joint PMF of some m discrete random variables. Therefore, for any such function, we can say, "Let $X_1, \ldots, X_m$ be discrete random variables with joint PMF p." This statement makes sense regardless of whether we explicitly give $X_1, \ldots, X_m$ and the sample space on which they are defined.

When we state definitions and results in the general multivariate case, we use sub-scripted variables (e.g., $X_1, \ldots, X_m$). However, in the bivariate case, we continue to use unsubscripted variables (e.g., X, Y). In the trivariate case, we also sometimes use un-subscripted variables (e.g., X, Y, Z) in place of subscripted variables (e.g., X_1, X_2, X_3).

EXAMPLE 6.6   *Joint Probability Mass Functions*

Automobile Insurance Policies   An automobile insurance policy reimburses a fam-ily's accident losses up to a maximum of two accidents per year. For a (high-risk) three-person family, let X, Y, and Z denote the number of accidents in a single year for the three people in the family. Assume that the joint PMF of X, Y, Z is

$$p_{X,Y,Z}(x, y, z) = c(x + 2y + z), \qquad x \in \{0, 1\}, \ y, z \in \{0, 1, 2\},$$

and $p_{X,Y,Z}(x, y, z) = 0$ otherwise. Determine the constant c.

Solution   According to Proposition 6.4(c), the values of a joint PMF must sum to 1, which allows us to determine the constant c. We have

$$1 = \sum_{(x,y,z)} \sum \sum p_{X,Y,Z}(x, y, z) = \sum_{x=0}^{1} \sum_{y=0}^{2} \sum_{z=0}^{2} c(x + 2y + z)$$

$$= c \sum_{x=0}^{1} \sum_{y=0}^{2} \sum_{z=0}^{2} (x + 2y + z) = 63c.$$

Consequently, $c = 1/63$, as required.   ■

Marginal Probability Mass Functions

Let X and Y be two discrete random variables with joint PMF $p_{X,Y}$. As we noted previously, in the joint context, p_X (the PMF of X) is called the marginal PMF of X and, likewise, p_Y (the PMF of Y) is called the marginal PMF of Y. Proposition 6.2 on page 263 shows how to obtain the marginal PMFs of X and Y from their joint PMF.

When considering more than two discrete random variables, we have many more marginal PMFs. For instance, suppose that X, Y, and Z are three discrete random variables defined on the same sample space. The joint PMF of these random variables is

$$p_{X,Y,Z}(x, y, z) = P(X = x, Y = y, Z = z).$$

In this case, there are $\binom{3}{1} = 3$ univariate marginal PMFs—namely, p_X (the PMF of X), p_Y (the PMF of Y), and p_Z (the PMF of Z). Additionally, there are $\binom{3}{2} = 3$ bivariate marginal PMFs—namely, $p_{X,Y}$ (the joint PMF of X and Y), $p_{X,Z}$ (the joint PMF of X and Z), and $p_{Y,Z}$ (the joint PMF of Y and Z).

In general, if $X_1, \ldots, X_m$ are m discrete random variables defined on the same sample space, there are $\binom{m}{j}$ j-variate marginal PMFs ($1 \le j \le m - 1$). Each such j-variate marginal PMF is obtained by summing the joint PMF of $X_1, \ldots, X_m$ on the other $m - j$ variables. More precisely, we have Proposition 6.5 whose proof we leave to you as Exercise 6.37.

◆◆◆   **Proposition 6.5   Obtaining Marginal PMFs from the Joint PMF**

Suppose that $X_1, \ldots, X_m$ are discrete random variables defined on the same sample space. Let j be a positive integer between 1 and m, inclusive, and let $k_1 < \cdots < k_j$ be integers between 1 and m, inclusive. Then the joint PMF of $X_{k_1}, \ldots, X_{k_j}$ is

$$p_{X_{k_1}, \ldots, X_{k_j}}(x_{k_1}, \ldots, x_{k_j}) = \sum \cdots \sum p_{X_1, \ldots, X_m}(x_1, \ldots, x_m), \qquad \textbf{(6.7)}$$

where the $(m - j)$-fold sum is taken over all $(x_1, \ldots, x_m) \in \mathcal{R}^m$ whose k_ith coordinate is fixed at x_{k_i}, for $i = 1, \ldots, j$.

EXAMPLE 6.7   *Obtaining Marginal PMFs from the Joint PMF*

Automobile Insurance Policies   Refer to Example 6.6 on page 273. Let X, Y, and Z denote the number of accidents in a single year for the three people in the family. We assume that the joint PMF of X, Y, Z is

$$p_{X,Y,Z}(x, y, z) = (x + 2y + z)/63, \qquad x \in \{0, 1\}, \ y, z \in \{0, 1, 2\},$$

and $p_{X,Y,Z}(x, y, z) = 0$ otherwise. Determine all marginal PMFs.

Solution   First we determine the univariate marginal PMFs. To obtain the marginal PMF of X, we apply Proposition 6.5. For $x = 0$ or 1, we have

$$p_X(x) = \sum_{(y,z)} \sum p_{X,Y,Z}(x, y, z) = \sum_{y=0}^{2} \sum_{z=0}^{2} (x + 2y + z)/63 = (x + 3)/7.$$

Thus

$$p_X(x) = (x + 3)/7, \qquad x \in \{0, 1\},$$

and $p_X(x) = 0$ otherwise. Similarly,

$$p_Y(y) = (4y + 3)/21, \qquad y \in \{0, 1, 2\},$$

and $p_Y(y) = 0$ otherwise. Similarly,

$$p_Z(z) = (2z + 5)/21, \qquad z \in \{0, 1, 2\},$$

and $p_Z(z) = 0$ otherwise.

Next we determine the bivariate marginal PMFs. To obtain the marginal PMF of X and Y, we again apply Proposition 6.5. For $x = 0$ or 1 and $y = 0$, 1, or 2, we have

$$p_{X,Y}(x, y) = \sum_z p_{X,Y,Z}(x, y, z) = \sum_{z=0}^{2} (x + 2y + z)/63 = (x + 2y + 1)/21.$$

Thus
$$p_{X,Y}(x, y) = (x + 2y + 1)/21, \qquad x \in \{0, 1\}, \ y \in \{0, 1, 2\},$$
and $p_{X,Y}(x, y) = 0$ otherwise. Similarly,
$$p_{X,Z}(x, z) = (x + z + 2)/21, \qquad x \in \{0, 1\}, \ z \in \{0, 1, 2\},$$
and $p_{X,Z}(x, z) = 0$ otherwise. Similarly,
$$p_{Y,Z}(y, z) = (4y + 2z + 1)/63, \qquad y, z \in \{0, 1, 2\},$$
and $p_{Y,Z}(y, z) = 0$ otherwise.   ■

FPF for the General Multivariate Discrete Case

As for the case of one discrete random variable and two discrete random variables, once
we have the joint PMF of m discrete random variables, we can—at least conceptually—
determine any probability involving those random variables. Specifically, we again have
a **fundamental probability formula** (FPF), which we present as Proposition 6.6.

◆◆◆   **Proposition 6.6   Fundamental Probability Formula: Multivariate Discrete Case**

*Suppose that $X_1, \ldots, X_m$ are discrete random variables defined on the same sample
space. Then, for any subset $A \subset \mathcal{R}^m$, we have*

$$P\big((X_1, \ldots, X_m) \in A\big) = \sum \cdots \sum_{(x_1, \ldots, x_m) \in A} p_{X_1, \ldots, X_m}(x_1, \ldots, x_m). \qquad (6.8)$$

*In words, the probability that an m-tuple of discrete random variables takes a value in a
specified subset of m-dimensional space can be obtained by summing the joint PMF of
the m random variables over that subset.*

EXAMPLE 6.8   *The Fundamental Probability Formula*

Automobile Insurance Policies   Refer to Example 6.6 on page 273. Find the probability that, during a single year, all the family's automobile-accident losses are reimbursed.

Solution   The total number of accidents for the family during a single year is $X + Y + Z$. All the
losses are reimbursed if and only if that total is at most two, that is, if and only if the
event $\{X + Y + Z \leq 2\}$ occurs. Applying the FPF to the joint PMF of X, Y, and Z,
given at the beginning of Example 6.7, we get

$$
\begin{aligned}
P(X + Y + Z \leq 2) &= \sum\sum\sum_{x+y+z \leq 2} p_{X,Y,Z}(x, y, z) \\
&= p_{X,Y,Z}(0, 0, 0) + p_{X,Y,Z}(0, 0, 1) + p_{X,Y,Z}(0, 0, 2) \\
&\quad + p_{X,Y,Z}(0, 1, 0) + p_{X,Y,Z}(0, 1, 1) + p_{X,Y,Z}(0, 2, 0) \\
&\quad + p_{X,Y,Z}(1, 0, 0) + p_{X,Y,Z}(1, 0, 1) + p_{X,Y,Z}(1, 1, 0) \\
&= 18/63 = 0.286.
\end{aligned}
$$

There is a 28.6% chance that, during a single year, all the family's automobile-accident
losses will be reimbursed.   ■

The Multinomial Distribution

One of the most important multivariate distributions is the *multinomial distribution.* Before discussing this distribution, we briefly review the binomial distribution, which we examined in Chapter 5.

The binomial distribution gives the probability distribution of the number of successes in a finite sequence of Bernoulli trials. Recall that, in Bernoulli trials, we have independent repetitions (trials) of a random experiment in which the result of each trial is classified as either a success or a failure, depending on whether or not a specified event occurs; the success probability (i.e., the probability that the specified event occurs on any particular trial) is assumed to remain the same from trial to trial and is denoted p.

The number of successes, X, in n Bernoulli trials has PMF

$$p_X(x) = \binom{n}{x} p^x (1-p)^{n-x}, \qquad x = 0, 1, \ldots, n, \qquad \textbf{(6.9)}$$

and $p_X(x) = 0$ otherwise. The random variable X is called a binomial random variable and is said to have the binomial distribution with parameters n and p. The "bi" in "binomial" reflects the fact that the result of each trial is classified as one of two mutually exclusive possibilities—success (the specified event occurs) or failure (the specified event doesn't occur).

The binomial setting can also be viewed in terms of a bivariate distribution. Indeed, in n Bernoulli trials, let X denote the number of successes and let Y denote the number of failures. Recalling that the failure probability is $q = 1 - p$, the joint PMF of X and Y is

$$p_{X,Y}(x, y) = \binom{n}{x, y} p^x q^y, \qquad \textbf{(6.10)}$$

if x and y are nonnegative integers whose sum is n, and $p_{X,Y}(x, y) = 0$ otherwise. Because $\binom{n}{x} = \binom{n}{x, n-x}$, the right side of Equation (6.10) can be obtained by simply replacing $1 - p$ by q and $n - x$ by y in the right side of Equation (6.9).

We can generalize the binomial distribution by considering repeated independent trials of a random experiment in which the result of each trial is classified as one of r mutually exclusive possibilities. Example 6.9 introduces this idea.

EXAMPLE 6.9   *Introduces the Multinomial Distribution*

Roulette   A U.S. roulette wheel contains 38 numbers of which 18 are red, 18 are black, and 2 are green. When the roulette wheel is spun, the ball is equally likely to land on any of the 38 numbers. Here the random experiment consists of a play at the roulette wheel; each play constitutes one trial and the trials are independent.

To analyze this random experiment with regard to the color of the number on which the roulette ball lands, the result of each trial is classified as one of three mutually exclusive possibilities: red, black, or green. The probabilities of these three events are 18/38, 18/38, and 2/38, respectively. In 10 plays at the roulette wheel, let X_1, X_2, and X_3 denote the number of times that red, black, and green, respectively, occur.
a) Determine $P(X_1 = 3, X_2 = 6, X_3 = 1)$.
b) Determine the joint PMF of X_1, X_2, and X_3.

Solution   a) The event $\{X_1 = 3,\ X_2 = 6,\ X_3 = 1\}$ is that the ball lands on red exactly three times, on black exactly six times, and on green exactly one time. Because the trials are independent, any particular outcome of the 10 trials for which the ball lands on red exactly three times, on black exactly six times, and on green exactly one time has probability $(18/38)^3(18/38)^6(2/38)^1$. Furthermore, the number of such outcomes equals the number of ways that we can choose 3 trials from the 10 trials for when red occurs, 6 trials from the remaining 7 trials for when black occurs, and 1 trial from the remaining 1 trial for when green occurs. According to the partitions rule— Proposition 3.5 on page 104—this number is $\binom{10}{3,\,6,\,1} = 10!/(3!\,6!\,1!) = 840$. Hence

$$P(X_1 = 3,\ X_2 = 6,\ X_3 = 1) = \binom{10}{3,\ 6,\ 1}\left(\frac{18}{38}\right)^3\left(\frac{18}{38}\right)^6\left(\frac{2}{38}\right)^1 = 0.053.$$

b) Let x_1, x_2, and x_3 be nonnegative integers whose sum is 10. We want to determine the probability of the event $\{X_1 = x_1,\ X_2 = x_2,\ X_3 = x_3\}$—that is, the probability of the event that the ball lands on red exactly x_1 times, on black exactly x_2 times, and on green exactly x_3 times. Arguing as in part (a), we find that

$$P(X_1 = x_1,\ X_2 = x_2,\ X_3 = x_3) = \binom{10}{x_1,\ x_2,\ x_3}\left(\frac{18}{38}\right)^{x_1}\left(\frac{18}{38}\right)^{x_2}\left(\frac{2}{38}\right)^{x_3}.$$

Consequently, the joint PMF of X_1, X_2, and X_3 is

$$p_{X_1,X_2,X_3}(x_1, x_2, x_3) = \binom{10}{x_1,\ x_2,\ x_3}\left(\frac{18}{38}\right)^{x_1}\left(\frac{18}{38}\right)^{x_2}\left(\frac{2}{38}\right)^{x_3},$$

if x_1, x_2, and x_3 are nonnegative integers whose sum is 10; and, otherwise, we have $p_{X_1,X_2,X_3}(x_1, x_2, x_3) = 0$.   ■

Proceeding as in Example 6.9, we obtain Proposition 6.7, whose proof we leave to you as Exercise 6.39.

◆◆◆   **Proposition 6.7   PMF of Multinomial Random Variables**

Let $E_1, \ldots, E_m$ be mutually exclusive and exhaustive events of a random experiment with probabilities $p_1, \ldots, p_m$, respectively. In n independent repetitions of this random experiment, let $X_1, \ldots, X_m$ denote the number of times that events $E_1, \ldots, E_m$, respectively, occur. Then the joint PMF of the random variables $X_1, \ldots, X_m$ is

$$p_{X_1,\ldots,X_m}(x_1, \ldots, x_m) = \binom{n}{x_1, \ldots, x_m}p_1^{x_1} \cdots p_m^{x_m}, \tag{6.11}$$

if $x_1, \ldots, x_m$ are nonnegative integers whose sum is n; and, otherwise, we have $p_{X_1,\ldots,X_m}(x_1, \ldots, x_m) = 0$. The random variables $X_1, \ldots, X_m$ are called **multinomial random variables** *and are said to have the* **multinomial distribution with parameters n and $p_1, \ldots, p_m$.**

Note: Some researchers and textbook authors prefer an alternative version for a multinomial distribution, one that is consistent with the standard version of the binomial distribution. You are asked to explore that alternate version in Exercise 6.40.

EXAMPLE 6.10   *The Multinomial Distribution*

Roulette   Refer to Example 6.9. Identify the joint distribution of the random variables X_1, X_2, and X_3, the number of times that the roulette ball lands on red, black, and green, respectively, in 10 plays at the roulette wheel.

Solution   Here the random experiment consists of one play at the roulette wheel, with three mutually exclusive and exhaustive events under consideration: E_1, E_2, and E_3, which are the events that the ball lands on a red number, black number, and green number, respectively. The probabilities, p_1, p_2, and p_3, of those three events are 18/38, 18/38, and 2/38, respectively. Successive plays at the roulette wheel are independent and, because we are considering 10 plays, the number of trials is 10. Thus the random variables X_1, X_2, X_3 have the multinomial distribution with parameters $n = 10$ and $p_1 = 18/38$, $p_2 = 18/38$, $p_3 = 2/38$, as we showed in Example 6.9(b).   ■

Consider a finite population in which each member is classified as either having or not having a specified attribute. As previously noted, for random sampling with replacement, the number of members obtained that have the specified attribute is a binomial random variable; specifically, it has a binomial distribution with parameters n and p, where n is the sample size and p is the proportion of the population that has the specified attribute.

We can generalize this idea to a finite population in which each member is classified as having one of m mutually exclusive and exhaustive attributes. For instance, regarding region of residence, each member of the U.S. population is classified as having one of four mutually exclusive and exhaustive attributes: Northeast, Midwest, South, or West. Sampling with replacement from such a population leads to a multinomial distribution, as Example 6.11 shows.

EXAMPLE 6.11   *The Multinomial Distribution*

Random Sampling   Consider a finite population in which each member is classified as having one of m mutually exclusive and exhaustive attributes—say, $a_1, \ldots, a_m$. For each $j = 1, \ldots, m$, let p_j denote the proportion of the population that has attribute a_j. Suppose that a random sample of size n is taken with replacement and, for $j = 1, \ldots, m$, let X_j denote the number of members selected that have attribute a_j. Determine the joint PMF of the random variables $X_1, \ldots, X_m$.

Solution   There are n trials, where each trial comprises the random selection of one member of the population. Because the sampling is with replacement, the trials are independent. For $j = 1, \ldots, m$, let E_j denote the event that on any particular trial, the member obtained has attribute a_j. Then $E_1, \ldots, E_m$ are mutually exclusive and exhaustive and have probabilities $p_1, \ldots, p_m$, respectively. Therefore, by Proposition 6.7, $X_1, \ldots, X_m$ have the multinomial distribution with parameters n and $p_1, \ldots, p_m$.   ■

The Multiple Hypergeometric Distribution

In Example 6.11, we considered random sampling with replacement from a finite population in which each member is classified as having one of m mutually exclusive and exhaustive attributes—say, $a_1, \ldots, a_m$. If, for each j, we let X_j denote the number of members sampled that have attribute a_j, then, as we demonstrated, $X_1, \ldots, X_m$ have the multinomial distribution with parameters n and $p_1, \ldots, p_m$, where n is the sample size and p_j is the proportion of the population that has attribute a_j.

An essential assumption is that sampling is with replacement, which implies that the results are independent from one trial (random selection of one member) to the next. If the sampling is without replacement, we no longer have a multinomial distribution; rather we have a **multiple hypergeometric distribution,** a generalization of the hypergeometric distribution discussed in Section 5.4, beginning on page 208. You are asked to explore the multiple hypergeometric distribution in the exercises for this section.

EXERCISES 6.2   Basic Exercises

6.21 Let $X_1, \ldots, X_m$ be discrete random variables defined on the same sample space. Determine the total number of marginal PMFs, including univariate and multivariate.

6.22 Refer to Example 6.7 on page 274. Provide the details for obtaining all marginal PMFs of the random variables X, Y, and Z.

6.23 Let S consist of the 10 decimal digits. A number X is chosen according to the discrete uniform distribution on S, next a number Y is chosen according to the discrete uniform distribution on S with X removed, and then a number Z is chosen according to the discrete uniform distribution on S with X and Y removed.
a) Obtain the joint PMF of X, Y, and Z. *Hint:* Use the general multiplication rule.
b) Without doing any computations, determine $P(X > Y > Z)$.
c) Determine $P(X > Y > Z)$ by using the FPF.
d) Determine all marginal PMFs of X, Y, and Z by using Proposition 6.5 on page 274.
e) Determine all marginal PMFs of X, Y, and Z by using symmetry arguments.

6.24 In Bernoulli trials with success probability p, let X, Y, and Z denote the number of trials up to and including the first, second, and third successes, respectively.
a) Without doing any computations, identify the three univariate marginal PMFs of X, Y, and Z. *Hint:* Refer to Proposition 5.13 on page 241.
b) Determine the joint PMF of X, Y, and Z.
c) Use an argument similar to that in part (b) to obtain the three bivariate marginal PMFs.
d) Use Proposition 6.5 on page 274 and the result of part (b) to obtain the three bivariate marginal PMFs. Compare your results with those in part (c).
e) Use Proposition 6.5 and the result of part (b) to obtain the three univariate marginal PMFs. Compare your results with those obtained in part (a). *Hint:* In each case, draw a picture of the region in the plane over which the double summation is taken. For the univariate PMFs of X and Y, iterate the sums so that the sum on z is taken first.

6.25 Among eight people, two are Greens, two are Democrats, and four are Republicans. Three of these eight people are selected at random with replacement. Let X, Y, and Z be the number of Greens, Democrats, and Republicans, respectively, in the sample.
a) Identify and obtain the joint PMF of X, Y, and Z.
b) Use the result of part (a) and the FPF to determine and interpret each of $P(X > Y > Z)$, $P(X \leq Y < Z)$, and $P(X \leq Z, Y \geq Z)$.
c) Without doing any computations, identify and obtain the univariate marginal PMFs of X, Y, and Z.

d) Without doing any computations, identify and obtain the PMF of each of the three random variables $X + Y$, $X + Z$, and $Y + Z$.

e) Without doing any computations, identify and obtain the joint PMF of $X + Y$ and Z.

f) Use the FPF and the result of part (a) to find $P(1 \leq X + Z \leq 2)$.

g) Solve part (f) by using the result of part (d).

h) Without doing any computations, obtain the PMF of $X + Y + Z$.

6.26 In Example 5.1 on page 176, we considered the number of siblings for each of 40 students in one of Professor Weiss's classes. That class contains 6 freshmen, 15 sophomores, 12 juniors, and 7 seniors. Suppose that a random sample of 5 students is taken with replacement. Let X_1, X_2, X_3, and X_4 denote the number of freshmen, sophomores, juniors, and seniors, respectively, obtained.

a) Identify and determine the joint PMF of X_1, X_2, X_3, and X_4.

Use part (a) and the FPF to determine the probability that the sample contains

b) no freshmen, 3 sophomores, 1 junior, and 1 senior.

c) at least 3 sophomores.

d) at least 3 sophomores and at least 1 junior.

6.27 A balanced die is tossed 18 times. Determine the probability that each face appears

a) exactly 3 times.

b) at most 3 times.

c) at least 3 times.

6.28 Items from an assembly line are classified as either good or defective. On average, the assembly line produces 3% defective items. Two-thirds of the defective items are salvageable and can be repaired, but the other defective items must be scrapped. If 50 items are sampled from a week's output, what is the probability that the sample contains 48 good items and 1 each of salvageable and scrapped items?

6.29 Let $X_1, \ldots, X_5$ have the multinomial distribution with parameters n and $p_1, \ldots, p_5$. Without doing any computations, identify and determine the PMFs of each of the following random variables. Explain your reasoning.

a) X_2 **b)** $X_3 + X_5$ **c)** $X_1 + X_4 + X_5$

Without doing any computations, identify and determine the joint PMFs of each of the following collections of random variables. Explain your reasoning.

d) $X_1 + X_2$ and $X_3 + X_4 + X_5$

e) X_1, $X_2 + X_3$, and $X_4 + X_5$

f) X_1, $X_2 + X_3 + X_4$, and X_5

6.30 Let q and r be two nonnegative real numbers whose sum is less than 1 and let n be a positive integer. Verify the identity

$$\binom{n}{x} = \frac{1}{(q+r)^{n-x}} \sum_{y=0}^{n-x} \binom{n}{x,\ y,\ n-x-y} q^y r^{n-x-y}, \qquad x = 0, 1, \ldots, n,$$

by using

a) the binomial theorem.

b) a probabilistic argument based on the multinomial distribution.

6.31 Multiple hypergeometric distribution: Consider a finite population of size N in which each member is classified as having one of m mutually exclusive and exhaustive attributes, say, $a_1, \ldots, a_m$. For each $j = 1, \ldots, m$, let p_j denote the proportion of the population that has attribute a_j. Suppose that a random sample of size n is taken without replacement.

For $j = 1, \ldots, m$, let X_j denote the number of members selected that have attribute a_j. Show that the joint PMF of the random variables $X_1, \ldots, X_m$ is

$$p_{X_1, \ldots, X_m}(x_1, \ldots, x_m) = \binom{Np_1}{x_1} \cdots \binom{Np_m}{x_m} \Big/ \binom{N}{n},$$

if $x_1, \ldots, x_m$ are nonnegative integers whose sum is n, and $p_{X_1, \ldots, X_m}(x_1, \ldots, x_m) = 0$ otherwise. The random variables $X_1, \ldots, X_m$ are called *multiple hypergeometric random variables* and are said to have the *multiple hypergeometric distribution with parameters N, n, and $p_1, \ldots, p_m$*. *Note:* For $j = 1, \ldots, m$, the quantity Np_j equals the number of members of the population that have attribute a_j.

In Exercises 6.32–6.34, refer to Exercise 6.31.

6.32 Solve Exercise 6.25 if the sampling is without replacement.

6.33 Solve Exercise 6.26 if the sampling is without replacement.

6.34 Urn I contains 10 red, 11 green, and 7 blue balls. Urn II contains 9 red, 15 green, and 12 blue balls. A random sample of 6 balls is to be taken from one of the urns. The urn from which the sample is taken will be decided by the toss of a coin having probability 0.4 of a head. If the coin comes up a head, the sample is taken from Urn I; otherwise, the sample is taken from Urn II. Find the joint PMF of the numbers of red, green, and blue balls obtained if the sampling is
a) with replacement.   **b)** without replacement.

Theory Exercises

6.35 Prove Proposition 6.4 on page 272 by showing that the joint PMF of m discrete random variables $X_1, \ldots, X_m$ satisfies properties (a), (b), and (c) of that proposition. *Hint:* For property (b), use the facts that the Cartesian product of a finite number of countable sets is countable (Exercise 1.41) and that a subset of a countable set is countable (Exercise 1.40).

6.36 Show that a function $p: \mathcal{R}^m \to \mathcal{R}$ that satisfies properties (a), (b), and (c) of Proposition 6.4 on page 272 is the joint PMF of some m discrete random variables. *Hint:* Consider the sample space $\Omega = \{(x_1, \ldots, x_m) \in \mathcal{R}^m : p(x_1, \ldots, x_m) \neq 0\}$ and refer to Proposition 2.3 on page 43.

6.37 Prove Proposition 6.5 on page 274.

6.38 Prove the multivariate version of the fundamental probability formula for discrete random variables, Proposition 6.6 on page 275.

6.39 Prove Proposition 6.7 on page 277 by deriving the joint PMF of multinomial random variables $X_1, \ldots, X_m$ with parameters n and $p_1, \ldots, p_m$.

Advanced Exercises

6.40 Alternative version of the multinomial distribution: Let $X_1, \ldots, X_m$ have the multinomial distribution with parameters n and $p_1, \ldots, p_m$. Obtain the joint PMF of $X_1, \ldots, X_{m-1}$ by using the following steps.
a) Let $x_1, \ldots, x_{m-1}$ be nonnegative integers whose sum is at most n and, additionally, let $x_m = n - x_1 - \cdots - x_{m-1}$. Explain why

$$\{X_1 = x_1, \ldots, X_{m-1} = x_{m-1}\} = \{X_1 = x_1, \ldots, X_{m-1} = x_{m-1}, X_m = x_m\}.$$

b) Show that the joint PMF of $X_1, \ldots, X_{m-1}$ is

$$p_{X_1, \ldots, X_{m-1}}(x_1, \ldots, x_{m-1}) = \binom{n}{x_1, \ldots, x_{m-1}, n - x_1 - \cdots - x_{m-1}}$$
$$\times p_1^{x_1} \cdots p_{m-1}^{x_{m-1}} (1 - p_1 - \cdots - p_{m-1})^{n - x_1 - \cdots - x_{m-1}},$$

if $x_1, \ldots, x_{m-1}$ are nonnegative integers whose sum is at most n and that, otherwise, $p_{X_1, \ldots, X_{m-1}}(x_1, \ldots, x_{m-1}) = 0$. This probability distribution, rather than the one given in Proposition 6.7 on page 277, is sometimes called the *multinomial distribution*.

c) Show that, when $m = 2$, the PMF in part (b) reduces to that of a binomial distribution.

d) Show that all marginals of the PMF in part (b) have the same form as that PMF.

6.41 Multinomial approximation to the multiple hypergeometric: Refer to Exercise 6.31. Suppose that $X_1, \ldots, X_m$ have the multiple hypergeometric distribution with parameters N, n, and $p_1, \ldots, p_m$.

a) Prove that

$$\lim_{N \to \infty} p_{X_1, \ldots, X_m}(x_1, \ldots, x_m) = \binom{n}{x_1, \ldots, x_m} p_1^{x_1} \cdots p_m^{x_m},$$

for nonnegative integers $x_1, \ldots, x_m$ whose sum is n.

b) Interpret the result in part (a) and explain why it is reasonable.

6.42 According to *Current Population Reports*, of U.S. residents, 19.0% live in the Northeast, 23.1% live in the Midwest, 35.5% live in the South, and 22.4% live in the West. A random sample of n U.S. residents is taken without replacement. Let X_1, X_2, X_3, and X_4 denote the number of people in the sample that live in the Northeast, Midwest, South, and West, respectively.

a) Identify the exact probability distribution of X_1, X_2, X_3, and X_4.

b) Under what circumstances is it reasonable to use a multinomial distribution to approximate the probability distribution in part (a)? In those circumstances, identify the appropriate multinomial distribution.

c) Determine the approximate probability that, in a random sample of five U.S. residents, two live in the South and one each lives in the Northeast, Midwest, and West.

6.43 A politician who is basing her campaign on a controversial issue claims that, among her constituents, 65% favor her position on the issue, 30% oppose, and the rest are indifferent. A random sample of 40 constituents is taken by the opposition. The results: 18 in favor, 17 opposed, and 5 indifferent. Should the opposition use these data to refute the politician's claim? *Hint:* If the politician's claim is true, roughly how many of the 40 constituents sampled would you expect to be in favor, opposed, and indifferent?

6.3   Conditional Probability Mass Functions

We have discussed joint discrete distributions and marginal discrete distributions. Now we turn our attention to *conditional discrete distributions*, where we consider the (conditional) distribution of one or more discrete random variables, given the values of one or more discrete random variables. As before, we concentrate on the bivariate case of two discrete random variables—say, X and Y. For each value of X, we have a conditional distribution of Y, and, for each value of Y, we have a conditional distribution of X.

For instance, let X and Y denote the age and price, respectively, of a randomly selected Corvette. We might be interested in the price distribution of all Corvettes—that is, in the distribution of the random variable Y. However, we might be even more concerned with the price distribution of all Corvettes of a particular age—that is, in the conditional distribution of the random variable Y, given a value of the random variable X.

Conditional Probability Mass Functions

We first recall the conditional probability rule: The conditional probability that event F occurs, given that event E occurs, can be obtained from the formula

$$P(F \mid E) = \frac{P(E \cap F)}{P(E)}, \qquad (6.12)$$

where $P(E) > 0$.

Suppose now that X and Y are two discrete random variables defined on the same sample space and let x be a value of X with positive probability. Applying Equation (6.12) with $E = \{X = x\}$ and $F = \{Y = y\}$, we obtain

$$P(Y = y \mid X = x) = \frac{P(X = x,\, Y = y)}{P(X = x)} = \frac{p_{X,Y}(x, y)}{p_X(x)}.$$

In view of the previous equation, we now define **conditional probability mass function.**

DEFINITION 6.3   Conditional Probability Mass Function

Let X and Y be two discrete random variables defined on the same sample space. If $p_X(x) > 0$, the **conditional probability mass function of Y given $X = x$**, denoted $p_{Y \mid X}$, is the real-valued function

$$p_{Y \mid X}(y \mid x) = \frac{p_{X,Y}(x, y)}{p_X(x)}, \qquad y \in \mathcal{R}.$$

If $p_X(x) = 0$, we define $p_{Y \mid X}(y \mid x) = 0$ for all $y \in \mathcal{R}$, but we don't refer to $p_{Y \mid X}$ as a conditional probability mass function.

EXAMPLE 6.12   *Conditional Probability Mass Functions*

Bedrooms and Bathrooms   Data supplied by *Realty 2000/Better Homes and Gardens* of Prescott, Arizona, yielded the contingency table in Table 6.1 on page 261 for number of bedrooms and number of bathrooms for 50 homes currently for sale. Suppose that one of these 50 homes is selected at random. Let X and Y denote the number of bedrooms and number of bathrooms, respectively, of the home obtained. For each number of bedrooms, determine and interpret the conditional PMF of the number of bathrooms.

Solution   To obtain the conditional PMFs, we use Table 6.2 on page 262, which provides the joint and marginal PMFs of X and Y. We want to find the conditional PMF of Y for a given value of X. To illustrate, we obtain the conditional PMF of Y given $X = 3$. From Definition 6.3 and Table 6.2,

$$p_{Y \mid X}(y \mid 3) = \frac{p_{X,Y}(3, y)}{p_X(3)} = \frac{p_{X,Y}(3, y)}{0.56}. \tag{6.13}$$

Applying Equation (6.13) with $y = 2$ yields

$$p_{Y \mid X}(2 \mid 3) = \frac{p_{X,Y}(3, 2)}{0.56} = \frac{0.28}{0.56} = 0.500.$$

Similarly, we can use Equation (6.13) to obtain the values of $p_{Y \mid X}(y \mid 3)$ for $y = 3, 4$, and 5. Doing so, we get the conditional PMF of Y given $X = 3$, as displayed in Table 6.5. It shows that, among the homes with three bedrooms, 50% have two bathrooms, 42.9% have three bathrooms, 7.1% have four bathrooms, and none have five bathrooms.

Table 6.5 Conditional PMF of Y given $X = 3$ (the distribution of the number of bathrooms for homes with three bedrooms)

y	2	3	4	5
$p_{Y \mid X}(y \mid 3)$	0.500	0.429	0.071	0.000

Similarly, we obtain the conditional PMF of Y given each of the other two possible values of X. Each conditional PMF of Y is displayed as a row of Table 6.6.

Table 6.6 Conditional PMF of Y for each possible value of X

<table>
<tr><th></th><th colspan="5">Bathrooms, y</th></tr>
<tr><th></th><th>2</th><th>3</th><th>4</th><th>5</th><th>Total</th></tr>
<tr><td rowspan="3">Bedrooms, x 2</td><td>1.000</td><td>0.000</td><td>0.000</td><td>0.000</td><td>1.000</td></tr>
<tr><td>0.500</td><td>0.429</td><td>0.071</td><td>0.000</td><td>1.000</td></tr>
<tr><td>0.105</td><td>0.579</td><td>0.263</td><td>0.053</td><td>1.000</td></tr>
<tr><td>$p_Y(y)$</td><td>0.380</td><td>0.460</td><td>0.140</td><td>0.020</td><td>1.000</td></tr>
</table>

Each conditional PMF of Y given $X = x$ can be obtained by dividing each entry in the x row of the joint PMF in Table 6.2 by the marginal probability at the end of that row. ∎

Referring to the rows of Table 6.6, we note that the conditional PMFs of Y given $X = x$ differ from each other and from the marginal PMF of Y. By contrast, Example 6.13 provides an instance where all conditional PMFs of Y given $X = x$ are identical to each other and to the marginal PMF of Y.

EXAMPLE 6.13   *Conditional Probability Mass Functions*

Lifetimes of Electrical Components   Consider two identical electrical components. Let X and Y denote the respective lifetimes of the two components observed at discrete time units (e.g., every hour). The joint PMF of X and Y is

$$p_{X,Y}(x, y) = p^2(1 - p)^{x+y-2}, \qquad x, y \in \mathcal{N},$$

and $p_{X,Y}(x, y) = 0$ otherwise, where $0 < p < 1$.
a) Determine and identify the conditional PMF of Y given $X = x$.
b) Compare the conditional PMFs of Y given $X = x$ to each other and to the marginal PMF of Y. Interpret the results.

Solution   In Example 6.2 on page 263, we found that the marginal distributions of X and Y are both geometric with parameter p. In particular, then, $p_X(x) = p(1 - p)^{x-1}$ if $x \in \mathcal{N}$, and $p_X(x) = 0$ otherwise.
a) For each $x \in \mathcal{N}$, we have

$$p_{Y \mid X}(y \mid x) = \frac{p_{X,Y}(x, y)}{p_X(x)} = \frac{p^2(1 - p)^{x+y-2}}{p(1 - p)^{x-1}} = p(1 - p)^{y-1},$$

for $y \in \mathcal{N}$, and $p_{Y \mid X}(y \mid x) = 0$ otherwise. Thus, for each $x \in \mathcal{N}$, $Y_{|X=x} \sim \mathcal{G}(p)$; that is, the conditional distribution of Y given $X = x$ is the geometric distribution with parameter p.
b) From part (a), we conclude that the conditional PMFs of Y given $X = x$ are identical to each other and to the marginal PMF of Y. Consequently, knowing the value of the random variable X yields no information about the distribution of Y. In other words, knowing the value of X doesn't affect the probability distribution of Y.   ∎

A Conditional PMF Is a PMF

Proposition 6.8 establishes the result that a conditional PMF is in fact a genuine probability mass function.

◆◆◆   **Proposition 6.8   A Conditional PMF Is a PMF**

Let X and Y be two discrete random variables defined on the same sample space and let x be a value of X with positive probability. Then, as a function of y, the conditional PMF $p_{Y \mid X}(y \mid x)$ is a probability mass function; that is, it satisfies properties (a)–(c) of Proposition 5.1 on page 187.

Proof   That properties (a) and (b) are satisfied is obvious. We can verify property (c) in several ways. One method is to invoke Proposition 4.1 on page 134, which implies that

$P(\cdot \mid X = x)$ is a probability measure. Another method—which is the one we use here—relies on the formula for obtaining a marginal PMF from a joint PMF. Specifically, applying Proposition 6.2 on page 263, we conclude that

$$\sum_y p_{Y \mid X}(y \mid x) = \sum_y \frac{p_{X,Y}(x, y)}{p_X(x)} = \frac{1}{p_X(x)} \sum_y p_{X,Y}(x, y) = \frac{1}{p_X(x)} \cdot p_X(x) = 1,$$

as required.   ◆

Proposition 6.8 implies that any properties of (unconditional) PMFs also hold for conditional PMFs. For instance, the following conditional version of the FPF holds:

$$P(Y \in A \mid X = x) = \sum_{y \in A} p_{Y \mid X}(y \mid x), \tag{6.14}$$

for each subset $A \subset \mathcal{R}$.

EXAMPLE 6.14   *Conditional Probability Mass Functions*

Bedrooms and Bathrooms   Refer to Example 6.12. Given that a randomly selected home (from among the 50 homes) has three bedrooms, what is the probability that it has at most three bathrooms?

Solution   We want to find $P(Y \leq 3 \mid X = 3)$. The conditional PMF of Y given $X = 3$ is displayed in the second row of Table 6.6 on page 284. Applying Equation (6.14)—the conditional version of the FPF—gives

$$P(Y \leq 3 \mid X = 3) = \sum_{y \leq 3} p_{Y \mid X}(y \mid 3) = p_{Y \mid X}(2 \mid 3) + p_{Y \mid X}(3 \mid 3)$$
$$= 0.500 + 0.429 = 0.929.$$

Thus 92.9% of the homes with three bedrooms have three or fewer bathrooms.   ■

From Definition 6.3,

$$p_{Y \mid X}(y \mid x) = \frac{p_{X,Y}(x, y)}{p_X(x)}.$$

Multiplying both sides of that equation by $p_X(x)$, we get

$$p_{X,Y}(x, y) = p_X(x) p_{Y \mid X}(y \mid x). \tag{6.15}$$

Equation (6.15) is the **general multiplication rule,** Proposition 4.2 on page 138, when applied to the joint PMF of two discrete random variables, X and Y. In Exercise 6.57, you are asked to show that Equation (6.15) holds for all real numbers x and y, not just those where $p_X(x) > 0$.

Conditional Joint Probability Mass Functions

We can also determine conditional PMFs in the general multivariate discrete case. The formulas for them are obtained in the same way as in the bivariate discrete case—namely, by applying the conditional probability rule. For instance, if X, Y, and Z are three discrete random variables defined on the same sample space,

$$p_{Y,Z|X}(y, z \mid x) = \frac{p_{X,Y,Z}(x, y, z)}{p_X(x)} \quad \text{and} \quad p_{Z|X,Y}(z \mid x, y) = \frac{p_{X,Y,Z}(x, y, z)}{p_{X,Y}(x, y)}.$$

Example 6.15 illustrates the use of conditional PMFs in a trivariate environment.

EXAMPLE 6.15 *Conditional Probability Mass Functions*

Automobile Insurance An automobile insurance policy reimburses a family's accident losses up to a maximum of two accidents per year. For a (high-risk) three person family, let X, Y, and Z denote the number of accidents in a single year for the three people in the family. Assume that the joint PMF of X, Y, Z is

$$p_{X,Y,Z}(x, y, z) = (x + 2y + z)/63, \qquad x \in \{0, 1\}, \ y, z \in \{0, 1, 2\},$$

and $p_{X,Y,Z}(x, y, z) = 0$ otherwise.
a) Determine the conditional joint PMF of Y and Z given $X = 0$.
b) Given that $X = 0$ during a particular year, find the probability that all the family's automobile-accident losses will be reimbursed.

Solution a) The conditional joint PMF of Y and Z given $X = 0$ is

$$p_{Y,Z|X}(y, z \mid 0) = \frac{p_{X,Y,Z}(0, y, z)}{p_X(0)} = \frac{(2y + z)/63}{p_X(0)}, \qquad y, z \in \{0, 1, 2\}.$$

From Example 6.7 on page 274, the marginal PMF of X is $p_X(x) = (x + 3)/7$ if $x = 0$ or 1, and $p_X(x) = 0$ otherwise. In particular, then, $p_X(0) = 3/7$. Therefore,

$$p_{Y,Z|X}(y, z \mid 0) = \frac{(2y + z)/63}{3/7} = (2y + z)/27, \qquad y, z \in \{0, 1, 2\}, \qquad \textbf{(6.16)}$$

and $p_{Y,Z|X}(y, z \mid 0) = 0$ otherwise.
b) Given that $X = 0$, the probability that all the family's automobile-accident losses will be reimbursed is $P(Y + Z \leq 2 \mid X = 0)$. Applying the conditional form of the FPF and Equation (6.16) gives

$$P(Y + Z \leq 2 \mid X = 0) = \sum_{y+z \leq 2} \sum p_{Y,Z|X}(y, z \mid 0)$$

$$= p_{Y,Z|X}(0, 0 \mid 0) + p_{Y,Z|X}(0, 1 \mid 0)$$
$$+ p_{Y,Z|X}(0, 2 \mid 0) + p_{Y,Z|X}(1, 0 \mid 0)$$
$$+ p_{Y,Z|X}(1, 1 \mid 0) + p_{Y,Z|X}(2, 0 \mid 0)$$
$$= 12/27 = 0.444.$$

Given that $X = 0$, there is a 44.4% chance that all the family's automobile-accident losses will be reimbursed. ∎

EXERCISES 6.3   **Basic Exercises**

Note: Several of the exercises in this section are continuations of exercises presented in Sections 6.1 and 6.2.

6.44 Refer to Example 6.12 on page 283.
a) Obtain the conditional PMF of the number of bedrooms for each number of bathrooms. Construct a table similar to Table 6.6 on page 284.
b) Given that a randomly selected home (from among the 50 homes) has two bathrooms, what is the probability that it has at least three bedrooms?
c) Interpret your answer in part (b) in terms of percentages.

6.45 In Exercise 6.2 on page 269, you considered the number of siblings, X, and the number of sisters, Y, for a randomly selected student from one of Professor Weiss's classes.
a) For each number of siblings, determine and interpret the conditional PMF of the number of sisters.
b) Construct a table like Table 6.6 on page 284 that displays the marginal PMF of the number of sisters and the conditional PMF of the number of sisters for each number of siblings.
c) What percentage of students in the class with two siblings have at least one sister?
d) For each number of sisters, determine and interpret the conditional PMF of the number of siblings.
e) Construct a table that displays the marginal PMF of the number of siblings and the conditional PMF of the number of siblings for each number of sisters.
f) What percentage of students in the class with one sister have at least two siblings?

6.46 Let X be a number chosen according to the discrete uniform distribution on the set S of 10 decimal digits and let Y be a number chosen according to the discrete uniform distribution on S with X removed.
a) Determine and identify the conditional PMF of Y given $X = x$.
b) Determine and identify the conditional PMF of X given $Y = y$.
c) Find $P(3 \leq X \leq 4 \mid Y = 2)$.

6.47 Let E and F be events of a sample space.
a) Determine the conditional PMF of I_F given $I_E = x$ for each possible value x of I_E. Construct a table similar to Table 6.6 on page 284 that displays these conditional PMFs and the marginal PMF of I_F.
b) In this special case of indicator random variables, what result guarantees that the values in each row of your table in part (a) sum to 1?

6.48 Let X and Y denote the smaller and larger of the two faces, respectively, when a balanced die is tossed twice.
a) Determine each conditional PMF of Y given $X = x$ and construct a table similar to Table 6.6 on page 284 that displays these conditional PMFs and the marginal PMF of Y.
b) Repeat part (a) for the conditional PMFs of X given $Y = y$ and the marginal PMF of X.

6.49 Consider a sequence of Bernoulli trials with success probability p. Let X denote the number of trials up to and including the first success and let Y denote the number of trials up to and including the second success.
a) Determine the conditional PMF of Y given $X = x$ without doing any computations. Explain your reasoning.
b) Determine the conditional PMF of Y given $X = x$ by referring to Exercise 6.9 on page 270 and applying Definition 6.3 on page 283. Compare your result with that in part (a).

c) Determine and identify the conditional PMF of X given $Y = y$ without doing any computations but, rather, by using a symmetry argument.

d) Determine the conditional PMF of X given $Y = y$ by referring to Exercise 6.9 on page 270 and applying Definition 6.3 on page 283. Compare your result with that in part (c).

e) Determine and identify the conditional PMF of $Y - X$ given $X = x$ without doing any computations. Explain your reasoning.

f) Determine the conditional PMF of $Y - X$ given $X = x$ by using the conditional probability rule and referring to Exercise 6.9 on page 270.

6.50 Arrivals at the emergency rooms of two hospitals occur at average hourly rates of λ and μ, respectively. The joint PMF of the numbers of arrivals during any particular hour is

$$p_{X,Y}(x, y) = e^{-(\lambda+\mu)} \frac{\lambda^x \mu^y}{x! y!}, \qquad x, y = 0, 1, 2, \ldots,$$

and $p_{X,Y}(x, y) = 0$ otherwise.

a) Determine and identify each conditional distribution of Y given $X = x$.

b) Determine and identify each conditional distribution of X given $Y = y$.

c) What is the relationship between each conditional distribution of Y given $X = x$ and the marginal distribution of Y?

d) What is the relationship between each conditional distribution of X given $Y = y$ and the marginal distribution of X?

6.51 Suppose that X and Y are discrete random variables such that their joint PDF equals the product of their marginal PDFs.

a) Obtain the conditional PMF of Y given $X = x$ and compare it to the marginal PMF of Y.

b) Obtain the conditional PMF of X given $Y = y$ and compare it to the marginal PMF of X.

6.52 Refer to the automobile insurance illustration of Example 6.15 on page 287.

a) Determine the conditional joint PMF of X and Z given $Y = 1$.

b) Given that $Y = 1$ during a particular year, what is the probability that all the family's automobile-accident losses will be reimbursed?

c) Determine the conditional PMF of Z given $X = 0$ and $Y = 1$.

d) Given that $X = 0$ and $Y = 1$ during a particular year, what is the probability that all the family's automobile-accident losses will be reimbursed?

6.53 Let X, Y, and Z have the multinomial distribution with parameters n and p, q, r.

a) Obtain and identify the conditional joint PMF of Y and Z given $X = x$.

b) Obtain the conditional PMF of Z given $X = x$ and $Y = y$.

6.54 Let $X_1, \ldots, X_m$ have the multinomial distribution with parameters n and $p_1, \ldots, p_m$.

a) Determine and identify the conditional distribution of $X_1, \ldots, X_{m-1}$ given $X_m = x_m$.

b) Now refer to the roulette illustration of Example 6.10 on page 278. Apply the result of part (a) to determine and identify the conditional distribution of the number of times that red and black occur in 10 plays at the roulette wheel, given that green occurs once.

6.55 Refer to Exercise 6.31 on page 280. Suppose that $X_1, \ldots, X_m$ have the multiple hypergeometric distribution with parameters N, n, and $p_1, \ldots, p_m$.

a) Determine and identify the conditional distribution of $X_1, \ldots, X_{m-1}$ given $X_m = x_m$.

b) Now refer to Exercise 6.33 on page 281. Apply the result of part (a) to determine and identify the conditional distribution of the number of freshmen, sophomores, and juniors obtained, given that exactly one senior is obtained.

Theory Exercises

6.56 Let X and Y be discrete random variables defined on the same sample space. Suppose that, among all values x of X with positive probability, the conditional PMF of Y given $X = x$ doesn't depend on x. Prove that the common conditional PMF of Y given $X = x$ is the marginal PMF of Y.

6.57 Prove that the general multiplication rule for discrete random variables—presented as Equation (6.15) on page 286—holds for all real numbers x and y.

Advanced Exercises

6.58 Let $X_1, \ldots, X_m$ have the multinomial distribution with parameters n and $p_1, \ldots, p_m$ and let k be a positive integer less than m.
a) Obtain the conditional distribution of $X_1, \ldots, X_k$ given $X_{k+1} = x_{k+1}, \ldots, X_m = x_m$, and identify that distribution.
b) Without doing any computations, explain the result of part (a).

6.59 A number is chosen at random from the first N positive integers—call it X. Subsequently, a number is chosen at random from the first X positive integers—call it Y.
a) Obtain and identify the conditional PMF of Y given $X = x$.
b) Obtain the conditional PMF of X given $Y = y$. *Hint:* Refer to Exercise 6.20 on page 271.
c) Show that $p_{X \mid Y}(x \mid y) \sim 1/(x \ln N)$ as $N \to \infty$, where "$\sim$" means that the limit of the ratio of the two sides is 1.

6.4   Independent Random Variables

In this section, we examine an essential concept in the study of random variables—namely, that of *independent random variables*. The discussion applies to all types of random variables, not just discrete random variables. We begin with the bivariate case of two random variables, X and Y, defined on the same sample space.

There are several ways to approach the concept of independence of two random variables. One is to say that X and Y are independent random variables if knowing the value of one of these random variables doesn't affect the probability distribution of the other random variable. In terms of conditional and marginal distributions, this condition is equivalent to requiring that, for each possible value of x, the conditional distribution of Y given $X = x$ is the same as the (unconditional) distribution of Y.

However, using the concept of independent events, which we introduced in Chapter 3, is easier and more general. Recall that events E and F are independent if knowing that one of the events occurs doesn't affect the probability of occurrence of the other event. Mathematically, E and F are independent if and only if they satisfy the special multiplication rule,

$$P(E \cap F) = P(E)P(F). \tag{6.17}$$

Now, let X and Y be two random variables defined on the same sample space. Our intuitive notion of independence demands that, for these two random variables to be independent, the events $\{X \in A\}$ and $\{Y \in B\}$ must be independent for every two subsets A and B of $\mathcal{R}$. Applying Equation (6.17) to the events $E = \{X \in A\}$ and $F = \{Y \in B\}$, we arrive at the definition of **independent random variables.**

DEFINITION 6.4   Independence of Two Random Variables

Random variables X and Y defined on the same sample space are said to be **independent random variables** if

$$P(X \in A, \, Y \in B) = P(X \in A)P(Y \in B) \qquad (6.18)$$

for all subsets A and B of real numbers.

Note: The multiplication property defining independence in this context and others is ubiquitous in probability theory and its applications. Over and over again independence reduces to the multiplication property in one form or another.

In Proposition 6.9, we state and prove the useful fact that functions of independent random variables are also independent random variables.

◆◆◆   **Proposition 6.9**

Suppose that X and Y are independent random variables and that g and h are real-valued functions defined on the range of X and the range of Y, respectively. Then $g(X)$ and $h(Y)$ are independent random variables.[†]

Proof   Let A and B be subsets of real numbers. Applying Definition 6.4 to the subsets $g^{-1}(A)$ and $h^{-1}(B)$ of real numbers, we get

$$\begin{aligned} P\big(g(X) \in A, \, h(Y) \in B\big) &= P\big(X \in g^{-1}(A), \, Y \in h^{-1}(B)\big) \\ &= P\big(X \in g^{-1}(A)\big)P\big(Y \in h^{-1}(B)\big) \\ &= P\big(g(X) \in A\big)P\big(h(Y) \in B\big). \end{aligned}$$

Hence $g(X)$ and $h(Y)$ are independent random variables.   ◆

EXAMPLE 6.16   *Functions of Independent Random Variables*

Suppose that X and Y are independent random variables. Explain why each of the following pairs of random variables are independent.
a) X^2 and Y b) $\sin X$ and $\cos Y$ c) e^X and $Y^3 - 3Y^2 + 4Y - 8$

[†]To be perfectly correct, we must also assume that the functions g and h are such that the compositions $g(X)$ and $h(Y)$ are random variables in the sense given in the footnote on page 177. These conditions are almost always satisfied in practice.

Solution   To verify the independence of each pair of random variables, we simply apply Proposition 6.9 with the following functions.
a) $g(x) = x^2$ and $h(y) = y$
b) $g(x) = \sin x$ and $h(y) = \cos y$
c) $g(x) = e^x$ and $h(y) = y^3 - 3y^2 + 4y - 8$ ∎

Independent Discrete Random Variables: Bivariate Case

As we already mentioned, the definition of independent random variables presented in Definition 6.4 applies to any two random variables, discrete or not. Because, in this chapter, we are concentrating on discrete random variables, we now specialize to that case. Doing so, we can provide an equivalent condition for independence in terms of PMFs. Specifically, we have Proposition 6.10.

◆◆◆   **Proposition 6.10**   **Independent Discrete Random Variables: Bivariate Case**

Let X and Y be two discrete random variables defined on the same sample space. Then X and Y are independent if and only if

$$p_{X,Y}(x, y) = p_X(x)p_Y(y), \qquad \textit{for all } x, y \in \mathcal{R}. \tag{6.19}$$

In words, two discrete random variables are independent if and only if their joint PMF equals the product of their marginal PMFs.

Proof   Suppose that X and Y are independent. Let x and y be any two real numbers and set $A = \{x\}$ and $B = \{y\}$. Then, by Definition 6.4,

$$p_{X,Y}(x, y) = P(X = x, Y = y) = P(X \in A, Y \in B)$$
$$= P(X \in A)P(Y \in B) = P(X = x)P(Y = y) = p_X(x)p_Y(y).$$

Conversely, suppose that Equation (6.19) holds. Let A and B be any two subsets of real numbers. Applying the FPF and Equation (6.19) gives

$$P(X \in A, Y \in B) = P\big((X, Y) \in A \times B\big) = \sum\sum_{(x,y) \in A \times B} p_{X,Y}(x, y)$$

$$= \sum\sum_{(x,y) \in A \times B} p_X(x)p_Y(y) = \sum_{x \in A}\left(\sum_{y \in B} p_X(x)p_Y(y)\right)$$

$$= \sum_{x \in A} p_X(x)\left(\sum_{y \in B} p_Y(y)\right) = \sum_{x \in A} p_X(x)P(Y \in B)$$

$$= P(Y \in B)\sum_{x \in A} p_X(x) = P(X \in A)P(Y \in B),$$

as required.   ◆

For two discrete random variables, X and Y, to be independent, Equation (6.19) must hold for all real numbers x and y. Consequently, if we can find one pair of real numbers for which Equation (6.19) fails, we can conclude that X and Y aren't independent.

EXAMPLE 6.17 *The Concept of Independent Random Variables*

Bedrooms and Bathrooms Refer to Example 6.1 on page 261. Determine whether X and Y are independent random variables.

Solution The joint and marginal PMFs of X and Y are displayed in Table 6.2 on page 262. Note, for instance, that $p_X(4) = 0.38$, $p_Y(3) = 0.46$, and $p_{X,Y}(4, 3) = 0.22$. Because $0.22 \neq 0.38 \cdot 0.46$, we have $p_{X,Y}(4, 3) \neq p_X(4)p_Y(3)$. So X and Y aren't independent random variables. ■

In general, when the joint and marginal PMFs of two discrete random variables are presented in a contingency table, independence means that the number in each cell equals the product of the two numbers at the end of the row and column in which that cell resides. If we can find a cell in the contingency table for which that isn't the case, the two random variables aren't independent.

EXAMPLE 6.18 *The Concept of Independent Random Variables*

Lifetimes of Electrical Components Consider two identical electrical components. Let X and Y denote the respective lifetimes of the two components observed at discrete time units (e.g., every hour). Assume that the joint PMF of X and Y is

$$p_{X,Y}(x, y) = p^2(1 - p)^{x+y-2}, \qquad x, y \in \mathcal{N},$$

and $p_{X,Y}(x, y) = 0$ otherwise, where $0 < p < 1$. Determine whether X and Y are independent random variables.

Solution As we demonstrated in Example 6.2 on page 263, the marginal PMFs of X and Y are both $\mathcal{G}(p)$. Thus, $p_X(x) = p(1 - p)^{x-1}$ if $x \in \mathcal{N}$, and $p_X(x) = 0$ otherwise; and $p_Y(y) = p(1 - p)^{y-1}$ if $y \in \mathcal{N}$, and $p_Y(y) = 0$ otherwise. As

$$p^2(1 - p)^{x+y-2} = p(1 - p)^{x-1} \cdot p(1 - p)^{y-1},$$

it follows that $p_{X,Y}(x, y) = p_X(x)p_Y(y)$ for all real numbers x and y. Therefore X and Y are independent random variables. ■

Example 6.19 involves the binomial and Poisson distributions. Although we state it in terms of emergency-room arrivals, it applies to many other situations.

EXAMPLE 6.19 *The Concept of Independent Random Variables*

Emergency Room Traffic The number of people, Z, arriving at an emergency room during a one-hour period has a Poisson distribution with parameter λ. Independent of other arrivals, an arrival is male with probability p and female with probability q, where, of course, $p + q = 1$. Let X and Y denote the number of males and females, respectively, that arrive at the emergency room during a one-hour period.

a) Show that X has a Poisson distribution with parameters $p\lambda$ and that Y has a Poisson distribution with parameter $q\lambda$.

b) Show that X and Y are independent random variables.

Solution   By assumption,

$$p_Z(z) = e^{-\lambda} \frac{\lambda^z}{z!}, \qquad z = 0, 1, 2, \ldots, \tag{6.20}$$

and $p_Z(z) = 0$ otherwise. Moreover, the statement of the problem implies that, for each $z \in \mathcal{N}$, the conditional PMF of X given $Z = z$ is binomial with parameters z and p; that is,

$$p_{X \mid Z}(x \mid z) = \binom{z}{x} p^x (1 - p)^{z-x}, \qquad x = 0, 1, \ldots, z, \tag{6.21}$$

and $p_{X \mid Z}(x \mid z) = 0$ otherwise. Likewise, for each $z \in \mathcal{N}$, the conditional PMF of Y given $Z = z$ is binomial with parameters z and q; that is,

$$p_{Y \mid Z}(y \mid z) = \binom{z}{y} q^y (1 - q)^{z-y}, \qquad y = 0, 1, \ldots, z, \tag{6.22}$$

and $p_{Y \mid Z}(y \mid z) = 0$ otherwise. Similarly, for each $z \in \mathcal{N}$, the conditional joint PMF of X and Y given $Z = z$ is multinomial with parameters z and p, q; that is,

$$p_{X,Y \mid Z}(x, y \mid z) = \binom{z}{x, \, y} p^x q^y, \tag{6.23}$$

if x and y are nonnegative integers whose sum is z; and $p_{X,Y \mid Z}(x, y \mid z) = 0$ otherwise.

a) To obtain the marginal PMF of X, we apply Proposition 6.2 on page 263, the general multiplication rule for discrete random variables, Equation (6.20), Equation (6.21), and the exponential series to obtain

$$p_X(x) = \sum_z p_{X,Z}(x, z) = \sum_z p_Z(z) p_{X \mid Z}(x \mid z)$$

$$= \sum_{z=x}^{\infty} e^{-\lambda} \frac{\lambda^z}{z!} \cdot \binom{z}{x} p^x (1 - p)^{z-x} = \frac{e^{-\lambda} (p\lambda)^x}{x!} \sum_{z=x}^{\infty} \frac{\big((1 - p)\lambda\big)^{z-x}}{(z - x)!}$$

$$= \frac{e^{-\lambda} (p\lambda)^x}{x!} \sum_{k=0}^{\infty} \frac{\big((1 - p)\lambda\big)^k}{k!} = \frac{e^{-\lambda} (p\lambda)^x}{x!} \cdot e^{(1-p)\lambda} = e^{-p\lambda} \frac{(p\lambda)^x}{x!},$$

for $x = 0, 1, 2, \ldots$. Hence $X \sim \mathcal{P}(p\lambda)$. Similarly, we find that $Y \sim \mathcal{P}(q\lambda)$.

b) Let x and y be nonnegative integers. Noting that $Z = X + Y$ and applying Proposition 6.5 on page 274, Equation (6.20), Equation (6.23), and the results of part (a), we get

$$p_{X,Y}(x, y) = \sum_z p_{X,Y,Z}(x, y, z) = \sum_z p_Z(z) p_{X,Y \mid Z}(x, y \mid z)$$

$$= p_Z(x + y) p_{X,Y \mid Z}(x, y \mid x + y) = e^{-\lambda} \frac{\lambda^{x+y}}{(x + y)!} \cdot \binom{x + y}{x, \, y} p^x q^y$$

$$= e^{-p\lambda} \frac{(p\lambda)^x}{x!} \cdot e^{-q\lambda} \frac{(q\lambda)^y}{y!} = p_X(x) p_Y(y).$$

Hence X and Y are independent random variables.   ∎

In Section 6.1, we learned that the joint PMF of two discrete random variables determines the marginal PMFs, but that the converse isn't true—generally, the marginal PMFs don't determine the joint PMF. However, for independent discrete random variables, the marginal PMFs do determine the joint PMF because, by Proposition 6.10, the joint PMF is the product of the marginal PMFs.

Independence and Conditional Distributions

As Proposition 6.11 shows, independence of discrete random variables also conforms to our intuitive notion in terms of conditional distributions.

◆◆◆ **Proposition 6.11   Independence and Conditional Distributions**

Discrete random variables X and Y are independent if and only if either of the following properties holds:
a) Each conditional PMF of Y given $X = x$ is identical to the PMF of Y.
b) Each conditional PMF of X given $Y = y$ is identical to the PMF of X.

Proof   We verify that condition (a) is equivalent to independence and leave verification of condition (b) for you. Suppose that X and Y are independent. Let x be a value of X with positive probability. Then, by Definition 6.3 on page 283 and Proposition 6.10 on page 292, we have, for each $y \in \mathcal{R}$,

$$p_{Y \mid X}(y \mid x) = \frac{p_{X,Y}(x, y)}{p_X(x)} = \frac{p_X(x)p_Y(y)}{p_X(x)} = p_Y(y).$$

Conversely, suppose that condition (a) holds. To establish that X and Y are independent, we show that

$$p_{X,Y}(x, y) = p_X(x)p_Y(y), \qquad \text{for all } x, y \in \mathcal{R}. \tag{6.24}$$

If $p_X(x) = 0$, both sides of Equation (6.24) equal 0, so we can assume that $p_X(x) > 0$. By assumption, the conditional PMF of Y given $X = x$ is identical to the PMF of Y; that is, $p_{Y \mid X}(y \mid x) = p_Y(y)$ for all $y \in \mathcal{R}$. Therefore, by the general multiplication rule for discrete random variables,

$$p_{X,Y}(x, y) = p_X(x)p_{Y \mid X}(y \mid x) = p_X(x)p_Y(y),$$

as required.   ◆

Note the following consequences of Proposition 6.11.

• If each conditional PMF of Y given $X = x$ is identical to the PMF of Y, then each conditional PMF of X given $Y = y$ is identical to the PMF of X; and vice versa.

• In terms of contingency tables, X and Y are independent random variables if and only if the rows giving the conditional PMFs of Y given $X = x$ are identical to each other and to the row giving the marginal PMF of Y.

• In terms of contingency tables, X and Y are independent random variables if and only if the columns giving the conditional PMFs of X given $Y = y$ are identical to each other and to the column giving the marginal PMF of X.

Independent Random Variables: Multivariate Case

We can extend the definition of independence for random variables to any number of random variables. Again, the multiplication property provides the appropriate condition.

DEFINITION 6.5 Independence of Random Variables

Random variables $X_1, \ldots, X_m$ defined on the same sample space are said to be **independent random variables** if

$$P(X_1 \in A_1, \ldots, X_m \in A_m) = P(X_1 \in A_1) \cdots P(X_m \in A_m) \qquad (6.25)$$

for all subsets $A_1, \ldots, A_m$ of real numbers. The random variables of an infinite collection are said to be *independent* if the random variables of each finite subcollection are independent.

If $X_1, \ldots, X_m$ are independent random variables, so are the random variables obtained from a subcollection of those random variables. For instance, if $X_1, \ldots, X_5$ are independent random variables, the random variables X_1, X_2, and X_4 are also independent. Indeed, if A_1, A_2, and A_4 are subsets of $\mathcal{R}$, we have

$$P(X_1 \in A_1, \ X_2 \in A_2, \ X_4 \in A_4)$$
$$= P(X_1 \in A_1, \ X_2 \in A_2, \ X_3 \in \mathcal{R}, \ X_4 \in A_4, X_5 \in \mathcal{R})$$
$$= P(X_1 \in A_1)P(X_2 \in A_2)P(X_3 \in \mathcal{R})P(X_4 \in A_4)P(X_5 \in \mathcal{R})$$
$$= P(X_1 \in A_1)P(X_2 \in A_2)P(X_4 \in A_4).$$

Next, we recall from Definition 4.5 on page 150 that events $E_1, \ldots, E_m$ are independent if each subcollection of the m events satisfies the multiplication property—that is, if for each positive integer j, where $2 \le j \le m$, we have

$$P(E_{k_1} \cap E_{k_2} \cap \cdots \cap E_{k_j}) = P(E_{k_1})P(E_{k_2}) \cdots P(E_{k_j}),$$

where $k_1, k_2, \ldots, k_j$ are distinct integers between 1 and m, inclusive. Proposition 6.12, whose proof we leave to you as Exercise 6.76, provides an equivalent condition for independent random variables in terms of independent events.

◆◆◆ **Proposition 6.12**

Random variables $X_1, \ldots, X_m$ defined on the same sample space are independent if and only if the events $\{X_1 \in A_1\}, \ldots, \{X_m \in A_m\}$ are independent for all subsets $A_1, \ldots, A_m$ of real numbers.

Proposition 6.9 on page 291 states that, if two random variables are independent, so are the two random variables obtained by taking a function of one of the random variables and a function of the other random variable. This property also holds for more than two independent random variables. In fact, in Proposition 6.13, we state without proof an even stronger result to the effect that functions of disjoint subcollections of independent random variables are also independent random variables.

◆◆◆ Proposition 6.13

Suppose that $X_1, \ldots, X_m$ are independent random variables and let $k_1, k_2, \ldots, k_j$ be positive integers with $k_1 < k_2 < \cdots < k_j = m$. Further suppose that f_1 is a real-valued function on the range of $(X_1, \ldots, X_{k_1})$, that f_2 is a real-valued function on the range of $(X_{k_1+1}, \ldots, X_{k_2})$, $\ldots$, and that f_j is a real-valued function on the range of $(X_{k_{j-1}+1}, \ldots, X_m)$. Then the random variables $f_1(X_1, \ldots, X_{k_1})$, $f_2(X_{k_1+1}, \ldots, X_{k_2})$, $\ldots, f_k(X_{k_{j-1}+1}, \ldots, X_m)$ are independent.

EXAMPLE 6.20   *Functions of Independent Random Variables*

Suppose that $X_1, \ldots, X_6$ are independent random variables. Explain why the following are independent random variables.
a) $\min\{X_1, X_2\}$ and $\max\{X_3, X_4, X_5, X_6\}$
b) $X_1^2 + X_2^2 + X_3^2$, $\sin X_4$, and $\cos(X_5 + X_6)$
c) $X_1 X_2 X_4$ and $X_3 X_5$

Solution   To verify the independence of the random variables in each of parts (a)–(c), we apply Proposition 6.13 with the following functions, where in each case, we are taking functions of disjoint subcollections of the independent random variables $X_1, \ldots, X_6$.
a) $f_1(x_1, x_2) = \min\{x_1, x_2\}$ and $f_2(x_3, x_4, x_5, x_6) = \max\{x_3, x_4, x_5, x_6\}$
b) $f_1(x_1, x_2, x_3) = x_1^2 + x_2^2 + x_3^2$, $f_2(x_4) = \sin x_4$, and $f_3(x_5, x_6) = \cos(x_5 + x_6)$
c) $f_1(x_1, x_2, x_4) = x_1 x_2 x_4$ and $f_2(x_3, x_5, x_6) = x_3 x_5$ (Here we are applying Proposition 6.13 to the independent random variables $X_1, X_2, X_4, X_3, X_5, X_6$.)   ∎

Independent Discrete Random Variables: Multivariate Case

As in the bivariate discrete case, a necessary and sufficient condition for several discrete random variables to be independent is that their joint PMF equals the product of their marginal PMFs. Specifically, we have Proposition 6.14, whose proof is left to you as Exercise 6.77.

◆◆◆ Proposition 6.14   Independent Discrete Random Variables

Let $X_1, \ldots, X_m$ be discrete random variables defined on the same sample space. Then $X_1, \ldots, X_m$ are independent if and only if

$$p_{X_1, \ldots, X_m}(x_1, \ldots, x_m) = p_{X_1}(x_1) \cdots p_{X_m}(x_m), \qquad \text{for all } x_1, \ldots, x_m \in \mathcal{R}.$$

The definition of independence for events—Definition 4.5 on page 150—can be stated more easily in terms of independence of the corresponding indicator random variables. We do so in Proposition 6.15, whose proof is left to you as Exercise 6.78.

◆◆◆ Proposition 6.15

Events $E_1, E_2, \ldots, E_m$ are independent if and only if their indicators $I_{E_1}, I_{E_2}, \ldots, I_{E_m}$ are independent random variables.

EXERCISES 6.4   Basic Exercises

Note: Several of the exercises in this section are continuations of exercises presented in Sections 6.1–6.3.

6.60 Let X and Y be discrete random variables defined on the same sample space. Suppose that the joint PMF of X and Y can be factored into a function of x alone and a function of y alone—that is, suppose that there exist two real-valued functions q and r of one real variable such that $p_{X,Y}(x, y) = q(x)r(y)$ for all real numbers x and y.
a) Show that X and Y are independent random variables.
b) Is it necessarily true that q and r are the marginal PMFs of X and Y, respectively? If not, under what conditions are they?
c) Generalize to the multivariate case of m discrete random variables $X_1, \ldots, X_m$.

6.61 In Exercise 6.2 on page 269, you were asked to determine the joint and marginal PMFs of the number of siblings (X) and the number of sisters (Y) of a randomly selected student from among 40 students in one of Professor Weiss's classes. Are X and Y independent random variables? Explain your answer.

6.62 Let S consist of the 10 decimal digits. Suppose that a number X is chosen according to the discrete uniform distribution on S and then a number Y is chosen according to the discrete uniform distribution on S with X removed. Are X and Y independent random variables? Explain your answer.

6.63 A balanced die is tossed twice. Let X and Y denote the smaller and larger of the two faces, respectively. Are X and Y independent random variables? Explain your answer.

6.64 Refer to Exercise 6.9 on page 270, where, in a sequence of Bernoulli trials with success probability p, we let X denote the number of trials up to and including the first success and we let Y denote the number of trials up to and including the second success.
a) Without doing any computations, explain why X and Y aren't independent.
b) Use Proposition 6.10 on page 292 to show that X and Y aren't independent.
c) Use Proposition 6.11 on page 295 to show that X and Y aren't independent.
d) Without doing any computations, explain why X and $Y - X$ are independent.
e) Use Proposition 6.10 to show that X and $Y - X$ are independent.
f) Use Proposition 6.11 to show that X and $Y - X$ are independent.

6.65 Suppose that $X_1, \ldots, X_m$ are random variables defined on the same sample space. Further suppose that some two of the m random variables aren't independent. Is it possible that $X_1, \ldots, X_m$ are independent random variables? Explain your answer.

6.66 Refer to Exercise 6.26 on page 280. There, X_1, X_2, X_3, and X_4 denote the number of freshmen, sophomores, juniors, and seniors, respectively, obtained when a random sample of 5 students is taken with replacement from the 40 students in one of Professor Weiss's classes in which there are 6 freshmen, 15 sophomores, 12 juniors, and 7 seniors.
a) Without doing any computations, explain why X_1 and X_2 aren't independent.
b) Deduce from part (a) that X_1, X_2, X_3, and X_4 aren't independent.

6.67 Repeat Exercise 6.66 if the sampling is without replacement.

6.68 Let $X_1, \ldots, X_m$ have the multinomial distribution with parameters n and $p_1, \ldots, p_m$. Without doing any computations, explain why these random variables aren't independent.

6.69 A point (X, Y) is chosen randomly from $\{(x, y) : x, y \in \{0, 1, \ldots, 9\}\}$. Are X and Y independent random variables? Explain your answer.

6.70 A point (X, Y) is chosen randomly from $\{(x, y) : x, y \in \{0, 1, \ldots, 9\} \text{ and } x \geq y\}$.
a) Without doing any computations, explain why X and Y aren't independent.
b) Use Proposition 6.10 on page 292 to show that X and Y aren't independent.
c) Use Proposition 6.11 on page 295 to show that X and Y aren't independent.

6.71 A university gives separate placement exams in mathematics and verbal skills to incoming students. To the nearest minute, the time it takes a student to complete the mathematics exam has a geometric distribution with parameter 0.020, and the time it takes a student to complete the verbal exam has a geometric distribution with parameter 0.025. Assume that completion times for the two exams are independent random variables.
a) Determine the probability that a student takes longer to complete the mathematics exam than the verbal exam.
b) Determine the probability that a student takes at most twice as long to complete the mathematics exam as the verbal exam.
c) Determine the PMF of the total time that a student takes to complete both exams.
d) If students are allowed a total of 2 hours to complete both exams, what percentage finish in time?

6.72 Suppose that $X_1, \ldots, X_7$ are independent random variables and that $t \in \mathcal{R}$. In each part, decide whether the random variables are necessarily independent and explain your answers.
a) $\sqrt{X_1^2 + X_2^2 + X_3^2}$ and $\sqrt{X_4^2 + X_5^2 + X_6^2}$ b) $\sin(X_1 X_2 X_3)$ and $\cos(X_3 X_5)$
c) $\sin(X_1 X_2 X_5)$ and $\cos(X_3 X_4)$ d) $X_1 - X_2$ and $X_1 + X_2$
e) $\min\{X_1, X_2, X_7\}$ and $\max\{X_3, X_4\}$ f) $\sum_{k=1}^5 X_k \sin(kt)$ and $\sum_{k=6}^7 X_k \sin(kt)$

6.73 At a large hospital, an average of 6.9 patients arrive at the hospital emergency room each hour. The number who arrive during any particular hour has a Poisson distribution. At a smaller hospital, an average of 2.6 patients arrive at the hospital emergency room each hour. The number who arrive during any particular hour has a Poisson distribution. Assume that the number of patients arriving at the two emergency rooms are independent.
a) Determine and identify the PMF of the total number of patients that arrive each hour at both emergency rooms combined. *Hint:* Use the FPF and the binomial theorem.
b) On average, how many patients per hour arrive at the two hospital emergency rooms combined? Explain your answer.
c) Use your answer from part (a) to determine the probability that, during a one-hour period, at most eight patients arrive at the two hospital emergency rooms combined.

6.74 The number of people entering a bank during a specified time interval has a Poisson distribution with parameter λ. Each person entering makes a deposit with probability p, independent of the other people entering and the number of people entering. Determine and identify the PMF of the number of people making a deposit during the specified time interval.

6.75 The number of people entering a bank during a specified time interval has a Poisson distribution with parameter λ. Each arriving person is classified by type as follows.

● Type 1: drives to the bank and walks in

● Type 2: drives to the bank and uses the drive-thru facility

● Type 3: walks to the bank

An arriving person has probability p_i of being Type i, independent of other arrivals and the number of arrivals. Let X_i denote the number of Type i people that arrive during the specified time interval. Show that X_1, X_2, and X_3 are independent Poisson random variables with parameters λp_1, λp_2, and λp_3, respectively.

Theory Exercises

6.76 Prove Proposition 6.12: Random variables $X_1, \ldots, X_m$ defined on the same sample space are independent if and only if the events $\{X_1 \in A_1\}, \ldots, \{X_m \in A_m\}$ are independent for all subsets $A_1, \ldots, A_m$ of real numbers.

6.77 Prove Proposition 6.14: Discrete random variables $X_1, \ldots, X_m$ defined on the same sample space are independent if and only if

$$p_{X_1, \ldots, X_m}(x_1, \ldots, x_m) = p_{X_1}(x_1) \cdots p_{X_m}(x_m)$$

for all $x_1, \ldots, x_m \in \mathcal{R}$.

6.78 Prove Proposition 6.15: Events $E_1, E_2, \ldots, E_m$ are independent if and only if their indicators $I_{E_1}, I_{E_2}, \ldots, I_{E_m}$ are independent (discrete) random variables.

Advanced Exercises

6.79 Let X and Y be independent random variables that have geometric distributions with parameters p and q, respectively.
a) Find and identify the PMF of the random variable $\min\{X, Y\}$. *Hint:* Use tail probabilities.
b) Find the PMF of the random variable $X + Y$.
c) Identify the PMF in part (b) if $p = q$. Interpret your answer.

6.80 Suppose that X and Y are discrete random variables that jointly take the values $(1, 0)$, $(0, 1)$, $(-1, 0)$, and $(0, -1)$ with probability $1/4$ each.
a) Show that the random variables X and Y aren't independent.
b) Show that the random variables $X + Y$ and $X - Y$ are independent.

6.81 Suppose that X has a Poisson distribution with parameter λ. Further suppose that, given $X = x$, the random variables $X_1, \ldots, X_m$ have the multinomial distribution with parameters x and $p_1, \ldots, p_m$. Show that $X_1, \ldots, X_m$ are independent Poisson random variables with parameters $\lambda p_1, \ldots, \lambda p_m$, respectively.

6.82 Let $X_1, \ldots, X_m$ be independent integer-valued random variables with the same probability distribution.
a) Determine the PMF of $\min\{X_1, \ldots, X_m\}$ in terms of tail probabilities.
b) Use part (a) to determine and identify the PMF of $\min\{X_1, \ldots, X_m\}$ in case the common probability distribution is geometric with parameter p.
c) Determine the PMF of $\max\{X_1, \ldots, X_m\}$ in terms of tail probabilities.
d) Use part (c) to determine the PMF of $\max\{X_1, \ldots, X_m\}$ in case the common probability distribution is geometric with parameter p. Does $\max\{X_1, \ldots, X_m\}$ have a geometric distribution? Justify your answer.

6.83 Constant random variable: A random variable X is said to be a *constant random variable* if there is a real number c such that $P(X = c) = 1$.
a) Determine the PMF of a constant random variable.
b) Let X be a discrete random variable. Show that X is independent of itself if and only if it is a constant random variable.
c) Let X be a discrete random variable. Show that X is independent of all other random variables defined on the same sample space if and only it is a constant random variable.

6.84 Conditional independence: Let X, Y, and Z be discrete random variables defined on the same sample space. We say that X and Y are *conditionally independent* given Z if, for all values z of Z with positive probability, the conditional joint PMF of X and Y given $Z = z$

equals the product of the conditional PMF of X given $Z = z$ and the conditional PMF of Y given $Z = z$.

a) State the condition for conditional independence symbolically.

Consider now a coin tossing game in which you win $1 if the coin comes up a head and lose $1 if the coin comes up a tail. Your initial fortune is $0, and the coin has probability p of a head. Let X_1, X_2, and X_3 be your fortune after tosses 1, 2, and 3, respectively. In each case, answer the question and explain your reasoning.

b) Are X_1 and X_3 independent?

c) Are X_1 and X_3 conditionally independent given X_2?

d) Are X_1 and X_2 independent?

e) Are X_1 and X_2 conditionally independent given X_3?

6.5   Functions of Two or More Discrete Random Variables

In Section 5.8, we discussed functions of one discrete random variable. Now we consider functions of two or more discrete random variables.

As usual, we begin with the bivariate case. Let X and Y be two discrete random variables defined on the same sample space and let g be a real-valued function of two variables defined on the range of (X, Y). Then we can obtain another random variable, Z, by composing g with (X, Y)—that is, by letting $Z = g(X, Y)$. Because X and Y are discrete random variables, Z is also a discrete random variable. (See Exercise 6.99.)

EXAMPLE 6.21   *Functions of Two Discrete Random Variables*

Let X and Y be two discrete random variables defined on the same sample space. Discuss the following functions of these two discrete random variables.

a) $X + Y$ b) $|X - Y|$ c) $3X - 4Y$

d) $\sqrt{X^2 + Y^2}$ e) $\max\{X, Y\}$ f) $\min\{X, Y\}$

Solution   a) The random variable $X + Y$ is the sum of the random variables X and Y, obtained by letting $g(x, y) = x + y$.

b) The random variable $|X - Y|$ is the magnitude of the difference between the random variables X and Y, obtained by letting $g(x, y) = |x - y|$.

c) The random variable $3X - 4Y$ is a particular linear combination of the random variables X and Y, obtained by letting $g(x, y) = 3x - 4y$.

d) The random variable $\sqrt{X^2 + Y^2}$ gives the distance from the origin of the random point (X, Y), obtained by letting $g(x, y) = \sqrt{x^2 + y^2}$.

e) The random variable $\max\{X, Y\}$ is the larger of the random variables X and Y, obtained by letting $g(x, y) = \max\{x, y\}$.

f) The random variable $\min\{X, Y\}$ is the smaller of the random variables X and Y, obtained by letting $g(x, y) = \min\{x, y\}$.   ∎

PMF of a Function of Two Discrete Random Variables

Several techniques are available for determining the PMF of a function of two (or more) discrete random variables. The most straightforward method, although not necessarily the easiest, is presented in Proposition 6.16.

◆◆◆ **Proposition 6.16   PMF of a Function of Two Discrete Random Variables**

Let X and Y be two discrete random variables defined on the same sample space and let g be a real-valued function of two variables defined on the range of (X, Y). Then the PMF of the random variable $Z = g(X, Y)$ is

$$p_Z(z) = \sum\sum_{(x,y)\in g^{-1}(\{z\})} p_{X,Y}(x, y), \tag{6.26}$$

for z in the range of Z, and $p_Z(z) = 0$ otherwise. In words, if z is in the range of Z, we obtain the probability that $Z = z$—that is, the probability that $g(X, Y) = z$—by summing the joint PMF of X and Y over all points (x, y) in the plane such that $g(x, y) = z$.

Proof   Let z be in the range of Z. From the FPF for two discrete random variables,

$$p_Z(z) = P(Z = z) = P\big(g(X, Y) = z\big)$$

$$= P\big((X, Y) \in g^{-1}(\{z\})\big) = \sum\sum_{(x,y)\in g^{-1}(\{z\})} p_{X,Y}(x, y),$$

as required.   ◆

Note: We can express Equation (6.26) in the alternate form

$$p_{g(X,Y)}(z) = \sum\sum_{g(x,y)=z} p_{X,Y}(x, y), \tag{6.27}$$

where $\sum\sum_{g(x,y)=z}$ indicates that the double sum is taken over all x and y such that $g(x, y) = z$.

EXAMPLE 6.22   *PMF of a Function of Two Discrete Random Variables*

Let X and Y be independent random variables, both having the geometric distribution with parameter p. Determine and identify the PMF of the random variable
a) $X + Y$.   b) $\min\{X, Y\}$.

Solution   We apply Proposition 6.16. In doing so, we first need to obtain the joint PMF of X and Y. By assumption, both X and Y have the geometric distribution with parameter p. Therefore,

$$p_X(x) = \begin{cases} p(1 - p)^{x-1}, & \text{if } x \in \mathcal{N}; \\ 0, & \text{otherwise.} \end{cases} \qquad p_Y(y) = \begin{cases} p(1 - p)^{y-1}, & \text{if } y \in \mathcal{N}; \\ 0, & \text{otherwise.} \end{cases}$$

Furthermore, because X and Y are independent, their joint PMF equals the product of their marginal PMFs, or

$$p_{X,Y}(x, y) = p(1 - p)^{x-1} \cdot p(1 - p)^{y-1} = p^2(1 - p)^{x+y-2}, \qquad \textbf{(6.28)}$$

if $x, y \in \mathcal{N}$, and $p_{X,Y}(x, y) = 0$ otherwise. We can now apply Proposition 6.16 to solve parts (a) and (b).

a) The range of $X + Y$ consists of all positive integers greater than or equal to 2. From Proposition 6.16 or, equivalently, from Equation (6.27),

$$p_{X+Y}(z) = \sum_{x+y=z}\sum p_{X,Y}(x, y).$$

Figure 6.3 shows the set over which the double summation is taken.

Figure 6.3

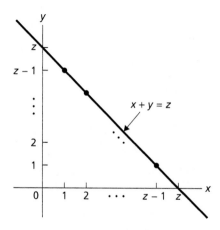

From Equation (6.28) and the preceding equation, we conclude that

$$p_{X+Y}(z) = \sum_{x+y=z}\sum p_{X,Y}(x, y) = \sum_{x=1}^{z-1}\left(\sum_{y=z-x}^{z-x} p^2(1 - p)^{x+y-2} \right)$$

$$= \sum_{x=1}^{z-1} p^2(1 - p)^{z-2} = p^2(1 - p)^{z-2} \sum_{x=1}^{z-1} 1 = (z - 1)p^2(1 - p)^{z-2}$$

$$= \binom{z - 1}{2 - 1} p^2(1 - p)^{z-2},$$

if $z = 2, 3, \ldots$, and $p_{X+Y}(z) = 0$ otherwise. Referring to Proposition 5.13 on page 241, we see that $X + Y \sim \mathcal{NB}(2, p)$; that is, $X + Y$ has the negative binomial distribution with parameters 2 and p.

b) The range of $\min\{X, Y\}$ consists of all positive integers. From Proposition 6.16 or, equivalently, from Equation (6.27),

$$p_{\min\{X,Y\}}(z) = \sum\sum_{\min\{x,y\}=z} p_{X,Y}(x, y).$$

Figure 6.4 shows the set over which the double summation is taken.

Figure 6.4

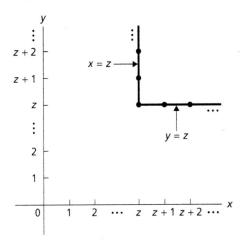

From Equation (6.28) and the preceding equation, we conclude that

$$p_{\min\{X,Y\}}(z) = \sum\sum_{\min\{x,y\}=z} p_{X,Y}(x, y)$$

$$= p_{X,Y}(z, z) + \sum_{y=z+1}^{\infty} p_{X,Y}(z, y) + \sum_{x=z+1}^{\infty} p_{X,Y}(x, z)$$

$$= p^2(1-p)^{2z-2} + \sum_{y=z+1}^{\infty} p^2(1-p)^{z+y-2} + \sum_{x=z+1}^{\infty} p^2(1-p)^{x+z-2}$$

$$= p^2(1-p)^{2z-2} + 2p^2(1-p)^{z-2} \sum_{w=z+1}^{\infty} (1-p)^{w}$$

$$= p^2(1-p)^{2z-2} + 2p^2(1-p)^{z-2} \cdot \frac{(1-p)^{z+1}}{1-(1-p)}$$

$$= (2p - p^2)\big(1 - (2p - p^2)\big)^{z-1},$$

if $z \in \mathcal{N}$, and $p_{\min\{X,Y\}}(z) = 0$ otherwise. Referring to Proposition 5.9 on page 229, we see that $\min\{X, Y\} \sim \mathcal{G}(2p - p^2)$; that is, $\min\{X, Y\}$ has the geometric distribution with parameter $2p - p^2$. ∎

As we mentioned earlier, although Proposition 6.16 is the most straightforward method for obtaining the PMF of a function of two discrete random variables, it isn't necessarily the easiest. For instance, we can solve Example 6.22(b) more quickly by using tail probabilities as follows. Let $Z = \min\{X, Y\}$ and let $z \in \mathcal{N}$. Note that $Z > z$ if and only if $X > z$ and $Y > z$—that is, $\{Z > z\} = \{X > z, Y > z\}$. Therefore, from the independence of X and Y and Proposition 5.10 on page 231,

$$P(Z > z) = P(X > z, Y > z) = P(X > z)P(Y > z) = (1 - p)^z(1 - p)^z$$
$$= (1 - p)^{2z} = \big((1 - p)^2\big)^z = \big(1 - (2p - p^2)\big)^z.$$

Again, from Proposition 5.10, the tail probabilities of Z are the same as those of a geometric random variable with parameter $2p - p^2$. And, as tail probabilities determine the PMF [see Equation (5.32) on page 230], we conclude that $Z \sim \mathcal{G}(2p - p^2)$. This result is the same as the one obtained in Example 6.22(b), but the method we just used is quicker and computationally less intensive.

PMF of a Function of Several Discrete Random Variables

The generalization of the earlier results of this section to the multivariate discrete case is clear. In particular, Proposition 6.17 is the multivariate version of Proposition 6.16.

◆◆◆   **Proposition 6.17   PMF of a Function of m Discrete Random Variables**

Let $X_1, \ldots, X_m$ be m discrete random variables defined on the same sample space and let g be a real-valued function of m variables defined on the range of $(X_1, \ldots, X_m)$. Then the PMF of the random variable $Z = g(X_1, \ldots, X_m)$ is

$$p_Z(z) = \sum \cdots \sum_{(x_1, \ldots, x_m) \in g^{-1}(\{z\})} p_{X_1, \ldots, X_m}(x_1, \ldots, x_m),$$

for z in the range of Z, and $p_Z(z) = 0$ otherwise. In words, if z is in the range of Z, we obtain the probability that $Z = z$—that is, the probability that $g(X_1, \ldots, X_m) = z$—by summing the joint PMF of $X_1, \ldots, X_m$ over all points $(x_1, \ldots, x_m)$ in m-dimensional Euclidean space such that $g(x_1, \ldots, x_m) = z$.

Again, in particular instances, use of a special technique, such as tail probabilities, to obtain the PMF of a function of m discrete random variables may very well be more efficient than Proposition 6.17.

EXERCISES 6.5   Basic Exercises

Note: Unless instructed otherwise, apply Proposition 6.16 on page 302 to determine all required PMFs.

6.85 In Example 6.1, we considered the number of bedrooms and number of bathrooms of a randomly selected home among 50 homes currently for sale. The joint PMF of those two random variables is provided in Table 6.2 on page 262. Determine the PMF of the
a) total number of bedrooms and bathrooms.

b) difference between the number of bedrooms and the number of bathrooms.

c) maximum of the number of bedrooms and the number of bathrooms.

d) minimum of the number of bedrooms and the number of bathrooms.

6.86 In Exercise 6.2 on page 269, you were asked to determine the joint PMF of the number of siblings, X, and the number of sisters, Y, of a randomly selected student from among 40 students in one of Professor Weiss's classes.

a) Determine the PMF of $X - Y$.

b) What does the random variable $X - Y$ represent?

6.87 A balanced die is tossed twice. Let X and Y denote the smaller and larger of the two faces, respectively.

a) Determine the PMF of $Y - X$. *Note:* You were asked to obtain the joint PMF of X and Y in Exercise 6.6 on page 270.

b) What does the random variable $Y - X$ represent?

6.88 Two people agree to meet at a specified place between 3:00 P.M. and 4:00 P.M. Suppose that you measure time to the nearest minute relative to 3:00 P.M. so that, for instance, time 40 represents 3:40 P.M. Further suppose that each person arrives according to the discrete uniform distribution on $\{0, 1, \ldots, 60\}$ and that the two arrival times are independent.

a) Determine the PMF of the number of minutes that the first person to arrive waits for the second person to arrive.

b) Use the result of part (a) to find the probability that the first person to arrive waits no longer than 10 minutes for the second person to arrive.

c) Determine the PMF of the time of the first person to arrive.

d) Determine the PMF of the time of the second person to arrive.

6.89 Suppose that X and Y are independent random variables, each having the discrete uniform distribution on $\{0, 1, \ldots, N\}$. Determine the PMF of the magnitude of the difference between X and Y by using

a) combinatorial analysis.   **b)** Proposition 6.16.

6.90 Let X and Y be as in Exercise 6.89. Determine the PMF of the

a) sum of X and Y.   **b)** minimum of X and Y.   **c)** maximum of X and Y.

6.91 Let X and Y be independent random variables that have geometric distributions with parameters p and q, respectively. Find the PMF of $\min\{X, Y\}$

a) directly, as in Example 6.22(b) on page 304.   **b)** by using tail probabilities.

c) Verify that your answers in parts (a) and (b) agree with that in Example 6.22(b) in the case where $p = q$.

Theory Exercises

6.92 Prove Proposition 6.17 on page 305, which provides the PMF of a real-valued function of $X_1, \ldots, X_m$ in terms of the joint PMF of $X_1, \ldots, X_m$.

6.93 Suppose that X and Y are discrete random variables defined on the same sample space. Use Proposition 6.16 on page 302 to derive the following formulas for the sum and difference of X and Y.

a) $p_{X+Y}(z) = \sum_x p_{X,Y}(x, z - x)$ **b)** $p_{Y-X}(z) = \sum_x p_{X,Y}(x, x + z)$

c) Specialize the formulas in parts (a) and (b) to the case where X and Y are independent random variables.

6.94 Derive the formulas in Exercise 6.93 by applying the law of partitions, Proposition 2.8 on page 68.

6.95 Suppose that X and Y are positive discrete random variables defined on the same sample space. Use Proposition 6.16 to derive the following formulas for the product and quotient of X and Y.

a) $p_{XY}(z) = \sum_x p_{X,Y}(x, z/x)$ **b)** $p_{Y/X}(z) = \sum_x p_{X,Y}(x, xz)$

c) Specialize the formulas in parts (a) and (b) to the case where X and Y are independent.

6.96 Derive the formulas in Exercise 6.95 by applying the law of partitions, Proposition 2.8 on page 68.

Advanced Exercises

6.97 Let X and Y be independent random variables with $X \sim \mathcal{B}(m, p)$ and $Y \sim \mathcal{B}(n, p)$.

a) Without doing any computations, identify the PMF of the random variable $X + Y$. Explain your reasoning.

b) Determine the PMF of $X + Y$ by applying Proposition 6.16. *Hint:* Use Vandermonde's identity, Exercise 3.48 on page 109.

c) Let $X_1, \ldots, X_m$ be independent random variables with $X_j \sim \mathcal{B}(n_j, p)$ for $j = 1, 2, \ldots, m$. Obtain and identify the PMF of the random variable $X_1 + \cdots + X_m$. *Hint:* Use part (b) and mathematical induction, but be sure to justify all your steps.

6.98 Let X and Y be independent random variables with $X \sim \mathcal{P}(\lambda)$ and $Y \sim \mathcal{P}(\mu)$.

a) Determine the PMF of $X + Y$. *Hint:* Use the binomial formula.

b) Let $X_1, \ldots, X_m$ be independent random variables with $X_j \sim \mathcal{P}(\lambda_j)$ for $j = 1, 2, \ldots, m$. Obtain and identify the PMF of the random variable $X_1 + \cdots + X_m$. *Hint:* Use part (a) and mathematical induction, but be sure to justify all your steps.

6.99 Refer to the definition of a discrete random variable as presented in Definition 5.2 on page 178. Let X and Y be discrete random variables defined on the same sample space and let g be a real-valued function of two real variables defined on the range of (X, Y). Prove that $g(X, Y)$ is a discrete random variable.

6.6   Sums of Discrete Random Variables

In Section 6.5, we introduced functions of discrete random variables. One of the most important functions in this regard is summation. In this section, we specialize the results of Section 6.5 to the case of sums of discrete random variables and, in particular, to sums of independent discrete random variables. As previously, we begin with the bivariate case of two discrete random variables. Our main result is Proposition 6.18.

◆◆◆   **Proposition 6.18   PMF of the Sum of Two Discrete Random Variables**

Let X and Y be discrete random variables defined on the same sample space. Then the PMF of the random variable $X + Y$ can be obtained from either of two formulas:

$$p_{X+Y}(z) = \sum_x p_{X,Y}(x, z - x), \qquad z \in \mathcal{R}, \tag{6.29}$$

or

$$p_{X+Y}(z) = \sum_y p_{X,Y}(z - y, y), \qquad z \in \mathcal{R}. \tag{6.30}$$

Proof   We verify Equation (6.29) and leave verification of Equation (6.30) to you as Exercise 6.113. Applying Proposition 6.16 on page 302 to the function $g(x, y) = x + y$ gives

$$p_{X+Y}(z) = \sum_{x+y=z}\sum p_{X,Y}(x, y) = \sum_x \left(\sum_{y=z-x}^{z-x} p_{X,Y}(x, y) \right) = \sum_x p_{X,Y}(x, z - x),$$

as required.   ◆

You can think of the formulas in Proposition 6.18 in terms of the law of partitions. For instance, consider Equation (6.29). As x ranges over the possible values of X, the events $\{X = x\}$ form a partition of the sample space. Moreover,

$$\{X = x\} \cap \{X + Y = z\} = \{X = x,\ X + Y = z\} = \{X = x,\ Y = z - x\}.$$

Therefore, by the law of partitions (Proposition 2.8 on page 68), we have

$$P(X + Y = z) = \sum_x P(X = x,\ Y = z - x),$$

which is Equation (6.29).

EXAMPLE 6.23   *PMF of the Sum of Two Discrete Random Variables*

Bedrooms and Bathrooms   Refer to Example 6.1 and the contingency table in Table 6.1 on page 261 for number of bedrooms and number of bathrooms for 50 homes currently for sale. Suppose that one of these 50 homes is selected at random. Determine the PMF of the total number of bedrooms and bathrooms of the home obtained.

Solution   Let X and Y denote the number of bedrooms and number of bathrooms, respectively, of the home obtained. We want to determine the PMF of the random variable $X + Y$, whose possible values are $4, 5, \ldots, 9$. From Equation (6.29) and Table 6.2 on page 262,

$$p_{X+Y}(4) = \sum_x p_{X,Y}(x, 4 - x)$$

$$= p_{X,Y}(2, 4 - 2) + p_{X,Y}(3, 4 - 3) + p_{X,Y}(4, 4 - 4)$$

$$= p_{X,Y}(2, 2) + p_{X,Y}(3, 1) + p_{X,Y}(4, 0)$$

$$= 0.06 + 0.00 + 0.00 = 0.06.$$

Similarly, we find $p_{X+Y}(z)$ for $z = 5, 6, \ldots, 9$, and thereby obtain the PMF of the random variable $X + Y$, as shown in Table 6.7.

Table 6.7 PMF of $X + Y$, the total number of bedrooms and bathrooms

z	4	5	6	7	8	9
$p_{X+Y}(z)$	0.06	0.28	0.28	0.26	0.10	0.02

Of course, we could also obtain the PMF of $X + Y$ by using Equation (6.30) instead of Equation (6.29).   ■

PMF of the Sum of Two Independent Discrete Random Variables

In this book, we concentrate on sums of random variables that are independent. For two independent discrete random variables, we have Proposition 6.19. We obtain it as a special case of Proposition 6.18 by using the fact that the joint PMF of two independent discrete random variables equals the product of their marginal PMFs.

◆◆◆ **Proposition 6.19   Sum of Two Independent Discrete Random Variables**

Let X and Y be independent discrete random variables. Then the PMF of the random variable X + Y can be obtained from either of two formulas:

$$p_{X+Y}(z) = \sum_x p_X(x) p_Y(z - x), \qquad z \in \mathcal{R}, \tag{6.31}$$

or

$$p_{X+Y}(z) = \sum_y p_X(z - y) p_Y(y), \qquad z \in \mathcal{R}. \tag{6.32}$$

Note: Each of the sums on the right of Equations (6.31) and (6.32) is called a **convolution** of the probability mass functions p_X and p_Y. Thus we can restate Proposition 6.19 as follows: *The PMF of the sum of two independent discrete random variables is the convolution of their marginal PMFs.*

As our first application of Proposition 6.19, we show that the sum of two independent binomial random variables with the same success-probability parameter is also a binomial random variable with that success-probability parameter.

EXAMPLE 6.24   *Sum of Two Independent Binomial Random Variables*

Suppose that X and Y are independent binomial random variables, both having the same success-probability parameter—say, $X \sim \mathcal{B}(m, p)$ and $Y \sim \mathcal{B}(n, p)$. Find and identify the PMF of the random variable $X + Y$.

Solution   We have

$$p_X(x) = \binom{m}{x} p^x (1 - p)^{m-x}, \qquad x = 0, 1, \ldots, m,$$

and $p_X(x) = 0$ otherwise. Likewise,

$$p_Y(y) = \binom{n}{y} p^y (1 - p)^{n-y}, \qquad y = 0, 1, \ldots, n,$$

and $p_Y(y) = 0$ otherwise.

To obtain the PMF of $X + Y$, we first note that the possible values of $X + Y$ are 0, 1, ..., $m + n$, to which we restrict our attention. From Proposition 6.19,

$$p_{X+Y}(z) = \sum_x p_X(x) p_Y(z - x).$$

To evaluate the sum on the right of this equation, we first note that $p_X(x) = 0$ if $x < 0$ and $p_Y(z - x) = 0$ if $x > z$. Therefore

$$p_{X+Y}(z) = \sum_x p_X(x) p_Y(z - x)$$

$$= \sum_{x=0}^{z} \binom{m}{x} p^x (1 - p)^{m-x} \cdot \binom{n}{z - x} p^{z-x} (1 - p)^{n-(z-x)}$$

$$= p^z (1 - p)^{m+n-z} \sum_{x=0}^{z} \binom{m}{x} \binom{n}{z - x} = \binom{m + n}{z} p^z (1 - p)^{m+n-z},$$

where the last equality follows from the combinatorial identity

$$\sum_{x=0}^{z} \binom{m}{x} \binom{n}{z - x} = \binom{m + n}{z}. \tag{6.33}$$

This identity—known as Vandermonde's identity—can be established as indicated in Exercise 3.48 on page 109. It also follows from the fact that the values of a hypergeometric PMF must sum to 1. In any case, the PMF of $X + Y$ is

$$p_{X+Y}(z) = \binom{m + n}{z} p^z (1 - p)^{m+n-z}, \qquad z = 0, 1, \ldots, m + n,$$

and $p_{X+Y}(z) = 0$ otherwise. Thus $X + Y \sim \mathcal{B}(m + n, p)$. In words, the sum of two independent binomial random variables with the same success-probability parameter is also a binomial random variable whose number-of-trials parameter is the sum of those of the binomial random variables in the sum and whose success-probability parameter is the common one of the binomial random variables in the sum.

Actually, we could have anticipated this result by arguing as follows. The random variable X gives the number of successes in m independent trials with success probability p, and the random variable Y gives the number of successes in n independent trials with success probability p. Because X and Y are independent, the random variable $X + Y$ gives the number of successes in $m + n$ independent trials with success probability p. Hence $X + Y$ has the binomial distribution with parameters $m + n$ and p. ∎

We have shown that the sum of two independent binomial random variables with the same success-probability parameter is also a binomial random variable. An analogous result holds for negative binomial random variables. Specifically, as you are asked to show in Exercise 6.116, if X and Y are independent negative binomial random variables with the same success-probability parameter, say, $X \sim \mathcal{NB}(r, p)$ and $Y \sim \mathcal{NB}(s, p)$, then $X + Y \sim \mathcal{NB}(r + s, p)$.

Another important result in this direction applies to Poisson random variables. If X and Y are independent Poisson random variables, say, $X \sim \mathcal{P}(\lambda)$ and $Y \sim \mathcal{P}(\mu)$, then $X + Y \sim \mathcal{P}(\lambda + \mu)$. Thus the sum of two independent Poisson random variables is also a Poisson random variable whose parameter is the sum of the parameters of the Poisson random variables in the sum. We leave the verification of this fact to you as Exercise 6.114.

PMF of the Sum of Several Independent Discrete Random Variables

Formulas analogous to those for the PMF of the sum of two discrete random variables are available in the multivariate case, although such formulas are generally difficult to implement. However, important special cases exist for which we can explicitly obtain the PMF of the sum of several independent discrete random variables. Three such cases of particular interest—the binomial, Poisson, and negative binomial—are given in Proposition 6.20. You should interpret each result in Proposition 6.20 in words, a task that we formally ask you to do in Exercise 6.106.

◆◆◆   **Proposition 6.20**

Let $X_1, \ldots, X_m$ be independent discrete random variables.
a) If $X_j \sim \mathcal{B}(n_j, p)$ for $1 \leq j \leq m$, then $X_1 + \cdots + X_m \sim \mathcal{B}(n_1 + \cdots + n_m, p)$.
b) If $X_j \sim \mathcal{P}(\lambda_j)$ for $1 \leq j \leq m$, then $X_1 + \cdots + X_m \sim \mathcal{P}(\lambda_1 + \cdots + \lambda_m)$.
c) If $X_j \sim \mathcal{NB}(r_j, p)$ for $1 \leq j \leq m$, then $X_1 + \cdots + X_m \sim \mathcal{NB}(r_1 + \cdots + r_m, p)$.

Proof   We prove part (a) and leave the proofs of the other two parts to you as Exercise 6.117. We use mathematical induction by assuming the truth of part (a) for $m - 1$ and then establishing it for m. Thus, suppose that $X_1, \ldots, X_m$ are independent binomial random variables with $X_j \sim \mathcal{B}(n_j, p)$ for $1 \leq j \leq m$. From Proposition 6.13 on page 297, $X_1 + \cdots + X_{m-1}$ and X_m are independent random variables. By the induction assumption, $X_1 + \cdots + X_{m-1} \sim \mathcal{B}(n_1 + \cdots + n_{m-1}, p)$. Referring now to Example 6.24 on page 309, we deduce that $(X_1 + \cdots + X_{m-1}) + X_m \sim \mathcal{B}\big((n_1 + \cdots + n_{m-1}) + n_m, p\big)$. In other words, $X_1 + \cdots + X_m \sim \mathcal{B}(n_1 + \cdots + n_m, p)$.   ◆

In Chapter 11, we present simpler and more efficient ways to deal with sums of independent random variables through the use of certain types of transforms—specifically, moment generating functions and characteristic functions. For instance, the proof of Proposition 6.20 becomes almost trivial by using moment generating functions.

Sum of a Random Number of Random Variables

Many applications of probability theory involve a sum of a random number of random variables. For example, insurance companies are interested in the total amount of money paid out on claims for specific types of policies. That total naturally decomposes into two parts: the number of claims—called the *claim frequency*—and the amount of each claim—called the *claim severity*. Both the claim frequency and the claim severity are random variables and, thus, the total amount of money paid out is a sum of a random number of random variables.

To obtain a probabilistic model for the sum of a random number of random variables, we let $X_1, X_2, \ldots$ be random variables and let N be a nonnegative integer-valued random variable independent of the X_js. We set $S_0 = 0$ and $S_n = X_1 + \cdots + X_n$, for $n \geq 1$. Then S_N is a random variable obtained as the sum of a random number of random variables. Proposition 6.21 provides a formula for the PMF of S_N in the case where the X_js are discrete random variables.

◆◆◆   **Proposition 6.21   Sum of a Random Number of Random Variables**

Let $X_1, X_2, \ldots$ be discrete random variables and let N be a nonnegative integer-valued random variable independent of the X_js. Then the PMF of S_N is

$$p_{S_N}(x) = \sum_{n=0}^{\infty} p_N(n) p_{S_n}(x), \qquad x \in \mathcal{R}. \tag{6.34}$$

Proof   From Proposition 6.2 on page 263 and the general multiplication rule,

$$p_{S_N}(x) = \sum_{n=0}^{\infty} p_{N, S_N}(n, x) = \sum_{n=0}^{\infty} p_N(n) p_{S_N \mid N}(x \mid n). \tag{6.35}$$

It is easy to see that $p_{S_N \mid N}(x \mid 0) = p_{S_0}(x)$. Moreover, because N and the X_js are independent, we can apply Proposition 6.11 on page 295 and Proposition 6.13 on page 297 to conclude that, for each $n \in \mathcal{N}$,

$$p_{S_N \mid N}(x \mid n) = P(S_N = x \mid N = n) = P(S_n = x \mid N = n)$$
$$= P(S_n = x) = p_{S_n}(x). \tag{6.36}$$

The required result now follows from Equations (6.35) and (6.36).   ◆

Frequently, the X_js are independent random variables and have the same probability distribution. In such cases, the distribution of S_N is called a **compound distribution.** Of particular importance in this regard is when N has a Poisson distribution; then the distribution of S_N is referred to as a **compound Poisson distribution.**

EXAMPLE 6.25   *A Compound Poisson Distribution*

Consider Bernoulli trials with success probability p. For each $j \in \mathcal{N}$, let $X_j = 1$ if the jth trial is a success and let $X_j = 0$ otherwise. Also, let N be independent of the X_js and have the Poisson distribution with parameter λ. Find and identify the PMF of the random variable S_N, the number of successes in a Poisson random number of Bernoulli trials.

Solution   Because Bernoulli trials are independent, $X_1, X_2, \ldots$ are independent discrete random variables. For each $n \in \mathcal{N}$, the sum $S_n = X_1 + \cdots + X_n$ gives the total number of successes in the first n Bernoulli trials and hence has the binomial distribution with parameters n and p. From Proposition 6.21, we conclude that

$$p_{S_N}(x) = \sum_{n=0}^{\infty} p_N(n) p_{S_n}(x) = \sum_{n=x}^{\infty} e^{-\lambda} \frac{\lambda^n}{n!} \cdot \binom{n}{x} p^x (1-p)^{n-x}$$

$$= \frac{e^{-\lambda}(p\lambda)^x}{x!} \sum_{n=x}^{\infty} \frac{\left((1-p)\lambda\right)^{n-x}}{(n-x)!} = \frac{e^{-\lambda}(p\lambda)^x}{x!} \sum_{k=0}^{\infty} \frac{\left((1-p)\lambda\right)^k}{k!}$$

$$= \frac{e^{-\lambda}(p\lambda)^x}{x!} \cdot e^{(1-p)\lambda} = e^{-p\lambda} \frac{(p\lambda)^x}{x!}.$$

for $x = 0, 1, 2, \ldots$. Thus $S_N \sim \mathcal{P}(p\lambda)$; that is, S_N has the Poisson distribution with parameter $p\lambda$. Compare this result with that obtained in Example 6.19(a) on page 294.   ■

EXERCISES 6.6   Basic Exercises

6.100 In Example 6.1, we considered the number of bedrooms and number of bathrooms of a randomly selected home among 50 homes currently for sale. The joint PMF of those two random variables is provided in Table 6.2 on page 262. Determine the PMF of the total number of bedrooms and bathrooms by applying Equation (6.30) on page 307.

6.101 A balanced die is tossed twice. Let X and Y denote the smaller and larger of the two faces, respectively. You were asked to obtain the joint PMF of X and Y in Exercise 6.6 on page 270.
a) Interpret the random variable $M = (X + Y)/2$.
b) Determine the PMF of M by first applying Proposition 6.18 on page 307 and then Corollary 5.1 on page 247.

6.102 Suppose that X and Y are independent random variables each having the discrete uniform distribution on $\{0, 1, \ldots, N\}$. Use Proposition 6.19 on page 309 to obtain the PMF of $X + Y$. Compare your result with that obtained in Exercise 6.90(a) on page 306.

6.103 Is the sum of two independent geometric random variables with the same success-probability parameter a geometric random variable? If not, what is its distribution?

6.104 During any particular week, the probability is $2^{-(n+1)}$ that an insurance agency receives n claims ($n = 0, 1, \ldots$). Furthermore, the numbers of claims received are independent from one week to the next. Determine the PMF of the total number of claims received in two consecutive weeks by using Proposition 6.19 on page 309.

6.105 In this exercise, you are asked to solve Exercise 6.104 without using Proposition 6.19, but by proceeding as follows. Let X and Y denote the number of claims received in weeks one and two, respectively.
a) Identify the common PMF of $X + 1$ and $Y + 1$.
b) Without doing any computations, use your result from part (a) to identify the PMF of the random variable $X + Y + 2$.
c) Use your result from part (b) and Corollary 5.1 on page 247 to obtain the PMF of $X + Y$, the total number of claims received in two consecutive weeks.

6.106 Interpret each of parts (a), (b), and (c) of Proposition 6.20 on page 311 in words.

6.107 Suppose that $X_1, X_2, \ldots, X_r$ are independent random variables each having the geometric distribution with parameter p. Without doing any computations, identify the probability distribution of $X_1 + X_2 + \cdots + X_r$.

6.108 Let X and Y be independent random variables with $X \sim \mathcal{P}(\lambda)$ and $Y \sim \mathcal{P}(\mu)$. Determine and identify the conditional PMF of X given $X + Y = z$.

6.109 Let X and Y be independent random variables with $X \sim \mathcal{NB}(r, p)$ and $Y \sim \mathcal{NB}(s, p)$.
a) Determine the conditional PMF of X given $X + Y = z$.
b) Without doing any computations, use the result of part (a) to obtain the combinatorial identity

$$\sum_{x=r}^{z-s} \binom{x-1}{r-1}\binom{z-x-1}{s-1} = \binom{z-1}{r+s-1},$$

for $z = r + s, r + s + 1, \ldots$.
c) The combinatorial identity in part (b) can be obtained without using probability. Use Proposition 6.19 and that identity to prove that $X + Y \sim \mathcal{NB}(r + s, p)$.

6.110 The number of customers waiting when the Ice Cream Shoppe opens at 10 A.M. has the discrete uniform distribution on $\{0, 1, \ldots, N\}$. The number of customers that arrive during the time that the shop is open has a Poisson distribution with parameter λ and is independent of the number of customers waiting when the shop opens. Determine the PMF of the total number of customers in a day.

6.111 Let X and Y be independent random variables with $X \sim \mathcal{B}(m, p)$ and $Y \sim \mathcal{B}(n, q)$, where p and q are real numbers between 0 and 1, exclusive.
a) Obtain a formula for the PMF of $X + Y$.
b) Without using the result in part (a), identify the PMF of $X + Y$ when $p = q$. Explain your reasoning.
c) Verify your answer to part (b) by substituting p for q in your result of part (a).

6.112 For a particular policy at an insurance company, the daily claim frequency has a Poisson distribution with parameter $\lambda = 4$; the claim severity, in dollars, has the discrete uniform distribution on $\{500, 501, \ldots, 1000\}$.
a) Assuming that claim amounts are independent from one claim to the next, what is the distribution of the total claim amount per day called?
b) Use a computer to simulate the total claim amount for each of 10 days. Construct a table that, for each day, shows the individual claim amounts and the total claim amount.

Theory Exercises

6.113 Complete the proof of Proposition 6.18 on page 307 by verifying Equation (6.30).

6.114 In this exercise, you are asked to prove that, if X and Y are independent random variables with $X \sim \mathcal{P}(\lambda)$ and $Y \sim \mathcal{P}(\mu)$, then $X + Y \sim \mathcal{P}(\lambda + \mu)$.
a) Interpret this result in words.
b) Use Proposition 6.19 on page 309 and the binomial theorem to establish the result.

6.115 Establish the following generalization of Vandermonde's identity (Exercise 3.48 on page 109): For any two real numbers a and b,

$$\sum_{j=0}^{k} \binom{a}{j}\binom{b}{k-j} = \binom{a+b}{k}.$$

Hint: Recall the binomial series—Equation (5.45) on page 241—and compare the coefficient of t^k on both sides of the equation $(1+t)^a(1+t)^b = (1+t)^{a+b}$.

6.116 In this exercise, you are asked to prove that, if X and Y are independent random variables with $X \sim \mathcal{NB}(r, p)$ and $Y \sim \mathcal{NB}(s, p)$, then $X + Y \sim \mathcal{NB}(r + s, p)$.
a) Interpret this result in words.
b) Provide an intuitive explanation of the result based on the interpretation of a negative binomial random variable as the number of trials up to and including a specified number of successes in Bernoulli trials.
c) Use Proposition 6.19 on page 309 and Exercise 6.115 to establish the required result. *Hint:* Use the alternative form of the PMF of a negative binomial random variable as presented in Equation (5.44) on page 241.

6.117 Complete the proof of Proposition 6.20 on page 311. *Note:* Use the results of Exercises 6.114 and 6.116.

Advanced Exercises

6.118 Prove that, if X, Y, and Z are independent discrete random variables, then

$$p_{X+Y+Z}(w) = \sum_x p_X(x) p_{Y+Z}(w - x), \qquad w \in \mathcal{R}.$$

6.119 Let $X_1, \ldots, X_m$ be discrete random variables defined on the same sample space.
a) Prove that

$$p_{X_1+\cdots+X_m}(w) = \sum_{(x_1,\ldots,x_{m-1})} \cdots \sum p_{X_1,\ldots,X_m}(x_1, \ldots, x_{m-1}, w - x_1 - \cdots - x_{m-1}), \qquad w \in \mathcal{R}.$$

b) Identify $m - 1$ formulas similar to the one in part (a) for the PMF of $X_1 + \cdots + X_m$.
c) Specialize parts (a) and (b) to obtain m formulas for the PMF of $X_1 + \cdots + X_m$ in case $X_1, \ldots, X_m$ are independent.
d) Specialize parts (a)–(c) to obtain formulas for the PMF of $X + Y + Z$ in the trivariate case of three discrete random variables X, Y, and Z.

6.120 Another method for obtaining the PMF of the sum of independent discrete random variables involves recursion. Specifically, prove that, if $X_1, X_2, \ldots$ are independent discrete random variables, then

$$p_{X_1+\cdots+X_m}(w) = \sum_x p_{X_1+\cdots+X_{m-1}}(x) p_{X_m}(w - x), \qquad w \in \mathcal{R},$$

for $m = 2, 3, \ldots$.

6.121 An automobile insurance policy reimburses a family's accident losses up to a maximum of two accidents per year. For a (high-risk) three-person family, let X, Y, and Z denote the number of accidents in a single year for the three people in the family. Assume that the joint PMF of X, Y, Z is

$$p_{X,Y,Z}(x, y, z) = (x + 2y + z)/63, \qquad x \in \{0, 1\}, \ y, z \in \{0, 1, 2\},$$

and $p_{X,Y,Z}(x, y, z) = 0$ otherwise. Determine the PMF of the total number of accidents that the family has in a single year. *Hint:* Use Exercise 6.119(d).

CHAPTER REVIEW

Summary

In this chapter, we discussed the simultaneous analysis of two or more discrete random variables defined on the same sample space. One of the primary objects used in such analyses is the joint probability mass function or, more briefly, joint PMF. The joint PMF gives the joint probability that the discrete random variables under consideration take any specified values. And once we have the joint PMF, we can, at least conceptually, use the fundamental probability formula (FPF) to find any probability for the discrete random variables under consideration.

Given the joint PMF of several random variables, we can obtain each individual PMF, called a marginal PMF. We do so by summing the joint PMF on the other variables. Although the joint PMF determines the marginal PMFs, the converse isn't true.

We introduced useful terminology for describing the number of random variables under consideration. The term univariate applies for one random variable, bivariate for two random variables, and trivariate for three random variables. Generally, we use the term multivariate when two or more random variables are being simultaneously analyzed.

Two widely applied multivariate distributions are the multinomial distribution and the multiple hypergeometric distribution. These two probability distributions are the multivariate versions of the binomial and hypergeometric distributions, respectively, discussed in Chapter 5.

A conditional probability mass function provides the PMF of one or more discrete random variables, given the values of other discrete random variables. A conditional PMF is a genuine PMF and hence all properties of (unconditional) PMFs also apply to conditional PMFs.

Next we examined the concept of independent random variables. For discrete random variables, independence is equivalent to the joint PMF being equal to the product of the marginal PMFs. Independence of random variables is one of the most important ideas in probability theory and its applications.

We discussed the process of taking a real-valued function of two or more discrete random variables to obtain a new discrete random variable. Given the joint PMF of the original random variables, we showed how to determine the PMF of the new random variable. Of particular significance in probability theory are sums of random variables, which, for the discrete case, we introduced and explored. Sums of independent discrete random variables—which play an especially important part in many applications—were also discussed, as were sums of a random number of random variables.

You Should Be Able To ...

1. obtain a joint PMF.

2. state the basic properties of a joint PMF.

3. determine the marginal PMFs, knowing the joint PMF.

4. state and apply the fundamental probability formula (FPF) for two or more discrete random variables.

5. identify and apply the multinomial distribution.

6. describe the multiple hypergeometric distribution.

7. define, obtain, and use conditional PMFs.

8. define and apply the concept of independent random variables.

9. state and apply basic properties of independent random variables.

10. state and apply basic properties of independent discrete random variables.

11. explain the relationship between independence of discrete random variables and conditional distributions.

12. obtain the PMF of a real-valued function of two or more discrete random variables, knowing the joint PMF of those random variables.

13. determine the PMF of the sum of two discrete random variables, knowing the joint PMF of those random variables.

14. state and apply the special formulas for obtaining the PMF of the sum of two independent discrete random variables, knowing the marginal PMFs of those random variables.

15. identify the distribution of the sum of independent binomial random variables with the same success-probability parameter.

16. identify the distribution of the sum of independent Poisson random variables.

17. identify the distribution of the sum of independent negative binomial random variables with the same success-probability parameter.

18. obtain the PMF of the sum of a random number of discrete random variables.

19. define and work with compound distributions.

Key Terms

bivariate distribution, *272*

compound distribution, *312*

compound Poisson
   distribution, *312*

conditional probability mass
   function, *283*

convolution, *309*

fundamental probability
   formula, *265, 275*

general multiplication rule, *286*

independent random
   variables, *291, 296*

joint probability histogram, *260*

joint probability mass
   function, *260, 272*

marginal probability mass
   function, *262*

multinomial distribution, *277*

multinomial random
   variables, *277*

multiple hypergeometric
   distribution, *279*

multivariate distribution, *272*

trivariate distribution, *272*

univariate distribution, *272*

Chapter Review Exercises

Basic Exercises

6.122 Each of 35 students in a class is enrolled in either the College of Arts and Sciences or the College of Engineering and is, of course, either female or male. The following contingency table cross classifies the students in the class by gender and college.

College

		Arts & Sciences	Engineering	Total
Gender	Female	6	9	15
	Male	8	12	20
	Total	14	21	35

One student is selected at random from the class. Let $X = 1$ if the student obtained is female and let $X = 2$ if the student obtained is male. Also, let $Y = 1$ if the student obtained is in the College of Arts and Sciences and let $Y = 2$ if the student obtained is in the College of Engineering.

a) Obtain the joint and marginal PMFs of the random variables X and Y.

b) Determine the number of contingency tables that agree with the previous table in the margins but not necessarily in the interior of the table—that is, find the number of ways of filling in the blanks in the following table with nonnegative integers.

College

		Arts & Sciences	Engineering	Total
Gender	Female			15
	Male			20
	Total	14	21	35

c) Explain why part (b) illustrates that marginal PMFs don't determine the joint PMF.

6.123 A small town in South Dakota has 250 families. The following contingency table cross classifies the number of televisions and number of radios owned by the 250 families.

Radios, y

		3	4	5	6	7	8	Total
Televisions, x	0	0	1	1	1	1	0	4
	1	0	6	15	18	2	0	41
	2	1	8	32	44	14	1	100
	3	0	6	37	35	9	1	88
	4	0	1	7	6	3	0	17
	Total	1	22	92	104	29	2	250

Suppose that one of these 250 families is selected at random. Let X and Y denote the number of televisions and the number of radios, respectively, owned by the family obtained.

a) Determine and interpret $p_{X,Y}(3, 6)$.

b) Determine the joint PMF and the marginal PMFs of the random variables X and Y. Display your results in a table similar to Table 6.2 on page 262.

c) Use random variable notation to express the event that the family obtained has twice as many radios as televisions, and then determine the probability of that event by using the FPF.

d) Use random variable notation to express the event that the family obtained has at least five more radios than televisions, and then determine the probability of that event by using the FPF.

6.124 Refer to Exercise 6.123.

a) For each number of televisions, determine and interpret the conditional PMF of the number of radios.

b) Construct a table like Table 6.6 on page 284 that displays the marginal PMF of the number of radios and the conditional PMF of the number of radios for each number of televisions.

c) What percentage of the town's families with two televisions have at most five radios?

d) For each number of radios, determine and interpret the conditional PMF of the number of televisions.

e) Construct a table that displays the marginal PMF of the number of televisions and the conditional PMF of the number of televisions for each number of radios.

f) What percentage of the town's families with five radios have at least three televisions?

6.125 A machine makes a random number X of parts in a day, where $X \sim \mathcal{B}(n, p)$. Each part works with probability α, independently of all other parts. Let Y be the number of working parts made in a day.

a) Obtain and identify the conditional PMF of Y given $X = x$.

b) Obtain the joint PMF of X and Y.

c) Obtain and identify the marginal PMF of Y.

d) Obtain the conditional PMF of X given $Y = y$.

e) Obtain and identify the conditional PMF of $X - Y$ given $Y = y$.

6.126 Jan and Jean play 18 holes of golf. On any given hole, Jan wins with probability p, Jean wins with probability q, and they tie with probability r. Assuming that the results on different holes are independent, determine and identify the joint PMF of the number of holes Jan wins, the number of holes Jean wins, and the number of holes tied.

6.127 Consider two identical electrical components. Let X and Y denote the respective lifetimes of the two components observed at discrete time units (e.g., every hour). Assume that the joint PMF of X and Y is $p_{X,Y}(x, y) = p^2(1 - p)^{x+y-2}$ if $x, y \in \mathcal{N}$, and $p_{X,Y}(x, y) = 0$ otherwise, where $0 < p < 1$. Use the FPF for two discrete random variables to determine the probability that one of the components lasts at least 3 time units longer than the other.

6.128 Refer to Exercise 6.127.

a) Obtain the PMF of the magnitude of the difference between the lifetimes of the two components.

b) Use part (a) and the FPF to solve Exercise 6.127.

6.129 An archer shoots an arrow into a square target 6 feet on a side whose center we call the origin. Assume that the archer will actually hit the target and is equally likely to hit any portion of the target. The archer scores 10 points if she hits the bull's-eye—a disk of radius 1 foot centered at the origin (region A); she scores 5 points if she hits the ring with inner radius 1 foot and outer radius 2 feet centered at the origin (region B); and she scores 0 points otherwise (region C). Suppose that the archer shoots independently at the target four times. Let X, Y, and Z denote the numbers of times that the archer hits regions A, B, and C, respectively.

a) Determine and identify the joint PMF of the random variables X, Y, and Z.

b) Determine and identify the univariate marginal PMFs of X, Y, and Z.

c) Determine, identify, and interpret the PMF of the random variable $X + Y$.

d) Express the total score, T, in four shots in terms of X, Y, and Z. Find the PMF of the random variable T.

6.130 Suppose that the random variables X_1, X_2, X_3, and X_4 have the multinomial distribution with parameters n and p_1, p_2, p_3, and p_4. Show that X_1, X_2, and $X_3 + X_4$ have the multinomial distribution with parameters n and p_1, p_2, and $p_3 + p_4$

a) by applying the multivariate form of the FPF (Proposition 6.6 on page 275) and the binomial formula.

b) without doing any computations but rather by using a probabilistic argument. *Hint:* What does $X_3 + X_4$ represent?

6.131 A point (X, Y) is randomly chosen from the unit square, $[0, 1] \times [0, 1]$. Let

$$U_{0.6} = \begin{cases} 0, & \text{if neither } X \text{ nor } Y \text{ is less than } 0.6; \\ 1, & \text{if either } X \text{ or } Y, \text{ but not both, is less than } 0.6; \\ 2, & \text{if both } X \text{ and } Y \text{ are less than } 0.6. \end{cases}$$

Define $U_{0.3}$ similarly by putting "0.3" in place of "0.6." Obtain the
a) PMF of $U_{0.6}$.
b) PMF of $U_{0.3}$.
c) joint PMF of $U_{0.3}$ and $U_{0.6}$.
d) conditional PMF of $U_{0.3}$ given $U_{0.6} = v$.
e) conditional PMF of $U_{0.6}$ given $U_{0.3} = u$.

6.132 Refer to the discussion of the multiple hypergeometric distribution in Exercise 6.31 on page 280. In a group of 100 voters, there are 50 Republicans, 30 Democrats, and 20 Independents. A committee of 10 people is randomly selected without replacement from the group to represent it at a state political convention.
a) What is the probability that the committee will be perfectly representative in the sense that the proportion of people on the committee in each political party is the same as that of the group as a whole?
b) What is the probability that the committee will be close to perfectly representative in the sense that the number of people on the committee in each political party is within 1 of that which would be perfect representation?

6.133 Refer to the discussion of the multiple hypergeometric distribution in Exercise 6.31 on page 280. The candy N&Ns is produced in seven colors: red, orange, yellow, green, blue, violet, and brown. A package contains 70 N&Ns: 10 red, 7 orange, 12 yellow, 15 green, 10 blue, 8 violet, and 8 brown. You randomly take 7 N&Ns from the package, without replacement. What is the probability that you get
a) all seven colors?   b) only one color?
c) 2 red, 2 yellow, 2 green, and 1 blue?

6.134 During a weekday, the number of accidents at a dangerous intersection is a Poisson random variable with parameter $\lambda = 2.4$, and each accident results in serious damage to at least one vehicle with probability $p = 0.6$. On weekends, the situation is the same except that $\lambda = 1.3$ and $p = 0.4$. What is the distribution of the total number of accidents in a week that result in serious damage to at least one vehicle?

6.135 Let X and Y be discrete random variables with joint PMF

$$p_{X,Y}(x, y) = \frac{x^y e^{-x}}{y!M}, \qquad x = 1, 2, \ldots, M, \quad y = 0, 1, \ldots,$$

and $p_{X,Y}(x, y) = 0$ otherwise.
a) Obtain and identify the marginal PMF of X.
b) Obtain the marginal PMF of Y.
c) Obtain and identify the conditional PMF of Y given $X = x$.
d) Are X and Y independent random variables? Explain your answer.

6.136 In a factory, each manufactured item is defective with probability p and, independently, inspected with probability q. If both defective status and inspection status are taken into account, there are four categories: nondefective-uninspected, nondefective-inspected, defective-uninspected, and defective-inspected. Determine the joint PMF of the numbers of items in the four categories for a random sample of n items.

6.137 Refer to Exercise 6.136.

a) Rework that exercise if the sample is taken without replacement from a lot of N manufactured items in which the proportions of items in the four categories are $(1 - p)(1 - q)$, $(1 - p)q$, $p(1 - q)$, and pq, respectively. *Hint:* Refer to Exercise 6.31 on page 280.

b) Under what condition is it reasonable to use the joint PMF in Exercise 6.136 as an approximation to the joint PMF in part (a) of this problem? Explain your answer.

6.138 In Exercise 6.136, find the conditional PMF of the number of undetected defectives when the number of detected defectives is k.

6.139 The number of people entering a bank during a specified time interval has a Poisson distribution with parameter λ. Each person entering makes a deposit with probability p, independent of the other people entering and the number of people entering.

a) Determine the conditional PMF of the number of people, N, entering the bank during the specified time interval, given the number of people, X, making a deposit.

b) Determine and identify the conditional PMF of $N - X$ given $X = x$.

c) Are the random variables X and $N - X$ independent? Explain your answer.

d) Interpret your result in part (c).

6.140 A car wash provides three options—basic wash, deluxe wash, and super-deluxe wash. The numbers of customers per hour requesting these options are independent Poisson random variables with parameters λ, μ, and ν, respectively. Determine the conditional joint distribution of the hourly numbers of customers requesting the three options, given the total number of customers during the hour.

6.141 Consider n Bernoulli trials with success probability p. Determine the conditional PMF of the trial on which the first success occurs, given that the number of successes is k.

6.142 Let S consist of the 10 decimal digits. Suppose that a number X is chosen according to the discrete uniform distribution on S and then a number Y is chosen according to the discrete uniform distribution on $\{X, X + 1, \ldots, 9\}$.

a) Determine the conditional PMF of X given $Y = y$ without first explicitly finding the marginal PMF of Y and the joint PDF of X and Y.

b) Evaluate the conditional PMF in part (a), given $Y = 6$.

6.143 Suppose that $X_1, X_2, \ldots$ are independent random variables, each having the discrete uniform distribution on the first N positive integers. Let m be a positive integer less than N and set $T = \min\{n \in \mathcal{N} : X_n > m\}$.

a) Show that $\{T = n\} = \bigcup_{k=m+1}^{N}\{X_1 \leq m, \ldots, X_{n-1} \leq m, X_n = k\}$.

b) Apply part (a) to determine and identify the PMF of the random variable T.

c) Obtain the result of part (b) without doing any computations by considering repeated Bernoulli trials with the appropriate random experiment and specified event.

6.144 Suppose that $X_1, X_2, \ldots$ are independent random variables each having the discrete uniform distribution on the first N positive integers. Let $U = \min\{n > 1 : X_n \geq X_1\}$. Determine the PMF of the random variable U. *Hint:* Use the law of total probability.

6.145 Suppose that $X_1, \ldots, X_m$ are independent Poisson random variables with $X_j \sim \mathcal{P}(\lambda_j)$ for $j = 1, 2, \ldots, m$. Determine and identify the conditional joint PMF of $X_1, \ldots, X_m$ given $X_1 + \cdots + X_m = z$.

6.146 Let $X_1, \ldots, X_m$ have the multinomial distribution with parameters n and $p_1, \ldots, p_m$.

a) Without doing any computations, identify the marginal PMF of X_m.

b) Determine and identify the conditional PMF of $X_1, \ldots, X_{m-1}$ given $X_m = x_m$.

6.147 A manufacturer of electronic components has two factories. Factory A produces m components per day and Factory B produces n components per day. At each factory, the probability of a defective component is p. Determine and identify

a) the probability distribution of the total number of defective components per day from the two factories combined.

b) the conditional probability distribution of the number of defective components per day from Factory A, given the combined total of defective components.

6.148 Repeat Exercise 6.147 if the manufacturer has three factories.

6.149 Let $X_1, \ldots, X_m$ be independent binomial random variables such that $X_j \sim \mathcal{B}(n_j, p)$ for $j = 1, 2, \ldots, m$.

a) Find and identify the conditional joint PMF of $X_1, \ldots, X_m$ given $X_1 + \cdots + X_m = z$. *Note:* Refer to Exercise 6.31 on page 280.

b) Use the result of part (a) to solve parts (b) of Exercises 6.147 and 6.148.

6.150 Let X denote the number of successes in n Bernoulli trials with success probability p. For j a positive integer between 1 and n, inclusive, let X_j be 1 or 0 depending on whether the jth trial is a success or a failure, respectively.

a) Identify the PMF of X_j.

b) Without doing any calculations, find and identify the conditional PMF of X_j given $X = x$.

c) Use the definition of conditional PMF to solve part (b).

Theory Exercises

6.151 In Bernoulli trials with success probability p, let X_1 denote the number of trials until the first success and, for $j \geq 2$, let X_j denote the number of trials from the $(j-1)$th to the jth success.

a) Prove that $X_1, X_2, \ldots$ are independent random variables, each having the geometric distribution with parameter p. *Hint:* Describe in detail what must happen for the event $\{X_1 = x_1, \ldots, X_m = x_m\}$ to occur. Also, refer to Exercise 6.60(c) on page 298.

b) Let X be the number of trials up to and including the rth success. Show that X can be expressed as the sum of r independent geometric random variables with common parameter p.

c) Use part (a) to obtain a relatively simple proof of the fact that, if X and Y are independent with $X \sim \mathcal{NB}(r, p)$ and $Y \sim \mathcal{NB}(s, p)$, then $X + Y \sim \mathcal{NB}(r + s, p)$.

Advanced Exercises

6.152 Consider Bernoulli trials with success probability p. Let X_1 denote the number of trials until the first success and, for $j \geq 2$, let X_j denote the number of trials from the $(j-1)$th to the jth success. Also, let N be independent of the X_js and have the Poisson distribution with parameter λ.

a) Is the distribution of S_N a compound distribution? *Hint:* Refer to Exercise 6.151.

b) Is the distribution of S_N a compound Poisson distribution?

c) Interpret the random variable S_N.

d) Determine an expression for the PMF of S_N.

e) Determine $P(S_N = 3)$.

6.153 Let $X_1, \ldots, X_m$ have the multinomial distribution with parameters n and $p_1, \ldots, p_m$.

a) Let k be a positive integer less than m. Without using Proposition 6.5 on page 274 or doing any computations, determine the joint PMF of the random variables $X_1, \ldots, X_k$.

b) Identify the conditional distribution of $X_1, \ldots, X_{m-2}$ given $X_{m-1} = x_{m-1}$ and $X_m = x_m$.

6.154 Let X be a nonnegative-integer valued random variable such that $0 < p_X(0) < 1$. Further suppose that X_1, X_2, ... are independent random variables, each having the same probability distribution as X. Set $T = \min\{n \in \mathcal{N} : X_n \neq 0\}$.
a) Without doing any computations, determine the PMF of T.
b) Interpret the random variable X_T.
c) Determine the joint PMF of X_T and T.
d) Use part (c) to obtain the PMF of X_T.
e) Are T and X_T independent random variables? Justify your answer.

6.155 Let X and Y be integer-valued random variables defined on the same sample space. Prove that, for $x, y \in \mathcal{Z}$,

$$p_{X,Y}(x, y) = P(X \leq x, Y \leq y) - P(X \leq x, Y \leq y - 1) - P(X \leq x - 1, Y \leq y)$$
$$+ P(X \leq x - 1, Y \leq y - 1).$$

6.156 Suppose that X_1, X_2, X_3, and X_4 are independent random variables each having the discrete uniform distribution on $\{0, 1, \ldots, 9\}$. According to Proposition 6.13, the random variables $U = \min\{X_1, X_2\}$ and $V = \max\{X_3, X_4\}$ are independent. Verify this result by showing that the joint PMF of U and V factors into the product of the marginal PMFs of U and V.

6.157 Suppose that X and Y are independent random variables, each being uniformly distributed on $\{0, 1, \ldots, N\}$. Determine the joint PMF of the random variables $U = \min\{X, Y\}$ and $V = \max\{X, Y\}$.

6.158 Poisson approximation to the multinomial distribution: Let $m > 1$ be a positive integer and let λ_1, ..., λ_{m-1} be positive real numbers. Suppose that, for each $n \in \mathcal{N}$, the random variables X_{1n}, ..., X_{mn} have the multinomial distribution with parameters n and p_{1n}, ..., p_{mn}, where, for $j = 1, \ldots, m - 1$, we have $np_{jn} \to \lambda_j$ as $n \to \infty$.
a) Prove that

$$\lim_{n \to \infty} P(X_{1n} = x_1, \ldots, X_{(m-1)n} = x_{m-1}) = e^{-(\lambda_1 + \cdots + \lambda_{m-1})} \frac{\lambda_1^{x_1} \cdots \lambda_{m-1}^{x_{m-1}}}{x_1! \cdots x_{m-1}!},$$

for all nonnegative integers $x_1, \ldots, x_{m-1}$.
b) Explain the meaning of the result in part (a) in terms of approximating a multinomial distribution by a multivariate Poisson distribution.
c) Show that, when $m = 2$, the result in part (a) reduces to the Poisson approximation to the binomial distribution, as presented in Exercise 5.86 on page 227.

Christiaan Huygens *1629–1695*

Christiaan Huygens was born on April 14, 1629, in The Hague, Netherlands, son of Constantijn Huygens, Dutch poet, musician, and statesman. The elder Huygens's status as a diplomat allowed Christiaan entrance to scientific society where he developed friendships with Mersenne and Descartes.

Prior to the age of 16, Huygens's education consisted of home tutoring. Subsequently, however, he entered the University of Leiden where he studied law and mathematics for 2 years before transferring to the College of Orange at Breda to complete his formal education.

Huygens began his publications with several papers in mathematics. Building on Pascal and Fermat's correspondences on probability, he published the first printed article on the calculus of probabilities, *De Ratiociniis in Ludo Aleae*. Included in this article was the introduction of the all-important concept of expected value.

Huygens was the first to propose the wave theory of light, known as Huygens's principle, which flew in the face of Newton's particle theory. Newton's reputation was so great that his rejection of the wave theory led to general disbelief in Huygens's principle. In the twentieth century, experiments were devised that showed that Newton's theory is implausible and that even Huygens's principle doesn't provide all the answers.

In 1665, Huygens moved to Paris under the patronage of Louis XIV. Nonetheless, Huygens made several visits to Holland, usually as a result of ill health. His return to The Hague in 1681 marked the end of his residence in France.

Huygens was elected a Fellow of the Royal Society in 1663. A mountain and a crater on the moon and a street in Paris are named for him. Huygens died at The Hague on July 8, 1695.

CHAPTER SEVEN

Expected Value of Discrete Random Variables

INTRODUCTION

In examining discrete random variables, so far we have concentrated on methods designed for obtaining probabilities. The primary object in this regard is the probability mass function (PMF).

In this chapter, we introduce and explore another essential quantity associated with random variables—namely, *expected value*. Roughly speaking, the expected value of a random variable is the value of the random variable that we would expect to observe "on average." Put another way, it is the long-run average of the observed values of the random variable in repeated independent trials.

Expected value is the key to understanding what happens in the long run when we repeatedly observe the value of a random variable. For instance, in setting life-insurance premiums, insurance companies need to know the life expectancy of policyholders of any particular age. Here the random variable under consideration is the remaining number of years a policyholder lives, starting at the particular age. The expected value of that random variable is the life expectancy of policyholders of that age.

We begin, in Section 7.1, by developing the formula used to define the expected value of a discrete random variable. Then, in Section 7.2, we present and apply the basic properties of expected value. In particular, we discuss the *fundamental expected-value formula (FEF)*, a formula that provides a method for obtaining the expected value of a real-valued function of one or more discrete random variables in terms of the joint PMF of those random variables.

We can use the expected value of an appropriate function of a random variable to measure the variation of the possible values of the random variable. We can also use the expected value of an appropriate function of two random variables to measure association between those random variables. We examine these and related ideas in Sections 7.3 and 7.4.

In Section 7.5, we apply the concept of expected value to conditional distributions and thereby obtain conditional expected values. Then we show how to determine the expected value of a discrete random variable from its various conditional expected values.

7.1 From Averages to Expected Values

In this section, we introduce the *expected value* of a discrete random variable. The concept of expected value is central to the theory of probability and its applications. To develop the formal definition of expected value, we first provide the most common interpretation of its meaning, the **long-run-average interpretation of expected value.**

Long-Run-Average Interpretation of Expected Value

The expected value of a random variable is the long-run-average value of the random variable in repeated independent observations.

Alternatively, we can express the long-run-average interpretation of expected value as follows. Let X be a random variable and let $\mathcal{E}(X)$ denote its expected value. For n independent repetitions of the random experiment, let $X_1, \ldots, X_n$ represent the n observed values of X. Then,

$$\frac{X_1 + \cdots + X_n}{n} \approx \mathcal{E}(X), \qquad \text{for large } n. \tag{7.1}$$

In words, for large n, the average value of X in n independent observations will approximately equal the expected value of X.

All attempts to use the long-run-average interpretation of expected value as a definition of expected value [by defining the expected value of a random variable X to be $\mathcal{E}(X) = \lim_{n \to \infty}(X_1 + \cdots + X_n)/n$] have failed. Nonetheless, that interpretation is invaluable for developmental purposes. Furthermore, as we demonstrate in Chapter 11, the theory of probability based on the Kolmogorov axioms can be applied to prove a mathematically precise version of Relation (7.1) as a theorem. That theorem is one of the most important results in probability theory; it is known as the *law of large numbers*.

The Expected Value of a Discrete Random Variable

We now use the long-run-average interpretation of expected value to arrive at the formal definition of the expected value of a discrete random variable whose range is finite—say, $\{x_1, \ldots, x_m\}$. Assume that the random experiment is repeated independently a large number of times—say, n times. Then, in view of Relation (7.1), we have

$$\mathcal{E}(X) \approx \frac{X_1 + \cdots + X_n}{n}$$

$$= \frac{1}{n} \sum_{k=1}^{m} n(\{X = x_k\}) \cdot x_k = \sum_{k=1}^{m} x_k \cdot \frac{n(\{X = x_k\})}{n}, \tag{7.2}$$

where, as usual, $n(E)$ denotes the number of times that an event E occurs in n repetitions of the random experiment. Because n is large, the frequentist interpretation of probability indicates that

$$\frac{n(\{X = x_k\})}{n} \approx P(X = x_k), \tag{7.3}$$

for each k. From Relations (7.2) and (7.3), we conclude that

$$\mathcal{E}(X) \approx \sum_{k=1}^{m} x_k P(X = x_k) = \sum_{k=1}^{m} x_k p_X(x_k) = \sum_{x} x p_X(x). \tag{7.4}$$

With Relation (7.4) in mind, we now present the definition of the **expected value** of a discrete random variable.

DEFINITION 7.1   Expected Value of a Discrete Random Variable

The **expected value** of a discrete random variable X, denoted $\mathcal{E}(X)$, is defined by

$$\mathcal{E}(X) = \sum_{x} x p_X(x). \tag{7.5}$$

In words, the expected value of a discrete random variable is a weighted average of its possible values, weighted by probability.

Note the following.

- Other commonly used terms for *expected value* are **expectation, mean,** and **first moment.**

- The notation μ_X, as well as $\mathcal{E}(X)$, is used to represent the expected value of a random variable X.

If the range of X is finite, the right side of Equation (7.5) is a finite sum of real numbers and hence is itself a real number. However, if the range of X is infinite, the right side of Equation (7.5) is an infinite series, which may or may not converge to a real number.

To avoid convergence problems, we require that the right side of Equation (7.5) converge absolutely, that is,

$$\sum_{x} |x| p_X(x) < \infty. \tag{7.6}$$

If Relation (7.6) holds, the sum on the right side of Equation (7.5) is a real number, regardless of whether it is a finite sum or an infinite series. We then say that the random variable X has **finite expectation** and define the expected value of X as in Definition 7.1. However, if Relation (7.6) fails to hold, we say that the random variable X doesn't have finite expectation and we don't define the expected value of X. For more on this issue, see Exercise 7.27.

EXAMPLE 7.1   *Expected Value*

Number of Siblings   Professor Weiss asked the 40 students in one of his classes to state how many siblings they have. The resulting data are shown in Table 7.1. Suppose that a student is selected at random from the class, and let X denote the number of siblings of the student obtained. Determine the expected value of the random variable X.

Table 7.1 Number of siblings for the 40 students

1	1	0	1	3	1	1	0	2	4
1	1	1	2	2	1	2	1	2	1
0	1	2	0	1	2	1	3	2	0
0	0	2	2	1	3	1	1	0	2

Solution   To apply Definition 7.1, we need the PMF of the random variable X, which was found in Example 5.4 on page 185. We repeat the PMF in the first two columns of Table 7.2.

Table 7.2 Table for calculating the expected value of X

Siblings x	Probability $p_X(x)$	$x p_X(x)$
0	0.200	0.000
1	0.425	0.425
2	0.275	0.550
3	0.075	0.225
4	0.025	0.100
	1.000	1.300

The third column of Table 7.2 gives the values of $x p_X(x)$. Adding those values, we get the expected value of X:

$$\mathcal{E}(X) = \sum_x x p_X(x) = 1.3.$$

The expected number of siblings of a randomly selected student is 1.3. ■

In Example 7.1, we obtained 1.3 as the expected number of siblings of a randomly selected student from one of Professor Weiss's classes. Thus, on average, we would expect the student selected to have 1.3 siblings. Obviously, a person can't have 1.3 siblings. The expected value of 1.3 simply indicates that, if we independently repeat the random experiment of selecting a student at random a large number of times, the average number of siblings of the students obtained will be about 1.3.

To illustrate, we used a computer to simulate 100 independent observations of the number of siblings, X, of a randomly selected student. The data are displayed in Table 7.3. The average value of the 100 observations is 1.26, which is quite close to the expected value of X, or 1.3. If we made, say, 1000 observations instead of 100, the average value of those 1000 observations would most likely be even closer to 1.3.

Table 7.3 One hundred independent observations of the random variable X, the number of siblings of a randomly selected student

2	2	1	3	1	1	0	1	1	2	0	1	2	1	1	0	1	0	2	2
1	2	0	1	2	4	0	2	1	2	2	1	2	0	0	2	2	1	2	0
1	1	1	1	1	1	0	3	0	2	0	0	0	0	1	1	1	3	1	2
2	3	1	1	0	2	0	1	2	1	1	1	2	2	2	2	0	1	1	1
1	2	4	1	2	1	1	1	1	2	1	0	0	0	3	2	0	3	2	2

Figure 7.1(a) shows a plot of the average number of siblings against the number of observations for the data in Table 7.3. The dashed line in the figure is at $\mathcal{E}(X) = 1.3$. Figure 7.1(b) shows a plot for a second simulation of 100 independent observations of the number of siblings, X, of a randomly selected student. Both plots suggest that, as the number of observations increases, the average number of siblings approaches the expected value, $\mathcal{E}(X) = 1.3$. In other words, the simulations seem to corroborate the long-run-average interpretation of expected value.

Figure 7.1 Plots of the average number of siblings against the number of observations for two simulations of 100 observations each

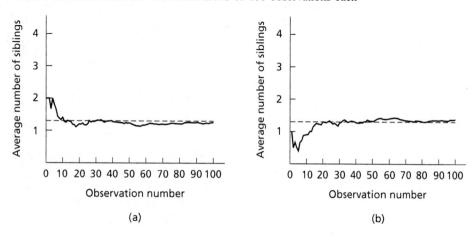

(a)   (b)

Expected Value as a Population Mean

Consider a finite population and a variable defined thereon. In statistics and related fields, the **population mean,** denoted μ, is defined to be the arithmetic mean of all possible observations of the variable for the entire population.

For instance, in Example 7.1, the population of interest consists of the 40 students in one of Professor Weiss's classes, and the variable of interest is "number of siblings." Referring to Table 7.1 on page 328, we see that the population mean number of siblings is

$$\mu = \frac{1 + 1 + 0 + 1 + \cdots + 1 + 0 + 2}{40} = 1.3.$$

We note that the population mean of 1.3 is the same as the expected value of the random variable X, the number of siblings of a randomly selected student. This result is no accident, as Proposition 7.1 shows.

◆◆◆   **Proposition 7.1   Expected Value as a Population Mean**

Consider a finite population and a variable defined on it. Suppose that a member is selected at random from the population and let X denote the value of the variable for the member obtained. Then the expected value of the random variable X equals the population mean.

Proof   Let N denote the size of the population. The sample space Ω for the random experiment of selecting one member at random from the population is the set whose elements are the N members of the population—say, $\Omega = \{\omega_1, \ldots, \omega_N\}$. The population mean, μ, which is the arithmetic mean of all possible observations of the variable for the entire population, can be expressed as

$$\mu = \frac{1}{N} \sum_{j=1}^{N} X(\omega_j). \tag{7.7}$$

Now, let $x_1, \ldots, x_m$ denote the possible values of the variable under consideration or, equivalently, the range of the random variable X. For each integer k between 1 and m, inclusive, set $E_k = \{X = x_k\} = \{\omega \in \Omega : X(\omega) = x_k\}$. Because a classical probability model is appropriate here, we have

$$p_X(x_k) = P(X = x_k) = \frac{N(\{X = x_k\})}{N(\Omega)} = \frac{N(E_k)}{N}, \tag{7.8}$$

for each k. From Equations (7.7) and (7.8), we conclude that

$$\mathcal{E}(X) = \sum_x x p_X(x) = \sum_{k=1}^{m} x_k p_X(x_k) = \sum_{k=1}^{m} x_k \cdot \frac{N(E_k)}{N}$$

$$= \frac{1}{N} \sum_{k=1}^{m} N(E_k) x_k = \frac{1}{N} \sum_{k=1}^{m} \sum_{\omega \in E_k} X(\omega) = \frac{1}{N} \sum_{j=1}^{N} X(\omega_j) = \mu.$$

Thus the expected value of the random variable X equals the population mean of the corresponding variable.   ◆

Expected Value for Some Families of Discrete Random Variables

For many families of discrete random variables, we can obtain formulas for the expected value in terms of the parameters. We do so by substituting the general form of the PMF for the family under consideration into the formula defining the expected value of a discrete random variable [i.e., Equation (7.5) on page 327].

To illustrate, we consider the indicator random variable, I_E, of an event E. Referring first to Definition 7.1 on page 327 and then to Proposition 5.12 on page 238, we find that

$$\mathcal{E}(I_E) = \sum_x x p_{I_E}(x) = 0 \cdot p_{I_E}(0) + 1 \cdot p_{I_E}(1) = 0 \cdot \big(1 - P(E)\big) + 1 \cdot P(E),$$

or

$$\mathcal{E}(I_E) = P(E). \tag{7.9}$$

In words, the expected value of the indicator random variable of an event equals the probability of that event. Although simple, Equation (7.9) is used often.

EXAMPLE 7.2   *Expected Value of a Binomial Random Variable*

Obtain the expected value of a binomial random variable with parameters n and p.

Solution   Let $X \sim \mathcal{B}(n, p)$. Recall that the PMF of X is

$$p_X(x) = \binom{n}{x} p^x (1 - p)^{n-x}, \qquad x = 0, 1, \ldots, n,$$

and $p_X(x) = 0$ otherwise. Substituting this PMF into Equation (7.5) gives

$$\mathcal{E}(X) = \sum_x x p_X(x) = \sum_{x=0}^{n} x \binom{n}{x} p^x (1 - p)^{n-x}$$

$$= \sum_{x=1}^{n} x \, \frac{n!}{x! \, (n - x)!} p^x (1 - p)^{n-x} = \sum_{x=1}^{n} \frac{n!}{(x - 1)! \, (n - x)!} p^x (1 - p)^{n-x}$$

$$= np \sum_{x=1}^{n} \frac{(n - 1)!}{(x - 1)! \, (n - x)!} p^{x-1} (1 - p)^{n-x}$$

$$= np \sum_{x=1}^{n} \binom{n - 1}{x - 1} p^{x-1} (1 - p)^{n-x} = np \sum_{j=0}^{n-1} \binom{n - 1}{j} p^j (1 - p)^{(n-1)-j} = np,$$

where the last equation follows from the binomial theorem or from the fact that the values of the PMF of a binomial random variable with parameters $n - 1$ and p sum to 1.

Thus, $\mathcal{E}(X) = np$; that is, the expected value of a binomial random variable is the product of its two parameters. We could have guessed this result by reasoning as follows. The random variable X gives the number of successes in n Bernoulli trials with success probability p. Because of the frequentist interpretation of probability, in the long run, $100p\%$ of the trials will result in success. Hence, in n trials, we would expect about $100p\%$ of the n trials, or np, to result in success.   ∎

EXAMPLE 7.3   *Expected Value of a Poisson Random Variable*

Obtain the expected value of a Poisson random variable with parameter λ.

Solution   Let $X \sim \mathcal{P}(\lambda)$. Recall that the PMF of X is

$$p_X(x) = e^{-\lambda} \frac{\lambda^x}{x!}, \qquad x = 0, 1, 2, \ldots,$$

and $p_X(x) = 0$ otherwise. Substituting this PMF into Equation (7.5) yields

$$\mathcal{E}(X) = \sum_x x p_X(x) = \sum_{x=0}^{\infty} x e^{-\lambda} \frac{\lambda^x}{x!} = e^{-\lambda} \sum_{x=1}^{\infty} x \frac{\lambda^x}{x!}$$

$$= \lambda e^{-\lambda} \sum_{x=1}^{\infty} \frac{\lambda^{x-1}}{(x-1)!} = \lambda e^{-\lambda} \sum_{k=0}^{\infty} \frac{\lambda^k}{k!} = \lambda e^{-\lambda} e^{\lambda} = \lambda,$$

where the penultimate equation follows from the exponential series or from the fact that the values of the PMF of a Poisson random variable with parameter λ sum to 1. Thus, $\mathcal{E}(X) = \lambda$; that is, the expected value of a Poisson random variable is its parameter. ∎

Using arguments similar to those in Examples 7.2 and 7.3, we can obtain simple formulas for the expected values of several other families of discrete random variables. Table 7.4 provides the expected values for some of the most important families of discrete random variables.

Table 7.4 Expected values for selected families of discrete random variables

Family	Parameter(s)	Expected value
Binomial	n and p	np
Hypergeometric	N, n, and p	np
Poisson	λ	λ
Geometric	p	$1/p$
Negative binomial	r and p	r/p
Indicator	E	$P(E)$

Let's take a moment to discuss the expected value of a geometric random variable. We provide a heuristic argument for this expected value and leave the mathematical derivation to you as Exercise 7.24. If $X \sim \mathcal{G}(p)$, we can interpret X as the number of trials up to and including the first success in a sequence of Bernoulli trials with success probability p. Because of the frequentist interpretation of probability, in the long run, $100p\%$ of the trials will result in success or, put another way, we expect about one success every $1/p$ trials. Thus the expected number of trials up to and including the first success is $1/p$; that is, $\mathcal{E}(X) = 1/p$, as indicated in Table 7.4.

EXAMPLE 7.4   *Expected Value*

The Lotto   An Arizona state lottery, called *Lotto,* is played as follows: A player specifies six numbers of his or her choice from the numbers 1–42; these six numbers constitute the player's "ticket" for which the player pays $1. In the lottery drawing, six winning numbers are chosen at random (without replacement) from the numbers 1–42. To win a prize, a *Lotto* ticket must contain three or more of the winning numbers. Suppose that the player buys one *Lotto* ticket per week.
a) Determine the expected number of weeks until the player wins a prize.
b) Determine the expected number of prizes that the player wins in a year.

Solution   In Example 5.15(b) on page 212, we showed that, if a player buys one ticket, the probability she wins a prize is 0.029 (to three decimal places). Let a success correspond to the player winning a prize during any given week. As lottery results from one week to the next are independent, we have Bernoulli trials with success probability $p = 0.029$.
a) Let X denote the number of weeks until the player wins a prize. Then X has the geometric distribution with parameter $p = 0.029$. Referring to Table 7.4, we see that $\mathcal{E}(X) = 1/p = 1/0.029 = 34.483$. The expected number of weeks until the player wins a prize is roughly 34.5.
b) Let Y denote the number of prizes that the player wins in a year. Then Y has the binomial distribution with parameters $n = 52$ and $p = 0.029$. Again, from Table 7.4, $\mathcal{E}(Y) = np = 52 \cdot 0.029 = 1.508$. The expected number of prizes that the player wins in a year is roughly 1.5.   ∎

EXERCISES 7.1   **Basic Exercises**

7.1 Suppose that two balanced dice are rolled. Let X be the sum of the two faces showing. Determine the expected value of the random variable X. *Note:* Exercise 5.24 on page 193 asks for the PMF of X.

7.2 An archer shoots an arrow into a square target 6 feet on a side whose center we call the origin. Assume that the archer will actually hit the target and is equally likely to hit any portion of the target. The archer scores 10 points if she hits the bull's eye—a disk of radius 1 foot centered at the origin (region A); she scores 5 points if she hits the ring with inner radius 1 foot and outer radius 2 feet centered at the origin (region B); and she scores 0 points otherwise (region C). Determine the expected score by the archer. *Note:* Exercise 5.27 on page 194 asks for the PMF of the score.

7.3 Urn I contains four red balls and one white ball; Urn II contains one red ball and four white balls. A coin with probability p of a head is tossed. If the result is a head, Urn I is chosen; otherwise, Urn II is chosen. Two balls are then drawn at random without replacement from the chosen urn. Let X be the number of red balls drawn. Obtain the mean of X in case
a) $p = 0.5$ (balanced coin).   **b)** $p = 0.4$ (biased coin).
Note: Exercise 5.28 on page 194 asks for the PMF of X in each case.

7.4 Of six men and five women applying for a job at Alpha, Inc., three are selected at random for interviews. Find the expected number of women in the interview pool by using
a) the PMF obtained in Exercise 5.29(a) on page 194.
b) an appropriate expected-value formula obtained in this section.

7.5 Expected value as center of gravity: Let X be a random variable with finite range, say, $\{x_1, \ldots, x_m\}$. Set $p_k = p_X(x_k)$ for $k = 1, 2, \ldots, m$. Think of the x-axis as a seesaw and each p_k as a mass placed at point x_k on the seesaw. The *center of gravity* of these masses is defined to be the point $\bar{x}$ on the x-axis at which a fulcrum could be placed to balance the seesaw. Relative to the center of gravity, the torque acting on the seesaw by the mass p_k is proportional to the product of that mass and the signed distance of the point x_k from $\bar{x}$. Show that the center of gravity equals the expected value of X; that is, $\bar{x} = \mathcal{E}(X)$. *Hint:* To balance, the total torque acting on the seesaw must be 0.

7.6 Consider a random experiment with a finite number of possible outcomes and let Ω be the sample space.
a) For a random variable X defined on Ω, show that $\mathcal{E}(X) = \sum_{\omega \in \Omega} X(\omega) P(\{\omega\})$.
b) Show that, if the outcomes are equally likely, the expression in part (a) reduces to the arithmetic average of the possible observations of the random variable X.

7.7 A variable of a finite population has mean 82.4.
a) Let X denote the value of the variable for a randomly selected member of the population. Find the expected value of X.
b) If a random sample of size n is taken from the population with replacement and the value of the variable is observed each time, roughly what will be the sum of the observed values?

7.8 As part of a screening exam for prospective insured, a physician conducts tests for acute problems. On average, the first positive test occurs with the 25th person tested by the physician.
a) Determine the probability that the first positive test occurs by the sixth person tested.
b) On average, how many people must the physician test until obtaining the fifth positive test?

7.9 Pinworm infestation, commonly found in children, can be treated with the drug pyrantel pamoate. According to the *Merck Manual*, the treatment is effective in 90% of cases. If 20 children with pinworm infestation are treated with pyrantel pamoate, what is the expected number cured?

7.10 According to the *Daily Racing Form*, the probability is about 0.67 that the favorite in a horse race will finish in the money (first, second, or third place). Determine the smallest number of races required so that the expected number of times that the favorite finishes in the money is at least 10.

7.11 Sickle cell anemia is an inherited blood disease that occurs primarily in blacks. In the United States, about 15 of every 10,000 black children have sickle cell anemia. The red blood cells of an affected person are abnormal; the result is severe chronic anemia, which causes headaches, shortness of breath, jaundice, increased risk of pneumococcal pneumonia and gallstones, and other severe problems. Sickle cell anemia occurs in children who inherit an abnormal type of hemoglobin, called hemoglobin S, from both parents. If hemoglobin S is inherited from only one parent, the person is said to have sickle cell trait and is generally free from symptoms. There is a 50% chance that a person who has sickle cell trait will pass hemoglobin S to an offspring. If two people who have sickle cell trait have five children, how many children should they expect to have sickle cell anemia?

7.12 An upper-level probability class has six undergraduate students and four graduate students. A random sample of three students is taken from the class. Determine the expected number of undergraduate students selected if the sampling is
a) without replacement.   b) with replacement.
c) Why are your results in parts (a) and (b) the same?

7.13 Refer to Example 5.14 on page 208, where X denotes the number of defective TVs obtained when 5 TVs are randomly selected without replacement from a lot of 100 TVs in which 6 are defective. Determine and interpret the expected value of X.

7.14 Suppose that a random sample of size n is taken from a population of size N in which $100p\%$ of the members have a specified attribute. Determine the expected number of the members sampled that have the specified attribute if the sampling is
a) without replacement.   **b)** with replacement.

7.15 Between 5:00 P.M. and 6:00 P.M., the number of cars that use the drive-up window at a fast-food restaurant has a Poisson distribution. Data show that, on average, 15 cars use the drive-up window during that hour.
a) What is the parameter of the Poisson distribution? Explain your answer.
b) Roughly what percentage of days are the number of cars that use the drive-up window between 5:00 P.M. and 6:00 P.M. within three, inclusive, of the average number?

7.16 Refer to Proposition 5.7 on page 220, which gives the Poisson approximation to the binomial distribution. What is the relationship between the mean of the binomial distribution and the mean of the approximating Poisson distribution?

7.17 The number of customers that enter the Downtown Coffee Shop in an hour has a Poisson distribution with mean 31. Thirty percent of the customers buy a café mocha with no whipped cream. What is the expected number of customers per hour who buy a café mocha with no whipped cream? State any assumptions that you make.

7.18 The number of eggs that the female grunge beetle lays in her nest has a $\mathcal{P}(4)$ distribution. Assuming that an empty nest can't be recognized, determine the expected number of eggs observed. *Note:* Exercise 5.85 on page 227 asks for the PMF of the number of eggs observed in a nest.

7.19 Use the formula for the expected value of a binomial random variable to obtain the expected value of an indicator random variable.

7.20 A baseball player has a batting average of .260. Suppose that you observe successive at-bats of the player and note for each at-bat whether the player gets a hit. Presuming that the assumption of Bernoulli trials is appropriate, how many at-bats, on average, does it take until the player gets his
a) first hit?   **b)** second hit?   **c)** tenth hit?

7.21 Simulation: Let X have the discrete uniform distribution on the set $\{0, 1, \ldots, 9\}$.
a) Use simulation to estimate the mean of the random variable X.
b) Use the definition of expected value to determine the mean of the random variable X. Compare your result with that in part (a).

7.22 Let X have the discrete uniform distribution on the set $\{1, 2, \ldots, N\}$, where N is a positive integer. Determine the expected value of the random variable X. *Hint:* Use the formula for the sum of the first N positive integers.

Theory Exercises

7.23 Let X be a hypergeometric random variable with parameters N, n, and p. Show that $\mathcal{E}(X) = np$ by using an argument similar to the one given for a binomial random variable in Example 7.2 on page 331.

7.24 Let X be a geometric random variable with parameter p. Prove that $\mathcal{E}(X) = 1/p$. *Hint:* Start by showing that

$$\mathcal{E}(X) = -p \sum_{x=0}^{\infty} \frac{d}{dp} (1 - p)^x.$$

7.25 Let X be a negative binomial random variable with parameters r and p.
a) Provide a heuristic argument indicating that $\mathcal{E}(X) = r/p$.
b) Apply the alternative form for the PMF of a negative binomial random variable, given in Equation (5.44) on page 241, to show that

$$\mathcal{E}(X) = r + p^r (p - 1) \sum_{j=0}^{\infty} \binom{-r}{j} \frac{d}{dp} (p - 1)^j.$$

Hint: Start by writing $\mathcal{E}(X) = \sum_x \big(r + (x - r)\big) p_X(x)$.
c) Use part (b) and the binomial series to conclude that $\mathcal{E}(X) = r/p$.
d) Deduce that, for a geometric random variable with parameter p, we have $\mathcal{E}(X) = 1/p$.

Advanced Exercises

7.26 Pascal distribution: Let α be a positive real number. A random variable X with PMF $p_X(x) = \alpha^x/(1 + \alpha)^{x+1}$ for $x = 0, 1, 2, \ldots$, and $p_X(x) = 0$ otherwise, is said to have the *Pascal distribution with parameter* α. (This Pascal distribution is the same as the one presented in Equation (5.50) on page 243 with $\alpha = 1$ and $p = 1/(1 + \alpha)$.) Determine the expected value of a Pascal random variable with parameter α.

7.27 Infinite expectation: Let X be a discrete random variable. Define

$$\mathcal{E}(X^+) = \sum_{x>0} x p_X(x) \quad \text{and} \quad \mathcal{E}(X^-) = -\sum_{x<0} x p_X(x).$$

a) Explain why the series defining $\mathcal{E}(X^+)$ is either a real number or equals ∞. In the latter case, we write $\mathcal{E}(X^+) = \infty$.
b) Explain why the series (without the minus sign) defining $\mathcal{E}(X^-)$ is either a real number or equals $-\infty$. In the latter case, we write $\mathcal{E}(X^-) = \infty$.
c) Prove that X has finite expectation if and only if both $\mathcal{E}(X^+)$ and $\mathcal{E}(X^-)$ are real numbers, in which case,

$$\mathcal{E}(X) = \mathcal{E}(X^+) - \mathcal{E}(X^-). \tag{$*$}$$

Suppose now that X doesn't have finite expectation, but that only one of $\mathcal{E}(X^+)$ and $\mathcal{E}(X^-)$ equals ∞. Then the expression on the right side of Equation ($*$) makes sense and equals either ∞ or $-\infty$. In this case, we say that X has *infinite expectation* and define the expected value of X by using Equation ($*$).
d) Construct a discrete random variable with infinite expectation. *Hint:* For which real numbers r does the series $\sum_{n=1}^{\infty} n^{-r}$ converge?

7.28 The St. Petersburg paradox: Consider the game where a person tosses a balanced coin until the first head appears. If the first head appears on the nth toss, the person wins 2^n dollars from the house. Let X denote the amount that the person wins.
a) Show that the random variable X doesn't have finite expectation.
b) Referring to Exercise 7.27, show that $\mathcal{E}(X) = \infty$.
c) Why is the result of part (b) called a paradox in this setting? *Hint:* What should the house charge as a fair price to play the game?

7.2   Basic Properties of Expected Value

In this section, we examine the basic properties of expected value. Our first result, which we call the **fundamental expected-value formula** (FEF), provides the foundation for obtaining properties of expected value. We introduce the FEF in Example 7.5.

EXAMPLE 7.5   *Introduces the Fundamental Expected-Value Formula*

Consider the discrete random variable X whose PMF is given in Table 5.18 on page 246. Determine $\mathcal{E}(X^2)$, the expected value of the random variable X^2.

Solution   We can apply either of two methods to determine $\mathcal{E}(X^2)$.

Method 1: Obtaining $\mathcal{E}(X^2)$ by using the definition of expected value.

With this method, we first determine the PMF of X^2 and then apply the definition of expected value (Definition 7.1 on page 327) to that random variable. The PMF of X^2 was obtained in Example 5.27 and is repeated here in the first two columns of Table 7.5. Applying Definition 7.1, we find from the third column of Table 7.5 that

$$\mathcal{E}(X^2) = \sum_y y p_{X^2}(y) = 3.15.$$

Table 7.5 Table for calculation of $\mathcal{E}(X^2)$ by using the definition of expected value

y	$p_{X^2}(y)$	$y p_{X^2}(y)$
0	0.20	0.00
1	0.35	0.35
4	0.25	1.00
9	0.20	1.80
	1.00	3.15

Method 2: Obtaining $\mathcal{E}(X^2)$ by using the PMF of X.

With this method, we first express the expected value of X^2 in terms of the PMF of X:

$$\mathcal{E}(X^2) = \sum_y y p_{X^2}(y) = \sum_y y P(X^2 = y)$$

$$= 0 \cdot P(X^2 = 0) + 1 \cdot P(X^2 = 1) + 4 \cdot P(X^2 = 4) + 9 \cdot P(X^2 = 9)$$

$$= 0 \cdot P(X = 0) + 1 \cdot P(X = \pm 1) + 4 \cdot P(X = \pm 2) + 9 \cdot P(X = \pm 3)$$

$$= 0 \cdot p_X(0) + 1 \cdot \big(p_X(-1) + p_X(1)\big)$$
$$\quad + 4 \cdot \big(p_X(-2) + p_X(2)\big) + 9 \cdot \big(p_X(-3) + p_X(3)\big)$$

$$= 0^2 \cdot p_X(0) + (-1)^2 \cdot p_X(-1) + 1^2 \cdot p_X(1)$$
$$\quad + (-2)^2 \cdot p_X(-2) + 2^2 \cdot p_X(2) + (-3)^2 \cdot p_X(-3) + 3^2 \cdot p_X(3).$$

Thus we see that

$$\mathcal{E}(X^2) = \sum_x x^2 p_X(x).$$

To evaluate the sum on the right side of the previous display, we use Table 7.6. The first two columns of that table repeat the PMF of X from Table 5.18; the third column provides the values of $x^2 p_X(x)$.

Table 7.6 Table for calculation of $\mathcal{E}(X^2)$ by using the PMF of X

x	$p_X(x)$	$x^2 p_X(x)$
-2	0.10	0.40
-1	0.20	0.20
0	0.20	0.00
1	0.15	0.15
2	0.15	0.60
3	0.20	1.80
	1.00	3.15

Summing the values in the third column of Table 7.6 gives

$$\mathcal{E}(X^2) = \sum_x x^2 p_X(x) = 3.15,$$

which is the same answer that we obtained by using Method 1.   ■

Let's take a moment to compare the two methods used in Example 7.5 for calculating the expected value of X^2. Method 1 appears quicker and easier than Method 2. But that appearance is misleading because, in Method 1, we relied on the work already done in Example 5.27, where we obtained the PMF of X^2 from the PMF of X. The advantage of Method 2 is that we don't need to determine the PMF of X^2 to obtain the expected value of X^2; rather, we can use the already known PMF of X.

If we define $g(x) = x^2$, we can express generically the computations done in Method 2 to obtain the expected value of X^2. Specifically, Proposition 5.14 on page 247 provides a formula for the PMF of a function of a discrete random variable, from which we get

$$\mathcal{E}(g(X)) = \sum_y y p_{g(X)}(y) = \sum_y y \sum_{x \in g^{-1}(\{y\})} p_X(x) = \sum_y \sum_{x \in g^{-1}(\{y\})} y p_X(x)$$

$$= \sum_y \sum_{x \in g^{-1}(\{y\})} g(x) p_X(x) = \sum_x g(x) p_X(x),$$

where $g^{-1}(\{y\}) = \{x : g(x) = y\}$.

This argument works for any discrete random variable X and any real-valued function g for which $g(X)$ has finite expectation. In fact, as Proposition 7.2 shows, an essentially identical argument provides the corresponding result in the multivariate case.

◆◆◆ **Proposition 7.2** **Fundamental Expected-Value Formula: Discrete Case**

Let $X_1, \ldots, X_m$ be discrete random variables defined on the same sample space and let g be a real-valued function of m variables defined on the range of $(X_1, \ldots, X_m)$. Then $g(X_1, \ldots, X_m)$ has finite expectation if and only if

$$\sum_{(x_1, \ldots, x_m)} \cdots \sum |g(x_1, \ldots, x_m)| \, p_{X_1, \ldots, X_m}(x_1, \ldots, x_m) < \infty. \qquad (7.10)$$

In that case,

$$\mathcal{E}\big(g(X_1, \ldots, X_m)\big) = \sum_{(x_1, \ldots, x_m)} \cdots \sum g(x_1, \ldots, x_m) \, p_{X_1, \ldots, X_m}(x_1, \ldots, x_m). \qquad (7.11)$$

In the univariate and bivariate cases, Equation (7.11) reduces to

$$\mathcal{E}\big(g(X)\big) = \sum_{x} g(x) p_X(x) \qquad (7.12)$$

and

$$\mathcal{E}\big(g(X, Y)\big) = \sum_{(x,y)} \sum g(x, y) p_{X,Y}(x, y), \qquad (7.13)$$

respectively.

Proof For convenience, we set $Z = g(X_1, \ldots, X_m)$. From Proposition 6.17 on page 305,

$$\sum_{z} |z| p_Z(z) = \sum_{z} |z| \sum_{(x_1, \ldots, x_m) \in g^{-1}(\{z\})} \cdots \sum p_{X_1, \ldots, X_m}(x_1, \ldots, x_m)$$

$$= \sum_{z} \sum_{(x_1, \ldots, x_m) \in g^{-1}(\{z\})} \cdots \sum |z| p_{X_1, \ldots, X_m}(x_1, \ldots, x_m)$$

$$= \sum_{z} \sum_{(x_1, \ldots, x_m) \in g^{-1}(\{z\})} \cdots \sum |g(x_1, \ldots, x_m)| \, p_{X_1, \ldots, X_m}(x_1, \ldots, x_m)$$

$$= \sum_{(x_1, \ldots, x_m)} \cdots \sum |g(x_1, \ldots, x_m)| \, p_{X_1, \ldots, X_m}(x_1, \ldots, x_m).$$

Hence $g(X_1, \ldots, X_m)$ has finite expectation if and only if Equation (7.10) holds.

Applying the same argument but without absolute values, we obtain Equation (7.11). Equations (7.12) and (7.13) are obvious special cases of Equation (7.11). ◆

Thus, in Example 7.5, Method 2, we used the FEF with $g(x) = x^2$ to obtain the expected value of X^2:

$$\mathcal{E}(X^2) = \mathcal{E}\big(g(X)\big) = \sum_{x} g(x) p_X(x) = \sum_{x} x^2 p_X(x) = 3.15.$$

Examples 7.6 and 7.7 provide more interesting applications of the FEF.

EXAMPLE 7.6   *The Fundamental Expected-Value Formula*

The Kelly Criterion   The *Kelly criterion* is a theory of optimal resource allocation (diversification) in favorable investment environments. It was first proposed by J. L. Kelly as an interpretation of information rate and has subsequently been applied to casino games and modern portfolio theory. The Kelly criterion is to maximize the expected value of the logarithm of investor capital, which ensures optimal long-run rate of capital growth.

Here we consider the simplest type of investment scenario. Successive investments are independent of one another and, for each investment, the investor, with probability p, gains k units per unit invested and, with probability $1 - p$, loses 1 unit per unit invested.

a) The investment is favorable to the investor if and only if the investor's expected gain per unit invested is positive. Show that this condition is equivalent to the condition that $k > (1 - p)/p$.

b) Use the FEF to obtain the expected value of the logarithm of investor capital after one investment.

c) Assume that the investment is favorable to the investor and apply the Kelly criterion to determine the investor's optimal long-run investment strategy.

Solution   a) Let X denote the investor's gain per unit invested. Then, by assumption, X is a discrete random variable that takes the value k with probability p and the value -1 with probability $1 - p$. Thus the PMF of X is

$$p_X(x) = \begin{cases} p, & \text{if } x = k; \\ 1 - p, & \text{if } x = -1; \\ 0, & \text{otherwise.} \end{cases}$$

The expected gain per unit invested is, therefore,

$$\mathcal{E}(X) = \sum_x x p_X(x) = kp + (-1)(1 - p) = pk - (1 - p).$$

Hence the investment is favorable to the investor if and only if $pk - (1 - p) > 0$ or, equivalently, $k > (1 - p)/p$.

b) Let b denote the amount of the investment, let w_0 denote the investor's capital before the investment, and let W denote the investor's capital after the investment. Then W is a discrete random variable that takes the value $w_0 + bk$ with probability p and the value $w_0 - b$ with probability $1 - p$. Thus the PMF of W is

$$p_W(w) = \begin{cases} p, & \text{if } w = w_0 + bk; \\ 1 - p, & \text{if } w = w_0 - b; \\ 0, & \text{otherwise.} \end{cases}$$

Applying the FEF with $g(w) = \ln w$, we find that the expected value of the logarithm of investor capital after one investment is

$$\mathcal{E}(\ln W) = \sum_w \ln w \cdot p_W(w) = p \ln(w_0 + bk) + (1 - p) \ln(w_0 - b). \qquad \textbf{(7.14)}$$

c) To maximize the expected value of the logarithm of investor capital, we need to maximize the right side of Equation (7.14) as a function of b. To do so, we let

$$G(b) = p \ln(w_0 + bk) + (1 - p) \ln(w_0 - b).$$

Taking the derivative with respect to b yields

$$G'(b) = \frac{pk}{w_0 + bk} - \frac{1-p}{w_0 - b}.$$

Setting $G'(b)$ equal to 0 and solving for b, we get

$$b = \left(\frac{pk - (1-p)}{k}\right) w_0. \qquad (7.15)$$

Use of calculus shows that this value of b indeed maximizes G. Consequently, Equation (7.15) provides the optimal investment strategy: The investor should always invest a fixed proportion of his capital—namely, the proportion $(pk - (1-p))/k$, known as the *optimal Kelly betting fraction*. Note that the numerator of the optimal Kelly betting fraction is the expected gain per unit invested. ∎

EXAMPLE 7.7 *The Fundamental Expected-Value Formula*

Automobile Insurance Policies An automobile insurance policy reimburses a family's accident losses up to a maximum of two accidents per year. For a (high-risk) three-person family, let X, Y, and Z denote the number of accidents in a single year for the three people in the family. We assume that the joint PMF of X, Y, Z is given by

$$p_{X,Y,Z}(x, y, z) = (x + 2y + z)/63, \qquad x \in \{0, 1\}, \ y, z \in \{0, 1, 2\},$$

and $p_{X,Y,Z}(x, y, z) = 0$ otherwise. Determine the expected number of unreimbursed accidents per year.

Solution Let U denote the number of unreimbursed accidents in a single year. Then

$$U = \begin{cases} 0, & \text{if } X + Y + Z \le 2; \\ X + Y + Z - 2, & \text{if } X + Y + Z > 2. \end{cases}$$

To obtain $\mathcal{E}(U)$, we apply the FEF with the function

$$g(x, y, z) = \begin{cases} 0, & \text{if } x + y + z \le 2; \\ x + y + z - 2, & \text{if } x + y + z > 2. \end{cases}$$

We have

$$\mathcal{E}(U) = \mathcal{E}\big(g(X, Y, Z)\big) = \sum\sum\sum_{(x,y,z)} g(x, y, z) p_{X,Y,Z}(x, y, z)$$

$$= \sum\sum\sum_{x+y+z \le 2} 0 \cdot p_{X,Y,Z}(x, y, z) + \sum\sum\sum_{x+y+z > 2} (x + y + z - 2) p_{X,Y,Z}(x, y, z)$$

$$= 0 + (0 + 1 + 2 - 2) p_{X,Y,Z}(0, 1, 2) + \cdots + (1 + 2 + 2 - 2) p_{X,Y,Z}(1, 2, 2)$$

$$= 76/63 = 1.206.$$

The expected number of unreimbursed accidents per year is roughly 1.2. ∎

Basic Properties of Expected Value

Now that we have the fundamental expected-value formula at our disposal, we can establish some of the basic and most widely used properties of expected value. *Although the properties of expected value given in this section are stated in terms of discrete random variables, they also hold for any types of random variables.*

Let's consider for a moment the simplest possible type of random variable, a constant random variable. A random variable X is said to be a **constant random variable** if there is a real number c such that $P(X = c) = 1$ or, equivalently, if its PMF is of the form $p_X(x) = 1$ if $x = c$, and $p_X(x) = 0$ otherwise. For such a random variable, we often write c instead of X. Note that, for a constant random variable X,

$$\mathcal{E}(X) = \sum_x x p_X(x) = c \cdot 1 = c,$$

or

$$\mathcal{E}(c) = c.$$

This relation makes sense: If a random variable equals a constant with probability 1, its observed value is always that constant and hence its expected value is that constant.

Proposition 7.3 shows how expected value behaves with regard to sums of random variables and constant multiples of random variables.

◆◆◆   **Proposition 7.3   Linearity Property of Expected Value**

Let X and Y be discrete random variables defined on the same sample space and having finite expectation, and let c be a real number. Then the following relations hold.
a) The random variable $X + Y$ has finite expectation and

$$\mathcal{E}(X + Y) = \mathcal{E}(X) + \mathcal{E}(Y). \tag{7.16}$$

> In words, the expected value of the sum of two random variables equals the sum of their expected values.

b) The random variable cX has finite expectation and

$$\mathcal{E}(cX) = c\,\mathcal{E}(X). \tag{7.17}$$

In words, the expected value of a constant times a random variable equals the constant times the expected value of the random variable.

Proof   a) Because X and Y have finite expectation, the triangle inequality and Proposition 6.2 on page 263 imply that

$$\sum_{(x,y)} |x + y| p_{X,Y}(x, y) \le \sum_{(x,y)} |x| p_{X,Y}(x, y) + \sum_{(x,y)} |y| p_{X,Y}(x, y)$$

$$= \sum_x |x| \left(\sum_y p_{X,Y}(x, y) \right) + \sum_y |y| \left(\sum_x p_{X,Y}(x, y) \right)$$

$$= \sum_x |x| p_X(x) + \sum_y |y| p_Y(y) < \infty.$$

Thus, from the FEF, $X + Y$ has finite expectation. To establish Equation (7.16), we apply the FEF first with $g(x, y) = x + y$ and then with $g(x, y) = x$ and $g(x, y) = y$:

$$\mathcal{E}(X + Y) = \sum_{(x,y)} \sum (x + y) p_{X,Y}(x, y)$$

$$= \sum_{(x,y)} \sum x p_{X,Y}(x, y) + \sum_{(x,y)} \sum y p_{X,Y}(x, y) = \mathcal{E}(X) + \mathcal{E}(Y).$$

b) Because X has finite expectation, we have

$$\sum_x |cx| p_X(x) = |c| \sum_x |x| p_X(x) < \infty,$$

and therefore cX has finite expectation. Applying the FEF with $g(x) = cx$, we get

$$\mathcal{E}(cX) = \sum_x (cx) p_X(x) = c \sum_x x p_X(x) = c \mathcal{E}(X),$$

as required.   ◆

The two properties of expected value given in Proposition 7.3 are collectively referred to as the **linearity property.** Together, they are equivalent to the property that

$$\mathcal{E}(aX + bY) = a \mathcal{E}(X) + b \mathcal{E}(Y), \tag{7.18}$$

for any two real numbers a and b. Also, a simple induction argument shows that we can extend the linearity property to any finite number of discrete random variables:

$$\mathcal{E}\left(\sum_{j=1}^{m} c_j X_j \right) = \sum_{j=1}^{m} c_j \mathcal{E}(X_j), \tag{7.19}$$

where $X_1, \ldots, X_m$ are discrete random variables defined on the same sample space and having finite expectation, and $c_1, \ldots, c_m$ are real numbers. See Exercise 7.45.

Our first application of the linearity property of expected value is to statistics—specifically, to the problem of estimating an unknown population mean. Although we state the result for finite populations, it also holds for infinite populations.

EXAMPLE 7.8   *The Linearity Property of Expected Value*

Estimating a Population Mean   Consider a finite population and a variable defined on it. Let μ denote the population mean—that is, the arithmetic mean of all possible observations of the variable for the entire population. Suppose that we don't know μ and that we want to estimate it. To do so, we take a random sample of n members from the population. Let $X_1, \ldots, X_n$ denote the values of the variable for the n members selected and set

$$\bar{X}_n = \frac{X_1 + \cdots + X_n}{n} = \frac{1}{n} \sum_{j=1}^{n} X_j. \tag{7.20}$$

The random variable $\bar{X}_n$ is called the **sample mean,** based on a sample of size n. Show that $\bar{X}_n$ is an **unbiased estimator** of μ—that is, $\mathcal{E}(\bar{X}_n) = \mu$.

Solution Let j be a positive integer between 1 and n, inclusive. Because the sample is random, the jth member obtained is equally likely to be any of the members of the population. Hence, from Proposition 7.1 on page 330, we have $\mathcal{E}(X_j) = \mu$. Consequently, by the linearity property of expected value,

$$\mathcal{E}(\bar{X}_n) = \mathcal{E}\left(\frac{1}{n}\sum_{j=1}^{n} X_j\right) = \frac{1}{n}\sum_{j=1}^{n}\mathcal{E}(X_j) = \frac{1}{n}\sum_{j=1}^{n}\mu = \frac{1}{n}\cdot n\mu = \mu.$$

This result holds regardless of whether the sampling is with or without replacement. ■

In Example 7.2 on page 331, we obtained a formula for the expected value of a binomial random variable. As a second application of the linearity property of expected value, we provide a simpler derivation of that formula.

EXAMPLE 7.9   *Expected Value of a Binomial Random Variable*

Use the linearity property of expected value to obtain the expected value of a binomial random variable with parameters n and p.

Solution Let $X \sim \mathcal{B}(n, p)$. Recall that X represents the number of successes in n Bernoulli trials with success probability p. For each $1 \leq j \leq n$, let $X_j = 1$ if the jth trial results in success and let $X_j = 0$ otherwise. Clearly, $X = \sum_{j=1}^{n} X_j$ and, consequently, by the linearity property of expected value,

$$\mathcal{E}(X) = \sum_{j=1}^{n}\mathcal{E}(X_j). \tag{7.21}$$

Each X_j is an indicator random variable—namely, the indicator random variable of the event E_j that the jth trial results in success. Because the expected value of the indicator random variable of an event equals the probability of that event, we have $\mathcal{E}(X_j) = P(E_j) = p$. Then, from Equation (7.21), we conclude that $\mathcal{E}(X) = np$. ■

Exercise 7.23 on page 335 asks for a formula for the expected value of a hypergeometric random variable. Using the linearity property of expected value, we can more easily obtain that formula, as shown in Example 7.10.

EXAMPLE 7.10   *Expected Value of a Hypergeometric Random Variable*

Use the linearity property of expected value to obtain the expected value of a hypergeometric random variable with parameters N, n, and p.

Solution Let $X \sim \mathcal{H}(N, n, p)$. Recall that X represents the number of members sampled that have a specified attribute when a random sample of size n is taken without replacement from a population of size N in which the proportion of members having the specified attribute is p. For each $1 \leq j \leq n$, let $X_j = 1$ if the jth member sampled has the specified attribute and let $X_j = 0$ otherwise. Clearly, $X = \sum_{j=1}^{n} X_j$ and, consequently, by the linearity property of expected value, we have

$$\mathcal{E}(X) = \sum_{j=1}^{n} \mathcal{E}(X_j). \tag{7.22}$$

Each X_j is an indicator random variable—namely, the indicator random variable of the event E_j that the jth member sampled has the specified attribute. Because the sampling is random, the jth member sampled is equally likely to be any of the N members of the population. Therefore the probability that the jth member sampled has the specified attribute equals the proportion of the population that has that attribute, which is p. Thus $\mathcal{E}(X_j) = P(E_j) = p$ and, in view of this result and Equation (7.22), we conclude that $\mathcal{E}(X) = np$. ∎

Proposition 7.4 provides some additional properties of expected value. Again, although we state and prove the proposition for discrete random variables, the properties presented hold for all types of random variables.

◆◆◆ **Proposition 7.4**

Let X and Y be discrete random variables defined on the same sample space and having finite expectation.
a) If a and b are real numbers, then $a + bX$ has finite expectation and

$$\mathcal{E}(a + bX) = a + b\,\mathcal{E}(X).$$

b) If $X \leq Y$, then $\mathcal{E}(X) \leq \mathcal{E}(Y)$.

Proof   a) This result follows easily from the linearity property of expected value and the fact that the expected value of a constant random variable is that constant.
b) Let $Z = Y - X$. Then, by assumption, the range of Z is a subset of the nonnegative real numbers. In particular, $p_Z(z) = 0$ for $z < 0$. Using this fact and the linearity property of expected value, we have

$$\mathcal{E}(Y) - \mathcal{E}(X) = \mathcal{E}(Y - X) = \mathcal{E}(Z) = \sum_{z} z p_Z(z) = \sum_{z \geq 0} z p_Z(z) \geq 0.$$

Hence $\mathcal{E}(X) \leq \mathcal{E}(Y)$. ◆

Note the following.

- Proposition 7.4(a) shows that expected value preserves location and scale change; that is, the expected value of a linear function of a random variable equals the linear function of the expected value.
- Proposition 7.4(b) is called the **monotonicity property** of expected value.

Expected Value of a Product of Random Variables

We have shown that the expected value of the sum of two or more random variables equals the sum of their expected values. What about products of random variables: Does the expected value of the product of two or more random variables equal the product of their expected values? In general, the answer is *no*.

For instance, let X and Y be discrete random variables with joint and marginal PMFs as presented in Table 7.7. Simple computations show that $\mathcal{E}(X) = \mathcal{E}(Y) = 1/3$ and that $\mathcal{E}(XY) = 0$. Consequently, $\mathcal{E}(XY) \neq \mathcal{E}(X)\mathcal{E}(Y)$.

Table 7.7 Random variables for which the expected value of their product doesn't equal the product of their expected values

		y		
		0	1	$p_X(x)$
x	0	$\frac{1}{3}$	$\frac{1}{3}$	$\frac{2}{3}$
	1	$\frac{1}{3}$	0	$\frac{1}{3}$
	$p_Y(y)$	$\frac{2}{3}$	$\frac{1}{3}$	1

However, for independent random variables, the expected value of their product does equal the product of their expected values. We state and prove this fact in Proposition 7.5.

◆◆◆ **Proposition 7.5**

Let X and Y be independent discrete random variables having finite expectation. Then the random variable XY has finite expectation and

$$\mathcal{E}(XY) = \mathcal{E}(X)\mathcal{E}(Y). \tag{7.23}$$

In words, for independent random variables, the expected value of their product equals the product of their expected values.

Proof By independence, the joint PMF of X and Y equals the product of their marginal PMFs. Therefore

$$\sum_{(x,y)}\sum |xy|\, p_{X,Y}(x, y) = \sum_{(x,y)}\sum |xy|\, p_X(x) p_Y(y)$$
$$= \sum_x |x|\, p_X(x) \sum_y |y|\, p_Y(y) < \infty,$$

where the inequality follows from the fact that both X and Y have finite expectation. Hence, from the FEF, the random variable XY has finite expectation. Applying the FEF with $g(x, y) = xy$, we have

$$\mathcal{E}(XY) = \sum_{(x,y)}\sum xy\, p_{X,Y}(x, y) = \sum_{(x,y)}\sum xy\, p_X(x) p_Y(y)$$
$$= \sum_x x p_X(x) \sum_y y p_Y(y) = \mathcal{E}(X)\mathcal{E}(Y),$$

as required.   ◆

Although independence is a sufficient condition for expected values to multiply [i.e., for Equation (7.23) to hold], it isn't necessary. For instance, let X and Y be discrete random variables with joint and marginal PMFs as presented in Table 7.8. Simple computations show that $\mathcal{E}(Y) = 0$ and that $\mathcal{E}(XY) = 0$. Consequently, $\mathcal{E}(XY) = \mathcal{E}(X)\,\mathcal{E}(Y)$. However, clearly, X and Y aren't independent random variables.

Table 7.8 Dependent random variables for which the expected value of their product equals the product of their expected values

		y		
	-1	0	1	$p_X(x)$
x $\quad 0$	$\frac{1}{4}$	0	$\frac{1}{4}$	$\frac{1}{2}$
$\quad\;\; 1$	0	$\frac{1}{2}$	0	$\frac{1}{2}$
$p_Y(y)$	$\frac{1}{4}$	$\frac{1}{2}$	$\frac{1}{4}$	1

We can use either mathematical induction or a direct argument to show that Proposition 7.5 holds for any finite number of independent discrete random variables. Specifically, as you are asked to prove in Exercise 7.46, if $X_1, \ldots, X_m$ are independent discrete random variables having finite expectation, then $\prod_{j=1}^{m} X_j$ has finite expectation and

$$\mathcal{E}\left(\prod_{j=1}^{m} X_j\right) = \prod_{j=1}^{m} \mathcal{E}(X_j). \tag{7.24}$$

Using Tail Probabilities to Obtain Expected Values

Most of the discrete random variables that we have considered and most that occur in practice are nonnegative-integer valued. For such a random variable, computing its expected value by using tail probabilities is sometimes easier. Proposition 7.6 provides the formula.

◆◆◆ **Proposition 7.6   Expected Value by Using Tail Probabilities**

Suppose that X is a nonnegative-integer valued random variable. Then X has finite expectation if and only if $\sum_{n=0}^{\infty} P(X > n) < \infty$, in which case

$$\mathcal{E}(X) = \sum_{n=0}^{\infty} P(X > n). \tag{7.25}$$

Proof   We have

$$\sum_{x} x p_X(x) = \sum_{x=0}^{\infty} x p_X(x) = \sum_{x=1}^{\infty} x p_X(x) = \sum_{x=1}^{\infty}\left(\sum_{y=1}^{x} 1\right) p_X(x).$$

In Figure 7.2, we have provided a graphical display of the set of points over which the preceding double sum is taken.

Figure 7.2

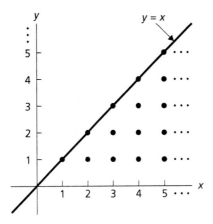

Referring to Figure 7.2 and applying the FPF, we get

$$\sum_{x=1}^{\infty}\left(\sum_{y=1}^{x}1\right)p_X(x) = \sum_{y=1}^{\infty}\left(\sum_{x=y}^{\infty}p_X(x)\right) = \sum_{y=1}^{\infty}P(X \ge y)$$

$$= \sum_{y=1}^{\infty}P(X > y - 1) = \sum_{n=0}^{\infty}P(X > n).$$

The required result now follows easily.   ◆

EXAMPLE 7.11   *Using Tail Probabilities to Obtain Expected Values*

Geometric Random Variables   Use tail probabilities to obtain the expected value of a geometric random variable with parameter p.

Solution   Let $X \sim \mathcal{G}(p)$. From Proposition 5.10 on page 231, $P(X > n) = (1 - p)^n$ for each nonnegative integer n. Referring to Proposition 7.6 and using a geometric formula, we get

$$\mathcal{E}(X) = \sum_{n=0}^{\infty}P(X > n) = \sum_{n=0}^{\infty}(1 - p)^n = \frac{1}{1 - (1 - p)} = \frac{1}{p},$$

as required. Compare this simple computation using tail probabilities to a direct computation using the definition of expected value, as required by Exercise 7.24 on page 336.   ■

EXERCISES 7.2   Basic Exercises

Note: The sum of the first n positive integers and the sum of the squares of the first n positive integers can be obtained from the formulas

$$\sum_{k=1}^{n} k = \frac{n(n+1)}{2} \qquad and \qquad \sum_{k=1}^{n} k^2 = \frac{n(n+1)(2n+1)}{6},$$

respectively. You may find these formulas useful in doing some of the exercises in this and other sections. Both of them can be proved by using mathematical induction.

7.29 Two balanced dice are rolled, one black and one gray. Let Y and Z be the faces showing on the black and gray dice, respectively, and let X be the sum of the two faces showing.
a) Determine the expected values of Y and Z.
b) Express X in terms of Y and Z and then use the linearity property of expected value and the result of part (a) to obtain the expected value of X.
c) Compare your answer in part (b) to that obtained in Exercise 7.1 on page 333. Which method for obtaining $\mathcal{E}(X)$ is easier? Explain your answer.

7.30 Let Y and Z be as in Exercise 7.29, and let W be the product of the two faces showing. Determine the expected value of W by using each of the following methods. Compare the work done in each case.
a) First find the PMF of W and then apply the definition of expected value (Definition 7.1 on page 327).
b) First express W in terms of Y and Z and then apply the FEF.
c) First express W in terms of Y and Z and then use the independence of Y and Z and the result of Exercise 7.29(a).

7.31 Let Y and Z be as in Exercise 7.29 and let U be the minimum of the two faces showing. Determine the expected value of U by using each of the following methods. Compare the work done in each case.
a) First find the PMF of U and then apply the definition of expected value.
b) First express U in terms of Y and Z and then apply the FEF.

7.32 Let Y and Z be as in Exercise 7.29, and let V be the maximum of the two faces showing. Determine the expected value of V by using each of the following methods. Compare the work done in each case.
a) First find the PMF of V and then apply the definition of expected value.
b) First express V in terms of Y and Z and then apply the FEF.

7.33 Consider n Bernoulli trials with success probability p. Let Y be 0 if the number of successes is even and let Y be 1 if the number of successes is odd. Determine the expected value of the random variable Y
a) by using the FEF.
b) without using the FEF, but by using the FPF.

7.34 An archer shoots an arrow into a square target 6 feet on a side whose center we call the origin. Assume that the archer will actually hit the target and is equally likely to hit any portion of the target. The archer scores 10 points if she hits the bull's eye—a disk of radius 1 foot centered at the origin (region A); she scores 5 points if she hits the ring with inner radius 1 foot and outer radius 2 feet centered at the origin (region B); and she scores 0 points otherwise (region C). Suppose that the archer shoots independently at the target four times.

Let X, Y, and Z denote the numbers of times that she hits regions A, B, and C, respectively. Also, let T denote the total score in the four shots.

a) Determine and identify the joint PMF of the random variables X, Y, and Z.

b) Express T in terms of X, Y, and Z, and then use the FEF and the result of part (a) to obtain $\mathcal{E}(T)$.

c) Determine and identify the univariate marginal PMFs of X, Y, and Z and, from them, find the expected values of X, Y, and Z.

d) Express T in terms of X, Y, and Z and then use the linearity property of expected value and the result of part (c) to obtain $\mathcal{E}(T)$.

e) Exercise 7.2 on page 333 asks for the expected score for one shot at the target. Use that result and the linearity property of expected value to obtain $\mathcal{E}(T)$.

f) Identify yet another way to obtain $\mathcal{E}(T)$.

7.35 N married couples are randomly seated at a rectangular table, the women on one side and the men on the other. What is the expected number of men who sit across from their wives? *Hint:* One way to solve this problem is to first determine the PMF of the number of men who sit across from their wives and then apply the definition of expected value. However, there is an easier way.

7.36 A random sample of size n is taken from a very large lot of items in which $100p_1\%$ have exactly one defect and $100p_2\%$ have two or more defects, where $0 < p_1 + p_2 < 1$. An item with exactly one defect costs \$1 to repair, whereas an item with two or more defects costs \$3 to repair. Determine the expected cost of repairing the defective items in the sample.

7.37 Determine the expected number of times a balanced die must be thrown to get all six possible numbers. *Hint:* For $1 \le k \le 6$, let X_k denote the number of throws from the appearance of the $(k-1)$th distinct number until the appearance of the kth distinct number.

7.38 Expected utility: One method for deciding among various investments involves the concept of *expected utility*. Economists describe the importance of various levels of wealth by using *utility functions*. For instance, in most cases, a single dollar is more important (has greater utility) for someone with little wealth than for someone with great wealth. Consider two investments—say, Investment A and Investment B. Measured in thousands of dollars, suppose that Investment A yields 0, 1, and 4 with probabilities 0.1, 0.5, and 0.4, respectively, and that Investment B yields 0, 1, and 16 with probabilities 0.5, 0.3, and 0.2, respectively. Let Y denote the yield of an investment. For the two investments, determine and compare

a) $\mathcal{E}(Y)$, the expected yield.

b) $\mathcal{E}(\sqrt{Y})$, the expected utility, using the utility function $u(y) = \sqrt{y}$. Interpret the utility function u.

c) $\mathcal{E}(Y^{3/2})$, the expected utility, using the utility function $v(y) = y^{3/2}$. Interpret the utility function v.

7.39 A lot contains 17 items, each of which is subject to inspection by two quality assurance engineers. Each engineer randomly and independently selects 4 items from the lot. Determine the expected number of items selected by

a) both engineers.   b) neither engineer.   c) exactly one engineer.

d) Without doing any computations, obtain the sum of the three expected values found in parts (a), (b), and (c). Explain your reasoning.

7.40 Consider two electrical components whose lifetimes observed at discrete time units (e.g., every hour) are independent geometric random variables with parameter p. Use tail probabilities to determine the expected time until

a) the first component to fail.   b) the second component to fail.

7.41 Two people agree to meet at a specified place between 3:00 P.M. and 4:00 P.M. Suppose that you measure time to the nearest minute relative to 3:00 P.M. so that, for instance, time 40 represents 3:40 P.M. Further suppose that each person arrives according to the discrete uniform distribution on $\{0, 1, \ldots, 60\}$ and that the two arrival times are independent. Determine the expected time
a) until the first arrival.   **b)** until the second arrival.
c) that the first person to arrive waits for the second person to arrive.

7.42 Show that $\mathcal{E}(X^2) \geq (\mathcal{E}(X))^2$. *Hint:* Consider the random variable $(X - \mathcal{E}(X))^2$.

7.43 Random variables X and Y defined on the same sample space are said to be *linearly uncorrelated* if $\mathcal{E}((X - \mathcal{E}(X))(Y - \mathcal{E}(Y))) = 0$.
a) Show that X and Y are linearly uncorrelated if and only if $\mathcal{E}(XY) = \mathcal{E}(X)\mathcal{E}(Y)$.
b) Are independent random variables (with finite expectations) linearly uncorrelated? Justify your answer.
c) Are linearly uncorrelated random variables independent? Justify your answer.

Theory Exercises

7.44 Let X have the negative binomial distribution with parameters r and p.
a) From Exercise 6.151 on page 322, we know that X can be expressed as the sum of r independent geometric random variables with common parameter p. Use this fact and Example 7.11 on page 348 to deduce that $\mathcal{E}(X) = r/p$.
b) Is the independence of the geometric random variables required for the result in part (a)? Explain your answer.

7.45 Let $X_1, \ldots, X_m$ be discrete random variables defined on the same sample space and having finite expectation, and let $c_1, \ldots, c_m$ be real numbers. Use Proposition 7.3 on page 342 and mathematical induction to prove that $\sum_{j=1}^{m} c_j X_j$ has finite expectation and that

$$\mathcal{E}\left(\sum_{j=1}^{m} c_j X_j\right) = \sum_{j=1}^{m} c_j \mathcal{E}(X_j).$$

7.46 Let $X_1, \ldots, X_m$ be independent discrete random variables having finite expectation.
a) Use Proposition 7.5 on page 346 and mathematical induction to prove that $\prod_{j=1}^{m} X_j$ has finite expectation and that

$$\mathcal{E}\left(\prod_{j=1}^{m} X_j\right) = \prod_{j=1}^{m} \mathcal{E}(X_j).$$

b) Use a direct argument to establish the results of part (a). *Hint:* Refer to Proposition 6.14 on page 297.

7.47 Let X be a random variable with finite expectation. Prove that $|\mathcal{E}(X)| \leq \mathcal{E}(|X|)$.

7.48 Bounded random variable: A random variable X is said to be *bounded* if there is a positive real number M such that $P(|X| \leq M) = 1$. Prove that a bounded discrete random variable, bounded by M, has finite expectation and that $|\mathcal{E}(X)| \leq M$. *Hint:* Refer to Exercise 7.47.

7.49 Markov's inequality: Let X be a nonnegative random variable with finite expectation. *Markov's inequality* is that $P(X \geq t) \leq \mathcal{E}(X)/t$ for all positive real numbers t.
a) Prove Markov's inequality for discrete random variables by applying the FPF.

b) Show that, for a nonnegative random variable X, we have $t I_{\{X \geq t\}} \leq X$ for all positive real numbers t.

c) Prove Markov's inequality by using part (b) and properties of expected value.

Advanced Exercises

7.50 A number X is chosen at random from the interval $(0, 1)$. Find the expected value of
a) the first digit of the decimal expansion of X.
b) the first digit of the decimal expansion of $\sqrt{X}$.
c) the second digit of the decimal expansion of $\sqrt{X}$.

7.51 Of the customers who enter a store, 60% are women and 40% are men. If 50 customers enter the store, what is the expected number of times that a woman is followed by a man? State any assumptions that you make.

7.52 Probability generating function: Let X be a nonnegative-integer valued random variable. The *probability generating function* of X, denoted P_X, is defined by $P_X(t) = \mathcal{E}(t^X)$. We use PGF as an abbreviation for "probability generating function."
a) Verify that P_X is well defined for $|t| \leq 1$ by showing that the random variable t^X has finite expectation for each t in that interval.
b) Prove that $P_X(t) = \sum_{n=0}^{\infty} p_X(n) t^n$.
c) Show that, if X has finite expectation, then $\mathcal{E}(X) = P_X'(1)$.

7.53 Refer to Exercise 7.52. For a random variable with each of the following probability distributions, determine the PGF in closed form and use the PGF to obtain the expected value.
a) Binomial with parameters n and p
b) Poisson with parameter λ
c) Geometric with parameter p
d) Negative binomial with parameters r and p
e) Discrete uniform on $\{1, 2, \ldots, N\}$

7.3   Variance of Discrete Random Variables

In this section, we apply the concept of expected value to obtain additional characteristics of random variables. Although we concentrate on discrete random variables, the ideas presented here apply to all types of random variables.

Recall that another term for *expected value* is *mean*. At this point, we frequently use the phrase "mean of X" instead of "expected value of X"; and we often say that "X has finite mean" rather than "X has finite expectation." Also recall that the symbol μ_X is an alternative to $\mathcal{E}(X)$.

To begin, we consider the *moments* of a random variable. Let r be a positive integer and let X be a random variable. We say that X has a **moment of order r** or that X has a **finite rth moment** if the random variable X^r has finite expectation. In that case, we define the **rth moment** of X to be the expected value of the random variable X^r:

$$r\text{th moment of } X = \mathcal{E}(X^r). \tag{7.26}$$

If X^r doesn't have finite expectation, we say that X doesn't have a moment of order r or that X doesn't have a finite rth moment.

Note the following.

- The first moment of a random variable X is its mean, $\mathcal{E}(X)$.
- If X has a moment of order r, it also has moments of all lower orders, as you are asked to show in Exercise 7.76.

Also of importance are **central moments.** Let X be a random variable with finite expectation. If $X - \mu_X$ has a finite rth moment, we define the **rth central moment** of X to be the rth moment of the random variable $X - \mu_X$:

$$r\text{th central moment of } X = \mathcal{E}\big((X - \mu_X)^r\big). \tag{7.27}$$

We leave it to you as Exercise 7.77 to show that X has an rth central moment if and only if it has an rth moment.

From the FEF, we can compute moments and central moments of a discrete random variable X as follows:

$$\mathcal{E}(X^r) = \sum_x x^r p_X(x) \tag{7.28}$$

and

$$\mathcal{E}\big((X - \mu_X)^r\big) = \sum_x (x - \mu_X)^r p_X(x). \tag{7.29}$$

Moments and central moments play an important role in both the theory and application of probability and in many other fields as well. For example, in physics, the first moment (mean, expected value) can be interpreted as a center of gravity and the second central moment can be interpreted as a moment of inertia. We concentrate on moments and central moments of order 2 or less.

Variance of a Discrete Random Variable

Recall that the mean of a random variable is the value of the random variable expected on average. So, in some sense, the mean of a random variable provides a typical or central value of the random variable. For that reason, the mean of a random variable is often referred to as a **measure of center.**

We can also use the concept of expected value to obtain other descriptive measures of random variables. One of the most important such measures is called the **variance,** which is the second central moment.

DEFINITION 7.2   Variance of a Random Variable

Let X be a random variable with finite second moment. Then the **variance** of X, denoted **Var(X)**, is defined by

$$\text{Var}(X) = \mathcal{E}\big((X - \mu_X)^2\big). \tag{7.30}$$

In words, the variance of a random variable is the expected value of the square of its deviation from its mean.

Note the following.

- The variance of a random variable is a nonnegative real number. Indeed, because $(X - \mu_X)^2 \geq 0$, the monotonicity property of expected value implies that

$$\text{Var}(X) = \mathcal{E}\big((X - \mu_X)^2\big) \geq \mathcal{E}(0) = 0.$$

- A random variable has a finite second moment if and only if it has a finite second central moment—that is, a **finite variance.** We generally use this latter term.
- Another commonly used notation for the variance of a random variable X is σ_X^2.
- The square root of the variance of a random variable X is called its **standard deviation,** denoted σ_X. In statistics and other applied fields, the standard deviation is used more commonly than the variance. One reason is that the units for the standard deviation are the same as those for the random variable, whereas the units for the variance are the square of those for the random variable.

The variance of a random variable provides a **measure of variation** of the possible values of the random variable. To show why, we first note that the random variable $X - \mu_X$ gives the deviation of X from its mean. So, roughly speaking, $\text{Var}(X) = \mathcal{E}\big((X - \mu_X)^2\big)$ tells us how far X is from its mean on average.[†]

Consequently, if $\text{Var}(X)$ is large, then, on average, X is far from its mean and, conversely, if $\text{Var}(X)$ is small, then, on average, X is close to its mean. In fact, in the extreme case of zero variance, we have the useful result presented in Proposition 7.7.

◆◆◆ **Proposition 7.7**

A random variable has zero variance if and only if it is a constant random variable. That is, $\text{Var}(X) = 0$ if and only if there is a constant c such that $P(X = c) = 1$.

Proof   For convenience, set $Y = (X - \mu_X)^2$. Suppose that X is a constant random variable—say, $P(X = c) = 1$. Then, $\mu_X = c$ and, so, $P(Y = 0) = P(X = \mu_X) = 1$. Therefore we have $\text{Var}(X) = \mathcal{E}(Y) = 0$.

Conversely, suppose that $\text{Var}(X) = 0$ or, equivalently, that $\mathcal{E}(Y) = 0$. Because Y is a nonnegative random variable, $p_Y(y) = 0$ for $y < 0$. Hence

$$0 = \mathcal{E}(Y) = \sum_y y p_Y(y) = \sum_{y \geq 0} y p_Y(y).$$

The only way that the sum on the right side of this equation can equal 0 is if $p_Y(y) = 0$ for all $y > 0$. Thus $p_Y(y) = 0$ unless $y = 0$ and hence $p_Y(0) = 1$. Consequently, $P(X = \mu_X) = P(Y = 0) = 1$, so X is a constant random variable.[‡]   ◆

[†] The quantity $\mathcal{E}(|X - \mu_X|)$—called the *mean absolute deviation*—is a more natural measure of how far X is from its mean on average. However, $\text{Var}(X) = \mathcal{E}\big((X - \mu_X)^2\big)$—or its square root σ_X, which has the same units as X—is much easier to work with mathematically and still provides a measure of how far X is from its mean on average.

[‡] In proving the converse in Proposition 7.7, we tacitly assumed that X is a discrete random variable. Nonetheless, the proposition holds generally, as we ask you to verify in Exercise 7.81.

An alternative formula for the variance in terms of (noncentral) moments is presented as Proposition 7.8. It is especially useful computationally when the mean of the random variable isn't an integer. For ease of reference, we call this alternative formula the **computing formula for the variance** and we call the formula in Definition 7.2 on page 353 the **defining formula for the variance.**

◆◆◆   **Proposition 7.8   Computing Formula for the Variance**

The variance of a random variable X can be obtained from the formula

$$\text{Var}(X) = \mathcal{E}(X^2) - (\mathcal{E}(X))^2. \tag{7.31}$$

In words, the variance of a random variable equals its second moment minus the square of its mean (first moment).

Proof   Applying the linearity property of expected value to the definition of the variance of a random variable and using the fact that the expected value of a constant random variable is the constant, we get

$$\text{Var}(X) = \mathcal{E}\big((X - \mu_X)^2\big) = \mathcal{E}\big(X^2 - 2\mu_X X + \mu_X^2\big) = \mathcal{E}(X^2) - 2\mu_X \mathcal{E}(X) + \mu_X^2$$
$$= \mathcal{E}(X^2) - 2\mu_X \mu_X + \mu_X^2 = \mathcal{E}(X^2) - \mu_X^2 = \mathcal{E}(X^2) - (\mathcal{E}(X))^2,$$

as required.   ◆

EXAMPLE 7.12   *The Variance of a Random Variable*

Number of Siblings   Refer to Example 7.1 on page 328 and obtain the variance of the random variable X by using the
a) defining formula.
b) computing formula.

Solution   The PMF of X is repeated here in the first two columns of Table 7.9.

Table 7.9

x	$p_X(x)$	$xp_X(x)$	$x - \mu_X$	$(x - \mu_X)^2$	$(x - \mu_X)^2 p_X(x)$	x^2	$x^2 p_X(x)$
0	0.200	0.000	−1.3	1.69	0.33800	0	0.000
1	0.425	0.425	−0.3	0.09	0.03825	1	0.425
2	0.275	0.550	0.7	0.49	0.13475	4	1.100
3	0.075	0.225	1.7	2.89	0.21675	9	0.675
4	0.025	0.100	2.7	7.29	0.18225	16	0.400
	1.000	1.300			0.91000		2.600

a) The third column shows that $\mu_X = 1.3$, as we found previously. From Definition 7.2, the FEF, and the sixth column, we obtain

$$\text{Var}(X) = \mathcal{E}\big((X - \mu_X)^2\big) = \sum_x (x - \mu_X)^2 p_X(x) = 0.91.$$

b) From Proposition 7.8, the FEF, and the third and eighth columns, we get

$$\text{Var}(X) = \mathcal{E}(X^2) - (\mathcal{E}(X))^2 = \sum_x x^2 p_X(x) - \left(\sum_x x p_X(x) \right)^2$$

$$= 2.6 - (1.3)^2 = 0.91.$$

Of course, the methods used in parts (a) and (b) both yield the same value for the variance of X. ∎

Variance as a Population Variance

For a variable of a finite population, the **population variance,** denoted σ^2, is defined to be the arithmetic mean of the squared deviations from the population mean of all possible observations of the variable for the entire population. Proposition 7.9, whose proof we leave to you as Exercise 7.79, shows the relationship between the population variance and the variance of the associated random variable.

◆◆◆ **Proposition 7.9   Variance as a Population Variance**

Consider a finite population and a variable defined on it. Suppose that a member is selected at random from the population and let X denote the value of the variable for the member obtained. Then the variance of the random variable X equals the population variance.

Variance for Some Families of Discrete Random Variables

For many families of discrete random variables, we can obtain formulas for the variance in terms of the parameters. We do so by substituting the general form of the PMF for the family under consideration into the defining or computing formula for the variance of a discrete random variable.

To illustrate, we consider the indicator random variable, I_E, of an event E. We have already shown that $\mu_{I_E} = P(E)$. Applying the defining formula for the variance and the FEF, we obtain

$$\text{Var}(I_E) = \mathcal{E}\big((I_E - \mu_{I_E})^2\big) = \sum_x (x - \mu_{I_E})^2 p_{I_E}(x)$$

$$= (0 - P(E))^2 \cdot (1 - P(E)) + (1 - P(E))^2 \cdot P(E)$$

or

$$\text{Var}(I_E) = P(E)(1 - P(E)). \tag{7.32}$$

Alternatively, we can obtain $\text{Var}(I_E)$ by first noting that $I_E^2 = I_E$ and then applying the computing formula for the variance. We leave the details to you as Exercise 7.69. In any case, Equation (7.32) shows that the variance of the indicator random variable of an event equals the probability of the event times the probability of the complement of the event. Although simple, that result is used often.

EXAMPLE 7.13   *Variance of a Binomial Random Variable*

Obtain the variance of a binomial random variable with parameters n and p.

Solution   Let $X \sim \mathcal{B}(n, p)$. Recall that the PMF of X is

$$p_X(x) = \binom{n}{x} p^x (1 - p)^{n-x}, \qquad x = 0, 1, \dots, n,$$

and $p_X(x) = 0$ otherwise. From Example 7.2, $\mathcal{E}(X) = np$. Hence, to obtain $\mathrm{Var}(X)$, we need only determine $\mathcal{E}(X^2)$ and then apply the computing formula for the variance. Using the FEF yields

$$\mathcal{E}(X^2) = \sum_x x^2 p_X(x) = \sum_{x=0}^{n} x^2 \binom{n}{x} p^x (1 - p)^{n-x}$$

$$= \sum_{x=0}^{n} \left((x^2 - x) + x\right) \binom{n}{x} p^x (1 - p)^{n-x}$$

$$= \sum_{x=0}^{n} (x^2 - x) \binom{n}{x} p^x (1 - p)^{n-x} + \sum_{x=0}^{n} x \binom{n}{x} p^x (1 - p)^{n-x}.$$

The second summation in the previous line is the mean of X, which equals np. For the first summation in that line, we have

$$\sum_{x=0}^{n} (x^2 - x) \binom{n}{x} p^x (1 - p)^{n-x} = \sum_{x=2}^{n} x(x - 1) \frac{n!}{x! \, (n - x)!} p^x (1 - p)^{n-x}$$

$$= n(n - 1) \sum_{x=2}^{n} \frac{(n - 2)!}{(x - 2)! \, (n - x)!} p^x (1 - p)^{n-x}$$

$$= n(n - 1) p^2 \sum_{x=2}^{n} \binom{n - 2}{x - 2} p^{x-2} (1 - p)^{n-x}$$

$$= n(n - 1) p^2 \sum_{j=0}^{n-2} \binom{n - 2}{j} p^j (1 - p)^{(n-2)-j}$$

$$= n(n - 1) p^2,$$

where the last equation follows from the binomial theorem or from the fact that the values of the PMF of a binomial random variable with parameters $n - 2$ and p sum to 1. Consequently, $\mathcal{E}(X^2) = n(n - 1)p^2 + np$ and hence

$$\mathrm{Var}(X) = \mathcal{E}(X^2) - \left(\mathcal{E}(X)\right)^2 = n(n - 1)p^2 + np - (np)^2 = np(1 - p).$$

Thus $\mathrm{Var}(X) = np(1 - p)$. That is, the variance of a binomial random variable is the product of the number of trials, the success probability, and the failure probability.   ■

The argument used in Example 7.13 to find the variance of a binomial random variable is somewhat tricky. In Section 7.4, we employ indicators and properties of the variance to obtain that variance more easily.

Using arguments similar to those in Example 7.13, we can obtain variance formulas for each of several other families of discrete random variables. Table 7.10 displays the variances for some of the most important families of discrete random variables.

Table 7.10 Variances for selected families of discrete random variables

Family	Parameter(s)	Variance
Binomial	n and p	$np(1-p)$
Hypergeometric	N, n, and p	$\left(\dfrac{N-n}{N-1}\right)np(1-p)$
Poisson	λ	λ
Geometric	p	$(1-p)/p^2$
Negative binomial	r and p	$r(1-p)/p^2$
Indicator	E	$P(E)(1-P(E))$

Some Basic Properties of Variance

We next examine how the variance of a random variable is affected by location or scale change. Keeping in mind that the variance measures the variation of the possible values of a random variable, we have the following.

- Because making a scale change (i.e., multiplying a random variable by a constant) in general affects the variation, doing so should affect the variance.
- Because making a location change (i.e., adding a constant to a random variable) doesn't affect the variation, doing so shouldn't affect the variance.

We make these statements precise in Proposition 7.10.

◆◆◆ **Proposition 7.10**

Let X be a random variable with finite variance and let c be a real number. Then
a) $\mathrm{Var}(cX) = c^2 \mathrm{Var}(X)$.
b) $\mathrm{Var}(X+c) = \mathrm{Var}(X)$.

Together, properties (a) and (b) are equivalent to

$$\mathrm{Var}(a+bX) = b^2 \mathrm{Var}(X) \tag{7.33}$$

for all real numbers a and b.

Proof　a) Because of the linearity property of expected value, we have $\mu_{cX} = c\mu_X$. Thus

$$\mathrm{Var}(cX) = \mathcal{E}\big((cX - \mu_{cX})^2\big) = \mathcal{E}\big((cX - c\mu_X)^2\big)$$
$$= \mathcal{E}\big(c^2(X - \mu_X)^2\big) = c^2\mathcal{E}\big((X - \mu_X)^2\big) = c^2\mathrm{Var}(X).$$

b) From Proposition 7.4(a) on page 345, $\mu_{X+c} = \mu_X + c$. Thus

$$\text{Var}(X + c) = \mathcal{E}\big([(X + c) - \mu_{X+c}]^2\big) = \mathcal{E}\big([(X + c) - (\mu_X + c)]^2\big)$$
$$= \mathcal{E}\big((X - \mu_X)^2\big) = \text{Var}(X).$$

The equivalence of Equation (7.33) to properties (a) and (b) together is straightforward and is left to you as Exercise 7.82.   ◆

From Proposition 7.10, we get the following important properties of the standard deviation of a random variable:

$$\sigma_{cX} = |c|\sigma_X \qquad \text{and} \qquad \sigma_{X+c} = \sigma_X, \tag{7.34}$$

for all real numbers c.

Standardized Random Variable

Properties of expected value and variance are particularly relevant to the concept of *standardizing* (or *normalizing*) a random variable. To begin, we present the definition of **standardized random variable.**

DEFINITION 7.3   Standardized Random Variable

Let X be a random variable with finite nonzero variance. Then the **standardized random variable** corresponding to X, denoted X^*, is defined by

$$X^* = \frac{X - \mathcal{E}(X)}{\sqrt{\text{Var}(X)}} = \frac{X - \mu_X}{\sigma_X}. \tag{7.35}$$

In words, the standardized random variable corresponding to X is obtained by subtracting from X its mean and then dividing by its standard deviation.

Note the following.

- The standardized random variable corresponding to a random variable X represents the number of standard deviations that X is from its mean. Consequently, a standardized random variable is unitless.
- The mean of a standardized random variable is 0:

$$\mathcal{E}(X^*) = \mathcal{E}\left(\frac{X - \mu_X}{\sigma_X}\right) = \frac{1}{\sigma_X}\mathcal{E}(X - \mu_X) = \frac{1}{\sigma_X}\big(\mathcal{E}(X) - \mu_X\big) = 0.$$

We ask you to supply the rationale for these steps in Exercise 7.71.

- The variance of a standardized random variable is 1:

$$\text{Var}(X^*) = \text{Var}\left(\frac{X - \mu_X}{\sigma_X}\right) = \frac{1}{\sigma_X^2}\text{Var}(X - \mu_X) = \frac{1}{\sigma_X^2}\text{Var}(X) = 1.$$

We ask you to supply the rationale for these steps in Exercise 7.71.

The concept of standardizing is of particular importance with respect to the normal distribution—which we introduce in Chapter 8—and to limit theorems such as the central limit theorem—which we introduce in Chapter 11.

Chebyshev's Inequality

Recall that the variance of a random variable is a measure of variation of the possible values of the random variable. A large variance indicates considerable variation among the possible values of the random variable, whereas a small variance suggests small variation among the possible values of the random variable.

Another way to interpret these properties of the variance is to look at the formula for the variance of a discrete random variable in terms of the PMF. From the FEF,

$$\text{Var}(X) = \mathcal{E}\big((X - \mu_X)^2\big) = \sum_x (x - \mu_X)^2 p_X(x). \qquad \textbf{(7.36)}$$

If $\text{Var}(X)$ is small, each term in the sum on the right of Equation (7.36) must be small, which in turn implies that values of X far from the mean must have small probability. More precisely, we have Proposition 7.11, known as **Chebyshev's inequality** in honor of its discoverer, Russian mathematician Pafnuty L. Chebyshev.

◆◆◆   **Proposition 7.11   Chebyshev's Inequality**

Let X be a random variable with finite variance. Then, for each positive real number t,

$$P(|X - \mu_X| \geq t) \leq \frac{\text{Var}(X)}{t^2}. \qquad \textbf{(7.37)}$$

In words, the probability that a random variable will deviate from its mean by at least some specified number is at most the variance of the random variable divided by the square of the specified number. (See Figure 7.3.)

Figure 7.3 Graphical representation of Chebyshev's inequality

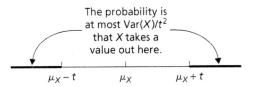

The probability is at most $\text{Var}(X)/t^2$ that X takes a value out here.

$\mu_X - t \qquad \mu_X \qquad \mu_X + t$

Proof   We present the proof of Chebyshev's inequality for discrete random variables but note that it holds for all types of random variables, as indicated in Exercise 7.80. We have

$$\text{Var}(X) = \sum_x (x - \mu_X)^2 p_X(x) \geq \sum_{|x - \mu_X| \geq t} (x - \mu_X)^2 p_X(x)$$

$$\geq t^2 \sum_{|x - \mu_X| \geq t} p_X(x) = t^2 P(|X - \mu_X| \geq t),$$

where the last equation follows from the FPF. Hence Relation (7.37) holds.   ◆

The power of Chebyshev's inequality lies in its generality—it holds for any random variable (with finite variance)—which makes Chebyshev's inequality a useful theoretical tool. Chebyshev's inequality also provides a method for obtaining bounds for probabilities of a random variable based only on its mean and variance. Although we can't expect such bounds to be sharp in individual cases, particular cases show that, in general, the bound given by Chebyshev's inequality can't be improved upon. Example 7.14 illustrates Chebyshev's inequality and addresses the two issues raised.

EXAMPLE 7.14   *Chebyshev's Inequality*

Number of Siblings
a) In Examples 7.1 and 7.12, we considered the number of siblings, X, of a randomly selected student from one of Professor Weiss's classes. Use Chebyshev's inequality to bound the probability that X will differ from its mean by at least two siblings. Then compare that bound to the actual probability.

b) Let Y be a random variable that takes the values 3 and 9, each with probability 1/18, and the value 6 with probability 8/9. Show that, for this random variable, equality holds in Chebyshev's inequality when $t = 3$.

Solution   a) As we discovered earlier, the mean and variance of X are 1.3 and 0.91, respectively. From Chebyshev's inequality,

$$P(|X - 1.3| \geq 2) \leq \frac{0.91}{2^2} = 0.228.$$

Because, in this case, we actually know the PMF of X (see the first two columns of Table 7.9 on page 355), we can compute the exact probability that X will differ from its mean by at least two siblings. Applying the FPF, we get

$$P(|X - 1.3| \geq 2) = \sum_{|x-1.3|\geq 2} p_X(x) = p_X(4) = 0.025.$$

So, in this case, the probability bound given by Chebyshev's inequality is almost 10 times the actual probability.

b) The random variable Y has PMF

$$p_Y(y) = \begin{cases} 1/18, & \text{if } y = 3 \text{ or } 9; \\ 8/9, & \text{if } y = 6; \\ 0, & \text{otherwise.} \end{cases}$$

Simple computations show that $\mathcal{E}(Y) = 6$ and $\text{Var}(Y) = 1$. Thus, by the FPF,

$$P(|Y - \mathcal{E}(Y)| \geq 3) = P(|Y - 6| \geq 3) = \sum_{|y-6|\geq 3} p_Y(y)$$

$$= p_Y(3) + p_Y(9) = \frac{1}{18} + \frac{1}{18} = \frac{1}{9} = \frac{\text{Var}(Y)}{3^2}.$$

So, in this case, the probability bound given by Chebyshev's inequality is exact. ∎

Chebyshev's inequality is commonly stated in a form alternative to that presented in Proposition 7.11. See Exercise 7.73.

EXERCISES 7.3   Basic Exercises

Note: Several exercises in this section are continuations of those in Section 7.1.

7.54 Suppose that two balanced dice are rolled. Let X be the sum of the two faces showing. Determine the variance of the random variable X by using the
a) defining formula.   **b)** computing formula.
Note: Exercise 7.1 on page 333 asks for the mean of X and also provides a reference for the PMF of X.

7.55 An archer shoots an arrow into a square target 6 feet on a side whose center we call the origin. Assume that the archer will actually hit the target and is equally likely to hit any portion of the target. The archer scores 10 points if she hits the bull's eye—a disk of radius 1 foot centered at the origin (region A); she scores 5 points if she hits the ring with inner radius 1 foot and outer radius 2 feet centered at the origin (region B); and she scores 0 points otherwise (region C). Determine the variance of the score by the archer. *Note:* Exercise 7.2 on page 333 asks for the mean of the score and also provides a reference for its PMF.

7.56 Urn I contains four red balls and one white ball; Urn II contains one red ball and four white balls. A coin with probability p of a head is tossed. If the result is a head, Urn I is chosen; otherwise, Urn II is chosen. Two balls are then drawn at random without replacement from the chosen urn. Let X be the number of red balls drawn. Obtain the variance of X in case
a) $p = 0.5$ (balanced coin).   **b)** $p = 0.4$ (biased coin).
Note: Exercise 7.3 on page 333 asks for the mean of X and provides a reference for the PMF of X in each case.

7.57 Of six men and five women applying for a job at Alpha, Inc., three are selected at random for interviews. Determine the variance of the number of women in the interview pool by using
a) the PMF obtained in Exercise 5.29(a) on page 194.
b) an appropriate variance formula obtained in this section.

7.58 A variable of a finite population has standard deviation 3.2. Let X denote the value of the variable for a randomly selected member of the population. Find the variance of X.

7.59 As part of a screening exam for prospective insured, a physician conducts tests for acute problems. On average, the first positive test occurs with the 25th person tested by the physician. Determine the variance of the number of people tested until
a) the first positive test.   **b)** the fifth positive test.

7.60 Pinworm infestation, commonly found in children, can be treated with the drug pyrantel pamoate. According to the *Merck Manual,* the treatment is effective in 90% of cases. If 20 children with pinworm infestation are treated with pyrantel pamoate, what is the variance of the number cured?

7.61 Refer to Exercise 7.11 on page 334, where sickle cell anemia is discussed. Two people with sickle cell trait have five children. What is the standard deviation of the number of those five children who have sickle cell anemia?

7.62 An upper-level probability class has six undergraduate students and four graduate students. A random sample of three students is taken from the class. Determine the variance of the number of undergraduate students selected if the sampling is
a) without replacement.   **b)** with replacement.
c) Why are your results in parts (a) and (b) different?

7.63 Refer to Example 5.14 on page 208, where X denotes the number of defective TVs obtained when 5 TVs are randomly selected without replacement from a lot of 100 TVs in which 6 are defective. Determine the variance of X.

7.64 Suppose that a random sample of size n is taken from a population of size N in which $100p\%$ of the members have a specified attribute. Determine the variance of the number of members sampled that have the specified attribute if the sampling is
a) without replacement.   **b)** with replacement.

7.65 Between 5:00 P.M. and 6:00 P.M., the number of cars that use the drive-up window at a fast-food restaurant has a Poisson distribution. Data show that, on average, 15 cars use the drive-up window during that hour. Find the variance of the number of cars that use the drive-up window between 5:00 P.M. and 6:00 P.M.

7.66 The number of customers that enter the Downtown Coffee Shop in an hour has a Poisson distribution with mean 31. Thirty percent of the customers buy a café mocha with no whipped cream. What is the variance of the number of customers per hour who buy a café mocha with no whipped cream? State any assumptions that you make.

7.67 The number of eggs that the female grunge beetle lays in her nest has a $\mathcal{P}(4)$ distribution. Assuming that an empty nest can't be recognized, determine the variance of the number of eggs observed. *Note:* Exercise 5.85 on page 227 asks for the PMF of the number of eggs observed in a nest.

7.68 A baseball player has a batting average of .260. Suppose that you observe successive at-bats of the player and note for each at-bat whether the player gets a hit. Presuming that the assumption of Bernoulli trials is appropriate, what is the variance of the number of at-bats until the player gets his
a) first hit?   **b)** second hit?   **c)** tenth hit?

7.69 Derive the formula for the variance of the indicator random variable of an event E, Equation (7.32) on page 356, by first noting that $I_E^2 = I_E$ and then applying the computing formula for the variance.

7.70 Use the formula for the variance of a binomial random variable to obtain the variance of an indicator random variable.

7.71 Let X be a random variable with finite nonzero variance. By using properties of expected value and variance, verify in detail that the mean and variance of X^*—the standardized random variable corresponding to X—are 0 and 1, respectively.

7.72 For each $n \in \mathcal{N}$, let X_n be a random variable that equals 0 with probability $1 - 1/n^2$ and equals $\pm n$ with probability $1/2n^2$ each.
a) Apply Chebyshev's inequality to obtain an upper bound for the probability that X_n is at least 3 in magnitude.
b) For $n \geq 3$, determine the exact probability that X_n is at least 3 in magnitude.
c) Compare your results in parts (a) and (b). How do they compare as $n \to \infty$?

7.73 Chebyshev's inequality is commonly stated in a form alternative to that given in Proposition 7.11 on page 360—namely,

$$P(|X - \mu_X| < k\sigma_X) \geq 1 - \frac{1}{k^2}. \qquad (*)$$

a) State Relation $(*)$ in words.

b) For $k = 2$ and $k = 3$, interpret Relation (∗) in terms of the chance that X will take a value within k standard deviations of its mean.

c) Derive Relation (∗) from Proposition 7.11.

7.74 Let X have the discrete uniform distribution on the set $\{1, 2, \ldots, N\}$, where N is a positive integer. Determine the variance of the random variable X. *Hint:* Use the formulas for the sum of and the sum of the squares of the first N positive integers.

7.75 In this exercise, you are to determine how to best estimate a random variable Y by a constant, c. A commonly used criterion for deciding on "best" is to minimize the *mean square error*—that is, to choose c so that $\mathcal{E}\big((Y - c)^2\big)$ is as small as possible.

a) Obtain the value of c that minimizes the mean square error. *Hint:* Use a calculus technique for finding extrema.

b) Find the minimum mean square error.

Theory Exercises

7.76 Show that, if X has a moment of order r, it also has moments of all lower orders. *Hint:* Begin by verifying that the inequality $|x|^{r-1} \le |x|^r + 1$ holds for all real numbers x and all positive integers r.

7.77 Show that $X - \mu_X$ has a finite rth moment (and hence that X has an rth central moment) if and only if X has a finite rth moment. *Hint:* Use Exercise 7.76 and the binomial theorem.

7.78 Moments of bounded random variables: A random variable X is said to be *bounded* if there is a positive real number M such that $P(|X| \le M) = 1$. Prove that a bounded discrete random variable, bounded by M, has finite moments of all orders and that $|\mathcal{E}(X^r)| \le M^r$ for each positive integer r.

7.79 Consider a finite population of size N and a variable defined on it.

a) Suppose that a member is selected at random from the population and let X denote the value of the variable for the member obtained. Show that the population variance σ^2, defined on page 356, can be expressed as $\sigma^2 = N^{-1} \sum_{j=1}^{N}(X(\omega_j) - \mu)^2$, where $\omega_1, \ldots, \omega_N$ denote the N members of the population.

b) Use part (a) to prove Proposition 7.9 on page 356—namely, that the variance of the random variable X equals the population variance.

7.80 Chebyshev's inequality: The proof of Chebyshev's inequality presented in Proposition 7.11 on page 360 applies only to discrete random variables. Use Markov's inequality (Exercise 7.49 on page 351) to obtain a proof of Chebyshev's inequality that applies to all types of random variables.

7.81 Proposition 7.7 on page 354 states that a random variable has zero variance if and only if it is a constant random variable. The proof presented presumes that X is a discrete random variable. Use Chebyshev's inequality—proved for all types of random variables in Exercise 7.80—to provide a proof of Proposition 7.7 that applies to all random variables. *Hint:* Start by verifying that $\{|X - \mu_X| > 0\} = \bigcup_{n=1}^{\infty}\{|X - \mu_X| \ge 1/n\}$.

7.82 Complete the proof of Proposition 7.10 on page 358 by establishing the equivalence of Equation (7.33) to properties (a) and (b) together.

Advanced Exercises

7.83 Pascal distribution: Determine the variance of a Pascal random variable with parameter α, as defined in Exercise 7.26 on page 336.

7.84 Probability generating function: In Exercise 7.52 on page 352, we introduced the probability generating function (PGF), denoted P_X, of a nonnegative-integer valued random variable X. Show that, if X has finite variance, $\text{Var}(X) = P_X''(1) + P_X'(1) - \left(P_X'(1)\right)^2$.

7.85 Refer to Exercise 7.84. For a random variable with each of the following probability distributions, use the PGF to obtain the variance. *Note:* In Exercise 7.53 on page 352, we asked for the PGFs of these probability distributions.
a) Binomial with parameters n and p **b)** Poisson with parameter λ
c) Geometric with parameter p **d)** Negative binomial with parameters r and p
e) Discrete uniform on $\{1, 2, \ldots, N\}$

7.4 Variance, Covariance, and Correlation

We now continue our discussion of the variance and introduce some related concepts—namely, the *covariance* and the *correlation coefficient*. In doing so, we need the facts that, if X and Y are discrete random variables defined on the same sample space and having finite variances, then XY has finite mean and $X + Y$ has finite variance. You are asked to verify those facts in Exercise 7.99.

Variance of a Sum; Covariance

Let X and Y be random variables defined on the same sample space and having finite variances. We want to obtain a formula for the variance of the sum of X and Y, that is, for $\text{Var}(X + Y)$. By the linearity property of expected value, $\mu_{X+Y} = \mu_X + \mu_Y$. Thus

$$\left((X + Y) - \mu_{X+Y}\right)^2 = \left((X + Y) - (\mu_X + \mu_Y)\right)^2 = \left((X - \mu_X) + (Y - \mu_Y)\right)^2$$
$$= (X - \mu_X)^2 + 2(X - \mu_X)(Y - \mu_Y) + (Y - \mu_Y)^2.$$

Taking expected values in the previous display, again applying the linearity property of expected value, and using the definition of the variance, we get

$$\text{Var}(X + Y) = \text{Var}(X) + 2\,\mathcal{E}\left((X - \mu_X)(Y - \mu_Y)\right) + \text{Var}(Y). \qquad \textbf{(7.38)}$$

The second term on the right of Equation (7.38) is so important that it's given a special name: the **covariance** of two random variables.

DEFINITION 7.4 Covariance of Two Random Variables

Let X and Y be random variables defined on the same sample space and having finite variances. The **covariance** of X and Y, denoted **Cov(X, Y)**, is

$$\text{Cov}(X, Y) = \mathcal{E}\left((X - \mu_X)(Y - \mu_Y)\right). \qquad \textbf{(7.39)}$$

In words, the covariance of two random variables is the expected value of the product of their deviations from their means.

Proposition 7.12, whose proof we leave to you as Exercise 7.105, provides the essential properties of covariance.

◆◆◆ Proposition 7.12   Basic Properties of Covariance

Let X, Y, and Z be random variables defined on the same sample space and having finite variances. Also, let a, b, and c be real numbers. Then
a) $\text{Cov}(X, X) = \text{Var}(X)$.
b) $\text{Cov}(X, Y) = \text{Cov}(Y, X)$.
c) $\text{Cov}(cX, Y) = c\,\text{Cov}(X, Y)$.
d) $\text{Cov}(X, Y + Z) = \text{Cov}(X, Y) + \text{Cov}(X, Z)$.
e) $\text{Cov}(a + bX, Y) = b\,\text{Cov}(X, Y)$.

Let $X_1, \ldots, X_m$ and $Y_1, \ldots, Y_n$ be random variables defined on the same sample space and having finite variances, and let $a_1, \ldots, a_m$ and $b_1, \ldots, b_n$ be real numbers. From Proposition 7.12(b)–(d), we get

$$\text{Cov}\left(\sum_{j=1}^{m} a_j X_j, \sum_{k=1}^{n} b_k Y_k\right) = \sum_{j=1}^{m}\sum_{k=1}^{n} a_j b_k\,\text{Cov}(X_j, Y_k). \tag{7.40}$$

Equation (7.40) is referred to as the **bilinearity property** of covariance. We leave its proof to you as Exercise 7.106.

We also leave to you as Exercise 7.107 verification of the formula

$$\text{Cov}(X, Y) = \mathcal{E}(XY) - \mathcal{E}(X)\,\mathcal{E}(Y), \tag{7.41}$$

called the **computing formula for the covariance.** In words, Equation (7.41) states that the covariance of two random variables equals the expected value of their product minus the product of their expected values.

We provide an interpretation of the covariance later in this section when we discuss the correlation coefficient. But, for now, let's return to the variance of the sum of random variables. From Equation (7.38) and Definition 7.4, we get the first equation in Proposition 7.13 and, from Proposition 7.12(a) and Equation (7.40), we get the second equation in Proposition 7.13.

◆◆◆ Proposition 7.13   Variance of a Sum of Random Variables

Let X and Y be random variables defined on the same sample space and with finite variances. Then

$$\text{Var}(X + Y) = \text{Var}(X) + \text{Var}(Y) + 2\,\text{Cov}(X, Y). \tag{7.42}$$

More generally, if $X_1, \ldots, X_m$ are random variables defined on the same sample space and with finite variances, then

$$\text{Var}\left(\sum_{k=1}^{m} X_k\right) = \sum_{k=1}^{m} \text{Var}(X_k) + 2\sum\sum_{i<j}\text{Cov}(X_i, X_j). \tag{7.43}$$

The double sum is taken over all $\binom{n}{2}$ pairs of integers i and j such that $1 \le i < j \le n$.

EXAMPLE 7.15   *Covariance and the Variance of a Sum*

Bedrooms and Bathrooms   Data supplied by *Realty 2000/Better Homes and Gardens* of Prescott, Arizona, resulted in the contingency table given in Table 6.1 on page 261 for number of bedrooms and number of bathrooms for 50 homes currently for sale. Suppose that one of these 50 homes is selected at random. Let X and Y denote the number of bedrooms and number of bathrooms, respectively, of the home obtained.
a) Determine $Cov(X, Y)$.
b) Find $Var(X + Y)$.

Solution   The joint and marginal PMFs of X and Y were obtained in Example 6.1 on page 261 and are repeated here in Table 7.11.

Table 7.11 Joint and marginal PMFs of X and Y

		Bathrooms, y				
		2	3	4	5	$p_X(x)$
Bedrooms, x	2	0.06	0.00	0.00	0.00	0.06
	3	0.28	0.24	0.04	0.00	0.56
	4	0.04	0.22	0.10	0.02	0.38
	$p_Y(y)$	0.38	0.46	0.14	0.02	1.00

a) We use the computing formula for the covariance to obtain $Cov(X, Y)$. We first calculate $\mathcal{E}(XY)$, $\mathcal{E}(X)$, and $\mathcal{E}(Y)$. Applying the FEF and Table 7.11 gives

$$\mathcal{E}(XY) = \sum_{(x,y)} \sum xy p_{X,Y}(x, y)$$

$$= 2 \cdot 2 \cdot 0.06 + 2 \cdot 3 \cdot 0.00 + \cdots + 4 \cdot 5 \cdot 0.02 = 9.52.$$

Similarly, $\mathcal{E}(X) = 3.32$ and $\mathcal{E}(Y) = 2.8$. Thus, from Equation (7.41),

$$Cov(X, Y) = \mathcal{E}(XY) - \mathcal{E}(X)\mathcal{E}(Y) = 9.52 - 3.32 \cdot 2.8 = 0.224.$$

b) Again applying the FEF and Table 7.11, we easily find that $Var(X) = 0.3376$ and $Var(Y) = 0.56$. Therefore, from Equation (7.42) and part (a),

$$Var(X + Y) = Var(X) + Var(Y) + 2\,Cov(X, Y)$$

$$= 0.3376 + 0.56 + 2 \cdot 0.224 = 1.3456.$$

We can also obtain $Var(X + Y)$ by using the definition of the variance and the PMF of $X + Y$, which we determined in Example 6.23 (see Table 6.7 on page 308). We leave it to you as Exercise 7.88 to show that this method gives the same value for $Var(X + Y)$ as the one we just found. ∎

Covariance and Variance for Independent Random Variables

For independent random variables, the formula for the variance of a sum of random variables simplifies considerably. We begin by recalling that the expected value of the product of two independent random variables equals the product of their expected values (Proposition 7.5 on page 346). Applying now the computing formula for the covariance, we get Proposition 7.14.

◆◆◆ **Proposition 7.14**

If X and Y are independent random variables, then $\text{Cov}(X, Y) = 0$.

The converse of Proposition 7.14 isn't true; that is, two random variables can have zero covariance without being independent. For instance, the random variables whose joint and marginal PMFs are given in Table 7.8 on page 347 aren't independent but, nevertheless, have zero covariance.

From Proposition 7.13 and Proposition 7.14, we get Proposition 7.15, which shows that, for independent random variables, the variance of the sum equals the sum of the variances. But, again, although independence is a sufficient condition for this property, it isn't necessary.

◆◆◆ **Proposition 7.15   Variance of a Sum of Independent Random Variables**

Let X and Y be independent random variables with finite variances. Then

$$\text{Var}(X + Y) = \text{Var}(X) + \text{Var}(Y). \tag{7.44}$$

More generally, if $X_1, \ldots, X_m$ are independent random variables with finite variances,

$$\text{Var}\left(\sum_{k=1}^{m} X_k\right) = \sum_{k=1}^{m} \text{Var}(X_k). \tag{7.45}$$

Applying Properties of Variance

We now apply properties of variance to obtain some practical results of interest. Our first application is to statistics. To begin, we define a **random sample from a distribution,** a generalization of a random sample with replacement from a finite population.

Let X be a random variable. For n independent repetitions of the random experiment, let $X_1, \ldots, X_n$ represent the n observed values of X. We say that $X_1, \ldots, X_n$ constitute a *random sample of size n from the distribution of X*. Note that $X_1, \ldots, X_n$ are independent random variables, all having the same probability distribution as X.

We also need to define *statistic* and *unbiased estimator.* Let X be a random variable and let $X_1, \ldots, X_n$ be a random sample of size n from the distribution of X. A **statistic** is any function of $X_1, \ldots, X_n$ whose numerical value can be determined by knowing only the values of the random sample. If θ is an unknown parameter of the distribution of X, then a statistic $T = g(X_1, \ldots, X_n)$ is said to be an **unbiased estimator** of θ if $\mathcal{E}(T) = \theta$ for all θ. Thus, on average, an unbiased estimator equals the parameter that it is estimating.

EXAMPLE 7.16   *Properties of the Variance*

Estimating a Mean   Consider a random variable X having finite variance. Suppose that we don't know the mean of X and that we want to estimate it. To do so, we take a random sample, $X_1, \ldots, X_n$, from the distribution of X and set

$$\bar{X}_n = \frac{X_1 + \cdots + X_n}{n} = \frac{1}{n} \sum_{j=1}^{n} X_j. \tag{7.46}$$

The random variable $\bar{X}_n$ is called the **sample mean** for a sample of size n.
a) Show that $\bar{X}_n$ is an unbiased estimator of the mean of X—that is, $\mathcal{E}(\bar{X}_n) = \mu_X$.
b) Show that $\text{Var}(\bar{X}_n) = \sigma_X^2/n$.

Solution   First, because X_j has the same probability distribution as X, it must have the same mean and variance as X. Thus

$$\mathcal{E}(X_j) = \mu_X \quad \text{and} \quad \text{Var}(X_j) = \sigma_X^2, \tag{7.47}$$

for $j = 1, 2, \ldots, n$.
a) Applying the linearity property of expected value and the first equality in Equation (7.47) yields

$$\mathcal{E}(\bar{X}_n) = \mathcal{E}\left(\frac{1}{n} \sum_{j=1}^{n} X_j\right) = \frac{1}{n} \sum_{j=1}^{n} \mathcal{E}(X_j) = \frac{1}{n} \sum_{j=1}^{n} \mu_X = \frac{1}{n} \cdot n\mu_X = \mu_X.$$

b) From the independence of $X_1, \ldots, X_n$, properties of variance, and the second equality in Equation (7.47), we get

$$\text{Var}(\bar{X}_n) = \text{Var}\left(\frac{1}{n} \sum_{j=1}^{n} X_j\right) = \frac{1}{n^2} \text{Var}\left(\sum_{j=1}^{n} X_j\right)$$

$$= \frac{1}{n^2} \sum_{j=1}^{n} \text{Var}(X_j) = \frac{1}{n^2} \sum_{j=1}^{n} \sigma_X^2 = \frac{1}{n^2} \cdot n\sigma_X^2 = \frac{\sigma_X^2}{n},$$

as required.   ◼

Note: Taking square roots in the equation of Example 7.16(b) gives

$$\sigma_{\bar{X}_n} = \frac{\sigma_X}{\sqrt{n}}. \tag{7.48}$$

In statistics, the quantity $\sigma_{\bar{X}_n}$ is referred to as the **standard error of the mean** because it determines the amount of sampling error to be expected when $\bar{X}_n$ is used to estimate μ_X.

In Example 7.13 on page 357, we obtained a formula for the variance of a binomial random variable. As a second application of the properties of variance, Example 7.17 provides a simpler derivation of that formula.

EXAMPLE 7.17   *Variance of a Binomial Random Variable*

Use properties of variance to obtain the variance of a binomial random variable with parameters n and p.

Solution   Let $X \sim \mathcal{B}(n, p)$. Recall that X represents the number of successes in n Bernoulli trials with success probability p. For each $1 \le k \le n$, let $X_k = 1$ if the kth trial results in success and let $X_k = 0$ otherwise. Clearly, $X_1, \ldots, X_n$ are independent random variables and $X = \sum_{k=1}^{n} X_k$. Consequently, by Proposition 7.15,

$$\text{Var}(X) = \sum_{k=1}^{n} \text{Var}(X_k). \tag{7.49}$$

Each X_k is an indicator random variable—namely, the indicator random variable of the event E_k that the kth trial results in success. Because the variance of the indicator random variable of an event equals the product of the probability of the event and the probability of the complement of the event,

$$\text{Var}(X_k) = P(E_k)\big(1 - P(E_k)\big) = p(1 - p), \qquad 1 \le k \le n.$$

From this equation and Equation (7.49), we conclude that $\text{Var}(X) = np(1 - p)$.   ∎

In Example 7.18, we apply properties of variance to obtain a formula for the variance of a hypergeometric random variable. This case is somewhat more challenging than the binomial case because the trials aren't independent.

EXAMPLE 7.18   *Variance of a Hypergeometric Random Variable*

Use properties of variance to obtain the variance of a hypergeometric random variable with parameters N, n, and p.

Solution   Let $X \sim \mathcal{H}(N, n, p)$. Recall that X represents the number of members sampled that have a specified attribute when a random sample of size n is taken without replacement from a population of size N in which the proportion of members having the specified attribute is p.

For each $1 \le k \le n$, let $X_k = 1$ if the kth member sampled has the specified attribute and let $X_k = 0$ otherwise. Clearly, $X = \sum_{k=1}^{n} X_k$ and, consequently, by Proposition 7.13 on page 366,

$$\text{Var}(X) = \text{Var}\left(\sum_{k=1}^{n} X_k\right) = \sum_{k=1}^{n} \text{Var}(X_k) + 2 \sum \sum_{i < j} \text{Cov}(X_i, X_j). \tag{7.50}$$

Each X_k is an indicator random variable—namely, the indicator random variable of the event E_k that the kth member sampled has the specified attribute. Because the sampling is random, the kth member sampled is equally likely to be any of the N members of the population. Therefore the probability that the kth member sampled has the specified attribute equals the proportion of the population that has that attribute, which is p. Thus

$$\text{Var}(X_k) = P(E_k)\big(1 - P(E_k)\big) = p(1 - p), \qquad 1 \le k \le n. \qquad \textbf{(7.51)}$$

To determine $\text{Cov}(X_i, X_j)$, we use the computing formula. We already know that $\mathcal{E}(X_k) = P(E_k) = p$ for each k. To obtain $\mathcal{E}(X_i X_j)$, we note that $X_i X_j$ equals 1 if both the ith member and jth member sampled have the specified attribute, and it equals 0 otherwise. Hence $X_i X_j$ is an indicator random variable—namely, of the event $E_i \cap E_j$. Its mean therefore is $P(E_i \cap E_j)$. From the general multiplication rule,

$$P(E_i \cap E_j) = P(E_i) P(E_j \mid E_i). \qquad \textbf{(7.52)}$$

Now, given that event E_i occurs—that is, that the ith member sampled has the specified attribute—the jth member sampled is equally likely to be any of the remaining $N - 1$ members of the population of which $Np - 1$ have the specified attribute. Therefore $P(E_j \mid E_i) = (Np - 1)/(N - 1)$ and, consequently, from Equation (7.52),

$$\mathcal{E}(X_i X_j) = P(E_i \cap E_j) = p \cdot \frac{Np - 1}{N - 1}.$$

Applying the computing formula for the covariance yields

$$\text{Cov}(X_i, X_j) = \mathcal{E}(X_i X_j) - \mathcal{E}(X_i)\, \mathcal{E}(X_j) = p \cdot \frac{Np - 1}{N - 1} - p \cdot p = -\frac{p(1 - p)}{N - 1}. \qquad \textbf{(7.53)}$$

From Equations (7.50), (7.51), and (7.53),

$$\text{Var}(X) = \sum_{k=1}^{n} p(1 - p) - 2 \sum \sum_{i<j} \frac{p(1 - p)}{N - 1}$$

$$= np(1 - p) - n(n - 1) \cdot \frac{p(1 - p)}{N - 1} = \left(\frac{N - n}{N - 1}\right) np(1 - p),$$

as required. ■

The Correlation Coefficient

We demonstrate shortly that the covariance is a measure of the linear association between two random variables. However, because the covariance depends on the unit of measurement, we consider a modified version of the covariance to obtain a measure of linear association that doesn't depend on the unit of measurement. Specifically, we employ standardized random variables—we use the covariance of X^* and Y^* instead of the covariance of X and Y. Applying Proposition 7.12 on page 366, we find that

$$\text{Cov}(X^*, Y^*) = \text{Cov}\left(\frac{X - \mu_X}{\sigma_X}, \frac{Y - \mu_Y}{\sigma_Y}\right) = \frac{\text{Cov}(X, Y)}{\sigma_X \sigma_Y} = \frac{\text{Cov}(X, Y)}{\sqrt{\text{Var}(X) \cdot \text{Var}(Y)}}.$$

This quantity—which doesn't depend on the unit of measurement—is called the **correlation coefficient** of X and Y.

DEFINITION 7.5   Correlation Coefficient

Let X and Y be random variables defined on the same sample space and with finite nonzero variances. Then the **correlation coefficient** of X and Y, denoted $\rho(X, Y)$, is defined to be the covariance of the standardized random variables X^* and Y^*. We have

$$\rho(X, Y) = \frac{\text{Cov}(X, Y)}{\sqrt{\text{Var}(X) \cdot \text{Var}(Y)}}. \tag{7.54}$$

In words, the correlation coefficient of two random variables equals their covariance divided by the square root of the product of their variances.

We present some of the most important properties of the correlation coefficient of two random variables in Proposition 7.16.

◆◆◆ **Proposition 7.16**

Let X and Y be random variables defined on the same sample space and having finite nonzero variances. Then the following hold.

a) $|\rho(X, Y)| \leq 1$.

b) If X and Y are independent, then $\rho(X, Y) = 0$.

c) $\rho(X, Y) = 1$ if and only if there are real numbers a and $b > 0$ such that $Y = a + bX$ with probability 1.

d) $\rho(X, Y) = -1$ if and only if there are real numbers a and $b < 0$ such that $Y = a + bX$ with probability 1.

Proof   a) From properties of variance, covariance, and standardized random variables, we have

$$\text{Var}(X^* \pm Y^*) = \text{Var}(X^*) + \text{Var}(Y^*) \pm 2\,\text{Cov}(X^*, Y^*)$$
$$= 1 + 1 \pm 2\rho(X, Y) = 2(1 \pm \rho(X, Y)). \tag{7.55}$$

As the left side of Equation (7.55) must be nonnegative, we have $|\rho(X, Y)| \leq 1$.

b) This property follows from the fact that the covariance of two independent random variables is 0 (Proposition 7.14 on page 368).

c) Suppose that $\rho(X, Y) = 1$. Then Equation (7.55) implies that $\text{Var}(X^* - Y^*) = 0$ and hence, by Proposition 7.7 on page 354, $X^* - Y^*$ is a constant random variable. Because X^* and Y^* both have mean 0, the constant must be 0; hence $P(Y^* = X^*) = 1$. Simple algebra now reveals that, with probability 1,

$$Y = \mu_Y - \frac{\sigma_Y}{\sigma_X}\mu_X + \frac{\sigma_Y}{\sigma_X}X.$$

This result shows that there are real numbers a and $b > 0$ such that $Y = a + bX$ with probability 1. Conversely, suppose that there are real numbers a and $b > 0$ such that $Y = a + bX$ with probability 1. Then $\text{Var}(Y) = \text{Var}(a + bX) = b^2\,\text{Var}(X)$ and

$$\text{Cov}(X, Y) = \text{Cov}(X, a + bX) = b\,\text{Cov}(X, X) = b\,\text{Var}(X).$$

Consequently,

$$\rho(X, Y) = \frac{\text{Cov}(X, Y)}{\sqrt{\text{Var}(X) \cdot \text{Var}(Y)}} = \frac{b\,\text{Var}(X)}{\sqrt{\text{Var}(X) \cdot b^2\,\text{Var}(X)}} = 1.$$

d) The proof of this part is similar to that of part (c). ◆

The properties presented in parts (b)–(d) of Proposition 7.16 help explain why the correlation coefficient is a measure of linear association between two random variables. For further insight into why the correlation coefficient is a measure of linear association between two random variables, see Exercise 7.104.

Here is some additional terminology that is used with the correlation coefficient of two random variables, X and Y.

- If $\rho(X, Y) > 0$, we say X and Y are **positively correlated random variables.**[†]

- If $\rho(X, Y) < 0$, we say X and Y are **negatively correlated random variables.**[†]

- If $\rho(X, Y) = 0$, we say X and Y are **uncorrelated random variables.**[†]

Proposition 7.16(b) shows that, if two random variables are independent, they are uncorrelated. The converse, however, isn't true. For instance, the random variables whose joint and marginal PMFs are given in Table 7.8 on page 347 aren't independent, but are nonetheless uncorrelated.

The correlation coefficient measures only *linear* association between two random variables. For instance, suppose that X has the discrete uniform distribution on the set $\{-1, 0, 1\}$. If we let $Y = X^2$, then, as you are asked to verify in Exercise 7.103, the random variables X and Y are uncorrelated. But, nonetheless, X and Y are associated, being in fact functionally related.

EXAMPLE 7.19   *The Correlation Coefficient*

Bedrooms and Bathrooms   Refer to Example 7.15 on page 367. Find and interpret the correlation coefficient of X and Y.

Solution   From Example 7.15, $\text{Cov}(X, Y) = 0.224$, $\text{Var}(X) = 0.3376$, and $\text{Var}(Y) = 0.56$. Thus

$$\rho(X, Y) = \frac{\text{Cov}(X, Y)}{\sqrt{\text{Var}(X) \cdot \text{Var}(Y)}} = \frac{0.224}{\sqrt{0.3376 \cdot 0.56}} = 0.515.$$

Not surprisingly, the number of bedrooms (X) and the number of bathrooms (Y) are positively correlated, although only mildly so. ■

[†]Correlation coefficients other than the one discussed here are also used. Because $\rho(X, Y)$ is a measure of the *linear* association between X and Y, it is often referred to as the **linear correlation coefficient.** And, to avoid confusion, the word "linearly" is often used to describe various types of correlation with respect to $\rho(X, Y)$. For instance, the term *positively linearly correlated* is often used in place of *positively correlated.*

EXAMPLE 7.20   *Correlation of Events*

Suppose that A and B are events with positive probability. Show that I_A and I_B are positively correlated, negatively correlated, or uncorrelated depending on whether $P(B \mid A)$ is greater than, less than, or equal to $P(B)$.

Solution   Recall that, for each event E, we have $\mathcal{E}(I_E) = P(E)$. Noting that $I_A I_B = I_{A \cap B}$ and applying the general multiplication rule, we conclude that

$$\text{Cov}(I_A, I_B) = \mathcal{E}(I_A I_B) - \mathcal{E}(I_A)\mathcal{E}(I_B) = \mathcal{E}(I_{A \cap B}) - \mathcal{E}(I_A)\mathcal{E}(I_B)$$
$$= P(A \cap B) - P(A)P(B) = P(A)P(B \mid A) - P(A)P(B)$$
$$= P(A)\big(P(B \mid A) - P(B)\big).$$

Thus I_A and I_B are positively correlated, negatively correlated, or uncorrelated depending on whether $P(B \mid A)$ is greater than, less than, or equal to $P(B)$. *Note:* Referring to the definitions given in Exercise 4.17 on page 136, we see that I_A and I_B are positively correlated, negatively correlated, or uncorrelated depending on whether events A and B are positively correlated, negatively correlated, or independent.   ∎

EXERCISES 7.4   **Basic Exercises**

Note: Several exercises in this section are continuations of those in Sections 7.1–7.3.

7.86 Two balanced dice are rolled, one black and one gray. Let Y and Z be the faces showing on the black and gray dice, respectively, and let X be the sum of the two faces showing.
a) Determine the variances of Y and Z.
b) Express X in terms of Y and Z, and then use properties of variance and the result of part (a) to obtain the variance of X.
c) Compare your answer in part (b) to that obtained in Exercise 7.54 on page 362. Which method for obtaining $\text{Var}(X)$ is easier? Explain your answer.

7.87 An archer shoots an arrow into a square target 6 feet on a side whose center we call the origin. Assume that the archer will actually hit the target and is equally likely to hit any portion of the target. The archer scores 10 points if she hits the bull's eye—a disk of radius 1 foot centered at the origin (region A); she scores 5 points if she hits the ring with inner radius 1 foot and outer radius 2 feet centered at the origin (region B); and she scores 0 points otherwise (region C). Suppose that the archer shoots independently at the target four times. Let X, Y, and Z denote the numbers of times that the archer hits regions A, B, and C, respectively. Also, let T denote the total score in the four shots.
a) Determine and identify the joint PMF of the random variables X, Y, and Z.
b) Express T in terms of X, Y, and Z and then use the FEF and the result of part (a) to obtain $\mathcal{E}(T^2)$.
c) Exercise 7.34 on page 349 asks for $\mathcal{E}(T)$. Use that result and part (b) to find $\text{Var}(T)$.
d) Exercise 7.55 on page 362 asks for the variance of the score for one shot at the target. Use that result and properties of variance to find $\text{Var}(T)$.
e) Identify yet another way to find $\text{Var}(T)$.

7.88 Refer to Example 7.15 on page 367.
a) Use the definition of the variance and the PMF of $X + Y$, which was found in Example 6.23 (see Table 6.7 on page 308), to determine $\text{Var}(X + Y)$.

b) Compare your answer here to that obtained in Example 7.15, where Proposition 7.13 was applied to determine $\text{Var}(X + Y)$.

7.89 N married couples are randomly seated at a rectangular table, the women on one side and the men on the other. What is the variance of the number of men who sit across from their wives? *Hint:* Use indicators.

7.90 In Exercise 7.89, let X_k be the indicator of the event that husband k sits across from his wife, where $1 \leq k \leq N$. Without doing any computations, explain why, for $i \neq j$, the random variables X_i and X_j are positively correlated.

7.91 Determine the variance of the number of times a balanced die must be thrown to get all six possible numbers. *Hint:* For $1 \leq k \leq 6$, let X_k denote the number of throws from the appearance of the $(k - 1)$th distinct number until the appearance of the kth distinct number.

7.92 Exercise 6.2 on page 269 asks for the joint PMF of the number of siblings (X) and number of sisters (Y) of a randomly selected student from one of Professor Weiss's classes.
a) Without doing any computations, decide whether the correlation coefficient between X and Y is positive or negative. Explain your answer.
b) Obtain and interpret the correlation coefficient of X and Y.

7.93 Covariance for multinomial random variables: Let $X_1, \ldots, X_m$ have the multinomial distribution with parameters n and $p_1, \ldots, p_m$. Recall that, for $1 \leq k \leq m$, the random variable X_k is the number of times that event E_k occurs in n independent repetitions of a random experiment in which $E_1, \ldots, E_m$ are mutually exclusive and exhaustive events with probabilities $p_1, \ldots, p_m$, respectively. For $1 \leq k \leq m$ and $1 \leq i \leq n$, let $X_{ik} = 1$ if event E_k occurs on trial i and let $X_{ik} = 0$ otherwise. Use the following steps to show that

$$\text{Cov}(X_k, X_\ell) = -np_k p_\ell, \qquad k \neq \ell. \tag{$*$}$$

a) Explain why $\text{Cov}(X_{ik}, X_{j\ell}) = 0$ for $i \neq j$ and all k and ℓ.
b) Show that $\text{Cov}(X_{ik}, X_{i\ell}) = -p_k p_\ell$ for all i and $k \neq \ell$.
c) Use parts (a) and (b) and Equation (7.40) on page 366 to deduce Equation $(*)$.
d) Without doing any computations, explain why the random variables X_k and X_ℓ are negatively correlated for $k \neq \ell$.
e) Obtain $\rho(X_k, X_\ell)$ for all k and ℓ.

7.94 Refer to Exercise 7.87. Express T in terms of $X, Y,$ and Z, and then use Proposition 7.13 on page 366, properties of variance, and Exercise 7.93 to find $\text{Var}(T)$.

7.95 A random sample of size n is taken from a very large lot of items in which $100p_1\%$ have exactly one defect and $100p_2\%$ have two or more defects, where $0 < p_1 + p_2 < 1$. An item with exactly one defect costs \$1 to repair, whereas an item with two or more defects costs \$3 to repair. Determine the variance of the cost of repairing the defective items in the sample. *Hint:* Use Exercise 7.93.

7.96 An automobile insurance policy reimburses a family's accident losses up to a maximum of two accidents per year. For a three-person family, let $X, Y,$ and Z denote the number of accidents in a single year for the three people in the family. Example 6.7 on page 274 provides the joint and marginal PMFs of $X, Y,$ and Z. Determine the mean and standard deviation of the total number of accidents for this family in a single year.

7.97 A balanced die is tossed twice. Let X and Y denote the smaller and larger of the two faces, respectively. Obtain and interpret the correlation coefficient of X and Y. *Note:* Exercise 6.6 on page 270 asks for the joint and marginal PMFs of X and Y.

7.98 An actuary is modeling the sizes of surgical claims and their associated hospital claims, measured in thousands of dollars. The first and second moments of a surgical claim are 5 and 27.4, respectively, the first and second moments of a hospital claim are 7 and 51.4, respectively, and the variance of the total of the surgical and hospital claims is 8. Let X and Y denote the sizes of a combined surgical and hospital claim before and after the application of a 20% surcharge on the hospital portion of the claim, respectively. Determine and interpret the covariance and correlation coefficient of X and Y.

7.99 Let X and Y be discrete random variables defined on the same sample space and having finite variances.
a) Prove that XY has finite mean. *Hint:* First show that $|xy| \le (x^2 + y^2)/2$.
b) Prove that $X + Y$ has finite variance. *Hint:* First show that $(x + y)^2 \le 2(x^2 + y^2)$.

7.100 Verify the computing formula for the covariance: $\mathrm{Cov}(X, Y) = \mathcal{E}(XY) - \mathcal{E}(X)\mathcal{E}(Y)$.

7.101 Consider a finite population of size N and a variable defined thereon. Let σ^2 denote the population variance—that is, the arithmetic mean of all possible square deviations of the variable from the population mean for the entire population. Suppose that a random sample of size n is taken with replacement and let $\bar{X}_n$ denote the sample mean. Using results that we have already obtained, explain why $\mathrm{Var}(\bar{X}_n) = \sigma^2/n$. *Note:* No calculations are required.

7.102 Use properties of covariance to show $\mathrm{Cov}(X^*, Y^*) = \mathrm{Cov}(X, Y)/\sqrt{\mathrm{Var}(X) \cdot \mathrm{Var}(Y)}$ and explain why that quantity is unitless.

7.103 Let X have the discrete uniform distribution on $\{-1, 0, 1\}$ and let $Y = X^2$.
a) Show that $\rho(X, Y) = 0$, although X and Y are functionally related and hence associated.
b) Why doesn't the result of part (a) conflict with the correlation coefficient's role as a measure of association?

7.104 In Exercise 7.75 on page 364, we discussed how to best estimate a random variable Y by a constant, c, where "best" means choosing c to minimize the mean square error, $\mathcal{E}((Y - c)^2)$. Now determine how to best estimate Y by a linear function of a random variable X, again using minimum mean square error as the criterion for "best."
a) Obtain the values of a and b that minimize the mean square error $\mathcal{E}([Y - (a + bX)]^2)$. *Hint:* Use a calculus technique for finding extrema of functions of two variables.
b) Find the minimum mean square error.
Use the result of part (b) to
c) deduce part (a) of Proposition 7.16 on page 372.
d) deduce parts (c) and (d) of Proposition 7.16.
e) explain why the correlation coefficient is a measure of linear association.

Theory Exercises

7.105 Establish the properties of covariance presented in Proposition 7.12 on page 366.

7.106 Prove the bilinearity property of covariance, Equation (7.40) on page 366.

7.107 Use properties of expected value to establish the computing formula for the covariance, Equation (7.41) on page 366.

7.108 Complete the proof of Proposition 7.13 by verifying Equation (7.43) on page 366.

Advanced Exercises

7.109 Sample variance: Let X be a random variable with finite variance. Suppose that you don't know the variance of X and want to estimate it. You take a random sample, $X_1, \ldots, X_n$, from the distribution of X and set $S_n^2 = (n-1)^{-1} \sum_{j=1}^{n} (X_j - \bar{X}_n)^2$. Show that the random variable S_n^2—which is called the *sample variance* based on a sample of size n—is an unbiased estimator of σ_X^2.

7.110 Let $X_1, \ldots, X_n$ be a random sample from the distribution of a random variable X whose third central moment equals 0. Prove that the sample mean $\bar{X}_n$ and the sample variance S_n^2 (defined in Exercise 7.109) are uncorrelated. Use the following steps.
a) Let $Y_j = X_j - \mu_X$, for $1 \le j \le n$, and let T_n^2 denote the sample variance of the Y_js. Show that $\operatorname{Cov}(\bar{X}_n, S_n^2) = \operatorname{Cov}(\bar{Y}_n, T_n^2)$.
b) Prove that $\operatorname{Cov}(\bar{Y}_n, T_n^2) = 0$.
c) Deduce that $\bar{X}_n$ and S_n^2 are uncorrelated.

7.111 Consider a finite population of size N and a variable defined thereon. Let σ^2 denote the population variance—that is, the arithmetic mean of all possible square deviations of the variable from the population mean for the entire population. Suppose that a random sample of size n is taken without replacement and let $\bar{X}_n$ denote the sample mean. Use the following steps to prove that

$$\operatorname{Var}(\bar{X}_n) = \left(\frac{N-n}{N-1} \right) \frac{\sigma^2}{n}. \qquad (**)$$

a) Let X denote the value of the variable for one randomly selected member from the population. For each $1 \le k \le N$, let $I_k = 1$ if the kth member of the population is in the sample, and let $I_k = 0$ otherwise. Show that $\bar{X}_n = n^{-1} \sum_{k=1}^{N} X(\omega_k) I_k$, where $\omega_1, \ldots, \omega_N$ denote the N members of the population.
b) Show that, for $1 \le k \le N$, we have $\mathcal{E}(I_k) = n/N$ and $\operatorname{Var}(I_k) = n(N-n)/N^2$ and that, for $i \ne j$,

$$\operatorname{Cov}(I_i, I_j) = -\frac{n(N-n)}{N^2(N-1)}.$$

c) Use parts (a) and (b) to conclude that

$$\operatorname{Var}(\bar{X}_n) = \frac{N-n}{nN^2} \sum_{k=1}^{N} \left(X(\omega_k) \right)^2 - \frac{2(N-n)}{nN^2(N-1)} \sum_{i<j} \sum X(\omega_i) X(\omega_j).$$

d) Obtain the required result.
e) Compare the expression for $\operatorname{Var}(\bar{X}_n)$ in Equation $(**)$, where the sampling is without replacement, to the one in Exercise 7.101, where the sampling is with replacement. How do the two expressions compare when the population size is large relative to the sample size? Why is your answer to that question not surprising?

7.5 Conditional Expectation

In this section, we introduce and apply the concept of *conditional expected value*, better known as *conditional expectation*. For instance, let X and Y denote the age and price, respectively, of a randomly selected Corvette. We might want to know the mean price of all Corvettes—that is, the expected value of the random variable Y. However, we probably would be more interested in knowing the mean price of all Corvettes of a particular age—that is, the conditional expectation of the random variable Y given a value of the random variable X.

Actually, **conditional expectation** is nothing new. It is simply expected value relative to a conditional distribution instead of an unconditional distribution.

DEFINITION 7.6   Conditional Expectation: Discrete Case

Let X and Y be discrete random variables defined on the same sample space. If $p_X(x) > 0$, we define the **conditional expectation of Y given $X = x$**, denoted $\mathcal{E}(Y \mid X = x)$, to be the expected value of Y relative to the conditional distribution of Y given $X = x$; that is,

$$\mathcal{E}(Y \mid X = x) = \sum_y y p_{Y \mid X}(y \mid x), \tag{7.56}$$

provided that $Y_{\mid X = x}$ has finite expectation.

Conditional expectation generalizes the concept of conditional probability. Indeed, suppose that A and B are events and that A has positive probability. Then

$$p_{I_B \mid I_A}(y \mid 1) = \begin{cases} P(B^c \mid A), & \text{if } y = 0; \\ P(B \mid A), & \text{if } y = 1; \\ 0, & \text{otherwise.} \end{cases}$$

Therefore, by Definition 7.6,

$$\mathcal{E}(I_B \mid I_A = 1) = \sum_y y p_{I_B \mid I_A}(y \mid 1) = P(B \mid A).$$

Similarly, we find that $\mathcal{E}(I_B \mid I_A = 0) = P(B \mid A^c)$.

EXAMPLE 7.21   *Conditional Expectation*

Bedrooms and Bathrooms   Refer to Example 7.15 on page 367. Determine and interpret the conditional expectation of Y given $X = x$ for each possible value x of X.

Solution   The conditional PMFs of Y given $X = x$ were obtained in Example 6.12 and are repeated in Table 7.12 at the top of the next page. From the "$x = 3$" row of Table 7.12, we get

$$\mathcal{E}(Y \mid X = 3) = \sum_y y p_{Y \mid X}(y \mid 3)$$

$$= 2 \cdot 0.500 + 3 \cdot 0.429 + 4 \cdot 0.071 + 5 \cdot 0.000$$

$$= 2.571.$$

Similarly, we obtain the other two conditional expectations of Y given $X = x$. We summarize our findings in Table 7.13. It shows that the expected number of bathrooms for a randomly selected home with two bedrooms is 2; with three bedrooms is 2.571; and with four bedrooms is 3.264. ∎

Table 7.12 Conditional PMF of Y for each possible value of X

Bathrooms, y

		2	3	4	5	Total
Bedrooms, x	2	1.000	0.000	0.000	0.000	1.000
	3	0.500	0.429	0.071	0.000	1.000
	4	0.105	0.579	0.263	0.053	1.000
	$p_Y(y)$	0.380	0.460	0.140	0.020	1.000

Table 7.13 Conditional expectations of Y given $X = x$

Bedrooms x	Conditional expectation $\mathcal{E}(Y \mid X = x)$
2	2.000
3	2.571
4	3.264

EXAMPLE 7.22   *Conditional Expectation*

Let X and Y be independent Poisson random variables with parameters λ and μ, respectively. For each nonnegative integer z, find and interpret $\mathcal{E}(X \mid X + Y = z)$.

Solution   For convenience, we set $Z = X + Y$. Then we want to determine $\mathcal{E}(X \mid Z = z)$. From Proposition 6.20(b) on page 311, the sum of two independent Poisson random variables is also a Poisson random variable whose parameter is the sum of the parameters of the Poisson random variables in the sum. Thus $Z \sim \mathcal{P}(\lambda + \mu)$.

To obtain $\mathcal{E}(X \mid Z = z)$, we must first find the conditional PMF of X given $Z = z$, that is, $p_{X \mid Z}(x \mid z)$. Now,

$$p_{Z,X}(z, x) = P(Z = z, X = x) = P(X + Y = z, X = x)$$
$$= P(Y = z - x, X = x) = p_{Y,X}(z - x, x) = p_Y(z - x)p_X(x),$$

where the last equation follows from the independence of X and Y. Consequently, for $x = 0, 1, \ldots, z$,

$$p_{X \mid Z}(x \mid z) = \frac{p_{Z,X}(z, x)}{p_Z(z)} = \frac{p_Y(z - x)p_X(x)}{p_Z(z)} = \frac{e^{-\mu} \dfrac{\mu^{z-x}}{(z-x)!} \cdot e^{-\lambda} \dfrac{\lambda^x}{x!}}{e^{-(\lambda+\mu)} \dfrac{(\lambda + \mu)^z}{z!}}$$

$$= \binom{z}{x} \frac{\lambda^x \mu^{z-x}}{(\lambda + \mu)^z} = \binom{z}{x} \left(\frac{\lambda}{\lambda + \mu}\right)^x \left(1 - \frac{\lambda}{\lambda + \mu}\right)^{z-x}.$$

Thus $X_{|Z=z} \sim \mathcal{B}\big(z, \lambda/(\lambda + \mu)\big)$; that is, the conditional distribution of X given $Z = z$ is the binomial distribution with parameters z and $\lambda/(\lambda + \mu)$. Because the expected value of a binomial random variable is the product of its two parameters, we can conclude without further computation that

$$\mathcal{E}(X \mid X + Y = z) = \mathcal{E}(X \mid Z = z) = z \cdot \frac{\lambda}{\lambda + \mu}.$$

Hence, for each nonnegative integer z, the conditional expectation of X, given that the sum of X and Y is z, equals $z\lambda/(\lambda + \mu)$. ∎

Law of Total Expectation

Sometimes we know the conditional expectation of a discrete random variable Y for each possible value of a discrete random variable X; that is, we know $\mathcal{E}(Y \mid X = x)$ for each x with $p_X(x) > 0$. From this information and the PMF of X, we can obtain the (unconditional) expectation of Y; that is, we can determine $\mathcal{E}(Y)$. To develop the necessary formula, we consider Example 7.23.

EXAMPLE 7.23   *Motivates the Law of Total Expectation*

Ages of Men and Women   Suppose that we know that the mean age of five men is 38 years and that the mean age of three women is 40 years.
a) Determine the mean age of all eight people.
b) Express the computations in part (a) in the language of probability.

Solution   For convenience, we represent the mean ages of the five men, three women, and all eight people by μ_{men}, μ_{women}, and μ_{all}, respectively. We know that $\mu_{\text{men}} = 38$ and $\mu_{\text{women}} = 40$, and we want to determine μ_{all}.

a) Let y_1, y_2, y_3, y_4, and y_5 denote the ages of the five men, and let y_6, y_7, and y_8 denote the ages of the three women. Then

$$\mu_{\text{men}} = \frac{y_1 + y_2 + y_3 + y_4 + y_5}{5} \quad \text{and} \quad \mu_{\text{women}} = \frac{y_6 + y_7 + y_8}{3}.$$

Therefore

$$\mu_{\text{all}} = \frac{y_1 + y_2 + y_3 + y_4 + y_5 + y_6 + y_7 + y_8}{8}$$

$$= \frac{(y_1 + y_2 + y_3 + y_4 + y_5) + (y_6 + y_7 + y_8)}{8} = \frac{5 \cdot \mu_{\text{men}} + 3 \cdot \mu_{\text{women}}}{8}$$

or

$$\mu_{\text{all}} = \mu_{\text{men}} \cdot \frac{5}{8} + \mu_{\text{women}} \cdot \frac{3}{8}. \tag{7.57}$$

From Equation (7.57) and the known values of μ_{men} and μ_{women}, we get

$$\mu_{\text{all}} = 38 \cdot \frac{5}{8} + 40 \cdot \frac{3}{8} = 38.75.$$

Thus the mean age of all eight people is 38.75 years.

b) To express the computations in part (a) in the language of probability, we consider the random experiment of selecting one person at random from the eight people. Let Y denote the age of the person selected and let $X = 1$ or $X = 2$, depending on whether the person selected is a man or a woman. Now,

$$\mathcal{E}(Y) = \mu_{\text{all}}, \qquad \mathcal{E}(Y \mid X = 1) = \mu_{\text{men}}, \qquad \mathcal{E}(Y \mid X = 2) = \mu_{\text{women}},$$

and

$$p_X(1) = \frac{5}{8} \quad \text{and} \quad p_X(2) = \frac{3}{8}.$$

Thus, in the language of probability, we express Equation (7.57) as

$$\mathcal{E}(Y) = \mathcal{E}(Y \mid X = 1)\, p_X(1) + \mathcal{E}(Y \mid X = 2)\, p_X(2), \tag{7.58}$$

as required. ∎

We now state and prove the **law of total expectation** (also known as the *double expectation formula*), a result that provides a method for computing the expected value of a discrete random variable in terms of its conditional expectations.

◆◆◆ **Proposition 7.17 Law of Total Expectation**

Let X and Y be two discrete random variables defined on the same sample space and suppose that Y has finite expectation. Then

$$\mathcal{E}(Y) = \sum_x \mathcal{E}(Y \mid X = x)\, p_X(x), \tag{7.59}$$

where the sum is taken over all x with $p_X(x) > 0$. In words, the expected value of Y is a weighted average of the conditional expectations of Y given $X = x$, weighted by the probabilities for X.

Proof Applying the FEF with $g(x, y) = y$, the general multiplication rule, and the definition of conditional expectation, we get

$$\mathcal{E}(Y) = \sum_{(x,y)} y p_{X,Y}(x, y) = \sum_{(x,y)} y p_X(x) p_{Y \mid X}(y \mid x)$$

$$= \sum_x p_X(x) \left(\sum_y y p_{Y \mid X}(y \mid x) \right) = \sum_x p_X(x) \mathcal{E}(Y \mid X = x),$$

as required. ◆

The law of total expectation generalizes the law of total probability, Proposition 4.3 on page 142. Indeed, suppose that $A_1, A_2, \ldots$ form a partition of the sample space and that B is an event. Let $X = \sum_n n I_{A_n}$; that is, $X = n$ if and only if event A_n occurs. We have $\mathcal{E}(I_B) = P(B)$, $\mathcal{E}(I_B \mid X = n) = P(B \mid A_n)$, and $p_X(n) = P(A_n)$. Hence the law of total expectation applied to the random variables I_B and X yields

$$P(B) = \sum_n P(B \mid A_n) P(A_n), \tag{7.60}$$

which is the law of total probability.

EXAMPLE 7.24   *The Law of Total Expectation*

Bedrooms and Bathrooms   Refer again to Example 7.15 on page 367. Use the law of total expectation to obtain $\mathcal{E}(Y)$, the expected number of bathrooms of a randomly selected home.

Solution   To obtain $\mathcal{E}(Y)$ with the law of total expectation, we need the conditional expectations of Y given $X = x$ and the PMF of X. The conditional expectations are provided in Table 7.13 on page 379 and the PMF of X can be found in Table 7.11 on page 367. Applying the law of total expectation gives

$$\mathcal{E}(Y) = \sum_x \mathcal{E}(Y \mid X = x)\, p_X(x)$$
$$= \mathcal{E}(Y \mid X = 2)\, p_X(2) + \mathcal{E}(Y \mid X = 3)\, p_X(3) + \mathcal{E}(Y \mid X = 4)\, p_X(4)$$
$$= 2.000 \cdot 0.06 + 2.571 \cdot 0.56 + 3.264 \cdot 0.38 = 2.80008.$$

Verify that, except for roundoff error, this value is the same as that obtained for $\mathcal{E}(Y)$ by using the definition of expected value and the PMF of Y (which can be found in Table 7.11). ∎

As Example 7.25 shows, we can sometimes apply the law of total expectation to efficiently obtain moments of a random variable.

EXAMPLE 7.25   *The Law of Total Expectation*

Geometric Random Variables   Use the law of total expectation to obtain the expected value of a geometric random variable with parameter p.

Solution   Let $X \sim \mathcal{G}(p)$. We recall that X represents the number of trials up to and including the first success in repeated Bernoulli trials with success probability p. We use the law of total expectation to obtain $\mathcal{E}(X)$ by conditioning on the outcome of the first trial.

To that end, we let Z be the indicator random variable of the event that the first trial is a success. Applying the law of total expectation yields

$$\mathcal{E}(X) = \mathcal{E}(X \mid Z = 0)\, p_Z(0) + \mathcal{E}(X \mid Z = 1)\, p_Z(1). \tag{7.61}$$

We know that $p_Z(0) = 1 - p$ and $p_Z(1) = p$. To use Equation (7.61), we still need to find $\mathcal{E}(X \mid Z = 0)$ and $\mathcal{E}(X \mid Z = 1)$.

Given that $Z = 0$ (i.e., the first trial is a failure), we use the fact that the trials are independent to conclude that the number of trials up to and including the first success has the same distribution as $1 + X$; that is, $X_{|Z=0}$ has the same distribution as $1 + X$. Hence, $\mathcal{E}(X \mid Z = 0) = \mathcal{E}(1 + X) = 1 + \mathcal{E}(X)$. Given that $Z = 1$ (i.e., the first trial is a success), we have $X = 1$; that is, $X_{|Z=1}$ has the same distribution as the constant random variable 1. Thus $\mathcal{E}(X \mid Z = 1) = \mathcal{E}(1) = 1$. Consequently, in view of Equation (7.61),

$$\mathcal{E}(X) = \big(1 + \mathcal{E}(X)\big) \cdot (1 - p) + 1 \cdot p.$$

Simple algebra now shows that $\mathcal{E}(X) = 1/p$. ∎

Alternative Form of the Law of Total Expectation

Noting that the conditional expectation of Y given $X = x$ is a function of x, we define the function ψ by $\psi(x) = \mathcal{E}(Y \mid X = x)$ if $p_X(x) > 0$, and $\psi(x) = 0$ otherwise. Then we define $\mathcal{E}(Y \mid X) = \psi(X)$. The random variable $\mathcal{E}(Y \mid X)$ is called the **conditional expectation of Y given X.** Applying the FEF yields

$$\mathcal{E}\big(\mathcal{E}(Y \mid X)\big) = \mathcal{E}\big(\psi(X)\big) = \sum_x \psi(x) p_X(x) = \sum_x \mathcal{E}(Y \mid X = x)\, p_X(x).$$

However, by the law of total expectation, the sum on the right equals the expected value of Y. Thus we have the alternative version of the law of total expectation as presented in Proposition 7.18.

◆◆◆   **Proposition 7.18   Law of Total Expectation: Alternative Form**

Let X and Y be two discrete random variables defined on the same sample space and suppose that Y has finite expectation. Then

$$\mathcal{E}(Y) = \mathcal{E}\big(\mathcal{E}(Y \mid X)\big). \tag{7.62}$$

In words, the expected value of Y equals the expected value of the conditional expectation of Y given X.

Note: Let B be an event. It's easy to see that $\mathcal{E}(I_B \mid X = x) = P(B \mid X = x)$. Therefore, applying Equation (7.62) with $Y = I_B$ and using the fact that $\mathcal{E}(I_B) = P(B)$, we get

$$P(B) = \mathcal{E}\big(P(B \mid X)\big),$$

which is the general form of the *law of total probability.*

EXAMPLE 7.26   *The Law of Total Expectation*

Random Number of Random Variables   Let $X_1, X_2, \ldots$ be independent random variables having the same probability distribution with common finite mean μ and common finite variance σ^2. And let N be a nonnegative-integer valued random variable independent of the X_js. Set $S_0 = 0$ and $S_n = X_1 + \cdots + X_n$ for $n \geq 1$. Then S_N is a random variable obtained as the sum of a random number of random variables. Use Proposition 7.18 to obtain the expected value of S_N.

Solution   We have

$$\mathcal{E}(S_N \mid N = 0) = \mathcal{E}(S_0 \mid N = 0) = \mathcal{E}(0 \mid N = 0) = 0.$$

And, for $n \geq 1$, we use the fact that the X_js are independent of N to conclude that

$$\mathcal{E}(S_N \mid N = n) = \mathcal{E}(S_n \mid N = n) = \mathcal{E}(S_n) = \mathcal{E}\left(\sum_{k=1}^n X_k\right) = \sum_{k=1}^n \mathcal{E}(X_k) = n\mu.$$

Hence $\mathcal{E}(S_N \mid N) = N\mu$ and, therefore, from Proposition 7.18,

$$\mathcal{E}(S_N) = \mathcal{E}\big(\mathcal{E}(S_N \mid N)\big) = \mathcal{E}(N\mu)$$

or

$$\mathcal{E}(S_N) = \mu\,\mathcal{E}(N). \tag{7.63}$$

We could have arrived at Equation (7.63) by reasoning as follows. On average, the number of summands in the sum S_N is $\mathcal{E}(N)$ and, on average, each summand equals μ. So, on average, S_N equals $\mathcal{E}(N)\,\mu$—that is, $\mathcal{E}(S_N) = \mu\,\mathcal{E}(N)$. ■

Conditional Variance

Recall that, by definition, $\text{Var}(Y) = \mathcal{E}\big([Y - \mathcal{E}(Y)]^2\big)$; that is, the variance of a random variable Y is defined to be the expected value of the square of the deviation of Y from its mean. Likewise, we define the **conditional variance** of Y given $X = x$ to be the conditional expected value given $X = x$ of the square of the deviation of Y from its conditional mean.

DEFINITION 7.7   Conditional Variance: Discrete Case

Let X and Y be discrete random variables defined on the same sample space. If $p_X(x) > 0$, we define the **conditional variance of Y given $X = x$**, denoted **$\text{Var}(Y \mid X = x)$**, to be the variance of Y relative to the conditional distribution of Y given $X = x$. That is,

$$\text{Var}(Y \mid X = x) = \mathcal{E}\Big(\big(Y - \mathcal{E}(Y \mid X = x)\big)^2 \mid X = x\Big), \tag{7.64}$$

provided that $Y_{\mid X = x}$ has finite variance.

Again, note that conditional variance is nothing new. It is simply the variance of a random variable relative to a conditional distribution instead of an unconditional distribution. Thus properties that hold for variance also hold for conditional variance. In particular, we have the conditional version of the computing formula for a variance:

$$\text{Var}(Y \mid X = x) = \mathcal{E}(Y^2 \mid X = x) - \big(\mathcal{E}(Y \mid X = x)\big)^2. \tag{7.65}$$

Compare this computing formula for conditional variance to the computing formula for the variance given in Proposition 7.8 on page 355.

EXAMPLE 7.27   *Conditional Variance*

Bedrooms and Bathrooms   Refer again to Example 7.15 on page 367. Determine the conditional variance of Y given $X = x$ for each possible value of X.

Solution We apply the computing formula, Equation (7.65). In Example 7.21, we obtained the conditional expectations of Y given $X = x$ for each possible value of X, as displayed in Table 7.13 on page 379. To find the conditional second moments of Y given $X = x$, we use the conditional form of the FEF as follows:

$$\mathcal{E}(Y^2 \mid X = x) = \sum_y y^2 p_{Y \mid X}(y \mid x).$$

The conditional PMFs of Y given $X = x$ are shown in Table 7.12 on page 379. From the "$x = 3$" row of that table, we get

$$\mathcal{E}(Y^2 \mid X = 3) = \sum_y y^2 p_{Y \mid X}(y \mid 3)$$

$$= 2^2 \cdot 0.500 + 3^2 \cdot 0.429 + 4^2 \cdot 0.071 + 5^2 \cdot 0.000 = 6.997.$$

From this result, Equation (7.65), and Table 7.13,

$$\text{Var}(Y \mid X = 3) = \mathcal{E}(Y^2 \mid X = 3) - \big(\mathcal{E}(Y \mid X = 3)\big)^2 = 6.997 - (2.571)^2 = 0.387.$$

Similarly, we obtain the other two conditional variances of Y given $X = x$. The results are summarized in Table 7.14. ■

Table 7.14 Conditional variances of Y given $X = x$, the variance of the number of bathrooms for homes with x bedrooms

Bedrooms x	Conditional variance $\text{Var}(Y \mid X = x)$
2	0.000
3	0.387
4	0.510

The Law of Total Variance

The law of total expectation provides a method for finding the expected value of a random variable by conditioning on another random variable. Likewise, we can obtain the variance of a random variable by conditioning on another random variable. Proposition 7.19 presents the appropriate formula for doing so, the **law of total variance.**

◆◆◆ **Proposition 7.19 Law of Total Variance**

Let X and Y be two discrete random variables defined on the same sample space and suppose that Y has finite variance. Then

$$\text{Var}(Y) = \mathcal{E}\big(\text{Var}(Y \mid X)\big) + \text{Var}\big(\mathcal{E}(Y \mid X)\big). \qquad (7.66)$$

In words, the variance of Y equals the expectation of the conditional variance of Y given X plus the variance of the conditional expectation of Y given X.

Proof   Applying the law of total expectation to the random variable $(Y - \mu_Y)^2$ yields

$$\text{Var}(Y) = \mathcal{E}\big((Y - \mu_Y)^2\big) = \mathcal{E}\Big(\mathcal{E}\big((Y - \mu_Y)^2 \mid X\big)\Big). \tag{7.67}$$

From properties of (conditional) expectation,

$$\mathcal{E}\big((Y - \mu_Y)^2 \mid X\big) = \mathcal{E}\big(Y^2 - 2\mu_Y Y + \mu_Y^2 \mid X\big)$$
$$= \mathcal{E}\big(Y^2 \mid X\big) - 2\mu_Y \mathcal{E}(Y \mid X) + \mu_Y^2.$$

Subtracting and adding $\big(\mathcal{E}(Y \mid X)\big)^2$ to this last expression and keeping Equation (7.65) in mind, we obtain

$$\mathcal{E}\big((Y - \mu_Y)^2 \mid X\big) = \text{Var}(Y \mid X) + \big(\mathcal{E}(Y \mid X)\big)^2 - 2\mu_Y \mathcal{E}(Y \mid X) + \mu_Y^2$$
$$= \text{Var}(Y \mid X) + \big(\mathcal{E}(Y \mid X) - \mu_Y\big)^2$$
$$= \text{Var}(Y \mid X) + \Big(\mathcal{E}(Y \mid X) - \mathcal{E}\big(\mathcal{E}(Y \mid X)\big)\Big)^2.$$

Taking expectations yields

$$\text{Var}(Y) = \mathcal{E}\big(\text{Var}(Y \mid X)\big) + \mathcal{E}\Big(\big(\mathcal{E}(Y \mid X) - \mathcal{E}(\mathcal{E}(Y \mid X))\big)^2\Big)$$
$$= \mathcal{E}\big(\text{Var}(Y \mid X)\big) + \text{Var}\big(\mathcal{E}(Y \mid X)\big),$$

as required.   ◆

EXAMPLE 7.28   *The Law of Total Variance*

Bedrooms and Bathrooms   Refer again to Example 7.15 on page 367. Use the law of total variance to obtain $\text{Var}(Y)$.

Solution   To find $\text{Var}(Y)$ by using the law of total variance, we need the conditional variances of Y given $X = x$, the conditional expectations of Y given $X = x$, and the PMF of X. These quantities were determined previously and are displayed in Table 7.15.

Table 7.15 Table for calculation of $\text{Var}(Y)$ by using the law of total variance

Bedrooms x	Probability $p_X(x)$	Conditional variance $\text{Var}(Y \mid X = x)$	Conditional expectation $\mathcal{E}(Y \mid X = x)$
2	0.06	0.000	2.000
3	0.56	0.387	2.571
4	0.38	0.510	3.264

From the FEF and Table 7.15,

$$\mathcal{E}\big(\text{Var}(Y \mid X)\big) = \sum_x \text{Var}(Y \mid X = x)\, p_X(x)$$
$$= 0.000 \cdot 0.06 + 0.387 \cdot 0.56 + 0.510 \cdot 0.38$$
$$= 0.411.$$

To determine $\text{Var}\big(\mathcal{E}(Y \mid X)\big)$, we use the computing formula for the variance, the FEF, Table 7.15, the law of total expectation, and the previously determined $\mathcal{E}(Y) = 2.8$:

$$
\begin{aligned}
\text{Var}\big(\mathcal{E}(Y \mid X)\big) &= \mathcal{E}\Big(\big(\mathcal{E}(Y \mid X)\big)^2\Big) - \Big(\mathcal{E}\big(\mathcal{E}(Y \mid X)\big)\Big)^2 \\
&= \sum_x \big(\mathcal{E}(Y \mid X = x)\big)^2 p_X(x) - \big(\mathcal{E}(Y)\big)^2 \\
&= (2.000)^2 \cdot 0.06 + (2.571)^2 \cdot 0.56 + (3.264)^2 \cdot 0.38 - (2.8)^2 \\
&= 0.150.
\end{aligned}
$$

Thus, by the law of total variance, $\text{Var}(Y) = 0.411 + 0.150 = 0.561$. Verify that, except for roundoff error, this value is the same as that obtained for $\text{Var}(Y)$ by using the definition of the variance and the PMF of Y (which can be found in Table 7.11 on page 367). ■

EXAMPLE 7.29   ***The Law of Total Variance***

Random Number of Random Variables   Refer to Example 7.26 on page 383. Use the law of total variance to obtain the variance of S_N.

Solution   To use the law of total variance, we need to determine $\text{Var}(S_N \mid N)$ and $\mathcal{E}(S_N \mid N)$. From Example 7.26, we have $\mathcal{E}(S_N \mid N) = N\mu$. To find $\text{Var}(S_N \mid N)$, we first note that

$$
\text{Var}(S_N \mid N = 0) = \text{Var}(S_0 \mid N = 0) = \text{Var}(0 \mid N = 0) = 0.
$$

And, for $n \geq 1$, we use the fact that the X_js are independent of each other and of N to conclude that

$$
\begin{aligned}
\text{Var}(S_N \mid N = n) &= \text{Var}(S_n \mid N = n) = \text{Var}(S_n) \\
&= \text{Var}\left(\sum_{k=1}^{n} X_k\right) = \sum_{k=1}^{n} \text{Var}(X_k) = n\sigma^2.
\end{aligned}
$$

Thus $\text{Var}(S_N \mid N) = N\sigma^2$. Applying the law of total variance, we have

$$
\text{Var}(S_N) = \mathcal{E}\big(\text{Var}(S_N \mid N)\big) + \text{Var}\big(\mathcal{E}(S_N \mid N)\big) = \mathcal{E}\big(N\sigma^2\big) + \text{Var}(N\mu)
$$

or

$$
\text{Var}(S_N) = \sigma^2 \mathcal{E}(N) + \mu^2 \, \text{Var}(N), \tag{7.68}
$$

as required. ■

Multidimensional Conditional Expectation

We can easily extend the concept and results of conditional expectation to the case where more than two random variables are involved. To illustrate, we consider three discrete random variables.

Let X, Y, and Z be discrete random variables defined on the same sample space. If $p_{X,Y}(x, y) > 0$, we define the conditional expectation of Z given $X = x$ and $Y = y$, denoted $\mathcal{E}(Z \mid X = x, Y = y)$, to be the expected value of Z relative to the conditional distribution of Z given $X = x$ and $Y = y$; that is,

$$\mathcal{E}(Z \mid X = x, Y = y) = \sum_{z} z p_{Z \mid X, Y}(z \mid x, y), \qquad (7.69)$$

provided that $Z_{\mid X=x, Y=y}$ has finite expectation.

Also, suppose that g is a real-valued function of two real variables defined on the range of (Y, Z). Then, by the conditional form of the FEF, $g(Y, Z)_{\mid X=x}$ has finite expectation if and only if $\sum \sum_{(y,z)} |g(y, z)| p_{Y,Z \mid X}(y, z \mid x) < \infty$, in which case

$$\mathcal{E}\big(g(Y, Z) \mid X = x\big) = \sum_{(y,z)} \sum g(y, z) p_{Y,Z \mid X}(y, z \mid x). \qquad (7.70)$$

Similar results hold in the general multivariate case.

EXERCISES 7.5   Basic Exercises

7.112 Suppose that X and Y are independent random variables and that Y has finite mean. Determine the conditional expectation of Y given $X = x$ for each possible value x of X
a) without doing any computations. Explain your reasoning.
b) by using Definition 7.6 on page 378.

7.113 Consider two identical electrical components. Let X and Y denote the respective lifetimes of the two components observed at discrete time units (e.g., every hour). Assume that the joint PMF of X and Y is $p_{X,Y}(x, y) = p^2(1 - p)^{x+y-2}$ if $x, y \in \mathcal{N}$, and $p_{X,Y}(x, y) = 0$ otherwise, where $0 < p < 1$. For each possible value x of X, determine $\mathcal{E}(Y \mid X = x)$. *Hint:* Refer to Exercise 7.112.

7.114 Why is the law of total expectation also referred to as the double expectation formula?

7.115 In Exercise 6.2 on page 269, we considered the number of siblings, X, and the number of sisters, Y, for a randomly selected student from one of Professor Weiss's classes. Exercise 6.45(a) on page 288 asks for the conditional PMF of the number of sisters for each number of siblings. For each number of siblings,
a) determine and interpret the conditional expectation of the number of sisters. Construct a table similar to Table 7.13 on page 379.
b) determine the conditional variance of the number of sisters. Construct a table similar to Table 7.14 on page 385.

7.116 Refer to Exercise 7.115 and note that Exercise 6.2 on page 269 asks for the joint and marginal PMFs of X and Y. For a randomly selected student,
a) use the law of total expectation to obtain the expected number of sisters.
b) use the definition of expected value to obtain the expected number of sisters. Compare your answer to that found in part (a).
c) use the law of total variance to obtain the variance of the number of sisters.
d) use the definition of variance to obtain the variance of the number of sisters. Compare your answer to that found in part (c).

7.117 Let X be a number chosen according to the discrete uniform distribution on the set S of 10 decimal digits and let Y be a number chosen according to the discrete uniform distribution on S with X removed.

a) Find the conditional expectation of Y given $X = x$. *Note:* Exercise 6.46(a) on page 288 asks for the conditional PMF of Y given $X = x$.
b) Find the conditional variance of Y given $X = x$.

7.118 Refer to Exercise 7.117 and note that Exercise 6.3 on page 269 asks for the joint PMF of X and Y and the marginal PMF of Y.
a) Use the law of total expectation to determine $\mathcal{E}(Y)$.
b) Determine $\mathcal{E}(Y)$ by using the definition of expected value. Compare your answer to the one obtained in part (a).
c) Use the law of total variance to determine $\text{Var}(Y)$.
d) Determine $\text{Var}(Y)$ by using the definition of variance. Compare your answer to the one obtained in part (c).

7.119 An automobile insurance company divides its policyholders into three groups, which constitute 45.8%, 32.6%, and 21.6% of the policyholders. The average claim amounts for the three groups are $1457, $2234, and $2516, respectively. Obtain the average claim amount among all policyholders.

7.120 The hourly number of customers entering a bank for the purpose of making a deposit has a Poisson distribution with parameter 25.8. Each such customer deposits an average of $574 with a standard deviation of $3167. For a 1-hour period,
a) find the expected total deposits by all entering customers.
b) find the standard deviation of deposit amounts by all entering customers.

7.121 Part of a homeowner's insurance policy covers one miscellaneous loss per year, which is known to have a 10% chance of occurring. If there is a miscellaneous loss, the probability is c/x that the loss amount is $100x$, for $x = 1, 2, \ldots, 5$, where c is a constant. These are the only loss amounts possible. If the deductible for a miscellaneous loss is $200, determine the net premium for this part of the policy—that is, the amount that the insurance company must charge to break even.

7.122 An automobile insurance company classifies its policyholders as either good or bad drivers, of which there are 75% and 25%, respectively. The mean claim amounts for good and bad drivers are $2000 and $3500, respectively, and the standard deviations of those claim amounts are $200 and $400, respectively. Determine the mean and standard deviation of the claim amounts for all policyholders of this company.

7.123 Let $X \sim \mathcal{G}(p)$. Assume as known that $\mathcal{E}(X) = 1/p$, as found in Example 7.25 on page 382. Determine $\text{Var}(X)$ by using each of the following methods.
a) Condition on the outcome of the first trial to obtain the second moment of X.
b) Use the law of total variance.

7.124 Let $X \sim \mathcal{NB}(r, p)$. Assume as known that the mean of a geometric random variable with parameter p is $1/p$, as found in Example 7.25 on page 382. Use mathematical induction on r to show that $\mathcal{E}(X) = r/p$ by conditioning on the outcome of the first trial.

7.125 Let $X \sim \mathcal{NB}(r, p)$. Assume as known that $\mathcal{E}(X) = r/p$, as found in Exercise 7.124, and that the variance of a geometric random variable with parameter p is $(1 - p)/p^2$, as found in Exercise 7.123. Show that $\text{Var}(X) = r(1 - p)/p^2$ by using mathematical induction on r with each of the following methods.
a) Condition on the outcome of the first trial to obtain the second moment of X.
b) Use the law of total variance.

7.126 Let $X_1, \ldots, X_n$ be a random sample from the distribution of a discrete random variable X with finite mean. For $1 \leq k \leq n$, determine $\mathcal{E}(X_k \mid \bar{X}_n = x)$
a) without doing any computations. Explain your reasoning.
b) by using properties of (conditional) expectation. *Hint:* Consider $\mathcal{E}(\bar{X}_n \mid \bar{X}_n = x)$.

Theory Exercises

7.127 Let X and Y be random variables defined on the same sample space.
a) Prove that, if Y has finite expectation, then so does $Y_{|X=x}$ for each possible value x of X.
b) Prove that, if Y has finite variance, then so does $Y_{|X=x}$ for each possible value x of X.

7.128 Let X and Y be discrete random variables defined on the same sample space and let x be a possible value of X. Suppose that g is a real-valued function of two real variables defined on the range of (X, Y) and such that $\sum_y |g(x, y)| p_{Y \mid X}(y \mid x) < \infty$.
a) Prove that $g(X, Y)_{|X=x}$ has finite expectation and that

$$\mathcal{E}(g(X, Y) \mid X = x) = \sum_y g(x, y) p_{Y \mid X}(y \mid x).$$

Hint: Refer to Equation (7.70) on page 388.
b) Let $\xi(x) = \mathcal{E}(g(X, Y) \mid X = x)$ and define $\mathcal{E}(g(X, Y) \mid X)$ to be $\xi(X)$. Let k be a function of x alone and let ℓ be a function of both x and y. Prove that

$$\mathcal{E}(k(X)\ell(X, Y) \mid X) = k(X)\mathcal{E}(\ell(X, Y) \mid X),$$

provided the expectation on the right makes sense.
c) Prove that $\mathcal{E}(g(X, Y)) = \mathcal{E}\big(\mathcal{E}(g(X, Y) \mid X)\big)$.

Advanced Exercises

7.129 Refer to Example 6.15 on page 287. Given that $X = 0$, determine the expected number of unreimbursed automobile-accident losses for the family.

7.130 Let $X_1, \ldots, X_m$ have the multinomial distribution with parameters n and $p_1, \ldots, p_m$ and let $k \neq \ell$ be integers between 1 and m, inclusive.
a) Use conditional expectation to determine $\mathcal{E}(X_k X_\ell)$. *Hint:* Refer to Exercise 7.128.
b) Use part (a) to determine $\text{Cov}(X_k, X_\ell)$. Compare your work with that done in Exercise 7.93 on page 375.

7.131 The number of customers who enter a certain store in a day is a random variable X having a Poisson distribution with parameter λ. Each customer makes a purchase with probability p, independent of all other customers. Let Y be the number of purchasing customers in a day. Determine $\rho(X, Y)$. *Hint:* Refer to Exercise 7.128.

7.132 In Exercise 7.75 (page 364) and Exercise 7.104 (page 376), we discussed how to best estimate a random variable Y by a constant c and by a linear function of a random variable X, respectively. In both cases, we use the minimum mean square as the criterion for "best." With that criterion for best, prove that the conditional expectation of Y given X is the best estimate among all functions of X. That is, prove that

$$\mathcal{E}\big((Y - \mathcal{E}(Y \mid X))^2\big) = \min_{h \in \mathcal{H}} \mathcal{E}\big((Y - h(X))^2\big), \qquad (*)$$

where $\mathcal{H}$ consists of all real-valued functions h defined on the range of X such that $h(X)$ has finite variance. Refer to Exercise 7.128 and use the following steps. Let $\psi(x) = \mathcal{E}(Y \mid X = x)$.

a) Show that $\mathcal{E}\Big((Y - \psi(X))k(X)\Big) = 0$ for each $k \in \mathcal{H}$.

b) Show that $\big(Y - h(X)\big)^2 \geq \big(Y - \psi(X)\big)^2 + 2\big(Y - \psi(X)\big)\big(\psi(X) - h(X)\big)$.

c) Use parts (a) and (b) to obtain Equation $(*)$.

7.133 Let X and Y be random variables defined on the same sample space and with finite nonzero variances. Suppose that $\mathcal{E}(Y \mid X) = a + bX$, where a and b are real numbers—that is, assume that the conditional expectation of Y given X is a linear function of X.

a) Show that $\big(\rho(X, Y)\big)^2 = \mathrm{Var}\big(\mathcal{E}(Y \mid X)\big) / \mathrm{Var}(Y)$. *Hint:* Refer to Exercise 7.128.

b) The terms $\mathrm{Var}\big(\mathcal{E}(Y \mid X)\big)$ and $\mathcal{E}\big(\mathrm{Var}(Y \mid X)\big)$ are called the *explained component* and *unexplained component* of $\mathrm{Var}(Y)$, respectively. Interpret the result of part (a) in terms of the proportion of the total variance of Y that is "explained." *Note:* The square of the correlation coefficient is called the *coefficient of determination*.

CHAPTER REVIEW

Summary

In this chapter, we introduced the concept of the expected value of a random variable. Conceptually, the expected value of a random variable is the long-run-average value of the random variable in repeated independent observations. This statement is the long-run-average interpretation of expected value.

Mathematically, the expected value of a random variable is a weighted average of its possible values, weighted by probability. In the discrete case, this weighted average is obtained by summing the possible values of the random variable against the probability mass function of the random variable—that is, by multiplying each possible value of the random variable by its probability and then adding those products. Other commonly used terms for expected value are expectation, mean, and first moment.

One of the most important results associated with the concept of expected value is the fundamental expected-value formula (FEF). This formula provides a method for obtaining the expected value of a function of one or more discrete random variables by using the (joint) PMF of those random variables. Applying the FEF, we showed that expected value has the linearity property; that is, the expected value of a linear combination of random variables equals the linear combination of the expected values. We also demonstrated that the expected value of a product of independent random variables equals the product of the expected values.

Next we discussed moments and central moments of a random variable. The rth moment of a random variable is the expected value of its rth power. The rth central moment of a random variable is the expected value of the rth power of the deviation of the random variable from its mean. Of particular importance is the second central moment, called the variance; its square root is called the standard deviation. The variance (or standard deviation) provides a measure of variation of the possible values of a random variable.

A random variable with finite variance is standardized by subtracting from it its mean and then dividing by its standard deviation. A standardized random variable always has mean 0 and variance 1.

We discussed several essential properties of variance, one of the most useful being that the variance of a sum of independent random variables equals the sum of the variances. Closely related to the variance of one random variable is the covariance of two random variables, which is the expected value of the product of the deviations of the random variables from their respective means. We obtained a general formula for the variance of a sum of random variables in terms of the variances and covariances of the summands.

The concept of covariance also leads to a measure of linear association between two random variables. That measure, called the correlation coefficient, is defined as the covariance of the standardized versions of the two random variables.

The conditional expectation of a discrete random variable is the expected value of that random variable relative to its conditional PMF given the value of another discrete random variable. An important application of conditional expectation is as a tool for obtaining the expected value of a random variable by conditioning on the value of another random variable. This process involves use of the law of total expectation.

Similarly, the variance of a random variable can be obtained by applying the law of total variance. This formula expresses the variance of a random variable in terms of its conditional variances and conditional expectations relative to another random variable.

You Should Be Able To ...

1. state and understand the long-run-average interpretation of expected value.
2. define, apply, and interpret the expected value of a discrete random variable.
3. relate the concepts of population mean and expected value.
4. state and apply the fundamental expected-value formula (FEF).
5. state, interpret, and apply the basic properties of expected value.
6. use tail probabilities to obtain the expected value of a nonnegative-integer valued random variable, when appropriate.
7. define and obtain moments and central moments of a discrete random variable.
8. define, apply, and interpret the variance.
9. relate the concepts of population variance and variance of a random variable.
10. state, interpret, and apply the basic properties of variance.
11. obtain and interpret the standardized version of a random variable.
12. state and apply Chebyshev's inequality.
13. define, apply, and interpret the covariance of two random variables.
14. state, interpret, and apply the basic properties of covariance.
15. obtain the variance of the sum of random variables from their variances and covariances.
16. define, apply, and interpret the correlation coefficient of two random variables.
17. define, apply, and interpret conditional expectation.
18. state and apply both forms of the law of total expectation.
19. define, apply, and interpret conditional variance.
20. state and apply the law of total variance.

Key Terms

bilinearity property, *366*

central moments, *353*

Chebyshev's inequality, *360*

computing formula for the
   covariance, *366*

computing formula for the
   variance, *355*

conditional expectation, *378, 383*

conditional variance, *384*

constant random variable, *342*

correlation coefficient, *372*

covariance, *365*

defining formula for the
   variance, *355*

expectation, *327*

expected value, *327*

finite expectation, *328*

finite *r*th moment, *352*

finite variance, *354*

first moment, *327*

fundamental expected-value
   formula, *339*

law of total expectation, *381, 383*

law of total variance, *385*

linear correlation coefficient, *373*

linearity property, *342, 343*

long-run-average interpretation of
   expected value, *326*

mean, *327*

measure of center, *353*

measure of variation, *354*

moment of order *r*, *352*

monotonicity property, *345*

negatively correlated random
   variables, *373*

population mean, *330*

population variance, *356*

positively correlated random
   variables, *373*

random sample from a
   distribution, *368*

*r*th central moment, *353*

*r*th moment, *352*

sample mean, *343, 369*

standard deviation, *354*

standard error of the mean, *369*

standardized random variable, *359*

statistic, *368*

unbiased estimator, *343, 368*

uncorrelated random
   variables, *373*

variance, *353*

Chapter Review Exercises

Basic Exercises

7.134 A random variable has mean 31.4. Roughly, what would you expect the average value of the random variable to be in 1000 independent repetitions of the random experiment? Explain your answer.

7.135 A probability distribution for the loss due to fire damage in warehouses is as follows.

Loss ($1000)	0	0.5	1	10	50	100
Probability	0.900	0.060	0.030	0.008	0.001	0.001

a) Determine the expected amount of a loss.

b) Determine the expected amount of a loss given that the loss is positive.

7.136 On average, a salesman makes a sale to 60% of the customers he contacts, and he contacts 20 customers per day. On each sale, he makes a commission of $35.

a) Find the mean and standard deviation of the salesman's total daily commission.

b) What assumption did you make in your solution to part (a)? Do you think that assumption is reasonable?

7.137 A baseball team schedules its opening game for April 1, but won't play if it rains. To offset any losses in revenue, the team insures against rain. The insurance policy pays $1 million for each day, up to 2 days, that the opening game is delayed. In the town where the team opens, the number of consecutive days of rain, beginning on April 1, has a Poisson distribution with mean 0.6 days. Determine the mean and standard deviation of the amount that the insurance company pays.

7.138 Suppose that the random variables X and Y represent the amount of return on two different investments. Further suppose that the mean of X equals the mean of Y but that the standard deviation of X is greater than the standard deviation of Y.
a) On average, is there a difference between the returns of the two investments?
b) Which investment is more conservative?

7.139 Suppose that X has the Poisson distribution with parameter λ.
a) For $n \in \mathcal{N}$, determine $\mathcal{E}(X(X-1)\cdots(X-n+1))$, called the *nth factorial moment*.
b) Use the result of part (a) to obtain the mean and variance of X.

7.140 Let X have the geometric distribution with parameter p, let m be a positive integer, and set $Y = \min\{X, m\}$. Determine the expected value of Y by using
a) the definition of expected value. *Note:* In Exercise 5.136(a) on page 249, we asked for the PMF of Y.
b) the FEF and the PMF of X.   c) tail probabilities of Y.

7.141 A university gives a placement exam in mathematics skills to incoming students. To the nearest minute, the time it takes a student to complete the exam has a geometric distribution with parameter 0.020.
a) On average, how long does it take a student to complete the exam?
b) Repeat part (a) if a 1-hour time limit is in force. *Hint:* Refer to Exercise 7.140.

7.142 A small town in South Dakota has 250 families. A contingency table for the number of televisions and number of radios owned by the 250 families is presented in Exercise 6.123 on page 318. Part (b) of that problem asks for the joint and marginal PMFs of the number of televisions (X) and the number of radios (Y) of a randomly selected family. And part (a) of Exercise 6.124 asks for the conditional PMF of the number of radios for each number of televisions.
a) For each number of televisions, determine and interpret the conditional expectation of the number of radios. Construct a table similar to Table 7.13 on page 379.
b) For each number of televisions, determine the conditional variance of the number of radios. Construct a table similar to Table 7.14 on page 385.

7.143 Refer to Exercise 7.142.
a) Use the law of total expectation to obtain the expected number of radios.
b) Use the definition of expected value to obtain the expected number of radios. Compare your answer to that found in part (a).
c) Use the law of total variance to obtain the variance of the number of radios.
d) Use the definition of variance to obtain the variance of the number of radios. Compare your answer to that found in part (c).

7.144 Two balanced dice are rolled, one black and one gray. Let Y and Z be the faces showing on the black and gray dice, respectively, and let X be the smaller of the two faces showing. Obtain the expected value of X
a) by first obtaining the PMF of X and then using the definition of expected value.
b) by first obtaining the joint PMF of Y and Z and then using the FEF.

7.145 Let X and Y be random variables such that $\mathcal{E}(X) = 19$, $\mathcal{E}(Y) = 25$, $\text{Var}(X) = 25$, $\text{Var}(Y) = 144$, and $\rho(X, Y) = -1/2$. Determine each of the following quantities.
a) $\text{Cov}(X, Y)$ b) $\text{Cov}(X + Y, X - Y)$ c) $\text{Var}(X + Y)$
d) $\text{Var}(X - Y)$ e) $\text{Var}(3X + 4Y)$ f) $\text{Cov}(3X + 4Y, -2X + 5Y)$

7.146 A company is drilling for water in an arid region. It costs \$500 to drill each well regardless of whether water is found, and chances are 2/3 that a well won't produce water.

A producing well can be sold for $4000. If the company drills until it finds three producing wells, what are the mean and standard deviation of its profit?

7.147 The owner of Billy Bob's Quick Stop doesn't know the probability distribution of the annual repair cost to the cash register, but he does know that the mean and standard deviation of the annual repair cost are $125 and $12, respectively. How much money should the owner put aside annually for cash register repair to ensure that the annual repair cost won't exceed that amount more than 5% of the time?

7.148 A number X is generated at random from the interval $(0, 1)$. However, the recording device gives only the first three digits of the decimal expansion of X. Determine the mean and standard deviation of the recorded value.

7.149 A computer company provides an insurance policy for one of its larger systems. If the system fails during the first year, the policy pays $5000. The benefit decreases by $1000 each year until it reaches $0. If the system hasn't failed at the beginning of a year, the probability that it fails during that year is 0.1. How much must the company charge for the insurance policy so that, on average, it nets $100 per policy?

7.150 An urn contains six red marbles and four blue marbles. Two marbles are drawn at random from the urn. Find and identify the PMF of the number of red marbles obtained assuming that they are drawn
a) with replacement.   **b)** without replacement.

Apply the results of parts (a) and (b) to find and compare the two means by using
c) the PMFs obtained there and the definition of expected value.
d) the identifications made there and Table 7.4 on page 332.

Apply the results of parts (a) and (b) to find and compare the two variances by using
e) the PMFs obtained there and the definition of variance.
f) the identifications made there and Table 7.10 on page 358.

7.151 A random sample of n people is taken from a very large population. The n people sampled will undergo blood tests for a certain disease. The blood test is unfailing, meaning that a person with the disease always tests positive and a person without the disease always tests negative. If the n people are tested separately, then n tests are required. Alternatively, all the individual blood samples could be pooled and tested as a whole. If the test is negative, one test suffices for the entire sample. If the test is positive, each person sampled is tested separately and, therefore, $n + 1$ tests are needed. Let p denote the proportion of people in the population who have the disease.
a) Determine the expected number of tests required under the alternative scheme.
b) As a function of n, for which values of p is the alternative scheme superior relative to the expected number of tests required?

7.152 Refer to Exercise 7.151. Let the n people in the sample be randomly divided into m groups of equal size n/m, following which the alternative scheme is applied to each group.
a) Determine the expected number of tests required under this scheme.
b) For the three schemes discussed in this and the preceding exercise, compare the expected number of tests required when $n = 100$, $p = 0.01$, and $m = 20$.

7.153 Suppose that n balls are randomly placed in n urns in such a way that each ball is equally likely to go in each urn. What is the expected number of empty urns?

7.154 Let X and Y be independent geometric random variables with parameters p and q, respectively. Use tail probabilities to determine the expected value of
a) the smaller of X and Y.   **b)** the larger of X and Y.
c) Specialize the results of parts (a) and (b) to the case where $q = p$.

7.155 One of the 16 cells in the following table is chosen at random. The random variable Y equals 1, 3, 4, or 6, depending on whether the chosen cell is in the first, second, third, or fourth column. The random variable X equals 1 if the chosen cell is in the first row, 2 if it's in either the second or the third row, and 3 if it's in the fourth row.

Y

		1	3	4	6
	1	2	4	5	7
	2	3	5	6	8
X	2	3	5	6	8
	3	4	6	7	9

The numbers in the upper margin are the values of Y, those in the left margin are the values of X, and those in the interior of the table are the values of $X + Y$. Hence this table is an "addition table." Show that the variances of the addition table add—that is, show that $\mathrm{Var}(X + Y) = \mathrm{Var}(X) + \mathrm{Var}(Y)$.

7.156 In a factory, each manufactured item is defective with probability p and, independently, inspected with probability q. If both defective status and inspection status are taken into account, there are four categories: nondefective-uninspected, nondefective-inspected, defective-uninspected, and defective-inspected. For a random sample of n items, determine the mean and variance of the number of undetected defectives, given that the number of detected defectives is k.

7.157 The heights of the 10 players on the traveling squad of a basketball team are $6'\,2''$, $6'\,6''$, $6'\,6''$, $6'\,9''$, $6'\,7''$, $6'\,7''$, $6'\,6''$, $6'\,10''$, $6'\,1''$, and $6'\,3''$. Let X be the height of a randomly selected player.
a) Determine the mean and standard deviation of the random variable X.
b) A decision is made to measure height in number of centimeters over 2 meters. Call this random variable Y. Find the mean and standard deviation of Y.
c) Obtain the correlation coefficient of X and Y.

7.158 Let $X \sim \mathcal{G}(p)$. Determine $\mathcal{E}(1/X)$ and compare it to $1/\mathcal{E}(X)$.

7.159 Bernoulli trials are performed until the first success occurs. Call this number of trials X. Next, X more Bernoulli trials are performed. Let Y be the number of successes in this second group of Bernoulli trials.
a) Obtain $\mathcal{E}(Y)$ by using a heuristic argument.
b) Obtain $\mathcal{E}(Y)$ by using a formal mathematical argument. *Hint:* Condition on X.
c) Determine $\mathrm{Var}(Y)$.

7.160 Let λ be a positive real number. Suppose that, for each $n \in \mathcal{N}$, the random variable X_n has the binomial distribution with parameters n and p_n, where $np_n \to \lambda$ as $n \to \infty$. According to Exercise 5.86 on page 227, the distribution of X_n converges to that of a Poisson random variable with parameter λ. Use this result and the formula for the variance of a binomial random variable to conjecture the variance of a Poisson random variable.

7.161 Suppose that a random sample of size n is taken without replacement from a population of size N in which the proportion of members with a specified attribute is p. Let X denote the number of members obtained that have the specified attribute. Suppose that N is large relative to n.

a) Make an educated guess about the approximate variance of X.

b) Verify your educated guess in part (a) by letting $N \to \infty$ in the formula for the exact variance of X.

7.162 Let X and Y be independent with $X \sim \mathcal{B}(m, p)$ and $Y \sim \mathcal{B}(n, p)$. Determine
a) $\mathcal{E}(X \mid X + Y = z)$.   **b)** $\text{Var}(X \mid X + Y = z)$.

Theory Exercises

7.163 Suppose that X and Y are discrete random variables defined on the same sample space and such that $P(X = Y) = 1$—that is, X and Y are equal with probability 1.
a) Show that X and Y have the same PMF.
b) Conclude from part (a) that X and Y have the same mean and variance.

7.164 Weak law of large numbers: The *weak law of large numbers* is one of the forms of the law of averages. It provides mathematical confirmation of the long-run-average interpretation of expected value and can be stated as follows. Let X be a random variable with finite mean. Suppose that the random experiment is independently repeated indefinitely and let $X_1, X_2, \ldots$ represent the observed values. Then, for each $\epsilon > 0$,

$$\lim_{n \to \infty} P\left(\left|\frac{X_1 + \cdots + X_n}{n} - \mathcal{E}(X)\right| < \epsilon\right) = 1.$$

a) Prove the weak law of large numbers in case X has finite variance. *Hint:* Use Chebyshev's inequality.

b) State the weak law of large numbers in the special case where X is the indicator random variable of an event E. Of which interpretation does this form of the weak law of large numbers provide mathematical confirmation?

Advanced Exercises

7.165 Pólya distribution: Let α and β be positive real numbers. A random variable X with PMF given by $p_X(0) = (1 + \alpha\beta)^{-1/\beta}$,

$$p_X(x) = \left(\frac{\alpha}{1 + \alpha\beta}\right)^x \frac{1(1 + \beta) \cdots (1 + (x - 1)\beta)}{x!} p_X(0), \qquad x \in \mathcal{N},$$

and $p_X(x) = 0$ otherwise, is said to have the *Pólya distribution with parameters α and β.* Find the expected value and variance of a Pólya random variable.

7.166 Suppose that you are collecting coupons, of which there are M types, and that each coupon you get is equally likely to be any one of the M types. Let $\mathcal{E}_M$ and $\mathcal{V}_M$ denote the expected value and variance, respectively, of the number of coupons that you must collect to obtain a complete set, that is, at least one of each type.
a) Determine $\mathcal{E}_M$.   **b)** Determine $\mathcal{V}_M$.
c) Prove that $\mathcal{E}_M \sim M \ln M$, as $M \to \infty$, where "$\sim$" means that the limit of the ratio of the two sides is 1.

7.167 An insurance company classifies its policyholders as either high risk or low risk; 10% are classified as high risk. For a given calendar year, the number of claims made by a policyholder has a Poisson distribution with parameter λ and is independent of the number of claims made by the policyholder in the previous calendar year. Furthermore, records show that $\lambda = 0.6$ for high-risk policyholders and $\lambda = 0.1$ for low-risk policyholders. Determine the expected number of claims in a calendar year by a policyholder who made one claim the preceding year.

7.168 A spelunker is in an underground cavern and doesn't know the way out. Three routes are possible. One route returns her to the cavern after 3 hours travel, another route returns her to the cavern after 4 hours travel, and the third route leads to the outside after 1 hour travel. Find the expected number of hours that the spelunker is underground if
a) each time she chooses a route at random.
b) each time she chooses a route at random from among the routes not already tried.

7.169 Bernoulli trials with success probability p are performed until two consecutive successes occur. What is the expected number of trials required?

7.170 Let $X_1, \ldots, X_m$ have the multinomial distribution with parameters n and $p_1, \ldots, p_m$. Find $\mathrm{Var}(X_1 + \cdots + X_m)$
a) by using Proposition 7.13 on page 366 and Exercise 7.93 on page 375.
b) without doing any computations.

Probability generating function: Recall from Exercise 7.52 on page 352 that the *probability generating function* (PGF) of a nonnegative-integer valued random variable X is defined by $P_X(t) = \mathcal{E}(t^X)$. In Exercises 7.171–7.175, you explore further properties of the PGF.

7.171 Let X be a nonnegative-integer valued random variable with finite nth moment. The *nth factorial moment* of X is defined as $\mathcal{E}(X(X-1) \cdots (X-n+1))$.
a) Show that $\mathcal{E}(X(X-1) \cdots (X-n+1)) = P_X^{(n)}(1)$.
b) If $X \sim \mathcal{P}(\lambda)$, we have $P_X(t) = e^{\lambda(t-1)}$, as determined in Exercise 7.53(b) on page 352. Use this result and part (a) to obtain the nth factorial moment of X and compare your answer with that obtained in Exercise 7.139(a).

7.172 Let X and Y be nonnegative-integer valued random variables.
a) Show that, if X and Y have the same probability distribution, they have the same PGF.
b) Show that, if X and Y have the same PGF, they have the same probability distribution.
c) Use part (b) to identify the distribution of a random variable whose PGF is $e^{4(t-1)}$.

7.173 Let X and Y be independent nonnegative-integer valued random variables.
a) Prove that $P_{X+Y} = P_X P_Y$ and interpret that result in words.
b) Extend the result of part (a) to the case of m independent nonnegative-integer valued random variables.
c) Suppose that $X_1, \ldots, X_m$ are independent nonnegative-integer valued random variables, all having the same probability distribution as a random variable X. Obtain a formula for the PGF of $X_1 + \cdots + X_m$ in terms of the PGF of X.

7.174 Use Exercises 7.172 and 7.173 and the results of Exercise 7.53 on page 352 to provide a simple proof of Proposition 6.20 on page 311.

7.175 Let $X_1, X_2, \ldots$ be independent nonnegative-integer valued random variables having the same probability distribution as a random variable X with mean μ and variance σ^2. And let N be a nonnegative-integer valued random variable independent of the X_js. Set $S_0 = 0$ and $S_n = X_1 + \cdots + X_n$ for $n \geq 1$. Then S_N is a random variable obtained as the sum of a random number of random variables.
a) Obtain the PGF of S_N in terms of the PGFs of N and X. *Hint:* Refer to Exercise 7.173(c), condition on N, and apply the law of total expectation.
b) Use the result of part (a) to obtain the mean and variance of S_N. Compare your answers with Equations (7.63) and (7.68) on pages 384 and 387, respectively.
c) Use the result of part (a) and Exercise 7.172(b) to solve Example 6.25 on page 312. *Note:* If $X \sim \mathcal{P}(\lambda)$, we have $P_X(t) = e^{\lambda(t-1)}$, as determined in Exercise 7.53(b) on page 352.

PART THREE

Continuous Random Variables

Carl Friedrich Gauss 1777–1855

Born on April 30, 1777, in Brunswick, Germany, the only son in a poor, semiliterate peasant family, Carl Friedrich Gauss taught himself to calculate before he could talk. At the age of 3, he pointed out an error in his father's calculations of wages. In addition to his arithmetic experimentation, he taught himself to read. At the age of 8, Gauss instantly solved the summing of all numbers from 1 to 100. His father was persuaded to allow him to stay in school and to study after school instead of working to help support the family.

Impressed by Gauss's brilliance, the Duke of Brunswick supported him monetarily from the ages of 14 to 30. This patronage permitted Gauss to pursue his studies exclusively. He conceived most of his mathematical discoveries by the time he was 17. Gauss was granted a doctorate in absentia from the university at Helmstedt; his doctoral thesis developed the concept of complex numbers and proved the fundamental theorem of algebra, which had previously been only partially established. Shortly thereafter, Gauss published his theory of numbers, which is considered one of the most brilliant achievements in mathematics.

Gauss made important discoveries in mathematics, physics, astronomy, and statistics. Two of his major contributions to probability and statistics were the development of the least-squares method and fundamental work with the normal distribution, often called the *Gaussian distribution* in his honor.

In 1807, Gauss accepted the directorship of the observatory at the University of Göttingen, which ended his dependence on the Duke of Brunswick. He remained there the rest of his life. In 1833, Gauss and a colleague, Wilhelm Weber, invented a working electric telegraph, 5 years before Samuel Morse. Gauss died in Göttingen in 1855.

CHAPTER EIGHT

Continuous Random Variables and Their Distributions

INTRODUCTION

In this chapter, we examine continuous random variables. We concentrate on the univariate case here and discuss the multivariate case in Chapter 9. To begin, in Section 8.1, we introduce continuous random variables and explain why the probability mass function (PMF) is useless for describing the probability distribution of such a random variable.

In Section 8.2, we present the *cumulative distribution function* (CDF), which provides all the information required to study the general theory of random variables. Although the concept of the CDF applies to any random variable, it's particularly useful for investigation of continuous random variables.

Armed with the concept of the CDF, we introduce and examine, in Section 8.3, the *probability density function* (PDF) of a continuous random variable. The PDF completely determines the probability distribution of a continuous random variable and is the analogue of the PMF for discrete random variables.

In Sections 8.4–8.6, we present some of the most important continuous random variables, including the *uniform random variable, exponential random variable, normal random variable, gamma random variable, beta random variable,* and *triangular random variable.* And, in Section 8.7—the final section of the chapter—we examine functions of a continuous random variable.

8.1 Introducing Continuous Random Variables

As we mentioned in Chapter 5, there are two main types of random variables—*discrete random variables* and *continuous random variables.* So far, our examination of random variables has been confined to discrete random variables.

A discrete random variable often involves a count of something, such as the number of siblings of a randomly selected student, the number of people waiting for a haircut in a barbershop, or the number of households in a sample that own a computer. In contrast, a **continuous random variable** typically involves a measurement of something, such as the height of a randomly selected person, the diameter of a bolt coming off a production line, or the life of a car battery.

For a discrete random variable X, the probabilities $P(X = x)$, $x \in \mathcal{R}$, completely determine the probability distribution of X in the sense that, once we know them, we can obtain any probability for X. Indeed, according to the fundamental probability formula (FPF) for discrete random variables, we have

$$P(X \in A) = \sum_{x \in A} P(X = x), \tag{8.1}$$

for all subsets A of $\mathcal{R}$.

In contrast, because a continuous random variable X typically involves measurement, the probability should be 0 that X equals any particular value x. Consequently, we have Definition 8.1.

DEFINITION 8.1   Continuous Random Variable

A random variable X is called a **continuous random variable** if

$$P(X = x) = 0 \qquad \text{for all } x \in \mathcal{R}. \tag{8.2}$$

There are several reasons why the adjective "continuous" is used to describe random variables that satisfy the condition of Definition 8.1. One reason is that the range of such a random variable forms a continuum of real numbers. Another reason will become apparent after we introduce the *cumulative distribution function* of a random variable in Section 8.2. For now, however, let's consider Examples 8.1 and 8.2, which provide two simple illustrations of continuous random variables.

EXAMPLE 8.1   *Continuous Random Variables*

Random Number Generators   Suppose that a number is selected at random from the interval (0, 1), a task that a basic random number generator aims to accomplish. Let X denote the number obtained. Show that X is a continuous random variable.

Solution   As we showed in Example 2.23 on page 57, the sample space for this random experiment consists of all real numbers between 0 and 1, so $\Omega = (0, 1) = \{x : 0 < x < 1\}$. Because we are selecting a number at random, a geometric probability model is appropriate. Thus, for each event E,

$$P(E) = \frac{|E|}{|\Omega|} = \frac{|E|}{1} = |E|, \qquad (8.3)$$

where $|E|$ denotes the length (in the extended sense) of the set E.

The random variable X is defined on Ω by $X(x) = x$; that is, X is the identity function on Ω. We want to show that X is a continuous random variable: $P(X = x) = 0$ for all $x \in \mathcal{R}$. Clearly, the range of X is the interval $(0, 1)$ and, consequently, we have $P(X = x) = 0$ if $x \notin (0, 1)$. For $x \in (0, 1)$, the event $\{X = x\}$ is that the randomly selected number is x—that is, $\{X = x\} = \{x\}$. As the length of a single point (i.e., a singleton set) is 0, we conclude from Equation (8.3) that

$$P(X = x) = |\{X = x\}| = |\{x\}| = 0. \qquad (8.4)$$

Hence X is a continuous random variable.   ▪

EXAMPLE 8.2   *Continuous Random Variables*

Bacteria on a Petri Dish   A petri dish is a small, shallow dish of thin glass or plastic, used especially for cultures in bacteriology. Suppose that a petri dish of unit radius, containing nutrients upon which bacteria can multiply, is smeared with a uniform suspension of bacteria. Subsequently, spots indicating colonies of bacteria will appear. Suppose that we observe the location of the center of the first spot to appear. Let Z denote the distance of the center of the first spot from the center of the petri dish. Show that Z is a continuous random variable.

Solution   As we demonstrated in Example 2.6 on page 28, we can take the sample space for this random experiment to be the unit disk: $\Omega = \{(x, y) : x^2 + y^2 < 1\}$. Because the dish is smeared with a uniform suspension of bacteria, use of a geometric probability model is reasonable. Specifically, we can think of the location of the center of the first spot (visible bacteria colony) as a point selected at random from the unit disk. Consequently, for each event E,

$$P(E) = \frac{|E|}{|\Omega|} = \frac{|E|}{\pi}, \qquad (8.5)$$

where $|E|$ denotes the area (in the extended sense) of the set E.

The random variable Z is defined on Ω by $Z(x, y) = \sqrt{x^2 + y^2}$. We want to show that Z is a continuous random variable: $P(Z = z) = 0$ for all $z \in \mathcal{R}$. Clearly, the range of Z is the interval $[0, 1)$ and hence $P(Z = z) = 0$ if $z \notin [0, 1)$.

Now let $z \in [0, 1)$. The event $\{Z = z\}$ is that the center of the first spot to appear is z units from the center of the petri dish or, equivalently, that the center of the first spot to appear lies on the circle $C = \{(x, y) : x^2 + y^2 = z^2\}$, as illustrated in Figure 8.1. Because the area of a circle (boundary of a disk) is 0, Equation (8.5) yields that $P(Z = z) = P(C) = |C|/\pi = 0$.

Figure 8.1 The event C that the center of the first spot to appear
is z units from the center of the petri dish

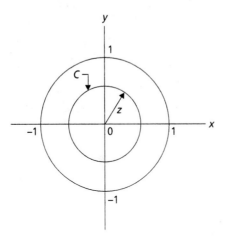

Here is another way to show that $P(Z = z) = 0$. Let $A_n = \{z \leq Z \leq z + 1/n\}$ for
each $n \in \mathcal{N}$. We note that A_n is the event that the center of the first spot to appear is
between z units and $z + 1/n$ units from the center of the petri dish, as shown in Figure 8.2.

Figure 8.2 The event A_n that the center of the first spot to appear
is between z units and $z + 1/n$ units from the center of the petri dish

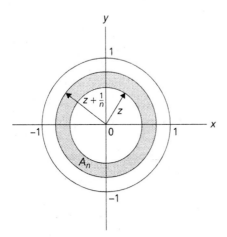

From Equation (8.5), for sufficiently large n,

$$P(A_n) = \frac{|A_n|}{\pi} = \frac{\pi(z + 1/n)^2 - \pi z^2}{\pi} = \frac{2z}{n} + \frac{1}{n^2}. \tag{8.6}$$

Because $\{Z = z\} \subset A_n$, Equation (8.6) implies that $P(Z = z) \leq 2z/n + 1/n^2$ for sufficiently large n. Letting $n \to \infty$ in this inequality yields $P(Z = z) = 0$. ∎

Describing the Probability Distribution of a Continuous Random Variable

As we mentioned earlier, the probabilities $P(X = x)$ completely determine the probability distribution of a discrete random variable X. However, for a continuous random variable X, the probabilities $P(X = x)$ always equal 0 and hence are totally useless in describing the probability distribution of X.

How then do we describe the probability distribution of a continuous random variable X? The answer is that, instead of considering, for each $x \in \mathcal{R}$, the probability that X equals x, we consider the probability that X is "near" x. First, however, we must examine another function associated with a random variable, the *cumulative distribution function,* which we do in Section 8.2.

Basic Exercises

8.1 On page 402, we stated that: "...because a continuous random variable X typically involves measurement, the probability should be 0 that X equals any particular value x." Explain in your own words why this statement makes sense.

8.2 A point is chosen at random from the interior of a sphere of radius r. Let Z denote the distance from the center of the sphere to the point chosen. Use both methods in Example 8.2 on page 403 to show that Z is a continuous random variable.

8.3 Let X denote the tangent of an angle chosen at random from the interval $(-\pi/2, \pi/2)$. Show that X is a continuous random variable.

8.4 Let X be a continuous random variable. Verify that each of the following functions of X is a continuous random variable.
a) $a + bX$, where a and $b \neq 0$ are real numbers
b) X^2 **c)** $\sin X$

8.5 A point is chosen at random from the interior of a triangle with base b and height h. Let Y denote the distance of the point chosen to the base of the triangle. Show that Y is a continuous random variable.

8.6 Let f be a nonnegative real-valued function of a real variable such that $\int_{-\infty}^{\infty} f(x)\, dx = 1$. Suppose that X is a random variable that satisfies $P(X \in A) = \int_A f(x)\, dx$ for all subsets A of $\mathcal{R}$. Show that X is a continuous random variable.

Theory Exercises

8.7 Let X be a continuous random variable.
a) Show that $P(X \in K) = 0$ for any countable set K.
b) Use the result of part (a) to conclude that a continuous random variable can't be a discrete random variable. *Hint:* Refer to Definition 5.2 on page 178.

Advanced Exercises

8.8 Refer to Example 8.1 on page 402, where X denotes a randomly chosen number from the interval $(0, 1)$. Determine the probability that the number chosen is a rational number.

8.9 Let X be a continuous random variable and let p be a polynomial of positive degree. Show that $p(X)$ is a continuous random variable.

8.10 Let X be a continuous random variable and let g be a real-valued function defined and one-to-one on the range of X. Show that $g(X)$ is a continuous random variable.

8.11 Let X be a continuous random variable and let g be a real-valued function defined on the range of X.
a) According to Exercise 8.10, a sufficient condition for $g(X)$ to be a continuous random variable is that g is one-to-one on the range of X. Is that condition necessary?
b) Is $g(X)$ necessarily a continuous random variable? Justify your answer.

8.12 Suppose that a number is chosen at random from the interval $(0, 1)$ and let X denote the number obtained. Define the random variable Y to be the smaller of X and 0.75; that is, $Y = \min\{X, 0.75\}$. Establish each of the following results.
a) The range of Y is a continuum of real numbers.
b) Y isn't a continuous random variable.
c) Y isn't a discrete random variable. *Hint:* Refer to Exercise 5.17 on page 183.

8.13 Let X be a random variable and define $F_X(x) = P(X \le x)$ for $x \in \mathcal{R}$.
a) Prove that $P(a < X \le b) = F_X(b) - F_X(a)$ for all real numbers a and b with $a < b$.
b) Prove that, if F_X is continuous on $\mathcal{R}$, then X is a continuous random variable. *Note:* In Section 8.2, we establish the converse of this result.

8.14 For each $n \in \mathcal{N}$, let the random variable X_n have the discrete uniform distribution on the set $\{0, 1/n, \ldots, (n-1)/n\}$.
a) Show for $0 < a < b < 1$ that $P(a < X_n \le b) = (\lfloor nb \rfloor - \lfloor na \rfloor)/n$, where $\lfloor x \rfloor$ denotes the *floor function*—the greatest integer smaller than or equal to x.
b) Deduce from part (a) that $\lim_{n \to \infty} P(a < X_n \le b) = b - a$ for $0 < a < b < 1$.
c) Based on the result of part (b), identify a continuous random variable X whose probability distribution is that of the limiting distribution of the discrete random variables $X_1, X_2, \ldots$.

8.2 Cumulative Distribution Functions

The probability mass function gives the probability that X equals x [i.e., $P(X = x)$] for each $x \in \mathcal{R}$. Alternatively, we can consider the function that provides the probability that X equals at most x [i.e., $P(X \le x)$] for each $x \in \mathcal{R}$. Because $P(X \le x)$ accumulates the probabilities for X up to and including x, the associated function is called the **cumulative distribution function.**

DEFINITION 8.2   Cumulative Distribution Function

Let X be a random variable. Then the **cumulative distribution function** of X, denoted F_X, is the real-valued function defined on $\mathcal{R}$ by

$$F_X(x) = P(X \le x), \qquad x \in \mathcal{R}.$$

We use **CDF** as an abbreviation for "cumulative distribution function."

Note the following.

- The CDF applies to any random variable, regardless of type (discrete, continuous, or otherwise).

- The CDF of a random variable is denoted by an uppercase F subscripted with the letter representing the random variable. Thus F_X denotes the CDF of the random variable X, F_Y denotes the CDF of the random variable Y, and so on.

- Other terms used for *cumulative distribution function* are **distribution function** and **cumulative distribution.**

- Although it's by no means obvious, the CDF of a random variable completely determines its probability distribution, regardless of type.

Examples 8.3–8.6 illustrate how to determine CDFs for various different types of random variables.

EXAMPLE 8.3 *Cumulative Distribution Functions*

Coin Tossing   Suppose that a balanced coin is tossed three times and let W denote the total number of heads in the three tosses. Determine the CDF of the random variable W.

Solution   As we know, W is a discrete random variable with PMF as given in Table 8.1.

Table 8.1 PMF of the random variable W

w	0	1	2	3
$p_W(w)$	0.125	0.375	0.375	0.125

From Table 8.1 and the FPF for discrete random variables, we can obtain the CDF of W. If $w < 0$, $F_W(w) = P(W \le w) = 0$; if $w \ge 3$, $F_W(w) = P(W \le w) = 1$; if $0 \le w < 1$,

$$F_W(w) = P(W \le w) = \sum_{t \le w} p_W(t) = p_W(0) = 0.125;$$

if $1 \le w < 2$,

$$F_W(w) = P(W \le w) = \sum_{t \le w} p_W(t) = p_W(0) + p_W(1) = 0.125 + 0.375 = 0.5;$$

and, if $2 \le w < 3$,

$$F_W(w) = P(W \le w) = \sum_{t \le w} p_W(t) = p_W(0) + p_W(1) + p_W(2)$$

$$= 0.125 + 0.375 + 0.375 = 0.875.$$

Thus the CDF of W is

$$F_W(w) = \begin{cases} 0, & \text{if } w < 0; \\ 0.125, & \text{if } 0 \leq w < 1; \\ 0.5, & \text{if } 1 \leq w < 2; \\ 0.875, & \text{if } 2 \leq w < 3; \\ 1, & \text{if } w \geq 3. \end{cases}$$

The positions of the weak and strict inequalities are important! A graph of the CDF of W is provided in Figure 8.3.

Figure 8.3 CDF of the random variable W

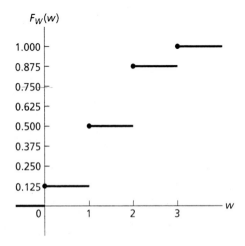

Note that the CDF of W is a step function. Either Figure 8.3 or the mathematical formulas for the PMF and CDF of W reveal that the jumps of this step function equal the values of the PMF. ∎

EXAMPLE 8.4   *Cumulative Distribution Functions*

Random Number Generators   Let X denote a number selected at random from the interval $(0, 1)$. Find the CDF of the random variable X.

Solution   From Example 8.1, X is a continuous random variable. The range of X is the interval $(0, 1)$. Therefore, if $x < 0$, we have $F_X(x) = P(X \leq x) = 0$; and, if $x \geq 1$, we have $F_X(x) = P(X \leq x) = 1$.

Now let $0 \leq x < 1$. The event $\{X \leq x\}$ is that the randomly selected number is between 0 and x. From Equation (8.3) on page 403,

$$F_X(x) = P(X \leq x) = \left|\{X \leq x\}\right| = \left|(0, x]\right| = x - 0 = x.$$

We conclude that the CDF of X is

$$F_X(x) = \begin{cases} 0, & \text{if } x < 0; \\ x, & \text{if } 0 \le x < 1; \\ 1, & \text{if } x \ge 1. \end{cases}$$

A graph of the CDF of X is provided in Figure 8.4.

Figure 8.4 CDF of the random variable X

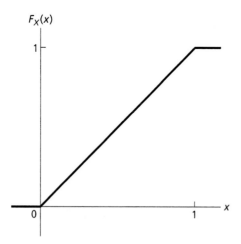

Note that the CDF of X is everywhere continuous. As you will soon see, this property characterizes continuous random variables. ∎

EXAMPLE 8.5 *Cumulative Distribution Functions*

Bacteria on a Petri Dish Refer to Example 8.2 on page 403, where Z denotes the distance of the center of the first spot (visible bacteria colony) from the center of a petri dish. Determine the CDF of the random variable Z.

Solution From Example 8.2, Z is a continuous random variable. The range of Z is the interval $[0, 1)$. Therefore, if $z < 0$, we have $F_Z(z) = P(Z \le z) = 0$; and, if $z \ge 1$, we have $F_Z(z) = P(Z \le z) = 1$.

Now let $0 \le z < 1$. The event $\{Z \le z\}$ is that the distance of the center of the first spot from the center of the petri dish is at most z. From Equation (8.5) on page 403,

$$F_Z(z) = P(Z \le z) = \frac{|\{Z \le z\}|}{\pi} = \frac{\pi z^2}{\pi} = z^2.$$

We conclude that the CDF of Z is

$$F_Z(z) = \begin{cases} 0, & \text{if } z < 0; \\ z^2, & \text{if } 0 \le z < 1; \\ 1, & \text{if } z \ge 1. \end{cases}$$

A graph of the CDF of Z is provided in Figure 8.5.

Figure 8.5 CDF of the random variable Z

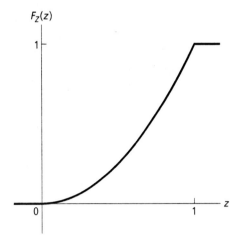

Note that the CDF of Z is everywhere continuous. As we said before, this property characterizes continuous random variables. ∎

EXAMPLE 8.6   *Cumulative Distribution Functions*

A Random Variable of Mixed Type   Let X denote a number selected at random from the interval $(0, 1)$, and set $Y = \min\{X, 0.75\}$; that is, Y is the smaller of the random number obtained and 0.75. Obtain the CDF of the random variable Y.

Solution   The range of Y is the interval $(0, 0.75]$. So, if $y < 0$, we have $F_Y(y) = P(Y \le y) = 0$; and, if $y \ge 0.75$, we have $F_Y(y) = P(Y \le y) = 1$.

Now let $0 \le y < 0.75$. The event $\{Y \le y\}$ is that the smaller of the randomly selected number and 0.75 is between 0 and y. However, because we are assuming that $0 \le y < 0.75$, it follows that $\{Y \le y\}$ is the event that the randomly selected number is between 0 and y. Therefore, by Equation (8.3) on page 403,

$$F_Y(y) = P(Y \le y) = P(X \le y) = \big|\{X \le y\}\big| = \big|(0, y]\big| = y - 0 = y.$$

We conclude that the CDF of Y is

$$F_Y(y) = \begin{cases} 0, & \text{if } y < 0; \\ y, & \text{if } 0 \le y < 0.75; \\ 1, & \text{if } y \ge 0.75. \end{cases}$$

A graph of the CDF of Y is provided in Figure 8.6. Exercise 8.12 on page 406 shows that Y is neither a discrete random variable nor a continuous random variable. ∎

Figure 8.6 CDF of the random variable Y

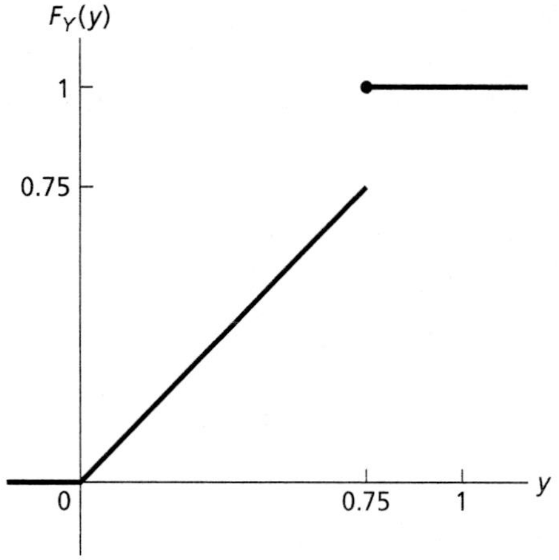

Basic Properties of a CDF

Figures 8.3–8.6 show that all four CDFs are nondecreasing and right-continuous. In fact, every CDF satisfies the four basic properties presented in Proposition 8.1.

◆◆◆ **Proposition 8.1   Basic Properties of a CDF**

The cumulative distribution function F_X of a random variable X satisfies the following four properties.

a) F_X is nondecreasing.
b) F_X is everywhere right-continuous.
c) $F_X(-\infty) \equiv \lim_{x \to -\infty} F_X(x) = 0$.
d) $F_X(\infty) \equiv \lim_{x \to \infty} F_X(x) = 1$.

Proof   a) Suppose that $x_1 < x_2$. Then we have $\{X \le x_1\} \subset \{X \le x_2\}$ and, so, by the domination principle (Proposition 2.6 on page 64),

$$F_X(x_1) = P(X \le x_1) \le P(X \le x_2) = F_X(x_2).$$

Thus F_X is nondecreasing.

b) Suppose that $x \in \mathcal{R}$ and let $\{x_n\}_{n=1}^{\infty}$ be any decreasing sequence of real numbers that converges to x. For each $n \in \mathcal{N}$, set $A_n = \{X \le x_n\}$. Then $A_1 \supset A_2 \supset \cdots$ and $\bigcap_{n=1}^{\infty} A_n = \{X \le x\}$. Applying the continuity property of probability (Proposition 2.11 on page 74) gives

$$\lim_{n \to \infty} F_X(x_n) = \lim_{n \to \infty} P(X \le x_n) = \lim_{n \to \infty} P(A_n)$$

$$= P\left(\bigcap_{n=1}^{\infty} A_n \right) = P(X \le x) = F_X(x).$$

Hence F_X is everywhere right-continuous.

c) Let $\{x_n\}_{n=1}^{\infty}$ be any decreasing sequence of real numbers such that $\lim_{n\to\infty} x_n = -\infty$. For each $n \in \mathcal{N}$, set $A_n = \{X \le x_n\}$. Then $A_1 \supset A_2 \supset \cdots$ and $\bigcap_{n=1}^{\infty} A_n = \emptyset$. Applying the continuity property of probability yields

$$\lim_{n\to\infty} F_X(x_n) = \lim_{n\to\infty} P(X \le x_n) = \lim_{n\to\infty} P(A_n) = P\left(\bigcap_{n=1}^{\infty} A_n\right) = P(\emptyset) = 0.$$

Therefore $\lim_{x\to -\infty} F_X(x) = 0$.

d) Let $\{x_n\}_{n=1}^{\infty}$ be any increasing sequence of real numbers such that $\lim_{n\to\infty} x_n = \infty$. For each $n \in \mathcal{N}$, set $A_n = \{X \le x_n\}$. Then $A_1 \subset A_2 \subset \cdots$ and $\bigcup_{n=1}^{\infty} A_n = \Omega$. Again, applying the continuity property of probability gives

$$\lim_{n\to\infty} F_X(x_n) = \lim_{n\to\infty} P(X \le x_n) = \lim_{n\to\infty} P(A_n) = P\left(\bigcup_{n=1}^{\infty} A_n\right) = P(\Omega) = 1.$$

Thus $\lim_{x\to\infty} F_X(x) = 1$. ◆

It can be shown that a function $F: \mathcal{R} \to \mathcal{R}$ that satisfies properties (a)–(d) of Proposition 8.1 is the CDF of some random variable. Therefore, for any such function, we can say, "Let X be a random variable with CDF F." This statement makes sense regardless of whether we explicitly give X and the sample space on which it is defined.

Now let X be any random variable. Proposition 8.1(b) shows that the CDF of X is everywhere right-continuous. Figures 8.3 and 8.6 indicate that the same can't be said about left-continuity. What we *can* say is that, for each $x \in \mathcal{R}$, the left-hand limit of F_X at x equals the probability that X is less than x; in symbols,

$$F_X(x-) = P(X < x). \tag{8.7}$$

Here we use the notation $F_X(x-)$ to denote the left-hand limit of F_X at x—that is, $F_X(x-) = \lim_{t\to x^-} F_X(t)$. The proof of Equation (8.7) is similar to that of Proposition 8.1(d) and is left for you as Exercise 8.35.

From the definition of a CDF and Equation (8.7), we obtain important relationships between the probabilities for a random variable and the values of its CDF. Specifically, we have Proposition 8.2.

◆◆◆ **Proposition 8.2**

Let X be a random variable. Then, for all real numbers $a < b$,
a) $P(a < X \le b) = F_X(b) - F_X(a)$.
b) $P(a \le X \le b) = F_X(b) - F_X(a-)$.
c) $P(a < X < b) = F_X(b-) - F_X(a)$.
d) $P(a \le X < b) = F_X(b-) - F_X(a-)$.

Proof   We prove property (c) and leave the others for you to prove as Exercise 8.36. To begin, we observe that $\{X < b\} = \{X \le a\} \cup \{a < X < b\}$. Noting that the two events on the right of this equation are mutually exclusive and referring to Equation (8.7), we have

$$F_X(b-) = P(X < b) = P(X \le a) + P(a < X < b) = F_X(a) + P(a < X < b).$$

Property (c) now follows easily. ◆

From Equation (8.7), we can also conclude that

$$P(X = x) = F_X(x) - F_X(x-), \qquad x \in \mathcal{R}. \tag{8.8}$$

Geometrically, this relation means that the probability that X takes the value x equals the jump (if any) of the CDF of X at x. To show why Equation (8.8) is true, we first observe that $\{X \le x\} = \{X < x\} \cup \{X = x\}$. Noting that the two events on the right of this equation are mutually exclusive and referring to Equation (8.7), we have

$$F_X(x) = P(X \le x) = P(X < x) + P(X = x) = F_X(x-) + P(X = x).$$

Hence Equation (8.8) holds. From Equation (8.8) and Proposition 8.1(b), we deduce the important fact presented in Proposition 8.3.

◆◆◆   **Proposition 8.3   Continuous Random Variables and Continuous CDFs**

A random variable is continuous if and only if its CDF is an everywhere continuous function.

Proof   We first recall from Proposition 8.1(b) that the CDF of any random variable is everywhere right-continuous. Thus a CDF is continuous at a point if and only if it is left-continuous at that point.

Suppose that X is a continuous random variable. Then, by definition, $P(X = x) = 0$ for all $x \in \mathcal{R}$. This fact and Equation (8.8) imply that F_X is left-continuous and hence continuous for all $x \in \mathcal{R}$. Conversely, suppose that the CDF of a random variable X is everywhere continuous. Then, in particular, it is everywhere left-continuous. From Equation (8.8), $P(X = x) = 0$ for all $x \in \mathcal{R}$. So X is a continuous random variable.   ◆

Note: Proposition 8.3 shows that, for a continuous random variable X, the right sides of all four equations in Proposition 8.2 equal $F_X(b) - F_X(a)$.

Proposition 8.3 provides another justification for using the adjective "continuous" for a random variable X that satisfies $P(X = x) = 0$ for all $x \in \mathcal{R}$—namely, the CDF of such a random variable is everywhere continuous.

EXERCISES 8.2   Basic Exercises

In Exercises 8.15–8.21, do the following.
a) Obtain and graph the CDF of the specified random variable.
b) Use the results of part (a) to verify both mathematically and graphically that the CDF of the specified random variable satisfies properties (a)–(d) of Proposition 8.1 on page 411.
c) Use Proposition 8.3 and the results of part (a) to decide whether the specified random variable is a continuous random variable.

8.15 A point is chosen at random from the interior of a sphere of radius r. Let Z denote the distance from the center of the sphere to the point chosen.

8.16 Let X denote the tangent of an angle chosen at random from the interval $(-\pi/2, \pi/2)$.

8.17 A point is chosen at random from the interior of a triangle with base b and height h. Let Y denote the distance from the point chosen to the base of the triangle.

8.18 Let X be the number of siblings of a randomly selected student from one of Professor Weiss's classes, as discussed in Example 5.1 on page 176. *Note:* Table 5.5 on page 186 gives the PMF of the random variable X.

8.19 Let Y be the indicator random variable of an event E.

8.20 According to *Vital Statistics of the United States*, published by the U.S. National Center for Health Statistics, chances are 80% that a person aged 20 will be alive at age 65. Of three people aged 20 selected at random, let X denote the number who live to be at least age 65.

8.21 Six men and five women apply for a job at Alpha, Inc. Three of the applicants are selected for interviews. Let X denote the number of women in the interview pool.

8.22 Let X be a discrete random variable.
a) Express F_X in terms of p_X.   **b)** Express p_X in terms of F_X.

8.23 Let X have the discrete uniform distribution on the set of the first N positive integers.
a) Without using the PMF of X, obtain and graph the CDF of X.
b) Use part (a) and Equation (8.8) on page 413 to obtain the PMF of X.

8.24 Let X have the geometric distribution with parameter p.
a) Use the FPF to obtain the CDF of X.
b) Use tail probabilities to obtain the CDF of X.
c) Graph the CDF of X.
d) Use the CDF of X and Equation (8.8) on page 413 to obtain the PMF of X.

8.25 Refer to Example 8.3 on page 407. Use the CDF obtained in that example to determine $P(a < W \le b)$, $P(a < W < b)$, $P(a \le W < b)$, and $P(a \le W \le b)$ for the specified values of a and b.
a) $a = 1, b = 2$ **b)** $a = 0.5, b = 2$ **c)** $a = 1, b = 2.75$ **d)** $a = 0.5, b = 2.75$
e) Repeat parts (a)–(d) by using the FPF and the PMF of W (Table 8.1 on page 407).

8.26 Refer to Example 8.4 on page 408. Use the CDF obtained in that example to determine $P(a < X \le b)$, $P(a < X < b)$, $P(a \le X < b)$, and $P(a \le X \le b)$ for the specified values of a and b.
a) $a = 0.2, b = 0.8$ **b)** $a = 0, b = 0.8$ **c)** $a = 0.2, b = 1.5$
d) $a = -1, b = 1.5$ **e)** $a = -2, b = -1$ **f)** $a = 1, b = 2$
g) For each of parts (a)–(f), why are the four probabilities identical?

8.27 Refer to Example 8.5 on page 409. Use the CDF obtained in that example to determine $P(a < Z \le b)$, $P(a < Z < b)$, $P(a \le Z < b)$, and $P(a \le Z \le b)$ for the specified values of a and b.
a) $a = 0.2, b = 0.8$ **b)** $a = 0, b = 0.8$ **c)** $a = 0.2, b = 1.5$
d) $a = -1, b = 1.5$ **e)** $a = -2, b = -1$ **f)** $a = 1, b = 2$
g) For each of parts (a)–(f), why are the four probabilities identical?

8.28 Refer to Example 8.6 on page 410. Use the CDF obtained in that example to determine $P(a < Y \le b)$, $P(a < Y < b)$, $P(a \le Y < b)$, and $P(a \le Y \le b)$ for the specified values of a and b.
a) $a = 0.2, b = 0.8$ **b)** $a = 0.2, b = 0.75$ **c)** $a = -1, b = 1.5$
d) $a = -1, b = 0.75$ **e)** $a = -2, b = -1$ **f)** $a = 0.75, b = 2$

8.29 Let f be a nonnegative real-valued function such that $\int_{-\infty}^{\infty} f(x)\,dx = 1$. Suppose that X is a random variable that satisfies $P(X \in A) = \int_A f(x)\,dx$ for all subsets A of $\mathcal{R}$.
a) Obtain F_X in terms of f.
b) What is the relationship between F_X' and f?

c) Let a and b be real numbers with $a < b$. Determine $P(a < X \le b)$, $P(a < X < b)$, $P(a \le X < b)$, and $P(a \le X \le b)$ in terms of f.

d) Why are all four answers in part (c) the same?

8.30 For each function f, let X be as in Exercise 8.29. Graph f and determine and graph F_X.

a) $f(x) = \lambda e^{-\lambda x}$ if $x > 0$ and $f(x) = 0$ otherwise, where λ is a positive real number.

b) $f(x) = 1/(b - a)$ if $a < x < b$ and $f(x) = 0$ otherwise, where a and b are real numbers with $a < b$.

c) $f(x) = b^{-1}(1 - |x|/b)$ if $-b < x < b$ and $f(x) = 0$ otherwise, where b is a positive real number.

8.31 Decide whether each function F is the CDF of a random variable by checking properties (a)–(d) of Proposition 8.1 on page 411. For each F that is the CDF of a random variable, classify the random variable as discrete, continuous, or mixed.

a) $F(x) = 0$ if $x < 0$ and $F(x) = 1$ if $x \ge 0$.

b) $F(x) = 0$ if $x < 0$, $F(x) = 1 - p$ if $0 \le x < 1$, and $F(x) = 1$ if $x \ge 1$. Here p is a real number with $0 < p < 1$.

c) $F(x) = 0$ if $x < 0$ and $F(x) = \lfloor x \rfloor$ if $x \ge 0$.

d) $F(x) = 0$ if $x < 0$ and $F(x) = \sum_{n=0}^{\lfloor x \rfloor} a_n$ for $x \ge 0$, where $\{a_n\}_{n=0}^{\infty}$ is a sequence of nonnegative real numbers whose sum is 1.

e) $F(x) = 0$ if $x < 0$ and $F(x) = 1 - e^{-\lambda x}\sum_{j=0}^{r-1}(\lambda x)^j/j!$ if $x \ge 0$. Here λ is a positive real number and r is a positive integer.

f) $F(x) = 0$ if $x < 0$ and $F(x) = x$ if $x \ge 0$.

g) $F(x) = 0$ if $x < -1$, $F(x) = \frac{1}{2} + \frac{3}{8}x$ if $-1 \le x < 1$, and $F(x) = 1$ if $x \ge 1$.

8.32 Let X be a random variable and let m be a real number. Determine the CDF of each of the following random variables in terms of the CDF of X.

a) $Y = \max\{X, m\}$ b) $Z = \min\{X, m\}$

8.33 Let $X_1, \dots, X_m$ be independent random variables, each having the same probability distribution as a random variable X. Determine the CDF of each of the following random variables in terms of the CDF of X.

a) $Y = \max\{X_1, \dots, X_m\}$ b) $Z = \min\{X_1, \dots, X_m\}$

8.34 A function F can be the CDF of a continuous random variable X and still have "flat spots." What is the probabilistic meaning of $F(x) = c$ for all $x \in [a, b]$, where c is a constant and a and b are real numbers with $a < b$?

Theory Exercises

8.35 Prove Equation (8.7) on page 412: If X is a random variable, then $F_X(x-) = P(X < x)$ for all $x \in \mathcal{R}$.

8.36 Prove parts (a), (b), and (d) of Proposition 8.2 on page 412.

Advanced Exercises

8.37 Let X be a random variable. Prove that the CDF of X has a countable number of discontinuities. *Hint:* For each $n \in \mathcal{N}$, consider the set $D_n = \{x \in \mathcal{R} : F_X(x) - F_X(x-) \ge 1/n\}$.

8.38 For each $n \in \mathcal{N}$, let the random variable X_n have the discrete uniform distribution on the set $\{0, 1/n, \dots, (n - 1)/n\}$.

a) Determine the CDF of X_n.

b) Use part (a) to find the limiting CDF—that is, the pointwise limit of the sequence of functions $F_{X_1}, F_{X_2}, \ldots$.

c) Based on the result of part (b), identify a continuous random variable X whose CDF is the limiting CDF of the discrete random variables $X_1, X_2, \ldots$.

8.39 For each $n \in \mathcal{N}$, let X_n denote a randomly selected number from the interval $(0, n)$.

a) Determine the CDF of X_n for each $n \in \mathcal{N}$.

b) Use part (a) to find the pointwise limit of the sequence of functions $F_{X_1}, F_{X_2}, \ldots$.

c) Is the function obtained in part (b) a CDF? Explain your answer.

d) Provide an intuitive explanation of the results of parts (b) and (c).

8.40 Let X be a random variable.

a) Prove that F_X can be written as a convex combination of a discrete CDF and a continuous CDF. That is, show that there exist discrete and continuous random variables X_d and X_c and a real number α, with $0 \le \alpha \le 1$, such that $F_X = \alpha F_{X_d} + (1 - \alpha) F_{X_c}$.

b) Apply the result of part (a) to the random variable in Example 8.6 on page 410.

c) Identify a discrete random variable and a continuous random variable whose CDFs are those in the convex combination of CDFs in part (b).

8.3 Probability Density Functions

For a discrete random variable X, the probabilities $P(X = x)$, $x \in \mathcal{R}$, which constitute the probability mass function p_X, completely determine the probability distribution of X in the sense that, once we know them, we can obtain any probability for X. Indeed, by the fundamental probability formula (FPF) for discrete random variables,

$$P(X \in A) = \sum_{x \in A} p_X(x), \qquad A \subset \mathcal{R}. \tag{8.9}$$

However, for a continuous random variable X, the probabilities $P(X = x)$, $x \in \mathcal{R}$, are all 0 and hence are of no use in describing the probability distribution of X. What we need for continuous random variables is a viable analogue of the probability mass function (PMF) for discrete random variables. At the end of Section 8.1, we suggested that we can get this analogue by considering, for each $x \in \mathcal{R}$, the probability that X is "near" x rather than the probability that X equals x.

Suppose then that X is a continuous random variable. To examine the probability that X is "near" x, we let Δx represent a small positive number and consider the probability that X takes a value between x and $x + \Delta x$, that is, $P(x \le X \le x + \Delta x)$. As X is a continuous random variable, we have from Propositions 8.2 and 8.3 that

$$P(x \le X \le x + \Delta x) = F_X(x + \Delta x) - F_X(x). \tag{8.10}$$

Assuming that F_X is differentiable at x, we have, in view of Equation (8.10), that

$$P(x \le X \le x + \Delta x) \approx F_X'(x) \Delta x. \tag{8.11}$$

The approximation is good because Δx is small. Relation (8.11) shows that F_X' gives the density (rate of change) of probability for X per unit.

We now provide a heuristic argument to show that F_X'—the derivative function of the CDF of X—gives the required analogue for continuous random variables of the PMF for discrete random variables. That is, once we know F_X', we can determine any probability for the random variable X. Results from measure theory imply that we need only verify this fact for probabilities involving bounded closed intervals.

So we want to show that, for all real numbers $a < b$, we can express the probability that X takes a value in the bounded closed interval $[a, b]$ in terms of F_X'. To do so, we consider the partition of $[a, b]$ given by $a = x_0 < x_1 < \cdots < x_n = b$, where n is a large positive integer and $x_j = a + j(b - a)/n$ for $0 \leq j \leq n$. Applying the additivity property of a probability measure, the fact that X is a continuous random variable, and Relation (8.11), we have

$$P(a \leq X \leq b) = \sum_{j=0}^{n-1} P(x_j \leq X \leq x_{j+1}) \approx \sum_{j=0}^{n-1} F_X'(x_j)\Delta x_j,$$

where $\Delta x_j = x_{j+1} - x_j = (b - a)/n$, for $0 \leq j \leq n - 1$. However, because n is large, the last term in the preceding display, which is a Riemann sum, approximately equals the Riemann integral of F_X' over the interval $[a, b]$. In other words,

$$P(a \leq X \leq b) \approx \int_a^b F_X'(x)\,dx. \tag{8.12}$$

Relation (8.12) suggests that, once we know F_X', we can obtain the probability that X takes a value in any specified bounded closed interval and, therefore, that F_X' completely determines the probability distribution of the random variable X.

The heuristic argument leading to Relation (8.12) presumes that F_X is everywhere differentiable. As that isn't always the case, we make the following definition.

DEFINITION 8.3   Probability Density Function

Let X be a continuous random variable. A nonnegative function f_X is said to be a **probability density function** of X if, for all real numbers $a < b$,

$$P(a \leq X \leq b) = \int_a^b f_X(x)\,dx. \tag{8.13}$$

We use **PDF** as an abbreviation for "probability density function."

Note the following.

- A PDF of a continuous random variable is denoted by a lowercase f subscripted with the letter representing the random variable. Therefore f_X denotes a PDF of the random variable X, f_Y denotes a PDF of the random variable Y, and so on.

- Another term used for *probability density function* is simply **density function**.

Recalling from calculus the area interpretation of integration, we deduce from Equation (8.13) that the probability a continuous random variable X (with a PDF) takes a value between a and b equals the area under the curve $y = f_X(x)$ between $x = a$ and $x = b$, as illustrated in Figure 8.7.

Figure 8.7 Probability for a continuous random variable as area under its PDF

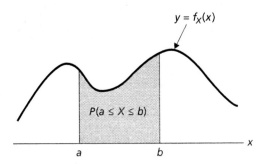

Several technical issues arise in the study of probability density functions. Here are two such issues.

- If we change the values of a PDF of a random variable at a finite number of points, the resulting function is still a PDF of the random variable (assuming, of course, that the changes are to nonnegative values). This result holds because the integral of a function isn't affected by changing the values of the function at a finite number of points. Thus, if g is defined on $\mathcal{R}$ and we write $f_X = g$, we mean that g is a nonnegative function that satisfies Equation (8.13) for all real numbers $a < b$. But, keep in mind that, if one such function exists, there are infinitely many such functions.

- Not all continuous random variables have PDFs; that is, for a continuous random variable X, there may not exist a function f_X satisfying Equation (8.13) for all real numbers $a < b$. However, all continuous random variables that we consider in this book and almost all continuous random variables in practice do have PDFs.[†]

Assuming that a continuous random variable does have a PDF, how do we find it? Our heuristic argument preceding Definition 8.3—specifically, Relation (8.12)—suggests that the likely candidate for a PDF is the derivative of the CDF. And, although, even in simple cases, the CDF may not be everywhere differentiable, a PDF is "essentially" the derivative of the CDF. In fact, in most applications, you can rely on Proposition 8.4.

[†] In more advanced courses in probability, a continuous random variable that has a PDF is called an *absolutely continuous random variable.*

◆◆◆ **Proposition 8.4   A PDF is the Derivative of the CDF**

Let X be a continuous random variable. If F_X' exists and is continuous except possibly at a finite number of points, then X has a PDF, which we can take to be

$$f_X(x) = \begin{cases} F_X'(x), & \text{if } F_X \text{ is differentiable at } x; \\ 0, & \text{otherwise.} \end{cases} \tag{8.14}$$

Proof   We give a proof in the case that F_X' exists and is continuous at all points. The proof in the case that F_X' is piecewise continuous follows easily and is left for you as Exercise 8.55. The general case requires concepts examined in a real analysis course and is therefore beyond the scope of this text.

From Proposition 8.1(a), F_X is nondecreasing; thus F_X' is a nonnegative function. Now let a and b be real numbers with $a < b$. According to the second fundamental theorem of calculus, if g is a continuous function on the interval $[a, b]$ and G is an antiderivative of g on $[a, b]$, then $\int_a^b g(x)\,dx = G(b) - G(a)$. Applying this result with $g = F_X'$ and $G = F_X$ and then using Proposition 8.2(b) on page 412 and Proposition 8.3 on page 413, we get

$$\int_a^b F_X'(x)\,dx = F_X(b) - F_X(a) = P(a \le X \le b).$$

We have now shown that F_X' is a nonnegative function that satisfies Equation (8.13) for all real numbers $a < b$. Hence F_X' is a PDF of the random variable X.   ◆

A Procedure for Finding a PDF

For a continuous random variable X, the conditions on F_X in Proposition 8.4 are sufficient for the existence of a PDF for X but aren't necessary. Nonetheless, when those conditions are satisfied—which usually is the case in practice—Proposition 8.4 provides a method for obtaining a PDF of X. We formalize this method in Procedure 8.1.

PROCEDURE 8.1   To Find a PDF for a Continuous Random Variable X

Assumptions

- F_X is everywhere continuous (i.e., X is a continuous random variable).
- F_X' exists and is continuous except possibly at a finite number of points.

Step 1 *Determine the CDF, F_X, of X.*

Step 2 *Obtain the derivative, F_X', of the CDF of X, where it exists.*

Step 3 *A PDF of X is given by $f_X(x) = F_X'(x)$ if F_X is differentiable at x, and $f_X(x) = 0$ otherwise.*

We apply Procedure 8.1 in Examples 8.7 and 8.8 to obtain PDFs of two continuous random variables considered previously in this chapter.

EXAMPLE 8.7 *Obtaining a PDF*

Random Number Generators Let X denote a number selected at random from the interval $(0, 1)$. Obtain a PDF of the random variable X.

Solution We apply Procedure 8.1.

Step 1 *Determine the CDF, F_X, of X.*

We did so in Example 8.4 and found that

$$F_X(x) = \begin{cases} 0, & \text{if } x < 0; \\ x, & \text{if } 0 \le x < 1; \\ 1, & \text{if } x \ge 1, \end{cases}$$

as shown in Figure 8.8(a). Observe that F_X is everywhere continuous.

Figure 8.8 (a) The CDF of X; (b) a PDF of X

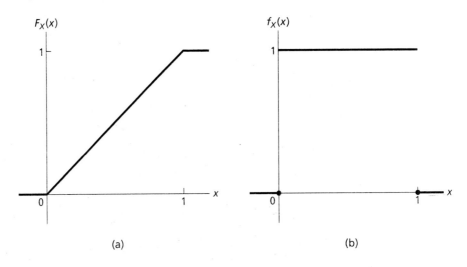

(a)   (b)

Step 2 *Obtain the derivative, F_X', of the CDF of X, where it exists.*

From Step 1, we find that

$$F_X'(x) = \begin{cases} 1, & \text{if } 0 < x < 1; \\ 0, & \text{if } x < 0 \text{ or } x > 1. \end{cases}$$

F_X is not differentiable at $x = 0$ or $x = 1$.

Step 3 *A PDF of X is given by $f_X(x) = F_X'(x)$ if F_X is differentiable at x, and $f_X(x) = 0$ otherwise.*

From Step 2, a PDF of X is

$$f_X(x) = \begin{cases} 1, & \text{if } 0 < x < 1; \\ 0, & \text{otherwise,} \end{cases}$$

as shown in Figure 8.8(b).

Note: That the PDF is constant on the range of X reflects the fact that we have a geometric probability model. Indeed, a number is being selected at random from the interval $(0, 1)$ and hence the number X obtained is no more or less likely to be near one number between 0 and 1 than another.   ▨

EXAMPLE 8.8   *Obtaining a PDF*

Bacteria on a Petri Dish   Refer to Example 8.2 on page 403, where Z denotes the distance of the center of the first spot (visible bacteria colony) from the center of a petri dish. Determine a PDF of the random variable Z.

Solution   We apply Procedure 8.1.

Step 1 *Determine the CDF, F_Z, of Z.*

We did so in Example 8.5 and found that

$$F_Z(z) = \begin{cases} 0, & \text{if } z < 0; \\ z^2, & \text{if } 0 \le z < 1; \\ 1, & \text{if } z \ge 1, \end{cases}$$

as shown in Figure 8.9(a). Observe that F_Z is everywhere continuous.

Figure 8.9 (a) The CDF of Z; (b) a PDF of Z

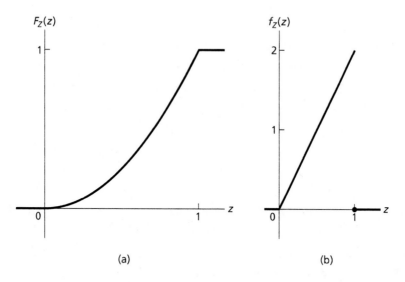

(a)   (b)

Step 2 *Obtain the derivative, F_Z', of the CDF of Z, where it exists.*

From Step 1, we find that

$$F_Z'(z) = \begin{cases} 2z, & \text{if } 0 \le z < 1; \\ 0, & \text{if } z < 0 \text{ or } z > 1. \end{cases}$$

F_Z is not differentiable at $z = 1$.

Step 3 *A PDF of Z is given by $f_Z(z) = F'_Z(z)$ if F_Z is differentiable at z, and $f_Z(z) = 0$ otherwise.*

From Step 2, a PDF of Z is

$$f_Z(z) = \begin{cases} 2z, & \text{if } 0 \le z < 1; \\ 0, & \text{otherwise,} \end{cases}$$

as shown in Figure 8.9(b).

Note: That the PDF is linear and increasing on the range of Z reflects the fact that we have a geometric probability model and that the rate of change of area with respect to radius is a linear function. In particular, the distance of the center of the first spot from the center of the petri dish is more likely to be large (close to 1) than small (close to 0). You can illustrate this fact visually by drawing two pairs of concentric circles of the same small width, one pair close to the origin and the other pair farther away. The pair farther away encloses more area—and hence more probability—than the pair closer in.   ■

Another Procedure for Finding a PDF

Proposition 8.4 on page 419 provides a basis for a procedure to find a PDF for a continuous random variable. An alternative and more general procedure is developed in Proposition 8.5.

◆◆◆ **Proposition 8.5   An Equivalent Condition for the Existence of a PDF**

A random variable X has a PDF if and only if there is a nonnegative function f defined on $\mathcal{R}$ such that

$$F_X(x) = \int_{-\infty}^{x} f(t)\, dt, \qquad x \in \mathcal{R}. \tag{8.15}$$

In this case, f is a PDF of X and

$$f(x) = F'_X(x) \tag{8.16}$$

at all points x where f is continuous.

Proof   Suppose that there is a nonnegative function f defined on $\mathcal{R}$ such that Equation (8.15) holds. Then F_X is everywhere continuous (i.e., X is a continuous random variable) and, for $a < b$,

$$P(a \le X \le b) = F_X(b) - F_X(a) = \int_{-\infty}^{b} f(t)\, dt - \int_{-\infty}^{a} f(t)\, dt = \int_{a}^{b} f(t)\, dt.$$

Hence, by Definition 8.3 on page 417, f is a PDF of X.

Conversely, suppose that X has a PDF f_X. Then, by Definition 8.3, f_X is a nonnegative function defined on $\mathcal{R}$ such that, for all $a < b$,

$$P(a \le X \le b) = \int_{a}^{b} f_X(t)\, dt. \tag{8.17}$$

Let $x \in \mathcal{R}$ and set $A_n = \{-n \le X \le x\}$ for each $n \in \mathcal{N}$. Then $A_1 \subset A_2 \subset \cdots$ and $\bigcup_{n=1}^{\infty} A_n = \{X \le x\}$. Applying the continuity property of a probability measure and Equation (8.17) yields

$$F_X(x) = P(X \le x) = P\left(\bigcup_{n=1}^{\infty} A_n\right) = \lim_{n \to \infty} P(A_n)$$

$$= \lim_{n \to \infty} P(-n \le X \le x) = \lim_{n \to \infty} \int_{-n}^{x} f_X(t)\, dt = \int_{-\infty}^{x} f_X(t)\, dt.$$

Thus f_X is a nonnegative function defined on $\mathcal{R}$ that satisfies Equation (8.15).

Equation (8.16) is a consequence of the first fundamental theorem of calculus. It shows again that a PDF of a continuous random variable—when it exists—is essentially the derivative of the CDF of the random variable.   ♦

Basic Properties of a PDF

Proposition 8.6 provides the basic properties of a PDF. Note the analogy with the basic properties of a PMF, as given in Proposition 5.1 on page 187.

◆◆◆ **Proposition 8.6   Basic Properties of a PDF**

A probability density function f_X of a continuous random variable X satisfies the following two properties.

a) $f_X(x) \ge 0$ for all $x \in \mathcal{R}$; that is, a PDF is a nonnegative function.

b) $\int_{-\infty}^{\infty} f_X(x)\, dx = 1$; that is, the integral of the values of a PDF equals 1.

Proof   The property in part (a) is part of the definition of a PDF, as given in Definition 8.3 on page 417. The property in part (b) is intuitively clear in that it reflects the fact that the probability that X takes on some real number equals 1. To prove it mathematically, we refer to the definition of a PDF and to Proposition 8.1 on page 411. Specifically, we have

$$\int_{-\infty}^{\infty} f_X(x)\, dx = \lim_{n \to \infty} \int_{-n}^{n} f_X(x)\, dx = \lim_{n \to \infty} P(-n \le X \le n)$$

$$= \lim_{n \to \infty} \left(F_X(n) - F_X(-n)\right) = 1 - 0 = 1,$$

as required.   ♦

A function $f \colon \mathcal{R} \to \mathcal{R}$ that satisfies properties (a) and (b) of Proposition 8.6 is a PDF of some continuous random variable. Therefore, for any such function, we can say, "Let X be a continuous random variable with PDF f." This statement makes sense regardless of whether we explicitly give X and the sample space on which it is defined.

We can use this fact to construct a PDF with a prescribed "variable form" (governing the shape of its graph). More precisely, suppose that $g \colon \mathcal{R} \to \mathcal{R}$ is a nonnegative function whose integral over $\mathcal{R}$ is finite and nonzero. We define the function f by

$$f(x) = cg(x), \quad \text{where} \quad c = \left(\int_{-\infty}^{\infty} g(x)\, dx\right)^{-1}.$$

Clearly, f satisfies properties (a) and (b) of Proposition 8.6 and hence is a PDF.

For example, suppose that we want a PDF whose variable form is x^3 for $0 < x < 1$, and 0 otherwise. This variable form is nonnegative, and its integral over $\mathcal{R}$ equals 1/4. Hence the corresponding PDF is $f(x) = 4x^3$ for $0 < x < 1$, and 0 otherwise.

Fundamental Probability Formula for a Continuous Random Variable

Once we have a PDF of a continuous random variable, we can obtain any probability involving that random variable. In other words, a PDF of a continuous random variable completely determines its probability distribution.

We refer to this result—which is a consequence of Definition 8.3 and a theorem in measure theory—as the **fundamental probability formula** (FPF) for continuous random variables and present it formally in Proposition 8.7. Note the analogy with the FPF for discrete random variables, Proposition 5.2 on page 190.

◆◆◆ **Proposition 8.7   Fundamental Probability Formula: Univariate Continuous Case**

Suppose that X is a continuous random variable having a PDF. Then, for any subset A of real numbers,

$$P(X \in A) = \int_A f_X(x)\,dx. \tag{8.18}$$

In words, the probability that a continuous random variable takes a value in a specified subset of real numbers can be obtained by integrating a PDF of the random variable over that subset of real numbers.[†]

We won't prove Proposition 8.7, as its rigorous verification requires the use of measure theory. However, the following is the basic idea of the proof.

- By definition of a PDF, Equation (8.18) holds if A is a bounded closed interval.

- Because X is a continuous random variable, it follows from the previous bulleted item that Equation (8.18) also holds if A is a bounded interval and, from this result it follows that Equation (8.18) holds if A is any interval.

- Each (measurable) subset A of $\mathcal{R}$ can be approximated as closely as we want by a union of intervals.

EXAMPLE 8.9   *Fundamental Probability Formula*

Bacteria on a Petri Dish   Refer to Example 8.2 on page 403. Obtain the probability that the center of the first spot (visible bacteria colony) will be
a) within 1/2 unit of the center of the petri dish.
b) either less than 1/4 unit or more than 3/4 unit from the center of the petri dish.

[†]Although, for technical reasons, Equation (8.18) actually isn't valid for all subsets of real numbers, it does hold for all subsets of real numbers that arise in practice.

Solution   Let Z denote the distance of the center of the first spot from the center of the petri dish. In Example 8.8, we obtained a PDF of Z:

$$f_Z(z) = \begin{cases} 2z, & \text{if } 0 \le z < 1; \\ 0, & \text{otherwise.} \end{cases}$$

Using this PDF and the FPF, we can obtain the required probabilities.

a) We want to determine $P(Z \le 1/2)$. Applying the FPF, we get

$$P(Z \le 1/2) = \int_{-\infty}^{1/2} f_Z(z)\, dz = \int_{-\infty}^{0} 0\, dz + \int_{0}^{1/2} 2z\, dz = 1/4.$$

The probability is 0.25 that the center of the first spot to appear will be within 1/2 unit of the center of the petri dish.

b) We need to find $P(Z < 1/4 \text{ or } Z > 3/4)$. Applying the FPF, we get

$$P(Z < 1/4 \text{ or } Z > 3/4) = \int_{-\infty}^{1/4} f_Z(z)\, dz + \int_{3/4}^{\infty} f_Z(z)\, dz$$

$$= \int_{0}^{1/4} 2z\, dz + \int_{3/4}^{1} 2z\, dz = 1/16 + 7/16 = 0.5.$$

The probability is 0.5 that the center of the first spot to appear will be either less than 1/4 unit or more than 3/4 unit from the center of the petri dish.   ∎

EXERCISES 8.3   **Basic Exercises**

8.41 A point is chosen at random from the interior of a sphere of radius r. Let Z denote the distance from the center of the sphere to the point chosen.

a) Determine a PDF of the random variable Z. *Note:* Exercise 8.15 on page 413 asks for the CDF of Z.

b) Interpret the PDF of Z obtained in part (a) with regard to which values of Z are more likely than others.

c) Use the PDF of Z obtained in part (a) to find the probability that the distance from the center of the sphere to the point chosen is at most $r/2$; at least $r/4$; between $r/4$ and $r/2$.

d) Use the CDF of Z obtained in Exercise 8.15 to determine the three probabilities required in part (c).

8.42 By referring to parts (c) and (d) of Exercise 8.41, answer the following question for a continuous random variable with a PDF: If you know (in closed form) the CDF of the random variable, is there any need to apply the FPF to the PDF to obtain the probability that the random variable takes a value in any specified interval? Explain your answer.

8.43 Let X denote the tangent of an angle chosen at random from the interval $(-\pi/2, \pi/2)$.

a) Determine a PDF of the random variable X. *Note:* Exercise 8.16 on page 413 asks for the CDF of X.

b) Interpret the PDF of X obtained in part (a) with regard to which values of X are more likely than others.

c) Apply the FPF to the PDF of X obtained in part (a) to find the probability that the tangent of the angle chosen is at most 1.

d) Find the probability required in part (c) by using the CDF of X obtained in Exercise 8.16.

8.44 A point is chosen at random from the interior of a triangle with base b and height h. Let Y denote the distance from the point chosen to the base of the triangle.
a) Determine a PDF of the random variable Y. *Note:* Exercise 8.17 on page 413 asks for the CDF of Y.
b) Interpret the PDF of Y obtained in part (a) with regard to which values of Y are more likely than others.

8.45 Refer to Example 8.2 on page 403. Let X and Y denote the x and y coordinates, respectively, of the center of the first spot (visible bacteria colony) to appear.
a) Obtain a PDF of X. *Hint:* First use calculus to express the CDF of X as an integral.
b) Interpret the PDF of X obtained in part (a) with regard to which values of X are more likely than others.
c) Without doing any calculations, obtain a PDF of Y. Explain your reasoning.

8.46 Let X be a continuous random variable with a PDF. Express a PDF of each of the following functions of X in terms of f_X.
a) $a + bX$, where a and $b \neq 0$ are real numbers   b) X^2

8.47 Let f be a nonnegative real-valued function such that $\int_{-\infty}^{\infty} f(x)\,dx = 1$. Suppose that X is a random variable that satisfies $P(X \in A) = \int_A f(x)\,dx$ for all subsets A of $\mathcal{R}$. Determine a PDF of X.

8.48 Construct a PDF from each of the following functions. Explain your reasoning.
a) $g(x) = x^3(1-x)^2$ for $0 < x < 1$ and $g(x) = 0$ otherwise
b) $g(x) = e^{-4x}$ for $0 < x < \infty$ and $g(x) = 0$ otherwise
c) $g(x) = x^2 e^{-6x}$ for $0 < x < \infty$ and $g(x) = 0$ otherwise
d) $g(x) = 1/(1 + x^2)$ for $-\infty < x < \infty$
e) $g(x) = \sin x$ for $0 < x < \pi$ and $g(x) = 0$ otherwise

8.49 Decide whether each function f is a PDF of a random variable by checking properties (a) and (b) of Proposition 8.6 on page 423. Obtain the CDF corresponding to each function f that is a PDF.
a) $f(x) = \lambda e^{-\lambda x}$ for $x > 0$ and $f(x) = 0$ otherwise, where λ is a positive real number
b) $f(x) = 1/(b-a)$ for $a < x < b$ and $f(x) = 0$ otherwise, where a and b are real numbers with $a < b$
c) $f(x) = x/4$ for $-1 < x < 3$ and $f(x) = 0$ otherwise
d) $f(x) = x^2$ for $-1 < x < 3$ and $f(x) = 0$ otherwise
e) $f(x) = b^{-1}(1 - |x|/b)$ for $-b < x < b$ and $f(x) = 0$ otherwise, where b is a positive real number

8.50 Provide an example of an unbounded PDF. What is the corresponding CDF? Is an unbounded CDF possible?

8.51 Give an example of a PDF that is positive for all $x \in \mathcal{R}$. Find the corresponding CDF.

8.52 Define $F: \mathcal{R} \to \mathcal{R}$ by

$$F(x) = \begin{cases} 0, & \text{if } x < 0; \\ x/2, & \text{if } 0 \le x < 1; \\ (x+2)/6, & \text{if } 1 \le x < 4; \\ 1, & \text{if } x \ge 4. \end{cases}$$

a) Is F the CDF of a continuous random variable? Explain your answer.
b) If your answer to part (a) is "yes," determine the corresponding PDF.

8.53 In Examples 8.4 and 8.7 on pages 408 and 420, respectively, we obtained the CDF and a PDF of a number X selected at random from the interval $(0, 1)$.
a) Determine the CDF and a PDF of a number Y selected at random from the interval $(2, 5)$.
b) Determine the CDF and a PDF of a number Z selected at random from the interval (a, b), where a and b are real numbers with $a < b$.

8.54 The loss, in hundreds of thousands of dollars, due to a fire in a commercial building has a PDF given by $f(x) = 0.005(20 - x)$ for $0 < x < 20$ and $f(x) = 0$ otherwise. Of fire losses that exceed $0.8 million, what percentage exceed $1.6 million?

Theory Exercises

8.55 Provide a proof of Proposition 8.4 on page 419 in case F_X' is piecewise continuous.

Advanced Exercises

8.56 Obtain a PDF with variable form $e^{-x^2/2}$ for $-\infty < x < \infty$. *Hint:* First evaluate the quantity $\left(\int_{-\infty}^{\infty} e^{-x^2/2} \, dx \right)^2$ by using polar coordinates.

8.57 Obtain a PDF with variable form $e^{-(x-\mu)^2/2\sigma^2}$, where μ and $\sigma > 0$ are real numbers.. *Hint:* Refer to Exercise 8.56.

8.58 Let X be a continuous random variable with a PDF and let g be a real-valued function defined, strictly monotone, and differentiable on the range of X. Obtain a PDF of the random variable $Y = g(X)$.

8.59 Show that a convex linear combination of PDFs is also a PDF; that is, show that, if $f_1, \ldots, f_n$ are PDFs and $\alpha_1, \ldots, \alpha_n$ are nonnegative real numbers whose sum is 1, then $f = \sum_{k=1}^{n} \alpha_k f_k$ is also a PDF.

8.60 Let $X_1, \ldots, X_m$ be independent random variables, each having the same probability distribution as a random variable X with a PDF. Determine the PDF of each of the following random variables in terms of the PDF of X.
a) $Y = \max\{X_1, \ldots, X_m\}$
b) $Z = \min\{X_1, \ldots, X_m\}$

8.61 Provide an example of a continuous random variable whose PDF has a countably infinite number of discontinuities.

8.62 Prove or give a counterexample: If f is a PDF, then $\lim_{x \to \infty} f(x) = 0$.

8.63 **Gini's index:** As a measure of how unevenly payroll is distributed among employees of a company, you can use *Gini's index*, defined by $G = 1 - 2 \int_0^1 F(x) \, dx$, where $F(x)$ is the proportion of total payroll earned by the lowest paid fraction x of employees.
a) Determine and interpret Gini's index in the extreme case where $F(x) = 0$ for $0 \le x < 1$ and $F(1) = 1$.
b) Determine and interpret Gini's index in the extreme case where $F(x) = x$ for $0 \le x \le 1$.

8.64 **Payroll density function:** Refer to Exercise 8.63. We say that a *payroll density function* exists if there is a nonnegative function f defined on $[0, 1]$ such that $F(x) = \int_0^x f(t) \, dt$ for $0 \le x \le 1$.
a) Is there a payroll density function in Exercise 8.63(a)? Justify your answer.
b) Is there a payroll density function in Exercise 8.63(b)? Justify your answer.
c) Determine Gini's index in the case of a payroll density function given by $f(x) = 3x^2$.

8.4   Uniform and Exponential Random Variables

In Sections 8.1–8.3, we examined continuous random variables and their probability density functions. In doing so, we considered examples of these concepts in a specific context (e.g., a randomly selected number from the unit interval, the distance of the center of the first visible bacteria colony from the center of a petri dish).

In the discussion of discrete random variables, we demonstrated that certain families of probability mass functions play a prominent role in probability theory and its applications. The same is true for probability density functions. In this section, we introduce two important families of probability density functions, the *uniform* and *exponential*.

Uniform Random Variables

We already encountered an example of a *uniform random variable* in the random-number experiment of Examples 8.1, 8.4, and 8.7. There we let X denote a number selected at random from the interval $(0, 1)$. As we showed in Example 8.7, a PDF of X is constant on the range of X—specifically, $f_X(x) = 1$ for $0 < x < 1$, and $f_X(x) = 0$ otherwise.

Random variables whose PDFs are constant on their ranges characterize uniform random variables, as further illustrated in Example 8.10.

EXAMPLE 8.10   *Introduces Uniform Random Variables*

Waiting for the Train   A commuter train arrives punctually at a station every half hour. Each morning, a commuter named John leaves his house and casually strolls to the train station. Let X denote the amount of time, in minutes, that John waits for the train from the time he reaches the train station. Obtain a PDF of the random variable X.

Solution   We can take the sample space for this random experiment to consist of all real numbers between 0 and 30: $\Omega = (0, 30) = \{ x : 0 < x < 30 \}$.[†] Because John arrives "randomly" at the train station, we can use a geometric probability model. Specifically, we can think of the amount of time that John waits for the train as a number selected at random from the interval $(0, 30)$. Consequently, for each event E,

$$P(E) = \frac{|E|}{|\Omega|} = \frac{|E|}{30}, \qquad (8.19)$$

where $|E|$ denotes the length (in the extended sense) of the set E.

The random variable X is defined on Ω by $X(x) = x$; that is, X is the identity function on Ω. We want to obtain a PDF of X. To do so, we apply Procedure 8.1 on page 419.

Step 1 *Determine the CDF, F_X, of X.*

The range of X is the interval $(0, 30)$. Hence, if $x < 0$, $F_X(x) = P(X \leq x) = 0$, and if $x \geq 30$, $F_X(x) = P(X \leq x) = 1$. Now let $0 \leq x < 30$. The event $\{X \leq x\}$ is that

---

[†]It may seem more natural to take the sample space to be $[0, 30)$. However, the sample space $(0, 30)$ will do and is consistent with the type of interval used for uniform random variables.

John waits for the train at most x minutes. From Equation (8.19),

$$F_X(x) = P(X \leq x) = \frac{|\{X \leq x\}|}{30} = \frac{|(0, x]|}{30} = \frac{x - 0}{30} = \frac{x}{30}.$$

We conclude that the CDF of X is

$$F_X(x) = \begin{cases} 0, & \text{if } x < 0; \\ x/30, & \text{if } 0 \leq x < 30; \\ 1, & \text{if } x \geq 30. \end{cases}$$

Step 2 *Obtain the derivative, F_X', of the CDF of X, where it exists.*

From Step 1, we find that

$$F_X'(x) = \begin{cases} 1/30, & \text{if } 0 < x < 30; \\ 0, & \text{if } x < 0 \text{ or } x > 30. \end{cases}$$

F_X is not differentiable at $x = 0$ or $x = 30$.

Step 3 *A PDF of X is given by $f_X(x) = F_X'(x)$ if F_X is differentiable at x, and $f_X(x) = 0$ otherwise.*

From Step 2, a PDF of X is

$$f_X(x) = \begin{cases} 1/30, & \text{if } 0 < x < 30; \\ 0, & \text{otherwise.} \end{cases}$$

That the PDF of X is constant on the range of X reflects the fact that we have a geometric probability model; in this case, a number is being selected at random from the interval $(0, 30)$. The constant PDF indicates that the amount of time that John waits for the train is equally likely to be anywhere between 0 and 30 minutes. ∎

With Example 8.10 in mind, we now define **uniform random variable,** following which we specify its probability density function.

DEFINITION 8.4   Uniform Random Variable

A continuous random variable X is called a **uniform random variable** if, for some finite interval (a, b) of real numbers, its value is equally likely to lie anywhere in that interval or, equivalently, its PDF is constant on that interval and 0 elsewhere. We say that X has the **uniform distribution on the interval (a, b)** or that X is **uniformly distributed on the interval (a, b)**. For convenience, we sometimes write $X \sim \mathcal{U}(a, b)$ to indicate that X has the uniform distribution on the interval (a, b).

Let $X \sim \mathcal{U}(a, b)$. To obtain a PDF of X, we first note that, by assumption, there is a constant c such that $f_X(x) = c$ for $a < x < b$ and $f_X(x) = 0$ otherwise. Because a PDF must integrate to 1, it follows easily that $c = 1/(b - a)$. Therefore a PDF of X is

$$f_X(x) = \frac{1}{b - a}, \qquad a < x < b, \tag{8.20}$$

and $f_X(x) = 0$ otherwise. See Figure 8.10 .

Figure 8.10 PDF of a uniform random variable on (a, b)

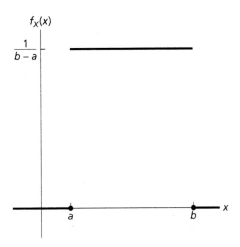

EXAMPLE 8.11   *Uniform Random Variables*

Waiting for the Train   Refer to Example 8.10, where X denotes the amount of time that a commuter, named John, waits for the train. Find the probability that John waits
a) between 10 and 15 minutes.
b) at least 10 minutes.

Solution   As we determined in Example 8.10, $X \sim \mathcal{U}(0, 30)$; that is, X is a uniform random variable on the interval $(0, 30)$. Its PDF is given by $f_X(x) = 1/30$ if $0 < x < 30$, and $f_X(x) = 0$ otherwise. We apply the FPF to obtain the required probabilities.
a) The probability that John waits between 10 and 15 minutes is

$$P(10 < X < 15) = \int_{10}^{15} f_X(x)\, dx = \int_{10}^{15} \frac{1}{30}\, dx = \frac{5}{30} = 0.167.$$

Chances are 16.7% that John will wait between 10 and 15 minutes for the train.
b) The probability that John waits at least 10 minutes is

$$P(X > 10) = \int_{10}^{\infty} f_X(x)\, dx = \int_{10}^{30} \frac{1}{30}\, dx + \int_{30}^{\infty} 0\, dx = \frac{20}{30} = 0.667.$$

Chances are 66.7% that John will wait at least 10 minutes for the train.   ■

Exponential Random Variables

Next we discuss the family of *exponential random variables.* As we now show, exponential random variables are the continuous analogue of geometric random variables, a family of discrete random variables that we presented in Section 5.6.

If we consider repeated Bernoulli trials with success probability p, the number of trials up to and including the first success is a geometric random variable with parameter p. Such a random variable has PMF and tail probabilities given by $p(1 - p)^{x-1}$ and $(1 - p)^x$, respectively, for $x = 1, 2, \dots$. We can think of a geometric random variable as giving the time of the first success in a sequence of Bernoulli trials, where time is quantized into integer values.

Let $n \in \mathcal{N}$ and suppose that Bernoulli trials with success probability p are performed at times $\frac{1}{n}, \frac{2}{n}, \dots$; that is, time is quantized into units of duration $\frac{1}{n}$. Let X denote the time of the first success. For a positive real number x, we set $k = \lfloor nx \rfloor$. Note that $\frac{k}{n} \le x < \frac{k+1}{n}$, as shown in Figure 8.11.

Figure 8.11 Graph showing the relationship between x and $k = \lfloor nx \rfloor$

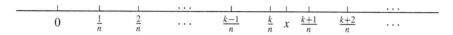

It follows that $\{X > x\} = \left\{ X > \frac{k}{n} \right\}$ and hence that

$$P(X > x) = P\left(X > \frac{k}{n} \right) = (1 - p)^k.$$

We now assume that n is large and that p is small. Referring to Relation (5.18) on page 219, noting that $x \approx \frac{k}{n}$, and using the preceding equation, we get

$$P(X > x) = (1 - p)^k \approx e^{-pk} = e^{-np\frac{k}{n}} \approx e^{-npx}. \tag{8.21}$$

Thus, as quantized time becomes finer and finer, approaching a continuum, the time of the first success has tail probabilities of the form e^{-npx}.

Therefore it's natural to consider continuous random variables whose tail probabilities take the form $e^{-\lambda x}$, where λ is a positive real number. We first show that the numbers $e^{-\lambda x}$, $x \ge 0$, are indeed the tail probabilities of some positive continuous random variable. Equivalently, we need to show that the function $F : \mathcal{R} \to \mathcal{R}$ defined by

$$F(x) = 1 - e^{-\lambda x}, \qquad x \ge 0, \tag{8.22}$$

and $F(x) = 0$ otherwise, is the CDF of a continuous random variable—that is, that F is everywhere continuous (see Proposition 8.3 on page 413) and satisfies properties (a)–(d) of Proposition 8.1 on page 411. We leave these verifications to you as Exercise 8.69.

We also observe that F' exists and is continuous except at $x = 0$. Noting that $F'(x) = \lambda e^{-\lambda x}$, for $x > 0$, we conclude from Proposition 8.4 on page 419 that a random variable with F as its CDF has a PDF given by $f(x) = \lambda e^{-\lambda x}$ if $x > 0$, and $f(x) = 0$ otherwise. A random variable with that PDF is called an **exponential random variable,** as indicated in Definition 8.5.

DEFINITION 8.5   Exponential Random Variable

A continuous random variable X is called an **exponential random variable** if it has a probability density function of the form

$$f_X(x) = \lambda\, e^{-\lambda x}, \qquad x > 0,$$

and $f_X(x) = 0$ otherwise, where λ is some positive real number. We say that X has an **exponential distribution with parameter λ** or that X is **exponentially distributed with parameter λ.** For convenience, we sometimes write $X \sim \mathcal{E}(\lambda)$ to indicate that X has an exponential distribution with parameter λ. See Figure 8.12.[†]

Figure 8.12 PDF of an exponential random variable with parameter λ

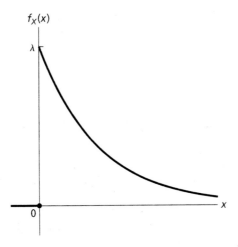

The exponential distribution is widely applied in many areas of study, including probability theory, statistics, stochastic processes, and operations research. It is often used as the distribution for the time required to complete a certain task or for the elapsed time between successive occurrences of a specified event.

[†]The PDF of an exponential random variable is frequently expressed in an alternative form—namely, as

$$f_X(x) = \frac{1}{\theta}\, e^{-x/\theta}, \qquad x > 0.$$

and $f_X(x) = 0$ otherwise, where θ is some positive real number. Using this form for the PDF of an exponential random variable has the advantages that (1) the expected value of an exponential random variable equals its parameter (instead of the reciprocal of its parameter) and (2) the parameter is a scale parameter. Using the form for the PDF of an exponential random variable given in Definition 8.5 has the advantage that the parameter can be considered a rate.

EXAMPLE 8.12 *Exponential Random Variables*

Emergency Room Traffic Desert Samaritan Hospital in Mesa, Arizona, keeps records of its emergency-room traffic. Those records indicate that, beginning at 6:00 P.M. on any given day, the elapsed time until the first patient arrives has an exponential distribution with parameter $\lambda = 6.9$, where time is measured in hours. Determine the probability that, beginning at 6:00 P.M. on any given day, the first patient arrives
a) between 6:15 P.M. and 6:30 P.M.
b) before 7:00 P.M.

Solution Let X denote the time, in hours, until the first patient arrives at the emergency room. By assumption, $X \sim \mathcal{E}(6.9)$; that is, X is an exponential random variable with parameter $\lambda = 6.9$. Its PDF therefore is $f_X(x) = 6.9\,e^{-6.9x}$ if $x > 0$, and $f_X(x) = 0$ otherwise. We apply the FPF to obtain the required probabilities.
a) The probability that the first patient arrives between 6:15 P.M. and 6:30 P.M. is

$$P(1/4 < X < 1/2) = \int_{1/4}^{1/2} f_X(x)\,dx = \int_{1/4}^{1/2} 6.9\,e^{-6.9x}\,dx = 0.146.$$

There is a 14.6% chance that the first patient will arrive between 6:15 P.M. and 6:30 P.M.
b) The probability that the first patient arrives before 7:00 P.M. is

$$P(X < 1) = \int_{-\infty}^{1} f_X(x)\,dx = \int_{-\infty}^{0} 0\,dx + \int_{0}^{1} 6.9\,e^{-6.9x}\,dx = 0.999.$$

The first patient will almost certainly arrive before 7:00 P.M. ■

Equation (8.22) on page 431 gives the CDF of an exponential random variable with parameter λ. From it, we easily get Proposition 8.8.

◆◆◆ **Proposition 8.8** **Tail Probabilities for an Exponential Random Variable**

Suppose that X has an exponential distribution with parameter λ. Then

$$P(X > x) = e^{-\lambda x}, \qquad x > 0. \tag{8.23}$$

Lack of Memory of Exponential Random Variables

In Section 5.6, we introduced the lack-of-memory property for positive-integer valued random variables and showed that the only such random variables with that property are geometric random variables. For positive continuous random variables, we express the **lack-of-memory property** in the form

$$P(X > s + t \mid X > s) = P(X > t), \qquad s, t \geq 0. \tag{8.24}$$

(Compare this equation for positive continuous random variables to the one for positive-integer valued random variables as given in Exercise 5.106 on page 237.)

On page 431, we showed that exponential random variables are the continuous analogue of geometric random variables. Therefore, because geometric random variables have the lack-of-memory property, it isn't surprising that exponential random variables also have the lack-of-memory property, as we now show.

Let $X \sim \mathcal{E}(\lambda)$. By the conditional probability rule and Proposition 8.8, we have, for $s, t \geq 0$,

$$P(X > s + t \mid X > s) = \frac{P(X > s + t, \, X > s)}{P(X > s)} = \frac{P(X > s + t)}{P(X > s)}$$

$$= \frac{e^{-\lambda(s+t)}}{e^{-\lambda s}} = e^{-\lambda t} = P(X > t).$$

Hence an exponential random variable has the lack-of-memory property.

The converse of this result is also true—that is, exponential random variables are the only positive continuous random variables with the lack-of-memory property. We leave the proof of this result to you as Exercise 8.85. In summary, we have Proposition 8.9.

◆◆◆ **Proposition 8.9   Lack-of-Memory Property**

A positive continuous random variable has the lack-of-memory property if and only if it has an exponential distribution.

An important interpretation of the lack-of-memory property arises when the elapsed time until a specified event occurs (e.g., the arrival of the first patient at an ER) has an exponential distribution. Then, given that the specified event hasn't occurred by a certain time, the remaining elapsed time until it occurs has that same exponential distribution.

EXAMPLE 8.13   *Exponential Random Variables*

Emergency Room Traffic   Refer to Example 8.12. Given that the first patient doesn't arrive by 6:15 P.M., determine the probability that he or she arrives by 6:45 P.M.

Solution   Let X denote the time, in hours, until the first patient arrives at the emergency room. By assumption, $X \sim \mathcal{E}(6.9)$. We want to find $P(X \leq 3/4 \mid X > 1/4)$. Applying the complementation rule, the lack-of-memory property of exponential random variables, and Proposition 8.8, we get

$$P(X \leq 3/4 \mid X > 1/4) = 1 - P(X > 3/4 \mid X > 1/4)$$

$$= 1 - P(X > 1/4 + 1/2 \mid X > 1/4)$$

$$= 1 - P(X > 1/2) = 1 - e^{-6.9/2} = 0.968,$$

a result that we could also obtain directly by using the conditional probability rule.   ∎

EXERCISES 8.4   **Basic Exercises**

8.65   Show that, if X is uniformly distributed on the interval (a, b), its CDF is $F_X(x) = 0$ if $x < a$, $F_X(x) = (x - a)/(b - a)$ if $a \leq x < b$, and $F_X(x) = 1$ if $x \geq b$. Graph this CDF.

8.66   Solve Example 8.11 on page 430 by using
a) the CDF of X. *Note:* Refer to Exercise 8.65.   **b)** Equation (8.19) on page 428.

8.67 Let $X \sim \mathcal{U}(0, 1)$ and let c and $d > 0$ be real numbers.
a) Without doing any calculations, make an educated guess about the distribution of the random variable $Y = c + dX$.
b) Use Procedure 8.1 on page 419 to determine and identify a PDF of $Y = c + dX$.
c) Suppose that a and b be real numbers with $a < b$. Use part (b) to find c and d such that $c + dX \sim \mathcal{U}(a, b)$.

8.68 Simulation: This exercise requires access to a computer or graphing calculator.
a) Use the result of Exercise 8.67(c) to simulate 10,000 values of a $\mathcal{U}(-2, 3)$ random variable by employing a basic random number generator—that is, a random number generator that simulates the selection of a number at random from the interval $(0, 1)$.
b) Roughly, what would you expect a histogram of the 10,000 values obtained in part (a) to look like?
c) Obtain a histogram of the 10,000 values obtained in part (a) and compare it to your expectation in part (b).

8.69 Define $F: \mathcal{R} \to \mathcal{R}$ by $F(x) = 1 - e^{-\lambda x}$ for $x \geq 0$ and $F(x) = 0$ otherwise. Show that F is the CDF of a continuous random variable—that is, that F is everywhere continuous and satisfies properties (a)–(d) of Proposition 8.1 on page 411.

8.70 Let X be exponentially distributed with parameter λ.
a) Use the PDF of X and the FPF to show that the CDF of X is given by

$$F_X(x) = \begin{cases} 0, & \text{if } x < 0; \\ 1 - e^{-\lambda x}, & \text{if } x \geq 0. \end{cases}$$

b) Graph the CDF of X.

8.71 Solve Example 8.12 on page 433 by using the CDF of X. *Note:* Refer to Exercise 8.70.

8.72 Let $X \sim \mathcal{E}(1)$ and let b be a positive real number.
a) Use Procedure 8.1 on page 419 to determine and identify a PDF of $Y = bX$.
b) Let λ be a positive real number. Use part (a) to find b such that $bX \sim \mathcal{E}(\lambda)$.

8.73 Let $X \sim \mathcal{U}(0, 1)$ and let λ be a positive real number.
a) Use Procedure 8.1 on page 419 to show that $-\ln X \sim \mathcal{E}(1)$.
b) Use part (a) and Exercise 8.72(b) to conclude that $-\lambda^{-1} \ln X \sim \mathcal{E}(\lambda)$.

8.74 Simulation: This exercise requires access to a computer or graphing calculator.
a) Use the result of Exercise 8.73(b) to simulate 10,000 values of an $\mathcal{E}(6.9)$ random variable by employing a basic random number generator, that is, a random number generator that simulates the selection of a number at random from the interval $(0, 1)$.
b) Roughly, what would you expect a histogram of the 10,000 values obtained in part (a) to look like?
c) Obtain a histogram of the 10,000 values obtained in part (a) and compare it to your expectation in part (b).

8.75 According to the text *Rhythms of Dialogue* by J. Jaffee and S. Feldstein (New York: Academic Press, 1970), the duration, in seconds, of a pause during a monologue has an exponential distribution with parameter 1.4. Determine the probability that a pause
a) lasts between 0.5 second and 1 second.
b) exceeds 1 second.
c) exceeds 3 seconds.
d) exceeds 3 seconds, given that it exceeds 2 seconds.

8.76 A trucker drives between fixed locations in Los Angeles and Phoenix. The duration, in hours, of a round trip has an exponential distribution with parameter 1/20. Determine the probability that a round trip
a) takes at most 15 hours.   b) takes between 15 and 25 hours.   c) exceeds 25 hours.
d) Without using the conditional probability rule, find the probability that a round trip takes at most 40 hours, given that it exceeds 15 hours.
e) Assuming that round-trip durations are independent from one trip to the next, find the probability that exactly two of five round trips take more than 25 hours.

8.77 Ten years ago at a certain insurance company, the size of claims under homeowner insurance policies had an exponential distribution. Furthermore, 25% of claims were less than $1000. Today, the size of claims still has an exponential distribution but, owing to inflation, every claim made today is twice the size of a similar claim made 10 years ago. Determine the probability that a claim made today is less than $1000.

8.78 Let X be a positive continuous random variable. Consider the following two mathematical relations:

$$P(X > s + t \mid X > s) = P(X > t), \qquad s, t \geq 0.$$

$$P(X > s + t \mid X > s) = P(X > s + t), \qquad s, t \geq 0.$$

a) Which of the two mathematical relations describes the lack-of-memory property?
b) Describe in words the other mathematical statement.
c) Is it possible for the second relation to hold?

8.79 Let $X \sim \mathcal{U}(0, 1)$ and let $0 < s < s + t < 1$.
a) If X had the lack-of-memory property, what would be $P(X > s + t \mid X > s)$?
b) Determine $P(X > s + t \mid X > s)$ and compare your answer to that in part (a).

8.80 Suppose that $X_1, \ldots, X_m$ are independent exponential random variables with parameters $\lambda_1, \ldots, \lambda_m$, respectively. Determine and identify the PDF of the random variable $X = \min\{X_1, \ldots, X_m\}$.

8.81 Suppose that the time T, in minutes, for a customer service representative to respond to 10 telephone inquiries is uniformly distributed on the interval $(8, 12)$. Let R denote the average rate, in customers per minute, at which the representative responds to inquiries. Determine the PDF of the random variable R.

8.82 Beginning at 6:00 P.M. on any given day, the number of patients, $N(t)$, that arrive at an emergency room within the first t hours has a Poisson distribution with parameter $6.9t$. Using this fact only, determine and identify the probability distribution of the elapsed time X until the first patient arrives.

8.83 Median of a random variable: A *median* of a random variable X is any number M such that $P(X \leq M) \geq 1/2$ and $P(X \geq M) \geq 1/2$. Every random variable has at least one median, but medians are not necessarily unique.
a) Show that M is a median of X if and only if $F_X(M-) \leq 1/2 \leq F_X(M)$.
b) Show that M is a median of X if and only if $P(X < M) \leq 1/2$ and $P(X > M) \leq 1/2$.
c) Determine the median(s) for an indicator random variable.
d) Suppose that X is a continuous random variable whose range is an interval and whose CDF is strictly increasing on its range. Show that X has a unique median given by the number M that satisfies $F_X(M) = 1/2$.
e) Suppose that $X \sim \mathcal{U}(a, b)$. Without doing any calculations, make an educated guess for the median of X.

f) Determine the median when $X \sim \mathcal{U}(a, b)$ and compare your result to the educated guess that you made in part (e).

g) Determine the median when $X \sim \mathcal{E}(\lambda)$.

8.84 An insurance policy reimburses dental expense, X, up to a maximum benefit of $250. The probability density function for X is $f_X(x) = ce^{-0.004x}$ for $x \geq 0$ and $f_X(x) = 0$ otherwise, where c is a constant. Calculate the median benefit for this policy. *Note:* Refer to Exercise 8.83.

Theory Exercises

8.85 Complete the proof of Proposition 8.9 on page 434 by showing that a positive continuous random variable has the lack-of-memory property only if it has an exponential distribution. Proceed by using the following steps. Suppose that X is a positive continuous random variable with the lack-of-memory property and let $G(x) = P(X > x)$.

a) Show that Equation (8.24) on page 433 is equivalent to $G(s + t) = G(s)G(t)$ for $s, t \geq 0$.

b) Use part (a) and mathematical induction to show that $G(m/n) = [G(1/n)]^m$ for all positive integers m and n.

c) Show that $G(1) < 1$. *Hint:* Set $n = 1$ in the result of part (b).

d) Show that $G(1) > 0$. *Hint:* Assume to the contrary that $G(1) = 0$ and set $m = n$ in the result of part (b).

e) Show that $G(r) = e^{-\lambda r}$ for all positive rational numbers r, where $\lambda = -\ln\big(G(1)\big)$.

f) Show that $G(x) = e^{-\lambda x}$ for all positive real numbers x.

g) Conclude that X is an exponential random variable with parameter λ.

Advanced Exercises

8.86 For a certain material, let $G(t)$ denote the probability that a thread of length t will not break under a specified load. Assuming that there is no breakage interaction between disjoint pieces of thread, identify the probability distribution of the length L at which the thread will break under the specified load.

8.87 A stick of length ℓ is cut at a random point, yielding two segments.

a) Find the probability that the shorter segment is less than half as long as the longer segment.

b) Determine the PDF of the ratio of the length of the longer segment to the length of the shorter segment.

8.5 Normal Random Variables

In this section, we introduce the most important distribution in probability theory and its applications—the *normal distribution*. The normal distribution was discovered in 1733 by Abraham De Moivre (1667–1754) in his investigation of approximating coin tossing probabilities. He named the PDF of his discovery the *exponential bell-shaped curve*.

In 1809, Carl Friedrich Gauss (1777–1855) firmly established the importance of the normal distribution by using it to predict the location of astronomical bodies. As a result, the normal distribution then became commonly known as the *Gaussian distribution*, a terminology that is still used.

Later, in the last half of the nineteenth century, researchers discovered that many variables have distributions that follow or are well-approximated by a Gaussian distribution. Roughly speaking, researchers found that it is quite usual, or "normal," for a variable to have a Gaussian distribution. Consequently, following the lead of noted British statistician Karl Pearson (1857–1936), the Gaussian distribution began to be referred to as the *normal distribution*.

Introducing Normal Random Variables

To introduce normal random variables, we begin by recalling binomial random variables, which we discussed in Section 5.3. In n Bernoulli trials with success probability p, the number of successes, X, is a binomial random variable with parameters n and p. Such a random variable has PMF given by

$$p_X(x) = \binom{n}{x} p^x (1-p)^{n-x}, \qquad x = 0, 1, \ldots, n, \tag{8.25}$$

and $p_X(x) = 0$ otherwise.

We now heuristically analyze the behavior of $p_X(k)$ when n is large. For convenience, we set $q = 1 - p$. First we examine values of k such that $k \approx np$ and hence such that $n - k \approx nq$. In our development, we use the following fact from calculus:

$$\ln(1+t) \approx t - \frac{t^2}{2}, \qquad \text{for } t \approx 0. \tag{8.26}$$

We also use **Stirling's formula,** which provides an approximation for large factorials:

$$m! \approx \sqrt{2\pi}\, m^{m+\frac{1}{2}} e^{-m}, \qquad \text{for large } m.^{\dagger} \tag{8.27}$$

Applying Stirling's formula, we find, after some algebraic manipulation, that

$$p_X(k) = \binom{n}{k} p^k q^{n-k} = \frac{n!}{k!\,(n-k)!} p^k q^{n-k}$$

$$\approx \frac{\sqrt{2\pi}\, n^{n+\frac{1}{2}} e^{-n}}{\sqrt{2\pi}\, k^{k+\frac{1}{2}} e^{-k} \sqrt{2\pi}\, (n-k)^{(n-k)+\frac{1}{2}} e^{-(n-k)}} p^k q^{n-k}$$

$$= \left(\frac{n}{2\pi k(n-k)} \right)^{1/2} \left(\frac{np}{k} \right)^k \left(\frac{nq}{n-k} \right)^{n-k}.$$

Because $k \approx np$ and $n - k \approx nq$, we conclude that

$$p_X(k) \approx \frac{1}{\sqrt{2\pi}\,\sqrt{npq}} \left(\frac{np}{k} \right)^k \left(\frac{nq}{n-k} \right)^{n-k}. \tag{8.28}$$

---

†More precisely, Stirling's formula states that $m! \sim \sqrt{2\pi}\, m^{m+\frac{1}{2}} e^{-m}$ as $m \to \infty$, where "$\sim$" means that the limit of the ratio of the two sides is 1.

Using Relation (8.26), we obtain

$$\ln\left(\frac{np}{k}\right) = \ln\left(1 + \frac{np}{k} - 1\right) \approx \left(\frac{np}{k} - 1\right) - \left(\frac{np}{k} - 1\right)^2 \Big/ 2$$

and

$$\ln\left(\frac{nq}{n-k}\right) = \ln\left(1 + \frac{nq}{n-k} - 1\right) \approx \left(\frac{nq}{n-k} - 1\right) - \left(\frac{nq}{n-k} - 1\right)^2 \Big/ 2.$$

From the preceding two relations and after some tedious, but simple, algebraic manipulation, we find that

$$\ln\left[\left(\frac{np}{k}\right)^k \left(\frac{nq}{n-k}\right)^{n-k}\right] = k\ln\left(\frac{np}{k}\right) + (n-k)\ln\left(\frac{nq}{n-k}\right)$$

$$\approx -\frac{(k-np)^2}{2} \cdot \frac{n}{k(n-k)} \approx -\frac{(k-np)^2}{2} \cdot \frac{n}{npnq}$$

$$= -\frac{(k-np)^2}{2npq},$$

or

$$\left(\frac{np}{k}\right)^k \left(\frac{nq}{n-k}\right)^{n-k} \approx e^{-(k-np)^2/2npq}.$$

From this last relation and Relation (8.28), we get

$$p_X(k) \approx \frac{1}{\sqrt{2\pi}\sqrt{npq}} e^{-(k-np)^2/2npq}, \tag{8.29}$$

for $k \approx np$. For values of k not near np, both sides of Relation (8.29) are close to 0 and hence to each other. Thus we conclude heuristically that Relation (8.29) holds for all k when n is large.

The function on the right of Relation (8.29) is of the form

$$f(x) = \frac{1}{\sqrt{2\pi}\sigma} e^{-(x-\mu)^2/2\sigma^2}, \tag{8.30}$$

where μ and $\sigma > 0$ are real constants. We claim that f is a PDF of a continuous random variable; that is, f is nonnegative and integrates to 1. That f is nonnegative is trivial. To show that f integrates to 1, we first make the substitution $y = (x - \mu)/\sigma$ to obtain

$$\int_{-\infty}^{\infty} f(x)\,dx = \int_{-\infty}^{\infty} \frac{1}{\sqrt{2\pi}\sigma} e^{-(x-\mu)^2/2\sigma^2}\,dx = \int_{-\infty}^{\infty} \frac{1}{\sqrt{2\pi}} e^{-y^2/2}\,dy.$$

It follows that

$$\left(\int_{-\infty}^{\infty} f(x)\,dx\right)^2 = \left(\int_{-\infty}^{\infty} \frac{1}{\sqrt{2\pi}} e^{-u^2/2}\,du\right)\left(\int_{-\infty}^{\infty} \frac{1}{\sqrt{2\pi}} e^{-v^2/2}\,dv\right)$$

$$= \frac{1}{2\pi} \int_{-\infty}^{\infty}\int_{-\infty}^{\infty} e^{-(u^2+v^2)/2}\,du\,dv.$$

Changing to polar coordinates and then making the substitution $w = r^2/2$, we get

$$\left(\int_{-\infty}^{\infty} f(x)\,dx \right)^2 = \frac{1}{2\pi} \int_0^{2\pi} \left(\int_0^{\infty} re^{-r^2/2}\,dr \right) d\theta$$

$$= \frac{1}{2\pi} \int_0^{2\pi} \left(\int_0^{\infty} e^{-w}\,dw \right) d\theta = \frac{1}{2\pi} \int_0^{2\pi} 1\,d\theta = 1.$$

Therefore $\int_{-\infty}^{\infty} f(x)\,dx = 1$.

We have shown that the function f in Equation (8.30) is a PDF of a continuous random variable. A random variable with that PDF is called a **normal random variable**, as indicated in Definition 8.6.

DEFINITION 8.6   Normal Random Variable

A continuous random variable X is called a **normal random variable** if it has a probability density function of the form

$$f_X(x) = \frac{1}{\sqrt{2\pi}\,\sigma} e^{-(x-\mu)^2/2\sigma^2}, \qquad -\infty < x < \infty,$$

where μ and $\sigma > 0$ are real constants. We say that X has a **normal distribution with parameters μ and σ^2** or that X is **normally distributed with parameters μ and σ^2.** We often write $X \sim \mathcal{N}(\mu, \sigma^2)$ to indicate that X has a normal distribution with parameters μ and σ^2. See Figure 8.13.

Figure 8.13 PDF of a normal random variable with parameters μ and σ^2

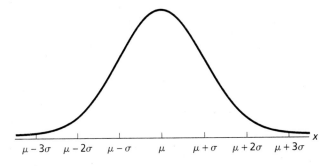

As indicated by Figure 8.13, a normal distribution is symmetric about and centered on its μ parameter; its spread depends on its σ parameter—the larger the σ parameter, the flatter and more spread out is the distribution. Figure 8.14 displays three normal distributions, that is, the PDFs of three normal random variables.

Figure 8.14 PDFs of three normal random variables

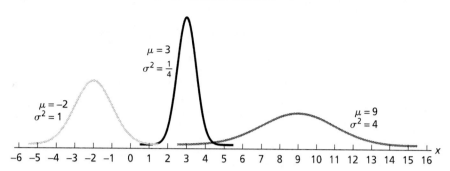

The parameter μ is an example of a **location parameter,** which provides the position of a PDF relative to the origin of the measurement scale; that is, it locates the PDF. The parameter σ is an example of a **scale parameter,** which relates to the physical units of the random variable; that is, it scales the PDF. Another important type of parameter is a **shape parameter,** which affects the shape of a PDF. We encounter location, scale, and shape parameters throughout the remainder of this chapter.

Obtaining Probabilities for a Normal Random Variable

Let $X \sim \mathcal{N}(\mu, \sigma^2)$. Referring to the FPF (Proposition 8.7 on page 424), we can, in theory, find any probability for the random variable X by integrating its PDF over the appropriate set:

$$P(X \in A) = \int_A \frac{1}{\sqrt{2\pi}\,\sigma}\, e^{-(x-\mu)^2/2\sigma^2}\, dx. \tag{8.31}$$

However, there is no simple formula for the antiderivative of a normal PDF or, equivalently, for its CDF. Thus we must resort to numerical methods in order to evaluate probabilities for a normally distributed random variable. We can simplify the process by standardizing, which converts any normal distribution into one particular normal distribution. Proposition 8.10 supplies the details.

◆◆◆ **Proposition 8.10** **Standardizing a Normal Random Variable**

Let X be normally distributed with parameters μ and σ^2. Then the random variable

$$Z = \frac{X - \mu}{\sigma}$$

has the normal distribution with parameters 0 and 1.

Proof For $z \in \mathcal{R}$, we have

$$F_Z(z) = P(Z \le z) = P\left(\frac{X - \mu}{\sigma} \le z\right) = P(X \le \mu + \sigma z)$$

$$= \int_{-\infty}^{\mu+\sigma z} \frac{1}{\sqrt{2\pi}\,\sigma}\, e^{-(x-\mu)^2/2\sigma^2}\, dx.$$

Making the change of variable $t = (x - \mu)/\sigma$, we find that

$$F_Z(z) = \int_{-\infty}^{z} \frac{1}{\sqrt{2\pi}} e^{-t^2/2} \, dt.$$

Thus, from Proposition 8.5 on page 422, a PDF of Z is

$$f_Z(z) = \frac{1}{\sqrt{2\pi}} e^{-z^2/2}, \qquad -\infty < z < \infty,$$

which is a PDF of a random variable with a normal distribution with parameters 0 and 1. Hence Z has the normal distribution with parameters 0 and 1.   ◆

Figure 8.15 graphically displays the standardizing of the three normal random variables whose PDFs are shown in Figure 8.14.

Figure 8.15 Standardizing the three normal random variables whose PDFs are shown in Figure 8.14

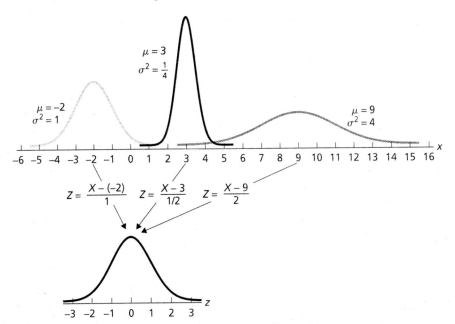

Because of the importance of the normal distribution with parameters 0 and 1, it is given a special name: the **standard normal distribution**. This and some further terminology and notation are presented in Definition 8.7.

DEFINITION 8.7   Standard Normal Distribution

A normal random variable with parameters 0 and 1 is called a **standard normal random variable** and is said to have the **standard normal distribution.** For a standard normal random variable, its PDF is denoted ϕ and its CDF is denoted Φ. Thus, for $-\infty < z < \infty$,

$$\phi(z) = \frac{1}{\sqrt{2\pi}} e^{-z^2/2} \tag{8.32}$$

and

$$\Phi(z) = \int_{-\infty}^{z} \frac{1}{\sqrt{2\pi}} e^{-t^2/2} \, dt = \int_{-\infty}^{z} \phi(t) \, dt. \tag{8.33}$$

By using Proposition 8.10, we can express normal probabilities in terms of Φ, the standard normal CDF. Specifically, if $X \sim \mathcal{N}(\mu, \sigma^2)$, we have

$$P(a < X < b) = P\left(\frac{a - \mu}{\sigma} < \frac{X - \mu}{\sigma} < \frac{b - \mu}{\sigma} \right)$$

$$= P\left(\frac{a - \mu}{\sigma} < Z < \frac{b - \mu}{\sigma} \right) = \Phi\left(\frac{b - \mu}{\sigma} \right) - \Phi\left(\frac{a - \mu}{\sigma} \right),$$

where the last equality follows from Proposition 8.2 on page 412 and the fact that Z is a continuous random variable. In summary, we have Proposition 8.11.

◆◆◆ **Proposition 8.11   Obtaining Probabilities for a Normal Random Variable**

If X is normally distributed with parameters μ and σ^2, then

$$P(a < X < b) = \Phi\left(\frac{b - \mu}{\sigma} \right) - \Phi\left(\frac{a - \mu}{\sigma} \right), \tag{8.34}$$

for $-\infty \leq a \leq b \leq \infty$, where Φ is the standard normal CDF.

From Proposition 8.11, if $X \sim \mathcal{N}(\mu, \sigma^2)$, the CDF of X is related to the standard normal CDF by

$$F_X(x) = \Phi\left(\frac{x - \mu}{\sigma} \right). \tag{8.35}$$

Differentiating Equation (8.35), we find that a PDF of X is related to the standard normal PDF by

$$f_X(x) = \frac{1}{\sigma} \phi\left(\frac{x - \mu}{\sigma} \right), \tag{8.36}$$

a fact that can be easily verified directly. Equation (8.36) shows how any normal PDF is obtained from the standard normal PDF by a location and scale change.

Approximate values of Φ (obtained by using numerical methods) are provided in Table I in the Appendix. From Table I, Proposition 8.11, and the relation

$$\Phi(-z) = 1 - \Phi(z), \qquad -\infty < z < \infty, \tag{8.37}$$

(see Exercise 8.113), we can determine probabilities for any normal random variable.

EXAMPLE 8.14   *Obtaining Probabilities for a Normal Random Variable*

Gestation Periods   The gestation periods of women are normally distributed with $\mu = 266$ days and $\sigma = 16$ days. Determine the probability that a gestation period is
a) less than 234 days.
b) between 265 and 295 days.

Solution   Let X denote gestation period for women. By assumption, $X \sim \mathcal{N}(266, 16^2)$.
a) We want $P(X < 234)$. From Proposition 8.11, Equation (8.37), and Table I,

$$P(X < 234) = P(-\infty < X < 234) = \Phi\left(\frac{234 - 266}{16}\right) - \Phi\left(\frac{-\infty - 266}{16}\right)$$

$$= \Phi(-2) - \Phi(-\infty) = 1 - \Phi(2) - 0 = 1 - 0.9772 = 0.0228.$$

Only 2.3% of human gestation periods are less than 234 days.
b) We want $P(265 < X < 295)$. From Proposition 8.11 and Equation (8.37),

$$P(265 < X < 295) = \Phi\left(\frac{295 - 266}{16}\right) - \Phi\left(\frac{265 - 266}{16}\right)$$

$$= \Phi(1.8125) - \Phi(-0.0625) = \Phi(1.8125) - 1 + \Phi(0.0625).$$

Table I provides only two decimal places for arguments of Φ. So, using that table, we approximate the required probability as

$$\Phi(1.81) - 1 + \Phi(0.06) = 0.9649 - 1 + 0.5239 = 0.4888.$$

Applying statistical software, we get the more accurate probability of 0.489963. Roughly half of all human gestation periods are between 265 and 295 days.   ∎

A Normal Approximation to the Binomial Distribution

In Section 5.5, we showed that a binomial distribution can be approximated by an appropriate Poisson distribution. Specifically, let X be a binomial random variable with parameters n and p. Then, according to Proposition 5.7 on page 220,

$$p_X(x) \approx e^{-np} \frac{(np)^x}{x!}, \qquad x = 0, 1, \ldots, n. \tag{8.38}$$

Thus the binomial distribution with parameters n and p can be approximated by the Poisson distribution with parameter np. The approximation is appropriate if n is large and p is small (i.e., close to 0). By interchanging the roles of success and failure, we can also apply the Poisson approximation to the binomial distribution when n is large and p is large (i.e., close to 1).

The question now is: How can we approximate a binomial distribution when n is large and p is moderate? Actually, we provided one method for doing that, when, at the beginning of this section, we gave a heuristic argument to introduce the normal distribution. Specifically, Relation (8.29) on page 439 shows that, for large n, the PMF of a binomial distribution with parameters n and p can be approximated by a PDF of a normal distribution with parameters np and $np(1 - p)$. Thus we have Proposition 8.12, due to Abraham De Moivre and Pierre Laplace.

◆◆◆ **Proposition 8.12** **Normal Approximation to the Binomial: Local Form**

Suppose that X has the binomial distribution with parameters n and p. Then

$$p_X(x) \approx \frac{1}{\sqrt{2\pi np(1 - p)}} e^{-(x-np)^2/2np(1-p)}, \qquad x = 0, 1, \ldots, n. \tag{8.39}$$

The approximation works well if n is large and p is moderate.

Proposition 8.12 is often referred to as the **local De Moivre–Laplace theorem.** It is called "local" because its statement involves "individual probabilities"—that is, probabilities of the form $P(X = x)$. In Chapter 11, we present another form of the normal approximation to the binomial distribution, known as the *integral De Moivre–Laplace theorem.* The local De Moivre–Laplace theorem is useful when we want to estimate only a few individual binomial probabilities, whereas the integral De Moivre–Laplace theorem is useful in any case.

EXAMPLE 8.15 *Normal Approximation to the Binomial*

Teen Pregnancy According to a recent issue of the periodical *Zero Population Growth,* the United States leads the industrialized world in teen pregnancy rates, with 40% of U.S. females getting pregnant at least once before reaching the age of 20. Use the normal approximation to find the probability that, from among 50 U.S. 20-year-old females, the number who have been pregnant at least once before the age of 20 will be
a) exactly 21.
b) between 19 and 21, inclusive.

Solution Let X denote the number of the 50 females who have been pregnant at least once before the age of 20. Then the random variable X has the binomial distribution with parameters $n = 50$ and $p = 0.4$. Thus

$$p_X(x) = \binom{50}{x}(0.4)^x(0.6)^{50-x}, \qquad x = 0, 1, \ldots, 50. \tag{8.40}$$

Note, however, that $n = 50$ is (relatively) large and that $p = 0.4$ is moderate. Thus, instead of applying Equation (8.40), we can use the normal approximation to obtain the required probabilities more easily. Specifically, noting that $np = 20$ and $np(1 - p) = 12$, we can, in view of Proposition 8.12, use the approximation

$$p_X(x) \approx \frac{1}{\sqrt{24\pi}} e^{-(x-20)^2/24}, \qquad x = 0, 1, \ldots, 50. \tag{8.41}$$

a) We want $P(X = 21)$. Applying Relation (8.41) gives

$$P(X = 21) = p_X(21) \approx \frac{1}{\sqrt{24\pi}} e^{-(21-20)^2/24} = 0.110.$$

There is about an 11.0% chance that exactly 21 of the 50 women have been pregnant at least once before the age of 20.

b) We want $P(19 \leq X \leq 21)$. Using the FPF and Relation (8.41) gives

$$P(19 \leq X \leq 21) = \sum_{19 \leq x \leq 21} p_X(x)$$

$$= p_X(19) + p_X(20) + p_X(21)$$

$$\approx \frac{1}{\sqrt{24\pi}} \left(e^{-(19-20)^2/24} + e^{-(20-20)^2/24} + e^{-(21-20)^2/24} \right)$$

$$= 0.336.$$

Chances are roughly 33.6% that between 19 and 21, inclusive, of the 50 women have been pregnant at least once before the age of 20. ■

Referring to Example 8.15, we now illustrate the accuracy of the normal approximation to the binomial distribution. We used a computer to obtain both the binomial distribution with parameters $n = 50$ and $p = 0.4$ and the values of the normal approximation $(1/\sqrt{24\pi})e^{-(x-20)^2/24}$, as displayed in Table 8.2. Note that we didn't list the probabilities that are 0 to four decimal places. Table 8.2 shows that, even for this moderate-size n, the normal approximates the binomial quite well.

Table 8.2 Binomial probabilities and normal approximation: $n = 50$ and $p = 0.4$

Successes x	Binomial probability	Normal approximation	Successes x	Binomial probability	Normal approximation
8	0.0002	0.0003	21	0.1091	0.1105
9	0.0005	0.0007	22	0.0959	0.0975
10	0.0014	0.0018	23	0.0778	0.0792
11	0.0035	0.0039	24	0.0584	0.0591
12	0.0076	0.0080	25	0.0405	0.0406
13	0.0147	0.0149	26	0.0259	0.0257
14	0.0260	0.0257	27	0.0154	0.0149
15	0.0415	0.0406	28	0.0084	0.0080
16	0.0606	0.0591	29	0.0043	0.0039
17	0.0808	0.0792	30	0.0020	0.0018
18	0.0987	0.0975	31	0.0009	0.0007
19	0.1109	0.1105	32	0.0003	0.0003
20	0.1146	0.1152	33	0.0001	0.0001

Figure 8.16 graphically portrays the information displayed in Table 8.2. The bars give a partial probability histogram of the binomial distribution with parameters $n = 50$ and $p = 0.4$, the curve is the PDF of the approximating normal distribution, and the dots show the values of the PDF for the integral values of x between 8 and 33, inclusive.

Figure 8.16 Partial probability histogram and approximating normal distribution for the binomial distribution with parameters $n = 50$ and $p = 0.4$

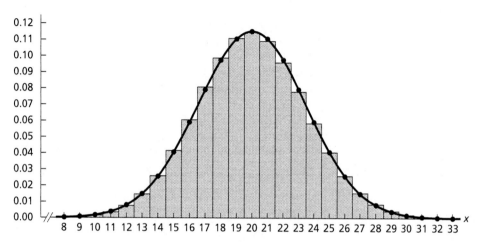

For large n and moderate p, you can't always use a computer to determine a required binomial distribution—sometimes n is so large that even a computer can't handle the computations needed to obtain a binomial distribution. Nonetheless, the normal approximation is still easy to apply.

EXERCISES 8.5   Basic Exercises

8.88 Provide all the algebraic details leading up to Relation (8.29) on page 439.

8.89 Let $X \sim \mathcal{N}(\mu, \sigma^2)$. Obtain the graph of the PDF of X, as shown in Figure 8.13 on page 440, by plotting several points and by verifying that
a) f_X is a nonnegative function.   **b)** $f_X(x) \downarrow 0$ as $x \to \pm\infty$.
c) f_X attains its maximum at $x = \mu$.   **d)** f_X has inflection points at $x = \mu \pm \sigma$.

8.90 Obtain the graphs of the normal PDFs with the following parameters. Interpret the center and spread of your graphs in terms of the parameters.
a) $\mu = 0$ and $\sigma^2 = 1$ **b)** $\mu = -3$ and $\sigma^2 = 1$ **c)** $\mu = -3$ and $\sigma^2 = 4$

8.91 Graph the standard normal PDF and CDF.

8.92 Which normal distribution has a wider spread, a $\mathcal{N}(1, 4)$ distribution or a $\mathcal{N}(4, 1)$ distribution? Explain your answer.

8.93 Consider two normal distributions, one with parameters -4 and 9 and the other with parameters 6 and 9. Answer true or false to each statement and explain your answers.
a) The two normal distributions have the same shape.
b) The two normal distributions are centered at the same place.

8.94 Let $X \sim \mathcal{N}(\mu, \sigma^2)$. Show that, for all $t > 0$,
a) $P(|X - \mu| \le t) = 2\Phi(t/\sigma) - 1.$ **b)** $P(|X - \mu| \ge t) = 2(1 - \Phi(t/\sigma)).$

8.95 Directly verify Equation (8.36) on page 443. That is, show that

$$\frac{1}{\sigma}\phi\left(\frac{x - \mu}{\sigma}\right) = \frac{1}{\sqrt{2\pi}\,\sigma}\,e^{-(x-\mu)^2/2\sigma^2}$$

directly from the definition of ϕ.

8.96 Let $X \sim \mathcal{N}(\mu, \sigma^2)$ and let $z > 0$.
a) Without evaluation, explain why a probability of the form $P(\mu - z\sigma \le X \le \mu + z\sigma)$ doesn't depend on either μ or σ.
b) Determine $P(\mu - z\sigma \le X \le \mu + z\sigma)$ for $z = 1, 2,$ and 3.

8.97 Two normal random variables, X and Y, have the same μ and σ^2 parameters. What can you say about the probability distributions of X and Y? Explain your answer.

8.98 Students in an introductory statistics course at the U.S. Air Force Academy participated in Nabisco's "Chips Ahoy! 1,000 Chips Challenge" by confirming that there were at least 1000 chips in every 18-ounce bag of cookies that they examined. As part of their assignment, they concluded that the number of chips per bag is approximately normally distributed. [Source: Brad Warner and Jim Rutledge, "Checking the Chips Ahoy! Guarantee," *Chance*, 1999, Vol. 12(1), pp. 10–14] Give two reasons why the number of chips in a bag couldn't be exactly normally distributed.

8.99 As reported in *Runner's World* magazine, the times of the finishers in the New York City 10 km run are normally distributed with $\mu = 61$ minutes and $\sigma = 9$ minutes. Let X be the time, in minutes, of a randomly selected finisher. Find
a) $P(X > 75)$.   **b)** $P(X < 50 \text{ or } X > 70)$.

8.100 In 1905, R. Pearl published the article "Biometrical Studies on Man. I. Variation and Correlation in Brain Weight" (*Biometrika*, Vol. 4, pp. 13–104). According to the study, brain weights of Swedish men are normally distributed with $\mu = 1.40$ kg and $\sigma = 0.11$ kg. Obtain the percentage of Swedish men who have brain weights
a) between 1.50 kg and 1.70 kg.   **b)** less than 1.6 kg.

8.101 Refer to Example 8.14 on page 444.
a) What percentage of pregnant women give birth before 300 days?
b) Among those women with a longer than average gestation, what percentage give birth within 300 days? *Note:* As we show in Chapter 10, the parameter μ of a normal random variable is its average (expected) value.
c) In a court case, the prosecuting attorney claims that the defendant is the father of a child who was born on July 6, 2002. The defendant can prove that he was out of town from September 1, 2001, to April 3, 2002. Can the defendant use this information to refute the prosecuting attorney's claim? Explain your answer.

8.102 When you put your money into a soft drink machine at the Student Union, a paper cup comes down, and some cola is put into the cup. You are supposed to get 8 oz of cola. However, the actual amount of cola dispensed is random, having a normal distribution with μ equal to the machine setting and $\sigma = 0.25$ oz. What should the machine setting be so that, in the long run, only 2% of the drinks will contain less than 8 oz?

8.103 A hardware manufacturer produces 10 mm bolts. The manufacturer knows that the diameters of the bolts produced vary somewhat from 10 mm and also from each other. But even if he is willing to accept some variation in bolt diameters, he can't tolerate too much variation—if the variation is too large, too many of the bolts produced will be unusable. The manufacturer has set the tolerance specifications for the 10 mm bolts at ±0.3 mm; that is, a bolt's diameter is considered satisfactory if it is between 9.7 mm and 10.3 mm. Furthermore, the manufacturer has decided that only 1 in 1000 bolts produced should be defective. Assuming that the diameters of bolts produced are normally distributed with $\mu = 10$ mm, what must σ be to insure that the manufacturer's production criteria are met?

8.104 Blood alcohol concentration (BAC) is the amount of alcohol in the bloodstream, measured in percentages. In many states, a driver is considered legally intoxicated if his or her BAC is 0.10% (i.e., 1 part alcohol per 1000 parts blood in the body) or higher. When a suspected DUI driver is stopped, police request that the person take a breathalyzer test to determine his or her BAC. Such tests are imperfect and exhibit a certain amount of measurement error. Suppose that the measured BAC is a normal random variable with μ equal to the person's actual BAC and $\sigma = 0.005\%$.

a) What is the probability that a driver with a BAC of 0.11% will pass the breathalyzer test?
b) What is the probability that a driver with a BAC of 0.095% will incorrectly be determined to be DUI?

8.105 Let $X \sim \mathcal{N}(0, 1/\alpha^2)$, where α is a positive real number. Determine the PDF of the random variable $1/X^2$. *Note:* The distribution of $1/X^2$ is called a *one-sided stable distribution of index 1/2.*

8.106 The diameters of ball bearings made by the Acme Ball Bearing Company are normally distributed with $\mu = 1.4$ cm and $\sigma = 0.025$ cm. The bearings are fully inspected and those that have diameters either less than 1.35 cm or greater than 1.48 cm are discarded. Determine the PDF of the diameters of the remaining ball bearings.

8.107 At a bottling plant, two machines are used for filling 16 oz bottles of soda. Machine I has an average fill (μ) of 16.21 oz with $\sigma = 0.14$ oz; Machine II has an average fill (μ) of 16.12 oz with $\sigma = 0.07$ oz. Both fill amounts are normally distributed. Machine I fills twice as many bottles per day as Machine II. What percentage of bottles that contain less than 15.96 oz of soda are filled by Machine I?

8.108 As reported by a spokesperson for Southwest Airlines, the no-show rate for reservations is 16%—that is, the probability is 0.16 that a person making a reservation will not take the flight. For a certain flight, 42 people have reservations. For each part, determine and compare the exact probability by using the appropriate binomial PMF and an approximate probability by using the normal approximation to the binomial as given in Proposition 8.12 on page 445. The probability that the number of people who don't take the flight is
a) exactly 5. **b)** between 9 and 12, inclusive. **c)** at least 1. **d)** at most 2.

8.109 According to *USA TODAY*, Anchorage, Alaska, is the city with the highest rate of cell phone ownership, with 56% of the residents owning cell phones. Of 500 randomly selected (without replacement) Anchorage residents, let X denote the number who own a cell phone.
a) Identify the exact probability distribution of the random variable X.
b) Identify the binomial distribution that should be used to approximate the probability distribution of X.
c) Identify the normal distribution that should be used to approximate the probability distribution of X.

Use the local De Moivre–Laplace theorem to approximate the probability that, of 500 randomly selected Anchorage residents, the number who own a cell phone is
d) exactly 280. **e)** between 278 and 280, inclusive.

8.110 The publication *Monitoring the Future* reports that 52.7% of all 12th graders in the United States have consumed alcohol in the past month. If 250 U.S. 12th graders are selected at random, find the probability that the number who have consumed alcohol in the past month is exactly one-half of those sampled. Use the normal approximation.

8.111 The second leading genetic cause of mental retardation is Fragile X Syndrome, named for the fragile appearance of the tip of the X chromosome in affected individuals. Worldwide, 1 in every 1500 males is affected, with no ethnic bias. For a sample of 10,000 males, use the normal approximation to the binomial distribution (Proposition 8.12 on page 445) to determine the probability that the number who have Fragile X Syndrome
a) exceeds 7.   b) is at most 10.
c) The probabilities in parts (a) and (b) were obtained in Exercise 5.80 on page 226 by using the Poisson approximation to the binomial distribution. Which estimates of the true binomial probabilities would you expect to be better, the ones using the normal approximation or those using the Poisson approximation? Explain your answer.

8.112 In Exercise 8.83 on page 436, we defined the *median* of a random variable. Show that, if $X \sim \mathcal{N}(\mu, \sigma^2)$, then the (unique) median of X equals μ.

Theory Exercises

8.113 Prove Equation (8.37) on page 444, that is, show that $\Phi(-z) = 1 - \Phi(z)$ for all $z \in \mathcal{R}$.

Advanced Exercises

8.114 This exercise establishes bounds and an asymptotic estimate for the tail probabilities of a random variable with the standard normal distribution.
a) Prove that

$$\left(\frac{1}{x} - \frac{1}{x^3}\right)\phi(x) \le 1 - \Phi(x) \le \frac{1}{x}\phi(x), \qquad x > 0.$$

Hint: First show that $\phi'(x) = -x\phi(x)$.
b) Deduce from part (a) that $1 - \Phi(x) \sim x^{-1}\phi(x)$ as $x \to \infty$, where "$\sim$" means that the limit of the ratio of the two sides is 1.

8.6 Other Important Continuous Random Variables

In this section, we discuss some other important families of continuous random variables. First, we consider *gamma random variables*. The gamma distribution is a generalization of the exponential distribution and has a wide range of applications in probability theory, statistics, actuarial science, operations research, and engineering—to name just a few.

The Gamma Function

Before we commence our discussion of gamma random variables, we need to examine the *gamma function*, a function that is used extensively in mathematics and its applications.

The **gamma function** is denoted Γ and is defined by

$$\Gamma(t) = \int_0^\infty x^{t-1} e^{-x}\, dx, \qquad t > 0. \tag{8.42}$$

A graph of the gamma function is provided in Figure 8.17. We ask you to verify the form of the graph in Exercise 8.115.

Figure 8.17 Graph of the gamma function, Γ

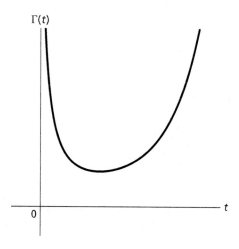

One of the most important properties of the gamma function is that

$$\Gamma(t + 1) = t\Gamma(t), \qquad t > 0. \tag{8.43}$$

This property is easily obtained by doing an integration by parts, as we ask you to do in Exercise 8.133.

We now show that the gamma function interpolates between the values of the factorials of positive integers and hence can be considered a generalization of factorials to all positive real numbers. Specifically, we have

$$\Gamma(n) = (n - 1)!, \qquad n = 1,\, 2,\, \ldots. \tag{8.44}$$

To prove Equation (8.44), we first substitute $t = 1$ into Equation (8.42) and get

$$\Gamma(1) = \int_0^\infty e^{-x}\, dx = 1 = 0!,$$

so Equation (8.44) holds for $n = 1$. Then we use mathematical induction and Equation (8.43) to deduce Equation (8.44). We leave the details to you as Exercise 8.134.

It is also helpful to have a formula for the values of the gamma function at half integers. First we observe that, by making the substitution $y = \sqrt{2x}$, we get

$$\Gamma\left(\tfrac{1}{2}\right) = \int_0^\infty x^{\frac{1}{2}-1} e^{-x}\, dx = \int_0^\infty x^{-\frac{1}{2}} e^{-x}\, dx = \int_0^\infty \sqrt{2}\, e^{-y^2/2}\, dy$$

$$= \sqrt{2}\sqrt{2\pi} \int_0^\infty \frac{1}{\sqrt{2\pi}} e^{-y^2/2}\, dy = 2\sqrt{\pi}\left(1 - \Phi(0)\right) = 2\sqrt{\pi}\left(1 - \tfrac{1}{2}\right),$$

or

$$\Gamma\left(\tfrac{1}{2}\right) = \sqrt{\pi}. \qquad (8.45)$$

Using mathematical induction and Equations (8.45) and (8.43), we get

$$\Gamma\left(n + \tfrac{1}{2}\right) = \frac{(2n)!}{n!\,2^{2n}}\sqrt{\pi}, \qquad n = 0, 1, 2, \ldots. \qquad (8.46)$$

We leave the details to you as Exercise 8.135.

Gamma Random Variables

As we now show, gamma random variables are the continuous analogue of negative binomial random variables, a family of discrete random variables that we presented in Section 5.7. If we consider repeated Bernoulli trials with success probability p, the number of trials up to and including the rth success is a negative binomial random variable with parameters r and p.

We can think of this negative binomial random variable as giving the time of the rth success in a sequence of Bernoulli trials, where time is quantized into integer values. In any case, such a random variable exceeds a positive integer m if and only if the number of successes in the first m trials is at most $r - 1$. Applying the FPF and the binomial PMF with parameters m and p gives the probability of this latter event as $\sum_{j=0}^{r-1} \binom{m}{j} p^j (1-p)^{m-j}$.

Let $n \in \mathcal{N}$ and suppose that Bernoulli trials with success probability p are performed at times $\frac{1}{n}, \frac{2}{n}, \ldots$; that is, time is quantized into units of duration $\frac{1}{n}$. Let X denote the time of the rth success. For a positive real number x, we set $k = \lfloor nx \rfloor$. Note that $\frac{k}{n} \leq x < \frac{k+1}{n}$, as shown in Figure 8.11 on page 431. It follows that $\{X > x\} = \left\{X > \frac{k}{n}\right\}$ and hence that

$$P(X > x) = P\left(X > \frac{k}{n}\right) = \sum_{j=0}^{r-1} \binom{k}{j} p^j (1-p)^{k-j}.$$

We now assume that n is large and that p is small. Referring to the Poisson approximation to the binomial distribution, specifically Relation (5.21) on page 220, noting that $x \approx \frac{k}{n}$, and using the previous equation, we get

$$P(X > x) \approx \sum_{j=0}^{r-1} e^{-kp} \frac{(kp)^j}{j!} \approx e^{-npx} \sum_{j=0}^{r-1} \frac{(npx)^j}{j!}. \qquad (8.47)$$

So, as quantized time becomes finer and finer, approaching a continuum, the time of the rth success has tail probabilities of the form given on the right of Relation (8.47).

Therefore it is natural to consider continuous random variables whose tail probabilities take the form

$$e^{-\lambda x} \sum_{j=0}^{r-1} \frac{(\lambda x)^j}{j!}, \qquad x > 0, \tag{8.48}$$

where λ is a positive real number and r is a positive integer. We first show that the numbers in Expression (8.48) are indeed the tail probabilities of some positive continuous random variable. Equivalently, we need to show that the function $F: \mathcal{R} \to \mathcal{R}$ defined by

$$F(x) = 1 - e^{-\lambda x} \sum_{j=0}^{r-1} \frac{(\lambda x)^j}{j!}, \qquad x \geq 0, \tag{8.49}$$

and $F(x) = 0$ otherwise, is the CDF of a continuous random variable—that is, that F is everywhere continuous (see Proposition 8.3 on page 413) and satisfies properties (a)–(d) of Proposition 8.1 on page 411. These results are easily verified and are left for you to do as Exercise 8.116.

We also observe that F' exists and is continuous except at $x = 0$. Routine calculations show that

$$F'(x) = \frac{\lambda^r}{(r-1)!} x^{r-1} e^{-\lambda x}, \qquad x > 0,$$

and $F'(x) = 0$, for $x < 0$. Consequently, by Proposition 8.4 on page 419, a random variable with F as its CDF has a PDF given by

$$f(x) = \frac{\lambda^r}{(r-1)!} x^{r-1} e^{-\lambda x} = \frac{\lambda^r}{\Gamma(r)} x^{r-1} e^{-\lambda x}, \qquad x > 0, \tag{8.50}$$

and $f(x) = 0$ otherwise.

We know that the function f in Equation (8.50) is a PDF when r is a positive integer. In fact, it's a PDF provided only that r is a positive real number. That is, for each fixed positive real number α, the function f defined by

$$f(x) = \frac{\lambda^\alpha}{\Gamma(\alpha)} x^{\alpha-1} e^{-\lambda x}, \qquad x > 0, \tag{8.51}$$

and $f(x) = 0$ otherwise, is a PDF. To verify this result, we need to show that f is nonnegative and integrates to 1. That f is nonnegative is trivial. To show that f integrates to 1, we make the substitution $y = \lambda x$ and use the definition of the gamma function [Equation (8.42) on page 451] to obtain

$$\int_{-\infty}^{\infty} f(x)\, dx = \int_0^\infty \frac{\lambda^\alpha}{\Gamma(\alpha)} x^{\alpha-1} e^{-\lambda x}\, dx = \frac{\lambda^\alpha}{\Gamma(\alpha)} \int_0^\infty \frac{y^{\alpha-1}}{\lambda^{\alpha-1}} e^{-y} \frac{1}{\lambda}\, dy$$

$$= \frac{1}{\Gamma(\alpha)} \int_0^\infty y^{\alpha-1} e^{-y}\, dy = \frac{1}{\Gamma(\alpha)} \Gamma(\alpha) = 1.$$

We have shown that the function f in Equation (8.51) is a PDF of a continuous random variable. A random variable with that PDF is called a **gamma random variable,** as indicated in Definition 8.8.

DEFINITION 8.8   Gamma Random Variable

A continuous random variable X is called a **gamma random variable** if it has a probability density function of the form

$$f_X(x) = \frac{\lambda^\alpha}{\Gamma(\alpha)} x^{\alpha-1} e^{-\lambda x}, \qquad x > 0,$$

and $f_X(x) = 0$ otherwise, where α and λ are positive real numbers. We say that X has a **gamma distribution with parameters α and λ** or that X is **gamma distributed with parameters α and λ.** For convenience, we sometimes write $X \sim \Gamma(\alpha, \lambda)$ to indicate that X has a gamma distribution with parameters α and λ.[†]

The overall shape of a gamma PDF depends on its α parameter; its λ parameter affects scale only. Refer to Figure 8.18 as you read the following details.

- If $\alpha = 1$, the gamma PDF is the exponential PDF with parameter λ and hence is decreasing and concave up.
- If $\alpha < 1$, the gamma PDF is decreasing and concave up and approaches infinity as x approaches 0 from the right.
- If $\alpha > 1$, the gamma PDF has a single hump.
- As α increases, the gamma PDF looks increasingly like a normal PDF.

Note: There is no simple formula for the CDF of a $\Gamma(\alpha, \lambda)$ random variable unless α is a positive integer, say, r. In that case, the CDF is given by Equation (8.49) on page 453.

EXAMPLE 8.16   *Gamma Random Variables*

Emergency Room Traffic   Desert Samaritan Hospital in Mesa, Arizona, keeps records of its emergency-room traffic. Those records indicate that, beginning at 6:00 P.M. on a given day, the elapsed time until the third patient arrives has the gamma distribution with parameters $\alpha = 3$ and $\lambda = 6.9$, where time is measured in hours. Determine the probability that, beginning at 6:00 P.M. on a given day, the third patient arrives
a) between 6:15 P.M. and 6:30 P.M.
b) before 7:00 P.M.

[†]As for an exponential random variable, the PDF of a gamma random variable is often expressed in an alternative form—namely, as

$$f_X(x) = \frac{1}{\theta^\alpha \Gamma(\alpha)} x^{\alpha-1} e^{-x/\theta}, \qquad x > 0.$$

and $f_X(x) = 0$ otherwise, where α and θ are positive real numbers.

Figure 8.18 PDFs of gamma random variables for various choices of the parameter α and $\lambda = 2$

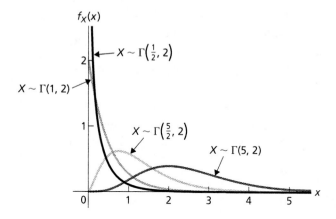

Solution   Let X denote the elapsed time, in hours, until the third patient arrives at the emergency room. By assumption, $X \sim \Gamma(3, 6.9)$. Its PDF therefore is

$$f_X(x) = \frac{6.9^3}{\Gamma(3)} x^{3-1} e^{-6.9x} = 164.2545 x^2 e^{-6.9x}, \qquad x > 0,$$

and $f_X(x) = 0$ otherwise. However, because $\alpha = 3$ is a positive integer, we can most easily obtain the required probabilities by using the CDF of X which is, by Equation (8.49) on page 453,

$$F_X(x) = 1 - e^{-6.9x} \sum_{j=0}^{2} \frac{(6.9x)^j}{j!}, \qquad x \geq 0. \tag{8.52}$$

In doing so, we note that X is a continuous random variable.

a) We want $P(1/4 < X < 1/2)$. From Proposition 8.2 on page 412 and Equation (8.52),

$$P(1/4 < X < 1/2) = P(1/4 < X \leq 1/2) = F_X(1/2) - F_X(1/4) = 0.420.$$

There is a 42.0% chance that the third patient will arrive between 6:15 P.M. and 6:30 P.M.

b) We want $P(X < 1)$. Again, from Equation (8.52),

$$P(X < 1) = P(X \leq 1) = F_X(1) = 0.968.$$

There is a 96.8% chance that the third patient will arrive before 7:00 P.M.   ■

As we said, if α isn't a positive integer, there is no simple formula for the CDF of a gamma random variable. In this case, we must resort to tables, computer software, or sophisticated calculators to obtain probabilities for a gamma random variable, in the same way that we do for a normal random variable.

The following important probability distributions are some of the special cases of the gamma distribution.

- *Exponential distribution.* The exponential distribution with parameter λ is also the gamma distribution with parameters 1 and λ.

- *Erlang distribution.* The **Erlang distribution** with parameters r and λ, where r is a positive integer, is also the gamma distribution with those two parameters. The Erlang distribution, named for Agner Krarup Erlang (1878–1929), is used extensively in queueing theory.

- *Chi-square distribution.* The **chi-square distribution** with ν degrees of freedom is also the gamma distribution with parameters $\nu/2$ and $1/2$. The chi-square distribution is of major importance in statistics.

Beta Random Variables

We next consider the family of *beta random variables.* The range of a beta random variable is the interval of real numbers between 0 and 1, which makes beta distributions particularly useful for modeling proportions, percentages, or probabilities. For instance, we might use a beta distribution to model

- the proportion of customers who are satisfied with their service each month,

- the percentage of defective items in a shipment,

- the percentage of data-entry errors for a particular task, or

- an unknown success probability in Bernoulli trials.

Beta distributions are also useful in the study of *order statistics,* where a finite number of random observations are placed in increasing order. Additionally, by using an appropriate linear function, we can often use a beta distribution to model a random variable whose range is a finite interval.

Before we begin our discussion of beta random variables, we need to examine the *beta function* which, like the gamma function, is widely used in mathematics and its applications. The **beta function** is defined by

$$B(s, t) = \int_0^1 x^{s-1}(1 - x)^{t-1}\, dx, \qquad s, t > 0. \tag{8.53}$$

It's related to the gamma function via the equation

$$B(s, t) = \frac{\Gamma(s)\Gamma(t)}{\Gamma(s + t)}, \qquad s, t > 0. \tag{8.54}$$

This identity can be proved by using calculus but, for our purposes, a probabilistic argument is superior, one that we give in Chapter 9.

We now present the definition of a **beta random variable.**

DEFINITION 8.9   Beta Random Variable

A continuous random variable X is called a **beta random variable** if it has a probability density function of the form

$$f_X(x) = \frac{1}{B(\alpha, \beta)} x^{\alpha-1}(1 - x)^{\beta-1}, \qquad 0 < x < 1,$$

and $f_X(x) = 0$ otherwise, where α and β are positive real numbers. We say that X has a **beta distribution with parameters α and β** or that X is **beta distributed with parameters α and β**.

Beta distributions are particularly useful because we can model an extensive variety of random phenomena by adjusting the two parameters appropriately. Indeed, the shape of a beta PDF varies considerably, depending on the values of its two parameters, α and β, as indicated by the following details.

- If $\alpha = \beta = 1$, the beta distribution reduces to the uniform distribution on the interval $(0, 1)$. Thus, in this case, the beta PDF is constant (on its range), as shown in Figure 8.19(a).

- If $\alpha < 1$, the beta PDF approaches infinity as x approaches 0 from the right, as shown in Figure 8.19(b), (c), and (f).

- If $\beta < 1$, the beta PDF approaches infinity as x approaches 1 from the left, as shown in Figure 8.19(b), (d), and (f).

- If $\alpha < 1$ and $\beta < 1$, the beta PDF is "U-shaped," as shown in Figure 8.19(b) and (f).

- If $\alpha < 1$ and $\beta \geq 1$, the beta PDF is decreasing, as shown in Figure 8.19(c).

- If $\alpha \geq 1$ and $\beta < 1$, the beta PDF is increasing, as shown in Figure 8.19(d).

- If $\alpha > 1$ and $\beta > 1$, the beta PDF has a single hump, as shown in Figure 8.19(e).

- If $\alpha = \beta$, the beta PDF is symmetric about $x = 1/2$, as shown in Figure 8.19(a) and (b); otherwise the beta PDF is skewed, as shown in Figure 8.19(c)–(f).

Note: There is no simple formula for the CDF of a beta random variable unless both parameters are integers. Specifically, if X has a beta distribution with parameters m and n, both positive integers, we have

$$F_X(x) = \sum_{j=m}^{m+n-1} \binom{m + n - 1}{j} x^j (1 - x)^{m+n-1-j}, \qquad 0 \leq x < 1. \qquad \textbf{(8.55)}$$

Of course, for all beta random variables, we have $F_X(x) = 0$ if $x < 0$ and $F_X(x) = 1$ if $x \geq 1$.

Figure 8.19 PDFs of beta random variables for various choices of α and β

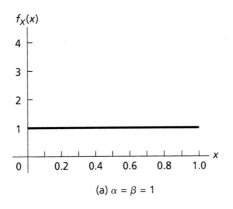

(a) $\alpha = \beta = 1$

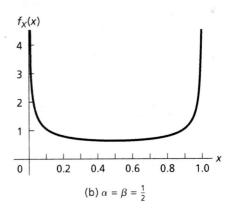

(b) $\alpha = \beta = \frac{1}{2}$

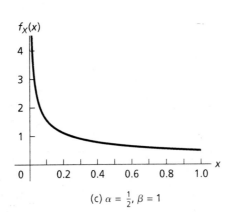

(c) $\alpha = \frac{1}{2}, \beta = 1$

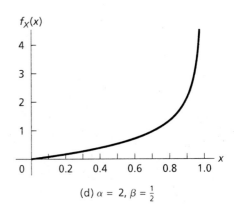

(d) $\alpha = 2, \beta = \frac{1}{2}$

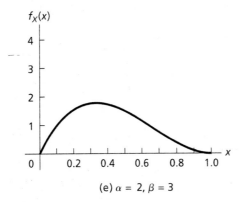

(e) $\alpha = 2, \beta = 3$

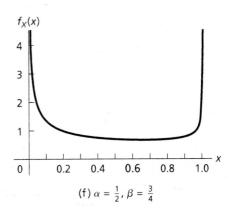

(f) $\alpha = \frac{1}{2}, \beta = \frac{3}{4}$

EXAMPLE 8.17   *Beta Random Variables*

Stockbrokers   A stockbroker has a large number of clients. The proportion of clients with whom she doesn't communicate during a given month has the beta distribution with parameters $\alpha = 6$ and $\beta = 2$. Determine the probability that, during a given month, the percentage of clients with whom the broker doesn't communicate will be
a) at least 70%.
b) between 80% and 90%.

Solution   Let X denote the proportion of clients with whom the broker doesn't communicate during a given month. By assumption, X has the beta distribution with parameters $\alpha = 6$ and $\beta = 2$. Referring to Definition 8.9 and Equation (8.54), we find that a PDF of X is

$$f_X(x) = \frac{1}{B(6, 2)} x^{6-1}(1-x)^{2-1} = 42x^5(1-x), \qquad 0 < x < 1,$$

and $f_X(x) = 0$ otherwise. However, because $\alpha = 6$ and $\beta = 2$ are both positive integers, we can most easily obtain the required probabilities by using the CDF of X which is, by Equation (8.55),

$$F_X(x) = \sum_{j=6}^{7} \binom{7}{j} x^j (1-x)^{7-j} = 7x^6 - 6x^7, \qquad 0 < x < 1. \qquad \textbf{(8.56)}$$

In doing so, we note that X is a continuous random variable.
a) From the complementation rule and Equation (8.56),

$$P(X \geq 0.7) = P(X > 0.7) = 1 - P(X \leq 0.7)$$
$$= 1 - F_X(0.7) = 1 - \left(7 \cdot (0.7)^6 - 6 \cdot (0.7)^7\right)$$
$$= 0.671.$$

Chances are 67.1% that, during a given month, the broker won't communicate with 70% or more of her clients.
b) From Proposition 8.2 on page 412 and Equation (8.56),

$$P(0.8 < X < 0.9) = P(0.8 < X \leq 0.9) = F_X(0.9) - F_X(0.8)$$
$$= \left(7 \cdot (0.9)^6 - 6 \cdot (0.9)^7\right) - \left(7 \cdot (0.8)^6 - 6 \cdot (0.8)^7\right)$$
$$= 0.274.$$

Chances are 27.4% that, during a given month, the broker won't communicate with between 80% and 90% of her clients.   ∎

Triangular Random Variables

Next we discuss the family of *triangular random variables*. These random variables obtain their name from the fact that their PDFs form isosceles triangles. Triangular random variables occur in a wide variety of applications, two particularly important ones being the spectral analysis of time series and electrical signals. We now present the definition of a **triangular random variable**.

DEFINITION 8.10   Triangular Random Variable

A continuous random variable X is called a **triangular random variable** if it has a probability density function of the form

$$f_X(x) = \frac{2}{b-a}\left(1 - \frac{|a+b-2x|}{b-a}\right), \qquad a < x < b,$$

and $f_X(x) = 0$ otherwise, where a and b are real constants with $a < b$. We say that X has a **triangular distribution on the interval** (a, b) or that X is **triangularly distributed on the interval** (a, b). For convenience, we sometimes write $X \sim \mathcal{T}(a, b)$ to indicate that X has a triangular distribution on the interval (a, b). See Figure 8.20.

Figure 8.20 PDF of a triangular random variable on the interval (a, b)

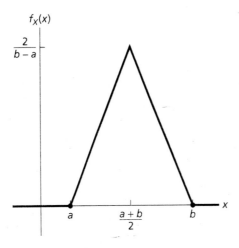

Note: In the case where $b > 0$ and $a = -b$—that is, X has a triangular distribution on the interval $(-b, b)$—we can simplify the expression for the PDF as follows:

$$f_X(x) = \frac{1}{b}\left(1 - \frac{|x|}{b}\right), \qquad -b < x < b, \tag{8.57}$$

and $f_X(x) = 0$ otherwise.

EXAMPLE 8.18   *Triangular Random Variables*

Arrival Time   A person plans to arrive at a specified place at noon. Let's assume that the time the person actually arrives has the triangular distribution on the interval $(-5, 5)$, where time is measured in minutes relative to noon. Determine the probability that the person will arrive within 2 minutes of noon.

Solution   Let X denote the time, relative to noon, that the person arrives at the specified place. By assumption, $X \sim T(-5, 5)$. From Equation (8.57), we can express the PDF of X as

$$f_X(x) = \frac{1}{5}\left(1 - \frac{|x|}{5}\right), \qquad -5 < x < 5, \qquad \textbf{(8.58)}$$

and $f_X(x) = 0$ otherwise.

The problem is to determine $P(|X| \le 2)$. Applying the FPF, Equation (8.58), and the symmetry of the PDF of X, we get

$$P(|X| \le 2) = \int_{|x| \le 2} f_X(x)\, dx = \int_{-2}^{2} \frac{1}{5}\left(1 - \frac{|x|}{5}\right) dx = \frac{2}{5}\int_{0}^{2}\left(1 - \frac{x}{5}\right) dx = 0.64.$$

Chances are 64% that the person will arrive within 2 minutes of noon.   ■

Other Continuous Random Variables

In addition to the families of continuous random variables that we have discussed so far, many others occur frequently in practice. They include the *Weibull, Pareto, lognormal,* and *Cauchy* families. We discuss some of them in the exercises for this section and others in Section 8.7. Later in the book, we encounter still other important families of continuous random variables.

EXERCISES 8.6   **Basic Exercises**

8.115 Apply calculus techniques to obtain a graph of the gamma function, as defined in Equation (8.42) on page 451. *Note:* You may assume that it's permissible to take the derivative of the gamma function by differentiating under the integral sign with respect to t.

8.116 Let λ be a positive real number and let r be a positive integer. Show that the function $F: \mathcal{R} \to \mathcal{R}$ defined by

$$F(x) = 1 - e^{-\lambda x}\sum_{j=0}^{r-1}\frac{(\lambda x)^j}{j!}, \qquad x \ge 0,$$

and $F(x) = 0$ otherwise, is the CDF of a continuous random variable. *Note:* You must show that F is everywhere continuous and satisfies properties (a)–(d) of Proposition 8.1 on page 411.

8.117 Determine a PDF of each of the following distributions.
a) Erlang distribution with parameters r and λ
b) Chi-square distribution with ν degrees of freedom

8.118 Let $X \sim \mathcal{N}(0, \sigma^2)$. Determine and identify the probability distribution of each of the following random variables.
a) X^2 **b)** X^2/σ^2

8.119 Suppose that X has the beta distribution with parameters m and n, both positive integers. Show that, for $0 < x < 1$, the identity $P(X \le x) = P(Y \ge m)$ holds, where Y has the binomial distribution with parameters $m + n - 1$ and x.

8.120 Let X have the beta distribution with parameters α and β. Determine a PDF of the random variable $X^{-1} - 1$.

8.121 Verify the properties of beta distributions that are given in the bulleted list on page 457.

8.122 Let X have the uniform distribution on the interval $(0, 1)$ and let α be a positive real number. Obtain and identify the probability distribution of the random variable $Y = X^{1/\alpha}$.

8.123 For a certain manufactured item, the proportion that require service during the first 5 years of use has the beta distribution with parameters $\alpha = 2$ and $\beta = 3$. Determine the probability that the percentage of these manufactured items that require service during the first 5 years of use is
a) at most 30%.
b) between 10% and 20%.

8.124 For this exercise, you'll need access to statistical software. In his article "A Juiced Analysis" (*Chance*, 2002, Vol. 15, No. 4, pp. 50–53), Scott M. Berry modeled the probability of a Barry Bonds home run by a beta distribution with parameters $\alpha = 51.3$ and $\beta = 539.75$.
a) Graph and interpret a PDF for the probability of a Barry Bonds home run.
b) Determine the probability that the probability of a Barry Bonds home run is between 0.07 and 0.10.

8.125 Show that any triangular distribution can be obtained from a particular triangular distribution, say, $\mathcal{T}(-1, 1)$, by a location and scale change.

8.126 Obtain and graph the CDF of a random variable with a triangular distribution on the interval (a, b).

8.127 According to the paper "Table for the Likelihood Solutions of Gamma Distribution and Its Medical Applications" (*Reports of Statistical Application Research (JUSE)*, 1952, Vol. 1, pp. 18–23) by M. Masuyama and Y. Kuroiwa, during the 30th week of pregnancy the sedimentation rate, X, has (approximately) the gamma distribution with parameters $\alpha = 5$ and $\lambda = 0.1$.
a) Specify the PDF and CDF of X.
b) Determine the probability that, during the 30th week of pregnancy, the sedimentation rate is at most 60.
c) Determine the probability that, during the 30th week of pregnancy, the sedimentation rate is between 40 and 50.

8.128 Pareto random variable: Let X have the exponential distribution with parameter α and let β be a positive real number.
a) Determine the CDF of the random variable $Y = \beta e^X$.
b) Determine a PDF of Y. *Note:* A random variable with this PDF is called a *Pareto random variable* and is said to have the *Pareto distribution with parameters α and β*.

8.129 Refer to Exercise 8.128. An actuary models the loss amounts of a certain policy by the Pareto distribution with parameters $\alpha = 2$ and $\beta = 3$, where loss amounts are measured in thousands of dollars.
a) Graph the PDF and CDF of the loss amounts.
b) Determine the probability that a loss amount exceeds $8000.
c) Determine the probability that a loss amount is between $4000 and $5000.

8.130 Weibull random variable: Let X have the exponential distribution with parameter α and let β be a positive real number.
a) Determine the CDF of the random variable $Y = X^{1/\beta}$.
b) Determine a PDF of Y. *Note:* A random variable with this PDF is called a *Weibull random variable* and is said to have the *Weibull distribution with parameters α and β*.
c) Graph the PDF and CDF of a Weibull random variable if $\beta < 1$; $\beta = 1$; $\beta > 1$.

8.131 Refer to Exercise 8.130. A Weibull distribution is often used to model the time at which a component or item fails. Suppose that the lifetime of a component has the Weibull distribution with parameters $\alpha = 2$ and $\beta = 3$, where time is measured in years.
a) Graph the PDF and CDF of component lifetime.
b) Determine the probability that the component lasts at most 6 months; between 6 months and 1 year.

8.132 Arcsine distribution: The beta distribution with $\alpha = \beta = 1/2$ is called the *arcsine distribution* and arises in the study of fair games. For instance, suppose that two people—say, A and B—play a game by tossing a balanced coin. If the coin comes up a head, A wins \$1 from B, whereas, if the coin comes up a tail, B wins \$1 from A. In a large number of tosses, the proportion of times, X, that A is ahead has approximately the arcsine distribution.
a) Without doing any calculations or referring to previous material, guess whether X is more likely to be near 0 or 1 or whether it is more likely to be near 1/2. Explain your reasoning.
b) Referring to Figure 8.19(b) on page 458, explain why X is more likely to be near 0 or 1 than near 1/2.
c) Determine the CDF of X.
d) Determine $P(0.4 \leq X \leq 0.6)$ and $P(X \leq 0.1 \text{ or } X \geq 0.9)$. Interpret your results.

Theory Exercises

8.133 Apply integration by parts to verify Equation (8.43) on page 451; that is, show that $\Gamma(t + 1) = t\Gamma(t)$ for all $t > 0$.

8.134 Complete the verification of Equation (8.44) on page 451 that $\Gamma(n) = (n - 1)!$ for each $n \in \mathcal{N}$.

8.135 On page 452, we showed that $\Gamma(1/2) = \sqrt{\pi}$. Use that result and Exercise 8.133 to show that
$$\Gamma\left(n + \tfrac{1}{2}\right) = \frac{(2n)!}{n!\,2^{2n}}\sqrt{\pi}, \qquad n = 0, 1, 2, \ldots.$$

Advanced Exercises

Hazard-rate functions: Let the positive continuous random variable T denote the lifetime of some item. Assume that T has a PDF and that $F_T'(t) = f_T(t)$ for all $t > 0$. The *hazard-rate function* (or *failure-rate function*) of T, denoted h_T, is then defined by
$$h_T(t) = \frac{f_T(t)}{1 - F_T(t)}, \qquad t > 0.$$
Exercises 8.136–8.140 examine hazard-rate functions.

8.136 Refer to the definition just given of a hazard-rate function.
a) Explain why $h_T(t)$ can be interpreted as the instantaneous risk of item failure at time t. *Hint:* Consider $P(T \leq t + \Delta t \mid T > t)$.
b) Prove that $F_T(t) = 1 - e^{-\int_0^t h_T(u)\,du}$ for all $t > 0$.
c) Show that $P(T > t + s \mid T > t) = e^{-\int_t^{t+s} h_T(u)\,du}$ for all $s, t > 0$.
d) Deduce that, if h_T is a decreasing function, the item improves with age—that is, for fixed s, $P(T > t + s \mid T > t)$ is an increasing function of t—whereas, if h_T is an increasing function, the item deteriorates with age—that is, for fixed s, $P(T > t + s \mid T > t)$ is a decreasing function of t.
e) What can be said of an item whose hazard-rate function is constant?

8.137 Assume that T is exponentially distributed with parameter λ.

a) Without doing any calculations, make an educated guess at the form of the hazard-rate function, h_T, of T.

b) Use the definition of the hazard-rate function to obtain and interpret h_T. *Note:* Refer to Exercise 8.136(e).

8.138 Let T have the Weibull distribution with parameters α and β, as defined in Exercise 8.130. Find the hazard-rate function of T and then apply the results of Exercises 8.136(d) and (e) to interpret it in case $\beta < 1$; $\beta = 1$; $\beta > 1$.

8.139 Use Exercise 8.136(b) to obtain a PDF of T if its hazard-rate function is

a) constant—say, $h_T(t) = \lambda$ where λ is a positive real number.

b) linear—say, $h_T(t) = \alpha + \beta t$ where α and β are positive real numbers.

8.140 Suppose that the hazard-rate function of Item II is twice that of Item I.

a) Without doing any calculations, guess at the relationship between the probability that, at age t years, Item II lasts (at least) an additional s years and the probability that, at age t years, Item I lasts (at least) an additional s years.

b) Use Exercise 8.136(c) to determine the relationship between the probabilities in part (a).

8.7 Functions of a Continuous Random Variable

As we noted in Section 5.8, it's quite common in probability theory and its applications to consider random variables obtained by taking functions of other random variables. There we examined functions of a discrete random variable. Now we investigate functions of a continuous random variable.

Recall that, if X is a random variable and g is a real-valued function defined on the range of X, the composition $g(X)$ is a random variable—specifically, a random variable obtained as the function g of the random variable X. In the discrete case—that is, when X is a discrete random variable—$g(X)$ is also a discrete random variable.

In the continuous case—that is, when X is a continuous random variable—$g(X)$ need not be a continuous random variable. For the most part, though, we concentrate on functions of continuous random variables that are also continuous random variables. Specifically, we mostly consider situations where X is a continuous random variable with a known PDF, and the function g is such that $g(X)$ is also a continuous random variable with a PDF. The problem then is to find a PDF of $g(X)$.

We employ two methods to solve such problems. The first is the *CDF method,* which is based on Procedure 8.1 on page 419, and applies when g is differentiable on the range of X. The second is the *transformation method,* which applies only when g is differentiable and strictly monotone (increasing or decreasing) on the range of X.

The CDF Method

Suppose that X is a continuous random variable with a known PDF and that g is a real-valued function defined and differentiable on the range of X. Let $Y = g(X)$. To obtain a PDF of Y by using the **CDF method,** we utilize the following five-step procedure.

PROCEDURE 8.2   CDF Method for Finding a PDF of $Y = g(X)$

Assumptions

- X is a continuous random variable with a known PDF.
- g is a real-valued function defined and differentiable on the range of X.

Step 1 *Identify a PDF of X.*

Step 2 *Identify the range of Y.*

Step 3 *For y in the range of Y, determine the CDF of Y in terms of that of X.*

Step 4 *For y in the range of Y, obtain the derivative of the CDF of Y, where it exists.*

Step 5 *A PDF of Y is given by $f_Y(y) = F'_Y(y)$, where the derivative exists, and $f_Y(y) = 0$ otherwise.*

EXAMPLE 8.19   *The CDF Method*

Squaring a Standard Normal   Let $X \sim \mathcal{N}(0, 1)$. Use the CDF method to determine and identify the probability distribution of the random variable X^2.

Solution   Here $Y = X^2$—that is, $g(x) = x^2$. We apply Procedure 8.2, noting first that the assumptions for its use are satisfied.

Step 1 *Identify a PDF of X.*

As $X \sim \mathcal{N}(0, 1)$, its PDF is $f_X(x) = \left(1/\sqrt{2\pi}\,\right)e^{-x^2/2}$ for $-\infty < x < \infty$.

Step 2 *Identify the range of Y.*

The range of X is $(-\infty, \infty)$, so the range of Y is $[0, \infty)$.

Step 3 *For y in the range of Y, determine the CDF of Y in terms of that of X.*

For y in the range of Y—that is, for $y \geq 0$—we have

$$F_Y(y) = P(Y \leq y) = P(X^2 \leq y)$$
$$= P(-\sqrt{y} \leq X \leq \sqrt{y}) = F_X(\sqrt{y}) - F_X(-\sqrt{y}).$$

Step 4 *For y in the range of Y, obtain the derivative of the CDF of Y, where it exists.*

From Step 3, the chain rule, and the symmetry about 0 of the standard normal PDF,

$$F'_Y(y) = F'_X(\sqrt{y}) \left(\frac{1}{2\sqrt{y}}\right) - F'_X(-\sqrt{y})\left(-\frac{1}{2\sqrt{y}}\right)$$

$$= \frac{1}{2\sqrt{y}} f_X(\sqrt{y}) + \frac{1}{2\sqrt{y}} f_X(-\sqrt{y}) = \frac{1}{2\sqrt{y}} f_X(\sqrt{y}) + \frac{1}{2\sqrt{y}} f_X(\sqrt{y})$$

$$= \frac{1}{\sqrt{y}} f_X(\sqrt{y}) = \frac{1}{\sqrt{y}} \frac{1}{\sqrt{2\pi}} e^{-(\sqrt{y})^2/2} = \frac{1}{\sqrt{2\pi y}} e^{-y/2}.$$

Step 5 *A PDF of Y is given by $f_Y(y) = F_Y'(y)$, where the derivative exists, and $f_Y(y) = 0$ otherwise.*

From the preceding steps, we conclude that a PDF of Y is

$$f_Y(y) = \frac{1}{\sqrt{2\pi y}} e^{-y/2}, \qquad y > 0, \tag{8.59}$$

and $f_Y(y) = 0$ otherwise.

We also want to identify the probability distribution of the random variable $Y = X^2$. Referring to Equation (8.59) and using the fact that $\Gamma(1/2) = \sqrt{\pi}$, we write

$$f_Y(y) = \frac{\left(\frac{1}{2}\right)^{\frac{1}{2}}}{\Gamma\left(\frac{1}{2}\right)} y^{\frac{1}{2}-1} e^{-\frac{1}{2}y}, \qquad y > 0,$$

and $f_Y(y) = 0$ otherwise. Thus $Y \sim \Gamma(1/2, 1/2)$. Equivalently, Y has the chi-square distribution with 1 degree of freedom. This result is so important that we summarize it as Proposition 8.13. ∎

♦♦♦ **Proposition 8.13   Square of a Standard Normal Random Variable**

If X has the standard normal distribution, then X^2 has the chi-square distribution with 1 degree of freedom.

The Transformation Method

The CDF method requires that we first find the CDF of the function of the random variable and then differentiate that CDF to obtain a PDF. The transformation method—although more restrictive than the CDF method—has the advantage of giving a direct formula for a PDF of the function of the random variable.

To state the transformation method, we first prove Proposition 8.14, which we refer to as the **univariate transformation theorem.** Compare it to its discrete analogue, Corollary 5.1 on page 247.

♦♦♦ **Proposition 8.14   Univariate Transformation Theorem**

Let X be a continuous random variable with a PDF and let g be a real-valued function defined, strictly monotone, and differentiable on the range of X. Then a PDF of the random variable $Y = g(X)$ is

$$f_Y(y) = \frac{1}{|g'(x)|} f_X(x), \tag{8.60}$$

for y in the range of Y, where x is the unique real number in the range of X such that $g(x) = y$; otherwise, $f_Y(y) = 0$.

Proof To prove the univariate transformation theorem, we apply the CDF method. We provide the proof when g is strictly decreasing on the range of X and leave the case of strictly increasing g for you to do as Exercise 8.168. In the proof, we use the notation g^{-1} for the inverse function of g, defined on the range of Y.

For y in the range of Y, we have

$$F_Y(y) = P(Y \le y) = P(g(X) \le y)$$
$$= P(X \ge g^{-1}(y)) = 1 - P(X < g^{-1}(y)) = 1 - F_X(g^{-1}(y)),$$

where the third equality follows from the fact that g is decreasing. Taking the derivative with respect to y and applying the chain rule, we obtain

$$f_Y(y) = F_Y'(y) = -F_X'(g^{-1}(y))\frac{d}{dy}g^{-1}(y) = -\frac{1}{g'(g^{-1}(y))}f_X(g^{-1}(y)), \quad \textbf{(8.61)}$$

where, in the last equality, we used the calculus result $(d/dy)g^{-1}(y) = 1/g'(g^{-1}(y))$. Because g is decreasing, g' is negative. Letting x denote the unique real number in the range of X such that $g(x) = y$, we conclude from Equation (8.61) that, for y in the range of Y, $f_Y(y) = f_X(x)/|g'(x)|$, as required.   ◆

Note: We can express Equation (8.60) in the alternate form

$$f_{g(X)}(y) = \frac{1}{|g'(g^{-1}(y))|}f_X(g^{-1}(y)). \quad \textbf{(8.62)}$$

In Procedure 8.3, we provide a five-step procedure for implementing the **transformation method** to obtain a PDF of the random variable $Y = g(X)$ under the conditions of the univariate transformation theorem.

PROCEDURE 8.3   Transformation Method for Finding a PDF of $Y = g(X)$

Assumptions

- X is a continuous random variable with a known PDF.
- g is a real-valued function defined, strictly monotone, and differentiable on the range of X.

Step 1 *Identify a PDF of X and the range of X, and check whether g satisfies the conditions of the univariate transformation theorem.*

Step 2 *Identify the range of Y.*

Step 3 *Apply Equation (8.60) in the univariate transformation theorem.*

Step 4 *For y in the range of Y, solve for x in the equation $y = g(x)$.*

Step 5 *Use Steps 2–4 to obtain a PDF of Y.*

Note that the transformation method can't be used to solve Example 8.19 because the function $g(x) = x^2$ is not monotone on the range of X, which is $(-\infty, \infty)$. But we can and do use the transformation method in Examples 8.20 and 8.21.

EXAMPLE 8.20   *The Transformation Method*

Linear Function of a Normal   Let $X \sim \mathcal{N}(\mu, \sigma^2)$ and let a and $b \neq 0$ be real numbers. Use the transformation method to obtain and identify the probability distribution of the random variable $a + bX$.

Solution   Here $Y = a + bX$—that is, $g(x) = a + bx$. We apply Procedure 8.3.

Step 1   *Identify a PDF of X and the range of X, and check whether g satisfies the conditions of the univariate transformation theorem.*

As $X \sim \mathcal{N}(\mu, \sigma^2)$, its PDF is $\left(1/\sqrt{2\pi}\sigma\right)e^{-(x-\mu)^2/2\sigma^2}$ for $-\infty < x < \infty$. In particular, the range of X is $(-\infty, \infty)$. We have $g'(x) = b$ for all x. Thus, if $b > 0$, g is strictly increasing, whereas, if $b < 0$, g is strictly decreasing. In either case, g satisfies the conditions of the univariate transformation theorem.

Step 2   *Identify the range of Y.*

The range of X is $(-\infty, \infty)$ and, as $Y = a + bX$, the range of Y is also $(-\infty, \infty)$.

Step 3   *Apply Equation (8.60) in the univariate transformation theorem.*

As $g'(x) = b$, the univariate transformation theorem applies to give

$$f_Y(y) = \frac{1}{|g'(x)|} f_X(x) = \frac{1}{|b|} \frac{1}{\sqrt{2\pi}\,\sigma} e^{-(x-\mu)^2/2\sigma^2},$$

where x is the unique real number such that $a + bx = y$.

Step 4   *For y in the range of Y, solve for x in the equation $y = g(x)$.*

We have $y = g(x) = a + bx$. Solving for x yields $x = (y - a)/b$.

Step 5   *Use Steps 2–4 to obtain a PDF of Y.*

Based on Steps 2–4, we obtain

$$f_Y(y) = \frac{1}{|b|} \frac{1}{\sqrt{2\pi}\,\sigma} e^{-(x-\mu)^2/2\sigma^2} = \frac{1}{|b|} \frac{1}{\sqrt{2\pi}\,\sigma} e^{-[(y-a)/b-\mu]^2/2\sigma^2}$$

or

$$f_Y(y) = \frac{1}{\sqrt{2\pi}\,(|b|\sigma)} e^{-[y-(a+b\mu)]^2/2b^2\sigma^2}, \qquad -\infty < y < \infty. \tag{8.63}$$

We also want to identify the probability distribution of $Y = a + bX$. In view of Equation (8.63) and Definition 8.6 on page 440, we conclude that $Y \sim \mathcal{N}(a + b\mu, b^2\sigma^2)$. This result is so important that we summarize it as Proposition 8.15.   ∎

◆◆◆   **Proposition 8.15**   **Linear Function of a Normal Random Variable**

If X has the normal distribution with parameters μ and σ^2 and a and $b \neq 0$ are real numbers, then $a + bX$ has the normal distribution with parameters $a + b\mu$ and $b^2\sigma^2$.

We can generalize Proposition 8.15 by considering a linear function of any continuous random variable X with a PDF. Let a and $b \neq 0$ be real numbers. Invoking the transformation method, we find that a PDF of $a + bX$ is

$$f_{a+bX}(y) = \frac{1}{|b|} f_X\left(\frac{y-a}{b}\right). \tag{8.64}$$

Consequently, applying a linear function to a continuous random variable corresponds to making a location and scale change to its PDF.

For comparison purposes, we use both the CDF method and the transformation method in Example 8.21.

EXAMPLE 8.21   *The CDF and Transformation Methods*

Tangent of a Uniform   Find a PDF of $\tan X$, where $X \sim \mathcal{U}(-\pi/2, \pi/2)$.

Solution   Here $Y = \tan X$—that is, $g(x) = \tan x$.

Method 1: CDF Method (Procedure 8.2 on page 465)

Step 1 *Identify a PDF of X.*
    Because $X \sim \mathcal{U}(-\pi/2, \pi/2)$, its PDF is $f_X(x) = 1/\pi$ for $-\pi/2 < x < \pi/2$, and $f_X(x) = 0$ otherwise.

Step 2 *Identify the range of Y.*
    The range of X is $(-\pi/2, \pi/2)$ and, as $Y = \tan X$, the range of Y is $(-\infty, \infty)$.

Step 3 *For y in the range of Y, determine the CDF of Y in terms of that of X.*
    For y in the range of Y—that is, for all real numbers y—we have

$$F_Y(y) = P(Y \leq y) = P(\tan X \leq y) = P(X \leq \arctan y) = F_X(\arctan y).$$

Step 4 *For y in the range of Y, obtain the derivative of the CDF of Y, where it exists.*
    From Step 3 and the chain rule,

$$F_Y'(y) = F_X'(\arctan y)\frac{d}{dy}\arctan y = \frac{1}{1+y^2}f_X(\arctan y) = \frac{1}{\pi(1+y^2)}.$$

Step 5 *A PDF of Y is given by $f_Y(y) = F_Y'(y)$, where the derivative exists, and $f_Y(y) = 0$ otherwise.*
    From the preceding steps, we conclude that a PDF of Y is

$$f_Y(y) = \frac{1}{\pi(1+y^2)}, \qquad -\infty < y < \infty.$$

Method 2: Transformation Method (Procedure 8.3 on page 467)

Step 1 *Identify a PDF of X and the range of X, and check whether g satisfies the conditions of the univariate transformation theorem.*
    Because $X \sim \mathcal{U}(-\pi/2, \pi/2)$, its PDF is $f_X(x) = 1/\pi$ for $-\pi/2 < x < \pi/2$, and $f_X(x) = 0$ otherwise. In particular, the range of X is $(-\pi/2, \pi/2)$. On the range of X,

we have $g'(x) = (d/dx) \tan x = \sec^2 x > 0$, which implies that g is strictly increasing. Hence g satisfies the conditions of the univariate transformation theorem.

Step 2 *Identify the range of Y.*

The range of X is $(-\pi/2, \pi/2)$ and, as $Y = \tan X$, the range of Y is $(-\infty, \infty)$.

Step 3 *Apply Equation (8.60) in the univariate transformation theorem.*

As $g'(x) = \sec^2 x$, the univariate transformation theorem applies to give

$$f_Y(y) = \frac{1}{|g'(x)|} f_X(x) = \frac{1}{\sec^2 x} \frac{1}{\pi},$$

where x is the unique real number such that $\tan x = y$.

Step 4 *For y in the range of Y, solve for x in the equation $y = g(x)$.*

We have $y = g(x) = \tan x$. Solving for x yields $x = \arctan y$.

Step 5 *Use Steps 2–4 to obtain a PDF of Y.*

Based on Steps 2–4, we obtain

$$f_Y(y) = \frac{1}{\sec^2 x} \frac{1}{\pi} = \frac{1}{\sec^2(\arctan y)} \frac{1}{\pi} = \frac{1}{\pi(1 + y^2)}, \qquad -\infty < y < \infty.$$

Of course, the CDF and transformation methods yield the same result. ∎

A random variable with the PDF found in Example 8.21 is called a **standard Cauchy random variable** and is said to have the **standard Cauchy distribution,** named in honor of the French mathematician Augustin-Louis Cauchy (1789–1857). More generally, we have the following definitions for a **Cauchy random variable** and **Cauchy distribution.**

DEFINITION 8.11 Cauchy Random Variable

A continuous random variable X is called a **Cauchy random variable** if it has a probability density function of the form

$$f_X(x) = \frac{\theta}{\pi\left(\theta^2 + (x - \eta)^2\right)} \qquad -\infty < x < \infty, \tag{8.65}$$

where η and $\theta > 0$ are real constants. We say that X has a **Cauchy distribution with parameters η and θ** or that X is **Cauchy distributed with parameters η and θ.** For convenience, we sometimes write $X \sim \mathcal{C}(\eta, \theta)$ to indicate that X has a Cauchy distribution with parameters η and θ.

Note: A $\mathcal{C}(0, 1)$ distribution—that is, the Cauchy distribution with parameters 0 and 1— is the standard Cauchy distribution.

Cauchy distributions are similar in shape to normal distributions, and the parameters η and θ for a Cauchy distribution play roles similar to the parameters μ and σ for a normal distribution. A Cauchy distribution is symmetric and centered on its η parameter, and its spread is determined by its θ parameter—the larger the θ parameter, the flatter and more spread out is the distribution. Thus, for a Cauchy distribution, η and θ are location and scale parameters, respectively. Figure 8.21 shows three Cauchy distributions, that is, the PDFs of three Cauchy random variables.

Figure 8.21 PDFs of three Cauchy random variables

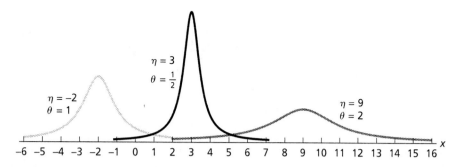

In Exercise 8.153, we ask you to verify that, if $X \sim \mathcal{C}(\eta, \theta)$, then

$$f_X(x) = \frac{1}{\theta}\, \psi\left(\frac{x - \eta}{\theta}\right), \tag{8.66}$$

where ψ is the standard Cauchy PDF. Equation (8.66) shows how any Cauchy PDF is obtained from the standard Cauchy PDF by a location and scale change.

Cauchy random variables have some unusual probabilistic properties. For instance, a Cauchy random variable doesn't have a mean (expected value). Nonetheless, Cauchy random variables are used in many applied areas such as anthropology, astronomy, physics, psychophysics, and statistics.

Simulation of Random Variables

With the advent of high-speed computers, simulation has become an indispensable tool in the sciences and related fields. By **simulation** of a random variable, we mean using a computer to generate observations of the random variable.

We can provide simulation methods for random variables based only on the use of a random number generator that simulates a $\mathcal{U}(0, 1)$ random variable, which we call a **basic random number generator.** Proposition 8.16 presents the necessary theory.

◆◆◆   **Proposition 8.16**

Let X be a continuous random variable whose range is an interval and whose CDF is strictly increasing on its range. Also, let $U \sim \mathcal{U}(0, 1)$. Then the following hold.

a) The random variable $F_X(X)$ is uniformly distributed on the interval $(0, 1)$.

b) The random variable $F_X^{-1}(U)$ has the same probability distribution as X.

Proof   We first recall that the CDF of U is given by $F_U(u) = u$, for $0 \leq u < 1$.

a) Let $V = F_X(X)$. Then, for $0 \leq v < 1$,

$$F_V(v) = P(V \leq v) = P\big(F_X(X) \leq v\big) = P\big(X \leq F_X^{-1}(v)\big) = F_X\big(F_X^{-1}(v)\big) = v.$$

Hence $V \sim \mathcal{U}(0, 1)$; that is, $F_X(X)$ is uniformly distributed on the interval $(0, 1)$.

b) Let $W = F_X^{-1}(U)$ and note that the ranges of W and X are identical. For w in the range of W,

$$F_W(w) = P(W \leq w) = P\big(F_X^{-1}(U) \leq w\big) = P\big(U \leq F_X(w)\big) = F_X(w).$$

Thus W and X have the same CDF and hence the same probability distribution.   ◆

Note: Almost all continuous random variables that arise in practice satisfy the conditions of Proposition 8.16. Nonetheless, as we ask you to show in Exercises 8.172 and 8.174, Proposition 8.16 holds provided only that X is a continuous random variable.

In Example 8.22, we apply Proposition 8.16 to exponential random variables. In doing so, we use the easily verified fact that, if the random variable U is uniformly distributed on the interval $(0, 1)$, so is the random variable $1 - U$. See Exercise 8.162.

EXAMPLE 8.22   *Simulation of Random Variables*

Exponential Random Variables   Let $X \sim \mathcal{E}(\lambda)$.

a) Show that $e^{-\lambda X} \sim \mathcal{U}(0, 1)$.

b) Show that, if $U \sim \mathcal{U}(0, 1)$, then $-\lambda^{-1} \ln U \sim \mathcal{E}(\lambda)$.

c) Explain how to use part (b) to simulate, say, 10,000 observations of an exponential random variable with parameter λ by using only a basic random number generator.

d) For a particular value of λ—say, $\lambda = 1/2$—conduct the simulation described in part (c) and provide a graph that portrays the results.

Solution   We recall that the CDF of X is $F_X(x) = 1 - e^{-\lambda x}$ for $x \geq 0$, and $F_X(x) = 0$ otherwise.

a) From Proposition 8.16(a), $F_X(X) \sim \mathcal{U}(0, 1)$, which here means $1 - e^{-\lambda X} \sim \mathcal{U}(0, 1)$. Therefore $e^{-\lambda X} = 1 - (1 - e^{-\lambda X}) \sim \mathcal{U}(0, 1)$.

b) Solving for x in the equation $u = 1 - e^{-\lambda x}$, we find that $x = -\lambda^{-1} \ln(1 - u)$; hence $F_X^{-1}(u) = -\lambda^{-1} \ln(1 - u)$. From Proposition 8.16(b), $F_X^{-1}(U)$ has the same distribution as X, which, in this case, means $-\lambda^{-1} \ln(1 - U) \sim \mathcal{E}(\lambda)$. As U and $1 - U$ have the same distribution, $-\lambda^{-1} \ln U \sim \mathcal{E}(\lambda)$.

c) Because of part (b), we can simulate 10,000 observations of an exponential random variable with parameter λ as follows. First we obtain 10,000 random numbers by using a basic random number generator. Then we apply the function $-\lambda^{-1} \ln u$ to each of those 10,000 random numbers.

d) We used a computer to carry out the procedure described in part (c) in the case where $\lambda = 1/2$. Figure 8.22 shows a histogram of the 10,000 simulated observations of an $\mathcal{E}(1/2)$ random variable. For comparison, we superimposed on the histogram the PDF of an $\mathcal{E}(1/2)$ random variable, which, as we know, is $(1/2)e^{-x/2}$ for $x > 0$.   ■

Figure 8.22 Histogram of 10,000 simulated observations of an $\mathcal{E}(1/2)$ random variable, with superimposed PDF

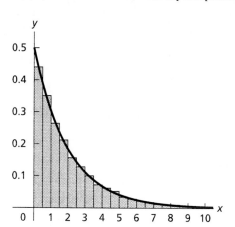

Continuous to Discrete

As we mentioned earlier, in the continuous case—that is, when X is a continuous random variable—$g(X)$ need not be a continuous random variable. Example 8.23 provides an instance where a function of a continuous random variable is a discrete random variable.

EXAMPLE 8.23 *Continuous to Discrete*

From Exponential to Geometric Let $X \sim \mathcal{E}(\lambda)$. Determine and identify the probability distribution of the random variable $\lceil X \rceil$, where $\lceil x \rceil$ denotes the *ceiling function,* the smallest integer greater than or equal to x.

Solution Let $Y = \lceil X \rceil$. A PDF of X is $f_X(x) = \lambda e^{-\lambda x}$ for $x > 0$, and $f_X(x) = 0$ otherwise. In particular, the range of X is $(0, \infty)$. Therefore the range of Y is $\mathcal{N}$, the positive integers. Thus Y is a discrete random variable whose PMF we now obtain. For $y \in \mathcal{N}$,

$$p_Y(y) = P(Y = y) = P(\lceil X \rceil = y) = P(y - 1 < X \le y)$$

$$= \int_{y-1}^{y} \lambda e^{-\lambda x}\, dx = e^{-\lambda(y-1)} - e^{-\lambda y} = e^{-\lambda(y-1)}(1 - e^{-\lambda})$$

$$= (1 - e^{-\lambda})\big(1 - (1 - e^{-\lambda})\big)^{y-1}.$$

Referring to Proposition 5.9 on page 229, we conclude that $Y \sim \mathcal{G}(1 - e^{-\lambda})$; that is, Y has the geometric distribution with parameter $1 - e^{-\lambda}$.

Thus, discretizing an exponential random variable yields a geometric random variable. This result is no surprise because, as we demonstrated earlier, the exponential distribution is the continuous analogue of the geometric distribution. ∎

EXERCISES 8.7 **Basic Exercises**

In Exercises 8.141–8.149, we cite exercises that appeared earlier in this chapter. Basically, their solutions involved the application of the CDF method, Procedure 8.2 on page 465. In each case, decide whether the transformation method, Procedure 8.3 on page 467, is appropriate; if it is, apply it.

8.141 Exercise 8.67(b) on page 435.

8.142 Exercise 8.72(a) on page 435.

8.143 Exercise 8.73(a) on page 435.

8.144 Exercise 8.105 on page 449.

8.145 Exercise 8.118 on page 461.

8.146 Exercise 8.120 on page 461.

8.147 Exercise 8.122 on page 462.

8.148 Exercise 8.128(b) on page 462.

8.149 Exercise 8.130(b) on page 462.

8.150 Solve Example 8.20 on page 468 by using the CDF method, Procedure 8.2 on page 465.

8.151 Use the transformation method to verify Equation (8.64): If X is a continuous random variable with a PDF and a and $b \neq 0$ are real numbers, then $f_{a+bX}(y) = |b|^{-1} f_X((y-a)/b)$.

8.152 An actuary modeled the lifetime of a device by the random variable $Y = 10X^{0.8}$, where X is exponentially distributed with parameter 1. Find a PDF of Y.

8.153 Let $X \sim \mathcal{C}(\eta, \theta)$ and let ψ denote the standard Cauchy PDF.
a) Verify Equation (8.66) on page 471: $f_X(x) = \theta^{-1} \psi((x-\eta)/\theta)$.
b) Use part (a) to conclude that $(X - \eta)/\theta$ has the standard Cauchy distribution.

8.154 Lognormal random variable: Let $X \sim \mathcal{N}(\mu, \sigma^2)$.
a) Determine a PDF of the random variable $Y = e^X$. *Note:* A random variable with this PDF is called a *lognormal random variable* and is said to have the *lognormal distribution with parameters μ and σ^2.*
b) Explain why the term "lognormal" is used for a random variable with a PDF as found in part (a).

8.155 Find a PDF of the random variable $Y = \sin X$ when
a) $X \sim \mathcal{U}(-\pi/2, \pi/2)$. **b)** $X \sim \mathcal{U}(-\pi, \pi)$.

8.156 A point is selected at random from the unit circle (i.e., boundary of the unit disk).
a) Determine a PDF of the x coordinate of the point chosen.
b) Without doing any calculations, determine a PDF of the y coordinate of the point chosen. Explain your reasoning.
c) Determine a PDF of the distance from the point chosen to the point $(1, 0)$.

8.157 Let X have the beta distribution with parameters α and β and set $Y = a + (b-a)X$, where a and b are real numbers with $a < b$.
a) Identify the range of Y. **b)** Obtain a PDF of Y.

8.158 Show that, if X has the standard Cauchy distribution, so does $1/X$.

8.159 Rayleigh random variable: A random variable R with PDF given by

$$f_R(r) = \frac{r}{\sigma^2} e^{-r^2/2\sigma^2}, \qquad r > 0,$$

and $f_R(r) = 0$ otherwise, is called a *Rayleigh random variable* and is said to have the *Rayleigh distribution with parameter* σ^2. Obtain and identify a PDF of the random variable R^2.

8.160 Let $Y = \sigma \sqrt{X}$, where X has the chi-square distribution with n degrees of freedom and σ is a positive real number.

a) Determine a PDF of the random variable Y.

b) Show that, if $n = 1$, then Y has the distribution of the absolute value of a $\mathcal{N}(0, \sigma^2)$ random variable.

c) Show that, if $n = 2$, then Y has the Rayleigh distribution with parameter σ^2, as defined in Exercise 8.159.

d) **Maxwell random variable:** A random variable S with PDF given by

$$f_S(s) = \frac{\sqrt{2/\pi}}{\sigma^3} s^2 e^{-s^2/2\sigma^2}, \qquad s > 0,$$

and $f_S(s) = 0$ otherwise, is called a *Maxwell random variable* and is said to have the *Maxwell distribution with parameter* σ^2. Show that, if $n = 3$, then Y has the Maxwell distribution with parameter σ^2.

8.161 Suppose that X has the standard Cauchy distribution. Determine and identify the probability distribution of the random variable $1/(1 + X^2)$.

8.162 Let U be uniformly distributed on the interval $(0, 1)$.

a) Without doing any computations, make an educated guess at the probability distribution of the random variable $1 - U$. Explain your reasoning.

b) Determine and identify the probability distribution of $1 - U$.

8.163 Let $X \sim \mathcal{U}(a, b)$ and let $U \sim \mathcal{U}(0, 1)$. Use Proposition 8.16 on page 471 to show that

a) $(X - a)/(b - a) \sim \mathcal{U}(0, 1)$.   b) $a + (b - a)U \sim \mathcal{U}(a, b)$.

8.164 Simulation: This exercise requires access to a computer or statistical software.

a) Explain how to use Exercise 8.163(b) to simulate 5000 observations of a uniform random variable on the interval (a, b) by using only a basic random number generator.

b) Conduct the simulation described in part (a) when $a = -0.5$ and $b = 0.5$.

c) Provide a graph of your results in part (b). Is the graph what you would expect? Explain.

8.165 Let $X \sim \mathcal{U}(0, 1)$ and let m and n be integers with $m < n$. Determine and identify the probability distribution of the random variable $\lfloor m + (n - m + 1)X \rfloor$, where $\lfloor x \rfloor$ denotes the *floor function*, the greatest integer smaller than or equal to x.

8.166 Simulation: This exercise requires access to a computer or statistical software.

a) Let m and n be integers with $m < n$. Explain how to simulate 10,000 observations of a discrete uniform random variable on the set $\{m, m + 1, \ldots, n\}$ by using only a basic random number generator. *Note:* Refer to Exercise 8.165.

b) Conduct the simulation described in part (a) when $m = 1$ and $n = 8$.

c) Provide a graph of your results in part (b). Is the graph what you would expect? Explain.

Theory Exercises

8.167 Show that Proposition 8.10 (page 441) is a special case of Proposition 8.15 (page 468).

8.168 Prove the univariate transformation theorem, Proposition 8.14 on page 466, in the case where g is strictly increasing on the range of X.

8.169 Let X be a continuous random variable with a PDF and let g be a real-valued function defined and differentiable on the range of X. Prove that a PDF of $Y = g(X)$ is

$$f_Y(y) = \frac{d}{dy} \int_{g^{-1}((-\infty,y])} f_X(x)\,dx,$$

for y in the range of Y, and $f_Y(y) = 0$ otherwise. Here $g^{-1}(A) = \{x : g(x) \in A\}$. Compare this result to Proposition 5.14 on page 247, which applies for discrete random variables X.

8.170 Use Exercise 8.169, the first fundamental theorem of calculus, and the chain rule to supply an alternative proof of the univariate transformation theorem, Proposition 8.14.

Advanced Exercises

8.171 Simulation: This exercise requires access to a computer or statistical software. Ten numbers are rounded to the nearest integer and then summed. Use simulation to approximate the probability that the sum of the rounded numbers will equal the rounded sum of the unrounded numbers. *Hint:* Refer to Exercise 8.164.

8.172 Probability integral transformation: Generalize Proposition 8.16(a) on page 471 by assuming only that X is a continuous random variable. That is, prove that, if X is a continuous random variable, then $F_X(X) \sim \mathcal{U}(0, 1)$. This result is known as the *probability integral transformation. Hint:* For each $0 < y < 1$, consider the set $\{x : F_X(x) = y\}$.

8.173 Is the converse of the probability integral transformation (Exercise 8.172) true—that is, does $F_X(X) \sim \mathcal{U}(0, 1)$ imply that X is a continuous random variable? Justify your answer.

8.174 Generalize Proposition 8.16(b) on page 471 by assuming only that X is a continuous random variable. That is, prove that, if X is a continuous random variable and $U \sim \mathcal{U}(0, 1)$, then $F_X^{-1}(U)$ has the same probability distribution as X, where, for $0 < y < 1$, we define $F_X^{-1}(y) = \min\{x : F_X(x) = y\}$.

CHAPTER REVIEW

Summary

In this chapter, we introduced continuous random variables and examined some of the concepts associated with them. Formally, we say that X is a continuous random variable if $P(X = x) = 0$ for all real numbers x.

Our initial task in analyzing continuous random variables was to find an analogue of the probability mass function (PMF), which is primary to the study of discrete random variables. To do so, we first investigated the cumulative distribution function (CDF), a function that applies to any type of random variable—discrete, continuous, or otherwise. The CDF of a random variable accumulates the probabilities of the random variable up to and including any specified value. Thus the CDF of a random variable X, denoted F_X, is defined by $F_X(x) = P(X \le x)$ for $x \in \mathcal{R}$.

Armed with the CDF, we found an analogue for continuous random variables of the PMF for discrete random variables. That analogue is the probability density function (PDF). Roughly speaking, a PDF of a continuous random variable gives the probability that the random variable takes a value near any specified value. A PDF of a continuous random variable X, denoted f_X, is essentially the derivative of the CDF.

More precisely, we say that a nonnegative function f_X is a PDF of X if and only if $P(a \leq X \leq b) = \int_a^b f_X(x)\, dx$ for all real numbers $a < b$. Equivalently, f_X is a PDF of X if and only if $F_X(x) = \int_{-\infty}^x f_X(t)\, dt$ for all $x \in \mathcal{R}$. From a PDF and the fundamental probability formula (FPF), we can obtain any probability for a continuous random variable under consideration.

We examined and applied the PDFs of several of the most widely used families of continuous random variables: the uniform, exponential, normal, gamma, beta, and triangular. Finally, we introduced the concept of taking functions of a continuous random variable to form other random variables. Of particular interest is the case where a function of a continuous random variable with a PDF is also a continuous random variable with a PDF. In those circumstances, we presented two methods for finding a PDF of the function of the random variable—the CDF method and the transformation method.

You Should Be Able To ...

1. define a continuous random variable.

2. determine whether a random variable is a continuous random variable.

3. define, state the basic properties of, and obtain the cumulative distribution function of a random variable.

4. use the cumulative distribution function to determine probabilities for a random variable.

5. define, state the basic properties of, and obtain a probability density function of a continuous random variable.

6. state and apply the fundamental probability formula for continuous random variables.

7. apply uniform random variables.

8. apply exponential random variables, including the lack-of-memory property.

9. apply normal random variables.

10. determine probabilities for a normal random variable by first converting to a standard normal random variable and then using Table I in the Appendix.

11. state, know when to use, and apply the local normal approximation to the binomial distribution.

12. define and state the basic properties of the gamma function.

13. apply gamma random variables.

14. define the beta function and express it in terms of the gamma function.

15. apply beta random variables.

16. apply triangular random variables.

17. obtain a PDF of a function of a continuous random variable by employing the CDF method, when appropriate.

18. obtain a PDF of a function of a continuous random variable by employing the transformation method, when appropriate.

19. determine the parameters for a normal random variable obtained as a linear function of another normal random variable.

20. apply Cauchy random variables.

21. apply basic simulation techniques for random variables.

Key Terms

basic random number

  generator, *471*

beta distribution, *457*

beta function, *456*

beta random variable, *457*

Cauchy distribution, *470*

Cauchy random variable, *470*

CDF method, *465*

chi-square distribution, *456*

continuous random variable, *402*

cumulative distribution, *407*

cumulative distribution

  function, *406*

density function, *417*

distribution function, *407*

Erlang distribution, *456*

exponential distribution, *432*

exponential random variable, *432*

fundamental probability

  formula, *424*

gamma distribution, *454*

gamma function, *451*

gamma random variable, *454*

lack-of-memory property, *433*

local De Moivre–Laplace

  theorem, *445*

location parameter, *441*

normal distribution, *440*

normal random variable, *440*

probability density function, *417*

scale parameter, *441*

shape parameter, *441*

simulation, *471*

standard Cauchy distribution, *470*

standard Cauchy random

  variable, *470*

standard normal distribution, *443*

standard normal random

  variable, *443*

Stirling's formula, *438*

transformation method, *467*

triangular distribution, *460*

triangular random variable, *460*

uniform distribution, *429*

uniform random variable, *429*

univariate transformation

  theorem, *466*

Chapter Review Exercises

Basic Exercises

8.175 Let X have the discrete uniform distribution on the set $\{1, 2, 3\}$. Your naive colleague, having been asked to identify the CDF of X, wrote

$$F_X(x) = \begin{cases} 1/3, & \text{if } x = 1; \\ 2/3, & \text{if } x = 2; \\ 1, & \text{if } x = 3. \end{cases}$$

Correct your colleague's errors.

8.176 A dart is thrown randomly at the unit square $\{(x, y) \in \mathcal{R}^2 : 0 \le x \le 1,\ 0 \le y \le 1\}$. Let W be the minimum of the x and y coordinates of the point at which the dart hits.

a) Determine the CDF of W.   **b)** Determine a PDF of W.

c) Identify the probability distribution of the random variable W.

8.177 Let X be uniformly distributed on the interval $(0, 1)$ and set

$$Y = \begin{cases} 0.2, & \text{if } X < 0.2; \\ X, & \text{if } 0.2 < X < 0.9; \\ 0.9, & \text{if } X > 0.9. \end{cases}$$

a) The piecewise definition of Y given in the previous display doesn't specify the value of Y when $X = 0.2$ or $X = 0.9$. Suppose that you let $Y = 83$ when $X = 0.9$ and that you let $Y = \sqrt{3\pi}$ when $X = 0.2$. Find the CDF of Y and show that the CDF would still be the same if the numbers 83 and $\sqrt{3\pi}$ were replaced by something else.

b) Is Y a discrete random variable? a continuous random variable? Explain your answers.

c) Let $S = \{ y : P(Y = y) > 0 \}$. Find the conditional probability distribution of Y, given the event $\{Y \in S\}$. Is $Y_{|\{Y \in S\}}$ a discrete random variable? a continuous random variable?

d) Find and identify the conditional probability distribution of Y, given the event $\{Y \notin S\}$. Is $Y_{|\{Y \notin S\}}$ a discrete random variable? a continuous random variable?

e) Suppose that the probability distribution of the random variable V is the same as that found in part (c) and that the probability distribution of the random variable W is the same as that found in part (d). Find numbers a and b such that $F_Y = aF_V + bF_W$.

f) For each of the random variables X, Y, V, and W, use its CDF to obtain the probability that it takes a value in the interval $[0.2, 0.9)$; $(0.2, 0.9)$; $(0.2, 0.9]$; $[0.2, 0.9]$.

8.178 Equation (5.32) on page 230 shows that, for an integer-valued random variable, its tail probabilities determine the PMF and hence the probability distribution. Do the tail probabilities, in general, determine the probability distribution of a random variable? Explain.

8.179 A gas station's storage tank has a capacity of m gallons. Once each week the storage tank is filled. The weekly demand has an exponential distribution with parameter λ.

a) Determine the CDF of weekly gas sales, in gallons.

b) Is the weekly sales, in gallons, a discrete random variable? a continuous random variable? Explain your answers.

c) What is the probability that the gas station runs out of gas before the end of the week?

8.180 Let X be a random variable with PDF given by $f_X(x) = c$ if $0 \le x \le 1/2$, $f_X(x) = 2c$ if $1/2 < x \le 1$, and $f_X(x) = 0$ otherwise.

a) Find all points of discontinuity of f_X.

b) Without actually obtaining F_X, find all of its points of discontinuity.

c) Obtain the constant c.

d) Explicitly obtain F_X and show from your result that F_X is everywhere continuous.

e) Graph f_X and F_X.

8.181 Let X have a PDF given by $f_X(x) = cx(5 - x)$ for $0 < x < 5$ and $f_X(x) = 0$ otherwise. Determine the value of c and explain your reasoning.

8.182 Let $g(x) = 1/x$ for $0 < x < 1$ and $g(x) = 0$ otherwise. Show that there is no constant c such that cg is a PDF.

8.183 Suppose that X has a PDF given by $f_X(x) = cx^2(1 - x)$ for $0 < x < 1$ and $f_X(x) = 0$ otherwise, where c is a constant.

a) Without determining c, identify the probability distribution of X.

b) Use part (a) and properties of the beta and gamma functions to obtain the value of c.

c) By using an appropriate property of PDFs, obtain the value of c.

d) Find $P(X \le 1/2)$.   **e)** Find $P(X \le 1/4 \mid X \le 1/2)$.

8.184 Suppose that a PDF of a continuous random variable X is known and that you want to find a PDF of the random variable X^2.

a) For which values of x would it make sense to write

$$f_{X^2}(x) = \frac{d}{dx} F_{X^2}(x) = \frac{d}{dx} P\left(-\sqrt{x} \le X \le \sqrt{x}\right) = \frac{d}{dx}\left[F_X\left(\sqrt{x}\right) - F_X\left(-\sqrt{x}\right)\right]?$$

b) Suppose that you decide to make $f_{X^2}(x)$ constant for all values of x for which it would be incorrect to proceed as in part (a). What must that constant be?

c) If f_X is an even function—that is, $f_X(-x) = f_X(x)$ for all $x \in \mathcal{R}$—how can the expression displayed in part (a) be simplified?

8.185 Refer to the petri-dish illustration of Example 8.2 on page 403. Let (Z, Θ) be the polar coordinates of the center of the first spot (visible bacteria colony) to appear.

a) In Example 8.8 on page 421, we determined that a PDF of Z is given by $f_Z(z) = 2z$ for $0 \leq z < 1$, and $f_Z(z) = 0$ otherwise. Identify the probability distribution of Z.

b) Determine a CDF of Θ, where the polar angle is taken to be in the interval $[0, 2\pi)$.

c) Identify the probability distribution of Θ.

8.186 The number $N(t)$ of phone calls arriving at a switchboard during the first t minutes after you start your stopwatch has a Poisson distribution with parameter $3.8t$. Let W_n be the time that elapses between when you start your stopwatch and when the nth phone call arrives.

a) On average, how many phone calls arrive during the first t minutes?

b) If it is known that $W_1 > t$, what can be said about $N(t)$? Similarly, what would $W_1 \leq t$ imply about $N(t)$?

c) Use part (b) and your knowledge of the Poisson distribution to find $P(W_1 > t)$.

d) Use part (c) to identify the probability distribution of W_1.

e) Determine and identify the probability distribution of W_3. *Hint:* See the note directly preceding Example 8.16 on page 454.

f) Determine and identify the probability distribution of W_n.

8.187 To obtain the area of a circle, its diameter is measured. However, owing to measurement error, the value obtained for the diameter is a random variable D. Determine a PDF of the computed area of the circle when

a) $D \sim \mathcal{U}(d - \epsilon, d + \epsilon)$.   b) $D \sim \mathcal{N}(d, \epsilon^2)$.

Here d denotes the actual diameter of the circle, and ϵ denotes a small positive real number.

8.188 Refer to Example 8.12 on page 433. There we let X denote the elapsed time in hours, starting at 6:00 P.M., until the first patient arrives at an emergency room. We noted that records indicate that $X \sim \mathcal{E}(6.9)$. Now let $N(t)$ denote the number of patients that arrive by time t, where t is measured in hours starting at 6:00 P.M.

a) Determine $P(N(t) = 0)$.

b) Without using the conditional probability rule, determine $P(N(2.5) = 0 \mid N(2) = 0)$.

c) Determine $P(N(2.5) - N(2) = 0 \mid N(2) = 0)$.

8.189 Let $U \sim \mathcal{U}(0, 1)$, let $b > 1$ be a real number, and let $X = -\log_b U$.

a) Use the CDF method to obtain and identify the probability distribution of X.

b) Use Example 8.22(b) on page 472 to obtain and identify the probability distribution of X.

8.190 Let Y denote the total annual claim amount for an insured person. Suppose that the annual number of claims, X, for an insured person satisfies $P(X = 0) = 1/2$, $P(X = 1) = 1/3$, and $P(X > 1) = 1/6$. Also suppose that, given $X = 1$, $Y \sim \mathcal{E}(1/5)$, and that, given $X > 1$, $Y \sim \mathcal{E}(1/8)$. Determine $P(4 < Y < 8)$.

8.191 Suppose that f is a PDF and that F is its corresponding CDF.

a) Show that F^n is a CDF for each $n \in \mathcal{N}$ and obtain the corresponding PDF.

b) Show that $1 - (1 - F)^n$ is a CDF for each $n \in \mathcal{N}$ and obtain the corresponding PDF.

8.192 At time $t > 0$, let $X(t)$ represent the (signed) amplitude of a signal emitted by a random signal generator. The waveform $X(t)$ is passed through a signal processor, resulting in an output $|X(t)|$. Determine a PDF of the output when

a) $X(t) \sim \mathcal{N}(0, \sigma^2 t)$.   b) $X(t) \sim \mathcal{U}(-t, t)$.

8.193 Suppose that the output of the signal processor considered in Problem 8.192 is given instead by $X^+(t) = \max\{X(t), 0\}$. Determine the CDF of the output when

a) $X(t) \sim \mathcal{N}(0, \sigma^2 t)$.   b) $X(t) \sim \mathcal{U}(-t, t)$.

c) Is the output in part (a) a continuous random variable? in part (b)?

8.194 In the kinetic theory of gasses, a molecule of an ideal gas has a velocity V with PDF given by $f_V(v) = av^2 e^{-bmv^2}$ for $v > 0$, where m is the mass of the molecule and a and b are positive real numbers.
a) Find a PDF of the kinetic energy $K = \frac{1}{2}mV^2$.
b) Identify the probability distribution of the random variable K.
c) Express a in terms of b and m.

8.195 According to Example 8.14 on page 444, human gestation periods, X, are normally distributed with $\mu = 266$ days and $\sigma = 16$ days.
a) Explain why the assertion that the probability distribution of X is normal is approximate rather than exact.
b) Explain why the approximation error is far too small to have any practical import.

8.196 An article by Scott M. Berry titled "Drive for Show and Putt for Dough" (*Chance*, 1999, Vol. 12(4), pp. 50–54) discussed driving distances of PGA players. Assuming that tee-shot distances of PGA players are normally distributed with $\mu = 272.2$ yards and $\sigma = 8.12$ yards, find the percentage of such tee shots that go
a) between 260 and 280 yards.   **b)** more than 300 yards.

8.197 Suppose that heights of men in their 20s are normally distributed with $\mu = 69$ inches and $\sigma = 2.5$ inches.
a) Find the probability that the height of a randomly chosen 20-year-old man is between 68 inches and 73 inches.
b) Find the number c such that 80% of men in their 20s have heights between $(69 - c)$ inches and $(69 + c)$ inches. Interpret your result.

8.198 In a large southwestern city, 47% of the adult population favors strict gun control laws. If a sample of size 100 is taken from this population, determine the probability that the number in favor will be
a) exactly 47.   **b)** 46, 47, or 48.
Use the normal approximation to the binomial, Proposition 8.12 on page 445.

8.199 The U.N. Children's Fund defines a household living in poverty as one with an income below 50% of the national median. Mexico leads the world with 26% of households living in poverty. Use the normal approximation to find the probability that, of 300 randomly chosen Mexican households, the number living in poverty is
a) exactly 26% of those sampled.
b) between 25% and 27%, inclusive, of those sampled.

8.200 Let X be a number obtained from a basic random number generator. Suppose that the device recording X only displays the first two digits of the decimal expansion of X. Call this random variable Y. Determine and identify the probability distribution of Y.

8.201 Suppose that $X \sim \Gamma(\alpha, \lambda)$. In Chapter 10, you will need to evaluate $\int_0^\infty x f_X(x)\, dx$ and $\int_0^\infty x^2 f_X(x)\, dx$. Evaluate these integrals without doing integration by parts, but rather by using properties of the gamma function and the fact that a PDF integrates to 1.

8.202 An insurance company models auto claim amounts, in thousands, with a random variable having probability density function $f_X(x) = xe^{-x}$ for $x > 0$ and $f_X(x) = 0$ otherwise.
a) Identify the probability distribution of the random variable X.
b) Obtain a formula for the CDF of X. *Hint:* See the note directly preceding Example 8.16 on page 454.
c) The insurance company expects to pay 100 claims if there is no deductible. How many claims should the company expect to pay if it decides to introduce a deductible of 1000?

Theory Exercises

Symmetric random variable: A random variable X is said to be a *symmetric random variable* if X and $-X$ have the same probability distribution. Exercises 8.203–8.207 examine symmetry of random variables and related topics.

8.203 Let X be a random variable.
a) Show that X is symmetric if and only if $P(X \le -x) = P(X \ge x)$ for all $x \in \mathcal{R}$.
b) Obtain an equivalent condition for symmetry of X in terms of its CDF.
c) For a symmetric random variable X, show that $P(|X| \le x) = 2F_X(x) - 1$ for all $x > 0$.
d) Show that 0 is a median of any symmetric random variable. *Note:* Refer to Exercise 8.83 on page 436 for the definition of median.

8.204 Suppose that X is a continuous random variable with a PDF. Show that X is symmetric if and only if f_X is a symmetric (even) function.

8.205 Let c be a real number. A random variable X is said to be *symmetric about c* if $X - c$ is a symmetric random variable. Suppose that X is a continuous random variable with a PDF. Obtain an equivalent condition for symmetry of X about c in terms of
a) the CDF of X.   b) a PDF of X.

8.206 Refer to Exercise 8.205. For each of the following random variables, decide whether it is symmetric about some real number c. If it is, specify c.
a) $X \sim \mathcal{U}(a, b)$ b) $X \sim \mathcal{E}(\lambda)$ c) $X \sim \mathcal{N}(\mu, \sigma^2)$ d) $X \sim \Gamma(\alpha, \lambda)$
e) $X \sim \text{Beta}(\alpha, \beta)$ f) $X \sim \mathcal{T}(a, b)$ g) $X \sim \mathcal{C}(\eta, \theta)$

8.207 Refer to Exercise 8.205.
a) Let X be symmetric about c. Show that c is a median of X. *Note:* Refer to Exercise 8.83 on page 436 for the definition of median.
b) Determine the median of a random variable with each of the following probability distributions: $\mathcal{U}(a, b)$; $\mathcal{N}(\mu, \sigma^2)$; $\mathcal{T}(a, b)$; and $\mathcal{C}(\eta, \theta)$.

Advanced Exercises

8.208 Suppose that F is the CDF of some random variable. Show that, for $h > 0$, each of the following functions is also the CDF of some random variable.

a) $G(x) = \dfrac{1}{h} \displaystyle\int_x^{x+h} F(t)\, dt$ b) $H(x) = \dfrac{1}{2h} \displaystyle\int_{x-h}^{x+h} F(t)\, dt$

8.209 Integral De Moivre–Laplace theorem: In this exercise, you are to explore the integral De Moivre–Laplace theorem, which provides an alternative (and generally more useful) method to the local De Moivre–Laplace theorem for approximating binomial probabilities by using a normal distribution.
a) Let $Y \sim \mathcal{N}(\mu, \sigma^2)$. Explain why $f_Y(y) \approx P(y - 1/2 \le Y \le y + 1/2)$ for each $y \in \mathcal{R}$. *Hint:* Construct a graph of the PDF of Y,
b) Let $X \sim \mathcal{B}(n, p)$. Use the local De Moivre–Laplace theorem (Proposition 8.12 on page 445) and the result of part (a) to show that, for integers a and b with $0 \le a \le b \le n$,

$$P(a \le X \le b) \approx \Phi\left(\frac{b + 1/2 - np}{\sqrt{np(1 - p)}} \right) - \Phi\left(\frac{a - 1/2 - np}{\sqrt{np(1 - p)}} \right).$$

This result is known as the *integral De Moivre–Laplace theorem.*
c) Use part (b) to approximate the probabilities required in Example 8.15 on page 445. Compare your results to those obtained in that example.

d) Suppose that you want to use a normal distribution to approximate the probability that a binomial random variable takes a value among a relatively large number of consecutive nonnegative integers. Explain why the integral De Moivre–Laplace theorem is preferable to the local De Moivre–Laplace theorem.

Quantiles of a random variable: Let $0 < p < 1$. A *pth quantile* (or *100pth percentile*) of a random variable X is any number ξ_p such that $P(X \le \xi_p) \ge p$ and $P(X \ge \xi_p) \ge 1 - p$. Note that the median of X, as defined in Exercise 8.83 on page 436, is also the 0.5th quantile (50th percentile). Every random variable has at least one *p*th quantile, but *p*th quantiles aren't necessarily unique. Exercises 8.210–8.217 examine quantiles.

8.210 Provide an example of a random variable X and a value p such that
a) X has a unique *p*th quantile.   **b)** X has more than one *p*th quantile.

8.211 Prove that ξ_p is a *p*th quantile of X if and only if either of the following holds.
a) $F_X(\xi_p-) \le p \le F_X(\xi_p)$ **b)** $P(X < \xi_p) \le p$ and $P(X > \xi_p) \le 1 - p$

8.212 Determine the *p*th quantile of an indicator random variable.

8.213 Suppose that X is a continuous random variable.
a) Show that, for each $0 < p < 1$, there exists an x such that $F_X(x) = p$.
b) Show that ξ_p is a *p*th quantile of X if and only if $F_X(\xi_p) = p$.

8.214 Suppose that X is a continuous random variable whose range is an interval and whose CDF is strictly increasing on its range. Show that X has a unique *p*th quantile given by $\xi_p = F_X^{-1}(p)$, where F_X^{-1} is the inverse function of F_X, defined on $(0, 1)$.

8.215 Referring to Exercise 8.214, find the *p*th quantile of each of the following random variables.
a) $X \sim \mathcal{U}(a, b)$ **b)** $X \sim \mathcal{E}(\lambda)$

8.216 Prove that any random variable X has at least one *p*th quantile. *Hint:* Consider the set $\{x : F_X(x) \ge p\}$.

8.217 An insurance company sells an automobile insurance policy that covers losses incurred by a policyholder, subject to a deductible of $100. Losses incurred follow an exponential distribution with parameter 1/300. Determine the 95th percentile of actual losses that exceed the deductible.

Pierre de Fermat *1601–1665*

Pierre Fermat was born in Beaumont-de-Lomagne, France, on August 17, 1601. Fermat's father was a wealthy leather merchant and second consul of Beaumont-de-Lomagne. In his youth, Fermat was probably educated at the local Franciscan monastery. Later he attended the University of Toulouse.

In the latter half of the 1620s, Fermat moved to Bordeaux, where he began serious mathematical research. From Bordeaux, Fermat went to Orléans where he studied law at the University. He received a degree in civil law and, by 1631, Fermat was a lawyer and government official in Toulouse. Owing to the office he held, he was permitted to change his name from Pierre Fermat to Pierre de Fermat.

Fermat's initial contact with the scientific community came through his study of free fall, although he generally had little interest in physical applications of mathematics. For several reasons, Fermat was out of touch with his scientific colleagues from 1643–1654. Nonetheless, it was during this period that Fermat worked on number theory. It is this theory for which he is best remembered—in particular, for his statement and claimed proof of *Fermat's last theorem*: $x^n + y^n = z^n$ has no nonzero integer solutions when $n \geq 3$. Fermat's "proof" of the theorem was never found; British mathematician Andrew Wiles's celebrated proof appeared only recently, in 1994.

In 1654, Fermat resumed his correspondence with Paris mathematicians when Blaise Pascal wrote to him to ask for confirmation about his ideas on probability. Their short correspondences—which included discussions of problems posed by a gambler, Chevalier de Méré, to Fermat—contained some basic combinatorial arguments and their applications to computing probabilities. Because of these correspondences, some regard Fermat and Pascal as joint founders of probability theory.

Together with René Descartes, Fermat is considered one of the two leading mathematicians of the first half of the seventeenth century. Fermat died on January 12, 1665, in Castres, France.

CHAPTER NINE

Jointly Continuous Random Variables

INTRODUCTION

In Chapter 8, we introduced continuous random variables. There we considered one continuous random variable at a time. However, we often need to examine two or more continuous random variables simultaneously. For instance, consider the continuous random variables height and weight. Although we can study each of those continuous random variables separately, examining them jointly to discover relationships between them is important.

In this chapter, we discuss the simultaneous analysis of two or more continuous random variables. We begin in Sections 9.1–9.3 by presenting joint cumulative distribution functions and joint probability density functions. In Section 9.4, we investigate marginal and conditional probability density functions.

In Section 9.5, we examine the concept of independence of random variables as it relates specifically to continuous random variables. In Section 9.6, we discuss the process of forming a continuous random variable as a function of two or more continuous random variables. As an important special case of this process, we consider forming a continuous random variable by taking the sum or quotient of continuous random variables.

In the final section of this chapter, Section 9.7, we present a formula that expresses a joint PDF of a transformation of continuous random variables in terms of a joint PDF of the continuous random variables being transformed.

9.1 Joint Cumulative Distribution Functions

In Chapter 8, we introduced the concept of the *probability density function* (PDF) of a continuous random variable. Specifically, a nonnegative function f_X is said to be a PDF of the continuous random variable X if, for all real numbers $a < b$,

$$P(a \le X \le b) = \int_a^b f_X(x)\,dx. \tag{9.1}$$

Intuitively, as we demonstrated in Chapter 8, a PDF of X gives the probability that X takes a value near x:

$$P(x \le X \le x + \Delta x) \approx f_X(x)\Delta x, \tag{9.2}$$

for Δx close to 0.

To analyze two or more continuous random variables simultaneously, we use the *joint probability density function.* For the most part, we concentrate on the bivariate case of two continuous random variables defined on the same sample space. The generalization to more than two continuous random variables is conceptually straightforward and we usually leave the details to you.

Two approaches are possible for introducing the joint PDF of two continuous random variables: by analogy with the univariate continuous case and by analogy with the bivariate discrete case. For the second approach, we need the concept of the *joint cumulative distribution function,* which is important in its own right. We present that concept now so as not to interrupt the flow of the discussion of joint PDFs.

Joint Cumulative Distribution Functions

Recall that the *cumulative distribution function* (CDF) of one random variable X, denoted F_X, is defined by $F_X(x) = P(X \le x)$. Similarly, the **joint cumulative distribution function** of two random variables is defined as follows.

DEFINITION 9.1   Joint CDF: Bivariate Case

Let X and Y be random variables defined on the same sample space. Then the **joint cumulative distribution function** of X and Y, denoted $F_{X,Y}$, is the real-valued function defined on $\mathcal{R}^2$ by

$$F_{X,Y}(x, y) = P(X \le x, Y \le y), \qquad x, y \in \mathcal{R}. \tag{9.3}$$

We use **joint CDF** as an abbreviation for "joint cumulative distribution function." See Figure 9.1.

Figure 9.1 $F_{X,Y}(x, y)$ equals the probability that the random point (X, Y) falls in the shaded region

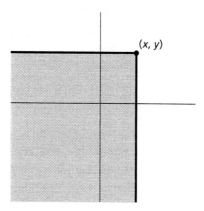

Note the following.

- The joint CDF applies to any two random variables (defined on the same sample space), regardless of type.

- The joint CDF of two random variables is denoted by an uppercase F subscripted with the two letters representing the random variables. For instance, $F_{X,Y}$ denotes the joint CDF of the random variables X and Y, and $F_{U,V}$ denotes the joint CDF of the random variables U and V.

- Although it's by no means obvious, the joint CDF of two random variables completely determines their joint probability distribution, regardless of type. This statement also holds for the general multivariate case.

EXAMPLE 9.1   *Joint Cumulative Distribution Functions*

Random Points in the Unit Square   Suppose that a point is selected at random from the unit square, $\{(x, y) \in \mathcal{R}^2 : 0 < x < 1, 0 < y < 1\}$. Let X and Y denote the x coordinate and y coordinate, respectively, of the point obtained. Determine the joint CDF of the random variables X and Y.

Solution   The sample space, Ω, for this random experiment is the unit square. Because we are selecting a point at random, use of a geometric probability model is appropriate. Thus, for each event E,

$$P(E) = \frac{|E|}{|\Omega|} = |E|, \qquad (9.4)$$

where $|E|$ denotes the area (in the extended sense) of the set E. The random variables X and Y are defined on Ω by $X(x, y) = x$ and $Y(x, y) = y$.

To find the joint CDF of X and Y, we begin by observing that the range of (X, Y) is the unit square. Therefore, if either x or y is negative, we have $F_{X,Y}(x, y) = 0$; and, if both x and y are 1 or greater, we have $F_{X,Y}(x, y) = 1$. We must still find $F_{X,Y}(x, y)$ in all other cases, of which we can consider three.

Case 1: $0 \le x < 1$ and $0 \le y < 1$

In this case, $\{X \le x, Y \le y\}$ is the event that the randomly selected point lies in the rectangle with vertices $(0, 0)$, $(x, 0)$, $(0, y)$, and (x, y), specifically, $(0, x] \times (0, y]$. By Equation (9.4),

$$F_{X,Y}(x, y) = P(X \le x, Y \le y) = |\{X \le x, Y \le y\}| = |(0, x] \times (0, y]| = xy.$$

Case 2: $0 \le x < 1$ and $y \ge 1$

In this case, $\{X \le x, Y \le y\}$ is the event that the randomly selected point lies in the rectangle with vertices $(0, 0)$, $(x, 0)$, $(0, 1)$, and $(x, 1)$, specifically, $(0, x] \times (0, 1)$. By Equation (9.4),

$$F_{X,Y}(x, y) = P(X \le x, Y \le y) = |\{X \le x, Y \le y\}| = |(0, x] \times (0, 1)| = x.$$

Case 3: $x \ge 1$ and $0 \le y < 1$

In this case, $\{X \le x, Y \le y\}$ is the event that the randomly selected point lies in the rectangle with vertices $(0, 0)$, $(1, 0)$, $(0, y)$, and $(1, y)$, specifically, $(0, 1) \times (0, y]$. By Equation (9.4),

$$F_{X,Y}(x, y) = P(X \le x, Y \le y) = |\{X \le x, Y \le y\}| = |(0, 1) \times (0, y]| = y.$$

We conclude that the joint CDF of the random variables X and Y is

$$F_{X,Y}(x, y) = \begin{cases} 0, & \text{if } x < 0 \text{ or } y < 0; \\ xy, & \text{if } 0 \le x < 1 \text{ and } 0 \le y < 1; \\ x, & \text{if } 0 \le x < 1 \text{ and } y \ge 1; \\ y, & \text{if } x \ge 1 \text{ and } 0 \le y < 1; \\ 1, & \text{if } x \ge 1 \text{ and } y \ge 1, \end{cases}$$

as required.   ■

Joint CDFs have properties similar to univariate CDFs. For our purposes, however, we need only Proposition 9.1, which is the two-dimensional analogue of Proposition 8.2(a) on page 412. Further properties of joint CDFs are considered in the exercises.

◆◆◆ Proposition 9.1

Let X and Y be two random variables defined on the same sample space. Then, for all real numbers a, b, c, and d, with $a < b$ and $c < d$,

$$\begin{aligned} P(a &< X \le b, c < Y \le d) \\ &= F_{X,Y}(b, d) - F_{X,Y}(b, c) - F_{X,Y}(a, d) + F_{X,Y}(a, c). \end{aligned} \tag{9.5}$$

If X and Y are both continuous random variables, Equation (9.5) holds, regardless of whether the inequalities are weak or strict.

Proof   Set $R = \{(x, y) : a < x \le b, c < y \le d\}$ and note that

$$P(a < X \le b, c < Y \le d) = P((X, Y) \in R). \tag{9.6}$$

For $(u, v) \in \mathcal{R}^2$, set $H_{u,v} = \{(x, y) : x \le u, y \le v\}$ and note that

$$F_{X,Y}(u, v) = P((X, Y) \in H_{u,v}). \tag{9.7}$$

We have $H_{b,d} = R \cup (H_{b,c} \cup H_{a,d})$, as you can easily see by drawing a picture. The sets R and $H_{b,c} \cup H_{a,d}$ are disjoint and, moreover, $H_{b,c} \cap H_{a,d} = H_{a,c}$. Applying the additivity property of a probability measure and the general addition rule, we get

$$\begin{aligned} P((X, Y) \in H_{b,d}) &= P((X, Y) \in R) + P((X, Y) \in H_{b,c}) \\ &\quad + P((X, Y) \in H_{a,d}) - P((X, Y) \in H_{a,c}). \end{aligned} \tag{9.8}$$

Referring now to Equations (9.6) and (9.7), we obtain the required result after simple rearrangement of Equation (9.8).   ◆

EXAMPLE 9.2   *Joint Cumulative Distribution Functions*

Random Points in the Unit Square   Let X and Y denote the x and y coordinates, respectively, of a point selected at random from the unit square. For $0 < a < b < 1$ and $0 < c < d < 1$, determine $P(a < X < b, c < Y < d)$ by using
a) the joint CDF obtained in Example 9.1.
b) the appropriate geometric probability model.

Solution   a) We first note that X and Y are continuous random variables. From Proposition 9.1 and the joint CDF of X and Y on page 488, we get

$$\begin{aligned} P(a < X < b, c < Y < d) &= P(a < X \le b, c < Y \le d) \\ &= F_{X,Y}(b, d) - F_{X,Y}(b, c) - F_{X,Y}(a, d) + F_{X,Y}(a, c) \\ &= bd - bc - ad + ac = (b - a)(d - c). \end{aligned}$$

b) We first note that $\{a < X < b, c < Y < d\}$ is the event that the randomly selected point lies in the rectangle with vertices (a, c), (b, c), (a, d), and (b, d)—specifically, the rectangle $(a, b) \times (c, d)$. Therefore, by Equation (9.4) on page 487,

$$\begin{aligned} P(a < X < b, c < Y < d) &= |\{a < X < b, c < Y < d\}| \\ &= |(a, b) \times (c, d)| = (b - a)(d - c). \end{aligned}$$

Of course, the answers obtained in parts (a) and (b) are the same.   ■

The definition of *joint cumulative distribution function* in the general multivariate case is a simple extension of that for the bivariate case. If $X_1, \ldots, X_m$ are m random variables defined on the same sample space, the **joint cumulative distribution function** of $X_1, \ldots, X_m$, denoted $F_{X_1,\ldots,X_m}$, is the real-valued function defined on $\mathcal{R}^m$ by

$$F_{X_1,\ldots,X_m}(x_1, \ldots, x_m) = P(X_1 \le x_1, \ldots, X_m \le x_m). \tag{9.9}$$

Several properties of multivariate joint cumulative distribution functions are considered in the exercises.

Marginal Cumulative Distribution Functions

Let X and Y be two random variables with joint CDF $F_{X,Y}$. In the joint context, each of F_X (the CDF of X) and F_Y (the CDF of Y) is called a **marginal cumulative distribution function,** or marginal CDF. Technically, however, the adjective "marginal" is redundant. To obtain a marginal CDF from the joint CDF, we use Proposition 9.2, whose proof we leave to you as Exercise 9.13.

◆◆◆ **Proposition 9.2   Obtaining Marginal CDFs From the Joint CDF**

Let X and Y be random variables defined on the same sample space. Then

$$F_X(x) = \lim_{y \to \infty} F_{X,Y}(x, y) \equiv F_{X,Y}(x, \infty), \qquad x \in \mathcal{R}, \tag{9.10}$$

and

$$F_Y(y) = \lim_{x \to \infty} F_{X,Y}(x, y) \equiv F_{X,Y}(\infty, y), \qquad y \in \mathcal{R}. \tag{9.11}$$

In words, we can obtain the (marginal) CDF of X by letting $y \to \infty$ in the joint CDF of X and Y and, likewise, we can obtain the (marginal) CDF of Y by letting $x \to \infty$ in the joint CDF of X and Y.

When analyzing more than two random variables, we have many more marginal CDFs. For instance, suppose that X, Y, and Z are three random variables defined on the same sample space. The joint CDF of these random variables is

$$F_{X,Y,Z}(x, y, z) = P(X \le x, Y \le y, Z \le z).$$

In this case, there are $\binom{3}{1} = 3$ univariate marginal CDFs—namely, F_X (the CDF of X), F_Y (the CDF of Y), and F_Z (the CDF of Z). Additionally, there are $\binom{3}{2} = 3$ bivariate marginal CDFs—namely, $F_{X,Y}$ (the joint CDF of X and Y), $F_{X,Z}$ (the joint CDF of X and Z), and $F_{Y,Z}$ (the joint CDF of Y and Z). Each univariate marginal CDF is obtained by letting the other two variables go to infinity in the joint CDF. For instance,

$$F_X(x) = \lim_{\substack{y \to \infty \\ z \to \infty}} F_{X,Y,Z}(x, y, z) \equiv F_{X,Y,Z}(x, \infty, \infty).$$

Similarly, each bivariate marginal CDF is obtained by letting the other variable go to infinity in the joint CDF. For instance,

$$F_X(x, y) = \lim_{z \to \infty} F_{X,Y,Z}(x, y, z) \equiv F_{X,Y,Z}(x, y, \infty).$$

Generally, if $X_1, \ldots, X_m$ are m random variables defined on the same sample space, there are $\binom{m}{j}$ j-variate marginal CDFs ($1 \le j \le m - 1$). Each j-variate marginal CDF is obtained by letting the other $m - j$ variables go to infinity in the joint CDF.

EXERCISES 9.1   Basic Exercises

9.1 Let X and Y be random variables defined on the same sample space. Set $U = a + bX$ and $V = c + dY$, where a, b, c, and d are real numbers with $b > 0$ and $d > 0$. Determine the joint CDF of U and V in terms of that of X and Y.

9.2 Suppose that X and Y are continuous random variables defined on the same sample space. Let $U = X^2$ and $V = Y^2$. Find the joint CDF of U and V in terms of that of X and Y.

9.3 Refer to Example 9.1 on page 487.
a) Use Proposition 9.2 to obtain the marginal CDF of X.
b) Use the result of part (a) to identify the (marginal) probability distribution of X.
c) Repeat parts (a) and (b) for the random variable Y.

9.4 Let X and Y be the x and y coordinates, respectively, of a point selected at random from the upper half of the unit disk—that is, from the set $\{(x, y) \in \mathcal{R}^2 : x^2 + y^2 < 1, \ y > 0\}$.
a) Determine the CDF of the random variable X without using the joint CDF of X and Y.
b) Determine the CDF of the random variable Y without using the joint CDF of X and Y.
c) Determine the joint CDF of the random variables X and Y.
d) Use the result of part (c) and Proposition 9.2 to obtain the marginal CDF of X. Compare your answer with that found in part (a).
e) Use the result of part (c) and Proposition 9.2 to obtain the marginal CDF of Y. Compare your answer with that found in part (b).

9.5 Let X and Y be the x and y coordinates, respectively, of a point selected at random from the four vertices of the unit square.
a) Determine the joint CDF of the random variables X and Y.
b) Use the result of part (a) to obtain the marginal CDF of X.
c) Use the result of part (b) to identify the (marginal) probability distribution of X.
d) Repeat parts (b) and (c) for the random variable Y.
e) What is the relationship between the joint CDF of X and Y and the marginal CDFs of X and Y?

9.6 Let X and Y be independent random variables, both having the Bernoulli distribution with parameter p.
a) Find the joint PMF of X and Y.
b) Use part (a) and Definition 9.1 on page 486 to obtain the joint CDF of X and Y.
c) Find the marginal CDFs of X and Y from the joint CDF of X and Y obtained in part (b).
d) Find the marginal CDFs of X and Y without using the joint CDF of X and Y.
e) Use the results of part (d) and the independence of X and Y to determine the joint CDF of X and Y. *Hint:* Apply the definition of independent random variables given in Definition 6.4 on page 291 with the sets $A = (-\infty, x]$ and $B = (-\infty, y]$.
f) Compare your answers in parts (b) and (e).

9.7 Let X and Y be independent exponential random variables with parameters λ and μ, respectively.
a) Determine the marginal CDFs of X and Y without using the joint CDF of X and Y.
b) Use part (a) and the independence of X and Y to obtain the joint CDF of X and Y.

9.8 The function $G: \mathcal{R} \to \mathcal{R}$ defined by $G(x) = 1 - e^{-x}$ for $x \geq 0$, and $G(x) = 0$ otherwise, is a univariate CDF—namely, the CDF of an exponential random variable with parameter $\lambda = 1$. Show that the function $F: \mathcal{R}^2 \to \mathcal{R}$ defined by $F(x, y) = 1 - e^{-(x+y)}$ for $x, y \geq 0$, and $F(x, y) = 0$ otherwise, can't be a joint CDF. *Hint:* Suppose that F is the joint CDF of random variables X and Y and compute an appropriate probability of the form $P(a < X \leq b, \ c < Y \leq d)$.

9.9 For random variables X and Y defined on the same sample space, let $U = \min\{X, Y\}$ and $V = \max\{X, Y\}$.
a) Determine the CDF of V in terms of the joint CDF of X and Y.

b) Determine the CDF of U in terms of the joint and marginal CDFs of X and Y.
c) Determine the CDF of U in terms of the joint CDF of X and Y.

9.10 In this exercise, you are to generalize the results of Exercise 9.9. Let $X_1, \ldots, X_m$ be random variables defined on the same sample space and set $U = \min\{X_1, \ldots, X_m\}$ and $V = \max\{X_1, \ldots, X_m\}$.
a) Determine the CDF of V in terms of the joint CDF of $X_1, \ldots, X_m$.
b) Determine the CDF of U in terms of the joint and marginal CDFs of $X_1, \ldots, X_m$. *Hint:* Use the inclusion–exclusion principle, Proposition 2.10 on page 73.

Theory Exercises

9.11 Suppose that X and Y are random variables defined on the same sample space. Prove that $F_{X,Y}$—the joint CDF of X and Y—satisfies the following properties.
a) $F_{X,Y}$ is nondecreasing in each variable separately.
b) $F_{X,Y}$ is everywhere right-continuous in each variable separately.
c) $F_{X,Y}(x, -\infty) \equiv \lim_{y \to -\infty} F_{X,Y}(x, y) = 0$
d) $F_{X,Y}(-\infty, y) \equiv \lim_{x \to -\infty} F_{X,Y}(x, y) = 0$
e) $F_{X,Y}(\infty, \infty) \equiv \lim_{x,y \to \infty} F_{X,Y}(x, y) = 1$.

9.12 Generalize Proposition 9.1 on page 488 to the case of three random variables, X_1, X_2, and X_3, defined on the same sample space. In other words, state and prove a formula for $P(a_1 < X_1 \le b_1, a_2 < X_2 \le b_2, a_3 < X_3 \le b_3)$ in terms of the joint CDF of X_1, X_2, and X_3, where a_j and b_j are real numbers with $a_j < b_j$ for $j = 1, 2, 3$.

9.13 Prove Proposition 9.2 on page 490, which gives formulas for obtaining marginal CDFs from the joint CDF of two random variables.

Advanced Exercises

9.14 Let X and Y be the x and y coordinates, respectively, of a point selected at random from the diagonal of the unit square—that is, from $\{ (x, y) \in \mathcal{R}^2 : y = x, \ 0 < x < 1 \}$.
a) Determine the CDF of the random variable X without using the joint CDF of X and Y.
b) Use the result of part (a) to identify the probability distribution of X.
c) Determine the joint CDF of the random variables X and Y.
d) Use the result of part (c) and Proposition 9.2 on page 490 to obtain the marginal CDF of X. Compare your answer with that found in part (a).
e) Without doing any further computations, determine the marginal CDF of Y and identify the probability distribution of Y.

9.15 Let X and Y be random variables defined on the same sample space.
a) Prove that, if X and Y are independent random variables, then their joint CDF equals the product of their marginal CDFs—that is, $F_{X,Y}(x, y) = F_X(x)F_Y(y)$ for all $x, y \in \mathcal{R}$. *Hint:* Refer to Definition 6.4 on page 291.
b) Prove that, if the joint CDF of X and Y equals the product of their marginal CDFs, then X and Y are independent random variables. *Hint:* Use the fact that the joint CDF of two random variables completely determines their joint probability distribution.

9.16 Generalize Exercise 9.15 to the multivariate case of m random variables defined on the same sample space. *Hint:* Refer to Definition 6.5 on page 296. Also, use the fact that the joint CDF of m random variables completely determines their joint probability distribution.

9.17 Let $X_1, \ldots, X_n$ be a random sample from the distribution of a random variable with CDF F—that is, $X_1, \ldots, X_n$ are independent and identically distributed random variables with common CDF F. Set $X = \min\{X_1, \ldots, X_n\}$ and $Y = \max\{X_1, \ldots, X_n\}$. Find the joint CDF of X and Y in terms of F. *Hint:* First obtain $P(X > x, Y \leq y)$.

9.18 Generalize Proposition 9.2 on page 490 to the multivariate case of m random variables by proving the following result. Suppose that $X_1, \ldots, X_m$ are random variables defined on the same sample space. Let j be a positive integer between 1 and m, inclusive, and let $k_1 < \cdots < k_j$ be integers between 1 and m, inclusive. Then the joint CDF of $X_{k_1}, \ldots, X_{k_j}$ can be obtained from the formula

$$F_{X_{k_1}, \ldots, X_{k_j}}(x_{k_1}, \ldots, x_{k_j}) = \lim_{\substack{x_i \to \infty \\ i \notin \{k_1, \ldots, k_j\}}} F_{X_1, \ldots, X_m}(x_1, \ldots, x_m),$$

for all $x_{k_1}, \ldots, x_{k_j} \in \mathcal{R}$.

9.2   Introducing Joint Probability Density Functions

As we noted at the beginning of Section 9.1, two approaches are possible for introducing joint probability density functions. One approach is by analogy with the discrete case.

For discrete random variables X and Y, the probabilities $P(X = x, Y = y)$, which constitute the joint probability mass function $p_{X,Y}$, completely determine the joint probability distribution of X and Y in the sense that, once we know them, we can obtain any joint probability for X and Y. Indeed, by the FPF for two discrete random variables (Proposition 6.3 on page 265),

$$P\big((X, Y) \in A\big) = \sum_{(x,y) \in A} \sum p_{X,Y}(x, y), \qquad A \subset \mathcal{R}^2. \tag{9.12}$$

For continuous random variables X and Y, the probabilities $P(X = x, Y = y)$ are all 0 and consequently are of no use in describing the joint probability distribution of X and Y. What we need for continuous random variables is a viable analogue of the joint PMF for discrete random variables. We can get this analogue by considering, for each $(x, y) \in \mathcal{R}^2$, the probability that X is "near" x and Y is "near" y, rather than the probability that X equals x and Y equals y.

Suppose then that X and Y are continuous random variables defined on the same sample space. To examine the probability that X is "near" x and Y is "near" y, we let Δx and Δy represent small positive numbers and consider the joint probability that X takes a value between x and $x + \Delta x$ and Y takes a value between y and $y + \Delta y$, that is, $P(x \leq X \leq x + \Delta x, y \leq Y \leq y + \Delta y)$. As X and Y are continuous random variables, we have from Proposition 9.1 on page 488 that

$$P(x \leq X \leq x + \Delta x, y \leq Y \leq y + \Delta y)$$
$$= F_{X,Y}(x + \Delta x, y + \Delta y) - F_{X,Y}(x + \Delta x, y) \tag{9.13}$$
$$- F_{X,Y}(x, y + \Delta y) + F_{X,Y}(x, y).$$

Assuming the existence of a mixed second-order partial derivative of $F_{X,Y}$, we have, in view of Equation (9.13), that

$$P(x \leq X \leq x + \Delta x, \ y \leq Y \leq y + \Delta y)$$

$$\approx \frac{\partial}{\partial y} F_{X,Y}(x + \Delta x, y)\Delta y - \frac{\partial}{\partial y} F_{X,Y}(x, y)\Delta y$$

$$= \left(\frac{\partial}{\partial y} F_{X,Y}(x + \Delta x, y) - \frac{\partial}{\partial y} F_{X,Y}(x, y) \right) \Delta y \approx \left(\frac{\partial^2}{\partial x \, \partial y} F_{X,Y}(x, y)\Delta x \right) \Delta y,$$

or

$$P(x \leq X \leq x + \Delta x, \ y \leq Y \leq y + \Delta y) \approx \frac{\partial^2}{\partial x \, \partial y} F_{X,Y}(x, y)\Delta x \Delta y. \qquad (9.14)$$

We now provide a heuristic argument to show that $\partial^2 F_{X,Y}/\partial x \partial y$—a mixed second-order partial derivative function of the joint CDF of X and Y—gives the required analogue for continuous random variables of the joint PMF for discrete random variables. That is, once we know $\partial^2 F_{X,Y}/\partial x \partial y$, we can determine any joint probability for X and Y. Results from measure theory imply that we need only verify this fact for joint probabilities involving bounded closed rectangles.

So we want to show that, for all real numbers $a < b$ and $c < d$, we can express the probability that (X, Y) takes a value in the bounded closed rectangle $[a, b] \times [c, d]$ in terms of $\partial^2 F_{X,Y}/\partial x \partial y$. To do so, we consider the partition of that rectangle determined by $a = x_0 < x_1 < \cdots < x_n = b$ and $c = y_0 < y_1 < \cdots < y_n = d$, where n is a large positive integer and $x_j = a + j(b - a)/n$ and $y_k = c + k(d - c)/n$ for $0 \leq j, k \leq n$. Applying the additivity property of a probability measure, the fact that X and Y are continuous random variables, and Relation (9.14), we get

$$P(a \leq X \leq b, \ c \leq Y \leq d) = \sum_{j=0}^{n-1} \sum_{k=0}^{n-1} P(x_j \leq X \leq x_{j+1}, \ y_k \leq Y \leq y_{k+1})$$

$$\approx \sum_{j=0}^{n-1} \sum_{k=0}^{n-1} \frac{\partial^2}{\partial x \, \partial y} F_{X,Y}(x_j, y_k)\Delta x_j \Delta y_k,$$

where we have let $\Delta x_j = x_{j+1} - x_j = (b - a)/n$ and $\Delta y_k = y_{k+1} - y_k = (d - c)/n$ for $0 \leq j, k \leq n - 1$. However, because n is large, the last term in the previous display, which is a double Riemann sum, approximately equals the double Riemann integral of $\partial^2 F_{X,Y}/\partial x \partial y$ over the rectangle $[a, b] \times [c, d]$. In other words,

$$P(a \leq X \leq b, \ c \leq Y \leq d) \approx \int_a^b \int_c^d \frac{\partial^2}{\partial x \, \partial y} F_{X,Y}(x, y) \, dx \, dy. \qquad (9.15)$$

Relation (9.15) suggests that, once we know $\partial^2 F_{X,Y}/\partial x \partial y$, we can obtain the probability that (X, Y) takes a value in any specified bounded closed rectangle and, therefore, that $\partial^2 F_{X,Y}/\partial x \partial y$ completely determines the joint probability distribution of X and Y.

The heuristic argument leading to Relation (9.15) presumes that a mixed second-order partial derivative of $F_{X,Y}$ exists everywhere. As that isn't always the case, we make the following definition of **joint probability density function** in the bivariate case. Observe the analogy between the definition of bivariate joint probability density function and that of (univariate) probability density function, as given in Definition 8.3 on page 417.

DEFINITION 9.2 Joint Probability Density Function: Bivariate Case

Let X and Y be random variables defined on the same sample space. A nonnegative function $f_{X,Y}$ is said to be a **joint probability density function** of X and Y if, for all real numbers a, b, c, and d, with $a < b$ and $c < d$,

$$P(a \le X \le b, \; c \le Y \le d) = \int_a^b \int_c^d f_{X,Y}(x, y) \, dx \, dy. \qquad (9.16)$$

We use **joint PDF** as an abbreviation for "joint probability density function."

Note: In view of Definition 9.2 and Relations (9.14) and (9.15), we write

$$P(x \le X \le x + \Delta x, \; y \le Y \le y + \Delta y) \approx f_{X,Y}(x, y) \Delta x \Delta y, \qquad (9.17)$$

for Δx and Δy close to 0.

Referring to Definition 9.2, we note the following.

- A joint PDF of two continuous random variables is denoted by a lowercase f subscripted with the letters that represent the random variables. For instance, $f_{X,Y}$ denotes a joint PDF of the random variables X and Y, and $f_{U,V}$ denotes a joint PDF of the random variables U and V.

- Another commonly used term for *joint probability density function* is simply **joint density function.**

Recalling from calculus the volume interpretation of double integration, we note from Equation (9.16) that the following statement holds: The joint probability that continuous random variables X and Y with a joint PDF take a value in the rectangle $[a, b] \times [c, d]$ equals the volume under the surface $z = f_{X,Y}(x, y)$ that lies above the rectangle. Figure 9.2 at the top of the next page illustrates this statement.

Obtaining a Joint Probability Density Function

Assuming that two continuous random variables have a joint PDF, how do we find it? Reference to Equation (9.16) and Relation (9.15) indicates that the likely candidate for a joint PDF is a mixed partial derivative of the joint CDF. And, although, even in simple cases, the mixed partial derivatives of the joint CDF may not exist everywhere, a joint PDF is "essentially" a mixed partial derivative of the joint CDF.

In fact, in most practical applications, you can rely on Proposition 9.3. We leave the proof of this proposition to you as Exercise 9.27. The proof is similar to that of its univariate counterpart, Proposition 8.4 on page 419.

Figure 9.2 Probability for two continuous random variables as volume under their joint PDF

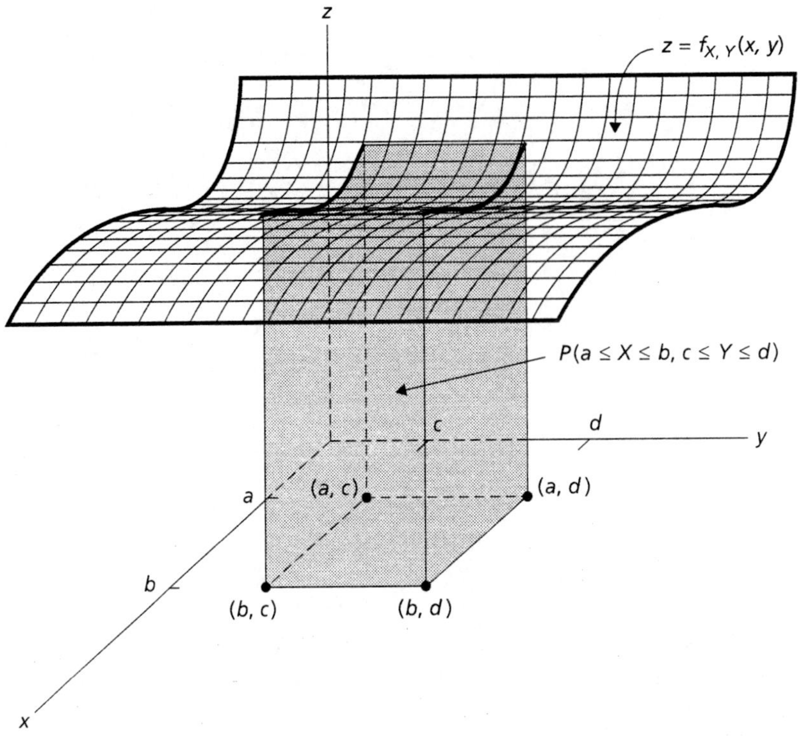

◆◆◆   **Proposition 9.3   Joint PDF is Mixed Partial Derivative of Joint CDF**

Let X and Y be continuous random variables defined on the same sample space. If the partial derivatives of $F_{X,Y}$, up to and including those of the second order, exist and are continuous except possibly at a finite number of (portions of) lines parallel to the coordinate axes, then X and Y have a joint PDF, which we can take to be

$$f_{X,Y}(x, y) = \frac{\partial^2}{\partial x\, \partial y} F_{X,Y}(x, y) = \frac{\partial^2}{\partial y\, \partial x} F_{X,Y}(x, y), \qquad (9.18)$$

at continuity points of the partials, and $f_{X,Y}(x, y) = 0$ otherwise.

EXAMPLE 9.3   *Obtaining a Joint PDF*

Random Points in the Unit Square   Let X and Y denote the x and y coordinates, respectively, of a point selected at random from the unit square. Find a joint PDF for the random variables X and Y.

Solution   In Example 9.1, we found the joint CDF of X and Y to be

$$F_{X,Y}(x, y) = \begin{cases} 0, & \text{if } x < 0 \text{ or } y < 0; \\ xy, & \text{if } 0 \le x < 1 \text{ and } 0 \le y < 1; \\ x, & \text{if } 0 \le x < 1 \text{ and } y \ge 1; \\ y, & \text{if } x \ge 1 \text{ and } 0 \le y < 1; \\ 1, & \text{if } x \ge 1 \text{ and } y \ge 1. \end{cases}$$

This result reveals that each second-order mixed partial derivative of $F_{X,Y}$ equals 1 if (x, y) lies in the interior of the unit square, doesn't exist on the boundary of the unit square, and equals 0 otherwise. Hence we conclude from Proposition 9.3 that a joint PDF of X and Y is

$$f_{X,Y}(x, y) = \begin{cases} 1, & \text{if } 0 < x < 1 \text{ and } 0 < y < 1; \\ 0, & \text{otherwise.} \end{cases}$$

That the joint PDF is constant on the range of (X, Y) reflects the fact that we have a geometric probability model. Indeed, a point is being selected at random from the unit square and hence the point (X, Y) obtained is no more or less likely to be near one point in the unit square than another.   ■

EXAMPLE 9.4   *Obtaining a Joint PDF*

Min and Max of a Random Sample   Let $X_1, \ldots, X_n$ be a random sample from a continuous distribution with CDF F and PDF f. That is, $X_1, \ldots, X_n$ are independent and identically distributed random variables with common CDF F and common PDF f. Set $X = \min\{X_1, \ldots, X_n\}$ and $Y = \max\{X_1, \ldots, X_n\}$.
a) Determine a joint PDF of X and Y by using a heuristic argument.
b) Justify the result of part (a) by using a rigorous mathematical derivation.
c) Apply the result for the joint PDF to the case of a random sample of size n from a uniform distribution on the interval $(0, 1)$.

Solution   a) Let $x < y$ and let Δx and Δy denote small positive numbers. Imagine the real line partitioned into five intervals, as depicted in Figure 9.3.

Figure 9.3 Partition of the real line into five intervals

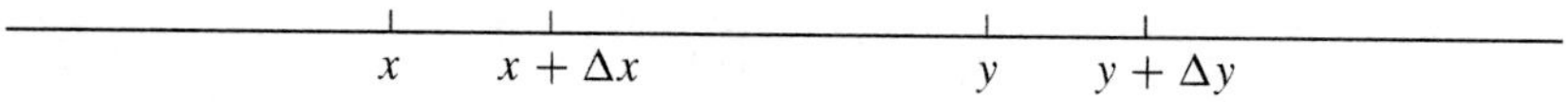

$$x \qquad x + \Delta x \qquad\qquad\qquad y \qquad y + \Delta y$$

We classify each of the n observations, $X_1, \ldots, X_n$, by the interval in which it falls. Because the n observations are independent, we can apply the multinomial distribution with parameters n and $p_1, \ldots, p_5$, where

$$p_1 = F(x), \qquad p_2 = f(x)\Delta x, \qquad p_3 = F(y) - F(x + \Delta x),$$

$$p_4 = f(y)\Delta y, \qquad \text{and} \qquad p_5 = 1 - F(y + \Delta y).$$

For the event $\{x \le X \le x + \Delta x, \ y \le Y \le y + \Delta y\}$ to occur, the number of observations that fall in the five intervals must be $0, 1, n - 2, 1$, and 0, respectively, where we have ignored higher order differentials. Referring now to the PMF of a multinomial distribution, as given in Proposition 6.7 on page 277, we conclude that

$$P(x \leq X \leq x + \Delta x, \ y \leq Y \leq y + \Delta y)$$

$$\approx \binom{n}{0, 1, n-2, 1, 0} \left(F(x)\right)^0 \left(f(x)\Delta x\right)^1 \left(F(y) - F(x + \Delta x)\right)^{n-2}$$

$$\times \left(f(y)\Delta y\right)^1 \left(1 - F(y + \Delta y)\right)^0$$

$$\approx \frac{n!}{0! \ 1! \ (n-2)! \ 1! \ 0!} f(x)\Delta x \left(F(y) - F(x)\right)^{n-2} f(y)\Delta y$$

$$= n(n-1)f(x)f(y)\left(F(y) - F(x)\right)^{n-2} \Delta x \Delta y.$$

Therefore, from Relation (9.17) on page 495,

$$f_{X,Y}(x, y) = n(n-1)f(x)f(y)\left(F(y) - F(x)\right)^{n-2}, \qquad x < y,$$

and $f_{X,Y}(x, y) = 0$ otherwise.

b) We apply Proposition 9.3 on page 496. Thus we first obtain the joint CDF of X and Y. We can more easily determine $P(X > x, \ Y \leq y)$ than $P(X \leq x, \ Y \leq y)$. Because $X_1, \ldots, X_n$ are independent and identically distributed random variables with common CDF F, we have, for $x < y$,

$$P(X > x, \ Y \leq y) = P(\min\{X_1, \ldots, X_n\} > x, \ \max\{X_1, \ldots, X_n\} \leq y)$$

$$= P(x < X_1 \leq y, \ldots, x < X_n \leq y)$$

$$= P(x < X_1 \leq y) \cdots P(x < X_n \leq y) = \left(F(y) - F(x)\right)^n.$$

Using this equation and the law of partitions gives

$$P(Y \leq y) = P(X > x, \ Y \leq y) + P(X \leq x, \ Y \leq y)$$

$$= \left(F(y) - F(x)\right)^n + F_{X,Y}(x, y).$$

Noting that $P(Y \leq y) = \left(F(y)\right)^n$, we obtain

$$F_{X,Y}(x, y) = \left(F(y)\right)^n - \left(F(y) - F(x)\right)^n, \qquad x < y. \tag{9.19}$$

We now apply Proposition 9.3 to Equation (9.19) and conclude that, for $x < y$,

$$f_{X,Y}(x, y) = \frac{\partial^2}{\partial y \, \partial x} F_{X,Y}(x, y) = \frac{\partial}{\partial y}\left(\frac{\partial}{\partial x} F_{X,Y}(x, y)\right)$$

$$= \frac{\partial}{\partial y}\left(nf(x)\left(F(y) - F(x)\right)^{n-1}\right)$$

$$= n(n-1)f(x)f(y)\left(F(y) - F(x)\right)^{n-2}.$$

In summary, then,

$$f_{X,Y}(x, y) = n(n-1)f(x)f(y)\left(F(y) - F(x)\right)^{n-2}, \qquad x < y, \tag{9.20}$$

and $f_{X,Y}(x, y) = 0$ otherwise. This result agrees with that obtained heuristically in part (a).

c) Here we assume a random sample of size n from a uniform distribution on the interval $(0, 1)$. For $0 < t < 1$, we have $F(t) = t$ and $f(t) = 1$; otherwise, $f(t) = 0$. In this case, Equation (9.20) becomes

$$f_{X,Y}(x, y) = n(n - 1)(y - x)^{n-2}, \qquad 0 < x < y < 1,$$

and $f_{X,Y}(x, y) = 0$ otherwise.   ■

Another Procedure for Finding a Joint PDF

Proposition 9.3 on page 496 provides a method for finding a joint PDF of two continuous random variables. An alternative and more general method is presented in Proposition 9.4, whose proof is left to you as Exercise 9.28.

◆◆◆ **Proposition 9.4   An Equivalent Condition for the Existence of a Joint PDF**

Random variables X and Y defined on the same sample space have a joint PDF if and only if there is a nonnegative function f defined on $\mathcal{R}^2$ such that

$$F_{X,Y}(x, y) = \int_{-\infty}^{x} \int_{-\infty}^{y} f(s, t)\, ds\, dt, \qquad x, y \in \mathcal{R}. \tag{9.21}$$

In this case, f is a joint PDF of X and Y, and

$$f(x, y) = \frac{\partial^2}{\partial x\, \partial y} F_{X,Y}(x, y) \tag{9.22}$$

at all points (x, y) where the mixed partial exists.

Multivariate Joint Probability Density Functions

The definitions and results that we presented for bivariate joint PDFs are easily generalized to the multivariate case of m continuous random variables. We ask you to supply these generalizations in Exercise 9.29.

EXERCISES 9.2   **Basic Exercises**

Note: Several of the exercises in this section are continuations of exercises presented in Section 9.1.

9.19 On page 493, we stated that "For continuous random variables X and Y, the probabilities $P(X = x, Y = y)$ are all 0 . . . ." Verify that statement.

9.20 Let X and Y be random variables defined on the same sample space. Set $U = a + bX$ and $V = c + dY$, where a, b, c, and d are real numbers with $b > 0$ and $d > 0$. Assuming that X and Y have a joint PDF, determine a joint PDF of U and V in terms of that of X and Y. *Note:* Exercise 9.1 on page 490 asks for the joint CDF of U and V in terms of that of X and Y.

9.21 Suppose that X and Y are continuous random variables with a joint PDF. Let $U = X^2$ and $V = Y^2$. Find a joint PDF of U and V in terms of that of X and Y. *Note:* Exercise 9.2 on page 491 asks for the joint CDF of U and V in terms of that of X and Y.

9.22 Let X and Y be the x and y coordinates of a point selected at random from the unit square. Determine $P(1/2 \le X \le 3/4, \ 1/4 \le Y \le 3/4)$ by using
a) Equation (9.4) on page 487.
b) the joint CDF of X and Y, obtained in Example 9.1 on page 487.
c) the joint PDF of X and Y, obtained in Example 9.3 on page 496.
d) Repeat parts (a)–(c) for $P(X > 0.6$ or $Y < 0.2)$.

9.23 Let X and Y be the x and y coordinates, respectively, of a point selected at random from the upper half of the unit disk—that is, from the set $\{ (x, y) \in \mathcal{R}^2 : x^2 + y^2 < 1, \ y > 0 \}$.
a) Determine a joint PDF of X and Y. *Note:* Exercise 9.4(c) on page 491 asks for the joint CDF of X and Y.
b) Explain why the joint PDF obtained in part (a) is a nonzero constant on the upper half of the unit disk and is zero elsewhere.

9.24 Let X and Y be independent exponential random variables with parameters λ and μ, respectively.
a) Obtain a joint PDF of X and Y. *Note:* Exercise 9.7(b) on page 491 asks for the joint CDF of X and Y.
b) What is the relationship between the joint PDF of X and Y found in part (a) and the (standard) individual PDFs of X and Y?
c) Does the relationship in part (b) surprise you? Explain your answer.

9.25 Let X and Y be the minimum and maximum, respectively, of a random sample of size n taken from an exponential distribution with parameter λ. Determine a joint PDF of X and Y.

9.26 Let X and Y have joint PDF given by $f(x, y) = x + y$ for $0 < x < 1$ and $0 < y < 1$, and $f(x, y) = 0$ otherwise.
a) Use the joint PDF to determine $P(1/4 < X < 3/4, \ 1/2 < Y < 1)$.
b) Determine the joint CDF of the random variables X and Y.
c) Use the joint CDF to determine $P(1/4 < X < 3/4, \ 1/2 < Y < 1)$. Compare your answer with that obtained in part (a).

Theory Exercises

9.27 Prove Proposition 9.3 on page 496 when the partial derivatives of $F_{X,Y}$, up to and including those of the second order, exist and are continuous everywhere.

9.28 Prove Proposition 9.4 on page 499, which provides an equivalent condition for the existence of a bivariate joint PDF.

9.29 Let $X_1, \ldots, X_m$ be random variables defined on the same sample space.
a) Define *joint probability density function* for these random variables. *Hint:* Refer to Definition 9.2 on page 495.
b) State the m-variate analogue of Proposition 9.3 on page 496.
c) State the m-variate analogue of Proposition 9.4 on page 499.

Advanced Exercises

9.30 Refer to Exercise 9.14 on page 492, where X and Y are the x and y coordinates, respectively, of a point selected at random from the diagonal of the unit square—that is, from $\{ (x, y) \in \mathcal{R}^2 : y = x, \ 0 < x < 1 \}$.
a) Show that X and Y are continuous random variables.
b) Show that X and Y can't possibly have a joint PDF.

9.31 Let X and Y be continuous random variables defined on the same sample space and having PDFs. Prove that X and Y are independent if and only if the function f defined on $\mathcal{R}^2$ by $f(x, y) = f_X(x) f_Y(y)$ is a joint PDF of X and Y. *Hint:* Refer to Exercise 9.15 on page 492 and to Proposition 9.4 on page 499.

9.32 Generalize Exercise 9.31 to the multivariate case of m continuous random variables defined on the same sample space and having PDFs. *Hint:* Refer to Exercise 9.16 on page 492 and assume as known the m-variate version of Proposition 9.4 on page 499.

Order statistics: Let $X_1, \ldots, X_n$ be a random sample from a continuous distribution with CDF F and PDF f. For each k between 1 and n, inclusive, the *kth order statistic,* denoted $X_{(k)}$, is defined to be the kth smallest of $X_1, \ldots, X_n$. Collectively, the random variables $X_{(1)}, \ldots, X_{(n)}$ are called the *order statistics* corresponding to $X_1, \ldots, X_n$. In Exercises 9.33–9.35, you are to examine some properties of order statistics.

9.33 Consider the kth order statistic.
a) Obtain the CDF of $X_{(k)}$. *Hint:* What does the event $\{X_{(k)} \le x\}$ mean in terms of the X_js?
b) Use part (a) to determine a PDF of $X_{(k)}$.
c) Use a heuristic argument similar to that in Example 9.4(a) on page 497 to determine a PDF of $X_{(k)}$. Compare your result with that obtained in part (b).
d) Obtain and identify the probability distribution of the kth order statistic for the case of a random sample of size n from a uniform distribution on the interval $(0, 1)$.
e) Obtain a PDF of the kth order statistic for the case of a random sample of size n from an exponential distribution with parameter λ.

9.34 Consider all n order statistics.
a) Determine a joint PDF of $X_{(1)}, \ldots, X_{(n)}$ by using a symmetry argument. *Hint:* Refer to Exercise 9.32.
b) Solve part (a) by using a heuristic argument similar to that in Example 9.4(a) on page 497.
c) Specialize the result of part (b) for the cases of a random sample of size n from a $\mathcal{U}(0, 1)$ distribution and from an $\mathcal{E}(\lambda)$ distribution.

9.35 Use a heuristic argument similar to that in Example 9.4(a) on page 497 to obtain a joint PDF of the random variables
a) $X_{(j)}$ and $X_{(k)}$, for $1 \le j < k \le n$.
b) $X_{(i)}$, $X_{(j)}$, and $X_{(k)}$, for $1 \le i < j < k \le n$.
c) Specialize the result of parts (a) and (b) for the cases of a random sample of size n from a $\mathcal{U}(0, 1)$ distribution and from an $\mathcal{E}(\lambda)$ distribution.

9.3 Properties of Joint Probability Density Functions

We now present some of the most important properties of joint probability density functions. Proposition 9.5—whose proof we leave to you as Exercise 9.50—provides the basic properties of a bivariate joint PDF. Note the analogy with the basic properties of a bivariate joint PMF, as given in Proposition 6.1 on page 262 and, as well, with the basic properties of a univariate PDF, as given in Proposition 8.6 on page 423.

◆◆◆   **Proposition 9.5   Basic Properties of a Joint PDF: Bivariate Case**

A joint probability density function $f_{X,Y}$ of two continuous random variables X and Y satisfies the following two properties.

a) $f_{X,Y}(x, y) \geq 0$ for all $(x, y) \in \mathcal{R}^2$—a joint PDF is a nonnegative function.

b) $\int_{-\infty}^{\infty} \int_{-\infty}^{\infty} f_{X,Y}(x, y)\, dx\, dy = 1$—the integral of the values of a joint PDF equals 1.

A function $f\colon \mathcal{R}^2 \to \mathcal{R}$ that satisfies properties (a) and (b) of Proposition 9.5 is a joint PDF of some pair of continuous random variables. Therefore, for any such function, we can say "Let X and Y be continuous random variables with joint PDF f." This statement makes sense regardless of whether we explicitly give X and Y and the sample space on which they are defined.

We can use this fact to construct a joint PDF with a prescribed "variable form." More precisely, suppose that $g\colon \mathcal{R}^2 \to \mathcal{R}$ is a nonnegative function whose integral over $\mathcal{R}^2$ is finite and nonzero. We define the function f as

$$f(x, y) = cg(x, y), \quad \text{where} \quad c = \left(\int_{-\infty}^{\infty} \int_{-\infty}^{\infty} g(x, y)\, dx\, dy \right)^{-1}.$$

We see that f satisfies properties (a) and (b) of Proposition 9.5 and hence is a joint PDF.

For example, suppose that we want a joint PDF whose variable form is $(y - x)^3$, for $0 < x < y < 1$, and 0 otherwise. This variable form is nonnegative and its integral over $\mathcal{R}^2$ equals 1/20. Hence the corresponding joint PDF is $f(x, y) = 20(y - x)^3$, for $0 < x < y < 1$, and 0 otherwise.

FPF for Two Continuous Random Variables

Once we have a joint PDF of two continuous random variables, we can obtain any probability involving those two random variables. Specifically, we have Proposition 9.6, which we again refer to as the **fundamental probability formula** (FPF). Note the analogy with the FPF for one continuous random variable (Proposition 8.7 on page 424) and the FPF for two discrete random variables (Proposition 6.3 on page 265).

◆◆◆   **Proposition 9.6   Fundamental Probability Formula: Bivariate Continuous Case**

Suppose that X and Y are continuous random variables with a joint PDF. Then, for any subset $A \subset \mathcal{R}^2$, we have

$$P\big((X, Y) \in A\big) = \iint_A f_{X,Y}(x, y)\, dx\, dy. \tag{9.23}$$

In words, the probability that a pair of continuous random variables takes a value in a specified subset of the plane can be obtained by integrating a joint PDF of the random variables over that subset.

We don't provide a proof of Proposition 9.6, as its rigorous verification requires the use of measure theory. However, referring to Definition 9.2 on page 495, we see that Equation (9.23) holds for all bounded closed rectangles. We can use this fact and measure theory to deduce that Equation (9.23) holds for all (measurable) subsets of $\mathcal{R}^2$.

EXAMPLE 9.5 *The Fundamental Probability Formula*

Lifetimes of Electrical Components Consider two electrical components, A and B, with respective lifetimes X and Y. Assume that a joint PDF of X and Y is

$$f_{X,Y}(x, y) = \lambda\mu\, e^{-(\lambda x + \mu y)}, \qquad x, y > 0, \tag{9.24}$$

and $f_{X,Y}(x, y) = 0$ otherwise, where λ and μ are positive constants.
a) Determine the probability that both components are functioning at time t.
b) Determine the probability that component A is the first to fail.
c) Determine the probability that component B is the first to fail.

Solution a) Both components functioning at time t means that $X > t$ and $Y > t$. From the FPF,

$$P(X > t, Y > t) = \iint\limits_{x>t,\, y>t} f_{X,Y}(x, y)\, dx\, dy.$$

The shaded and crosshatched region in Figure 9.4 at the top of the next page shows the set over which the double integral is taken. In view of that figure and Equation (9.24), we can evaluate the double integral as follows:

$$\iint\limits_{x>t,\, y>t} f_{X,Y}(x, y)\, dx\, dy = \int_t^\infty \left(\int_t^\infty \lambda\mu\, e^{-(\lambda x + \mu y)}\, dy \right) dx$$

$$= \lambda\mu \int_t^\infty e^{-\lambda x} \left(\int_t^\infty e^{-\mu y}\, dy \right) dx$$

$$= \lambda\mu \int_t^\infty e^{-\lambda x} \mu^{-1} e^{-\mu t}\, dx = \lambda e^{-\mu t} \int_t^\infty e^{-\lambda x}\, dx$$

$$= e^{-\mu t} e^{-\lambda t} = e^{-(\lambda + \mu)t}.$$

The probability is $e^{-(\lambda + \mu)t}$ that both components are functioning at time t.
b) The event that component A is the first to fail is $\{X < Y\}$. Applying the FPF gives

$$P(X < Y) = \iint\limits_{x<y} f_{X,Y}(x, y)\, dx\, dy.$$

The shaded region in Figure 9.5 shows the set over which the double integral is taken. The crosshatched portion of the shaded region indicates where the integrand is nonzero. In view of that figure, we can evaluate the double integral as follows:

$$\iint\limits_{x<y} f_{X,Y}(x, y)\, dx\, dy = \int_0^\infty \left(\int_x^\infty \lambda\mu\, e^{-(\lambda x + \mu y)}\, dy \right) dx$$

$$= \lambda\mu \int_0^\infty e^{-\lambda x} \left(\int_x^\infty e^{-\mu y}\, dy \right) dx$$

$$= \lambda\mu \int_0^\infty e^{-\lambda x} \mu^{-1} e^{-\mu x}\, dx = \frac{\lambda}{\lambda + \mu}.$$

Hence the probability that component A is the first to fail is $\lambda/(\lambda + \mu)$.

Figure 9.4

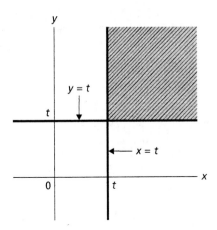

Figure 9.5

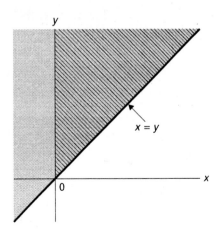

c) The event that component B is the first to fail is $\{X > Y\}$. We can obtain the probability of that event in several ways. One way is to apply the FPF as in part (b). A second way is to apply the complementation rule, the result of part (b), and the fact that $P(Y = X) = 0$. The easiest way, however, is to interchange the roles of X and Y and conclude from part (b) that $P(X > Y) = \mu/(\lambda + \mu)$. See Exercise 9.48.   ■

EXAMPLE 9.6   *The Fundamental Probability Formula*

Range of a Random Sample   Let $X_1, \ldots, X_n$ be a random sample from a continuous distribution with CDF F and PDF f. The *range* of the random sample is defined to be $R = Y - X$, where $X = \min\{X_1, \ldots, X_n\}$ and $Y = \max\{X_1, \ldots, X_n\}$.
a) Determine a PDF of the random variable R.
b) Apply the result for the PDF to the special case of a random sample of size n from a uniform distribution on the interval $(0, 1)$. Identify the probability distribution of R in this special case.

Solution   We first recall from Example 9.4 on page 497 that a joint PDF of X and Y is

$$f_{X,Y}(x, y) = n(n - 1)f(x)f(y)\left(F(y) - F(x)\right)^{n-2}, \qquad x < y, \tag{9.25}$$

and $f_{X,Y}(x, y) = 0$ otherwise.
a) To obtain a PDF of R, we first find its CDF. From the FPF, we have, for $r > 0$,

$$F_R(r) = P(R \le r) = P(Y - X \le r) = \iint\limits_{y-x \le r} f_{X,Y}(x, y)\, dx\, dy.$$

The shaded region in Figure 9.6 shows the set over which the double integral is taken. The crosshatched region depicts that portion of the shaded region where the integrand is possibly nonzero.

Figure 9.6

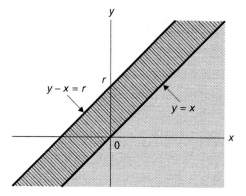

Referring to Figure 9.6, making the substitution $u = y - x$, and then interchanging the order of integration, we get that, for $r > 0$,

$$F_R(r) = \iint\limits_{y-x \leq r} f_{X,Y}(x, y)\, dx\, dy$$

$$= \int_{-\infty}^{\infty} \left(\int_{x}^{x+r} n(n-1) f(x) f(y) \left(F(y) - F(x) \right)^{n-2} dy \right) dx$$

$$= \int_{-\infty}^{\infty} \left(\int_{0}^{r} n(n-1) f(x) f(u+x) \left(F(u+x) - F(x) \right)^{n-2} du \right) dx$$

$$= \int_{0}^{r} \left(\int_{-\infty}^{\infty} n(n-1) f(x) f(u+x) \left(F(u+x) - F(x) \right)^{n-2} dx \right) du.$$

Applying Proposition 8.5 on page 422, we conclude that

$$f_R(r) = n(n-1) \int_{-\infty}^{\infty} f(x) f(r+x) \left(F(r+x) - F(x) \right)^{n-2} dx, \qquad \textbf{(9.26)}$$

for $r > 0$, and $f_R(r) = 0$ otherwise.

b) In this case, R takes values only between 0 and 1. Also, for $0 < t < 1$, we have $F(t) = t$ and $f(t) = 1$; otherwise, $f(t) = 0$. Applying Equation (9.26) yields

$$f_R(r) = n(n-1) \int_{0}^{1-r} \left((r+x) - x \right)^{n-2} dx$$

$$= n(n-1) \int_{0}^{1-r} r^{n-2} dx = n(n-1)(1-r)r^{n-2},$$

for $0 < r < 1$, and $f_R(r) = 0$ otherwise. Recalling the definition of a beta random variable (Definition 8.9 on page 457), we note that the "variable form" of the PDF here is that of a beta distribution with parameters $n - 1$ and 2. This characteristic implies that R must have that distribution. (Why?) We can also see explicitly that R has the beta distribution with parameters $n - 1$ and 2 by observing that

$$f_R(r) = n(n-1)(1-r)r^{n-2} = \frac{\Gamma(n+1)}{\Gamma(n-1)\Gamma(2)} r^{n-2}(1-r)$$

$$= \frac{\Gamma((n-1)+2)}{\Gamma(n-1)\Gamma(2)} r^{(n-1)-1}(1-r)^{2-1} = \frac{1}{B(n-1,2)} r^{(n-1)-1}(1-r)^{2-1},$$

for $0 < r < 1$, and $f_R(r) = 0$ otherwise.   ■

Properties of Multivariate Joint PDFs

The properties that we presented for bivariate joint PDFs—specifically, Propositions 9.5 and 9.6 on page 502—are easily generalized to the multivariate case of m continuous random variables. We ask you to supply these generalizations in Exercises 9.51 and 9.52.

EXERCISES 9.3   **Basic Exercises**

9.36 Show that, if X and Y are random variables having a joint PDF, each must be a continuous random variable. *Hint:* Use Proposition 9.6 on page 502 to obtain $P(X = x)$ for each $x \in \mathcal{R}$.

9.37 Suppose that f and g are univariate PDFs. Show that the function h defined on $\mathcal{R}^2$ by $h(x, y) = f(x)g(y)$ is a bivariate PDF; that is, it satisfies properties (a) and (b) of Proposition 9.5 on page 502.

9.38 Construct a joint PDF from each of the following functions, when possible. Explain your reasoning.
a) $g(x, y) = e^{-\lambda x}$ for $0 < y < x$, and $g(x, y) = 0$ otherwise
b) $g(x, y) = (y - x)^\alpha$ for $0 < x < y < 1$, and $g(x, y) = 0$ otherwise, where $\alpha \in \mathcal{R}$
c) $g(x, y) = x - y$ for $x, y \in (0, 1)$, and $g(x, y) = 0$ otherwise
d) $g(x, y) = x + y$ for $x, y \in (0, 1)$, and $g(x, y) = 0$ otherwise
e) $g(x, y) = x + y$ for $x > 0$ and $y > 0$, and $g(x, y) = 0$ otherwise

9.39 The lifetimes, in months, of two components of a machine have joint density function given by $f_{X,Y}(x, y) = c(50 - x - y)$ for $0 < x < 50 - y < 50$, and $f_{X,Y}(x, y) = 0$ otherwise, where $c = 6/125,000$. Which of the following provides the probability that both components are still functioning 20 months from now?

a) $c \int_0^{20} \left(\int_0^{20} (50 - x - y)\, dy \right) dx$ b) $c \int_{20}^{30} \left(\int_{20}^{50-x} (50 - x - y)\, dy \right) dx$

c) $c \int_{20}^{30} \left(\int_{20}^{50-x-y} (50 - x - y)\, dy \right) dx$ d) $c \int_{20}^{50} \left(\int_{20}^{50-x} (50 - x - y)\, dy \right) dx$

9.40 Bivariate uniform random variables: Let S be a subset of $\mathcal{R}^2$ with finite nonzero area. Suppose that a point is selected at random (i.e., uniformly) from S, and let X and Y be the x and y coordinates, respectively, of the point obtained.
a) Show that a joint PDF of X and Y is given by $f_{X,Y}(x, y) = 1/|S|$ for $(x, y) \in S$, and $f_{X,Y}(x, y) = 0$ otherwise. *Hint:* Use Proposition 9.5 on page 502.
b) Show that $P((X, Y) \in A) = |A \cap S|/|S|$ for each $A \subset \mathcal{R}^2$.

9.41 Let X and Y denote the x and y coordinates, respectively, of a point selected at random from the unit square.
a) Use Exercise 9.40(a) to obtain a joint PDF of X and Y.

b) Compare the work entailed in part (a) to that in Example 9.3 on page 496 (which also depended on the work done in Example 9.1 on page 487).

c) Find the probability that the magnitude of the difference of the x and y coordinates of the point obtained is at most 1/4 by using the joint PDF of X and Y; by using Exercise 9.40(b).

9.42 Let X and Y be the x and y coordinates, respectively, of a point selected at random from the upper half of the unit disk—that is, from the set $\{(x, y) \in \mathcal{R}^2 : x^2 + y^2 < 1,\ y > 0\}$.

a) Use Exercise 9.40(a) to obtain a joint PDF of X and Y.

b) Compare the work entailed in part (a) to that in Exercise 9.23 on page 500 (which also depended on the work done in Exercise 9.4(c) on page 491).

c) Find the probability that the point obtained lies in the triangle with vertices $(-1, 0), (0, 1)$, and $(1, 0)$ by using the joint PDF of X and Y; by using Exercise 9.40(b).

d) Determine the probability that the x coordinate of the point obtained is at least 1/2 unit from the origin.

9.43 Multivariate uniform random variables: Let S be a subset of $\mathcal{R}^m$ with finite nonzero m-dimensional volume. Suppose that a point is selected at random (i.e., uniformly) from S and, for $1 \leq j \leq m$, let X_j denote the x_j coordinate of the point obtained.

a) Determine a joint PDF of $X_1, \ldots, X_m$.

b) Determine a simple formula for $P\big((X_1, \ldots, X_m) \in A\big)$, where $A \subset \mathcal{R}^m$.

9.44 A point is selected a random from the unit cube, $\{(x, y, z) \in \mathcal{R}^3 : 0 < x, y, z < 1\}$. Let X, Y, and Z be the x, y, and z coordinates, respectively, of the point obtained.

a) Use the result of Exercise 9.43(a) to determine a joint PDF of X, Y and Z.

b) Use the result of part (a) to find the probability that $Z = \max\{X, Y, Z\}$—that is, that Z is the largest among X, Y and Z.

c) Use a symmetry argument to solve part (b).

d) Find the probability that the point obtained lies in the sphere of radius 1/4 centered at the point $(1/2, 1/2, 1/2)$.

e) Find the probability that, for the point obtained, the sum of the x and y coordinates exceeds the z coordinate.

9.45 A company is reviewing tornado damage claims under a farm insurance policy. Let X be the portion of a claim representing damage to the house and let Y be the portion of the same claim representing damage to the rest of the property. A joint density function of X and Y is $f_{X,Y}(x, y) = 6(1 - x - y)$ in the triangle with vertices $(0, 0)$, $(1, 0)$, and $(0, 1)$, and $f_{X,Y}(x, y) = 0$ otherwise.

a) Determine the probability that the portion of a claim representing damage to the house exceeds the portion of the same claim representing damage to the rest of the property.

b) Determine the probability that the portion of a claim representing damage to the house is less than 0.2.

9.46 In Example 9.5 on page 503, we considered two electrical components, A and B, with respective lifetimes X and Y whose joint PDF is $f_{X,Y}(x, y) = \lambda\mu e^{-(\lambda x + \mu y)}$ for $x > 0$ and $y > 0$, and $f_{X,Y}(x, y) = 0$ otherwise. Suppose that components A and B constitute an electrical unit.

a) Find a PDF of this electrical unit's lifetime if it's a parallel system—that is, if it functions when at least one of the components is working. *Hint:* First obtain the CDF of the electrical unit's lifetime by using the FPF.

b) Find a PDF of this electrical unit's lifetime if it's a series system—that is, if it functions only when both components are working. Identify the lifetime distribution in this case.

c) Determine the probability that exactly one of the two components is working at time t.

9.47 Suppose that X and Y are continuous random variables with a joint PDF. Use the FPF to show that $P(Y = X) = 0$.

9.48 Solve part (c) of Example 9.5 on page 503 by
a) applying the FPF as done in the solution to part (b) of that example.
b) using the complementation rule, the result of part (b) of that example, and Exercise 9.47.
c) interchanging the roles of X and Y and applying the result of part (b) of that example.

9.49 Let $X_1, \ldots, X_n$ be a random sample from an exponential distribution with parameter λ. Set $X = \min\{X_1, \ldots, X_n\}$ and $Y = \max\{X_1, \ldots, X_n\}$.
a) Obtain a joint PDF of X and Y. *Note:* A general formula for such a joint PDF is provided by Equation (9.20) on page 498.
b) Determine $P(Y > X + 1)$ by using your result from part (a).
c) Obtain a PDF for the range R of the random sample. *Note:* A general formula for such a PDF is provided by Equation (9.26) on page 505.
d) Determine $P(Y > X + 1)$ by using your result from part (c).

Theory Exercises

9.50 Prove Proposition 9.5 on page 502, which provides the basic properties of a bivariate joint PDF.

9.51 State the m-variate analogue of Proposition 9.5 on page 502, thus providing the basic properties of joint PDF in the general multivariate case.

9.52 State the m-variate analogue of Proposition 9.6 on page 502, thus providing the fundamental probability formula (FPF) in the general multivariate case.

9.53 Let X and Y be random variables with a joint PDF. Obtain a formula for a PDF of
a) $X + Y$.   **b)** $X - Y$.
Hint: In each case, first find the CDF by using the FPF.

Advanced Exercises

9.54 We know from Exercise 9.47 that, if X and Y are continuous random variables with a joint PDF, then $P(Y = X) = 0$. Does this result hold for any two continuous random variables defined on the same sample space? Justify your answer.

9.55 Let X and Y be random variables defined on the same sample space. Exercise 9.36 states that, if X and Y have a joint PDF, then both X and Y are continuous random variables. Is the converse of this result true? Justify your answer.

9.56 Three gas stations are independently and randomly placed along 1 mile of highway.
a) Determine the probability that no two of the gas stations are less than $\frac{1}{3}$ mile apart. *Hint:* Refer to Exercise 9.34 on page 501.
b) Suppose that you want to generalize part (a) to determine the probability that no two of the gas stations are less than d mile apart. What restriction on d is appropriate?
c) Determine the probability that no two of the gas stations are less than d mile apart for appropriate values of d.

9.57 Generalize the results of Exercise 9.56 as follows. Suppose that n gas stations are independently and randomly placed along 1 mile of highway.
a) Suppose that you want to determine the probability that no two of the gas stations are less than d mile apart. What restriction on d is appropriate?
b) Determine the probability that no two of the gas stations are less than d mile apart for appropriate values of d.

Mixed bivariate distributions: Frequently, in applications, we deal jointly with one continuous random variable, X, and one discrete random variable, Y. A nonnegative function $h_{X,Y}$ is called a *joint PDF/PMF* of X and Y if, for all real numbers $a < b$ and y,

$$P(a \leq X \leq b, \ Y = y) = \int_a^b h_{X,Y}(x, y)\, dx. \qquad (*)$$

In Exercises 9.58–9.62, you are to examine the concept of a joint PDF/PMF.

9.58 Proposition 9.5 on page 502 provides the basic properties of a bivariate joint PDF. Prove the analogue of that proposition for a bivariate joint PDF/PMF: A joint PDF/PMF $h_{X,Y}$ of random variables X and Y satisfies

a) $h_{X,Y}(x, y) \geq 0$ for all $(x, y) \in \mathcal{R}^2$.   **b)** $\int_{-\infty}^{\infty} \sum_y h_{X,Y}(x, y)\, dx = 1$.

Note: A function $h: \mathcal{R}^2 \to \mathcal{R}$ that satisfies properties (a) and (b) is a joint PDF/PMF of some pair of random variables, one continuous and the other discrete.

9.59 For a certain emergency room, let Y denote the number of patients who arrive during a 1-hour period and let X denote the average rate at which they arrive. Suppose that X and Y have joint PDF/PMF given by $h_{X,Y}(x, y) = \lambda e^{-(1+\lambda)x}(x^y/y!)$ for $x > 0$ and $y = 0, 1, 2, \ldots$, and $h_{X,Y}(x, y) = 0$ otherwise.
a) Verify that $h_{X,Y}$ satisfies properties (a) and (b) of Exercise 9.58.
b) Obtain the PMF of Y. *Hint:* Use the fact that $P(Y = y) = P(-\infty < X < \infty, \ Y = y)$ and that the PDF of a gamma random variable integrates to 1.
c) Obtain a PDF of X. *Hint:* First determine the CDF of X by using the law of partitions, Proposition 2.8 on page 68.

Exercises 9.60–9.62 provide the basis for common ways of specifying a joint PDF/PMF for two random variables.

9.60 Suppose that f is a univariate PDF and that p is a univariate PMF. Show that the function h defined by $h(x, y) = f(x)p(y)$ is a joint PDF/PMF, that is, it satisfies properties (a) and (b) of Exercise 9.58.

9.61 Suppose that f is a univariate PDF and that, for each $x \in \mathcal{R}$, p_x is a univariate PMF. Show that the function h defined by $h(x, y) = f(x)p_x(y)$ is a joint PDF/PMF.

9.62 Suppose that p is a univariate PMF and that, for each $y \in \mathcal{R}$, f_y is a univariate PDF. Show that the function h defined by $h(x, y) = f_y(x)p(y)$ is a joint PDF/PMF.

9.4   Marginal and Conditional Probability Density Functions

In this section, we continue our examination of the simultaneous analysis of two or more continuous random variables by considering the concepts of marginal and conditional probability density functions. As before, we concentrate on the bivariate case of two continuous random variables and mostly leave the details of the general multivariate case to you as exercises.

Marginal Probability Density Functions

Let X and Y be continuous random variables with joint PDF $f_{X,Y}$. When we analyze the two random variables together, we use a joint PDF. In this context, we refer to the individual PDFs of X and Y as **marginal probability density functions** or, more simply, as *marginal PDFs*.

Recall that, in the discrete bivariate case, each marginal PMF is obtained by summing the joint PMF on the unwanted variable, as presented in Proposition 6.2 on page 263. Likewise, in the continuous case, each marginal PDF is obtained by integrating the joint PMF on the unwanted variable, as shown in Proposition 9.7.

◆◆◆ **Proposition 9.7   Obtaining Marginal PDFs From a Joint PDF**

Let X and Y be continuous random variables with a joint PDF. Then

$$f_X(x) = \int_{-\infty}^{\infty} f_{X,Y}(x, y)\, dy, \qquad x \in \mathcal{R}, \tag{9.27}$$

and

$$f_Y(y) = \int_{-\infty}^{\infty} f_{X,Y}(x, y)\, dx, \qquad y \in \mathcal{R}. \tag{9.28}$$

In words, we can obtain a (marginal) PDF of X by integrating on y a joint PDF of X and Y and, likewise, we can obtain a (marginal) PDF of Y by integrating on x a joint PDF of X and Y.

Proof We verify Equation (9.27); the verification of Equation (9.28) is similar. From the FPF for two continuous random variables,

$$F_X(x) = P(X \le x) = P(-\infty < X \le x, -\infty < Y < \infty)$$

$$= \iint_{\substack{-\infty < s \le x \\ -\infty < y < \infty}} f_{X,Y}(s, y)\, ds\, dy = \int_{-\infty}^{x} \left(\int_{-\infty}^{\infty} f_{X,Y}(s, y)\, dy \right) ds.$$

Applying Proposition 8.5 on page 422, we conclude that Equation (9.27) holds.   ◆

EXAMPLE 9.7   *Obtaining Marginal PDFs From a Joint PDF*

Random Points in the Unit Square Let X and Y denote the x and y coordinates, respectively, of a point selected at random from the unit square. Find and identify marginal PDFs of X and Y.

Solution From Example 9.3 on page 496, we know that a joint PDF of X and Y is $f_{X,Y}(x, y) = 1$ if (x, y) is in the unit square, and $f_{X,Y}(x, y) = 0$ otherwise. To obtain a marginal PDF of X, we first observe that the range of X is the interval $(0, 1)$. Next we apply Proposition 9.7 by integrating on y the joint PDF of X and Y. Figure 9.7 shows the interval of integration for a fixed x in the range of X. The solid portion of the interval indicates where the joint PDF of X and Y is nonzero.

Figure 9.7

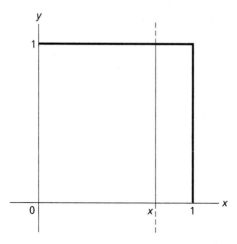

From Equation (9.27) and Figure 9.7,

$$f_X(x) = \int_{-\infty}^{\infty} f_{X,Y}(x, y)\, dy = \int_0^1 1\, dy = 1,$$

for $0 < x < 1$.   Consequently, $f_X(x) = 1$ if $0 < x < 1$, and $f_X(x) = 0$ otherwise. So, $X \sim \mathcal{U}(0, 1)$. Similarly, $Y \sim \mathcal{U}(0, 1)$.   ∎

In Example 9.7, the random point (X, Y) is uniformly distributed over the unit square. We found that the marginal distributions of X and Y are also uniformly distributed. Example 9.8 shows that marginals of bivariate uniform distributions aren't always uniform.

EXAMPLE 9.8   *Obtaining Marginal PDFs From a Joint PDF*

Bacteria on a Petri Dish   A petri dish is a small, shallow dish of thin glass or plastic, used especially for cultures in bacteriology. Suppose that a petri dish of unit radius, containing nutrients upon which bacteria can multiply, is smeared with a uniform suspension of bacteria. Subsequently, spots indicating colonies of bacteria will appear. Let X and Y denote the x and y coordinates, respectively, of the center of the first spot to appear.
a) Determine a joint PDF of X and Y.
b) Use part (a) to find marginal PDFs of X and Y.

Solution   Because the dish is smeared with a uniform suspension of bacteria, we use a geometric probability model. Specifically, we can think of the location of the center of the first spot as a point selected at random from the unit disk.
a) We see that (X, Y) is uniformly distributed over the unit disk. Hence a joint PDF of X and Y is $f_{X,Y}(x, y) = 1/\pi$ if (x, y) is in the unit disk, and $f_{X,Y}(x, y) = 0$ otherwise. (Why $1/\pi$?)

b) To obtain a marginal PDF of X, we first observe that the range of X is the interval $(-1, 1)$. Next we apply Proposition 9.7 by integrating on y the joint PDF of X and Y. Figure 9.8 shows the interval of integration for a fixed x in the range of X. The solid portion of the interval indicates where the joint PDF of X and Y is nonzero.

Figure 9.8

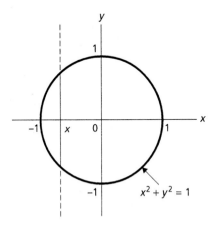

Noting that the equation of the boundary of the unit disk is $x^2 + y^2 = 1$, we find, in view of Equation (9.27) and Figure 9.8, that, for $-1 < x < 1$,

$$f_X(x) = \int_{-\infty}^{\infty} f_{X,Y}(x, y)\, dy = \int_{-\sqrt{1-x^2}}^{\sqrt{1-x^2}} \frac{1}{\pi}\, dy = \frac{2}{\pi}\sqrt{1 - x^2}.$$

Thus $f_X(x) = (2/\pi)\sqrt{1 - x^2}$ if $-1 < x < 1$, and $f_X(x) = 0$ otherwise. By symmetry, Y has that same PDF. Observe that the common distribution of X and Y isn't uniform. In fact, for each of these random variables, values near 0 are the most likely, with the likelihood decreasing as values move away from 0 in either direction.   ■

EXAMPLE 9.9   *Obtaining Marginal PDFs*

Min and Max of a Random Sample   Let $X_1, \ldots, X_n$ be a random sample from a continuous distribution with CDF F and PDF f. Set $X = \min\{X_1, \ldots, X_n\}$ and $Y = \max\{X_1, \ldots, X_n\}$.

a) Determine a PDF of X directly.

b) Determine a PDF of X by using a joint PDF of X and Y.

c) Apply the result for the PDF of X to the special case of a random sample of size n from a uniform distribution on the interval $(0, 1)$. Identify the probability distribution of X in this special case.

Solution   a) We first find the CDF of X. From the complementation rule and the independence of the X_js, we have

$$F_X(x) = P(X \le x) = 1 - P(X > x) = 1 - P(\min\{X_1, \ldots, X_n\} > x)$$
$$= 1 - P(X_1 > x, \ldots, X_n > x) = 1 - P(X_1 > x) \cdots P(X_n > x)$$
$$= 1 - \bigl(1 - P(X_1 \le x)\bigr) \cdots \bigl(1 - P(X_n \le x)\bigr) = 1 - \bigl(1 - F(x)\bigr)^n.$$

Taking the derivative, we get a PDF of X:

$$f_X(x) = F_X'(x) = 0 - n\bigl(1 - F(x)\bigr)^{n-1}\bigl(-F'(x)\bigr) = nf(x)\bigl(1 - F(x)\bigr)^{n-1}.$$

b) From Example 9.4—specifically, Equation (9.20) on page 498—we know that

$$f_{X,Y}(x, y) = n(n - 1)f(x)f(y)\bigl(F(y) - F(x)\bigr)^{n-2}, \qquad x < y,$$

and $f_{X,Y}(x, y) = 0$ otherwise. Next we apply Proposition 9.7 by integrating on y the joint PDF of X and Y. Figure 9.9 shows the interval of integration for a fixed x. The solid portion of the interval is the part where the joint PDF of X and Y may be nonzero.

Figure 9.9

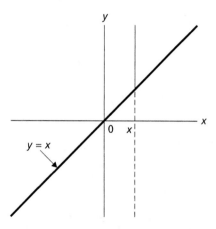

Applying Equation (9.27), Figure 9.9, and the substitution $u = F(y) - F(x)$ yields

$$f_X(x) = \int_{-\infty}^{\infty} f_{X,Y}(x, y)\,dy = \int_{x}^{\infty} n(n - 1)f(x)f(y)\bigl(F(y) - F(x)\bigr)^{n-2}\,dy$$
$$= n(n - 1)f(x)\int_{0}^{1-F(x)} u^{n-2}\,du = nf(x)\bigl(1 - F(x)\bigr)^{n-1}.$$

This result agrees with the one found in part (a).

c) For a $\mathcal{U}(0, 1)$ distribution, we have, for $0 < x < 1$, that $F(x) = x$ and $f(x) = 1$; otherwise, $f(x) = 0$. Hence, in this case, we find from either part (a) or part (b) that a PDF of X is $f_X(x) = n(1 - x)^{n-1}$ for $0 < x < 1$, and $f_X(x) = 0$ otherwise. It follows that X has the beta distribution with parameters 1 and n, as you are asked to verify in Exercise 9.67.   ∎

Conditional Probability Density Functions

We now turn to conditional continuous distributions, where we consider the conditional probability distribution of one or more continuous random variables, given the values of one or more continuous random variables. In the bivariate case of two continuous random variables X and Y, there is a conditional probability distribution of Y for each given value of X and a conditional probability distribution of X for each given value of Y.

For instance, let X and Y denote the height and weight, respectively, of a randomly selected woman. We might be interested in the weight distribution of all women—that is, in the probability distribution of the random variable Y. However, we might be even more concerned with the weight distribution of all women of a particular height—that is, in the conditional probability distribution of the random variable Y for a given value of the random variable X.

We want to define the *conditional probability density function* of one continuous random variable, given the value of another continuous random variable. One approach for making this definition is by analogy with the discrete case. According to Definition 6.3 on page 283, for discrete random variables X and Y, the conditional probability mass function of Y given $X = x$, denoted $p_{Y \mid X}$, is defined by

$$p_{Y \mid X}(y \mid x) = \frac{p_{X,Y}(x, y)}{p_X(x)}, \qquad y \in \mathcal{R}, \tag{9.29}$$

provided that $p_X(x) > 0$. Equation (9.29) is simply the conditional probability rule applied to the events $\{X = x\}$ and $\{Y = y\}$:

$$P(Y = y \mid X = x) = \frac{P(X = x, Y = y)}{P(X = x)}, \qquad y \in \mathcal{R}, \tag{9.30}$$

provided that the event $\{X = x\}$ has positive probability.

For continuous random variables X and Y, the event $\{X = x\}$ always has zero probability and hence Equation (9.30) never makes sense. Consequently, to obtain a viable analogue of the conditional PMF for discrete random variables that applies to continuous random variables, we consider the conditional probability that Y is "near" y given that X is "near" x, rather than the conditional probability that Y equals y given that X equals x.

Suppose then that X and Y are continuous random variables with a joint PDF. To examine the conditional probability that Y is "near" y, given that X is "near" x, we let Δx and Δy represent small positive numbers and consider the conditional probability that Y takes a value between y and $y + \Delta y$, given that X takes a value between x and $x + \Delta x$—that is, $P(y \le Y \le y + \Delta y \mid x \le X \le x + \Delta x)$. Applying first the conditional probability rule and then Relations (9.17) and (9.2) on pages 495 and 486, respectively, we get

$$f_{Y \mid X}(y \mid x)\Delta y \approx P(y \le Y \le y + \Delta y \mid x \le X \le x + \Delta x)$$

$$= \frac{P(x \le X \le x + \Delta x, \ y \le Y \le y + \Delta y)}{P(x \le X \le x + \Delta x)} \approx \frac{f_{X,Y}(x, y)\Delta x \Delta y}{f_X(x)\Delta x}.$$

In other words,

$$f_{Y|X}(y\,|\,x) \approx \frac{f_{X,Y}(x,\,y)}{f_X(x)}. \qquad (9.31)$$

In view of Relation (9.31), we have the following definition of **conditional probability density function.**

DEFINITION 9.3   Conditional Probability Density Function

Let X and Y be two continuous random variables defined on the same sample space and having a joint PDF. If $f_X(x) > 0$, the **conditional probability density function of Y given $X = x$**, denoted $f_{Y|X}$, is the real-valued function

$$f_{Y|X}(y\,|\,x) = \frac{f_{X,Y}(x,\,y)}{f_X(x)}, \qquad y \in \mathcal{R}. \qquad (9.32)$$

If $f_X(x) = 0$, we define $f_{Y|X}(y\,|\,x) = 0$ for all $y \in \mathcal{R}$, but we don't refer to $f_{Y|X}$ as a conditional probability density function.

Note: Observe the analogy between the definition of conditional PDF for continuous random variables, as presented in Equation (9.32), and the definition of conditional PMF for discrete random variables, as presented in Equation (9.29). The former can be obtained from the latter by formal substitution of PDFs for PMFs.

EXAMPLE 9.10   *Obtaining Conditional PDFs*

Bacteria on a Petri Dish   Refer to Example 9.8 on page 511. Determine and identify a conditional PDF of Y for each possible value of X.

Solution   To obtain a conditional PDF of Y for each possible value of X, we need a joint PDF of X and Y and a marginal PDF of X. From Example 9.8, $f_{X,Y}(x, y) = 1/\pi$ for (x, y) in the unit disk, and $f_{X,Y}(x, y) = 0$ otherwise; and $f_X(x) = (2/\pi)\sqrt{1 - x^2}$ for $x \in (-1, 1)$, and $f_X(x) = 0$ otherwise. In particular, the range of X is the interval $(-1, 1)$.

Let $x \in (-1, 1)$. Observe that the point (x, y) is in the unit disk if and only if y is in the interval $\left(-\sqrt{1 - x^2}, \sqrt{1 - x^2}\right)$. Applying Definition 9.3 yields

$$f_{Y|X}(y\,|\,x) = \frac{f_{X,Y}(x,\,y)}{f_X(x)} = \frac{1/\pi}{(2/\pi)\sqrt{1 - x^2}} = \frac{1}{2\sqrt{1 - x^2}},$$

for $-\sqrt{1 - x^2} < y < \sqrt{1 - x^2}$, and $f_{Y|X}(y\,|\,x) = 0$ otherwise. So, given that $X = x$, Y has the uniform distribution on the interval $\left(-\sqrt{1 - x^2}, \sqrt{1 - x^2}\right)$. We express this result symbolically as $Y_{|X=x} \sim \mathcal{U}\left(-\sqrt{1 - x^2}, \sqrt{1 - x^2}\right)$. ∎

We saw in Example 9.10 that the conditional distributions of Y given $X = x$ are all uniform distributions. This result is a special case of the fact that all conditional distributions of a bivariate uniform distribution are uniform distributions. See Exercise 9.81.

In Example 9.10, the conditional PDFs of Y given $X = x$ vary, depending on x, and also differ from the marginal PDF of Y (which was determined in Example 9.8). Example 9.11 provides an instance where all conditional PDFs of Y given $X = x$ are identical to each other and to the marginal PDF of Y, a property whose important consequences will be seen in Section 9.5.

EXAMPLE 9.11   *Obtaining Conditional PDFs*

Lifetimes of Electrical Components   In Example 9.5, we considered two electrical components whose lifetimes, X and Y, have joint PDF $f_{X,Y}(x, y) = \lambda\mu\, e^{-(\lambda x + \mu y)}$ for $x > 0$ and $y > 0$, and $f_{X,Y}(x, y) = 0$ otherwise.

a) Determine and identify a conditional PDF of Y for each possible value of X.
b) Compare the conditional PDFs of Y given $X = x$ to each other and to the marginal PDF of Y. Interpret the results.

Solution   To begin, we find marginal PDFs of X and Y. For $x > 0$,

$$f_X(x) = \int_{-\infty}^{\infty} f_{X,Y}(x, y)\, dy = \int_0^{\infty} \lambda\mu\, e^{-(\lambda x + \mu y)}\, dy$$

$$= \lambda\mu\, e^{-\lambda x} \int_0^{\infty} e^{-\mu y}\, dy = \lambda e^{-\lambda x},$$

and $f_X(x) = 0$ otherwise. Thus $X \sim \mathcal{E}(\lambda)$; that is, X has the exponential distribution with parameter λ. Similarly, we find that $Y \sim \mathcal{E}(\mu)$.

a) Applying Definition 9.3, we get, for $x > 0$,

$$f_{Y\,|\,X}(y\,|\,x) = \frac{f_{X,Y}(x, y)}{f_X(x)} = \frac{\lambda\mu\, e^{-(\lambda x + \mu y)}}{\lambda e^{-\lambda x}} = \mu e^{-\mu y}, \qquad y > 0,$$

and $f_{Y\,|\,X}(y\,|\,x) = 0$ otherwise. Thus, for each $x > 0$, we have $Y_{|X=x} \sim \mathcal{E}(\mu)$—that is, the conditional distribution of Y given $X = x$ is the exponential distribution with parameter μ.

b) From part (a), we conclude that the conditional PDFs of Y given $X = x$ are identical to each other and to the marginal PDF of Y. Thus, in this case, knowing the value of the random variable X imparts no information about the distribution of Y; in other words, knowing the value of X doesn't affect the probability distribution of Y.   ■

A Conditional PDF Is a PDF

A conditional probability density function is in fact a genuine probability density function, as stated in Proposition 9.8. We leave the proof to you as Exercise 9.77.

◆◆◆   **Proposition 9.8   A Conditional PDF Is a PDF**

Let X and Y be continuous random variables with a joint PDF and let x be a possible value of X—that is, $f_X(x) > 0$. Then a conditional PDF of Y given $X = x$ is a probability density function; that is, as a function of y, $f_{Y\,|\,X}(y\,|\,x)$ satisfies properties (a) and (b) of Proposition 8.6 on page 423.

Proposition 9.8 implies that all properties of (unconditional) PDFs also hold for conditional PDFs. For instance, the following conditional version of the FPF holds:

$$P(Y \in A \mid X = x) = \int_A f_{Y \mid X}(y \mid x) \, dy \tag{9.33}$$

for each subset $A \subset \mathcal{R}$.

EXAMPLE 9.12   *Conditional Probability Density Functions*

Min and Max of a Random Sample   Suppose that $X_1, \ldots, X_n$ are a random sample from a $\mathcal{U}(0, 1)$ distribution. Set $X = \min\{X_1, \ldots, X_n\}$ and $Y = \max\{X_1, \ldots, X_n\}$. Find $P\left(\frac{1}{4} < Y < \frac{3}{4} \mid X = \frac{1}{2}\right)$.

Solution   We first obtain a conditional PDF of Y given $X = x$. From Example 9.4(c) on page 499, a joint PDF of X and Y is $f_{X,Y}(x, y) = n(n - 1)(y - x)^{n-2}$ for $0 < x < y < 1$, and $f_{X,Y}(x, y) = 0$ otherwise. And, from Example 9.9(c) on page 513, a marginal PDF of X is $f_X(x) = n(1 - x)^{n-1}$ for $0 < x < 1$, and $f_X(x) = 0$ otherwise. Therefore, by Definition 9.3, for each $0 < x < 1$,

$$f_{Y \mid X}(y \mid x) = \frac{f_{X,Y}(x, y)}{f_X(x)} = \frac{n(n-1)(y-x)^{n-2}}{n(1-x)^{n-1}} = \frac{(n-1)(y-x)^{n-2}}{(1-x)^{n-1}},$$

for $x < y < 1$, and $f_{Y \mid X}(y \mid x) = 0$ otherwise. Hence, from Equation (9.33),

$$P\left(\tfrac{1}{4} < Y < \tfrac{3}{4} \mid X = \tfrac{1}{2}\right) = \int_{\frac{1}{4}}^{\frac{3}{4}} f_{Y \mid X}\left(y \mid \tfrac{1}{2}\right) dy = \int_{\frac{1}{2}}^{\frac{3}{4}} \frac{(n-1)\left(y - \tfrac{1}{2}\right)^{n-2}}{\left(1 - \tfrac{1}{2}\right)^{n-1}} \, dy$$

$$= (n - 1)2^{n-1} \int_0^{\frac{1}{4}} u^{n-2} \, du = \frac{1}{2^{n-1}},$$

as required.   ■

Equation (9.32) on page 515 defines the conditional PDF of Y given $X = x$. Multiplying both sides of that equation by $f_X(x)$ yields

$$f_{X,Y}(x, y) = f_X(x) f_{Y \mid X}(y \mid x). \tag{9.34}$$

Equation (9.34) holds for all real numbers x and y, not only for those where $f_X(x) > 0$. More precisely, the function f defined on $\mathcal{R}^2$ by $f(x, y) = f_X(x) f_{Y \mid X}(y \mid x)$ is a joint PDF of X and Y. This result is the **general multiplication rule** for the joint PDF of two continuous random variables, X and Y. We leave the details to you in Exercise 9.78.

EXAMPLE 9.13   *The General Multiplication Rule*

Regression Analysis   Let X have the standard normal distribution and let $-1 < \rho < 1$. Suppose that, for each $x \in \mathcal{R}$, the conditional distribution of Y given $X = x$ is a normal distribution with parameters ρx and $1 - \rho^2$; symbolically, $Y_{|X=x} \sim \mathcal{N}(\rho x, 1 - \rho^2)$.
a) Determine a joint PDF of X and Y.
b) Determine and identify a marginal PDF of Y.

Solution   By assumption,

$$f_X(x) = \frac{1}{\sqrt{2\pi}} e^{-x^2/2}, \qquad x \in \mathcal{R},$$

and

$$f_{Y|X}(y \,|\, x) = \frac{1}{\sqrt{2\pi}\sqrt{1-\rho^2}} e^{-(y-\rho x)^2/2(1-\rho^2)}, \qquad x, y \in \mathcal{R}.$$

a) Using the general multiplication rule, Equation (9.34), we have

$$f_{X,Y}(x, y) = f_X(x)f_{Y|X}(y \,|\, x) = \frac{1}{\sqrt{2\pi}} e^{-x^2/2} \cdot \frac{1}{\sqrt{2\pi}\sqrt{1-\rho^2}} e^{-(y-\rho x)^2/2(1-\rho^2)},$$

or

$$f_{X,Y}(x, y) = \frac{1}{2\pi\sqrt{1-\rho^2}} e^{-(x^2-2\rho xy+y^2)/2(1-\rho^2)}. \tag{9.35}$$

b) Applying Proposition 9.7 on page 510, Equation (9.35), and the technique of completing the square, we obtain

$$f_Y(y) = \int_{-\infty}^{\infty} f_{X,Y}(x, y)\, dx = \int_{-\infty}^{\infty} \frac{1}{2\pi\sqrt{1-\rho^2}} e^{-(x^2-2\rho xy+y^2)/2(1-\rho^2)}\, dx$$

$$= \frac{1}{2\pi\sqrt{1-\rho^2}} \int_{-\infty}^{\infty} e^{-((x-\rho y)^2+y^2-\rho^2 y^2)/2(1-\rho^2)}\, dx$$

$$= \frac{1}{2\pi\sqrt{1-\rho^2}} e^{-y^2/2} \int_{-\infty}^{\infty} e^{-(x-\rho y)^2/2(1-\rho^2)}\, dx$$

$$= \frac{1}{\sqrt{2\pi}} e^{-y^2/2} \int_{-\infty}^{\infty} \frac{1}{\sqrt{2\pi}\sqrt{1-\rho^2}} e^{-(x-\rho y)^2/2(1-\rho^2)}\, dx.$$

The integrand in the previous expression is the PDF of an $\mathcal{N}(\rho y, 1-\rho^2)$ distribution and therefore its integral must equal 1, as is the case for any PDF. Thus

$$f_Y(y) = \frac{1}{\sqrt{2\pi}} e^{-y^2/2}, \qquad y \in \mathcal{R}.$$

Hence $Y \sim \mathcal{N}(0, 1)$; that is, Y has the standard normal distribution.   ∎

Note: The joint PDF in Equation (9.35) is that of a particular *bivariate normal distribution*. We examine bivariate normal distributions in detail in Chapter 10. As you will learn there, ρ is the correlation coefficient of the bivariate normal random variables.

Multivariate Marginal and Conditional PDFs

The definitions and results that we presented for bivariate PDFs are easily extended to the general multivariate case. To begin, we explain how to obtain marginal PDFs. In the case of two continuous random variables with a joint PDF, there are two marginal PDFs.

However, when considering more than two continuous random variables, there are many more marginal PDFs.

For instance, suppose that X, Y, and Z are three continuous random variables with a joint PDF. In this case, there are $\binom{3}{1} = 3$ univariate marginal PDFs—namely, f_X (the PDF of X), f_Y (the PDF of Y), and f_Z (the PDF of Z). Additionally, there are $\binom{3}{2} = 3$ bivariate marginal PDFs—namely, $f_{X,Y}$ (the joint PDF of X and Y), $f_{X,Z}$ (the joint PDF of X and Z), and $f_{Y,Z}$ (the joint PDF of Y and Z). Each marginal PDF is obtained by integrating the joint PDF of X, Y, and Z on the unwanted variable(s). To illustrate, the univariate marginal PDF of Y and the bivariate marginal PDF of Y and Z are obtained as follows:

$$f_Y(y) = \int_{-\infty}^{\infty} \int_{-\infty}^{\infty} f_{X,Y,Z}(x, y, z)\, dx\, dz$$

and

$$f_{Y,Z}(y, z) = \int_{-\infty}^{\infty} f_{X,Y,Z}(x, y, z)\, dx.$$

In general, if $X_1, \ldots, X_m$ are m continuous random variables with a joint PDF, there are $\binom{m}{j}$ j-variate marginal PDFs ($1 \le j \le m - 1$). Each such j-variate marginal PDF is obtained by integrating the joint PDF of $X_1, \ldots, X_m$ on the other $m - j$ variables.

We can also define conditional PDFs in the general multivariate case. The formulas for these conditional PDFs are analogous to those in the bivariate case. For instance, if X, Y, and Z are continuous random variables with a joint PDF, then

$$f_{Y,Z \mid X}(y, z \mid x) = \frac{f_{X,Y,Z}(x, y, z)}{f_X(x)}$$

and

$$f_{Z \mid X,Y}(z \mid x, y) = \frac{f_{X,Y,Z}(x, y, z)}{f_{X,Y}(x, y)}.$$

From the previous formula and the general multiplication rule for the joint PDF of two continuous random variables—Equation (9.34) on page 517—we easily obtain the general multiplication rule in the trivariate case:

$$f_{X,Y,Z}(x, y, z) = f_X(x) f_{Y \mid X}(y \mid x) f_{Z \mid X,Y}(z \mid x, y). \tag{9.36}$$

Exercise 9.80 asks you to state the general multiplication rule for the joint PDF of m continuous random variables.

EXERCISES 9.4   Basic Exercises

9.63 Let X and Y denote the x and y coordinates, respectively, of a point selected at random from the unit square. In Example 9.7 on page 510, we obtained marginal PDFs of X and Y.
a) Obtain and identify all conditional PDFs.
b) Compare the conditional PDFs of Y given $X = x$ to each other and to the marginal PDF of Y. Interpret the results.

9.64 For each part, determine $P(X > 0.9 \mid Y = 0.8)$ for the specified joint PDF of a random point (X, Y) in the unit square.
a) $f_{X,Y}(x, y) = 1$ **b)** $f_{X,Y}(x, y) = x + y$ **c)** $f_{X,Y}(x, y) = \frac{3}{2}(x^2 + y^2)$

9.65 Let X and Y be the x and y coordinates, respectively, of a point selected at random from the upper half of the unit disk—that is, from the set $\{(x, y) \in \mathcal{R}^2 : x^2 + y^2 < 1, y > 0\}$. From Exercise 9.42 on page 507, a joint PDF of X and Y is $f_{X,Y}(x, y) = 2/\pi$ for (x, y) in the upper half of the unit disk, and $f_{X,Y}(x, y) = 0$ otherwise. Obtain a
a) marginal PDF of X.
b) marginal PDF of Y.
c) conditional PDF of Y given $X = x$.
d) conditional PDF of X given $Y = y$.

9.66 Let X and Y have joint probability density function given by $f_{X,Y}(x, y) = \frac{3}{8}(|x| + |y|)$ if $x^2 + y^2 < 1$, and $f_{X,Y}(x, y) = 0$ otherwise.
a) Obtain a marginal PDF of X.
b) Without doing further computations, obtain a marginal PDF of Y.
c) Find $P(|Y| > 1/2 \mid X = 0)$.

9.67 Let $X_1, \ldots, X_n$ be a random sample from a uniform distribution on the interval $(0, 1)$, and set $X = \min\{X_1, \ldots, X_n\}$. In Example 9.9(c) on page 513, we showed that a PDF of X is $f_X(x) = n(1 - x)^{n-1}$ for $0 < x < 1$, and $f_X(x) = 0$ otherwise. Verify that X has the beta distribution with parameters 1 and n.

9.68 Refer to Example 9.12 on page 517, where $X_1, \ldots, X_n$ is a random sample from a $\mathcal{U}(0, 1)$ distribution, $X = \min\{X_1, \ldots, X_n\}$, and $Y = \max\{X_1, \ldots, X_n\}$.
a) Is a conditional PDF used to determine $P(Y \leq 0.7 \mid X > 0.1)$? Explain your answer.
b) Determine $P(Y \leq 0.7 \mid X > 0.1)$.

9.69 Let T be the interior of the triangle with vertices $(0, 0)$, $(1, 0)$, and $(1, 1)$. Consider the following two methods for selecting a point (X, Y) from T.
    I. Select the point randomly (i.e., uniformly) from T.
    II. Select the x coordinate of the point randomly from the interval $(0, 1)$ and then select the y coordinate of the point randomly from the interval $(0, x)$.
a) Without doing any calculations, decide whether, probabilistically, a difference exists in the two methods. Explain your answer.
b) Determine and compare joint PDFs for both methods I and II.
c) Determine and compare marginal PDFs for both methods I and II. Identify the marginal distributions when possible.
d) Determine and compare conditional PDFs for both methods I and II. Identify the conditional distributions when possible.

9.70 Refer to Example 9.9 on page 512.
a) Determine a PDF of Y directly.
b) Determine a PDF of Y by using a joint PDF of X and Y, which can be found in Equation (9.20) on page 498.
c) Apply the results for the PDF of Y to the special case of a random sample of size n from a uniform distribution on the interval $(0, 1)$. Identify the probability distribution of Y in this special case.

9.71 Mechanical or electrical units often consist of several components, each of which is subject to failure. A unit is said to be a *parallel system* if it functions when at least one of the components is working. Consider a parallel system of n components, $C_1, \ldots, C_n$, with respective lifetimes $X_1, \ldots, X_n$, where $X_1, \ldots, X_n$ are independent exponential random variables with common parameter λ.
a) Explain how the lifetimes of the components can be considered a random sample of size n from an appropriate probability distribution.
b) Use the result of Example 9.9 on page 512 to obtain a PDF for the time, X, at which the first component failure occurs.

c) Use the result of Exercise 9.70(a) to obtain a PDF for system lifetime, Y.

d) Apply the definition of a conditional PDF and your result from part (b) to obtain a conditional PDF of system lifetime, given that the time at which the first component failure occurs is x, where $x > 0$. *Hint:* Refer to Equation (9.20) on page 498.

e) Obtain the result of part (d) by using the lack-of-memory property of the exponential distribution and the result of part (c). *Hint:* Consider a parallel system consisting of $n - 1$ components.

f) Apply the definition of a conditional PDF and your result from part (c) to obtain a conditional PDF of the time at which the first component failure occurs, given that the system lifetime is y, where $y > 0$.

9.72 This exercise requires results from Exercise 9.71. Suppose that $\lambda = 1$ and that there are 10 components.

a) Given that the time of first component failure is 1, what is the probability that system lifetime exceeds 4?

b) Given that system lifetime is 4, what is the probability that the time of first component failure exceeds 1?

9.73 Once a fire is reported to a fire insurance company, the company makes an initial estimate, X, of the amount it will pay to the claimant for the fire loss. When the claim is finally settled, the company pays an amount Y to the claimant. The company has determined that X and Y have joint density function

$$f_{X,Y}(x, y) = \frac{2}{x^2(x - 1)} y^{-(2x-1)/(x-1)}, \qquad x > 1, \ y > 1,$$

and $f_{X,Y}(x, y) = 0$ otherwise. Given that the initial estimate by the company is 2, what is the probability that the final settlement amount is between 1 and 3?

9.74 An automobile insurance policy will pay for damage to both the policyholder's car and the other driver's car in the event that the policyholder is responsible for an accident. The payment for damage to the policyholder's car, X, has a marginal density function of 1 for $0 < x < 1$. Given that $X = x$, the payment for damage to the other driver's car, Y, has a conditional density of 1 for $x < y < x + 1$. If the policyholder is responsible for an accident, what is the probability that the payment for damage to the other driver's car will exceed 0.5?

9.75 Suppose that the conditional distribution of Y given $\Lambda = \lambda$ is exponential with parameter λ and that Λ has a gamma distribution with parameters α and β. *Note:* These assumptions could also be stated as follows: Let $Y \sim \mathcal{E}(\lambda)$, where λ varies as a random variable Λ having a $\Gamma(\alpha, \beta)$ distribution.

a) Determine a marginal PDF of Y.

b) Determine and identify a conditional PDF of Λ given $Y = y$.

9.76 A number X is selected at random from the interval $(0, 1)$, next a number Y is selected at random from the interval $(0, X)$, and then a number Z is selected at random from the interval $(0, Y)$. Determine a

a) joint PDF of the random variables X, Y, and Z.

b) marginal PDF of Z.

c) conditional PDF of X and Y given $Z = z$.

Theory Exercises

9.77 Prove Proposition 9.8 on page 516, which shows that a conditional probability density function is in fact a genuine probability density function.

9.78 General multiplication rule for two continuous random variables: Let X and Y be continuous random variables with a joint PDF. Prove that the function f defined on $\mathcal{R}^2$ by $f(x, y) = f_X(x) f_{Y|X}(y|x)$ is a joint PDF of X and Y. Use the following steps.
a) Let $A = \{x : f_X(x) > 0\}$. Show that $P(X \in A^c \cap (-\infty, x], Y \leq y) = 0$ for all x and y.
b) Verify that $F_{X,Y}(x, y) = P(X \in A \cap (-\infty, x], Y \leq y)$ for all x and y.
c) Prove that, for all x and y, $F_{X,Y}(x, y) = \int_{-\infty}^{x} \int_{\infty}^{y} f_X(s) f_{Y|X}(t|s) \, ds \, dt$.
d) Conclude that the function f defined on $\mathcal{R}^2$ by $f(x, y) = f_X(x) f_{Y|X}(y|x)$ is a joint PDF of X and Y.

9.79 Law of total probability for two continuous random variables: Let X and Y be continuous random variables with a joint PDF.

a) Prove that $f_Y(y) = \int_{-\infty}^{\infty} f_{Y|X}(y|x) f_X(x) \, dx$. Interpret this result.

b) Prove that $P(Y \in A) = \int_{-\infty}^{\infty} P(Y \in A \mid X = x) f_X(x) \, dx$, for all $A \subset \mathcal{R}$.

9.80 Multivariate general multiplication rule: State the general multiplication rule for the joint PDF of m continuous random variables, $X_1, \ldots, X_m$.

Advanced Exercises

9.81 Bivariate uniform random variables: Suppose that a point is selected at random from a subset S of $\mathcal{R}^2$ with finite nonzero area. Let X and Y be the x and y coordinates, respectively, of the point obtained. Exercise 9.40 shows that a joint PDF of X and Y is $f_{X,Y}(x, y) = 1/|S|$ for $(x, y) \in S$, and $f_{X,Y}(x, y) = 0$ otherwise. For $x \in \mathcal{R}$, let $S_x = \{y \in \mathcal{R} : (x, y) \in S\}$ and, for $y \in \mathcal{R}$, let $S^y = \{x \in \mathcal{R} : (x, y) \in S\}$.
a) Determine marginal PDFs of X and Y.
b) Show that all conditional distributions of X and Y are (univariate) uniform distributions.
c) Are all marginal distributions of X and Y necessarily (univariate) uniform distributions?
d) Provide a sufficient condition for the marginals of X and Y to be uniform distributions.

9.82 Proposition 9.7 on page 510 implies the following result: If X and Y are continuous random variables with a joint PDF, then each of X and Y is a continuous random variable with a PDF. Is the converse of this result true? Justify your answer.

Mixed bivariate distributions: On page 509, we introduced the concept of mixed bivariate distributions, where we simultaneously consider a continuous random variable X and a discrete random variable Y. Exercises 9.83–9.87 continue the examination of mixed bivariate distributions.

9.83 Let X and Y be random variables with joint PDF/PMF $h_{X,Y}$, as defined on page 509.
a) Prove that a (marginal) PDF of X is given by $f_X(x) = \sum_y h_{X,Y}(x, y)$. *Hint:* First find the CDF of X by using the law of partitions, Proposition 2.8 on page 68.
b) Prove that the (marginal) PMF of Y is given by $p_Y(y) = \int_{-\infty}^{\infty} h_{X,Y}(x, y) \, dx$. *Hint:* First express $P(Y = y)$ as a joint probability involving X and Y.

9.84 Let X and Y have joint PDF/PMF given by $h_{X,Y}(x, y) = \frac{1}{2} e^{-x}$ for $x > 0$ and $y \in \{0, 1\}$, and $h_{X,Y}(x, y) = 0$ otherwise. Use Exercise 9.83 to obtain and identify the probability distributions of X and Y.

9.85 Let X and Y be random variables with joint PDF/PMF $h_{X,Y}$.
a) Define the conditional PMF of Y given $X = x$.
b) Define the conditional PDF of X given $Y = y$.

9.86 Let Y denote the number of patients who arrive at an emergency room during a 1-hour period and let X denote the rate at which they arrive. A joint PDF/PMF of X and Y is given by $h_{X,Y}(x, y) = \lambda e^{-(1+\lambda)x}(x^y/y!)$ for $x > 0$ and $y = 0, 1, 2, \ldots$, and $h_{X,Y}(x, y) = 0$ otherwise.

a) Use Exercise 9.83 to obtain and identify a marginal PDF of X and the marginal PMF of Y.
   Note: In identifying the probability distribution of Y, refer to Exercise 7.26 on page 336.

b) Use Exercise 9.85 and part (a) to obtain and identify the conditional PDF of X given $Y = y$ and the conditional PMF of Y given $X = x$.

9.87 The annual number of automobile accidents, Y, of a driver has a Poisson distribution with parameter λ, where λ varies as a random variable X having a gamma distribution with parameters α and β.

a) Determine a joint PDF/PMF of X and Y.

b) Determine and identify the marginal PMF of Y. *Note:* In identifying the probability distribution of Y, refer to Equation (5.50) on page 243.

c) Determine and identify a conditional PDF of X given $Y = y$.

9.5   Independent Continuous Random Variables

In Section 6.4, we introduced the concept of independent random variables. The discussion there applies to all types of random variables—discrete, continuous, or otherwise. In this section, we specialize to the situation of continuous random variables with a joint PDF. As usual, we first consider the bivariate case.

To begin, we recall Definition 6.4: Two random variables X and Y defined on the same sample space are said to be *independent random variables* if

$$P(X \in A, Y \in B) = P(X \in A)P(Y \in B) \tag{9.37}$$

for all subsets A and B of real numbers. In the continuous case, we can provide an equivalent condition for independence in terms of PDFs, as presented in Proposition 9.9.

◆◆◆   **Proposition 9.9   Independent Continuous Random Variables: Bivariate Case**

Let X and Y be continuous random variables with a joint PDF. Then X and Y are independent if and only if the function f defined on $\mathcal{R}^2$ by $f(x, y) = f_X(x)f_Y(y)$ is a joint PDF of X and Y. We write this condition symbolically as

$$f_{X,Y}(x, y) = f_X(x)f_Y(y), \qquad x, y \in \mathcal{R}. \tag{9.38}$$

In words, two continuous random variables with a joint PDF are independent if and only if their joint PDF equals the product of their marginal PDFs.

Proof   Suppose that X and Y are independent random variables. Then, for each $x, y \in \mathcal{R}$,

$$F_{X,Y}(x, y) = P(X \le x, Y \le y) = P(X \le x)P(Y \le y)$$

$$= \left(\int_{-\infty}^{x} f_X(s)\,ds \right) \left(\int_{-\infty}^{y} f_Y(t)\,dt \right) = \int_{-\infty}^{x} \int_{-\infty}^{y} f_X(s)f_Y(t)\,ds\,dt.$$

Consequently, from Proposition 9.4 on page 499, the function f defined on $\mathcal{R}^2$ by $f(x, y) = f_X(x)f_Y(y)$ is a joint PDF of X and Y.

Conversely, suppose that the function f defined on $\mathcal{R}^2$ by $f(x, y) = f_X(x) f_Y(y)$ is a joint PDF of X and Y. Let A and B be any two subsets of real numbers. Applying the FPFs for bivariate and univariate continuous random variables, we get

$$
\begin{aligned}
P(X \in A, Y \in B) = P\left((X, Y) \in A \times B\right) &= \iint\limits_{A \times B} f(x, y)\, dx\, dy \\
&= \iint\limits_{A \times B} f_X(x) f_Y(y)\, dx\, dy = \int_B \left(\int_A f_X(x) f_Y(y)\, dx \right) dy \\
&= \int_B f_Y(y) \left(\int_A f_X(x)\, dx \right) dy = P(X \in A) \int_B f_Y(y)\, dy \\
&= P(X \in A) P(Y \in B).
\end{aligned}
$$

Thus X and Y are independent random variables. ◆

EXAMPLE 9.14   *The Concept of Independent Random Variables*

Bacteria on a Petri Dish   Refer to Example 9.8 on page 511, where X and Y denote the x and y coordinates, respectively, of the center of the first spot (visible bacteria colony) to appear. Determine whether X and Y are independent random variables.

Solution   From Example 9.8, joint and marginal PDFs of X and Y are

$$
f_{X,Y}(x, y) = \begin{cases} \dfrac{1}{\pi}, & \text{if } (x, y) \text{ is in the unit disk;} \\[2mm] 0, & \text{otherwise;} \end{cases} \tag{9.39}
$$

$$
f_X(x) = \begin{cases} \dfrac{2}{\pi}\sqrt{1 - x^2}, & \text{if } -1 < x < 1; \\[2mm] 0, & \text{otherwise;} \end{cases} \qquad f_Y(y) = \begin{cases} \dfrac{2}{\pi}\sqrt{1 - y^2}, & \text{if } -1 < y < 1; \\[2mm] 0, & \text{otherwise.} \end{cases}
$$

Let f be the function defined on $\mathcal{R}^2$ by $f(x, y) = f_X(x) f_Y(y)$. Then,

$$
f(x, y) = \begin{cases} \dfrac{4}{\pi^2}\sqrt{(1 - x^2)(1 - y^2)}, & \text{if } -1 < x < 1 \text{ and } -1 < y < 1; \\[2mm] 0, & \text{otherwise.} \end{cases}
$$

We claim that f isn't a joint PDF of X and Y. One way to verify this claim is to note that, from Equation (9.39), the probability is 0 that the random point (X, Y) falls in the shaded region in Figure 9.10. If f were a joint PDF of X and Y, its double integral over that region would also be 0, which it clearly isn't.

We have shown that the function f defined on $\mathcal{R}^2$ by $f(x, y) = f_X(x) f_Y(y)$ isn't a joint PDF of X and Y. Consequently, by Proposition 9.9, X and Y aren't independent random variables. Other ways to show that X and Y aren't independent random variables are considered in Exercise 9.88. ■

Figure 9.10

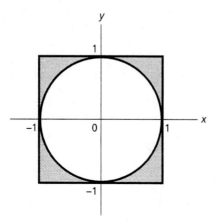

EXAMPLE 9.15   *The Concept of Independent Random Variables*

Lifetimes of Electrical Components   Consider two electrical components, A and B, with respective lifetimes X and Y. Assume that a joint PDF of X and Y is given by

$$f_{X,Y}(x, y) = \begin{cases} \lambda\mu\, e^{-(\lambda x + \mu y)}, & \text{if } x, y > 0; \\ 0, & \text{otherwise;} \end{cases} \qquad (9.40)$$

where λ and μ are positive constants. Determine whether X and Y are independent.

Solution   From Example 9.11 on page 516, marginal PDFs of X and Y are exponential with parameters λ and μ, respectively. Thus,

$$f_X(x) = \begin{cases} \lambda e^{-\lambda x}, & \text{if } x > 0; \\ 0, & \text{otherwise;} \end{cases} \qquad \text{and} \qquad f_Y(y) = \begin{cases} \mu e^{-\mu y}, & \text{if } y > 0; \\ 0, & \text{otherwise.} \end{cases}$$

Therefore,

$$f_X(x) f_Y(y) = \begin{cases} \lambda e^{-\lambda x}\mu e^{-\mu y}, & \text{if } x, y > 0; \\ 0, & \text{otherwise;} \end{cases} = \begin{cases} \lambda\mu\, e^{-(\lambda x + \mu y)}, & \text{if } x, y > 0; \\ 0, & \text{otherwise.} \end{cases}$$

Referring to Equation (9.40), we now conclude that the function f defined on $\mathcal{R}^2$ by $f(x, y) = f_X(x) f_Y(y)$ is a joint PDF of X and Y. Hence, by Proposition 9.9, X and Y are independent random variables.   ■

Sometimes we either are told or know from the context of a problem that two continuous random variables, X and Y, are independent. In that case, we can, in view of Proposition 9.9, obtain a joint PDF of X and Y by simply multiplying marginal PDFs of X and Y. Consider, for instance, Example 9.16.

EXAMPLE 9.16   *The Concept of Independent Random Variables*

Meeting for an Appointment   Two people agree to meet at a specified place at noon. Let's assume that the times that the two people actually arrive are independent random variables, each having a triangular distribution on the interval $(-5, 5)$, where time is measured in minutes relative to noon. Determine the probability that the first person to arrive will wait more than 5 minutes before the other person arrives.

Solution   Let X and Y denote the times that the two people arrive. By assumption, X and Y are both $T(-5, 5)$ random variables. Thus, from Equation (8.57) on page 460,

$$f_X(x) = \frac{1}{5}\left(1 - \frac{|x|}{5}\right), \qquad -5 < x < 5,$$

and $f_X(x) = 0$ otherwise; and, likewise, for Y.

Because X and Y are independent random variables, a joint PDF of X and Y can be obtained as the product of the two marginal PDFs. Thus,

$$f_{X,Y}(x, y) = \frac{1}{25}\left(1 - \frac{|x|}{5}\right)\left(1 - \frac{|y|}{5}\right), \qquad -5 < x < 5, \; -5 < y < 5,$$

and $f_{X,Y}(x, y) = 0$ otherwise.

We want to find the probability that the first person to arrive will wait more than 5 minutes before the other person arrives, $P(|X - Y| > 5)$. From the FPF,

$$P(|X - Y| > 5) = \iint\limits_{|x-y|>5} f_{X,Y}(x, y)\, dx\, dy.$$

The shaded region in Figure 9.11 shows the set over which the double integral is taken. The crosshatched portion of the shaded region indicates where the integrand is nonzero.

Figure 9.11

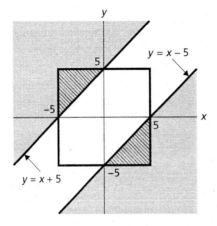

Using the symmetry of the joint PDF of X and Y and making the substitution $u = 1 + y/5$, we obtain

$$P(|X - Y| > 5) = \iint_{|x-y|>5} f_{X,Y}(x, y)\,dx\,dy$$

$$= 2 \int_0^5 \left(\int_{-5}^{x-5} \frac{1}{25} \left(1 - \frac{x}{5}\right) \left(1 + \frac{y}{5}\right) dy \right) dx$$

$$= \frac{2}{5} \int_0^5 \left(1 - \frac{x}{5}\right) \left(\int_0^{x/5} u\,du \right) dx = \frac{1}{5} \int_0^5 \left(1 - \frac{x}{5}\right) \frac{x^2}{25}\,dx$$

$$= \frac{1}{625} \int_0^5 \left(5x^2 - x^3\right) dx = \frac{1}{12} = 0.083.$$

Chances are only 8.3% that the first person to arrive at the specified place will wait more than 5 minutes before the other person arrives. ■

Independence and Conditional Distributions

As Proposition 9.10 shows, independence of continuous random variables also conforms to our intuitive notion in terms of conditional distributions.

◆◆◆ **Proposition 9.10   Independence and Conditional Distributions**

Let X and Y be continuous random variables with a joint PDF. Then X and Y are independent if and only if either of the following properties holds:
a) Each conditional PDF of Y given $X = x$ is a PDF of Y.
b) Each conditional PDF of X given $Y = y$ is a PDF of X.

Proof We verify that condition (a) is equivalent to independence and leave verification of condition (b) for you. Suppose that X and Y are independent. Then the function f defined on $\mathcal{R}^2$ by $f(x, y) = f_X(x) f_Y(y)$ is a joint PDF of X and Y. Thus, if $f_X(x) > 0$,

$$f_{Y|X}(y \mid x) = \frac{f_{X,Y}(x, y)}{f_X(x)} = \frac{f_X(x) f_Y(y)}{f_X(x)} = f_Y(y).$$

Hence $f_{Y|X}$ is a PDF of Y.

Conversely, suppose that condition (a) holds. To establish that X and Y are independent, we show that the function f defined on $\mathcal{R}^2$ by $f(x, y) = f_X(x) f_Y(y)$ is a joint PDF of X and Y. Applying the general multiplication rule, Equation (9.34) on page 517, we obtain

$$f_{X,Y}(x, y) = f_X(x) f_{Y|X}(y \mid x) = \begin{cases} f_X(x) f_Y(y), & \text{if } f_X(x) > 0; \\ 0, & \text{if } f_X(x) = 0. \end{cases} = f_X(x) f_Y(y),$$

for all x and y. Consequently, f is a joint PDF of X and Y. ◆

Note: From Proposition 9.10, if each conditional PDF of Y given $X = x$ is a PDF of Y, then each conditional PDF of X given $Y = y$ must be a PDF of X; and vice versa.

Multivariate Independent Continuous Random Variables

As in the bivariate continuous case, a necessary and sufficient condition for several continuous random variables with a joint PDF to be independent is that their joint PDF equals the product of their marginal PDFs. We leave the precise statement of this result and its proof to you as Exercise 9.108.

EXERCISES 9.5   **Basic Exercises**

9.88 In the petri-dish illustration of Example 9.8 on page 511, let X and Y denote the x and y coordinates, respectively, of the center of the first spot (visible bacteria colony) to appear. We showed, in Example 9.14 on page 524, that the random variables X and Y aren't independent. Here you are asked to provide three other arguments to establish that result.
a) Argue heuristically that X and Y aren't independent by considering the possible values of Y among different specified values of X.
b) Use the results of Examples 9.8(b) and 9.10 (pages 512 and 515, respectively) and Proposition 9.10 (page 527) to show that X and Y aren't independent.
c) Use Proposition 9.4 on page 499—specifically, Equation (9.22)—to show that X and Y aren't independent. *Hint:* Assume that the product of the marginals of X and Y is a joint PDF of X and Y. Obtain a contradiction.

9.89 Provide another verification that the random variables X and Y in Example 9.15 on page 525 are independent. *Hint:* Refer to Example 9.11 on page 516.

9.90 Solve Example 9.16 on page 526 if the arrival times are independent uniform random variables on the interval $(-5, 5)$, where time is measured in minutes relative to noon.

9.91 Let X and Y be the x and y coordinates, respectively, of a point selected at random from the unit square. Determine whether X and Y are independent random variables.

9.92 Let X and Y be the x and y coordinates, respectively, of a point selected at random from the upper half of the unit disk, that is, from the set $\{(x, y) \in \mathcal{R}^2 : x^2 + y^2 < 1, y > 0\}$. Determine whether X and Y are independent random variables. *Note:* Exercise 9.65 on page 520 provides a joint PDF of X and Y and asks for marginal and conditional PDFs.

9.93 Refer to the regression analysis illustration of Example 9.13 on page 517. Use Proposition 9.10 on page 527 to determine necessary and sufficient conditions for X and Y to be independent random variables.

9.94 Let X and Y be independent random variables, each uniform on the interval $(-1, 1)$. Find the probability that the roots of the random quadratic equation $x^2 + Xx + Y = 0$ are real.

9.95 Let X and Y be continuous random variables with a joint PDF. Suppose that there are nonnegative functions g and h defined on $\mathcal{R}$ such that $f_{X,Y}(x, y) = g(x)h(y)$ for all $x, y \in \mathcal{R}$. Show that X and Y are independent by proceeding as follows.
a) Obtain a marginal PDF of X in terms of g and h.
b) Obtain a marginal PDF of Y in terms of g and h.
c) Explain why $\left(\int_{-\infty}^{\infty} g(x)\,dx\right)\left(\int_{-\infty}^{\infty} h(y)\,dy\right) = 1$.
d) Verify that X and Y are independent random variables.

9.96 In Exercise 9.95, is it necessarily true that g is a marginal PDF of X and h is a marginal PDF of Y? If not, find conditions when that is the case.

9.97 Suppose that X_1 and X_2 are independent random variables with common PDF f. Set $X = \min\{X_1, X_2\}$ and $Y = \max\{X_1, X_2\}$.

a) Show that a joint PDF of X and Y is given by $f_{X,Y}(x, y) = 2f(x)f(y)$ for $x < y$, and $f_{X,Y}(x, y) = 0$ otherwise. *Hint:* Refer to Equation (9.20) on page 498.

b) Does Exercise 9.95 and part (a) of this exercise imply that X and Y are independent? Explain your answer.

c) Are X and Y independent? Explain your answer.

9.98 Consider two continuous random variables with a joint PDF. From Proposition 9.7 on page 510, you know that a joint PDF determines marginal PDFs. Show that the converse of that result isn't true; that is, in general, marginal PDFs don't determine a joint PDF. Proceed as follows. Let X and Y be dependent continuous random variables with a joint PDF.

a) Verify that the function f defined on $\mathcal{R}^2$ by $f(x, y) = f_X(x)f_Y(y)$ is a joint PDF of two continuous random variables—say, U and V—by showing that f satisfies properties (a) and (b) of Proposition 9.5 on page 502.

b) Determine marginal PDFs of U and V.

c) Show that U and V are independent random variables.

d) Conclude that marginal PDFs don't necessarily determine a joint PDF.

9.99 Let X and Y be continuous random variables with a joint PDF. Exercise 9.98 shows that marginal PDFs of X and Y don't necessarily determine a joint PDF of X and Y.

a) Explain why a marginal PDF of X and all conditional PDFs of Y given $X = x$ determine a joint PDF of X and Y.

b) State a condition under which marginal PDFs of X and Y do necessarily determine a joint PDF of X and Y.

9.100 Consider two electrical components with respective lifetimes X_1 and X_2, which are independent and have the same PDF—say, f. Let X and Y denote the times at which the first and second component failures occur, respectively. The random variable $R = Y - X$ is called the residual lifetime of the component remaining in service.

a) Determine a joint PDF of X and R.

b) Determine marginal PDFs of X and R.

c) Show that, if the common probability distribution of X_1 and X_2 is exponential, then X and R are independent random variables.

d) In general, are X and R independent random variables? Justify your answer.

9.101 Two insurers provide bids on an insurance policy to a large company. The bids must be between 2000 and 2200. The company decides to accept the lower bid if the two bids differ by 20 or more; otherwise, the company will consider the two bids further. Assume that the two bids are independent and are both uniformly distributed on the interval from 2000 to 2200. Determine the probability that the company considers the two bids further.

9.102 The waiting times, in years, for the first claims from a good driver and a bad driver are independent and follow exponential distributions with parameters $1/6$ and $1/3$, respectively.

a) What is the probability that the first claim from a good driver will be filed within 3 years and the first claim from a bad driver will be filed within 2 years?

b) What is the probability that the first claim from a good driver precedes the first claim from a bad driver?

c) Given that the first claim from a good driver occurs in 4 years, what is the probability that the first claim from a bad driver exceeds 4 years?

9.103 Let X, Y, and Z be independent random variables with common CDF F and PDF f.
a) Without doing any computations, explain why $P(X < Y < Z) = \frac{1}{6}$.
b) Use the independence assumption and the FPF to show that $P(X < Y < Z) = \frac{1}{6}$.
c) Let $X_1, \ldots, X_n$ be a random sample from a distribution with CDF F and PDF f. If $i_1, i_2, \ldots, i_n$ is a permutation of the first n positive integers, find $P(X_{i_1} < X_{i_2} < \cdots < X_{i_n})$.

9.104 In this exercise, you are asked to provide a method for simulating the random (uniform) selection of a point from a rectangle—say, $R = (a, b) \times (c, d)$.
a) Suppose that X and Y are the x and y coordinates, respectively, of a point selected at random from R. Determine joint and marginal PDFs of X and Y, and show that X and Y are independent random variables.
b) Use the result of part (a) to explain how a basic random number generator can be employed to simulate the random selection of a point from R. *Hint:* Refer to Proposition 8.16(b) on page 471.

9.105 Three gas stations are independently and randomly (i.e., uniformly) placed along a mile of highway.
a) Determine the probability that no two of the gas stations are less than $\frac{1}{3}$ mile apart.
b) Suppose that you want to generalize part (a) to determine the probability that no two of the gas stations are less than d mile apart. What restriction on d is appropriate?
c) Determine the probability that no two of the gas stations are less than d mile apart for appropriate values of d.

9.106 Let $X_1, \ldots, X_n$ be a random sample from a continuous distribution with CDF F and PDF f. For $1 \leq k \leq n$, obtain $P(X_k = \min\{X_1, \ldots, X_n\})$—the probability that the minimum of the random sample is X_k—by using
a) the FPF.   **b)** a symmetry argument.

9.107 Let $X_1, \ldots, X_n$ be a random sample from a continuous distribution with CDF F and PDF f. For $1 \leq k \leq n$, obtain $P(X_k = \max\{X_1, \ldots, X_n\})$—the probability that the maximum of the random sample is X_k—by using
a) the FPF.   **b)** a symmetry argument.

Theory Exercises

9.108 State and prove the analogue of Proposition 9.9 on page 523 for the multivariate case of m continuous random variables with a joint PDF.

9.109 Suppose that X and Y are random variables, each having a PDF.
a) Prove that, if X and Y are independent, they have a joint PDF.
b) If the independence assumption in part (a) is removed, do X and Y necessarily have a joint PDF? Justify your answer.
c) Is the converse of part (a) true?

9.110 Suppose that X and Y are independent random variables, each having a PDF. Use the FPF to obtain a PDF of the random variable
a) $X + Y$.   **b)** $X - Y$.

Advanced Exercises

9.111 Suppose that a point is selected at random (i.e., uniformly) from a subset S of $\mathcal{R}^2$ with finite nonzero area. Let X and Y be the x and y coordinates, respectively, of the point obtained. Prove that, if $S = A \times B$, where A and B are subsets of $\mathcal{R}$, then X and Y are independent random variables. *Note:* Refer to Exercises 9.40 and 9.81 on pages 506 and 522, respectively.

Mixed bivariate distributions: On pages 509 and 522, we introduced and applied the concept of mixed bivariate distributions, where we simultaneously consider a continuous random variable X and a discrete random variable Y. Exercises 9.112–9.114 continue the examination of mixed bivariate distributions.

9.112 Suppose that X and Y are random variables with joint PDF/PMF $h_{X,Y}$, as defined on page 509. Prove that X and Y are independent if and only if the function h defined on $\mathcal{R}^2$ by $h(x, y) = f_X(x)p_Y(y)$ is a joint PDF/PMF of X and Y.

9.113 Let Y denote the number of patients who arrive at an emergency room during a 1-hour period and let X denote the rate at which they arrive. A joint PDF/PMF of X and Y is given by $h_{X,Y}(x, y) = \lambda\, e^{-(1+\lambda)x}(x^y/y!)$ for $x > 0$ and $y = 0, 1, 2, \ldots$, and $h_{X,Y}(x, y) = 0$ otherwise. Determine whether X and Y are independent random variables.

9.114 Suppose that X and Y have joint PDF/PMF given by $h_{X,Y}(x, y) = \frac{1}{2}e^{-x}$ for $x > 0$ and $y \in \{0, 1\}$, and $h_{X,Y}(x, y) = 0$ otherwise. Decide whether X and Y are independent.

9.6 Functions of Two or More Continuous Random Variables

In Section 8.7, we discussed functions of one continuous random variable. We now examine functions of two or more continuous random variables. Let $X_1, \ldots, X_m$ be random variables defined on the same sample space and let g be a real-valued function of m variables defined on the range of $(X_1, \ldots, X_m)$. Then we can obtain another random variable, X, by composing g with $(X_1, \ldots, X_m)$—that is, by letting $X = g(X_1, \ldots, X_m)$.

When $X_1, \ldots, X_m$ are continuous random variables, $g(X_1, \ldots, X_m)$ need not be a continuous random variable. For the most part, though, we concentrate here on functions of continuous random variables that are also continuous random variables. Specifically, we mostly consider situations in which $X_1, \ldots, X_m$ are continuous random variables with a known joint PDF and the function g is such that $g(X_1, \ldots, X_m)$ is also a continuous random variable with a PDF. The task then is to find a PDF of $g(X_1, \ldots, X_m)$.

Generally, we can obtain a PDF of $g(X_1, \ldots, X_m)$ by first finding its CDF and then applying either Proposition 8.4 (page 419) or Proposition 8.5 (page 422). We've already used this technique—for instance, in Example 9.6(a) on page 504 and Example 9.9(a) on page 513. Example 9.17(a) and Example 9.18 further illustrate that technique.

EXAMPLE 9.17 *Functions of Two or More Continuous Random Variables*

Component Analysis Mechanical or electrical units often consist of several components, each of which is subject to failure. A unit is said to be a *series system* if it functions only when all the components are working. Consider a series system of n components, $C_1, \ldots, C_n$, with respective lifetimes $X_1, \ldots, X_n$, where $X_1, \ldots, X_n$ are independent exponential random variables with parameters $\lambda_1, \ldots, \lambda_n$, respectively.
a) Determine and identify the distribution of the lifetime of this series system.
b) Find the probability that a specified component is the first to fail.

Solution   By assumption, $X_j \sim \mathcal{E}(\lambda_j)$ for $1 \le j \le n$. Because a series system functions only when all the components are working, it fails at the time of the first failure among the n components. Thus, the lifetime of the series system is $X = \min\{X_1, \ldots, X_n\}$.

a) We first obtain the CDF of X. As it is easier to compute $P(X > x)$ than $P(X \le x)$, we proceed as follows. From the complementation rule, the independence of the X_js, and Proposition 8.8 on page 433, we have, for $x > 0$,

$$F_X(x) = P(X \le x) = 1 - P(X > x) = 1 - P(\min\{X_1, \ldots, X_n\} > x)$$
$$= 1 - P(X_1 > x, \ldots, X_n > x) = 1 - P(X_1 > x) \cdots P(X_n > x)$$
$$= 1 - e^{-\lambda_1 x} \cdots e^{-\lambda_n x} = 1 - e^{-(\lambda_1 + \cdots + \lambda_n)x}.$$

Differentiating F_X, we find that $f_X(x) = (\lambda_1 + \cdots + \lambda_n) e^{-(\lambda_1 + \cdots + \lambda_n)x}$ for $x > 0$, and $f_X(x) = 0$ otherwise. Thus, $X \sim \mathcal{E}(\lambda_1 + \cdots + \lambda_n)$. In other words, the lifetime of this series system has the exponential distribution with parameter $\lambda_1 + \cdots + \lambda_n$.

b) The probability that a specified component—say, component C_k—is the first to fail is $P(X = X_k)$ or, equivalently, $P(X_k < \min_{j \ne k}\{X_j\})$. Using the same reasoning as in part (a), we conclude that $\min_{j \ne k}\{X_j\}$ has the exponential distribution with parameter $\sum_{j \ne k} \lambda_j$. Furthermore, by Proposition 6.13 on page 297, we know that X_k and $\min_{j \ne k}\{X_j\}$ are independent random variables. Referring to the result of Example 9.5(b) on page 503, we conclude that

$$P\left(X_k < \min_{j \ne k}\{X_j\}\right) = \frac{\lambda_k}{\lambda_k + \sum_{j \ne k} \lambda_j} = \frac{\lambda_k}{\sum_{j=1}^{n} \lambda_j}.$$

The probability is $\lambda_k / (\lambda_1 + \cdots + \lambda_n)$ that component C_k is the first to fail.   ∎

By using the same reasoning as that in Example 9.17, we get Proposition 9.11, which is especially important in the theory of stochastic processes and operations research.

◆◆◆   Proposition 9.11   Minimum of Independent Exponentials

Suppose that $X_1, \ldots, X_m$ are independent exponential random variables with parameters $\lambda_1, \ldots, \lambda_m$, respectively. Then the following hold.

a) *We have*

$$\min\{X_1, \ldots, X_m\} \sim \mathcal{E}(\lambda_1 + \cdots + \lambda_m). \tag{9.41}$$

In words, the minimum of m independent exponential random variables is also an exponential random variable with parameter equal to the sum of the parameters of the component exponential random variables.

b) *For each k, with $1 \le k \le m$, we have*

$$P\left(X_k = \min\{X_1, \ldots, X_m\}\right) = \frac{\lambda_k}{\lambda_1 + \cdots + \lambda_m}. \tag{9.42}$$

The probability is $\lambda_k / (\lambda_1 + \cdots + \lambda_m)$ that X_k is the minimum of $X_1, \ldots, X_m$.

EXAMPLE 9.18   *Functions of Two or More Continuous Random Variables*

Speed of Gas Molecules   The three velocity components of a gas molecule are independent and identically distributed normal random variables with common parameters 0 and σ^2. Determine a PDF of the speed of a gas molecule.

Solution   Let X, Y, and Z denote the velocity components of a gas molecule. Then its speed is $S = \sqrt{X^2 + Y^2 + Z^2}$. The task is to obtain a PDF of the random variable S. By independence and the normality assumption, a joint PDF of X, Y, and Z is

$$f_{X,Y,Z}(x, y, z) = f_X(x) f_Y(y) f_Z(z)$$

$$= \frac{1}{\sqrt{2\pi}\sigma} e^{-x^2/2\sigma^2} \frac{1}{\sqrt{2\pi}\sigma} e^{-y^2/2\sigma^2} \frac{1}{\sqrt{2\pi}\sigma} e^{-z^2/2\sigma^2}$$

$$= \frac{1}{(2\pi)^{3/2}\sigma^3} e^{-(x^2+y^2+z^2)/2\sigma^2},$$

for all $x, y, z \in \mathcal{R}$. Applying the FPF and changing to spherical coordinates, we get, for $s > 0$,

$$F_S(s) = P(S \le s) = P\left(\sqrt{X^2 + Y^2 + Z^2} \le s\right)$$

$$= \iiint\limits_{\sqrt{x^2+y^2+z^2}\le s} f_{X,Y,Z}(x, y, z)\, dx\, dy\, dz$$

$$= \iiint\limits_{\sqrt{x^2+y^2+z^2}\le s} \frac{1}{(2\pi)^{3/2}\sigma^3} e^{-(x^2+y^2+z^2)/2\sigma^2}\, dx\, dy\, dz$$

$$= \int_0^s \int_0^{2\pi} \int_0^{\pi} \frac{1}{(2\pi)^{3/2}\sigma^3} e^{-\rho^2/2\sigma^2} \rho^2 \sin\phi\, d\rho\, d\theta\, d\phi$$

$$= \int_0^s \frac{1}{(2\pi)^{3/2}\sigma^3} \rho^2 e^{-\rho^2/2\sigma^2} \left(\int_0^{\pi} \sin\phi \left(\int_0^{2\pi} d\theta\right) d\phi\right) d\rho$$

$$= \int_0^s \frac{2 \cdot 2\pi}{(2\pi)^{3/2}\sigma^3} \rho^2 e^{-\rho^2/2\sigma^2}\, d\rho = \int_0^s \frac{\sqrt{2/\pi}}{\sigma^3} \rho^2 e^{-\rho^2/2\sigma^2}\, d\rho,$$

Therefore, by Proposition 8.5 on page 422,

$$f_S(s) = \frac{\sqrt{2/\pi}}{\sigma^3} s^2 e^{-s^2/2\sigma^2}, \qquad s > 0, \qquad\qquad \textbf{(9.43)}$$

and $f_S(s) = 0$ otherwise. Later in this section, we present a more efficient method for obtaining a PDF of S.   ■

Note: A random variable with a PDF as given in Equation (9.43) is said to have a **Maxwell distribution**, named in honor of James Clerk Maxwell (1831–1879), the British physicist who did fundamental research on the kinetic theory of gases.

Sums of Continuous Random Variables

Because of their wide use in applications, sums are one of the most important functions of random variables. For that reason, it's useful to obtain special formulas for a PDF of the sum of continuous random variables. Here we concentrate on the bivariate case. The main result is Proposition 9.12.

◆◆◆ **Proposition 9.12   PDF of the Sum of Two Continuous Random Variables**

Let X and Y be continuous random variables with a joint PDF. Then a PDF of the random variable $X + Y$ can be obtained from either of the two formulas

$$f_{X+Y}(z) = \int_{-\infty}^{\infty} f_{X,Y}(x, z - x)\, dx, \qquad z \in \mathcal{R}, \tag{9.44}$$

or

$$f_{X+Y}(z) = \int_{-\infty}^{\infty} f_{X,Y}(z - y, y)\, dy, \qquad z \in \mathcal{R}. \tag{9.45}$$

If, in addition, X and Y are independent random variables, Equations (9.44) and (9.45) take the form

$$f_{X+Y}(z) = \int_{-\infty}^{\infty} f_X(x) f_Y(z - x)\, dx, \qquad z \in \mathcal{R}, \tag{9.46}$$

or

$$f_{X+Y}(z) = \int_{-\infty}^{\infty} f_X(z - y) f_Y(y)\, dy, \qquad z \in \mathcal{R}, \tag{9.47}$$

respectively.

Proof   To derive Equation (9.44), we proceed as usual by first finding the CDF of $X + Y$. Applying the FPF, we have, for $z \in \mathcal{R}$,

$$F_{X+Y}(z) = P(X + Y \le z) = \iint_{x+y \le z} f_{X,Y}(x, y)\, dx\, dy.$$

The shaded region in Figure 9.12 shows the set over which the double integral is taken. Referring to that figure, making the substitution $u = y + x$, and interchanging the order of integration, we get

$$F_{X+Y}(z) = \iint_{x+y \le z} f_{X,Y}(x, y)\, dx\, dy = \int_{-\infty}^{\infty} \left(\int_{-\infty}^{z-x} f_{X,Y}(x, y)\, dy \right) dx$$

$$= \int_{-\infty}^{\infty} \left(\int_{-\infty}^{z} f_{X,Y}(x, u - x)\, du \right) dx$$

$$= \int_{-\infty}^{z} \left(\int_{-\infty}^{\infty} f_{X,Y}(x, u - x)\, dx \right) du.$$

From Proposition 8.5 on page 422, we now conclude that Equation (9.44) holds.

Equation (9.45) is obtained by using a similar argument. And Equations (9.46) and (9.47) follow from Equations (9.44) and (9.45), respectively, by applying the fact that a joint PDF of independent random variables is the product of the marginal PDFs.   ◆

Figure 9.12

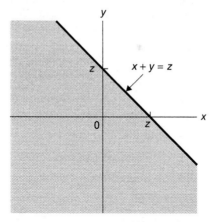

Each of the integrals on the right of Equations (9.46) and (9.47) is called a **convolution** of the probability density functions f_X and f_Y. Therefore we can restate those two results as follows: *A PDF of the sum of two independent continuous random variables is the convolution of their marginal PDFs.*

EXAMPLE 9.19   *Sum of Two Independent Uniform Random Variables*

Let X and Y be independent random variables, both uniformly distributed on the interval $(0, 1)$. Determine and identify the probability distribution of $X + Y$.

Solution   We know that

$$f_X(x) = \begin{cases} 1, & \text{if } 0 < x < 1; \\ 0, & \text{otherwise}; \end{cases} \quad \text{and} \quad f_Y(y) = \begin{cases} 1, & \text{if } 0 < y < 1; \\ 0, & \text{otherwise}. \end{cases}$$

To obtain a PDF of $X + Y$, we first note that its range is the interval $(0, 2)$, so we concentrate on values of z between 0 and 2. From Equation (9.46),

$$f_{X+Y}(z) = \int_{-\infty}^{\infty} f_X(x) f_Y(z - x) \, dx.$$

The integrand is nonzero if and only if $0 < x < 1$ and $0 < z - x < 1$—that is, if and only if $0 < x < 1$ and $z - 1 < x < z$. Therefore

$$f_{X+Y}(z) = \int_{\max\{0, z-1\}}^{\min\{1, z\}} 1 \cdot 1 \, dx = \min\{1, z\} - \max\{0, z - 1\}.$$

To obtain an explicit expression for $f_{X+Y}(z)$, we note that

$$\min\{1, z\} = \begin{cases} z, & \text{if } z \le 1; \\ 1, & \text{if } z > 1; \end{cases} \quad \text{and} \quad \max\{0, z - 1\} = \begin{cases} 0, & \text{if } z \le 1; \\ z - 1, & \text{if } z > 1. \end{cases}$$

So, $f_{X+Y}(z) = z$ for $0 < z \leq 1$, $f_{X+Y}(z) = 2 - z$ for $1 < z < 2$, and $f_{X+Y}(z) = 0$ otherwise. Equivalently, $f_{X+Y}(z) = 1 - |1 - z|$ for $0 < z < 2$, and $f_{X+Y}(z) = 0$ otherwise. Referring to Definition 8.10 on page 460, we see that $X + Y \sim \mathcal{T}(0, 2)$. Thus the sum of two independent uniform random variables on the interval $(0, 1)$ has the triangular distribution on the interval $(0, 2)$.   ∎

As Example 9.19 shows, distribution type isn't necessarily preserved by sums, even for independent random variables. However, there are cases in which sums do preserve distribution type. We consider one of the most important such cases in Example 9.20.

EXAMPLE 9.20   *Sum of Two Independent Gamma Random Variables*

Suppose that X and Y are independent gamma random variables, both having the same second parameter—say, $X \sim \Gamma(\alpha, \lambda)$ and $Y \sim \Gamma(\beta, \lambda)$.
a) Determine and identify the probability distribution of the random variable $X + Y$.
b) Use part (a) to prove that $B(s, t) = \Gamma(s)\Gamma(t)/\Gamma(s + t)$.

Solution   For positive x and y,

$$f_X(x) = \frac{\lambda^\alpha}{\Gamma(\alpha)} x^{\alpha-1} e^{-\lambda x} \qquad \text{and} \qquad f_Y(y) = \frac{\lambda^\beta}{\Gamma(\beta)} y^{\beta-1} e^{-\lambda y},$$

and $f_X(x) = f_Y(y) = 0$ otherwise.
a) As the ranges of both X and Y are the positive real numbers, so is the range of $X + Y$. Applying Proposition 9.12, we get, for $z > 0$,

$$f_{X+Y}(z) = \int_{-\infty}^{\infty} f_X(x) f_Y(z - x)\, dx$$

$$= \int_0^z \frac{\lambda^\alpha}{\Gamma(\alpha)} x^{\alpha-1} e^{-\lambda x} \frac{\lambda^\beta}{\Gamma(\beta)} (z - x)^{\beta-1} e^{-\lambda(z-x)}\, dx$$

$$= \frac{\lambda^{\alpha+\beta}}{\Gamma(\alpha)\Gamma(\beta)} e^{-\lambda z} \int_0^z x^{\alpha-1} (z - x)^{\beta-1}\, dx.$$

Making the substitution $u = x/z$, we obtain

$$f_{X+Y}(z) = \frac{\lambda^{\alpha+\beta}}{\Gamma(\alpha)\Gamma(\beta)} e^{-\lambda z} \int_0^1 (uz)^{\alpha-1} (z - uz)^{\beta-1} z\, du$$

$$= \frac{\lambda^{\alpha+\beta}}{\Gamma(\alpha)\Gamma(\beta)} z^{\alpha+\beta-1} e^{-\lambda z} \int_0^1 u^{\alpha-1} (1 - u)^{\beta-1}\, du$$

$$= B(\alpha, \beta) \frac{\lambda^{\alpha+\beta}}{\Gamma(\alpha)\Gamma(\beta)} z^{\alpha+\beta-1} e^{-\lambda z} \int_0^1 \frac{1}{B(\alpha, \beta)} u^{\alpha-1} (1 - u)^{\beta-1}\, du$$

$$= B(\alpha, \beta) \frac{\lambda^{\alpha+\beta}}{\Gamma(\alpha)\Gamma(\beta)} z^{\alpha+\beta-1} e^{-\lambda z},$$

where the last equation follows from the fact that the integral of the PDF of a beta random variable must equal 1. The "variable form" of the expression on the right of the preceding display, $z^{\alpha+\beta-1}e^{-\lambda z}$, is that of a gamma PDF with parameters $\alpha + \beta$ and λ, so we conclude that $X + Y \sim \Gamma(\alpha + \beta, \lambda)$. And, from that fact, it follows that

$$B(\alpha, \beta)\frac{\lambda^{\alpha+\beta}}{\Gamma(\alpha)\Gamma(\beta)} = \frac{\lambda^{\alpha+\beta}}{\Gamma(\alpha+\beta)}.$$

b) Letting $\alpha = s$ and $\beta = t$ in the preceding equation and performing some simple algebra, we get $B(s, t) = \Gamma(s)\Gamma(t)/\Gamma(s+t)$.   ∎

Using mathematical induction, Proposition 6.13 on page 297, and the results of Example 9.20(a), we obtain Proposition 9.13. We leave the proof is to you as Exercise 9.136.

◆◆◆   **Proposition 9.13   Sum of Independent Gamma Random Variables**

Let $X_1, \ldots, X_m$ be independent random variables with $X_j \sim \Gamma(\alpha_j, \lambda)$ for $1 \le j \le m$. Then

$$X_1 + \cdots + X_m \sim \Gamma(\alpha_1 + \cdots + \alpha_m, \lambda).$$

In words, the sum of independent gamma random variables with the same second parameter is also a gamma random variable whose first parameter is the sum of those of the gamma random variables in the sum and whose second parameter is the common parameter of the gamma random variables in the sum.

Proposition 9.13 has two important consequences. In stating them, we recall that (1) the exponential distribution with parameter λ is also the gamma distribution with parameters 1 and λ, and (2) the chi-square distribution with ν degrees of freedom is also the gamma distribution with parameters $\nu/2$ and $1/2$.

- *Sum of Independent Exponential Random Variables:* If $X_1, \ldots, X_m$ are independent exponential random variables with common parameter λ, the random variable $X_1 + \cdots + X_m$ has the gamma distribution with parameters m and λ or, equivalently, the Erlang distribution with those two parameters.

- *Sum of Independent Chi-Square Random Variables:* If $X_1, \ldots, X_m$ are independent chi-square random variables with degrees of freedom $\nu_1, \ldots, \nu_m$, respectively, the random variable $X_1 + \cdots + X_m$ has the chi-square distribution with $\nu_1 + \cdots + \nu_m$ degrees of freedom.

Using the second bulleted item and the fact that the square of a standard normal random variable is a chi-square random variable with 1 degree of freedom (Proposition 8.13 on page 466), we can more easily obtain the result for the distribution of the speed of a gas molecule, considered in Example 9.18.

EXAMPLE 9.21   *Sum of Independent Random Variables*

Speed of Gas Molecules   Recall from Example 9.18 on page 533 that the three velocity components of a gas molecule are independent and identically distributed normal random variables with common parameters 0 and σ^2. Find a PDF of the speed of a gas molecule.

Solution   Let X, Y, and Z denote the velocity components of a gas molecule. Then its speed is $S = \sqrt{X^2 + Y^2 + Z^2}$. The task is to obtain a PDF of the random variable S.

By assumption, $X \sim \mathcal{N}(0, \sigma^2)$. Applying, in turn, Proposition 8.10 on page 441 and Proposition 8.13 on page 466, we deduce that X^2/σ^2 has the chi-square distribution with 1 degree of freedom, as do both Y^2/σ^2 and Z^2/σ^2. Because X, Y, and Z are independent, so are X^2/σ^2, Y^2/σ^2, and Z^2/σ^2 (Proposition 6.13 on page 297). Applying the second bulleted item, we conclude that $W = (X^2 + Y^2 + Z^2)/\sigma^2$ has the chi-square distribution with three degrees of freedom:

$$f_W(w) = \frac{\left(\frac{1}{2}\right)^{3/2}}{\Gamma(3/2)} w^{3/2-1} e^{-w/2} = \frac{1}{\sqrt{2\pi}} w^{1/2} e^{-w/2}, \qquad w > 0,$$

and $f_W(w) = 0$ otherwise.

We note that $S = \sigma \sqrt{W}$. Applying the univariate transformation theorem (Proposition 8.14 on page 466) with $g(w) = \sigma \sqrt{w}$, we get, for $s > 0$,

$$f_S(s) = \frac{1}{|g'(w)|} f_W(w) = \frac{1}{\frac{1}{2}\sigma w^{-1/2}} \frac{1}{\sqrt{2\pi}} w^{1/2} e^{-w/2} = \frac{\sqrt{2/\pi}}{\sigma} w\, e^{-w/2},$$

where w is the unique number in the range of W such that $g(w) = s$. Solving for w in the equation $\sigma \sqrt{w} = s$ yields $w = s^2/\sigma^2$. Therefore,

$$f_S(s) = \frac{\sqrt{2/\pi}}{\sigma^3} s^2 e^{-s^2/2\sigma^2}, \qquad s > 0,$$

and $f_S(s) = 0$ otherwise. This result, of course, agrees with that obtained in Example 9.18 but is less computationally intensive. ∎

Our next task is to show that the sum of independent normal random variables is also a normal random variable and to identify the parameters of the sum in terms of those of the summands. To begin, we establish Lemma 9.1.

◆◆◆ **Lemma 9.1**

If X and Y are independent random variables with $X \sim \mathcal{N}(0, 1)$ and $Y \sim \mathcal{N}(0, \sigma^2)$, then $X + Y \sim \mathcal{N}(0, 1 + \sigma^2)$.

Proof   For $x, y \in \mathcal{R}$, we have

$$f_X(x) = \frac{1}{\sqrt{2\pi}} e^{-x^2/2} \qquad \text{and} \qquad f_Y(y) = \frac{1}{\sqrt{2\pi}\,\sigma} e^{-y^2/2\sigma^2}.$$

Applying Proposition 9.12, we obtain, for $z \in \mathcal{R}$,

$$f_{X+Y}(z) = \int_{-\infty}^{\infty} f_X(x) f_Y(z - x)\, dx = \int_{-\infty}^{\infty} \frac{1}{\sqrt{2\pi}} e^{-x^2/2} \frac{1}{\sqrt{2\pi}\,\sigma} e^{-(z-x)^2/2\sigma^2}\, dx$$

$$= \frac{1}{2\pi\sigma} \int_{-\infty}^{\infty} e^{-\left(\sigma^2 x^2 + (z-x)^2\right)/2\sigma^2}\, dx.$$

Using simple algebra and the technique of completing the square, we find that

$$\left(\sigma^2 x^2 + (z-x)^2\right)\bigg/2\sigma^2 = \left(x - \frac{z}{1+\sigma^2}\right)^2\bigg/\left(2\sigma^2/(1+\sigma^2)\right) + z^2\bigg/\left(2(1+\sigma^2)\right).$$

Therefore, for $z \in \mathcal{R}$,

$$f_{X+Y}(z) = \frac{1}{2\pi\sigma} e^{-z^2/2(1+\sigma^2)} \int_{-\infty}^{\infty} e^{-\left(x - \frac{z}{1+\sigma^2}\right)^2\big/\left(2\sigma^2/(1+\sigma^2)\right)} \, dx$$

$$= \frac{1}{2\pi\sigma} e^{-z^2/2(1+\sigma^2)} \sqrt{2\pi} \left(\sigma/\sqrt{1+\sigma^2}\right)$$

$$\times \int_{-\infty}^{\infty} \frac{1}{\sqrt{2\pi}\left(\sigma/\sqrt{1+\sigma^2}\right)} e^{-\left(x - \frac{z}{1+\sigma^2}\right)^2\big/\left(2\sigma^2/(1+\sigma^2)\right)} \, dx$$

$$= \frac{1}{\sqrt{2\pi}\sqrt{1+\sigma^2}} e^{-z^2/2(1+\sigma^2)},$$

where the last equation follows from the fact that the integral of the PDF of a normal random variable is 1. Consequently, $X + Y \sim \mathcal{N}(0, 1+\sigma^2)$. $\blacklozenge$

With Lemma 9.1 in mind, we can now show that, in general, the sum of two independent normal random variables is also normal. We do so in Example 9.22.

EXAMPLE 9.22   ***Sum of Two Independent Normal Random Variables***

Suppose that X and Y are independent normal random variables—say, $X \sim \mathcal{N}(\mu, \sigma^2)$ and $Y \sim \mathcal{N}(\nu, \tau^2)$. Determine and identify the probability distribution of $X + Y$.

Solution   Applying, in turn, Proposition 8.10 on page 441 and Proposition 8.15 on page 468, gives

$$\frac{X - \mu}{\sigma} \sim \mathcal{N}(0, 1) \qquad \text{and} \qquad \frac{\tau}{\sigma}\left(\frac{Y - \nu}{\tau}\right) \sim \mathcal{N}(0, \tau^2/\sigma^2).$$

Therefore, by Lemma 9.1,

$$W = \frac{(X+Y) - (\mu + \nu)}{\sigma} = \frac{X - \mu}{\sigma} + \frac{\tau}{\sigma}\left(\frac{Y - \nu}{\tau}\right) \sim \mathcal{N}(0, 1 + \tau^2/\sigma^2).$$

Again applying Proposition 8.15, we conclude that

$$X + Y = (\mu + \nu) + \sigma W \sim \mathcal{N}\left(\mu + \nu, \sigma^2(1 + \tau^2/\sigma^2)\right).$$

Thus $X + Y \sim \mathcal{N}(\mu + \nu, \sigma^2 + \tau^2)$. $\blacksquare$

We can use mathematical induction, Proposition 6.13 on page 297, and the results of Example 9.22 to show that the sum of a finite number of independent normal random variables is also a normal random variable whose first and second parameters are the sums of those of the normal random variables in the sum. More generally, in view of Proposition 8.15 on page 468, we have Proposition 9.14, whose proof is left to you as Exercise 9.137.

◆◆◆ **Proposition 9.14   Sum of Independent Normal Random Variables**

Let $X_1, \ldots, X_m$ be independent random variables with $X_j \sim \mathcal{N}(\mu_j, \sigma_j^2)$ for $1 \le j \le m$, and let $a, b_1, \ldots, b_m$ be real numbers, where not all the b_js equal 0. Then

$$a + b_1 X_1 + \cdots + b_m X_m \sim \mathcal{N}\left(a + b_1\mu_1 + \cdots + b_m\mu_m, b_1^2\sigma_1^2 + \cdots + b_m^2\sigma_m^2\right).$$

In particular, $X_1 + \cdots + X_m \sim \mathcal{N}\left(\mu_1 + \cdots + \mu_m, \sigma_1^2 + \cdots + \sigma_m^2\right)$.

Quotients of Continuous Random Variables

Quotients of continuous random variables occur frequently in applications, especially in statistics where they often arise as sampling distributions. Proposition 9.15 provides a general formula for a PDF of the quotient of two continuous random variables.

◆◆◆ **Proposition 9.15   PDF of the Quotient of Two Continuous Random Variables**

Let X and Y be continuous random variables with a joint PDF. Then a PDF of the random variable Y/X is given by

$$f_{Y/X}(z) = \int_{-\infty}^{\infty} |x| f_{X,Y}(x, xz)\, dx, \qquad z \in \mathcal{R}. \tag{9.48}$$

If, in addition, X and Y are independent random variables, then Equation (9.48) takes the form

$$f_{Y/X}(z) = \int_{-\infty}^{\infty} |x| f_X(x) f_Y(xz)\, dx, \qquad z \in \mathcal{R}. \tag{9.49}$$

Proof   Applying the FPF yields

$$F_{Y/X}(z) = P(Y/X \le z) = \iint\limits_{y/x \le z} f_{X,Y}(x, y)\, dx\, dy, \qquad z \in \mathcal{R}.$$

The form of the region $\{(x, y) : y/x \le z\}$ depends on the sign of z. In each case, the shaded region in either Figure 9.13(a) or Figure 9.13(b) shows the set over which the double integral is taken. Referring to Figure 9.13, making the substitution $u = y/x$, and interchanging the order of integration, we get

$$F_{Y/X}(z) = \int_{-\infty}^{0}\left(\int_{zx}^{\infty} f_{X,Y}(x, y)\, dy\right)dx + \int_{0}^{\infty}\left(\int_{-\infty}^{zx} f_{X,Y}(x, y)\, dy\right)dx$$

$$= \int_{-\infty}^{0}\left(\int_{z}^{-\infty} x f_{X,Y}(x, xu)\, du\right)dx + \int_{0}^{\infty}\left(\int_{-\infty}^{z} x f_{X,Y}(x, xu)\, du\right)dx$$

$$= \int_{-\infty}^{0}\left(\int_{-\infty}^{z} (-x) f_{X,Y}(x, xu)\, du\right)dx + \int_{0}^{\infty}\left(\int_{-\infty}^{z} x f_{X,Y}(x, xu)\, du\right)dx$$

$$= \int_{-\infty}^{\infty}\left(\int_{-\infty}^{z} |x| f_{X,Y}(x, xu)\, du\right)dx = \int_{-\infty}^{z}\left(\int_{-\infty}^{\infty} |x| f_{X,Y}(x, xu)\, dx\right)du.$$

Figure 9.13

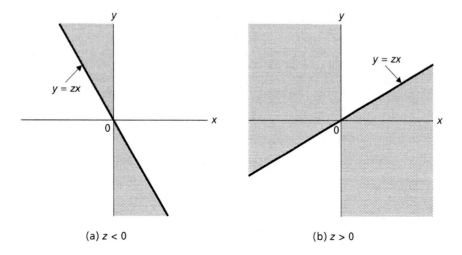

(a) $z < 0$ (b) $z > 0$

Thus, from Proposition 8.5 on page 422, Equation (9.48) holds. Equation (9.49) follows from Equation (9.48) and the fact that a joint PDF of independent random variables is the product of the marginal PDFs. ◆

EXAMPLE 9.23 *Quotient of Two Independent Normal Random Variables*

Let X and Y be independent normal random variables, both having parameters 0 and σ^2. Find and identify the probability distribution of the random variable Y/X.

Solution   By assumption, for $x, y \in \mathcal{R}$,

$$f_X(x) = \frac{1}{\sqrt{2\pi}\sigma} e^{-x^2/2\sigma^2} \qquad \text{and} \qquad f_Y(y) = \frac{1}{\sqrt{2\pi}\sigma} e^{-y^2/2\sigma^2}.$$

Applying Equation (9.49), using the symmetry of the integrand, and making the substitution $u = (1 + z^2)x^2/2\sigma^2$, we get, for $z \in \mathcal{R}$,

$$f_{Y/X}(z) = \int_{-\infty}^{\infty} |x| f_X(x) f_Y(xz)\, dx = \int_{-\infty}^{\infty} |x| \frac{1}{\sqrt{2\pi}\sigma} e^{-x^2/2\sigma^2} \frac{1}{\sqrt{2\pi}\sigma} e^{-x^2z^2/2\sigma^2}\, dx$$

$$= \frac{2}{2\pi\sigma^2} \int_{0}^{\infty} x\, e^{-(1+z^2)x^2/2\sigma^2}\, dx = \frac{1}{\pi} \frac{1}{1+z^2} \int_{0}^{\infty} e^{-u}\, du = \frac{1}{\pi(1+z^2)}.$$

Thus Y/X has the standard Cauchy distribution. ■

EXERCISES 9.6   **Basic Exercises**

9.115 This exercise considers the two-dimensional analogue of the Maxwell distribution discussed in Examples 9.18 and 9.21. Suppose that X and Y are independent normal random variables, both having parameters 0 and σ^2. Show that the random variable $R = \sqrt{X^2 + Y^2}$ has the Rayleigh distribution—introduced in Exercise 8.159 on page 475—by using an argument similar to that in
a) Example 9.18 on page 533.
b) Example 9.21 on page 537.

9.116 Midrange of a random sample: Let $X_1, \ldots, X_n$ be a random sample from a continuous distribution with CDF F and PDF f. The *midrange* of the random sample is defined to be $M = (X + Y)/2$, where $X = \min\{X_1, \ldots, X_n\}$ and $Y = \max\{X_1, \ldots, X_n\}$.
a) Show that a PDF of the random variable M is given by

$$f_M(m) = 2n(n-1) \int_{-\infty}^{m} f(x) f(2m - x) \left(F(2m - x) - F(x)\right)^{n-2} dx,$$

for $-\infty < m < \infty$. *Note:* A joint PDF of X and Y can be found in Equation (9.20) on page 498.
b) Apply the result of part (a) to the special case of a random sample of size n from a uniform distribution on the interval $(0, 1)$.

9.117 Mechanical or electrical units often consist of several components, each of which is subject to failure. A unit is said to be a *parallel system* if it functions when at least one of the components is working. Consider a parallel system of n components, $C_1, \ldots, C_n$, with respective lifetimes $X_1, \ldots, X_n$, where $X_1, \ldots, X_n$ are independent exponential random variables with parameters $\lambda_1, \ldots, \lambda_n$, respectively. Determine the probability distribution of the lifetime of this parallel system.

9.118 A critical component in a complex system has a built-in spare. When the original component fails, the spare comes on immediately and the system continues to function. When the spare fails, the system goes down. Assuming that the lifetimes of both the original component and the spare are exponentially distributed with parameter λ, find and identify the CDF of the proportion of time until system failure that the original component is working.

9.119 Bilateral exponential random variable: Let X and Y be independent random variables both having the exponential distribution with parameter λ. Determine and graph a PDF of $Z = X - Y$. *Note:* Any random variable with this PDF is called a *bilateral exponential random variable* and is said to have the *bilateral exponential distribution with parameter λ.*

9.120 Let X and Y be random variables with joint PDF given by $f_{X,Y}(x, y) = e^{-(x+y)}$ for $x > 0$ and $y > 0$, and $f_{X,Y}(x, y) = 0$ otherwise. An insurance policy is written to reimburse $X + Y$. Determine the probability that the reimbursement is at most 1 by using
a) the FPF.
b) Proposition 9.13 on page 537 and Equation (8.49) on page 453.

9.121 An insurance company sells two types of auto insurance policies: basic and deluxe. The time, in days, until the next basic policy claim is an exponential random variable with parameter 1/2. The time, in days, until the next deluxe policy claim is an independent exponential random variable with parameter 1/3. Determine the probability that the next claim will be a deluxe policy claim by using
a) the FPF.   **b)** Proposition 9.11 on page 532.

9.122 A device containing two key components fails when and only when both components fail. The lifetimes, T_1 and T_2, of these components are independent with common density function $f(t) = e^{-t}$ for $t > 0$, and $f(t) = 0$ otherwise. The cost of operating the device until failure is $2T_1 + T_2$. Obtain a PDF of this cost.

9.123 Let X and Y be independent random variables with PDFs given by $f_X(x) = f_Y(y) = 0$ for negative x and y, and $f_X(x) = ax^\alpha e^{-\lambda x}$ and $f_Y(y) = by^\beta e^{-\lambda y}$ for positive x and y, where a and b are constants. Without doing any calculations, identify the probability distribution of the random variable $X + Y$.

9.124 Let X and Y be independent random variables with $X \sim \mathcal{N}(2, 1)$ and $Y \sim \mathcal{N}(3, 2)$. Identify the probability distribution of each of the following random variables.
a) $X + Y$ **b)** $X - Y$ **c)** $2 - 3X + 4Y$

9.125 Let $X_1, \ldots, X_n$ be a random sample from a $\mathcal{N}(\mu, \sigma^2)$ distribution.
a) Determine the standardized version of the random variable $X_1 + \cdots + X_n$ and identify its probability distribution.
b) Identify the probability distribution of the sample mean, $\bar{X}_n$.

9.126 Two instruments are used to measure the height, h, of a tower. The error made by the less accurate instrument is normally distributed with parameters 0 and $(0.0056h)^2$. The error made by the more accurate instrument is normally distributed with parameters 0 and $(0.0044h)^2$. If the two measurements are independent random variables, what is the probability that their average value is within $0.005h$ of the height of the tower?

9.127 A company manufactures a brand of light bulb with a lifetime, in months, that is normally distributed with parameters 3 and 1. A consumer buys a number of these bulbs with the intention of replacing them successively as they burn out. The light bulbs have independent lifetimes. What is the smallest number of bulbs to be purchased so that the succession of light bulbs produces light for at least 40 months with probability at least 0.9772?

9.128 In Exercise 8.154 on page 474, we presented the definition of a lognormal random variable. Let $X_1, \ldots, X_m$ be independent lognormal random variables and let $a_1, \ldots, a_m$ be nonzero real numbers.
a) Show that $\prod_{j=1}^m X_j^{a_j}$ is lognormal.
b) Deduce from part (a) that the product of a finite number of independent lognormal random variables is a lognormal random variable.

9.129 A company uses batteries of two types, A and B. The lifetimes of type A and type B batteries are exponential random variables with parameters λ and μ, respectively. For a life test, n batteries of type A and m batteries of type B are used. What is the probability that the first battery to fail is of type A?

9.130 An insurance company that offers earthquake insurance models annual premiums and annual claims by exponential distributions with parameters $\frac{1}{2}$ and 1, respectively. Premiums and claims are independent. Obtain a density function for the ratio of claims to premiums.

9.131 Simulation: This exercise requires access to a computer or graphing calculator.
a) Use a basic random number generator to conduct two simulations of 10,000 observations each of a $\mathcal{U}(0, 1)$ random variable.
b) Add the first observations from the two simulations, the second observations from the two simulations, and so on.
c) Roughly, what would you expect a histogram of the 10,000 sums found in part (b) to look like? Explain your answer.
d) Obtain a histogram of the 10,000 sums found in part (b).

9.132 Simulation: This exercise requires access to a computer or graphing calculator.
a) Use a normal random number generator (i.e., a random number generator that simulates observations of a normal random variable) to obtain 10,000 observations each of a $\mathcal{N}(-2, 9)$ random variable and a $\mathcal{N}(3, 16)$ random variable.
b) Add the first observations from the two simulations, the second observations from the two simulations, and so on.
c) Roughly, what would you expect a histogram of the 10,000 sums found in part (b) to look like? Explain your answer.
d) Obtain a histogram of the 10,000 sums found in part (b).

9.133 Let X and Y be independent random variables with $X \sim \Gamma(\alpha, \lambda)$ and $Y \sim \Gamma(\beta, \lambda)$.
a) Show that the random variable Y/X has PDF given by

$$f_{Y/X}(z) = \frac{1}{B(\alpha, \beta)} \frac{z^{\beta-1}}{(z+1)^{\alpha+\beta}}, \qquad z > 0,$$

and $f_{Y/X}(z) = 0$ otherwise.
b) Determine and identify a PDF of $Z = X/(X + Y)$. *Hint:* Express Z as a function of Y/X.
c) Apply part (b) to obtain the result required in Exercise 9.118.

9.134 Recall from Example 9.19 that the sum of two independent uniform random variables on the interval $(0, 1)$ has the triangular distribution on the interval $(0, 2)$. Let $W, X, Y,$ and Z be independent random variables, all uniformly distributed on the interval $(0, 1)$.
a) Determine a PDF of $W + X + Y$. *Hint:* Write $W + X + Y = (W + X) + Y$.
b) Determine a PDF of $W + X + Y + Z$.
c) Graph PDFs of W, $W + X$, $W + X + Y$, and $W + X + Y + Z$. Observe how the shape of the distribution changes as the number of random variables in the sum increases.

Theory Exercises

9.135 Let X and Y be continuous random variables with a joint PDF.
a) Derive a formula for a PDF of XY.
b) Specialize the result of part (a) to the case where X and Y are independent.
c) Obtain a PDF of the product of two independent $\mathcal{U}(0, 1)$ random variables.

9.136 Prove Proposition 9.13 on page 537.

9.137 Prove Proposition 9.14 on page 540.

Advanced Exercises

9.138 Let $X, Y,$ and Z have joint PDF given by $f_{X,Y,Z}(x, y, z) = 6/(1 + x + y + z)^4$ for positive $x, y,$ and z, and $f_{X,Y,Z}(x, y, z) = 0$ otherwise. Obtain a PDF of $X + Y + Z$.

9.139 Let X and Y have the bivariate normal distribution with joint PDF given by Equation (9.35) on page 518. Find and identify a PDF of each of the following random variables.
a) $X + Y$ b) Y/X

9.140 The times between arrivals of customers at a service counter are independent exponential random variables with parameter λ. Determine and identify the probability distribution of the random variable $N(t)$, the number of customers that arrive by time t.

9.141 Student's t-distribution: Let Z and Y be independent random variables, where Z has the standard normal distribution and Y has the chi-square distribution with ν degrees of freedom. Determine a PDF of the random variable $T = Z/\sqrt{Y/\nu}$. *Note:* The random variable T is said to have the *Student's t-distribution with ν degrees of freedom.*

9.142 Let X and Y be independent standard Cauchy random variables.
a) Show that the random variable $(X + Y)/2$ is also a standard Cauchy random variable.
b) Generalize part (a) by showing that, for each real number c, where $0 \le c \le 1$, the random variable $cX + (1 - c)Y$ is a standard Cauchy random variable.
c) Prove that, if $X_1, \ldots, X_n$ are independent standard Cauchy random variables, then $(X_1 + \cdots + X_n)/n$ is also a standard Cauchy random variable.

In Exercises 9.143–9.145, you are to consider several analogues of Proposition 9.12 on page 534 for the sum of three or more continuous random variables with a joint PDF.

9.143 Prove that, if X, Y, and Z are independent continuous random variables, each with a PDF, then

$$f_{X+Y+Z}(w) = \int_{-\infty}^{\infty} f_X(x) f_{Y+Z}(w - x)\, dx, \qquad w \in \mathcal{R}.$$

9.144 Let $X_1, \ldots, X_m$ be continuous random variables with a joint PDF.
a) Prove that, for $w \in \mathcal{R}$,

$$f_{X_1 + \cdots + X_m}(w)$$
$$= \int_{-\infty}^{\infty} \cdots \int_{-\infty}^{\infty} f_{X_1, \ldots, X_m}(x_1, \ldots, x_{m-1}, w - x_1 - \cdots - x_{m-1})\, dx_1 \cdots dx_{m-1}.$$

Hint: Apply the same technique as in the proof of Equation (9.44) on page 534.
b) Identify $m - 1$ formulas similar to the one in part (a) for a PDF of $X_1 + \cdots + X_m$.
c) Specialize parts (a) and (b) to obtain m formulas for a PDF of $X_1 + \cdots + X_m$ in case $X_1, \ldots, X_m$ are independent.
d) Specialize parts (a)–(c) to obtain formulas for a PDF of $X + Y + Z$ in the trivariate case of three continuous random variables X, Y, and Z with a joint PDF.

9.145 Another method for obtaining a PDF of the sum of independent continuous random variables involves recursion. Specifically, prove that, if $X_1, X_2, \ldots$ are independent continuous random variables, each with a PDF, then

$$f_{X_1 + \cdots + X_m}(w) = \int_{-\infty}^{\infty} f_{X_1 + \cdots + X_{m-1}}(x) f_{X_m}(w - x)\, dx, \qquad w \in \mathcal{R},$$

for $m = 2, 3, \ldots$.

9.7 Multivariate Transformation Theorem

In Section 8.7, we presented two methods for obtaining a PDF of a function of one continuous random variable. One of those methods—the transformation method—is based on the univariate transformation theorem (Proposition 8.14 on page 466).

In this section, we generalize the univariate transformation theorem to the multivariate case and present several examples illustrating that generalization. As usual, we concentrate on the bivariate case, with the understanding that similar results and applications apply to the general multivariate case. We begin with Proposition 9.16, the **bivariate transformation theorem**.

◆◆◆ **Proposition 9.16   Bivariate Transformation Theorem**

Let X and Y be continuous random variables with a joint PDF. Suppose that g and h are real-valued functions of two real variables defined on the range of (X, Y) and that they and their first partial derivatives are continuous on it. Further suppose that the two-dimensional transformation defined on the range of (X, Y) by $u = g(x, y)$ and $v = h(x, y)$ is one-to-one. Then a joint PDF of the random variables $U = g(X, Y)$ and $V = h(X, Y)$ is

$$f_{U,V}(u, v) = \frac{1}{|J(x, y)|} f_{X,Y}(x, y), \tag{9.50}$$

for (u, v) in the range of (U, V), where (x, y) is the unique point in the range of (X, Y) such that $g(x, y) = u$ and $h(x, y) = v$, and $J(x, y)$ is the Jacobian determinant of the transformation:

$$J(x, y) = \begin{vmatrix} \dfrac{\partial g}{\partial x}(x, y) & \dfrac{\partial g}{\partial y}(x, y) \\[2ex] \dfrac{\partial h}{\partial x}(x, y) & \dfrac{\partial h}{\partial y}(x, y) \end{vmatrix}.$$

Proof   Applying the FPF, we have, for (u, v) in the range of (U, V),

$$F_{U,V}(u, v) = P(U \le u, V \le v)$$

$$= P\big(g(X, Y) \le u, h(X, Y) \le v\big) = \iint\limits_{\substack{g(x,y) \le u \\ h(x,y) \le v}} f_{X,Y}(x, y)\, dx\, dy.$$

Making the transformation $s = g(x, y)$ and $t = h(x, y)$ and applying from calculus the change of variable formula for double integrals, we get

$$F_{U,V}(u, v) = \int_{-\infty}^{u} \int_{-\infty}^{v} \frac{1}{|J(x, y)|} f_{X,Y}(x, y)\, ds\, dt,$$

where, for each point (s, t), the point (x, y) is the unique point in the range of (X, Y) such that $g(x, y) = s$ and $h(x, y) = t$. Referring now to Proposition 9.4 on page 499, we conclude that Equation (9.50) holds.   ◆

It's difficult to overemphasize the importance of the bivariate transformation theorem. We can use it to efficiently derive results, obtain univariate distributions, and, of course, determine bivariate distributions.

In Example 9.24, we use the bivariate transformation theorem to provide an alternative derivation of a formula for a PDF of the sum of two continuous random variables with a joint PDF. In doing so, we introduce the technique of using a "dummy variable" for the purpose of applying the bivariate transformation theorem to obtain a univariate PDF.

EXAMPLE 9.24   *The Bivariate Transformation Theorem*

PDF of the Sum of Two Continuous Random Variables   Let X and Y be continuous random variables with a joint PDF. Apply the bivariate transformation theorem to obtain a PDF of the random variable $X + Y$.

Solution Set $U = X$ (the "dummy variable") and $V = X + Y$. The Jacobian determinant of the transformation $u = x$ and $v = x + y$ is

$$J(x, y) = \begin{vmatrix} 1 & 0 \\ 1 & 1 \end{vmatrix} = 1.$$

Solving the equations $u = x$ and $v = x + y$ for x and y, we obtain the inverse transformation $x = u$ and $y = v - u$. Therefore, by the bivariate transformation theorem, a joint PDF of U and V is

$$f_{U,V}(u, v) = \frac{1}{|J(x, y)|} f_{X,Y}(x, y) = \frac{1}{|1|} f_{X,Y}(u, v - u) = f_{X,Y}(u, v - u).$$

Hence a joint PDF of X and $X + Y$ can be expressed in terms of a joint PDF of X and Y as $f_{X,X+Y}(u, v) = f_{X,Y}(u, v - u)$. To obtain a PDF of $X + Y$, we apply Proposition 9.7 on page 510:

$$f_{X+Y}(v) = \int_{-\infty}^{\infty} f_{X,X+Y}(u, v) \, du = \int_{-\infty}^{\infty} f_{X,Y}(u, v - u) \, du.$$

This result agrees with Equation (9.44) on page 534. ■

EXAMPLE 9.25 *The Bivariate Transformation Theorem*

Quality Assurance For purposes of quality assurance, an expensive item produced at a manufacturing plant is independently inspected by two engineers. The amount of time it takes each engineer to inspect the item has the exponential distribution with parameter λ. Let S denote the proportion of the total inspection time attributed to the first engineer, and let T denote the total inspection time by both engineers.
a) Use the bivariate transformation theorem to obtain a joint PDF of S and T.
b) Use the result of part (a) to obtain and identify a marginal PDF of S.
c) Use the result of part (a) to obtain and identify a marginal PDF of T.
d) Show that S and T are independent random variables.

Solution Let X and Y denote the inspection times for the first and second engineers, respectively. By assumption, X and Y are independent exponential random variables, both with parameter λ. Hence a joint PDF of X and Y is given by

$$f_{X,Y}(x, y) = \lambda e^{-\lambda x} \cdot \lambda e^{-\lambda y} = \lambda^2 e^{-\lambda(x+y)}, \qquad x > 0, \ y > 0, \qquad \textbf{(9.51)}$$

and $f_{X,Y}(x, y) = 0$ otherwise. We note that $S = X/(X + Y)$ and $T = X + Y$ and, furthermore, that the range of S is the interval $(0, 1)$ and the range of T is the interval $(0, \infty)$.
a) The Jacobian determinant of the transformation $s = x/(x + y)$ and $t = x + y$ is

$$J(x, y) = \begin{vmatrix} \dfrac{y}{(x + y)^2} & -\dfrac{x}{(x + y)^2} \\ 1 & 1 \end{vmatrix} = \frac{y}{(x + y)^2} + \frac{x}{(x + y)^2} = \frac{1}{x + y}. \qquad \textbf{(9.52)}$$

Solving the equations $s = x/(x + y)$ and $t = x + y$ for x and y, we obtain the inverse transformation $x = st$ and $y = (1 - s)t$. Applying the bivariate transformation theorem and keeping Equations (9.51) and (9.52) in mind, we obtain a joint PDF of S and T:

$$f_{S,T}(s,t) = \frac{1}{|J(x,y)|} f_{X,Y}(x,y) = (x+y)\lambda^2 e^{-\lambda(x+y)},$$

or

$$f_{S,T}(s,t) = \lambda^2 t e^{-\lambda t}, \qquad 0 < s < 1, \, t > 0, \tag{9.53}$$

and $f_{S,T}(s,t) = 0$ otherwise.

b) To obtain a PDF of S, we apply Proposition 9.7 on page 510 to Equation (9.53). For $0 < s < 1$, we have

$$f_S(s) = \int_{-\infty}^{\infty} f_{S,T}(s,t)\,dt = \int_0^{\infty} \lambda^2 t e^{-\lambda t}\,dt = \Gamma(2) \int_0^{\infty} \frac{\lambda^2}{\Gamma(2)} t^{2-1} e^{-\lambda t}\,dt.$$

Using the facts that $\Gamma(2) = 1$ and that a PDF of a gamma random variable integrates to 1, we conclude that

$$f_S(s) = 1, \qquad 0 < s < 1, \tag{9.54}$$

and $f_S(s) = 0$ otherwise. Hence, $S \sim \mathcal{U}(0,1)$.

c) To obtain a PDF of T, we again apply Proposition 9.7 to Equation (9.53). For $t > 0$, we have

$$f_T(t) = \int_{-\infty}^{\infty} f_{S,T}(s,t)\,ds = \int_0^1 \lambda^2 t e^{-\lambda t}\,ds = \lambda^2 t e^{-\lambda t}.$$

Thus,

$$f_T(t) = \frac{\lambda^2}{\Gamma(2)} t^{2-1} e^{-\lambda t}, \qquad t > 0, \tag{9.55}$$

and $f_T(t) = 0$ otherwise. Hence, $T \sim \Gamma(2, \lambda)$.

d) From Equations (9.53), (9.54), and (9.55), we see that the function f defined on $\mathcal{R}^2$ by $f(s,t) = f_S(s) f_T(t)$ is a joint PDF of S and T. Therefore, S and T are independent random variables. ∎

The Box–Müller Transformation

Next we use the bivariate transformation theorem to derive a result that provides a basis for simulating independent standard normal random variables by using only a basic random number generator, that is, one that simulates a $\mathcal{U}(0,1)$ random variable.

Theoretically, in view of Proposition 6.9 on page 291 and Proposition 8.16(b) on page 471, we can simulate independent standard normal random variables by using the fact that, if U and V are independent $\mathcal{U}(0,1)$ random variables, then $\Phi^{-1}(U)$ and $\Phi^{-1}(V)$ are independent standard normal random variables. However, because there is no simple formula for Φ^{-1}, we need to take a different approach. We do so in Example 9.26.

EXAMPLE 9.26   *The Bivariate Transformation Theorem*

Box–Müller Transformation   Let X and Y be independent standard normal random variables. Express X and Y in terms of independent $\mathcal{U}(0, 1)$ random variables by using the following steps.

a) Define the random variables R and Θ implicitly by $X = R\cos\Theta$ and $Y = R\sin\Theta$. Obtain PDFs of R and Θ and show that they are independent random variables.

b) Show that R and Θ can be expressed in the form $R = \sqrt{-2\ln U}$ and $\Theta = 2\pi V$, where U and V are independent $\mathcal{U}(0, 1)$ random variables.

c) Deduce from parts (a) and (b) that the random variables X and Y can be expressed in the form $X = \sqrt{-2\ln U}\cos(2\pi V)$ and $Y = \sqrt{-2\ln U}\sin(2\pi V)$, where U and V are independent $\mathcal{U}(0, 1)$ random variables.

Solution   By assumption, X and Y are independent standard normal random variables. So a joint PDF of X and Y is

$$f_{X,Y}(x, y) = f_X(x) f_Y(y) = \frac{1}{\sqrt{2\pi}} e^{-x^2/2} \cdot \frac{1}{\sqrt{2\pi}} e^{-y^2/2} = \frac{1}{2\pi} e^{-(x^2+y^2)/2},$$

for all $x, y \in \mathcal{R}$.

a) The Jacobian determinant of the inverse transformation, $x = r\cos\theta$ and $y = r\sin\theta$, is

$$J(r, \theta) = \begin{vmatrix} \cos\theta & -r\sin\theta \\ \sin\theta & r\cos\theta \end{vmatrix} = r\cos^2\theta + r\sin^2\theta = r.$$

From calculus, the Jacobians of a transformation and its inverse are reciprocals of each other. Therefore, applying the bivariate transformation theorem, we get

$$f_{R,\Theta}(r, \theta) = \frac{1}{|J(x, y)|} f_{X,Y}(x, y) = |J(r, \theta)| f_{X,Y}(x, y) = r\frac{1}{2\pi} e^{-(x^2+y^2)/2}.$$

Noting that $x^2 + y^2 = r^2$, we now obtain

$$f_{R,\Theta}(r, \theta) = \frac{1}{2\pi} r e^{-r^2/2}, \qquad r > 0, \quad 0 < \theta < 2\pi, \tag{9.56}$$

and $f_{R,\Theta}(r, \theta) = 0$ otherwise.

We next obtain (marginal) PDFs of R and Θ. For R,

$$f_R(r) = \int_{-\infty}^{\infty} f_{R,\Theta}(r, \theta)\, d\theta = \int_0^{2\pi} \frac{1}{2\pi} r e^{-r^2/2}\, d\theta.$$

Thus,

$$f_R(r) = r e^{-r^2/2}, \qquad r > 0, \tag{9.57}$$

and $f_R(r) = 0$ otherwise. For Θ,

$$f_\Theta(\theta) = \int_{-\infty}^{\infty} f_{R,\Theta}(r, \theta)\, dr = \int_0^{\infty} \frac{1}{2\pi} r e^{-r^2/2}\, dr = \frac{1}{2\pi} \int_0^{\infty} e^{-u}\, du.$$

Thus,

$$f_\Theta(\theta) = \frac{1}{2\pi}, \qquad 0 < \theta < 2\pi, \tag{9.58}$$

and $f_\Theta(\theta) = 0$ otherwise.

Referring to Equations (9.56)–(9.58), we see that the function f defined on $\mathcal{R}^2$ by $f(r, \theta) = f_R(r) f_\Theta(\theta)$ is a joint PDF of R and Θ. Hence, R and Θ are independent random variables. Also, from Equation (9.58), we see that $\Theta \sim \mathcal{U}(0, 2\pi)$. The distribution of the random variable R, whose PDF is given in Equation (9.57), is called a **Rayleigh distribution,** in honor of British mathematician and physicist John William Strutt Rayleigh (1842–1919).

b) From Equations (9.57) and (9.58), we find that the CDFs of R and Θ are

$$F_R(r) = \begin{cases} 0, & \text{if } r < 0; \\ 1 - e^{-r^2/2}, & \text{if } r \geq 0; \end{cases} \quad \text{and} \quad F_\Theta(\theta) = \begin{cases} 0, & \text{if } \theta < 0; \\ \theta/2\pi, & \text{if } 0 \leq \theta < 2\pi; \\ 1, & \text{if } \theta \geq 2\pi. \end{cases}$$

We now apply Proposition 8.16(a) on page 471 and Proposition 6.9 on page 291 to deduce that $1 - e^{-R^2/2}$ and $\Theta/2\pi$ are independent $\mathcal{U}(0, 1)$ random variables. And, because 1 minus a $\mathcal{U}(0, 1)$ random variable is also a $\mathcal{U}(0, 1)$ random variable, $U = e^{-R^2/2}$ and $V = \Theta/2\pi$ are independent $\mathcal{U}(0, 1)$ random variables. Solving for R and Θ in terms of U and V, we now conclude that, if U and V are independent $\mathcal{U}(0, 1)$ random variables, the random variables $R = \sqrt{-2 \ln U}$ and $\Theta = 2\pi V$ are independent and have the Rayleigh and $\mathcal{U}(0, 2\pi)$ distributions, respectively.

c) From parts (a) and (b), the random variables X and Y can be expressed in the form $X = \sqrt{-2 \ln U} \cos(2\pi V)$ and $Y = \sqrt{-2 \ln U} \sin(2\pi V)$, where U and V are independent $\mathcal{U}(0, 1)$ random variables. ∎

In view of the results of Example 9.26(c), we know that, if U and V are independent $\mathcal{U}(0, 1)$ random variables, then the random variables

$$\begin{aligned} X &= \sqrt{-2 \ln U} \cos(2\pi V) \\ Y &= \sqrt{-2 \ln U} \sin(2\pi V) \end{aligned} \tag{9.59}$$

are independent standard normal random variables. Consequently, to simulate an observation (x, y) of two independent standard normal random variables, we first obtain two numbers, say, u and v, from a basic random number generator, and then calculate $x = \sqrt{-2 \ln u} \cos(2\pi v)$ and $y = \sqrt{-2 \ln u} \sin(2\pi v)$.

The transformation given in Equations (9.59) is due to G. E. P. Box and M. E. Müller, and is described in their paper "A Note on the Generation of Random Normal Deviates" (*Annals of Mathematical Statistics*, 1958, Vol. 29, pp. 610–611). Hence the transformation is now referred to as the **Box–Müller transformation.** In Chapter 10, you are asked to show how the Box–Müller transformation can be applied to simulate observations of bivariate normal random variables.

The Multivariate Transformation Theorem

The bivariate transformation theorem—Proposition 9.16 on page 546—can easily be extended to the general multivariate case of m continuous random variables with a joint PDF. That result is called the **multivariate transformation theorem.** We leave its statement and proof to you as Exercise 9.160.

EXERCISES 9.7   Basic Exercises

9.146 Suppose that, in Example 9.25 on page 547, the times it takes the first and second engineers to inspect the item have $\Gamma(\alpha, \lambda)$ and $\Gamma(\beta, \lambda)$ distributions, respectively.
a) Use the bivariate transformation theorem to obtain a joint PDF of S and T.
b) Use the result of part (a) to obtain and identify a marginal PDF of S.
c) Use the result of part (a) to obtain and identify a marginal PDF of T.
d) Are S and T independent random variables? Justify your answer.

9.147 Suppose that, as in Example 9.25 on page 547, the times it takes the first and second engineers to inspect the item are both exponentially distributed but that the parameters for the exponential distributions differ; say that they are λ and μ, respectively, where $\lambda \neq \mu$.
a) Use the bivariate transformation theorem to obtain a joint PDF of S and T.
b) Use the result of part (a) to obtain a marginal PDF of S.
c) Use the result of part (a) to obtain a marginal PDF of T.
d) Are S and T independent random variables? Justify your answer.

9.148 Let X and Y be continuous random variables with a joint PDF. Apply the bivariate transformation theorem to obtain a PDF of the random variable Y/X. Compare your result with Equation (9.48) on page 540.

9.149 In the petri-dish illustration of Example 9.8 on page 511, let R and Θ denote the polar coordinates of the center of the first spot (visible bacterial colony) to appear.
a) Obtain a joint PDF of R and Θ.
b) Obtain and identify marginal PDFs of R and Θ.
c) Decide whether R and Θ are independent random variables.

9.150 Let U and V be independent $\mathcal{U}(0, 1)$ random variables and let a, b, c, and d be real numbers with $a < b$ and $c < d$.
a) Obtain and identify a joint PDF of $X = a + (b - a)U$ and $Y = c + (d - c)V$.
b) Apply your result from part (a) to explain how to simulate the random (uniform) selection of a point from the rectangle $(a, b) \times (c, d)$ by using a basic random number generator.

9.151 Range and midrange of a random sample: Let $X_1, \ldots, X_n$ be a random sample from a continuous distribution with CDF F and PDF f. The *range* and *midrange* of the random sample are defined to be $R = Y - X$ and $M = (X + Y)/2$, respectively, where $X = \min\{X_1, \ldots, X_n\}$ and $Y = \max\{X_1, \ldots, X_n\}$.
a) Apply the bivariate transformation theorem to obtain a joint PDF of R and M. *Note:* Equation (9.20) on page 498 provides a joint PDF of X and Y.
b) Use the result of part (a) to obtain a marginal PDF of R. Compare your answer to that found in Example 9.6 on page 504.
c) Use the result of part (a) to obtain a marginal PDF of M. Compare your answer to that presented in Exercise 9.116 on page 542.
d) Are R and M independent random variables? Justify your answer.

9.152 Let X and Y be independent $\mathcal{N}(0, \sigma^2)$ random variables. Show that the random variables $X^2 + Y^2$ and Y/X are independent.

9.153 Let X and Y be independent random variables with $X \sim \Gamma(\alpha, \lambda)$ and $Y \sim \Gamma(\beta, \lambda)$. Are $X + Y$ and Y/X independent? Justify your answer.

9.154 Suppose that X and Y are continuous random variables with a joint PDF.
a) Use the bivariate transformation theorem to obtain a joint PDF of the random variables $X + Y$ and $X - Y$.

b) Suppose that X and Y are independent $\mathcal{N}(\mu, \sigma^2)$ random variables. Use the result of part (a) to decide whether $X + Y$ and $X - Y$ are independent.

c) Suppose that X and Y are independent $\mathcal{U}(0, 1)$ random variables. Use the result of part (a) to decide whether $X + Y$ and $X - Y$ are independent.

9.155 Let the location of a particle moving in two dimensions be described by a Cartesian coordinate system whose origin is the location of the particle at time $t = 0$. A common model for this process is that the x and y coordinates of the location of the particle at time $t > 0$ are independent $\mathcal{N}(0, \sigma^2 t)$ random variables. Let $R(t)$ and $\Theta(t)$ denote the polar coordinates of the position of the particle at time $t > 0$. For fixed $t > 0$,

a) find a joint PDF of $R(t)$ and $\Theta(t)$.

b) obtain and identify marginal PDFs of $R(t)$ and $\Theta(t)$.

c) decide whether $R(t)$ and $\Theta(t)$ are independent random variables.

9.156 Let T denote the interior of the triangle with vertices $(0, 0)$, $(1, 0)$, and $(1, 1)$. Show how to simulate the random (uniform) selection of a point from T by using a basic random number generator. Proceed as follows. Let X and Y be the x and y coordinates, respectively, of a point selected at random from T.

a) Explain why a marginal PDF of X and all conditional PDFs of Y given $X = x$ determine a joint PDF of X and Y.

b) Determine a joint PDF of X and Y, a marginal PDF of X, and conditional PDFs of Y given $X = x$.

c) Let U and V be independent $\mathcal{U}(0, 1)$ random variables. Use your results from part (b) and Proposition 8.16(b) on page 471 to show that $\sqrt{U}$ has the same probability distribution as X and that $\sqrt{U} V_{|\sqrt{U}=x}$ has the same probability distribution as $Y_{|X=x}$.

d) Use parts (a) and (c) to conclude that $\sqrt{U}$ and $\sqrt{U} V$ have the same joint probability distribution as X and Y.

e) Use the bivariate transformation theorem to give another verification that $\sqrt{U}$ and $\sqrt{U} V$ have the same joint probability distribution as X and Y.

f) Use part (d) to explain how to simulate the random selection of a point from T by using a basic random number generator.

g) Generalize the approach applied in this exercise to explain when and how a basic random number generator can be used to simulate a specified bivariate continuous probability distribution.

9.157 Explain how the Box–Müller transformation can be used to simulate observations of a normal random variable with parameters μ and σ^2.

9.158 Simulation: Here, we assume that you have done Exercise 9.157 and that you have access to a computer or graphing calculator. The gestation periods of women are (approximately) normally distributed with $\mu = 266$ days and $\sigma = 16$ days.

a) Use a basic random number generator to simulate 10,000 human gestation periods.

b) Roughly, what would you expect a histogram of the 10,000 observations found in part (a) to look like?

c) Obtain a histogram of the 10,000 observations found in part (a).

9.159 Let Z_1 and Z_2 be independent standard normal random variables and let ρ be a real number such that $|\rho| < 1$.

a) Obtain a joint PDF of the random variables $X = Z_1$ and $Y = \rho Z_1 + \sqrt{1 - \rho^2} Z_2$.

b) Apply your result from part (a) and the Box–Müller transformation to explain how to use a basic random number generator to simulate random variables having the bivariate normal distribution with joint PDF given by Equation (9.35) on page 518.

Theory Exercises

9.160 Multivariate transformation theorem: State and prove the multivariate transformation theorem, the generalization of the bivariate transformation theorem to the case of m continuous random variables with a joint PDF.

9.161 Suppose that $X_1, \ldots, X_n$ are continuous random variables with a joint PDF. Let a_{ij} $(1 \le i \le n, 1 \le j \le n)$ be real numbers such that the $n \times n$ matrix $\mathbf{A} = [a_{ij}]$ is nonsingular. Define $Y_i = \sum_{j=1}^{n} a_{ij} X_j$ for $i = 1, 2, \ldots, n$. Use the multivariate transformation theorem to obtain a joint PDF of $Y_1, \ldots, Y_n$ in terms of the matrix $\mathbf{A}$ and a joint PDF of $X_1, \ldots, X_n$.

Advanced Exercises

9.162 Let $X_1, \ldots, X_n$ be a random sample from an exponential distribution with parameter λ, and set $Y_k = \sum_{j=1}^{k} X_j$ for $1 \le k \le n$.
a) Use the multivariate transformation theorem (or Exercise 9.161) to obtain a joint PDF of the random variables $Y_1, \ldots, Y_n$.
b) Let $Y_1, \ldots, Y_n$ be random variables with joint PDF $f_{Y_1, \ldots, Y_n}(y_1, \ldots, y_n) = \lambda^n e^{-\lambda y_n}$ for $0 < y_1 < \cdots < y_n$, and $f_{Y_1, \ldots, Y_n}(y_1, \ldots, y_n) = 0$ otherwise. Now let $X_1 = Y_1$ and $X_k = Y_k - Y_{k-1}$ for $2 \le k \le n$. Without doing any calculations obtain a joint PDF of the random variables $X_1, \ldots, X_n$.

9.163 Let the location of a particle moving in three dimensions be described by a Cartesian coordinate system whose origin is the location of the particle at time $t = 0$. A common model for this process is that the x, y, and z coordinates of the location of the particle at time $t > 0$ are independent $\mathcal{N}(0, \sigma^2 t)$ random variables. Let $\mathrm{P}(t)$, $\Phi(t)$, and $\Theta(t)$ denote the spherical coordinates of the position of the particle at time $t > 0$. For fixed $t > 0$,
a) find a joint PDF of $\mathrm{P}(t)$, $\Phi(t)$, and $\Theta(t)$.
b) obtain and identify marginal PDFs of $\mathrm{P}(t)$, $\Phi(t)$, and $\Theta(t)$.
c) decide whether $\mathrm{P}(t)$, $\Phi(t)$, and $\Theta(t)$ are independent random variables.

9.164 Student's t-distribution: Let Z and Y be independent random variables, where Z has the standard normal distribution and Y has the chi-square distribution with v degrees of freedom.
a) Use the bivariate transformation theorem to show that a PDF of $T = Z/\sqrt{Y/v}$ is given by

$$f_T(t) = \frac{\Gamma\left((v+1)/2\right)}{\sqrt{v\pi}\,\Gamma(v/2)} \left(1 + t^2/v\right)^{-(v+1)/2}, \qquad \infty < t < \infty.$$

Note: A random variable with this PDF is said to have the *Student's t-distribution with v degrees of freedom.*
b) Show that the Student's t-distribution with 1 degree of freedom is the same as the standard Cauchy distribution.

9.165 F-distribution: Let X_1 and X_2 be independent chi-square random variables with v_1 and v_2 degrees of freedom, respectively. Show that a PDF of $W = (X_1/v_1)/(X_2/v_2)$ is given by

$$f_W(w) = \frac{(v_1/v_2)^{v_1/2}\Gamma\left((v_1 + v_2)/2\right)}{\Gamma(v_1/2)\Gamma(v_2/2)} \frac{w^{v_1/2-1}}{\left(1 + (v_1/v_2)w\right)^{(v_1+v_2)/2}}, \qquad w > 0,$$

and $f_W(w) = 0$ otherwise. *Note:* A random variable with this PDF is said to have the *F-distribution with v_1 and v_2 degrees of freedom.*

CHAPTER REVIEW

Summary

In this chapter, we discussed the simultaneous analysis of two or more continuous random variables defined on the same sample space. One of the primary objects used in such analyses is the joint probability density function or, more simply, joint PDF. As a preliminary to our discussion of joint PDFs, we examined the joint cumulative distribution function (joint CDF), a function that applies to any types of random variables—discrete, continuous, or otherwise.

Roughly speaking, a joint PDF of continuous random variables gives the probability that those random variables take values near any specified values. A joint PDF is essentially the mixed partial derivative of the joint CDF. Using a joint PDF and the fundamental probability formula (FPF), we showed that we can, at least conceptually, obtain any probability for the continuous random variables under consideration.

Given a joint PDF of several continuous random variables, we can obtain each individual PDF, called a marginal probability density function (marginal PDF). This task is accomplished by integrating the joint PDF on the other variables. Although a joint PDF determines the marginal PDFs, the converse isn't true—knowing only the marginal PDFs generally isn't sufficient information to determine the joint PDF.

A conditional probability density function (conditional PDF) provides a PDF of one or more continuous random variables, given the values of other continuous random variables. A conditional PDF is a genuine PDF and hence all properties of (unconditional) PDFs also hold for conditional PDFs.

We examined the concept of independence, specialized to continuous random variables. For continuous random variables, independence is equivalent to the condition that the joint PDF equals the product of the marginal PDFs or, more precisely, that the product of the marginal PDFs is a joint PDF.

We discussed the process of taking a real-valued function of two or more continuous random variables with a joint PDF, concentrating on the case in which the function is such that the resulting random variable is also a continuous random variable with a PDF. Given a joint PDF of the original random variables, we demonstrated how to determine a PDF of the function of those random variables.

Of particular significance in probability theory are sums and quotients of random variables which, for the continuous case, we introduced and explored. Sums of independent continuous random variables—which play an essential part in many applications—were also investigated.

Finally, we examined the all-important multivariate transformation theorem, which can be applied to efficiently derive various results, obtain PDFs of univariate distributions, and, of course, determine joint PDFs of multivariate distributions.

You Should Be Able To ...

1. obtain the joint CDF of random variables defined on the same sample space.

2. define, state the basic properties of, and obtain a joint PDF of continuous random variables.

3. use the joint CDF of continuous random variables to obtain their joint PDF.

4. state and apply the fundamental probability formula (FPF) for continuous random variables.

5. determine marginal PDFs, knowing a joint PDF.

6. define, obtain, and use conditional PDFs.

7. state and apply basic properties of independent continuous random variables.

8. explain the relationship between independence of continuous random variables and conditional distributions.

9. determine a PDF of a real-valued function of two or more continuous random variables, knowing a joint PDF of those random variables.

10. determine a PDF of the sum of two continuous random variables, knowing a joint PDF of those random variables.

11. state and apply the special formulas for obtaining a PDF of the sum of two independent continuous random variables, knowing marginal PDFs of those random variables.

12. identify the distribution of the sum of independent gamma random variables with the same second parameter.

13. identify the distribution of the sum of independent normal random variables.

14. determine a PDF of the quotient of two continuous random variables, knowing a joint PDF of those random variables.

15. state and apply the bivariate and multivariate transformation theorems.

Key Terms

bivariate transformation theorem, *546*

Box-Müller transformation, *550*

conditional probability density function, *515*

convolution, *535*

fundamental probability formula, *502*

general multiplication rule, *517*

joint cumulative distribution function, *486, 489*

joint density function, *495*

joint probability density function, *495*

marginal cumulative distribution function, *490*

marginal probability density functions, *510*

Maxwell distribution, *533*

multivariate transformation theorem, *550*

Rayleigh distribution, *550*

Chapter Review Exercises

Basic Exercises

9.166 Let X and Y be random variables defined on the same sample space. Show that
a) $P(X > x, Y > y) = 1 - F_X(x) - F_Y(y) + F_{X,Y}(x, y)$.
b) $P(X > x, Y > y) = 1 - F_{X,Y}(x, \infty) - F_{X,Y}(\infty, y) + F_{X,Y}(x, y)$.

9.167 Let X and Y denote the x and y coordinates, respectively, of a point selected at random from the unit square. The joint CDF of X and Y was found in Example 9.1 on page 487; the marginal probability distributions of X and Y are both uniform on the interval $(0, 1)$, as

obtained in Exercise 9.3 on page 491; and a joint PDF of X and Y was found in Example 9.3 on page 496. Determine $P(X > 0.3, Y > 0.4)$ by using
a) Proposition 9.1.   **b)** Exercise 9.166(a).   **c)** Exercise 9.166(b).   **d)** the FPF.

9.168 Let X and Y denote the x and y coordinates, respectively, of a point selected at random from the rectangle $(a, b) \times (c, d) = \{ (x, y) \in \mathcal{R}^2 : a < x < b, c < y < d \}$.
a) Determine the joint CDF of the random variables X and Y.
b) Use part (a) to obtain the marginal CDFs of X and Y.
c) Use part (b) to identify the probability distributions of X and Y.
d) What is the relationship between the joint CDF and the marginal CDFs of X and Y?
e) Are X and Y independent random variables? *Hint:* Apply Proposition 9.4 on page 499 to your answer from part (d).

9.169 Obtain a joint PDF of the random variables X and Y in Exercise 9.168 by using
a) the CDF obtained in that exercise.
b) a symmetry argument and basic properties of a joint PDF.

9.170 By considering an appropriate joint probability, show that the function $F: \mathcal{R}^2 \to \mathcal{R}$ defined by $F(x, y) = 0$ for $x + y < 0$ and $F(x, y) = 1$ for $x + y \geq 0$ can't be a joint CDF of two random variables.

9.171 Can the function g defined on $\mathcal{R}^2$ by $g(x, y) = e^{-2(x^2 + xy + y^2)/3}$ be the "variable form" of a joint PDF for two random variables? If not, why not? If so, obtain the corresponding joint PDF. *Hint:* Complete the square in the exponent and use the fact that the PDF of a normal random variable integrates to 1.

9.172 An insurance company insures a large number of drivers. Let X and Y denote the company's losses under collision insurance and liability insurance, respectively. A joint PDF of X and Y is $f_{X,Y}(x, y) = (2x + 2 - y)/4$ for $0 < x < 1$ and $0 < y < 2$, and $f_{X,Y}(x, y) = 0$ otherwise. Determine the probability that the total loss is at least 1.

9.173 Let X and Y be continuous random variables with joint PDF given by $f_{X,Y}(x, y) = e^{-y}$ for $0 < x < y$, and $f_{X,Y}(x, y) = 0$ otherwise.
a) Obtain and identify marginal PDFs of X and Y.
b) Decide whether X and Y are independent.
c) Determine a PDF of $X + Y$.
d) Obtain and identify a conditional PDF of X given $Y = y$.
e) Calculate $P(X > 3 \mid Y < 5)$.
f) Calculate $P(X > 3 \mid Y = 5)$.

9.174 A bus arrives at a bus stop at time X. Independently, a potential passenger named Jane arrives at the bus stop at time Y. Jane will wait at most 5 minutes for the bus. Assume that X and Y are continuous random variables, each with a PDF.
a) Express the probability that Jane catches the bus in terms of marginal PDFs of X and Y.
b) Evaluate the expression that you obtained in part (a) if the bus arrives uniformly between 7:00 A.M. and 7:30 A.M. and Jane arrives uniformly between 7:05 A.M. and 7:20 A.M.

9.175 Let $X_1, \ldots, X_n$ be a random sample from a distribution with CDF F. For $k = 0, 1, \ldots, n$, let A_k denote the event that exactly k of the n observations are in the interval $(a, b]$.
a) Determine $P(A_n)$ by using the formula for $P(X > x, Y \leq y)$ obtained in the solution to Example 9.4(b) on page 498, where $X = \min\{X_1, \ldots, X_n\}$ and $Y = \max\{X_1, \ldots, X_n\}$.
b) Determine $P(A_k)$ in terms of F for $k = 0, 1, \ldots, n$.
c) Provide a table of values for $P(A_k)$ in case $n = 4$, $a = 0.2$, $b = 0.9$, and the common distribution of the X_js is $\mathcal{U}(0, 1)$.
d) Repeat part (c) if the common distribution of the X_js is $\mathcal{E}(1)$.

9.176 Let X, Y, and Z be the x, y, and z coordinates, respectively, of a point selected at random from the unit sphere, $S = \{(x, y, z) \in \mathcal{R}^3 : x^2 + y^2 + z^2 < 1\}$.

a) Without finding a joint PDF of X, Y, and Z, determine a formula for the probability that the point obtained lands in the set E, where $E \subset \mathcal{R}^3$.

b) Obtain a joint PDF of X, Y, and Z.

c) Use your result from part (b) and the FPF to solve part (a).

d) Obtain the three univariate marginal PDFs.

e) Obtain the three bivariate marginal PDFs.

f) Determine and identify a conditional PDF of Z given $X = x$ and $Y = y$.

g) Determine and identify a conditional joint PDF of Y and Z given $X = x$.

h) Are X, Y, and Z independent random variables? Justify your answer.

9.177 In Exercise 9.176, determine and identify a PDF of the distance of the point obtained from the origin.

9.178 A device consisting of two components fails if either component fails. The joint density function of the lifetimes of the components, measured in hours, equals $f(s, t)$ for $0 < s < 1$ and $0 < t < 1$, and equals 0 otherwise. Which of the following provide the probability that the device fails during the first half hour of operation?

a) $\int_0^{0.5} \left(\int_0^{0.5} f(s, t)\, ds \right) dt$ **b)** $\int_0^1 \left(\int_0^{0.5} f(s, t)\, ds \right) dt$

c) $\int_{0.5}^1 \left(\int_{0.5}^1 f(s, t)\, ds \right) dt$

d) $\int_0^{0.5} \left(\int_0^1 f(s, t)\, ds \right) dt + \int_0^1 \left(\int_0^{0.5} f(s, t)\, ds \right) dt$

e) $\int_0^{0.5} \left(\int_{0.5}^1 f(s, t)\, ds \right) dt + \int_0^1 \left(\int_0^{0.5} f(s, t)\, ds \right) dt$

9.179 A company offers a basic life insurance policy to its employees, as well as a supplemental life insurance policy. To purchase the supplemental policy, an employee must first purchase the basic policy. Let X denote the proportion of employees who purchase the basic policy and let Y denote the proportion of employees who purchase the supplemental policy. Suppose that X and Y have joint density function $f_{X,Y}(x, y) = 2(x + y)$ on the region where the density is positive. Given that 10% of the employees buy the basic policy, what is the probability that less than 5% buy the supplemental policy?

9.180 Let X and Y denote the x and y coordinates, respectively, of a point selected at random from the line segment $D = \{(x, y) : y = x, 0 < x < 1\}$.

a) Determine the joint CDF of X and Y.

b) Determine the marginal CDFs of X and Y.

c) Find and identify marginal PDFs of X and Y.

d) Show that X and Y don't have a joint PDF.

9.181 In a small metropolitan area, annual losses due to storm, fire, and theft are independent exponentially distributed random variables with respective parameters 1, $\frac{2}{3}$, and $\frac{5}{12}$. Determine the probability that the maximum of these losses exceeds 3.

9.182 Three people agree to meet at a specified place at noon. Assume that their arrival times are independent $\mathcal{U}(-5, 5)$ random variables, where time is measured in minutes relative to noon. Determine the probability that

a) all three people arrive before noon.

b) exactly one of the three people arrives after 12:03 P.M.

c) the first two people who arrive wait at least 3 minutes for the last person to arrive.

9.183 Let $C = \{ (x, y) \in \mathcal{R}^2 : 0 < y < f(x), x \in \mathcal{R} \}$, where f is a univariate PDF. Denote by X and Y the x and y coordinates, respectively, of a point selected at random from C. Determine a

a) joint PDF of X and Y.   b) marginal PDF of X.   c) marginal PDF of Y.
d) conditional PDF of Y given $X = x$.
e) conditional PDF of X given $Y = y$.

9.184 A random variable has a PDF given by $(2/\pi)/(e^{-x} + e^x)$ for all $x \in \mathcal{R}$. Determine the probability that, in two independent observations of this random variable, neither observation exceeds 1.

9.185 An insurance policy is written to cover a loss with density $(3/8)x^2$ for $0 < x < 2$, and 0 otherwise. The time, in hours, to process a claim of size x is uniformly distributed on the interval from x to $2x$. Determine the probability that the processing time of a randomly chosen claim on this policy is between 1 and 3 hours.

9.186 Let $X \sim \mathcal{U}(-h, h)$, where h is a positive real number. Also, let Y be a continuous random variable with a PDF, independent of X.

a) Use the FPF to show that $F_{X+Y}(z) = \dfrac{1}{2h} \displaystyle\int_{z-h}^{z+h} F_Y(u)\, du$.

b) Use part (a) to show that $f_{X+Y}(z) = \dfrac{1}{2h} \displaystyle\int_{z-h}^{z+h} f_Y(u)\, du$.

c) Use Proposition 9.12 on page 534 to obtain the result of part (b).
d) Use part (c) to obtain the result of part (a).

9.187 The time a certain device takes to warm up after it's turned on has the uniform distribution on the interval from 1 to 3 minutes. Once the device warms up, its lifetime, in hours, has an exponential distribution with parameter 0.1, independent of warm-up time. Determine a PDF for the elapsed time, T, from when the device is turned on until it fails.

9.188 Let the location of a particle moving in two dimensions be described by a Cartesian coordinate system whose origin is the location of the particle at time $t = 0$. A common model for this process is that the x and y coordinates at time $t > 0$ are independent $\mathcal{N}(0, \sigma^2 t)$ random variables. Let $\Theta(t)$ be the angle in radians between $-\pi/2$ and $\pi/2$ that the line through the origin and the particle's position at time $t > 0$ makes with the positive x-axis. Determine and identify the probability distribution of the random variable $\Theta(t)$.

9.189 In a chemical process, a fraction X of the product consists of impurities. Of these impurities, a fraction Y is harmful and the remaining impurities don't affect product quality.

a) Obtain an expression for a PDF of the fraction of harmful impurities in the product if X and Y are independent random variables.
b) Evaluate and interpret your expression in part (a) if $X \sim \mathcal{U}(0, 0.1)$ and $Y \sim \mathcal{U}(0, 0.5)$.

9.190 Let X denote the amount of an item stocked at the beginning of a day and let Y denote the amount of the item sold during the day. A joint PDF of X and Y is $f_{X,Y}(x, y) = 8xy$ for $0 < y < x < 1$, and $f_{X,Y}(x, y) = 0$ otherwise. Find a PDF of the amount of the item

a) stocked at the beginning of a day.
b) sold during the day.
c) remaining at the end of the day.
d) Determine the probability that the amount of the item remaining at the end of the day is less than one third of the amount stocked at the beginning of the day.

9.191 Claims filed under automobile insurance policies at a certain insurance company follow a normal distribution with parameters 19.4 and 25. What is the probability that the average of 25 randomly selected claims exceeds 20?

9.192 For Company A, there is a 60% chance that no claim is made during the coming year. If one or more claims are made, the total claim amount is (approximately) normally distributed with parameters 10 and 4. For Company B, there is a 70% chance that no claim is made during the coming year. If one or more claims are made, the total claim amount is (approximately) normally distributed with parameters 9 and 4. Total claim amounts of the two companies are independent. What is the probability that, in the coming year, Company B's total claim amount will exceed Company A's total claim amount?

9.193 When a current of I amperes flows through a resistance of R ohms, the power generated is $W = I^2 R$ watts. Assume that I and R are independent random variables, that I has the beta distribution with parameters 2 and 2, and that R has the beta distribution with parameters 2 and 1. Obtain a PDF of W
a) without using the bivariate transformation theorem.
b) by using the bivariate transformation theorem.

9.194 The stiffness, S, of a rectangular wooden beam is proportional to its width, W, times the cube of its depth, D. The proportionality constant, c, depends on the units used and the type of wood. Suppose that W and D are independent random variables, each with a PDF. Show that a PDF of S is given by

$$f_S(s) = \int_0^\infty \frac{1}{ct^3} f_W\left(\frac{s}{ct^3}\right) f_D(t)\, dt, \qquad s > 0,$$

and $f_S(s) = 0$ otherwise.

9.195 Let $X_1, \ldots, X_n$ be independent random variables, each having a $\mathcal{N}(\mu, \sigma^2)$ distribution, and let m be a positive integer less than n. Determine a joint PDF of the random variables $\sum_{k=1}^n X_k$ and $\sum_{k=1}^m X_k$.

Theory Exercises

9.196 Let X and Y be continuous random variables with a joint PDF. Prove that

$$P(X \in A,\ Y \in B) = \int_A P(Y \in B \mid X = x) f_X(x)\, dx,$$

for all subsets A and B of $\mathcal{R}$.

9.197 Bayes's rule for continuous random variables: Let X and Y be continuous random variables with a joint PDF.
a) Prove and interpret *Bayes's rule* in this context:

$$f_{X \mid Y}(x \mid y) = \frac{f_X(x) f_{Y \mid X}(y \mid x)}{\int_{-\infty}^\infty f_X(s) f_{Y \mid X}(y \mid s)\, ds}.$$

b) Suppose that the (prior) distribution of X is exponential with parameter λ. Further suppose that, given $X = x$, the distribution of Y is exponential with parameter λ on the interval (x, ∞); that is, for each $x > 0$, we have $f_{Y \mid X}(y \mid x) = \lambda e^{-\lambda(y-x)}$ if $y > x$, and $f_{Y \mid X}(y \mid x) = 0$ otherwise. Use Bayes's rule to obtain and identify the (posterior) distribution of X under the condition that $Y = y$.

9.198 Suppose that X and Y are random variables defined on the same sample space. Define $G(x, y) = \max\{F_X(x) + F_Y(y) - 1, 0\}$ and $H(x, y) = \min\{F_X(x), F_Y(y)\}$. It can be shown that both G and H are bivariate CDFs.

a) Prove that G has marginal CDFs F_X and F_Y. That is, if U and V are random variables having G as their joint CDF, then $F_U = F_X$ and $F_V = F_Y$.
b) Prove that H has marginal CDFs F_X and F_Y.
c) Prove that $G(x, y) \le F_{X,Y}(x, y) \le H(x, y)$ for all x and y. *Note:* This result provides upper and lower bounds on $F_{X,Y}$ in terms of F_X and F_Y even when $F_{X,Y}$ is unknown.

Advanced Exercises

9.199 Let $X \sim \mathcal{U}(0, 1)$ and set $Y = 1 - X$.
a) Determine the joint CDF of X and Y.
b) Use part (a) to obtain the marginal CDFs of X and Y.
c) Use part (b) to obtain and identify marginal PDFs of X and Y.
d) Do X and Y have a joint PDF? Explain your answer.

9.200 Let X and Y be random variables with joint CDF given by

$$F_{X,Y}(x, y) = \begin{cases} 0, & \text{if } x < 0 \text{ or } y < 0; \\ (1/2)(1 - e^{-x}), & \text{if } x \ge 0 \text{ and } 0 \le y < 1; \\ 1 - e^{-x}, & \text{if } x \ge 0 \text{ and } y \ge 1. \end{cases}$$

a) Determine the marginal CDFs of X and Y.
b) Use part (a) to identify the marginal probability distributions of X and Y.
c) Can X and Y have a joint PDF? Explain your answer.
d) Are X and Y independent random variables? *Hint:* Refer to Exercise 9.15 on page 492.
e) Find $P(3 < X < 5, Y = 0)$.

9.201 Suppose that X is uniformly distributed on the interval $(0, 3)$. Set $Y = X - \lfloor X \rfloor$ so that Y is the "fractional part" of X.
a) Obtain and identify the probability distribution of Y. *Hint:* First find the CDF of Y.
b) Graph the range of (X, Y).
c) Determine the joint CDF of X and Y.

9.202 Refer to Exercise 9.201.
a) Obtain and identify the probability distribution of the random variable $Z = X - Y$.
b) From Exercise 9.201, both X and Y are continuous random variables with PDFs. Use part (a) to explain why the random variables X and Y can't have a joint PDF.
c) Use the result of Exercise 9.201(b) to explain why the random variables X and Y can't have a joint PDF.

9.203 Sample median: Suppose that a random sample of size $2n + 1$ is taken from a continuous distribution with CDF F and PDF f. The *sample median* is defined to be the middle observation when the $2n + 1$ observations are arranged in increasing order.
a) Determine a PDF of the sample median. *Hint:* Refer to Exercise 9.33 on page 501.
b) Apply the result of part (a) to determine the probability that the sample median of a random sample of size 5 from a $\mathcal{U}(0, 1)$ distribution will be within 1/4 of the median of the distribution, which is 1/2.

9.204 The two employees of a small company are covered by an insurance policy that will reimburse the company for no more than one loss per employee per year. The policy reimburses the full amount of losses up to an annual company-wide maximum of $8000. Chances are 40% that an employee will incur a loss in a year, independent of the other employee's losses. Furthermore, the amount of each loss is uniformly distributed between $1000 and $5000. Given that "one" of the employees has incurred a loss in excess of $2000, find the probability that losses will exceed reimbursements. Solve this problem when "one" means
a) a specified one.   b) at least one.   c) exactly one.

9.205 Farlie–Morgenstern family of CDFs: Let G and H be univariate CDFs and let α be a real number with $|\alpha| \leq 1$. Define $F_\alpha : \mathcal{R}^2 \to \mathcal{R}$ by

$$F_\alpha(x, y) = G(x)H(y)\big[1 + \alpha\big(1 - G(x)\big)\big(1 - H(y)\big)\big].$$

It can be shown that F_α is a bivariate CDF. The collection $\{\, F_\alpha : -1 \leq \alpha \leq 1\,\}$, is called the *Farlie–Morgenstern family* of bivariate CDFs corresponding to the univariate CDFs G and H.
a) Determine the marginal CDFs of F_α.
b) Use the results of part (a) to construct an uncountably infinite number of bivariate CDFs whose marginals are both CDFs for a uniform distribution on the interval $(0, 1)$.
c) Obtain a joint PDF corresponding to each bivariate CDF in part (b).

9.206 Refer to Exercise 9.205. Assume now that G and H have corresponding PDFs—say, g and h, respectively.
a) Determine a joint PDF corresponding to F_α.
b) Use part (a) to find marginal PDFs.
c) Use the result of Exercise 9.205(a) to find marginal PDFs.
d) For what value(s) of α are the corresponding random variables independent? Explain your answer.
e) Use the result of part (a) to obtain a joint PDF corresponding to F_α when both G and H are CDFs for a uniform distribution on the interval $(0, 1)$. Compare your answer to that found in Exercise 9.205(c).

9.207 Bayes's rule for mixed bivariate distributions: On pages 509 and 522, we discussed mixed bivariate distributions, where X is a continuous random variable and Y is a discrete random variable. Suppose that X and Y have joint PDF/PMF $h_{X,Y}$.
a) Prove and interpret *Bayes's rule* in this context:

$$f_{X|Y}(x \mid y) = \frac{f_X(x)p_{Y|X}(y \mid x)}{\int_{-\infty}^{\infty} f_X(s)p_{Y|X}(y \mid s)\,ds}.$$

b) A company has n employees. The probability is p that an employee will take at least one sick day during any particular year, independent of the other employees. However, p varies as a random variable Π having a beta distribution with parameters α and β. Determine and identify the (posterior) probability distribution of Π given that the number of employees who take at least one sick day during the year is k.

9.208 Suppose that $X_1, X_2, \ldots$ are independent random variables, each having the uniform distribution on the interval $(0, 1)$. For each $n \in \mathcal{N}$, let $U_n = \min\{X_1, \ldots, X_n\}$ and let $V_n = \max\{X_1, \ldots, X_n\}$. The joint CDF of U_n and V_n, denoted F_n, can be obtained from Equation (9.19) on page 498.
a) Intuitively, what happens to the joint probability distribution of U_n and V_n as $n \to \infty$?
b) Find $\lim_{n \to \infty} F_n(x, y)$ for each $(x, y) \in \mathcal{R}^2$.
c) Consider random variables X and Y such that $P(X = 0) = 1$ and $P(Y = 1) = 1$. Determine the joint CDF of X and Y.
d) Do your results for parts (b) and (c) seem to confirm your answer to part (a)?

Pafnuty Chebyshev *1821–1894*

Pafnuty Chebyshev was born in Okatovo, Russia, on May 16, 1821. He founded the St. Petersburg mathematical school, sometimes referred to as the Chebyshev school.

In 1847, Chebyshev was appointed Assistant Professor of Mathematics at the University of St. Petersburg; in 1850, Extraordinary Professor; and, in 1860, Full Professor. He was voted Adjunct of the Petersburg Academy of Sciences with the Chair of Applied Mathematics in 1853, and, in 1859, Ordinary Academician, again with the Chair of Applied Mathematics. In 1860, Chebyshev became a Correspondent of the Institut de France and, in 1874, a Foreign Associate of that institute and of the Royal Society as well.

Chebyshev is well known for his development of the basic probability inequality now referred to as *Chebyshev's inequality*. He used that inequality to obtain a simple but mathematically precise proof of a general form of the weak law of large numbers.

In 1845, Bertrand conjectured that there is always at least one prime between n and $2n$, a result that Chebyshev proved in 1850. Chebyshev also worked on the *prime number theorem*, which states that $\pi(n)/(n/\ln n) \to 1$, as $n \to \infty$, where $\pi(n)$ denotes the number of primes not exceeding n. He proved that, if $\pi(n)/(n/\ln n)$ has a limit, that limit must be 1. That the limit actually exists was proved independently by Hadamard and de la Vallée Poussin two years after Chebyshev's death in 1894.

Chebyshev's mathematical writings include treatises on probability theory, the theory of integrals (including the beta function), quadratic forms, orthogonal functions, and formulas for volumes. He also worked on theoretical mechanics, contributed significantly to ballistics, invented hinge-lever gears, and built a calculating machine.

Chebyshev became widely known for his book *Teoria sravneny* (Theory of Congruences), published in 1849 and used in Russian universities for many years. He retired from the University of St. Petersburg in 1882, but held "open house" once a week. Chebyshev died on December 8, 1894, in St. Petersburg, Russia.

CHAPTER TEN

Expected Value of Continuous Random Variables

INTRODUCTION

So far in our discussion of continuous random variables, we have concentrated on methods for obtaining probabilities. As you know, the primary object in this regard is the probability density function.

In this chapter, we discuss expected value of continuous random variables and related concepts. As we demonstrated in Chapter 7, roughly speaking, the expected value of a random variable is the value of the random variable we would expect to observe "on average." Put another way, it's the long-run average value of the random variable in repeated independent observations.

We begin, in Section 10.1, by developing and illustrating the formula used to define the expected value of a continuous random variable. In Section 10.2, we present and apply the basic properties of expected value. In particular, we discuss the *fundamental expected-value formula (FEF)*, a formula that provides a method for obtaining the expected value of a real-valued function of one or more continuous random variables.

As for discrete random variables, we can use the expected value of an appropriate function of a continuous random variable to measure the variation of the possible values of the continuous random variable. We can also use the expected value of an appropriate function of two continuous random variables to measure association between those random variables. We examine these and related ideas in Section 10.3.

In Section 10.4, we apply the concept of expected value to conditional distributions and thereby obtain conditional expected values. Then we show how the expected value of a continuous random variable can be determined from its various conditional expected values.

In Section 10.5, we examine the *bivariate normal distribution*, one of the most important families of bivariate continuous distributions. As part of the discussion, we show how to interpret the parameters of a bivariate normal distribution in terms of expected values, variances, and correlation.

10.1 Expected Value of a Continuous Random Variable

In this section, we define and illustrate the concept of the *expected value* of a continuous random variable. *We assume throughout this chapter that any continuous random variable being discussed has a probability density function (PDF).*

To define the expected value of a continuous random variable, we first recall the definition of the expected value of a discrete random variable, as presented in Definition 7.1 on page 327: The expected value of a discrete random variable X, denoted $\mathcal{E}(X)$, is defined by

$$\mathcal{E}(X) = \sum_x x p_X(x) = \sum_x x P(X = x). \tag{10.1}$$

For a continuous random variable X, the probabilities $P(X = x)$ are all 0 and, consequently, Equation (10.1) won't work as a definition for the expected value of X. However, Equation (10.1) shows that the expected value of a discrete random variable is a weighted average of its possible values—weighted by probability—and it's this idea that we use to develop the definition of the expected value of a continuous random variable.

For simplicity, we assume that the range of the continuous random variable X is bounded—say, a subset of $[a, b]$. For a large positive integer n, we consider the partition of that interval into n equal-length subintervals: $a = x_0 < x_1 < \cdots < x_n = b$, where $x_j = a + j(b - a)/n$ for $0 \le j \le n$. Then, in view of Relation (9.2) on page 486, a weighted average of the possible values of X, weighted by probability, is given approximately by

$$\sum_{j=0}^{n-1} x_j P(x_j \le X \le x_{j+1}) \approx \sum_{j=0}^{n-1} x_j f_X(x_j) \Delta x_j, \tag{10.2}$$

where $\Delta x_j = x_{j+1} - x_j = (b - a)/n$ for $0 \le j \le n - 1$.

Because n is large, the right side of Relation (10.2), which is a Riemann sum, approximately equals the Riemann integral of $x f_X(x)$ over the interval $[a, b]$. Therefore,

$$\mathcal{E}(X) \approx \int_a^b x f_X(x)\, dx = \int_{-\infty}^{\infty} x f_X(x)\, dx. \tag{10.3}$$

Consequently, in view of Relation (10.3), we have provided a basis for the definition of the **expected value** of a continuous random variable as presented in Definition 10.1.

DEFINITION 10.1   Expected Value of a Continuous Random Variable

The **expected value** of a continuous random variable X, denoted $\mathcal{E}(X)$, is defined by

$$\mathcal{E}(X) = \int_{-\infty}^{\infty} x f_X(x)\, dx, \qquad (10.4)$$

where f_X is a probability density function of X.

Note: As you know, the continuous analogue of summation is integration, and the continuous analogue of the probability mass function is the probability density function. These analogues display themselves vividly in the definitions of the expected value of a discrete random variable [Equation (10.1)] and the expected value of a continuous random variable [Equation (10.4)].

The following comments apply to any type of random variables, in particular, to continuous random variables.

- Other commonly used terms for *expected value* are **expectation, mean,** and **first moment.**
- The notation μ_X, as well as $\mathcal{E}(X)$, is used to represent the expected value of a random variable X.
- We interpret the expected value of a random variable as the long-run-average value of the random variable in repeated independent observations—the *long-run-average interpretation of expected value.*

To ensure that the integral on the right of Equation (10.4) makes sense, we require that the integral converges absolutely—that is,

$$\int_{-\infty}^{\infty} |x| f_X(x)\, dx < \infty. \qquad (10.5)$$

If Relation (10.5) holds, the integral on the right of Equation (10.4) converges to a real number. We then say that the random variable X has **finite expectation** and define the expected value of X as in Definition 10.1. However, if Relation (10.5) fails to hold, we say that the random variable X doesn't have finite expectation and don't define the expected value of X. For more on this issue, see Exercise 10.21.

EXAMPLE 10.1   *Expected Value*

Bacteria on a Petri Dish   A petri dish of unit radius, containing nutrients upon which bacteria can multiply, is smeared with a uniform suspension of bacteria. Subsequently, spots indicating colonies of bacteria will appear. Suppose that we observe the location of the center of the first spot to appear. Let Z denote the distance of the center of the first spot from the center of the petri dish. Determine and interpret the expected value of the random variable Z.

Solution   From Example 8.8, a PDF of Z is $f_Z(z) = 2z$ for $0 < z < 1$, and $f_Z(z) = 0$ otherwise. Obviously, Z has finite expectation. Applying Definition 10.1 yields

$$\mathcal{E}(Z) = \int_{-\infty}^{\infty} z f_Z(z)\, dz = \int_0^1 z \cdot 2z\, dz = \int_0^1 2z^2\, dz = 2/3.$$

The expected distance of the center of the first spot from the center of the petri dish is 2/3 unit. On average, the center of the first spot to appear is 2/3 unit from the center of the petri dish.   ■

Expected Value for Some Families of Continuous Random Variables

For many families of continuous random variables, we can obtain formulas for the expected value in terms of the parameters. We do so by substituting the general form of a PDF for a family into the formula defining the expected value of a continuous random variable [i.e., Equation (10.4) on page 565]. In Examples 10.2–10.4, we obtain the expected value for the exponential, normal, and gamma families of random variables.

EXAMPLE 10.2   *Expected Value of an Exponential Random Variable*

Obtain the expected value of an exponential random variable with parameter λ.

Solution   Let $X \sim \mathcal{E}(\lambda)$ and recall that a PDF of X is $f_X(x) = \lambda e^{-\lambda x}$ for $x > 0$, and $f_X(x) = 0$ otherwise. Substituting this PDF into Equation (10.4), we obtain

$$\mathcal{E}(X) = \int_{-\infty}^{\infty} x f_X(x)\, dx = \int_0^{\infty} x \lambda e^{-\lambda x}\, dx = \lambda \int_0^{\infty} x e^{-\lambda x}\, dx.$$

To evaluate the integral on the right, we could use integration by parts. Alternatively, by observing that the "variable form" of the integrand is that of a gamma PDF with parameters 2 and λ and using the fact that a PDF must integrate to 1, we get

$$\mathcal{E}(X) = \lambda \int_0^{\infty} x e^{-\lambda x}\, dx = \lambda \frac{\Gamma(2)}{\lambda^2} \int_0^{\infty} \frac{\lambda^2}{\Gamma(2)} x^{2-1} e^{-\lambda x}\, dx = \lambda \frac{\Gamma(2)}{\lambda^2} \cdot 1 = \frac{1}{\lambda}.$$

Consequently, the expected value of an exponential random variable is the reciprocal of its parameter.   ■

We can apply the result of Example 10.2 to the emergency-room situation in Example 8.12. We recall that, beginning at 6:00 P.M. on any given day, the elapsed time until the first patient arrives has an exponential distribution with parameter $\lambda = 6.9$, where time is measured in hours. Therefore, from Example 10.2, the expected time until the first patient arrives is $\frac{1}{6.9}$ hour or, roughly, 8.7 minutes.

EXAMPLE 10.3   *Expected Value of a Normal Random Variable*

Obtain the expected value of a normal random variable with parameters μ and σ^2.

Solution   Let $X \sim \mathcal{N}(\mu, \sigma^2)$ and recall that a PDF of X is $f_X(x) = (1/\sqrt{2\pi}\sigma)e^{-(x-\mu)^2/2\sigma^2}$ for all $x \in \mathcal{R}$. Substituting this PDF into Equation (10.4) and making the substitution $y = (x - \mu)/\sigma$ gives

$$\mathcal{E}(X) = \int_{-\infty}^{\infty} x f_X(x)\, dx = \int_{-\infty}^{\infty} x \frac{1}{\sqrt{2\pi}\,\sigma} e^{-(x-\mu)^2/2\sigma^2}\, dx$$

$$= \frac{1}{\sqrt{2\pi}\sigma} \int_{-\infty}^{\infty} x e^{-(x-\mu)^2/2\sigma^2}\, dx = \frac{1}{\sqrt{2\pi}} \int_{-\infty}^{\infty} (\mu + \sigma y) e^{-y^2/2}\, dy$$

$$= \mu \int_{-\infty}^{\infty} \frac{1}{\sqrt{2\pi}} e^{-y^2/2}\, dy + \frac{\sigma}{\sqrt{2\pi}} \int_{-\infty}^{\infty} y e^{-y^2/2}\, dy$$

$$= \mu \cdot 1 + \frac{\sigma}{\sqrt{2\pi}} \cdot 0 = \mu,$$

where, in the penultimate equation, we used the facts that the integral of a standard normal PDF is 1 and the integral of an odd function is 0. Thus the expected value of a normal random variable is its first parameter, μ. ∎

We can apply the result of Example 10.3 to the gestation-period illustration in Example 8.14. It states that gestation periods of women are normally distributed with parameters $\mu = 266$ days and $\sigma = 16$ days. Therefore, from Example 10.3, the expected gestation period of a woman is 266 days.

EXAMPLE 10.4   *Expected Value of a Gamma Random Variable*

Obtain the expected value of a gamma random variable with parameters α and λ.

Solution   Let $X \sim \Gamma(\alpha, \lambda)$ and recall that a PDF of X is $f_X(x) = (\lambda^\alpha / \Gamma(\alpha))x^{\alpha-1}e^{-\lambda x}$ for $x > 0$, and $f_X(x) = 0$ otherwise. Substituting this PDF into Equation (10.4), using the fact that the integral of a gamma PDF is 1, and applying Equation (8.43) on page 451 gives

$$\mathcal{E}(X) = \int_{-\infty}^{\infty} x f_X(x)\, dx = \int_{0}^{\infty} x \frac{\lambda^\alpha}{\Gamma(\alpha)} x^{\alpha-1} e^{-\lambda x}\, dx$$

$$= \frac{\lambda^\alpha}{\Gamma(\alpha)} \frac{\Gamma(\alpha+1)}{\lambda^{\alpha+1}} \int_{0}^{\infty} \frac{\lambda^{\alpha+1}}{\Gamma(\alpha+1)} x^{(\alpha+1)-1} e^{-\lambda x}\, dx$$

$$= \frac{\lambda^\alpha}{\Gamma(\alpha)} \frac{\Gamma(\alpha+1)}{\lambda^{\alpha+1}} \cdot 1 = \frac{\alpha}{\lambda}.$$

Thus the expected value of a gamma random variable equals the quotient of its first and second parameters, respectively. ∎

We can apply the result of Example 10.4 to the emergency-room situation in Example 8.16. We recall that, beginning at 6:00 P.M. on any given day, the time until the third patient arrives has a gamma distribution with parameters $\alpha = 3$ and $\lambda = 6.9$, where time is measured in hours. Therefore, from Example 10.4, the expected time until the third patient arrives is $\frac{3}{6.9}$ hour or, roughly, 26.1 minutes.

As we showed in Section 8.6 (see page 456), several important probability distributions are special cases of the gamma distribution. For such distributions, we can apply the result of Example 10.4 to obtain expected values.

- *Exponential distribution.* The exponential distribution with parameter λ is also the gamma distribution with parameters 1 and λ. Hence the expected value of an exponential random variable with parameter λ is $1/\lambda$, a result that we found directly in Example 10.2.

- *Erlang distribution.* The Erlang distribution with parameters r and λ, where r is a positive integer, is also the gamma distribution with those two parameters. So the expected value of an Erlang random variable with parameters r and λ is r/λ.

- *Chi-square distribution.* The chi-square distribution with ν degrees of freedom is also the gamma distribution with parameters $\nu/2$ and $1/2$. So the expected value of a chi-square random variable with ν degrees of freedom is $(\nu/2)/(1/2) = \nu$. Thus the number of degrees of freedom of a chi-square random variable also gives the expected value.

Using arguments similar to those in Examples 10.2–10.4, we can obtain the expected values for several other families of continuous random variables, as shown in Table 10.1.

Table 10.1 Expected values for selected families of continuous random variables

Family	Parameter(s)	Expected value
Uniform	a and b	$(a+b)/2$
Exponential	λ	$1/\lambda$
Normal	μ and σ^2	μ
Gamma	α and λ	α/λ
Erlang	r and λ	r/λ
Chi-square	ν	ν
Beta	α and β	$\alpha/(\alpha+\beta)$
Triangular	a and b	$(a+b)/2$

We conclude this section by showing that a Cauchy random variable doesn't have finite expectation and hence doesn't have a mean.

EXAMPLE 10.5   *Expected Value*

Cauchy Random Variables   Show that a Cauchy random variable doesn't have finite expectation.

Solution We provide a verification for a standard Cauchy random variable and leave the general case for you to do as Exercise 10.15. Assume that X has the standard Cauchy distribution—that is, $f_X(x) = 1/\pi(1 + x^2)$ for $-\infty < x < \infty$. Using the fact that f_X is an even function and making the substitution $u = 1 + x^2$, we get

$$\int_{-\infty}^{\infty} |x| f_X(x)\, dx = 2 \int_{0}^{\infty} x f_X(x)\, dx = 2 \int_{0}^{\infty} \frac{x}{\pi(1 + x^2)}\, dx$$

$$= \frac{1}{\pi} \int_{1}^{\infty} \frac{du}{u} = \frac{1}{\pi} \lim_{n \to \infty} \int_{1}^{n} \frac{du}{u} = \frac{1}{\pi} \lim_{n \to \infty} \ln n = \infty.$$

Thus X doesn't have finite expectation. ∎

EXERCISES 10.1 Basic Exercises

10.1 A point is chosen at random from the interior of a sphere of radius r. On average, how far would you expect the point to be from the center of the sphere? *Note:* In Exercise 8.41 on page 425, we asked for a PDF of the distance from the point chosen to the center of the sphere.

10.2 Let X denote the tangent of an angle chosen at random from the interval $(-\pi/2, \pi/2)$. Does X have finite expectation? *Note:* In Exercise 8.43 on page 425, we asked for a PDF of X.

10.3 A point is chosen at random from the interior of a triangle with base b and height h. Let Y denote the distance from the point chosen to the base of the triangle. Determine and interpret the expected value of Y. *Note:* In Exercise 8.44 on page 426, we asked for a PDF of Y.

10.4 Refer to the petri-dish illustration of Example 10.1 on page 565. Let X and Y denote the x and y coordinates, respectively, of the first spot (visible bacteria colony) to appear.
a) Without doing any computations, guess the value of $\mathcal{E}(X)$. Explain your reasoning.
b) In Example 9.8 on page 511, we obtained a PDF of X. Use that PDF to find $\mathcal{E}(X)$. Compare your answer to that found in part (a).
c) Without doing further computations, find $\mathcal{E}(Y)$. Explain your reasoning.

10.5 A commuter train arrives punctually at a station every half hour. Each morning, a commuter named John leaves his house and casually strolls to the train station. Let X denote the amount of time, in minutes, that John waits for the train from the time he reaches the train station. In Example 8.10 on page 428, we found that a PDF of X is $f_X(x) = 1/30$ for $0 < x < 30$ and $f_X(x) = 0$ otherwise. Use that result and the definition of expected value to obtain and interpret the expected value of the random variable X.

10.6 Suppose that the random variable X has the uniform distribution on the interval (a, b).
a) Show that $\mathcal{E}(X) = (a + b)/2$.
b) Apply your result from part (a) to obtain a quick solution of Exercise 10.5.

10.7 According to the text *Rhythms of Dialogue* by J. Jaffee and S. Feldstein (New York: Academic Press, 1970), the duration, in seconds, of a pause during a monologue has an exponential distribution with parameter 1.4. Determine the expected duration of a pause during a monologue.

10.8 As reported in *Runner's World* magazine, the times of the finishers in the New York City 10 km run are normally distributed with $\mu = 61$ minutes and $\sigma = 9$ minutes. On average, how long does it take a finisher to complete the race?

10.9 Suppose that the random variable X has the beta distribution with parameters α and β. Show that $\mathcal{E}(X) = \alpha/(\alpha + \beta)$.

10.10 For a certain manufactured item, the proportion of the annual production that requires service during the first 5 years of use has the beta distribution with parameters $\alpha = 2$ and $\beta = 3$. Determine and interpret the expected proportion of the annual production of these manufactured items that require service during the first 5 years of use.

10.11 Because the expected value, μ_X, of a random variable X is often considered the "center" of the probability distribution of X, it's tempting to think that $P(X \geq \mu_X) = 1/2$. By using beta random variables with appropriate parameters, provide examples where
a) $P(X \geq \mu_X) = 1/2$.   **b)** $P(X \geq \mu_X) > 1/2$.   **c)** $P(X \geq \mu_X) < 1/2$.

10.12 Suppose that the random variable X has the triangular distribution on the interval (a, b). Show that $\mathcal{E}(X) = (a + b)/2$.

10.13 An insurance company's monthly claims are modeled by a continuous, positive random variable X, whose PDF is proportional to $(1 + x)^{-4}$ for $x > 0$. Determine the company's expected monthly claims.

10.14 A company agrees to accept the highest of four sealed bids on a property that it owns. The four bids are regarded as independent random variables with common CDF given by $F(x) = (1 + \sin \pi x)/2$ for $3/2 \leq x < 5/2$. Which expression represents the expected value of the accepted bid?

a) $\dfrac{\pi}{2} \displaystyle\int_{3/2}^{5/2} x \cos \pi x \, dx$ **b)** $\dfrac{1}{16} \displaystyle\int_{3/2}^{5/2} x(1 + \sin \pi x)^4 \, dx$

c) $\dfrac{\pi}{4} \displaystyle\int_{3/2}^{5/2} \cos \pi x (1 + \sin \pi x)^3 \, dx$ **d)** $\dfrac{\pi}{4} \displaystyle\int_{3/2}^{5/2} x \cos \pi x (1 + \sin \pi x)^3 \, dx$

10.15 Let X have the Cauchy distribution with parameters η and θ. Show that X doesn't have finite expectation.

10.16 Let T denote the interior of the triangle with vertices $(0, 0)$, $(2, 0)$, and $(2, 1)$. Suppose that X and Y have joint PDF given by $f_{X,Y}(x, y) = 2xy$ for $(x, y) \in T$ and $f(x, y) = 0$ otherwise. Determine $\mathcal{E}(X)$ and $\mathcal{E}(Y)$.

10.17 Let T_1 be the time between a car accident and reporting a claim to the insurance company. Let T_2 be the time between the report of the claim and payment of the claim. A joint density function of T_1 and T_2 is constant over the region $0 < t_1 < 6$, $0 < t_2 < 6$, $t_1 + t_2 < 10$, and zero otherwise. Determine the expected time between a car accident and payment of the claim. *Hint:* Use Proposition 9.12 on page 534 to obtain a PDF of the time between a car accident and payment of the claim.

10.18 A number X is chosen at random from the interval $(0, 1)$. Then another number Y is chosen at random from the interval $(0, X)$.
a) Without doing any calculations, guess the value of $\mathcal{E}(Y)$.
b) Calculate $\mathcal{E}(Y)$. *Note:* First obtain a PDF of Y and then apply Definition 10.1 on page 565.

Theory Exercises

10.19 Bounded random variable: A random variable X is said to be *bounded* if there is a positive real number M such that $P(|X| \leq M) = 1$. Prove that a bounded continuous random variable (with a PDF), bounded by M, has finite expectation and that $|\mathcal{E}(X)| \leq M$. *Hint:* First show that you can assume that $f_X(x) = 0$ for $|x| > M$.

10.20 Symmetric PDF: Suppose that a PDF of the random variable X is symmetric about c; that is, $f_X(c + x) = f_X(c - x)$ for all $x \in \mathcal{R}$.
a) Show that, if X has finite expectation, then $\mathcal{E}(X) = c$.
b) Apply part (a) to obtain the expected value of a random variable with each of the following distributions: $\mathcal{U}(a, b)$; $\mathcal{N}(\mu, \sigma^2)$; $\mathcal{T}(a, b)$.
c) Apply part (a) to obtain the expected value of a random variable having a PDF that is an even function (i.e., symmetric about 0).

Advanced Exercises

10.21 Infinite expectation: Let X be a continuous random variable. Define

$$\mathcal{E}(X^+) = \int_0^\infty x f_X(x)\, dx \qquad \text{and} \qquad \mathcal{E}(X^-) = -\int_{-\infty}^0 x f_X(x)\, dx.$$

a) Explain why the integral defining $\mathcal{E}(X^+)$ is either a real number or equals ∞. In the latter case, we write $\mathcal{E}(X^+) = \infty$.
b) Explain why the integral (without the minus sign) defining $\mathcal{E}(X^-)$ is either a real number or equals $-\infty$. In the latter case, we write $\mathcal{E}(X^-) = \infty$.
c) Prove that X has finite expectation if and only if both $\mathcal{E}(X^+)$ and $\mathcal{E}(X^-)$ are real numbers, in which case,

$$\mathcal{E}(X) = \mathcal{E}(X^+) - \mathcal{E}(X^-). \tag{$*$}$$

Suppose now that X doesn't have finite expectation, but that only one of $\mathcal{E}(X^+)$ and $\mathcal{E}(X^-)$ equals ∞. Then the expression on the right side of Equation ($*$) makes sense and equals either ∞ or $-\infty$. In this case, we say that X has *infinite expectation* and define the expected value of X by using Equation ($*$).
d) Construct a continuous random variable with infinite expectation.
e) In Example 10.5 on page 568, we showed that a standard Cauchy random variable doesn't have finite expectation. Does it have infinite expectation? Explain your answer.

10.22 Student's *t*-distribution: Let T have the Student's t-distribution with ν degrees of freedom. According to Exercise 9.164 on page 553, a PDF of T is given by

$$f_T(t) = \frac{\Gamma\left((\nu + 1)/2\right)}{\sqrt{\nu\pi}\,\Gamma(\nu/2)} \left(1 + t^2/\nu\right)^{-(\nu+1)/2}, \qquad \infty < t < \infty.$$

a) Determine when T has finite expectation.
b) Obtain $\mathcal{E}(T)$ when T has finite expectation.

10.2   Basic Properties of Expected Value

In this section, we examine the basic properties of expected value for continuous random variables. Our first result—the continuous version of the **fundamental expected-value formula**—supplies the foundation for obtaining such properties and also provides a useful way for finding the expected value of a real-valued function of one or more continuous random variables.

Before presenting the fundamental expected-value formula, we comment on the concept of *finite expectation*. We already defined it for discrete random variables—Relation (7.6) on page 327—and continuous random variables—Relation (10.5) on page 565. In fact, using advanced mathematics, we can define the concept of finite expectation for any type of random variable.

We now present the fundamental expected-value formula (FEF) for continuous random variables. Note the analogy with the discrete version of the FEF provided in Proposition 7.2 on page 339.

◆◆◆ **Proposition 10.1   Fundamental Expected-Value Formula: Continuous Case**

Let $X_1, \ldots, X_m$ be continuous random variables with a joint PDF and let g be a real-valued function of m variables defined on the range of $(X_1, \ldots, X_m)$. Then $g(X_1, \ldots, X_m)$ has finite expectation if and only if

$$\int_{-\infty}^{\infty} \cdots \int_{-\infty}^{\infty} |g(x_1, \ldots, x_m)|\, f_{X_1, \ldots, X_m}(x_1, \ldots, x_m)\, dx_1 \cdots dx_m < \infty. \qquad \textbf{(10.6)}$$

In that case,

$$\begin{aligned}
\mathcal{E}\big(g(X_1, \ldots, X_m)\big) \\
= \int_{-\infty}^{\infty} \cdots \int_{-\infty}^{\infty} g(x_1, \ldots, x_m)\, f_{X_1, \ldots, X_m}(x_1, \ldots, x_m)\, dx_1 \cdots dx_m.
\end{aligned} \qquad \textbf{(10.7)}$$

In the univariate and bivariate cases, Equation (10.7) reduces to

$$\mathcal{E}\big(g(X)\big) = \int_{-\infty}^{\infty} g(x) f_X(x)\, dx \qquad \textbf{(10.8)}$$

and

$$\mathcal{E}\big(g(X, Y)\big) = \int_{-\infty}^{\infty} \int_{-\infty}^{\infty} g(x, y) f_{X,Y}(x, y)\, dx\, dy, \qquad \textbf{(10.9)}$$

respectively.

Proof   Without the tools of measure theory, we cannot give a complete proof of this proposition. Nonetheless, we can and do provide the basic idea of the proof. We concentrate on the verification of Equation (10.7).

From results in measure theory, we need only verify that Equation (10.7) holds when g is any finite linear combination of indicators of sets in $\mathcal{R}^m$. The reason is that any real-valued (measurable) function on $\mathcal{R}^m$ can be approximated arbitrarily closely by such functions.

Let's first assume that g is the indicator of a set in $\mathcal{R}^m$—say, $g = I_A$. We observe that $g(X_1, \ldots, X_m) = I_A(X_1, \ldots, X_m) = I_{\{(X_1, \ldots, X_m) \in A\}}$. Therefore, by Equation (7.9) on page 331 and the FPF for continuous random variables,

$$\mathcal{E}\big(g(X_1, \ldots, X_m)\big)$$
$$= \mathcal{E}\big(I_{\{(X_1,\ldots,X_m)\in A\}}\big) = P\big(X_1, \ldots, X_m\big) \in A\big)$$
$$= \int \cdots \int_A f_{X_1,\ldots,X_m}(x_1, \ldots, x_m) \, dx_1 \cdots dx_m$$
$$= \int_{-\infty}^{\infty} \cdots \int_{-\infty}^{\infty} I_A(x_1, \ldots, x_m) \, f_{X_1,\ldots,X_m}(x_1, \ldots, x_m) \, dx_1 \cdots dx_m$$
$$= \int_{-\infty}^{\infty} \cdots \int_{-\infty}^{\infty} g(x_1, \ldots, x_m) \, f_{X_1,\ldots,X_m}(x_1, \ldots, x_m) \, dx_1 \cdots dx_m.$$

Hence Equation (10.7) holds when g is the indicator of a set in $\mathcal{R}^m$.

We still need to show that Equation (10.7) holds when g is a finite linear combination of indicators of sets in $\mathcal{R}^m$. Let $g = \sum_{j=1}^{n} c_j g_j$, where n is a positive integer, $c_1, \ldots, c_n$ are real numbers, and each g_j is the indicator of a set in $\mathcal{R}^m$. Each random variable $g_j(X_1, \ldots, X_m)$ is discrete; in fact, it is an indicator random variable. Applying the linearity property of expected value for discrete random variables given in Equation (7.19) on page 343, the fact that Equation (10.7) holds for indicators, and the linearity property of multiple integrals, we conclude that

$$\mathcal{E}\big(g(X_1, \ldots, X_m)\big)$$
$$= \mathcal{E}\left(\sum_{j=1}^{n} c_j g_j(X_1, \ldots, X_m)\right) = \sum_{j=1}^{n} c_j \mathcal{E}\big(g_j(X_1, \ldots, X_m)\big)$$
$$= \sum_{j=1}^{n} c_j \int_{-\infty}^{\infty} \cdots \int_{-\infty}^{\infty} g_j(x_1, \ldots, x_m) \, f_{X_1,\ldots,X_m}(x_1, \ldots, x_m) \, dx_1 \cdots dx_m$$
$$= \int_{-\infty}^{\infty} \cdots \int_{-\infty}^{\infty} \left(\sum_{j=1}^{n} c_j g_j(x_1, \ldots, x_m)\right) f_{X_1,\ldots,X_m}(x_1, \ldots, x_m) \, dx_1 \cdots dx_m$$
$$= \int_{-\infty}^{\infty} \cdots \int_{-\infty}^{\infty} g(x_1, \ldots, x_m) \, f_{X_1,\ldots,X_m}(x_1, \ldots, x_m) \, dx_1 \cdots dx_m,$$

as required.   ◆

EXAMPLE 10.6   *The Fundamental Expected-Value Formula*

Group Health Insurance   A group health insurance policy of a small business pays 100% of employee medical bills up to a maximum of $1 million per policy year. The total annual medical bills, X, in millions of dollars, incurred by the employees has PDF given by $f_X(x) = x(4 - x)/9$ for $0 < x < 3$, and $f_X(x) = 0$ otherwise. Determine the expected annual payout by the insurance company.

Solution   The annual payout, in millions of dollars, by the insurance company is $Y = \min\{X, 1\}$. To obtain the expected value of Y, we apply the FEF—specifically, Equation (10.8) on page 572, with $g(x) = \min\{x, 1\}$:

$$\mathcal{E}(Y) = \mathcal{E}\big(g(X)\big) = \int_{-\infty}^{\infty} g(x) f_X(x)\, dx = \int_0^3 \min\{x, 1\}\, \frac{x(4-x)}{9}\, dx$$

$$= \frac{1}{9} \int_0^1 x \cdot x(4-x)\, dx + \frac{1}{9} \int_1^3 1 \cdot x(4-x)\, dx$$

$$= \frac{1}{9} \int_0^1 (4x^2 - x^3)\, dx + \frac{1}{9} \int_1^3 (4x - x^2)\, dx = \frac{101}{108} = 0.935.$$

On average, the insurance company pays $0.935 million per policy year.   ∎

In Example 10.7, we compare an expected value determination with and without use of the fundamental expected-value formula.

EXAMPLE 10.7   *The Fundamental Expected-Value Formula*

Speed of Gas Molecules   The three velocity components of a gas molecule are independent and identically distributed normal random variables with common parameters 0 and σ^2. Determine the expected speed of a gas molecule
a) directly from the definition of expected value, Definition 10.1 on page 565.
b) by applying the FEF, Proposition 10.1.
c) Compare the methods used in parts (a) and (b).

Solution   The speed of a gas molecule is $S = \sqrt{X^2 + Y^2 + Z^2}$, where X, Y, and Z denote the velocity components, which, by assumption, are independent normal random variables, each having parameters 0 and σ^2. The task is to determine the expected value of S.
a) From Example 9.18 on page 533, a PDF of S is $f_S(s) = \big(\sqrt{2/\pi}/\sigma^3\big) s^2 e^{-s^2/2\sigma^2}$ for $s > 0$, and $f_S(s) = 0$ otherwise. Applying Definition 10.1, making the substitution $u = s^2/2\sigma^2$, and using the fact that a gamma PDF integrates to 1, we get

$$\mathcal{E}(S) = \int_{-\infty}^{\infty} s f_S(s)\, ds = \int_0^{\infty} s \frac{\sqrt{2/\pi}}{\sigma^3} s^2 e^{-s^2/2\sigma^2}\, ds$$

$$= \frac{\sqrt{2/\pi}}{\sigma^3} \int_0^{\infty} s^3 e^{-s^2/2\sigma^2}\, ds = \frac{\sqrt{2/\pi}}{\sigma} \int_0^{\infty} 2\sigma^2 u e^{-u}\, du \qquad \textbf{(10.10)}$$

$$= 2\sqrt{2/\pi}\, \sigma \frac{\Gamma(2)}{1^2} \int_0^{\infty} \frac{1^2}{\Gamma(2)} u^{2-1} e^{-u}\, du = 2\sqrt{2/\pi}\, \sigma.$$

The expected speed of a gas molecule is $2\sqrt{2/\pi}\, \sigma$.

b) To use the FEF to find $\mathcal{E}(S)$, we first note that a joint PDF of X, Y, and Z is

$$f_{X,Y,Z}(x, y, z) = f_X(x)f_Y(y)f_Z(z)$$

$$= \frac{1}{\sqrt{2\pi}\,\sigma} e^{-x^2/2\sigma^2} \frac{1}{\sqrt{2\pi}\,\sigma} e^{-y^2/2\sigma^2} \frac{1}{\sqrt{2\pi}\,\sigma} e^{-z^2/2\sigma^2}$$

$$= \frac{1}{(2\pi)^{3/2}\sigma^3} e^{-(x^2+y^2+z^2)/2\sigma^2},$$

for all x, y, $z \in \mathcal{R}$. Applying the FEF for continuous random variables and changing to spherical coordinates, we get

$$\mathcal{E}(S) = \mathcal{E}\left(\sqrt{X^2 + Y^2 + Z^2}\right)$$

$$= \int_{-\infty}^{\infty} \int_{-\infty}^{\infty} \int_{-\infty}^{\infty} \sqrt{x^2 + y^2 + z^2}\, f_{X,Y,Z}(x, y, z)\, dx\, dy\, dz$$

$$= \int_{-\infty}^{\infty} \int_{-\infty}^{\infty} \int_{-\infty}^{\infty} \sqrt{x^2 + y^2 + z^2}\, \frac{1}{(2\pi)^{3/2}\sigma^3} e^{-(x^2+y^2+z^2)/2\sigma^2}\, dx\, dy\, dz$$

$$= \frac{1}{(2\pi)^{3/2}\sigma^3} \int_0^{\infty} \int_0^{2\pi} \int_0^{\pi} \rho e^{-\rho^2/2\sigma^2} \rho^2 \sin\phi\, d\rho\, d\theta\, d\phi$$

$$= \frac{1}{(2\pi)^{3/2}\sigma^3} \int_0^{\infty} \rho^3 e^{-\rho^2/2\sigma^2} \left(\int_0^{\pi} \sin\phi \left(\int_0^{2\pi} d\theta\right) d\phi\right) d\rho$$

$$= \frac{2 \cdot 2\pi}{(2\pi)^{3/2}\sigma^3} \int_0^{\infty} \rho^3 e^{-\rho^2/2\sigma^2}\, d\rho = \frac{\sqrt{2/\pi}}{\sigma^3} \int_0^{\infty} \rho^3 e^{-\rho^2/2\sigma^2}\, d\rho.$$

This last integral is the same as the first one in the second row of Equation (10.10) on the previous page. Therefore, $\mathcal{E}(S) = 2\sqrt{2/\pi}\,\sigma$. As in part (a), we see that the expected speed of a gas molecule is $2\sqrt{2/\pi}\,\sigma$.

c) The method used in part (a) to find $\mathcal{E}(S)$ seems easier than that in part (b). However, keep in mind the work that was required to determine a PDF of S, as shown in Example 9.18. ∎

Basic Properties of Expected Value

We can now establish some of the basic and most widely used properties of expected value for continuous random variables. Actually, in Propositions 10.2–10.4, we provide only the statements of these properties. We leave the proof of Proposition 10.3 to you as Exercise 10.44. You can obtain the proofs of Propositions 10.2 and 10.4 by mimicking those in Section 7.2 for discrete random variables; simply replace sums by integrals and PMFs by PDFs. And, as we mentioned there, *these three propositions hold for all types of random variables, not just for discrete and continuous random variables.*

◆◆◆ **Proposition 10.2   Linearity Property of Expected Value**

Let X and Y be continuous random variables with a joint PDF and finite expectation. Also, let c be a real number. Then the following relations hold.
a) The random variable $X + Y$ has finite expectation and

$$\mathcal{E}(X + Y) = \mathcal{E}(X) + \mathcal{E}(Y).\tag{10.11}$$

In words, the expected value of the sum of two random variables equals the sum of their expected values.
b) The random variable cX has finite expectation and

$$\mathcal{E}(cX) = c\,\mathcal{E}(X).\tag{10.12}$$

In words, the expected value of a constant times a random variable equals the constant times the expected value of the random variable.

The two properties of expected value given in Proposition 10.2 are collectively referred to as the **linearity property.** Together, they are equivalent to the property that

$$\mathcal{E}(aX + bY) = a\,\mathcal{E}(X) + b\,\mathcal{E}(Y),\tag{10.13}$$

for any two real numbers a and b. Also, a simple induction argument shows that we can extend the linearity property to any finite number of continuous random variables:

$$\mathcal{E}\left(\sum_{j=1}^{m} c_j X_j\right) = \sum_{j=1}^{m} c_j\,\mathcal{E}(X_j),\tag{10.14}$$

where $X_1, \ldots, X_m$ are continuous random variables with a joint PDF and finite expectation, and $c_1, \ldots, c_m$ are real numbers.

◆◆◆ **Proposition 10.3**

Let X and Y be continuous random variables with a joint PDF and finite expectation.
a) If a and b are real numbers, then $a + bX$ has finite expectation and

$$\mathcal{E}(a + bX) = a + b\,\mathcal{E}(X).\tag{10.15}$$

b) If $X \leq Y$, then $\mathcal{E}(X) \leq \mathcal{E}(Y)$.

As you have seen, the expected value of the sum of two or more continuous random variables equals the sum of their expected values. However, the expected value of the product of two or more continuous random variables generally doesn't equal the product of their expected values. (See, for instance, Exercise 10.36.) Nonetheless, for independent continuous random variables, that property is true.

◆◆◆ **Proposition 10.4**

Let X and Y be independent continuous random variables having finite expectation. Then XY has finite expectation and

$$\mathcal{E}(XY) = \mathcal{E}(X)\,\mathcal{E}(Y).\tag{10.16}$$

In words, for independent random variables, the expected value of their product equals the product of their expected values.

Although independence is a sufficient condition for the expected value of a product of random variables to equal the product of their expected values, it isn't necessary. In other words, it's possible to have $\mathcal{E}(XY) = \mathcal{E}(X)\mathcal{E}(Y)$ for dependent random variables. See, for instance, Exercise 10.37.

We can use either mathematical induction or a direct argument to show that Proposition 10.4 holds for any finite number of independent continuous random variables. Specifically, if $X_1, \ldots, X_m$ are independent continuous random variables having finite expectation, then $\prod_{j=1}^{m} X_j$ has finite expectation and

$$\mathcal{E}\left(\prod_{j=1}^{m} X_j\right) = \prod_{j=1}^{m} \mathcal{E}(X_j). \tag{10.17}$$

Using Tail Probabilities to Obtain Expected Values

Many of the continuous random variables that we have considered and many that occur in practice are nonnegative. For such a random variable, computing its expected value by using tail probabilities is sometimes easier. Proposition 10.5 provides the formula.

◆◆◆ **Proposition 10.5   Using Tail Probabilities to Obtain Expected Value**

Suppose that X is a nonnegative continuous random variable. Then X has finite expectation if and only if $\int_0^\infty P(X > x)\,dx < \infty$, in which case,

$$\mathcal{E}(X) = \int_0^\infty P(X > x)\,dx. \tag{10.18}$$

Proof   We first note that

$$\int_{-\infty}^{\infty} x f_X(x)\,dx = \int_0^\infty x f_X(x)\,dx = \int_0^\infty \left(\int_0^x 1\,dy\right) f_X(x)\,dx.$$

Figure 10.1 shows the set over which the double integral is taken.

Figure 10.1

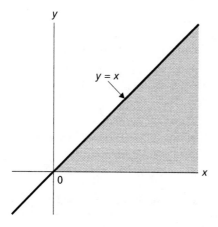

Referring to Figure 10.1, interchanging the order of integration, and applying the FPF, we get

$$\int_0^\infty \left(\int_0^x 1 \, dy \right) f_X(x) \, dx = \int_0^\infty \left(\int_y^\infty f_X(x) \, dx \right) dy = \int_0^\infty P(X > y) \, dy.$$

The required result now follows easily.   ◆

Note the following.

- Proposition 10.5 holds for any type of random variable—discrete, continuous, or otherwise.
- A version of Proposition 10.5 exists for random variables without the restriction of nonnegativity. See Exercise 10.45.

EXAMPLE 10.8   *Using Tail Probabilities to Obtain Expected Values*

Exponential Random Variables   Use tail probabilities to obtain the expected value of an exponential random variable with parameter λ.

Solution   Let $X \sim \mathcal{E}(\lambda)$. From Proposition 8.8 on page 433, $P(X > x) = e^{-\lambda x}$ for $x > 0$. Thus, by Proposition 10.5,

$$\mathcal{E}(X) = \int_0^\infty P(X > x) \, dx = \int_0^\infty e^{-\lambda x} \, dx = \frac{1}{\lambda}.$$

Compare this simple computation using tail probabilities to a direct computation using the definition of expected value, as required by Example 10.2 on page 566.   ■

EXERCISES 10.2   **Basic Exercises**

10.23 The speed of a gas molecule is $S = \sqrt{X^2 + Y^2 + Z^2}$, where X, Y, and Z denote the velocity components, which are independent $\mathcal{N}(0, \sigma^2)$ random variables. In Example 10.7 on page 574, we presented two ways to determine the expected value of S. Solve that problem a third way as follows:
a) Without doing any computations, explain why $W = (X^2 + Y^2 + Z^2)/\sigma^2$ has the chi-square distribution with three degrees of freedom.
b) Use your result from part (a) and the FEF to obtain $\mathcal{E}(S)$.

10.24 Let Z have the standard normal distribution. Obtain the expected value of $|Z|$
a) by first obtaining a PDF of $|Z|$ and then applying the definition of expected value.
b) by using the FEF.

10.25 Let T denote the interior of the triangle with vertices $(0, 0)$, $(2, 0)$, and $(2, 1)$. Let X and Y have joint PDF given by $f_{X,Y}(x, y) = 2xy$ for $(x, y) \in T$, and $f(x, y) = 0$ otherwise. In Exercise 10.16 on page 570, we asked for $\mathcal{E}(X)$ and $\mathcal{E}(Y)$, requiring you first to obtain marginal PDFs of X and Y. Apply the FEF to determine $\mathcal{E}(X)$ and $\mathcal{E}(Y)$ without first obtaining marginal PDFs of X and Y.

10.26 Let T_1 be the time between a car accident and reporting a claim to the insurance company. Let T_2 be the time between the report of the claim and payment of the claim.

A joint density function of T_1 and T_2 is constant over the region $0 < t_1 < 6$, $0 < t_2 < 6$, $t_1 + t_2 < 10$, and zero otherwise. In Exercise 10.17 on page 570, we asked for the expected time between a car accident and payment of the claim, $\mathcal{E}(T_1 + T_2)$, requiring you first to obtain a PDF of $T_1 + T_2$. Determine $\mathcal{E}(T_1 + T_2)$ without first obtaining a PDF of $T_1 + T_2$.

10.27 An insurance policy reimburses a loss up to a benefit limit of 10. The policyholder's loss follows a distribution with density function $f(y) = 2/y^3$ for $y > 1$, and $f(y) = 0$ otherwise. What is the expected value of the benefit paid under the insurance policy?

10.28 Let $X \sim \mathcal{U}(1, 2)$.
a) Determine $\mathcal{E}(1/X)$ by first obtaining a PDF of $1/X$.
b) Determine $\mathcal{E}(1/X)$ by using the FEF.
c) Compare $\mathcal{E}(1/X)$ and $1/\mathcal{E}(X)$. Conclude that, in general, $\mathcal{E}(1/X) \neq 1/\mathcal{E}(X)$.

10.29 A piece of equipment, whose lifetime is exponentially distributed with mean 10 years, is being insured against failure. The insurance will pay \$$x$ if failure occurs during the first year, \$$0.5x$ if failure occurs during the second or third year, and nothing if failure occurs after the first 3 years. At what level must x be set if the expected payment made under this insurance is to be \$1000?

10.30 Let $X_1, \ldots, X_n$ be a random sample from a uniform distribution on the interval $(0, 1)$. Set $X = \min\{X_1, \ldots, X_n\}$, $Y = \max\{X_1, \ldots, X_n\}$, $R = Y - X$ (the range of the random sample), and $M = (X + Y)/2$ (the midrange of the random sample). In Example 9.4(c) on page 497, we obtained a joint PDF of X and Y. Use that joint PDF and the FEF to determine
a) $\mathcal{E}(X)$.   **b)** $\mathcal{E}(Y)$.   **c)** $\mathcal{E}(R)$.   **d)** $\mathcal{E}(M)$.

10.31 Refer to Exercise 10.30.
a) In Example 9.9(c) on page 512, we showed that X has the beta distribution with parameters 1 and n. Use that fact to obtain $\mathcal{E}(X)$.
b) In Exercise 9.70(c) on page 520, we asked you to obtain the marginal distribution of Y, which is the beta distribution with parameters n and 1. Use that fact to obtain $\mathcal{E}(Y)$.
c) Use properties of expected value and your results from parts (a) and (b) to obtain $\mathcal{E}(R)$.
d) Use properties of expected value and your results from parts (a) and (b) to obtain $\mathcal{E}(M)$.

10.32 Let X and Y be random losses, in thousands of dollars, with joint density function $f_{X,Y}(x, y) = 2x$ for $0 < x < 1$ and $0 < y < 1$, and $f_{X,Y}(x, y) = 0$ otherwise. An insurance policy with a deductible of 1 is written to cover the loss $X + Y$. What is the expected payment under this policy?

10.33 For purposes of quality assurance, an expensive item produced at a manufacturing plant is independently inspected by two engineers.
a) Assume that the amount of time it takes each engineer to inspect the item has the exponential distribution with parameter λ. Use the result of Example 9.25(b) on page 547 to show that, on average, 50% of the total inspection time is attributed to the first engineer.
b) Now assume only that the probability distribution of the amount of time it takes to inspect the item is the same for both engineers. Show that the result obtained in part (a) of this problem still holds.

10.34 A stick of length ℓ is randomly broken into two pieces.
a) Determine and identify a PDF of the length, X, of the shorter piece.
b) Use your result from part (a) and the FEF to find the expected value of the ratio, R, of the shorter piece to the longer piece.
c) Use your result from part (a) to obtain a PDF of R.

d) Use your result from part (c) and the definition of expected value to find the expected value of the ratio, R, of the shorter piece to the longer piece.

e) Use your result from part (a) and the FEF to show that the ratio of the longer piece to the shorter piece doesn't have finite expectation.

10.35 Simulation: Refer to Exercise 10.34.

a) Use a basic random number generator and the long-run-average interpretation of expected value to estimate the expected value of the ratio, R, of the shorter piece to the longer piece. *Hint:* The distribution of R doesn't depend on the value of ℓ.

b) Compare your estimate from part (a) with the exact value obtained in Exercise 10.34.

10.36 Let X and Y be uniform on the interior of the triangle with vertices at $(0, 0)$, $(1, 0)$, and $(1, 1)$.

a) Show that $\mathcal{E}(XY) \neq \mathcal{E}(X)\mathcal{E}(Y)$.

b) Can we conclude from part (a) that X and Y aren't independent random variables? Justify your answer.

10.37 Let X and Y be uniform on the interior of the triangle with vertices at $(0, 0)$, $(1, -1)$, and $(1, 1)$.

a) Show that $\mathcal{E}(XY) = \mathcal{E}(X)\mathcal{E}(Y)$.

b) Can we conclude from part (a) that X and Y are independent random variables? Justify your answer.

c) Are X and Y independent random variables? Justify your answer.

10.38 Claim amounts for wind damage to insured homes are independent random variables with common density function $f(x) = 3/x^4$ for $x > 1$ and $f(x) = 0$ otherwise, where x is the amount of a claim in thousands. Suppose that three such claims will be made. Determine the expected value of the largest of the three claims

a) by first finding a PDF of the maximum of the three claim amounts and then applying the definition of expected value.

b) by using the FEF and a joint PDF of the three claim amounts.

10.39 Two people agree to meet at a specified place at a specified time. The times that the two people actually arrive are independent and identically distributed random variables with PDF f. Show that the expected amount of time that the first person to arrive waits for the second person to arrive equals $2 \int_{-\infty}^{\infty} f(y) \left(\int_{-\infty}^{y} (y - x) f(x)\, dx \right) dy$ by using

a) a joint PDF of the two arrival times.

b) a joint PDF of the arrival times of the first and second persons to arrive. *Hint:* Refer to Example 9.4 on page 497.

c) marginal PDFs of the arrival times of the first and second persons to arrive.

d) a PDF of the amount of time that the first person to arrive waits for the second person to arrive. *Hint:* Refer to Example 9.6 on page 504.

10.40 Refer to Exercise 10.39. Determine the expected amount of time that the first person to arrive waits for the second person to arrive in case the common distribution of the two arrival times is

a) uniform on the interval from 0 to 2ℓ.

b) triangular on the interval from 0 to 2ℓ.

10.41 Refer to Exercise 10.39. Suppose that the common distribution of the two arrival times is exponentially distributed with mean ℓ.

a) Without doing any computations, identify the probability distribution of the amount of time that the first person to arrive waits for the second person to arrive.

b) Use part (a) to determine the expected amount of time that the first person to arrive waits for the second person to arrive.

c) Use Exercise 10.39 to determine the expected amount of time that the first person to arrive waits for the second person to arrive.

d) Compare your answers in parts (b) and (c).

10.42 Let X be a nonnegative random variable that represents the time to failure of some device. In a planned replacement policy, the device is replaced at a predetermined time t_r or upon failure, whichever comes first.

a) Express the expected time at which the device is replaced in terms of F_X and t_r.

b) Suppose that it costs c_1 to replace a device at time t_r and that it costs c_2 to replace it upon failure. Express the expected cost of replacement in terms of c_1, c_2, F_X, and t_r.

c) Apply your result from part (b) to obtain the expected cost of replacement when X has the exponential distribution with mean 1000 hours, $t_r = 950$ hours, $c_1 = \$50$, and $c_2 = \$80$.

d) Apply your result from part (b) to obtain the expected cost of replacement when X has mean 1000 hours but fails uniformly between 900 hours and 1100 hours. Assume the same values for t_r, c_1, and c_2 as in part (c).

10.43 A certain type of electronic component has probability p of failing immediately when turned on. If a component doesn't fail immediately, it's life distribution has CDF F. Let X denote the time to failure of a component.

a) Determine a CDF of the random variable X in terms of p and F.

b) Use your result from part (a) to express the expected lifetime of a component in terms of p and F.

c) Suppose that the life distribution of a component that doesn't fail immediately is exponential with parameter λ. Determine the expected lifetime of such a component.

Theory Exercises

10.44 Prove Proposition 10.3 on page 576. *Hint:* For part (b), first show that you can assume without loss of generality that $f_{Y-X}(z) = 0$ for $z < 0$.

10.45 Let X be a continuous random variable with a PDF.

a) Generalize Proposition 10.5 on page 577 by proving that X has finite expectation if and only if $\int_0^\infty P(|X| > x)\,dx < \infty$, in which case

$$\mathcal{E}(X) = \int_0^\infty P(X > x)\,dx - \int_0^\infty P(X < -x)\,dx.$$

Note: This result holds for any random variable, regardless of type.

b) Use part (a) to obtain the expected value of a $\mathcal{U}(-1, 3)$ random variable.

10.46 Symmetric random variable: A random variable X is said to be *symmetric* if X and $-X$ have the same probability distribution.

a) Show that, if X is a symmetric random variable with finite expectation, then $\mathcal{E}(X) = 0$.

b) A random variable X is said to be *symmetric about* c if $X - c$ is a symmetric random variable. Show that, if X is symmetric about c and has finite expectation, then $\mathcal{E}(X) = c$.

c) Show that, if X has a PDF f_X that is symmetric about c—that is, $f_X(c + x) = f_X(c - x)$ for all $x \in \mathcal{R}$—then X is symmetric about c.

d) Use parts (b) and (c) to obtain the mean of a random variable with each of the following probability distributions: $\mathcal{U}(a, b)$; $\mathcal{N}(\mu, \sigma^2)$; and $\mathcal{T}(a, b)$.

Advanced Exercises

Order statistics: Let $X_1, \ldots, X_n$ be a random sample from a continuous distribution with CDF F and PDF f. For each k between 1 and n, inclusive, the *kth order statistic*, denoted $X_{(k)}$, is defined to be the kth smallest of $X_1, \ldots, X_n$. Collectively, the random variables $X_{(1)}, \ldots, X_{(n)}$ are called the *order statistics* corresponding to $X_1, \ldots, X_n$. In Exercises 10.47–10.50, we ask you to examine and apply properties of the means of order statistics for uniform and exponential distributions.

10.47 Determine the mean of the kth order statistic for a random sample of size n from a uniform distribution on the interval $(0, 1)$. *Note:* In Exercise 9.33(d) on page 501, we asked for a PDF of the kth order statistic in this case.

10.48 Consider a random sample of size n from an exponential distribution with parameter λ.
a) Show that $\mathcal{E}(X_{(1)}) = 1/n\lambda$.
b) Show that, for $2 \leq j \leq n$, the random variable $X_{(j)} - X_{(j-1)}$ has the exponential distribution with parameter $(n - j + 1)\lambda$. *Note:* In Exercise 9.35(c) on page 501, we asked for a joint PDF of the jth and kth order statistics in this case.
c) Use parts (a) and (b) to show that $\mathcal{E}(X_{(k)}) = (1/\lambda) \sum_{j=n-k+1}^{n} 1/j$ for $1 \leq k \leq n$. *Hint:* Write $X_{(k)}$ as a telescoping sum.

10.49 Mechanical or electrical units often consist of several components, each of which is subject to failure. Consider a system of n components, $C_1, \ldots, C_n$, with respective lifetimes $X_1, \ldots, X_n$, where $X_1, \ldots, X_n$ are independent exponential random variables with common parameter λ. Refer to Exercise 10.48 and solve the following problems.
a) A unit is said to be a *series system* if it functions only when all the components are working. Determine the expected lifetime of the unit if it is a series system.
b) A unit is said to be a *parallel system* if it functions when at least one of the components is working. Determine the expected lifetime of the unit if it is a parallel system.
c) If the unit is a parallel system, show that, for large n, its expected lifetime behaves like $\lambda^{-1} \ln n$.

10.50 By considering an appropriate parallel system of n components and using the lack-of-memory property of the exponential distribution, obtain the result of Exercise 10.48(b) without doing any computations.

10.3   Variance, Covariance, and Correlation

We next discuss the *variance, covariance,* and *correlation* of continuous random variables. These quantities are all defined in terms of expected value in the same way as for the discrete case. Consequently, we need only summarize the fundamental definitions and results. The proofs are, for the most part, identical to those in the discrete case. Again, *these definitions and results hold for all types of random variables, not just for discrete and continuous random variables.*

To begin, review the definitions of *moments* and *central moments* given at the beginning of Section 7.3 on pages 352 and 353. In Equations (7.28) and (7.29) on page 353, we provided formulas for the rth moment and rth central moment of a discrete random variable. They are a consequence of the FEF for discrete random variables. Similarly, by applying the FEF for continuous random variables, we obtain the following formulas for the rth moment and rth central moment, respectively, of a continuous random variable:

$$\mathcal{E}(X^r) = \int_{-\infty}^{\infty} x^r f_X(x)\, dx \tag{10.19}$$

and

$$\mathcal{E}((X - \mu_X)^r) = \int_{-\infty}^{\infty} (x - \mu_X)^r f_X(x)\, dx. \tag{10.20}$$

The most important moment is the first moment, which is just the expected value, or the mean. The most important central moment is the second central moment, or the variance. Thus the **variance** of a random variable X, denoted $\mathrm{Var}(X)$ or σ_X^2, is defined by

$$\mathrm{Var}(X) = \mathcal{E}((X - \mu_X)^2). \tag{10.21}$$

As we demonstrated in Proposition 7.8 on page 355, the variance can also be obtained by using the *computing formula*

$$\mathrm{Var}(X) = \mathcal{E}(X^2) - (\mathcal{E}(X))^2. \tag{10.22}$$

Also recall that the square root of the variance of a random variable X is called the **standard deviation** of X and is denoted σ_X.

EXAMPLE 10.9   *The Variance of a Random Variable*

Bacteria on a Petri Dish   Refer to Example 10.1 on page 565, where Z denotes the distance of the center of the first spot (visible bacteria colony) from the center of the petri dish. Obtain the variance of the random variable Z by using the
a) defining formula, Equation (10.21).
b) computing formula, Equation (10.22).

Solution   From Example 8.8, a PDF of the random variable Z is $f_Z(z) = 2z$ for $0 < z < 1$, and $f_Z(z) = 0$ otherwise. In Example 10.1, we found that $\mathcal{E}(Z) = 2/3$.
a) Applying Equation (10.21) and the FEF, we obtain

$$\mathrm{Var}(Z) = \mathcal{E}((Z - \mu_Z)^2) = \int_{-\infty}^{\infty} (z - \mu_Z)^2 f_Z(z)\, dz$$

$$= \int_0^1 (z - 2/3)^2 \cdot 2z\, dz = 2 \int_0^1 \left(z^3 - \frac{4}{3} z^2 + \frac{4}{9} z \right) dz = \frac{1}{18}.$$

b) To use the computing formula for the variance, we first obtain the second moment of Z. Applying the FEF, we get

$$\mathcal{E}(Z^2) = \int_{-\infty}^{\infty} z^2 f_Z(z)\, dz = \int_0^1 z^2 \cdot 2z\, dz = 2 \int_0^1 z^3\, dz = \frac{1}{2}.$$

As $\mathcal{E}(Z) = 2/3$, we apply Equation (10.22) and obtain

$$\mathrm{Var}(Z) = \mathcal{E}(Z^2) - (\mathcal{E}(Z))^2 = \frac{1}{2} - \left(\frac{2}{3} \right)^2 = \frac{1}{2} - \frac{4}{9} = \frac{1}{18}. \qquad \blacksquare$$

EXAMPLE 10.10   *The Standard Deviation of a Random Variable*

Group Health Insurance   A group health insurance policy of a small business pays 100% of employee medical bills up to a maximum of $1 million per policy year. The total annual medical bills, X, in millions of dollars, incurred by the employees has PDF given by $f_X(x) = x(4 - x)/9$ for $0 < x < 3$, and $f_X(x) = 0$ otherwise. Determine the standard deviation of the annual payout by the insurance company.

Solution   The annual payout, in millions of dollars, by the insurance company is $Y = \min\{X, 1\}$. In Example 10.6 on page 573, we found that $\mathcal{E}(Y) = 101/108$. To obtain the second moment of Y, we apply the FEF—specifically, Equation (10.8) on page 572, with $g(x) = (\min\{x, 1\})^2$:

$$\mathcal{E}(Y^2) = \mathcal{E}(g(X)) = \int_{-\infty}^{\infty} g(x) f_X(x)\, dx = \int_0^3 (\min\{x, 1\})^2\, \frac{x(4 - x)}{9}\, dx$$

$$= \frac{1}{9} \int_0^1 x^2 \cdot x(4 - x)\, dx + \frac{1}{9} \int_1^3 1^2 \cdot x(4 - x)\, dx$$

$$= \frac{1}{9} \int_0^1 (4x^3 - x^4)\, dx + \frac{1}{9} \int_1^3 (4x - x^2)\, dx = \frac{122}{135}.$$

Therefore,

$$\sigma_Y = \sqrt{\mathrm{Var}(Y)} = \sqrt{\mathcal{E}(Y^2) - (\mathcal{E}(Y))^2} = \sqrt{\frac{122}{135} - \left(\frac{101}{108}\right)^2} = 0.171.$$

The standard deviation of the annual payout is $0.171 million.   ■

EXAMPLE 10.11   *Moments of a Random Variable*

Min and Max of a Random Sample   Suppose that $X_1, \ldots, X_n$ are a random sample from a $\mathcal{U}(0, 1)$ distribution; that is, $X_1, \ldots, X_n$ are independent random variables all having a uniform distribution on the interval $(0, 1)$. Set $X = \min\{X_1, \ldots, X_n\}$ and $Y = \max\{X_1, \ldots, X_n\}$.
a) Determine the moments of X and Y.
b) Use part (a) to obtain the mean and variance of X and Y.

Solution   In Example 9.4(c) on page 497, we showed that a joint PDF of X and Y is

$$f_{X,Y}(x, y) = n(n - 1)(y - x)^{n-2}, \qquad 0 < x < y < 1,$$

and $f_{X,Y}(x, y) = 0$ otherwise.
a) In this case, we can obtain all moments of X and Y simultaneously. To do so, we use the bivariate form of the FEF—Equation (10.9) on page 572—and the joint PDF of X and Y. Alternatively, we could first find marginal PDFs of X and Y and then use the univariate form of the FEF—Equation (10.8) on page 572.

To determine the rth moment of Y, we apply the bivariate form of the FEF with $g(x, y) = y^r$. Doing so and making the substitution $u = y - x$, we get

$$\mathcal{E}(Y^r) = \int_{-\infty}^{\infty} \int_{-\infty}^{\infty} y^r f_{X,Y}(x, y) \, dx \, dy$$

$$= n(n-1) \int_0^1 \left(\int_0^y (y-x)^{n-2} \, dx \right) y^r \, dy$$

$$= n(n-1) \int_0^1 \left(\int_0^y u^{n-2} \, du \right) y^r \, dy = n \int_0^1 y^{r+n-1} \, dy,$$

or

$$\mathcal{E}(Y^r) = \frac{n}{r+n}. \tag{10.23}$$

To determine the rth moment of X, we apply the bivariate form of the FEF with $g(x, y) = x^r$. Doing so, making the substitution $u = y - x$, and using the fact that a beta PDF integrates to 1, we get

$$\mathcal{E}(X^r) = \int_{-\infty}^{\infty} \int_{-\infty}^{\infty} x^r f_{X,Y}(x, y) \, dx \, dy$$

$$= n(n-1) \int_0^1 \left(\int_x^1 (y-x)^{n-2} \, dy \right) x^r \, dx$$

$$= n(n-1) \int_0^1 \left(\int_0^{1-x} u^{n-2} \, du \right) x^r \, dx = n \int_0^1 x^r (1-x)^{n-1} \, dx$$

$$= n B(r+1, n) = n \frac{\Gamma(r+1)\Gamma(n)}{\Gamma(r+1+n)} = n \frac{r! \, (n-1)!}{(r+n)!},$$

or

$$\mathcal{E}(X^r) = \frac{r! \, n!}{(r+n)!}. \tag{10.24}$$

b) Applying Equation (10.24) with $r = 1$ and $r = 2$, we obtain $\mathcal{E}(X) = 1/(n+1)$ and $\mathcal{E}(X^2) = 2/(n+2)(n+1)$. From these results and the computing formula for the variance, we get $\text{Var}(X) = n/(n+2)(n+1)^2$. In summary, for future reference,

$$\mathcal{E}(X) = \frac{1}{n+1} \quad \text{and} \quad \text{Var}(X) = \frac{n}{(n+2)(n+1)^2}. \tag{10.25}$$

Also, from Equation (10.23), $\mathcal{E}(Y) = n/(n+1)$ and $\mathcal{E}(Y^2) = n/(n+2)$. These results and the computing formula for the variance give $\text{Var}(Y) = n/(n+2)(n+1)^2$. In summary, for future reference,

$$\mathcal{E}(Y) = \frac{n}{n+1} \quad \text{and} \quad \text{Var}(Y) = \frac{n}{(n+2)(n+1)^2}. \tag{10.26}$$

Later in this section, we apply Equations (10.25) and (10.26) to obtain the covariance and correlation of X and Y. ∎

Variance for Some Families of Continuous Random Variables

For many families of continuous random variables, we can obtain formulas for the variance in terms of the parameters. We do so by substituting the general form of the PDF for the family under consideration into the defining or computing formula for the variance of a continuous random variable.

In Examples 10.12 and 10.13, we obtain the variance for the normal and gamma families of random variables. Before proceeding, we recall a useful property of the variance as stated and proved in Proposition 7.10 on page 358: If X is a random variable with finite variance and a and b are real numbers, then

$$\mathrm{Var}(a + bX) = b^2\,\mathrm{Var}(X).\qquad(10.27)$$

EXAMPLE 10.12   *Variance of a Normal Random Variable*

Obtain the variance of a normal random variable with parameters μ and σ^2.

Solution   Let $X \sim \mathcal{N}(\mu, \sigma^2)$. It's easiest to work with the standardized version of X—namely, the random variable $Z = (X - \mu)/\sigma$—which, from Proposition 8.10 on page 441, has the standard normal distribution, that is, $Z \sim \mathcal{N}(0, 1)$.

As we demonstrated in Example 10.3 on page 567, the expected value of a normal random variable equals its first parameter; hence $\mathcal{E}(Z) = 0$. Applying the FEF, noting that $z^2\phi(z)$ is an even function, making the change of variable $u = z^2/2$, and using the fact that a gamma PDF integrates to 1, we get

$$\mathrm{Var}(Z) = \mathcal{E}(Z^2) - (\mathcal{E}(Z))^2 = \mathcal{E}(Z^2) = \int_{-\infty}^{\infty} z^2\phi(z)\,dz = 2\int_0^{\infty} z^2\phi(z)\,dz$$

$$= \frac{2}{\sqrt{2\pi}}\int_0^{\infty} z^2 e^{-z^2/2}\,dz = \frac{2}{\sqrt{2\pi}}\int_0^{\infty} \sqrt{2u}\,e^{-u}\,du$$

$$= \frac{2}{\sqrt{\pi}}\,\frac{\Gamma(3/2)}{1^{3/2}}\int_0^{\infty} \frac{1^{3/2}}{\Gamma(3/2)}\,u^{3/2-1}e^{-u}\,du = \frac{2}{\sqrt{\pi}}\,\frac{\sqrt{\pi}}{2}\cdot 1 = 1.$$

Therefore $\mathrm{Var}(Z) = 1$ and consequently, by Equation (10.27),

$$\mathrm{Var}(X) = \mathrm{Var}(\mu + \sigma Z) = \sigma^2\,\mathrm{Var}(Z) = \sigma^2.$$

Thus the variance of a normal random variable is its second parameter.   ∎

We can apply the result obtained in Example 10.12 to the gestation-period illustration in Example 8.14. It states that gestation periods of women are normally distributed with parameters $\mu = 266$ days and $\sigma = 16$ days. So, from Example 10.12, human gestation periods have a variance of 256 days^2 or, equivalently, a standard deviation of 16 days.

EXAMPLE 10.13   *Variance of a Gamma Random Variable*

Obtain the variance of a gamma random variable with parameters α and λ.

Solution   Let $X \sim \Gamma(\alpha, \lambda)$ and recall that a PDF of X is $f_X(x) = (\lambda^\alpha / \Gamma(\alpha))x^{\alpha-1}e^{-\lambda x}$ for $x > 0$, and $f_X(x) = 0$ otherwise. As we demonstrated in Example 10.4, $\mathcal{E}(X) = \alpha/\lambda$. To obtain the second moment of X, we apply the FEF and use the fact that the integral of a gamma PDF equals 1:

$$\mathcal{E}(X^2) = \int_{-\infty}^{\infty} x^2 f_X(x)\, dx = \int_0^\infty x^2 \frac{\lambda^\alpha}{\Gamma(\alpha)} x^{\alpha-1} e^{-\lambda x}\, dx$$

$$= \frac{\lambda^\alpha}{\Gamma(\alpha)} \frac{\Gamma(\alpha+2)}{\lambda^{\alpha+2}} \int_0^\infty \frac{\lambda^{\alpha+2}}{\Gamma(\alpha+2)} x^{(\alpha+2)-1} e^{-\lambda x}\, dx$$

$$= \frac{\lambda^\alpha}{\Gamma(\alpha)} \frac{\Gamma(\alpha+2)}{\lambda^{\alpha+2}} \cdot 1 = \frac{(\alpha+1)\alpha}{\lambda^2}.$$

Hence

$$\text{Var}(X) = \mathcal{E}(X^2) - (\mathcal{E}(X))^2 = \frac{(\alpha+1)\alpha}{\lambda^2} - \left(\frac{\alpha}{\lambda}\right)^2 = \frac{\alpha}{\lambda^2}.$$

Thus the variance of a gamma random variable equals the quotient of its first parameter and square of its second parameter. ∎

We can apply the result obtained in Example 10.13 to the emergency-room situation in Example 8.16 on page 454. Beginning at 6:00 P.M. on a given day, the elapsed time until the third patient arrives has a gamma distribution with parameters $\alpha = 3$ and $\lambda = 6.9$, where time is measured in hours. Therefore, from Example 10.13, the time until the third patient arrives has variance $3/6.9^2$ hour^2 or, equivalently, a standard deviation of roughly 15.1 minutes.

As we showed in Section 8.6 (see page 456), several important probability distributions are special cases of the gamma distribution. For such distributions, we can apply the result of Example 10.13 to obtain variances.

- *Exponential distribution.* The exponential distribution with parameter λ is also the gamma distribution with parameters 1 and λ. Hence the variance of an exponential random variable with parameter λ is $1/\lambda^2$.

- *Erlang distribution.* The Erlang distribution with parameters r and λ, where r is a positive integer, is also the gamma distribution with those two parameters. So the variance of an Erlang random variable with parameters r and λ is r/λ^2.

- *Chi-square distribution.* The chi-square distribution with ν degrees of freedom is also the gamma distribution with parameters $\nu/2$ and $1/2$. Hence the variance of a chi-square random variable with ν degrees of freedom is $(\nu/2)/(1/2)^2 = 2\nu$, twice the number of degrees of freedom.

Using arguments similar to those in Examples 10.12 and 10.13, we can obtain the variances for several other families of continuous random variables, as shown in Table 10.2.

Table 10.2 Variances for selected families of continuous random variables

Family	Parameter(s)	Variance
Uniform	a and b	$(b-a)^2/12$
Exponential	λ	$1/\lambda^2$
Normal	μ and σ^2	σ^2
Gamma	α and λ	α/λ^2
Erlang	r and λ	r/λ^2
Chi-square	ν	2ν
Beta	α and β	$\alpha\beta/(\alpha+\beta)^2(\alpha+\beta+1)$
Triangular	a and b	$(b-a)^2/24$

Covariance; Variance of a Sum

In Section 7.4, we introduced the concept of the *covariance* of two random variables. The definitions and results presented there apply not only to discrete random variables but to any types of random variables, in particular, to continuous random variables.

Recall that, if X and Y are random variables defined on the same sample space and having finite variances, the **covariance** of X and Y, denoted $\mathrm{Cov}(X, Y)$, is defined by

$$\mathrm{Cov}(X, Y) = \mathcal{E}\big((X - \mu_X)(Y - \mu_Y)\big). \qquad (10.28)$$

As we observed in Section 7.4 on page 366, we can also obtain the covariance by using the *computing formula*

$$\mathrm{Cov}(X, Y) = \mathcal{E}(XY) - \mathcal{E}(X)\,\mathcal{E}(Y). \qquad (10.29)$$

Note that $\mathrm{Cov}(X, Y) = \mathrm{Cov}(Y, X)$ and $\mathrm{Cov}(X, X) = \mathrm{Var}(X)$. These and other basic properties of covariance are given in Proposition 7.12 on page 366. Review those properties and the **bilinearity property** of covariance: If $X_1, \ldots, X_m$ and $Y_1, \ldots, Y_n$ are random variables defined on the same sample space and having finite variances and if $a_1, \ldots, a_m$ and $b_1, \ldots, b_n$ are real numbers, then

$$\mathrm{Cov}\left(\sum_{j=1}^{m} a_j X_j, \sum_{k=1}^{n} b_k Y_k\right) = \sum_{j=1}^{m}\sum_{k=1}^{n} a_j b_k \, \mathrm{Cov}(X_j, Y_k). \qquad (10.30)$$

From Equation (10.30), we can get the formula for the variance of the sum of random variables. Specifically, if $X_1, \ldots, X_m$ are random variables defined on the same sample space and having finite variances, then

$$\mathrm{Var}\left(\sum_{k=1}^{m} X_k\right) = \mathrm{Cov}\left(\sum_{i=1}^{m} X_i, \sum_{j=1}^{m} X_j\right) = \sum_{i=1}^{m}\sum_{j=1}^{m} \mathrm{Cov}(X_i, X_j)$$

$$= \sum_{k=1}^{m} \mathrm{Cov}(X_k, X_k) + \sum\sum_{i \neq j} \mathrm{Cov}(X_i, X_j).$$

Now using $\text{Cov}(X, X) = \text{Var}(X)$ and $\text{Cov}(X, Y) = \text{Cov}(Y, X)$, we obtain

$$\text{Var}\left(\sum_{k=1}^{m} X_k\right) = \sum_{k=1}^{m} \text{Var}(X_k) + 2 \sum \sum_{i<j} \text{Cov}(X_i, X_j). \qquad \textbf{(10.31)}$$

In particular, if X and Y are random variables defined on the same sample space and having finite variances, then

$$\text{Var}(X + Y) = \text{Var}(X) + \text{Var}(Y) + 2\,\text{Cov}(X, Y). \qquad \textbf{(10.32)}$$

EXAMPLE 10.14   *Covariance and the Variance of a Sum*

Min, Max, Range and Midrange of a Random Sample   Let $X_1, \ldots, X_n$ be a random sample from a $\mathcal{U}(0, 1)$ distribution and let us denote $X = \min\{X_1, \ldots, X_n\}$ and $Y = \max\{X_1, \ldots, X_n\}$. Determine
a) $\text{Cov}(X, Y)$.
b) the mean and variance of the range of the random sample.
c) the mean and variance of the midrange of the random sample.

Solution   We first recall from Example 9.4(c) on page 497 that a joint PDF of X and Y is

$$f_{X,Y}(x, y) = n(n - 1)(y - x)^{n-2}, \qquad 0 < x < y < 1,$$

and $f_{X,Y}(x, y) = 0$ otherwise.
a) From Example 10.11—specifically, Equations (10.25) and (10.26) on page 585—we
   know that $\mathcal{E}(X) = 1/(n + 1)$ and $\mathcal{E}(Y) = n/(n + 1)$. Hence, to obtain $\text{Cov}(X, Y)$,
   we need only find $\mathcal{E}(XY)$. Applying the bivariate form of the FEF with $g(x, y) = xy$
   and making the substitution $u = y - x$, we get

$$\mathcal{E}(XY) = \int_{-\infty}^{\infty} \int_{-\infty}^{\infty} xy f_{X,Y}(x, y)\, dx\, dy$$

$$= n(n - 1) \int_0^1 \left(\int_0^y x(y - x)^{n-2}\, dx\right) y\, dy$$

$$= n(n - 1) \int_0^1 \left(\int_0^y (y - u)u^{n-2}\, du\right) y\, dy$$

$$= n(n - 1) \int_0^1 \left[y\frac{u^{n-1}}{n - 1} - \frac{u^n}{n}\right]_0^y y\, dy = \int_0^1 y^{n+1}\, dy = \frac{1}{n + 2}.$$

Consequently,

$$\text{Cov}(X, Y) = \mathcal{E}(XY) - \mathcal{E}(X)\,\mathcal{E}(Y) = \frac{1}{n + 2} - \frac{1}{n + 1}\frac{n}{n + 1}$$

or

$$\text{Cov}(X, Y) = \frac{1}{(n + 2)(n + 1)^2}. \qquad \textbf{(10.33)}$$

b) The range of the random sample is defined by $R = Y - X$. We could obtain the mean and variance of R directly by using the PDF of R that we found in Example 9.6(b) on page 504. Here, however, we use the moments and covariance of X and Y and the properties of mean and variance. We have

$$\mathcal{E}(R) = \mathcal{E}(Y - X) = \mathcal{E}(Y) - \mathcal{E}(X) = \frac{n}{n+1} - \frac{1}{n+1} = \frac{n-1}{n+1}.$$

From Equations (10.25), (10.26), and (10.33),

$$\begin{aligned}
\mathrm{Var}(R) &= \mathrm{Var}(Y - X) = \mathrm{Var}\big(Y + (-X)\big) \\
&= \mathrm{Var}(Y) + \mathrm{Var}(-X) + 2\,\mathrm{Cov}(Y, -X) \\
&= \mathrm{Var}(Y) + \mathrm{Var}(X) - 2\,\mathrm{Cov}(X, Y) \\
&= \frac{n}{(n+2)(n+1)^2} + \frac{n}{(n+2)(n+1)^2} - 2 \cdot \frac{1}{(n+2)(n+1)^2} \\
&= \frac{2(n-1)}{(n+2)(n+1)^2}.
\end{aligned}$$

c) The midrange of the random sample is defined by $M = (X + Y)/2$. We could obtain the mean and variance of M directly by using a PDF of M [Exercise 9.116(b) on page 542 asks for that PDF]. However, it is easier to use the moments and covariance of X and Y and the properties of mean and variance. We have

$$\mathcal{E}(M) = \mathcal{E}\left(\frac{X+Y}{2}\right) = \frac{1}{2}\big(\mathcal{E}(X) + \mathcal{E}(Y)\big) = \frac{1}{2}\left(\frac{1}{n+1} + \frac{n}{n+1}\right) = \frac{1}{2}.$$

From Equations (10.25), (10.26), and (10.33), we find that

$$\begin{aligned}
\mathrm{Var}(M) &= \mathrm{Var}\left(\frac{X+Y}{2}\right) = \frac{1}{4}\big(\mathrm{Var}(X+Y)\big) \\
&= \frac{1}{4}\big(\mathrm{Var}(X) + \mathrm{Var}(Y) + 2\,\mathrm{Cov}(X, Y)\big) = \frac{1}{2(n+2)(n+1)},
\end{aligned}$$

as required.   ■

For independent random variables, the formula for the variance of a sum simplifies considerably. From the computing formula for the covariance and Proposition 10.4 on page 576, the covariance of two independent random variables equals 0. Hence, for such random variables, Equation (10.31) reduces to

$$\mathrm{Var}\left(\sum_{k=1}^{m} X_k\right) = \sum_{k=1}^{m} \mathrm{Var}(X_k). \tag{10.34}$$

Thus the variance of a sum of independent random variables equals the sum of their variances, as we noted in Proposition 7.15 on page 368. In particular, then, if X and Y are independent random variables with finite variances,

$$\mathrm{Var}(X + Y) = \mathrm{Var}(X) + \mathrm{Var}(Y). \tag{10.35}$$

EXAMPLE 10.15   *Variance and Covariance*

Let Z_1 and Z_2 be independent random variables with zero means and unit variances. Also, let μ_1 and μ_2 be real numbers, σ_1 and σ_2 be positive real numbers, and $-1 \leq \rho \leq 1$. Define the random variables X and Y by

$$X = \mu_1 + aZ_1 + bZ_2$$
$$Y = \mu_2 + cZ_1 + dZ_2,$$

where a, b, c, and d are any four real numbers that satisfy

$$a^2 + b^2 = \sigma_1^2, \qquad ac + bd = \rho\sigma_1\sigma_2, \qquad c^2 + d^2 = \sigma_2^2.$$

For instance, we could let $a = \sigma_1$, $b = 0$, $c = \rho\sigma_2$, and $d = \sqrt{1 - \rho^2}\,\sigma_2$.

Apply properties of expected value, variance, and covariance to show that $\mathcal{E}(X) = \mu_1$, $\text{Var}(X) = \sigma_1^2$, $\mathcal{E}(Y) = \mu_2$, $\text{Var}(Y) = \sigma_2^2$, and $\text{Cov}(X, Y) = \rho\sigma_1\sigma_2$.

Solution   By assumption, $\mathcal{E}(Z_1) = \mathcal{E}(Z_2) = 0$ and $\text{Var}(Z_1) = \text{Var}(Z_2) = 1$. Applying properties of expected value, we get

$$\mathcal{E}(X) = \mathcal{E}(\mu_1 + aZ_1 + bZ_2) = \mu_1 + a\mathcal{E}(Z_1) + b\mathcal{E}(Z_2)$$
$$= \mu_1 + a \cdot 0 + b \cdot 0 = \mu_1,$$

and, applying properties of variance, we get

$$\text{Var}(X) = \text{Var}(\mu_1 + aZ_1 + bZ_2) = a^2 \, \text{Var}(Z_1) + b^2 \, \text{Var}(Z_2)$$
$$= a^2 \cdot 1 + b^2 \cdot 1 = a^2 + b^2 = \sigma_1^2.$$

Similarly, $\mathcal{E}(Y) = \mu_2$ and $\text{Var}(Y) = \sigma_2^2$.

Applying properties of covariance and using the fact that the covariance of independent random variables is 0 gives

$$\text{Cov}(X, Y) = \text{Cov}(\mu_1 + aZ_1 + bZ_2, \mu_2 + cZ_1 + dZ_2)$$
$$= ac \, \text{Cov}(Z_1, Z_1) + ad \, \text{Cov}(Z_1, Z_2)$$
$$+ bc \, \text{Cov}(Z_2, Z_1) + bd \, \text{Cov}(Z_2, Z_2)$$
$$= ac \, \text{Var}(Z_1) + ad \cdot 0 + bc \cdot 0 + bd \, \text{Var}(Z_2)$$
$$= ac + bd = \rho\sigma_1\sigma_2, \qquad \blacksquare$$

The Correlation Coefficient

As a measure of linear association between two random variables, we introduced, in Section 7.4 on page 371, the *correlation coefficient*, a concept that applies to all types of random variables. Recall that, if X and Y are two random variables defined on the same sample space and having finite variances, the **correlation coefficient** of X and Y, denoted $\rho(X, Y)$, is defined to be the covariance of the standardized random variables X^* and Y^*. Applying basic properties of covariance gives

$$\rho(X, Y) = \frac{\text{Cov}(X, Y)}{\sqrt{\text{Var}(X) \cdot \text{Var}(Y)}}. \tag{10.36}$$

We presented several important properties of the correlation coefficient in Proposition 7.16 on page 372, which you should now review. One of those properties states that the correlation coefficient of independent random variables equals 0. Random variables that have zero correlation are said to be **uncorrelated random variables.** Although independent random variables are uncorrelated, uncorrelated random variables need not be independent. See, for instance, Exercise 10.52.

EXAMPLE 10.16   *The Correlation Coefficient*

Obtain the correlation coefficient of the random variables X and Y from Example 10.15.

Solution   From Example 10.15, we have $\text{Var}(X) = \sigma_1^2$, $\text{Var}(Y) = \sigma_2^2$, and $\text{Cov}(X, Y) = \rho\sigma_1\sigma_2$. Therefore,

$$\rho(X, Y) = \frac{\text{Cov}(X, Y)}{\sqrt{\text{Var}(X) \cdot \text{Var}(Y)}} = \frac{\rho\sigma_1\sigma_2}{\sqrt{\sigma_1^2\sigma_2^2}} = \frac{\rho\sigma_1\sigma_2}{\sigma_1\sigma_2} = \rho.$$

This result shows that we can construct random variables with any specified correlation coefficient, ρ, where, of course, $|\rho| \leq 1$.   ■

EXERCISES 10.3   **Basic Exercises**

10.51 Suppose that a tax of 20% is introduced on all items associated with the maintenance and repair of cars (i.e., everything is made 20% more expensive). How does that increase translate into the percentage increase in
a) the mean annual cost?
b) the variance of the annual cost?
c) the standard deviation of the annual cost?

10.52 Refer to the petri-dish illustration of Example 10.1 on page 565. Let X and Y denote the x and y coordinates, respectively, of the first spot (visible bacteria colony) to appear.
a) In Example 9.8(b) on page 511, we obtained a PDF of X. Use that PDF to find $\text{Var}(X)$. *Note:* From Exercise 10.4 on page 569, $\mathcal{E}(X) = 0$.
b) Without doing further computations, find $\text{Var}(Y)$. Explain your reasoning.
c) Obtain $\rho(X, Y)$.
d) From part (c), the random variables X and Y are uncorrelated. Does this result imply that X and Y are independent? Explain your answer.
e) Are X and Y independent random variables? Justify your answer.

10.53 A commuter train arrives punctually at a station every half hour. Each morning, a commuter named John leaves his house and casually strolls to the train station. Let X denote the amount of time, in minutes, that John waits for the train from the time he reaches the train station. In Example 8.10 on page 428, we found that a PDF of X is $f_X(x) = 1/30$ for $0 < x < 30$, and $f_X(x) = 0$ otherwise. Use that result to obtain the variance of the random variable X. *Note:* From Exercise 10.5 on page 569, $\mathcal{E}(X) = 15$.

10.54 Suppose that the random variable X has the uniform distribution on the interval (a, b).
a) Show that $\text{Var}(X) = (b - a)^2/12$.
b) Apply your result from part (a) to obtain a quick solution of Exercise 10.53.

10.55 According to the text *Rhythms of Dialogue* by J. Jaffee and S. Feldstein (New York: Academic Press, 1970), the duration, in seconds, of a pause during a monologue has an exponential distribution with parameter 1.4. Determine the standard deviation of the duration of a pause during a monologue.

10.56 As reported in *Runner's World* magazine, the times of the finishers in the New York City 10 km run are normally distributed with $\mu = 61$ minutes and $\sigma = 9$ minutes. What is the standard deviation of the finishing times?

10.57 Suppose that the random variable X has the beta distribution with parameters α and β. Show that $\text{Var}(X) = \alpha\beta/(\alpha + \beta)^2(\alpha + \beta + 1)$.

10.58 For a certain manufactured item, the proportion of the annual production that requires service during the first 5 years of use has the beta distribution with parameters $\alpha = 2$ and $\beta = 3$. Determine the standard deviation of the proportion of the annual production of these manufactured items that require service during the first 5 years of use.

10.59 Let T_1 be the time between a car accident and reporting a claim to the insurance company. Let T_2 be the time between the report of the claim and payment of the claim. A joint density function of T_1 and T_2 is constant over the region $0 < t_1 < 6$, $0 < t_2 < 6$, $t_1 + t_2 < 10$, and zero otherwise. Determine the standard deviation of the time between a car accident and payment of the claim
a) by first obtaining a PDF of the time between a car accident and payment of the claim.
b) without first obtaining a PDF of the time between a car accident and payment of the claim.

10.60 A number X is chosen at random from the interval $(0, 1)$. Then another number Y is chosen at random from the interval $(0, X)$. Obtain $\text{Var}(Y)$ by first finding a PDF of Y and then applying the definition of the variance of a random variable.

10.61 For purposes of quality assurance, an expensive item produced at a manufacturing plant is independently inspected by two engineers. Let X and Y denote the proportions of the total inspection time attributed to the first and second engineers, respectively.
a) Use properties of variance to show that $\text{Var}(Y) = \text{Var}(X)$.
b) Use part (a) and properties of variance to show that $\text{Cov}(X, Y) = -\text{Var}(X)$.
c) Use properties of covariance to show that $\text{Cov}(X, Y) = -\text{Var}(X)$.
d) Conclude from parts (a) and (b) that $\rho(X, Y) = -1$.
e) Obtain the result of part (d) by using properties of the correlation coefficient.

10.62 The warranty on a machine whose mean lifetime is 3 years specifies that the machine will be replaced at failure or at age 4 years, whichever occurs first. Determine the standard deviation of the age of the machine at the time of replacement if the distribution of the lifetime of the machine is
a) uniform.   **b)** triangular.   **c)** exponential.

10.63 A joint density function of X and Y is given by $f_{X,Y}(x, y) = cx$ for $0 < x < 1$ and $0 < y < 1$, and $f_{X,Y}(x, y) = 0$ otherwise. Here c is a constant. Obtain $\text{Cov}(X, Y)$
a) by using the joint PDF of X and Y.
b) without doing any computations.

10.64 The owner of an automobile insures it against damage by purchasing an insurance policy with a deductible of \$250. In the event that the automobile is damaged, repair costs (in dollars) can be modeled by a uniform random variable on the interval $(0, 1500)$. Determine the standard deviation of the insurance payment in the event that the automobile is damaged.

10.65 A stock market analyst knows from experience that the mean daily sales revenues for Companies A and B are both 100. Furthermore, a daily sales revenue above 100 for

Company A is always accompanied by a daily sales revenue below 100 for Company B, and vice versa. Let X and Y denote the daily sales revenues for Companies A and B, respectively. Which of the following relationships is correct? Justify your answer. *Hint:* The monotonicity property of expected value holds for strict inequality.

a) $\text{Var}(X + Y) = \text{Var}(X) + \text{Var}(Y)$

b) $\text{Var}(X + Y) < \text{Var}(X) + \text{Var}(Y)$

c) $\text{Var}(X + Y) > \text{Var}(X) + \text{Var}(Y)$

10.66 As stated on page 590, the covariance of two independent random variables equals 0. Is the converse of this statement true? Justify your answer.

10.67 Refer to Examples 10.11 and 10.14 on pages 584 and 589. Let $X_1, \ldots, X_n$ be a random sample from a $\mathcal{U}(0, 1)$ distribution. Set $X = \min\{X_1, \ldots, X_n\}$ and $Y = \max\{X_1, \ldots, X_n\}$.

a) Determine $\rho(X, Y)$.

b) From part (a), the correlation between the minimum and maximum of a random sample from a $\mathcal{U}(0, 1)$ distribution is the reciprocal of the sample size. In particular, those two random variables are asymptotically uncorrelated—their correlation approaches 0 as the sample size increases without bound. Can you explain why this fact makes sense?

c) Determine $\rho(R, M)$, where R and M are the range and midrange of the random sample, respectively.

d) From part (c), R and M are uncorrelated random variables. Does that imply they are independent random variables?

e) Are R and M independent random variables? Justify your answer.

10.68 Let $\bar{X}_n$ denote the sample mean for a random sample of size n from a distribution with mean μ and variance σ^2. Show that

a) $\mathcal{E}(\bar{X}_n) = \mu$.   **b)** $\text{Var}(\bar{X}_n) = \sigma^2/n$.   **c)** $\text{Cov}(\bar{X}_n, X_j - \bar{X}_n) = 0$ for $1 \le j \le n$.

d) Interpret the results in parts (a)–(c).

10.69 Let $X \sim \Gamma(\alpha, \lambda)$. Show that $\mathcal{E}(X^n) = \alpha(\alpha + 1) \cdots (\alpha + n - 1)/\lambda^n$ for $n = 1, 2, \ldots$.

10.70 Let $X \sim \mathcal{N}(\mu, \sigma^2)$.

a) Show that all odd central moments of X equal 0.

b) Show that the even central moments of X are given by the formula

$$\mathcal{E}\big((X - \mu)^{2n}\big) = \frac{(2n)!}{n!\, 2^n} \sigma^{2n}, \qquad n = 1, 2, \ldots.$$

Hint: Use Equation (8.46) on page 452.

10.71 Let X and Y be random variables defined on the same sample space and having finite variances, and let a and b be real numbers.

a) Show that $\text{Cov}(aX + bY, aX - bY) = a^2 \text{Var}(X) - b^2 \text{Var}(Y)$.

b) Conclude that, if X and Y have equal variances, then $X + Y$ and $X - Y$ are uncorrelated random variables.

Theory Exercises

10.72 Symmetric random variable: Recall that a random variable X is said to be *symmetric* if X and $-X$ have the same probability distribution.

a) Show that all odd moments of a symmetric random variable equal 0.

b) Recall that a random variable X is said to be *symmetric about c* if $X - c$ is a symmetric random variable. Show that all odd central moments of a random variable symmetric about c are 0.

10.73 Moments by tail probabilities: Generalize Proposition 10.5 on page 577 by proving the following result: Suppose that X is a nonnegative continuous random variable. Then X has finite nth moment if and only if $\int_0^\infty x^{n-1} P(X > x) \, dx < \infty$, in which case

$$\mathcal{E}(X^n) = n \int_0^\infty x^{n-1} P(X > x) \, dx.$$

Note: This result holds for any nonnegative random variable X, regardless of type.

10.74 Weak law of large numbers: Let $X_1, X_2, \ldots$ be independent and identically distributed random variables with finite variance and let μ denote the common mean of the X_js.

a) Prove the *weak law of large numbers:* For each $\epsilon > 0$,

$$\lim_{n \to \infty} P\left(\left|\frac{X_1 + \cdots + X_n}{n} - \mu\right| < \epsilon\right) = 1.$$

Hint: Use Chebyshev's inequality, Proposition 7.11 on page 360.

b) Interpret the weak law of large numbers.

c) How does the weak law of large numbers relate to the long-run-average interpretation of expected value?

Advanced Exercises

10.75 Best linear unbiased estimator: Let $X_1, \ldots, X_n$ be independent random variables with common mean μ and with variances $\sigma_1^2, \ldots, \sigma_n^2$, respectively. Now consider the linear estimator of μ defined by $Y = \sum_{j=1}^n c_j X_j$, where $c_1, \ldots, c_n$ are real numbers.

a) Show that Y is an unbiased estimator of μ if and only if $\sum_{j=1}^n c_j = 1$.

b) Among the linear unbiased estimators of μ, find the one with minimum variance—the so-called *best linear unbiased estimator* of μ.

c) Why are minimum variance and unbiasedness desirable properties for an estimator?

d) Let $X_1, \ldots, X_n$ be a random sample from a distribution with finite variance. Apply your result from part (b) to identify Y in this case.

10.76 Student's t-distribution: Let T have the Student's t-distribution with ν degrees of freedom. According to Exercise 9.164 on page 553, a PDF of T is given by

$$f_T(t) = \frac{\Gamma\left((\nu+1)/2\right)}{\sqrt{\nu\pi}\,\Gamma(\nu/2)} \left(1 + t^2/\nu\right)^{-(\nu+1)/2}, \qquad \infty < t < \infty.$$

a) Determine when T has finite variance.

b) Obtain $\mathrm{Var}(T)$ when T has finite variance.

10.4   Conditional Expectation

In this section, we introduce and apply the concept of *conditional expected value*—better known as *conditional expectation*—for continuous random variables. For instance, let X and Y denote the height and weight, respectively, of a randomly selected woman. We might be interested in the mean weight of all women—that is, the expected value of the random variable Y. However, we might be even more interested in the mean weight of all women of a particular height—that is, the conditional expectation of the random variable Y, given a value of the random variable X.

Actually, **conditional expectation** is nothing new. It's simply expected value relative to a conditional distribution instead of an unconditional distribution.

DEFINITION 10.2   Conditional Expectation: Continuous Case

Suppose that X and Y are continuous random variables with a joint PDF. If $f_X(x) > 0$, we define the **conditional expectation of Y given $X = x$**, denoted $\mathcal{E}(Y \mid X = x)$, to be the expected value of Y relative to the conditional distribution of Y given $X = x$. Thus, if $f_X(x) > 0$, we define

$$\mathcal{E}(Y \mid X = x) = \int_{-\infty}^{\infty} y f_{Y \mid X}(y \mid x)\, dy, \tag{10.37}$$

provided, of course, that $Y_{\mid X=x}$ has finite expectation. If $f_X(x) = 0$, we define $\mathcal{E}(Y \mid X = x) = 0$ but don't refer to $\mathcal{E}(Y \mid X = x)$ as a conditional expectation.

Because conditional expectation is simply expected value relative to a conditional distribution, all properties that hold for (unconditional) expectation also hold for conditional expectation. For instance, the conditional form of the univariate FEF is

$$\mathcal{E}\big(g(Y) \mid X = x\big) = \int_{-\infty}^{\infty} g(y) f_{Y \mid X}(y \mid x)\, dy. \tag{10.38}$$

Conditional variance is also a simple extension of (unconditional) variance. Recall that, by definition, $\mathrm{Var}(Y) = \mathcal{E}\big((Y - \mathcal{E}(Y))^2\big)$; that is, the variance of a random variable Y is defined to be the expected value of the square of the deviation of Y from its mean. Likewise, we define the **conditional variance** of Y given $X = x$ to be the conditional expected value given $X = x$ of the square of the deviation of Y from its conditional mean.

DEFINITION 10.3   Conditional Variance: Continuous Case

Suppose that X and Y are continuous random variables with a joint PDF. If $f_X(x) > 0$, we define the **conditional variance of Y given $X = x$**, denoted **$\mathrm{Var}(Y \mid X = x)$**, to be the variance of Y relative to the conditional distribution of Y given $X = x$. Thus, if $f_X(x) > 0$, we define

$$\mathrm{Var}(Y \mid X = x) = \mathcal{E}\Big((Y - \mathcal{E}(Y \mid X = x))^2 \mid X = x\Big), \tag{10.39}$$

provided, of course, that $Y_{\mid X=x}$ has finite variance. If $f_X(x) = 0$, we define $\mathrm{Var}(Y \mid X = x) = 0$ but don't refer to $\mathrm{Var}(Y \mid X = x)$ as a conditional variance.

Again, conditional variance is nothing new. It's simply the variance of a random variable relative to a conditional distribution instead of an unconditional distribution. Consequently, properties that hold for variance also hold for conditional variance. In particular, we have the following conditional version of the *computing formula* for a variance:

$$\text{Var}(Y \mid X = x) = \mathcal{E}(Y^2 \mid X = x) - \big(\mathcal{E}(Y \mid X = x)\big)^2. \qquad \textbf{(10.40)}$$

Compare this computing formula for conditional variance to the computing formula for (unconditional) variance as presented in Equation (10.22) on page 583.

EXAMPLE 10.17   *Conditional Expectation and Variance*

Bacteria on a Petri Dish   Refer to Example 10.1 on page 565. Let X and Y denote the x and y coordinates, respectively, of the center of the first spot to appear. Determine the conditional expectation and conditional variance of Y for each possible value of X.

Solution   The possible values of X are $-1 < x < 1$, so we restrict discussion to those values. In Example 9.10 on page 515, we found that a conditional PDF of Y given $X = x$ is

$$f_{Y \mid X}(y \mid x) = \frac{1}{2\sqrt{1 - x^2}}, \qquad -\sqrt{1 - x^2} < y < \sqrt{1 - x^2},$$

and $f_{Y \mid X}(y \mid x) = 0$ otherwise. Therefore, $Y_{\mid X=x} \sim \mathcal{U}\big(-\sqrt{1 - x^2}, \sqrt{1 - x^2}\big)$. From Table 10.1 on page 568, the expected value of a uniform random variable on the interval (a, b) is $(a + b)/2$. Consequently, for $-1 < x < 1$,

$$\mathcal{E}(Y \mid X = x) = \frac{\big(-\sqrt{1 - x^2}\big) + \sqrt{1 - x^2}}{2} = 0,$$

a result that we could have easily guessed because of symmetry.

From Table 10.2 on page 588, the variance of a uniform random variable on the interval (a, b) is $(b - a)^2/12$. Consequently, for $-1 < x < 1$,

$$\text{Var}(Y \mid X = x) = \frac{\big(\sqrt{1 - x^2} - \big(-\sqrt{1 - x^2}\big)\big)^2}{12} = \frac{1 - x^2}{3},$$

as required.   ∎

EXAMPLE 10.18   *Conditional Expectation and Variance*

Min and Max of a Random Sample   Let $X_1, \ldots, X_n$ be a random sample from a $\mathcal{U}(0, 1)$ distribution, and set $X = \min\{X_1, \ldots, X_n\}$ and $Y = \max\{X_1, \ldots, X_n\}$. Find
a) $\mathcal{E}(Y \mid X = x)$.   b) $\text{Var}(Y \mid X = x)$.

Solution   The possible values of X are $0 < x < 1$, so we restrict discussion to those values. In Example 9.12 on page 517, we found that a conditional PDF of Y given $X = x$ is

$$f_{Y \mid X}(y \mid x) = \frac{(n - 1)(y - x)^{n-2}}{(1 - x)^{n-1}}, \qquad x < y < 1,$$

and $f_{Y \mid X}(y \mid x) = 0$ otherwise.

a) Applying Definition 10.2 and making the substitution $u = y - x$, we get

$$\mathcal{E}(Y \mid X = x) = \int_{-\infty}^{\infty} y f_{Y \mid X}(y \mid x)\, dy = \int_{x}^{1} y \frac{(n-1)(y-x)^{n-2}}{(1-x)^{n-1}}\, dy$$

$$= \frac{n-1}{(1-x)^{n-1}} \int_{0}^{1-x} (u+x) u^{n-2}\, du \tag{10.41}$$

$$= (n-1) \left(\frac{1-x}{n} + \frac{x}{n-1} \right)$$

or

$$\mathcal{E}(Y \mid X = x) = 1 - \frac{1-x}{n}. \tag{10.42}$$

b) To obtain the conditional variance, we use the computing formula in Equation (10.40) and the result just found for the conditional mean, Equation (10.42). First we apply the conditional form of the FEF, Equation (10.38), with $g(y) = y^2$, and make the substitution $u = y - x$ to get

$$\mathcal{E}(Y^2 \mid X = x) = \int_{-\infty}^{\infty} y^2 f_{Y \mid X}(y \mid x)\, dy = \int_{x}^{1} y^2 \frac{(n-1)(y-x)^{n-2}}{(1-x)^{n-1}}\, dy$$

$$= \frac{n-1}{(1-x)^{n-1}} \int_{0}^{1-x} (u+x)^2 u^{n-2}\, du$$

$$= (n-1) \left(\frac{(1-x)^2}{n+1} + \frac{2x(1-x)}{n} + \frac{x^2}{n-1} \right).$$

Hence, from Equations (10.40) and (10.41),

$$\mathrm{Var}(Y \mid X = x) = \mathcal{E}(Y^2 \mid X = x) - \left(\mathcal{E}(Y \mid X = x) \right)^2$$

$$= (n-1) \left(\frac{(1-x)^2}{n+1} + \frac{2x(1-x)}{n} + \frac{x^2}{n-1} \right)$$

$$- \left((n-1) \left(\frac{1-x}{n} + \frac{x}{n-1} \right) \right)^2.$$

After some algebra, we find that

$$\mathrm{Var}(Y \mid X = x) = \frac{n-1}{n^2(n+1)} (1-x)^2, \tag{10.43}$$

as required.   ■

Laws of Total Expectation and Total Variance

Sometimes we have the conditional expectation of a continuous random variable Y for each possible value of a continuous random variable X; that is, we know $\mathcal{E}(Y \mid X = x)$

for each x with $f_X(x) > 0$. From this information and a PDF of X, we can obtain the (unconditional) expectation of Y; that is, we can determine $\mathcal{E}(Y)$. The formula for doing this calculation is called the **law of total expectation.**

Before stating and proving the law of total expectation for continuous random variables, it's useful to consider another concept. Let X and Y be continuous random variables with a joint PDF, and let ψ be the real-valued function defined on $\mathcal{R}$ by $\psi(x) = \mathcal{E}(Y \mid X = x)$. Then we define the **conditional expectation of Y given X,** denoted $\mathcal{E}(Y \mid X)$, to be the random variable $\psi(X)$. So, by definition, $\mathcal{E}(Y \mid X) = \psi(X)$.

◆◆◆   **Proposition 10.6   Law of Total Expectation**

Let X and Y be two continuous random variables with a joint PDF and suppose that Y has finite expectation. Then

$$\mathcal{E}(Y) = \int_{-\infty}^{\infty} \mathcal{E}(Y \mid X = x)\, f_X(x)\, dx. \tag{10.44}$$

In terms of the conditional expectation of Y given X, we can express Equation (10.44) in the compact form

$$\mathcal{E}(Y) = \mathcal{E}\big(\mathcal{E}(Y \mid X)\big). \tag{10.45}$$

In words, the expected value of Y equals the expected value of the conditional expectation of Y given X.

Proof   Applying the bivariate form of the FEF with $g(x, y) = y$, the general multiplication rule, and the definition of conditional expectation, we get

$$\mathcal{E}(Y) = \int_{-\infty}^{\infty} \int_{-\infty}^{\infty} y f_{X,Y}(x, y)\, dx\, dy = \int_{-\infty}^{\infty} \int_{-\infty}^{\infty} y f_X(x) f_{Y \mid X}(y \mid x)\, dx\, dy$$

$$= \int_{-\infty}^{\infty} f_X(x) \left(\int_{-\infty}^{\infty} y f_{Y \mid X}(y \mid x)\, dy \right) dx = \int_{-\infty}^{\infty} f_X(x) \mathcal{E}(Y \mid X = x)\, dx,$$

which establishes Equation (10.44).

Applying the univariate form of the FEF with $g(x) = \psi(x) = \mathcal{E}(Y \mid X = x)$ and then referring to Equation (10.44), we get

$$\mathcal{E}\big(\mathcal{E}(Y \mid X)\big) = \mathcal{E}\big(\psi(X)\big) = \int_{-\infty}^{\infty} \psi(x) f_X(x)\, dx$$

$$= \int_{-\infty}^{\infty} \mathcal{E}(Y \mid X = x)\, f_X(x)\, dx = \mathcal{E}(Y),$$

which establishes Equation (10.45).   ◆

We also have a formula, analogous to the law of total expectation, for obtaining variance by conditioning. This formula is aptly called the **law of total variance.** Its proof in the continuous case is identical to that for the discrete case, as given in Proposition 7.19 on page 385.

◆◆◆ **Proposition 10.7   Law of Total Variance**

Let X and Y be two continuous random variables with a joint PDF and suppose that Y has finite variance. Then

$$\text{Var}(Y) = \mathcal{E}\big(\text{Var}(Y \mid X)\big) + \text{Var}\big(\mathcal{E}(Y \mid X)\big). \tag{10.46}$$

In words, the variance of Y equals the expectation of the conditional variance of Y given X plus the variance of the conditional expectation of Y given X.

EXAMPLE 10.19   *Laws of Total Expectation and Total Variance*

Min and Max of a Random Sample   Let $X_1, \ldots, X_n$ be a random sample from a $\mathcal{U}(0, 1)$ distribution, and set $X = \min\{X_1, \ldots, X_n\}$ and $Y = \max\{X_1, \ldots, X_n\}$. Referring to the results obtained in Examples 10.11 and 10.18 on pages 584 and 597, respectively,

a) use the law of total expectation to determine $\mathcal{E}(Y)$.

b) use the law of total variance to determine $\text{Var}(Y)$.

c) compare the values for $\mathcal{E}(Y)$ and $\text{Var}(Y)$ obtained in parts (a) and (b), respectively, to those found in Example 10.11.

Solution   a) Applying the law of total expectation and properties of expected value, we obtain, by referring to Equations (10.42) and (10.25) on pages 598 and 585, respectively, that

$$\mathcal{E}(Y) = \mathcal{E}\big(\mathcal{E}(Y \mid X)\big) = \mathcal{E}\left(1 - \frac{1 - X}{n}\right)$$

$$= 1 - \frac{1 - \mathcal{E}(X)}{n} = 1 - \frac{1 - 1/(n+1)}{n} = \frac{n}{n+1}.$$

b) Referring again to Equations (10.42) and (10.25) and using properties of variance, we get

$$\text{Var}\big(\mathcal{E}(Y \mid X)\big) = \text{Var}\left(1 - \frac{1 - X}{n}\right) = \frac{1}{n^2}\text{Var}(X) = \frac{1}{n^2}\frac{n}{(n+2)(n+1)^2},$$

or

$$\text{Var}\big(\mathcal{E}(Y \mid X)\big) = \frac{1}{n(n+2)(n+1)^2}. \tag{10.47}$$

Next we refer to Equation (10.43) and Equation (10.24) on pages 598 and 585, respectively, to obtain

$$\mathcal{E}\big(\text{Var}(Y \mid X)\big) = \mathcal{E}\left(\frac{n-1}{n^2(n+1)}(1 - X)^2\right)$$

$$= \frac{n-1}{n^2(n+1)}\mathcal{E}\big((1-X)^2\big) = \frac{n-1}{n^2(n+1)}\left(1 - 2\mathcal{E}(X) + \mathcal{E}\big(X^2\big)\right)$$

$$= \frac{n-1}{n^2(n+1)}\left(1 - \frac{2}{n+1} + \frac{2}{(n+2)(n+1)}\right) = \frac{n-1}{n(n+1)(n+2)}.$$

Applying the law of total variance and using the preceding result and Equation (10.47), we find that

$$\text{Var}(Y) = \mathcal{E}\big(\text{Var}(Y \mid X)\big) + \text{Var}\big(\mathcal{E}(Y \mid X)\big)$$

$$= \frac{n-1}{n(n+1)(n+2)} + \frac{1}{n(n+2)(n+1)^2} = \frac{n}{(n+2)(n+1)^2}.$$

c) The values for $\mathcal{E}(Y)$ and $\text{Var}(Y)$ that we obtained in parts (a) and (b), respectively, agree with those found directly in Example 10.11, as presented in Equation (10.26) on page 585.   ∎

Prediction

In applications, we frequently need to predict the value of a random variable, based on known values of other random variables. Here we consider the simplest case of predicting the value of a random variable Y, based on the known value of one random variable X. Mathematically, the solution to this problem requires the selection of a real-valued function h defined on the range of X so that when we observe the value x of X, we predict $h(x)$ for the value of Y.

We want our predicted value for Y to be as close as possible to the actual value of Y. Consequently, to begin, we must choose a criterion for "closest." One commonly used criterion is to minimize the **mean square error**—that is, we select the function h so that $\mathcal{E}\big([Y - h(X)]^2\big)$ is as small as possible. As we prove shortly, the function h defined by $h(x) = \mathcal{E}(Y \mid X = x)$ does the trick.

To begin, we need to extend the definition of conditional expectation. We do so in the continuous case; the extension in the discrete case is similar. Suppose that X and Y are continuous random variables with a joint PDF and that g is a real-valued function of two real variables defined on the range of (X, Y). If $f_X(x) > 0$, we define the **conditional expectation of $g(X, Y)$ given $X = x$,** denoted $\mathcal{E}\big(g(X, Y) \mid X = x\big)$, to be

$$\mathcal{E}\big(g(X, Y) \mid X = x\big) = \int_{-\infty}^{\infty} g(x, y) f_{Y \mid X}(y \mid x) \, dy. \tag{10.48}$$

If $f_X(x) = 0$, we define $\mathcal{E}\big(g(X, Y) \mid X = x\big) = 0$ but don't refer to $\mathcal{E}\big(g(X, Y) \mid X = x\big)$ as a conditional expectation.

When $g(x, y) = y$, the extended definition of conditional expectation, as presented in Equation (10.48), is consistent with Definition 10.2 on page 596. More generally, if g is a function of y alone, the extended definition of conditional expectation agrees with the conditional form of the FEF in Equation (10.38) on page 596. You should verify that the extended definition of conditional expectation satisfies the usual properties of expectation: linearity, monotonicity, and so forth.

Let ξ denote the real-valued function defined on $\mathcal{R}$ by $\xi(x) = \mathcal{E}\big(g(X, Y) \mid X = x\big)$. Then, not surprisingly, we define the **conditional expectation of $g(X, Y)$ given X,** denoted $\mathcal{E}\big(g(X, Y) \mid X\big)$, to be the random variable $\xi(X)$.

A useful property is that, if k is a function of x alone and ℓ is a function of both x and y, then

$$\mathcal{E}\big(k(X)\ell(X, Y) \mid X\big) = k(X)\mathcal{E}\big(\ell(X, Y) \mid X\big). \tag{10.49}$$

To verify Equation (10.49) in the continuous case, we apply Equation (10.48), first with $g(x, y) = k(x)\ell(x, y)$ and then with $g(x, y) = \ell(x, y)$, as follows:

$$\mathcal{E}\big(k(X)\ell(X, Y) \mid X = x\big) = \int_{-\infty}^{\infty} k(x)\ell(x, y) f_{Y \mid X}(y \mid x)\, dy$$

$$= k(x) \int_{-\infty}^{\infty} \ell(x, y) f_{Y \mid X}(y \mid x)\, dy$$

$$= k(x)\mathcal{E}\big(\ell(X, Y) \mid X = x\big).$$

The proof in the discrete case is virtually identical.

The law of total expectation holds for the extended form of conditional expectation. That is, we have

$$\mathcal{E}\big(g(X, Y)\big) = \mathcal{E}\Big(\mathcal{E}\big(g(X, Y) \mid X\big)\Big). \tag{10.50}$$

We prove Equation (10.50) in the continuous case; the proof in the discrete case is essentially the same. Letting $\xi(x) = \mathcal{E}\big(g(X, Y) \mid X = x\big)$ and applying in turn the univariate form of the FEF, Equation (10.48), the general multiplication rule, and the bivariate form of the FEF gives

$$\mathcal{E}\Big(\mathcal{E}\big(g(X, Y) \mid X\big)\Big) = \mathcal{E}\big(\xi(X)\big) = \int_{-\infty}^{\infty} \xi(x) f_X(x)\, dx$$

$$= \int_{-\infty}^{\infty} \mathcal{E}\big(g(X, Y) \mid X = x\big) f_X(x)\, dx$$

$$= \int_{-\infty}^{\infty} \left(\int_{-\infty}^{\infty} g(x, y) f_{Y \mid X}(y \mid x)\, dy \right) f_X(x)\, dx$$

$$= \int_{-\infty}^{\infty} \int_{-\infty}^{\infty} g(x, y) f_{X,Y}(x, y)\, dx\, dy = \mathcal{E}\big(g(X, Y)\big).$$

In Proposition 10.8, we prove the **prediction theorem,** which shows that $\mathcal{E}(Y \mid X)$ is the optimal predictor of Y based on observing X.

◆◆◆ **Proposition 10.8   Prediction Theorem**

Let X and Y be random variables defined on the same sample space and suppose that Y has finite variance. Then, with minimum mean square error as the criterion, the best predictor of Y based on observing X is $\mathcal{E}(Y \mid X)$, the conditional expectation of Y given X.

Proof   The prediction theorem states that

$$\mathcal{E}\Big((Y - \mathcal{E}(Y \mid X))^2\Big) = \min_{h \in \mathcal{H}} \mathcal{E}\Big((Y - h(X))^2\Big), \tag{10.51}$$

where $\mathcal{H}$ consists of all real-valued functions h defined on the range of X such that $h(X)$ has finite variance.

For convenience, we set $\psi(x) = \mathcal{E}(Y \mid X = x)$. We first show that $Y - \psi(X)$ is uncorrelated with any function of X or, more to the point, that $\mathcal{E}([Y - \psi(X)]k(X)) = 0$ for each $k \in \mathcal{H}$. To verify this fact, we first apply Equations (10.50) and (10.49) to get

$$\mathcal{E}(Yk(X)) = \mathcal{E}\left(\mathcal{E}(Yk(X) \mid X)\right) = \mathcal{E}(k(X)\mathcal{E}(Y \mid X)) = \mathcal{E}(k(X)\psi(X)).$$

Therefore, by the linearity property of expected value,

$$\mathcal{E}\left((Y - \psi(X))k(X)\right) = \mathcal{E}(Yk(X)) - \mathcal{E}(\psi(X)k(X))$$
$$= \mathcal{E}(k(X)\psi(X)) - \mathcal{E}(\psi(X)k(X)) = 0.$$

Now, let $h \in \mathcal{H}$. We have

$$(Y - h(X))^2 = \left((Y - \psi(X)) + (\psi(X) - h(X))\right)^2$$
$$= (Y - \psi(X))^2 + 2(Y - \psi(X))(\psi(X) - h(X)) + (\psi(X) - h(X))^2$$
$$\geq (Y - \psi(X))^2 + 2(Y - \psi(X))(\psi(X) - h(X)).$$

As $\psi(X) - h(X)$ is a function of X alone, it is uncorrelated with $Y - \psi(X)$. Applying the monotonicity and linearity properties of expected value to the previous display yields

$$\mathcal{E}\left((Y - h(X))^2\right) \geq \mathcal{E}\left((Y - \psi(X))^2\right) + 2\mathcal{E}\left((Y - \psi(X))(\psi(X) - h(X))\right)$$
$$= \mathcal{E}\left((Y - \psi(X))^2\right).$$

Consequently, Equation (10.51) holds. ◆

EXAMPLE 10.20 *The Prediction Theorem*

Min and Max of a Random Sample Let $X_1, \ldots, X_n$ be a random sample from a $\mathcal{U}(0, 1)$ distribution, and set $X = \min\{X_1, \ldots, X_n\}$ and $Y = \max\{X_1, \ldots, X_n\}$.

a) Determine the best predictor of the maximum of the random sample, based on knowing the minimum of the random sample.

b) Compare the predicted value of the maximum of the random sample to the expected value of the maximum of the random sample.

Solution a) From Equation (10.42), $\mathcal{E}(Y \mid X) = 1 - (1 - X)/n$, which, by the prediction theorem, is the best predictor of Y based on observing X. Therefore, knowing that the minimum of the random sample is x, the predicted maximum of the random sample is $1 - (1 - x)/n$.

b) From Equation (10.26), the expected maximum of the random sample is $n/(n + 1)$. Simple algebra shows that the predicted maximum, $1 - (1 - x)/n$, is less than, equal to, or greater than the expected maximum, $n/(n + 1)$, depending on whether x is less than, equal to, or greater than $1/(n + 1)$, respectively. However, from Equation (10.25), we know that $1/(n + 1)$ is the expected minimum of the random sample. Hence the predicted maximum of the random sample is less than, equal to, or greater than the expected maximum, depending on whether x is less than, equal to, or greater than the expected minimum of the random sample, respectively. ∎

We emphasize that the results obtained in this section for continuous random variables hold in general:

> The law of total expectation, the law of total variance, and the prediction theorem are valid for any types or combination of types of random variables.

Conditional Expectation: Multivariate Case

Conditional expectation, conditional variance, and related concepts can be easily extended to the multivariate case. For instance, suppose that X, Y, and Z are continuous random variables with a joint PDF. Then the **conditional expectation of Z given $X = x$ and $Y = y$,** denoted $\mathcal{E}(Z \mid X = x, \, Y = y)$, is defined to be the expected value of Z relative to the conditional distribution of Z given $X = x$ and $Y = y$. Thus, if $f_{X,Y}(x, y) > 0$,

$$\mathcal{E}(Z \mid X = x, \, Y = y) = \int_{-\infty}^{\infty} z f_{Z \mid X,Y}(z \mid x, y) \, dz,$$

provided, of course, that $Z_{\mid X=x, Y=y}$ has finite expectation. If $f_{X,Y}(x, y) = 0$, we define $\mathcal{E}(Z \mid X = x, \, Y = y) = 0$ but don't refer to $\mathcal{E}(Z \mid X = x, \, Y = y)$ as a conditional expectation. The law of total expectation remains valid in the multivariate case. For instance, as you are asked to show in Exercise 10.92, $\mathcal{E}(Z) = \mathcal{E}\big(\mathcal{E}(Z \mid X, Y)\big)$.

EXERCISES 10.4   **Basic Exercises**

10.77 A number X is chosen at random from the interval $(0, 1)$. Then another number Y is chosen at random from the interval $(0, X)$.
a) Determine the conditional expectation of Y and the conditional variance of Y for each possible value of X.
b) In Exercises 10.18 and 10.60 on pages 570 and 593, we asked you to find $\mathcal{E}(Y)$ and Var(Y), respectively, by first obtaining a PDF of Y. Use the laws of total expectation and total variance to find $\mathcal{E}(Y)$ and Var(Y).

10.78 The stock prices of two companies at the end of any specified year are modeled with random variables X and Y that follow a distribution with joint density function $f_{X,Y}(x, y) = 2x$ for $x < y < x + 1$ and $0 < x < 1$, and $f_{X,Y}(x, y) = 0$ otherwise. Determine $\mathcal{E}(Y \mid X = x)$ and Var$(Y \mid X = x)$ for all $x \in \mathcal{R}$.

10.79 Let T denote the interior of the triangle with vertices at $(0, 0)$, $(2, 0)$ and $(2, 1)$. Let X and Y have joint PDF given by $f_{X,Y}(x, y) = 2xy$ for $(x, y) \in T$, and $f_{X,Y}(x, y) = 0$ otherwise. Use the law of total expectation to determine $\mathcal{E}(X)$.

10.80 Consider two electrical components, A and B, with respective lifetimes X and Y. Assume that a joint PDF of X and Y is given by $f_{X,Y}(x, y) = \lambda \mu \, e^{-(\lambda x + \mu y)}$ for positive x and y, and $f_{X,Y}(x, y) = 0$ otherwise, where λ and μ are positive constants.
a) Determine $\mathcal{E}(Y \mid X = x)$ and Var$(Y \mid X = x)$ for each possible value x of X.
b) Referring to part (a), why don't $\mathcal{E}(Y \mid X = x)$ and Var$(Y \mid X = x)$ depend on x?
c) Use the law of total expectation to obtain $\mathcal{E}(Y)$.
d) Use the law of total variance to obtain Var(Y).

10.81 Let X and Y be independent random variables having PDFs and finite variances.
a) Without doing any computations, obtain $\mathcal{E}(Y \mid X = x)$ for each possible value x of X. Explain your reasoning.
b) Use Definition 10.2 on page 596 to obtain $\mathcal{E}(Y \mid X = x)$ for each possible value x of X.
c) Without doing any computations, obtain $\mathrm{Var}(Y \mid X = x)$ for each possible value x of X. Explain your reasoning.
d) Use Definition 10.3 on page 596 to obtain $\mathrm{Var}(Y \mid X = x)$ for each possible value x of X.
e) Use part (b) to show that $\mathcal{E}(\mathcal{E}(Y \mid X)) = \mathcal{E}(Y)$, thus verifying the law of total expectation in this case.
f) Use part (d) to show that $\mathcal{E}(\mathrm{Var}(Y \mid X)) + \mathrm{Var}(\mathcal{E}(Y \mid X)) = \mathrm{Var}(Y)$, thus verifying the law of total variance in this case.

10.82 Let X have the standard normal distribution and let $-1 < \rho < 1$. Suppose that, for each $x \in \mathcal{R}$, the conditional distribution of Y given $X = x$ is a normal distribution with parameters ρx and $1 - \rho^2$.
a) Without doing any computations, obtain $\mathcal{E}(Y \mid X)$ and $\mathrm{Var}(Y \mid X)$.
b) Use your results from part (a) and the laws of total expectation and total variance to determine $\mathcal{E}(Y)$ and $\mathrm{Var}(Y)$.
c) Show that your results from part (b) are consistent with those that would be obtained by using the PDF of Y found in Example 9.13(b) on page 517.

10.83 A bus is scheduled to arrive at a bus stop within the next hour. Starting from now, the amount of time, in hours, until the bus actually arrives has a beta distribution with parameters 2 and 1. The number of passengers who have arrived at the bus stop t hours from now has a Poisson distribution with parameter λt, independent of when the bus arrives. Determine the mean and variance of the number of passengers at the bus stop when the bus arrives.

10.84 Suppose that X and Y are independent random variables.
a) What do you think is the best predictor of Y, based on observing X?
b) Use the prediction theorem to determine the best predictor of Y, based on observing X.

10.85 Suppose that Y is some function of X—say, $Y = g(X)$.
a) What do you think is the best predictor of Y, based on observing X?
b) Use the prediction theorem to determine the best predictor of Y, based on observing X.

10.86 Show that the minimum mean square error for prediction equals $\mathcal{E}(\mathrm{Var}(Y \mid X))$.

10.87 Fix a number $c \in (0, 1)$ and divide the interval $(0, 1)$ randomly into two subintervals.
a) What is the expected length of the subinterval that contains c?
b) What value of c maximizes the expected length of the subinterval that contains c?

10.88 In a textile mill, the number of defects per yard of fabric has a Poisson distribution with mean λ. However, λ varies from loom to loom and can be thought of as a random variable Λ whose probability distribution is concentrated on the interval $(0, 3)$. Determine the mean and variance of the number of defects per yard when
a) $\Lambda \sim \mathcal{U}(0, 3)$.
b) $\Lambda \sim \mathcal{T}(0, 3)$.

10.89 Two types of devices—say, Type 1 and Type 2—are used at a factory. The lifetime of a Type k device has mean μ_k and variance σ_k^2. A bin contains $100p\%$ Type 1 devices and the rest Type 2 devices. If one device is selected at random from the bin, what are the mean and variance of its lifetime?

10.90 Metro Insurance Company issued insurance policies to 32 independent risks. For each policy, the probability of a claim is 1/6. Given that there is a claim, the benefit amount has a beta distribution with parameters 1 and 2. Determine the mean and variance of the total benefits paid by expressing the total claim amount as a sum of
a) a random number of random variables.
b) 32 random variables.

10.91 An insurance company insures a car worth $15,000 for 1 year under a policy with a $1,000 deductible. During the policy year, there is a 4% chance of partial damage to the car and a 2% chance of a total loss. If there is partial damage to the car, the amount of damage, in thousands of dollars, follows a distribution with density function $f(x) = 0.5003\, e^{-x/2}$ for $0 < x < 15$, and $f(x) = 0$ otherwise. Find the expected claim payment.

Theory Exercises

10.92 Multivariate law of total expectation: The law of total expectation remains valid in the multivariate case. Here you are to establish that result for continuous random variables with a joint PDF.
a) Suppose that X, Y, and Z are continuous random variables with a joint PDF. Prove that $\mathcal{E}(Z) = \mathcal{E}\big(\mathcal{E}(Z \mid X, Y)\big)$.
b) Generalize part (a) to the multivariate case of m random variables with a joint PDF.

10.93 Law of total expectation for partitions: As we mentioned on page 604, the law of total expectation is valid for any types or combination of types of random variables.
a) Let Y be a random variable with finite expectation and let A_1, A_2, ... be a partition of the sample space. Prove that $\mathcal{E}(Y) = \sum_n \mathcal{E}(Y \mid A_n)\, P(A_n)$, where $\mathcal{E}(Y \mid A_n)$ is the expectation of Y relative to the conditional distribution of Y, given that event A_n occurs.
b) Use part (a) to derive the law of total probability, Proposition 4.3 on page 142.

Advanced Exercises

10.94 An electronic component has probability p of failing immediately when turned on. If it doesn't fail immediately, it's life distribution has CDF F. Express the expected lifetime of the component in terms of p and F.

10.95 Best linear predictor: Let X and Y be two random variables defined on the same sample space and having finite variances. A *linear predictor* of Y based on observing X is a random variable of the form $a + bX$, where a and b are real numbers.
a) With minimum mean square error as a criterion, show that the best linear predictor of Y based on observing X is $\mu_Y + \rho\sigma_Y(X - \mu_X)/\sigma_X$, where $\rho = \rho(X, Y)$. *Hint:* Apply calculus techniques.
b) Show that the minimum mean square error for linear prediction is $(1 - \rho^2)\sigma_Y^2$.

10.96 Let $X_1, \ldots, X_n$ be a random sample from a uniform distribution on the interval $(0, 1)$. Set $X = \min\{X_1, \ldots, X_n\}$ and $Y = \max\{X_1, \ldots, X_n\}$.
a) Referring to Exercise 10.95, Equations (10.25) and (10.26) on page 585, and Equation (10.33) on page 589, obtain the best linear predictor of Y based on observing X.
b) Compare the predictor that you obtained in part (a) to the best predictor found in Example 10.20(a) on page 603.
c) Why are the predictors in parts (a) and (b) identical?
d) From Exercises 10.86 and 10.95(b), the minimum mean square errors for prediction and linear prediction are $\mathcal{E}\big(\mathrm{Var}(Y \mid X)\big)$ and $(1 - \rho^2)\sigma_Y^2$, respectively. Why must these two quantities be equal in the present case? Verify explicitly that they are indeed equal.

10.97 A very large company allows each employee a maximum of 100 hours of sick leave per year. For a randomly selected employee, let X and Y denote the number of hundreds of hours of sick leave taken last year and this year, respectively. A joint PDF of X and Y is

$$f_{X,Y}(x, y) = 2 - \frac{6}{5}x - \frac{4}{5}y, \qquad 0 < x < 1,\ 0 < y < 1.$$

a) Determine the best predictor of an employee's sick leave this year, based on the amount of sick leave he took last year.

b) Determine the best linear predictor of an employee's sick leave this year, based on the amount of sick leave he took last year. *Note:* Refer to Exercise 10.95.

c) Compare the results obtained in parts (a) and (b). Why are they different?

d) Predict this year's sick leave for an employee who took 10 hours of sick leave last year by using (i) the best predictor and (ii) the best linear predictor. Compare your answers.

Mixed bivariate distributions: On pages 509 and 522, we introduced and applied the concept of mixed bivariate distributions, where we simultaneously consider a continuous random variable X and a discrete random variable Y. In Exercises 10.98–10.100, you are to consider conditional expectation and the law of total expectation for mixed bivariate distributions.

10.98 Let X and Y be random variables with joint PDF/PMF $h_{X,Y}$.

a) Define $\mathcal{E}(Y \mid X = x)$ and $\mathcal{E}(X \mid Y = y)$.

b) Prove that the law of total expectation holds for mixed bivariate distributions. That is, show that $\mathcal{E}(Y) = \mathcal{E}\big(\mathcal{E}(Y \mid X)\big)$ and $\mathcal{E}(X) = \mathcal{E}\big(\mathcal{E}(X \mid Y)\big)$.

10.99 Let Y denote the number of patients who arrive at an emergency room during a 1-hour period and let X denote the rate at which they arrive. A joint PDF/PMF of X and Y is given by $h_{X,Y}(x, y) = \lambda\, e^{-(1+\lambda)x}(x^y/y!)$ for $x > 0$ and $y = 0, 1, 2, \ldots$, and $h_{X,Y}(x, y) = 0$ otherwise.

a) Referring to Exercise 10.98, determine $\mathcal{E}(Y \mid X = x)$ and $\mathcal{E}(X \mid Y = y)$.

b) Use the law of total expectation to obtain $\mathcal{E}(Y)$ and $\mathcal{E}(X)$.

10.100 The annual number of automobile accidents, Y, of a driver has a Poisson distribution with parameter λ, where λ varies as a random variable X having a gamma distribution with parameters α and β.

a) Determine the conditional expectation of Y given $X = x$ for each possible value x of X.

b) Use the result of part (a) and the law of total expectation to determine $\mathcal{E}(Y)$.

c) Determine $\mathcal{E}(Y)$ by first finding the PMF of Y.

d) Determine the conditional expectation of X given $Y = y$ for each possible value y of Y.

e) Use the result of part (d) and the law of total expectation to determine $\mathcal{E}(X)$.

10.5   The Bivariate Normal Distribution

In this section, we discuss one of the most important bivariate probability distributions— the *bivariate normal distribution,* which generalizes the univariate normal distribution to two dimensions. The generalization is nontrivial, however, because it introduces another parameter—namely, the correlation coefficient.

First we recall from Examples 10.3 and 10.12, respectively, that the mean and variance of a normal random variable equal its first and second parameters, respectively.

Thus, if $W \sim \mathcal{N}(\mu, \sigma^2)$, then $\mathcal{E}(W) = \mu$ and $\text{Var}(W) = \sigma^2$, or $\mu_W = \mu$ and $\sigma_W^2 = \sigma^2$. Consequently, we can write $W \sim \mathcal{N}(\mu_W, \sigma_W^2)$. Using this notation, we have

$$f_W(w) = \frac{1}{\sqrt{2\pi}\,\sigma_W} e^{-(w-\mu_W)^2/2\sigma_W^2} = \frac{1}{\sqrt{2\pi}\,\sigma_W} e^{-\frac{1}{2}\left(\frac{w-\mu_W}{\sigma_W}\right)^2}, \tag{10.52}$$

for $-\infty < w < \infty$.

Introducing the Bivariate Normal Distribution

To develop the bivariate normal distribution, we begin with the simplest case—namely, that of two independent normal random variables, X and Y. As X and Y are independent, their joint PDF equals the product of their marginal PDFs, or

$$f_{X,Y}(x, y) = f_X(x) f_Y(y) = \frac{1}{\sqrt{2\pi}\,\sigma_X} e^{-\frac{1}{2}\left(\frac{x-\mu_X}{\sigma_X}\right)^2} \cdot \frac{1}{\sqrt{2\pi}\,\sigma_Y} e^{-\frac{1}{2}\left(\frac{y-\mu_Y}{\sigma_Y}\right)^2}$$

or

$$f_{X,Y}(x, y) = \frac{1}{2\pi\sigma_X\sigma_Y} e^{-\frac{1}{2}\left\{\left(\frac{x-\mu_X}{\sigma_X}\right)^2 + \left(\frac{y-\mu_Y}{\sigma_Y}\right)^2\right\}}, \tag{10.53}$$

for $-\infty < x < \infty$ and $-\infty < y < \infty$. The joint distribution of X and Y, whose PDF is given in Equation (10.53), is an example of a bivariate normal distribution. We note that, because X and Y are independent, they are uncorrelated—that is, $\rho(X, Y) = 0$.

We also want to consider correlated normal random variables. The question then is: What is the form of the joint PDF of a bivariate normal distribution for which the correlation between the two random variables equals some specified number ρ, where $-1 < \rho < 1$?

To answer this question, we first refer to Examples 10.15 and 10.16 on pages 591 and 592, respectively. There we showed the following: Suppose that Z_1 and Z_2 are independent random variables with zero means and unit variances, and that μ_1 and μ_2 are real numbers, σ_1 and σ_2 are positive real numbers, and $-1 \le \rho \le 1$. Furthermore suppose that a, b, c, and d are any four real numbers that satisfy

$$a^2 + b^2 = \sigma_1^2, \qquad ac + bd = \rho\sigma_1\sigma_2, \qquad c^2 + d^2 = \sigma_2^2, \tag{10.54}$$

and let the random variables X and Y be defined by

$$\begin{aligned} X &= \mu_1 + aZ_1 + bZ_2 \\ Y &= \mu_2 + cZ_1 + dZ_2. \end{aligned} \tag{10.55}$$

Then $\mathcal{E}(X) = \mu_1$, $\text{Var}(X) = \sigma_1^2$, $\mathcal{E}(Y) = \mu_2$, $\text{Var}(Y) = \sigma_2^2$, and $\rho(X, Y) = \rho$. Consequently, we can obtain random variables with arbitrarily specified means, variances, and correlation coefficient by appropriately transforming independent random variables with zero means and unit variances.

If in Equations (10.55), Z_1 and Z_2 are not only independent but also are normally distributed, then, from Proposition 9.14 on page 540, X and Y are normally distributed. To obtain a joint PDF of X and Y—assuming that Z_1 and Z_2 are independent standard normal random variables—we apply the bivariate transformation theorem. In doing so, we assume that $-1 < \rho < 1$. (Do you know why?)

As Z_1 and Z_2 are independent standard normal random variables, their joint PDF is

$$f_{Z_1,Z_2}(z_1, z_2) = f_{Z_1}(z_1)f_{Z_2}(z_2) = \frac{1}{\sqrt{2\pi}}e^{-z_1^2/2} \cdot \frac{1}{\sqrt{2\pi}}e^{-z_2^2/2}$$

or

$$f_{Z_1,Z_2}(z_1, z_2) = \frac{1}{2\pi}e^{-\frac{1}{2}(z_1^2+z_2^2)}, \qquad \textbf{(10.56)}$$

for $-\infty < z_1 < \infty$ and $-\infty < z_2 < \infty$. The Jacobian determinant of the transformation in Equations (10.55) is

$$J(z_1, z_2) = \begin{vmatrix} a & b \\ c & d \end{vmatrix} = ad - bc.$$

By referring to Equations (10.54) and performing some simple algebra, we find that $(ad - bc)^2 = \sigma_1^2\sigma_2^2(1 - \rho^2)$. Thus

$$|J(z_1, z_2)| = \sigma_1\sigma_2\sqrt{1 - \rho^2}. \qquad \textbf{(10.57)}$$

Solving the equations $x = \mu_1 + az_1 + bz_2$ and $y = \mu_2 + cz_1 + dz_2$ for z_1 and z_2 yields the inverse transformation

$$z_1 = \big((x - \mu_1)d - (y - \mu_2)b\big)/(ad - bc)$$
$$z_2 = \big(-(x - \mu_1)c + (y - \mu_2)a\big)/(ad - bc).$$

From these two equations and Equations (10.54), we find that

$$z_1^2 + z_2^2 = \frac{1}{1 - \rho^2}$$
$$\times \left\{ \left(\frac{x - \mu_1}{\sigma_1}\right)^2 - 2\rho\left(\frac{x - \mu_1}{\sigma_1}\right)\left(\frac{y - \mu_2}{\sigma_2}\right) + \left(\frac{y - \mu_2}{\sigma_2}\right)^2 \right\}. \qquad \textbf{(10.58)}$$

Applying the bivariate transformation theorem (Proposition 9.16 on page 546) and keeping Equations (10.56), (10.57), and (10.58) in mind, we get

$$f_{X,Y}(x, y) = \frac{1}{|J(z_1, z_2)|}f_{Z_1,Z_2}(z_1, z_2) = \frac{1}{\sigma_1\sigma_2\sqrt{1 - \rho^2}}\frac{1}{2\pi}e^{-\frac{1}{2}(z_1^2+z_2^2)}$$

or

$$f_{X,Y}(x, y) = \frac{1}{2\pi\sigma_1\sigma_2\sqrt{1 - \rho^2}}e^{-\frac{1}{2(1-\rho^2)}\left\{\left(\frac{x-\mu_1}{\sigma_1}\right)^2-2\rho\left(\frac{x-\mu_1}{\sigma_1}\right)\left(\frac{y-\mu_2}{\sigma_2}\right)+\left(\frac{y-\mu_2}{\sigma_2}\right)^2\right\}},$$

for $-\infty < x < \infty$ and $-\infty < y < \infty$. Based on this equation, we now define **bivariate normal random variables**.

> **DEFINITION 10.4   Bivariate Normal Random Variables**
>
> Continuous random variables X and Y are called **bivariate normal random variables** if they have a joint probability density function of the form
>
> $$f_{X,Y}(x, y) = \frac{1}{2\pi\sigma_X\sigma_Y\sqrt{1-\rho^2}}\, e^{-\frac{1}{2}Q(x,y)}, \qquad -\infty < x, y < \infty,$$
>
> where $-1 < \rho < 1$, and
>
> $$Q(x, y) = \frac{1}{1-\rho^2}\left[\left(\frac{x-\mu_X}{\sigma_X}\right)^2 - 2\rho\left(\frac{x-\mu_X}{\sigma_X}\right)\left(\frac{y-\mu_Y}{\sigma_Y}\right) + \left(\frac{y-\mu_Y}{\sigma_Y}\right)^2\right].$$
>
> We say that X and Y have the **bivariate normal distribution with parameters** $\mu_X, \sigma_X^2, \mu_Y, \sigma_Y^2,$ **and** ρ. To indicate that X and Y have that bivariate normal distribution, we often write $(X, Y) \sim \mathcal{BVN}(\mu_X, \sigma_X^2, \mu_Y, \sigma_Y^2, \rho)$.

We can obtain the marginal distributions for bivariate normal random variables by individually integrating the joint PDF in Definition 10.4 on each of the variables (Proposition 9.7 on page 510). However, there is an easier way. Indeed, because a joint distribution determines the marginal distributions, the discussion leading to Definition 10.4 shows the following about bivariate normal random variables, X and Y, with parameters $\mu_X, \sigma_X^2, \mu_Y, \sigma_Y^2,$ and ρ.

- $X \sim \mathcal{N}(\mu_X, \sigma_X^2)$: The marginal distribution of X is a normal distribution with mean μ_X and variance σ_X^2.

- $Y \sim \mathcal{N}(\mu_Y, \sigma_Y^2)$: The marginal distribution of Y is a normal distribution with mean μ_Y and variance σ_Y^2.

- $\rho(X, Y) = \rho$: The correlation between X and Y is ρ.

Thus the marginal distributions of bivariate normal random variables, X and Y, are both normal and the parameters $\mu_X, \sigma_X^2, \mu_Y, \sigma_Y^2,$ and ρ are indeed the mean and variance of X, the mean and variance of Y, and the correlation coefficient of X and Y, respectively.

Recall that, in the univariate case, the normal distribution with mean 0 and variance 1—an $\mathcal{N}(0, 1)$ distribution—is called the *standard normal distribution*. Similarly, in the bivariate case, a bivariate normal distribution whose marginals both have mean 0 and variance 1—a $\mathcal{BVN}(0, 1, 0, 1, \rho)$ distribution—is called a **standard bivariate normal distribution.** In the univariate case, there is only one standard normal distribution, whereas in the bivariate case, there are infinitely many standard bivariate normal distributions—one for each value of ρ between -1 and 1.

For a bivariate normal distribution, μ_X and μ_Y are location parameters and σ_X and σ_Y are scale parameters. Consequently, we can get the basic picture of bivariate normal distributions by just looking at the effect of the correlation coefficient, ρ. Figure 10.2 shows standard bivariate normal distributions for five different values of ρ.

Figure 10.2 Selected standard bivariate normal distributions

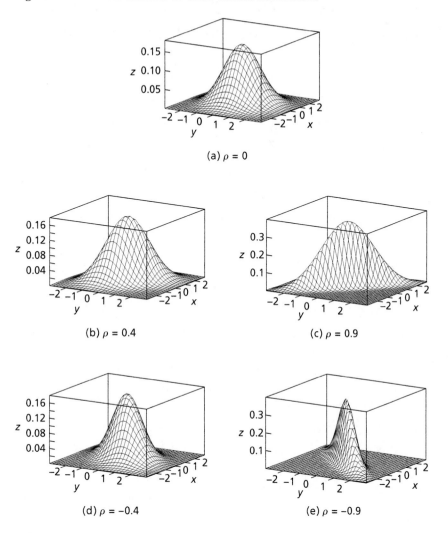

(a) $\rho = 0$

(b) $\rho = 0.4$

(c) $\rho = 0.9$

(d) $\rho = -0.4$

(e) $\rho = -0.9$

EXAMPLE   10.21   *The Bivariate Normal Distribution*

Heights of Mothers and Sons   For a family with sons, let X and Y denote the heights, in inches, of the mother and eldest son, respectively. Data from the *National Health and Nutrition Survey* and research by K. Pearson and A. Lee ("On the Laws of Inheritance in Man. I. Inheritance of Physical Characters," *Biometrika*, 1903, Vol. 2, pp. 357–462) show that it's reasonable to presume that X and Y have a bivariate normal distribution. Specifically, we can assume that $(X, Y) \sim \mathcal{BVN}(63.7, 2.7^2, 69.1, 2.9^2, 0.5)$.

a) Determine and graph a joint PDF of this bivariate normal distribution.

b) Identify and interpret the marginal distribution of X.

c) Identify and interpret the marginal distribution of Y.

d) Identify and interpret the correlation coefficient of X and Y.

Solution   Because $(X, Y) \sim \mathcal{BVN}(63.7, 2.7^2, 69.1, 2.9^2, 0.5)$, we have

$$\mu_X = 63.7, \quad \sigma_X = 2.7, \quad \mu_Y = 69.1, \quad \sigma_Y = 2.9, \quad \rho = 0.5. \qquad \textbf{(10.59)}$$

a) From Equations (10.59),

$$2\sigma_X\sigma_Y\sqrt{1 - \rho^2} = 2 \cdot 2.7 \cdot 2.9\sqrt{1 - 0.5^2} = 7.83\sqrt{3},$$

$$1 - \rho^2 = 1 - 0.5^2 = 3/4,$$

$$2\rho = 2 \cdot 0.5 = 1.$$

Therefore, by Definition 10.4 and Equations (10.59), a joint PDF of mother and eldest son heights is

$$f_{X,Y}(x, y) = \frac{1}{7.83\sqrt{3}\,\pi}\, e^{-\frac{2}{3}\left\{\left(\frac{x-63.7}{2.7}\right)^2 - \left(\frac{x-63.7}{2.7}\right)\left(\frac{y-69.1}{2.9}\right) + \left(\frac{y-69.1}{2.9}\right)^2\right\}}.$$

A graph of this bivariate normal PDF is shown in Figure 10.3.

Figure 10.3 Graph of joint PDF of a bivariate normal distribution with parameters $\mu_X = 63.7$, $\sigma_X^2 = 2.7^2$, $\mu_Y = 69.1$, $\sigma_Y^2 = 2.9^2$, and $\rho = 0.5$

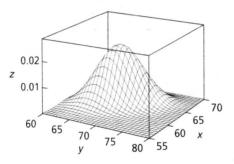

b) Because (X, Y) is bivariate normal, we know that the marginal distribution of X is $\mathcal{N}(\mu_X, \sigma_X^2) = \mathcal{N}(63.7, 2.7^2)$. Hence the heights of mothers (with sons) are normally distributed with mean 63.7 inches and standard deviation 2.7 inches.

c) Because (X, Y) is bivariate normal, we know that the marginal distribution of Y is $\mathcal{N}(\mu_Y, \sigma_Y^2) = \mathcal{N}(69.1, 2.9^2)$. Hence the heights of eldest sons are normally distributed with mean 69.1 inches and standard deviation 2.9 inches.

d) We know that $\rho(X, Y) = \rho = 0.5$. Hence the correlation between the heights of mothers and their eldest sons is 0.5. We see that a moderate positive correlation exists between the height of a mother and that of her eldest son. In particular, eldest sons of tall mothers tend to be taller than eldest sons of short mothers.   ∎

Independence and Uncorrelation for Bivariate Normal Random Variables

Recall that two random variables are said to be *uncorrelated* if their correlation coefficient is 0. As you know, although independent random variables (with finite variances) are uncorrelated, uncorrelated random variables aren't necessarily independent. For bivariate normal random variables, however, independence and uncorrelatedness are equivalent, as Proposition 10.9 shows.

◆◆◆ **Proposition 10.9**

Bivariate normal random variables are independent if and only if they are uncorrelated.

Proof   We already know that independent random variables are uncorrelated—bivariate normal or not. Consequently, we need only show that uncorrelated bivariate normal random variables are independent.

Let X and Y be uncorrelated bivariate normal random variables. From Definition 10.4, if $\rho = \rho(X, Y) = 0$, then a joint PDF of X and Y reduces to

$$f_{X,Y}(x, y) = \frac{1}{2\pi\sigma_X\sigma_Y} e^{-\frac{1}{2}\left\{\left(\frac{x-\mu_X}{\sigma_X}\right)^2 + \left(\frac{y-\mu_Y}{\sigma_Y}\right)^2\right\}}. \tag{10.60}$$

However, from the argument in the paragraph containing Equation (10.53) on page 608, the joint PDF in Equation (10.60) is that of independent normal random variables. Hence, as a joint distribution determines the marginal distributions, we conclude that X and Y are independent random variables.   ◆

Note: The assumption of bivariate normal is essential in Proposition 10.9. It's not even enough that X and Y have a joint PDF and that the marginal distributions of X and Y are both normal. Exercise 10.113 provides an example of two normally distributed random variables that are uncorrelated but dependent, and even have a joint PDF.

Conditional Distributions of Bivariate Normal Random Variables

Let X and Y be bivariate normal random variables: $(X, Y) \sim \mathcal{BVN}(\mu_X, \sigma_X^2, \mu_Y, \sigma_Y^2, \rho)$. As we have shown, the marginal distributions of X and Y are both normal—specifically, $X \sim \mathcal{N}(\mu_X, \sigma_X^2)$ and $Y \sim \mathcal{N}(\mu_Y, \sigma_Y^2)$. We now investigate the conditional distributions. By symmetry, we need look at only one of the two families of conditional distributions—say, $Y_{|X=x}$ for $-\infty < x < \infty$.

To obtain a conditional PDF of Y given $X = x$, we need a joint PDF of X and Y and a marginal PDF of X, both of which we know. A joint PDF of X and Y is

$$f_{X,Y}(x, y) = \frac{1}{2\pi\sigma_X\sigma_Y\sqrt{1-\rho^2}} e^{-\frac{1}{2}Q(x,y)}, \qquad -\infty < x, y < \infty,$$

where

$$Q(x, y) = \frac{1}{1-\rho^2}\left\{\left(\frac{x-\mu_X}{\sigma_X}\right)^2 - 2\rho\left(\frac{x-\mu_X}{\sigma_X}\right)\left(\frac{y-\mu_Y}{\sigma_Y}\right) + \left(\frac{y-\mu_Y}{\sigma_Y}\right)^2\right\},$$

and a marginal PDF of X is

$$f_X(x) = \frac{1}{\sqrt{2\pi}\,\sigma_X}\,e^{-(x-\mu_X)^2/2\sigma_X^2} = \frac{1}{\sqrt{2\pi}\,\sigma_X}\,e^{-\frac{1}{2}\left(\frac{x-\mu_X}{\sigma_X}\right)^2}.$$

Hence a conditional PDF of Y given $X = x$ is

$$f_{Y\mid X}(y\mid x) = \frac{f_{X,Y}(x,y)}{f_X(x)} = \frac{\dfrac{1}{2\pi\sigma_X\sigma_Y\sqrt{1-\rho^2}}\,e^{-\frac{1}{2}Q(x,y)}}{\dfrac{1}{\sqrt{2\pi}\,\sigma_X}\,e^{-\frac{1}{2}\left(\frac{x-\mu_X}{\sigma_X}\right)^2}},$$

or

$$f_{Y\mid X}(y\mid x) = \frac{1}{\sqrt{2\pi}\,\sigma_Y\sqrt{1-\rho^2}}\,e^{-\frac{1}{2}\left\{Q(x,y)-\left(\frac{x-\mu_X}{\sigma_X}\right)^2\right\}}. \tag{10.61}$$

Making the simplifying substitutions $u = (x - \mu_X)/\sigma_X$ and $v = (y - \mu_Y)/\sigma_Y$ gives

$$\begin{aligned}
Q(x,y) - \left(\frac{x-\mu_X}{\sigma_X}\right)^2 &= \frac{1}{1-\rho^2}\left(u^2 - 2\rho uv + v^2\right) - u^2 \\
&= \frac{1}{1-\rho^2}\left(v^2 - 2\rho uv + \rho^2 u^2\right) = \frac{1}{1-\rho^2}(v - \rho u)^2 \\
&= \frac{1}{1-\rho^2}\left(\frac{y-\mu_Y}{\sigma_Y} - \rho\frac{x-\mu_X}{\sigma_X}\right)^2 \\
&= \frac{1}{\sigma_Y^2(1-\rho^2)}\left[y - \left(\mu_Y + \rho\frac{\sigma_Y}{\sigma_X}(x-\mu_X)\right)\right]^2.
\end{aligned}$$

Substituting this last expression into Equation (10.61) yields

$$f_{Y\mid X}(y\mid x) = \frac{1}{\sqrt{2\pi}\,\sigma_Y\sqrt{1-\rho^2}}\,e^{-\left[y-\left(\mu_Y+\rho\frac{\sigma_Y}{\sigma_X}(x-\mu_X)\right)\right]^2/2\sigma_Y^2(1-\rho^2)}. \tag{10.62}$$

Equation (10.62) shows that

$$Y_{\mid X=x} \sim \mathcal{N}\left(\mu_Y + \rho\frac{\sigma_Y}{\sigma_X}(x-\mu_X),\ \sigma_Y^2(1-\rho^2)\right), \tag{10.63}$$

that is, the conditional distribution of Y given $X = x$ is the normal distribution with mean $\mu_Y + \rho\frac{\sigma_Y}{\sigma_X}(x - \mu_X)$ and variance $\sigma_Y^2(1 - \rho^2)$. In particular, then,

$$\mathcal{E}(Y \mid X = x) = \mu_Y + \rho\frac{\sigma_Y}{\sigma_X}(x - \mu_X). \tag{10.64}$$

Another important consequence of Equation (10.63) is that the conditional variance of Y given $X = x$ doesn't depend on x: $\text{Var}(Y \mid X = x) = \sigma_Y^2(1 - \rho^2)$. Figure 10.4 displays the conditional distributions of Y given $X = x$ for bivariate normal random variables.

Figure 10.4 Conditional distributions of Y given $X = x$ for bivariate normal random variables

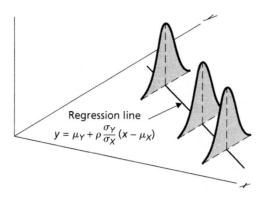

Regression line
$$y = \mu_Y + \rho \frac{\sigma_Y}{\sigma_X}(x - \mu_X)$$

Prediction in the Bivariate Normal Setting

Referring to Equation (10.64) and applying the prediction theorem (Proposition 10.8 on page 602), we conclude that the best possible predictor of Y, given that X equals x, is the quantity $\mu_Y + \rho \frac{\sigma_Y}{\sigma_X}(x - \mu_X)$. The equation $y = \mu_Y + \rho \frac{\sigma_Y}{\sigma_X}(x - \mu_X)$ is called the **regression equation,** and the straight line it represents is called the **regression line,** as illustrated in Figure 10.4.

EXAMPLE 10.22   *Prediction in the Bivariate Normal Setting*

Heights of Mothers and Sons   Refer to Example 10.21 on page 611.
a) Determine the conditional distribution of Y given $X = x$.
b) Obtain the regression equation for predicting the height of an eldest son based on the height of his mother.
c) Apply part (b) to predict the height of an eldest son whose mother is 62 inches tall.

Solution   Because $(X, Y) \sim \mathcal{BVN}(63.7, 2.7^2, 69.1, 2.9^2, 0.5)$, we have

$$\mu_X = 63.7, \quad \sigma_X = 2.7, \quad \mu_Y = 69.1, \quad \sigma_Y = 2.9, \quad \rho = 0.5. \tag{10.65}$$

a) From Equations (10.63) and (10.65), we conclude that the conditional distribution of Y given $X = x$ is the normal distribution with mean

$$\mu_Y + \rho \frac{\sigma_Y}{\sigma_X}(x - \mu_X) = 69.1 + 0.5 \cdot \frac{2.9}{2.7}(x - 63.7) \approx 34.9 + 0.537x \tag{10.66}$$

and variance $\sigma_Y^2(1 - \rho^2) = 2.9^2(1 - 0.5^2) \approx 6.31$.
b) From Relation (10.66), the regression equation for predicting the height of an eldest son, based on the height of his mother, is given approximately by $y = 34.9 + 0.537x$.
c) To use a mother's height to predict the height of her eldest son, we use the regression equation found in part (b). For a mother who is 62 inches tall, we predict the height of her eldest son to be roughly $34.9 + 0.537 \cdot 62$, or about 68.2 inches.   ∎

Bivariate Normal Random Variables in Summary

In Proposition 10.10, we summarize the results that we have obtained for bivariate normal random variables.

◆◆◆ **Proposition 10.10   Bivariate Normal Random Variables**

Let X and Y be bivariate normal random variables: $(X, Y) \sim \mathcal{BVN}(\mu_X, \sigma_X^2, \mu_Y, \sigma_Y^2, \rho)$. Then the following properties hold.

a) $X \sim \mathcal{N}(\mu_X, \sigma_X^2)$; that is, X is normally distributed with mean μ_X and variance σ_X^2.

b) $Y \sim \mathcal{N}(\mu_Y, \sigma_Y^2)$; that is, Y is normally distributed with mean μ_Y and variance σ_Y^2.

c) $\rho(X, Y) = \rho$; that is, the correlation between X and Y is ρ.

d) We have

$$Y_{|X=x} \sim \mathcal{N}\left(\mu_Y + \rho \frac{\sigma_Y}{\sigma_X}(x - \mu_X), \; \sigma_Y^2(1 - \rho^2)\right).$$

That is, given that $X = x$, Y is normally distributed with mean $\mu_Y + \rho \frac{\sigma_Y}{\sigma_X}(x - \mu_X)$ and variance $\sigma_Y^2(1 - \rho^2)$.

e) We have

$$X_{|Y=y} \sim \mathcal{N}\left(\mu_X + \rho \frac{\sigma_X}{\sigma_Y}(y - \mu_Y), \; \sigma_X^2(1 - \rho^2)\right).$$

That is, given that $Y = y$, X is normally distributed with mean $\mu_X + \rho \frac{\sigma_X}{\sigma_Y}(y - \mu_Y)$ and variance $\sigma_X^2(1 - \rho^2)$.

Beyond Bivariate Normal Random Variables

In this section, we examined bivariate normal random variables, the two-dimensional analogue of univariate normal random variables. The concept of normal random variables can be extended to dimensions higher than two, which we do in Section 12.3.

EXERCISES 10.5   **Basic Exercises**

10.101 Let $(X, Y) \sim \mathcal{BVN}(-2, 4, 3, 2, -0.7)$. Determine the mean and standard deviation of the random variable $W = -2X + 4Y - 12$.

10.102 Let f_1 and f_2 be PDFs of $\mathcal{BVN}(0, 1, 0, 1, \rho_1)$ and $\mathcal{BVN}(0, 1, 0, 1, \rho_2)$ random variables, respectively, where $\rho_1 \neq \rho_2$. Define $f = (f_1 + f_2)/2$.
a) Show that f is a bivariate PDF—that is, that f satisfies properties (a) and (b) of Proposition 9.5 on page 502.
b) Note that f isn't the joint PDF of bivariate normal random variables. Show that, nonetheless, the marginals of f are both standard normal PDFs.

10.103 Let X and Y be bivariate normal random variables.
a) Show that the level curves (contour lines) of the joint PDF of X and Y are ellipses.
b) Under what conditions on X and Y are the axes of the level-curve ellipses parallel to the coordinate axes? Justify your answer.

10.104 For a certain population, the blood pressure $(X, Y) = $ (systolic, diastolic) can be modeled as having a $\mathcal{BVN}(120, 400, 82, 225, 0.6)$ distribution.
a) Obtain a joint PDF of X and Y.
b) Identify and interpret the marginal distribution of X.
c) Identify and interpret the marginal distribution of Y.
d) Identify and interpret the correlation coefficient of X and Y.
e) Obtain, identify, and interpret the conditional distribution of Y given $X = x$ (for $x > 0$).
f) Find the predicted diastolic pressure for an individual whose systolic pressure is 123.
g) Find the probability that an individual with a diastolic pressure of 86 will have a systolic pressure exceeding 130.

10.105 Scientists concerned with environmental issues have long been interested in the sources of acid rain. Nitrates are a constituent of acid rain with arsenic as an accompanying element. Research by J. D. Scudlark and T. M. Church, published in the paper "The Atmospheric Deposition of Arsenic and Association with Acid Precipitation" (*Atmospheric Environment*, 1989, Vol. 22, pp. 937–943), suggests that the variables nitrate concentration (X), in micromoles per liter, and arsenic concentration (Y), in nanomoles per liter, have the parameters $\mu_X = 49.8$, $\sigma_X = 30.7$, $\mu_Y = 2.26$, $\sigma_Y = 1.18$, and $\rho = 0.732$. Suppose that you model the joint distribution of X and Y with a bivariate normal distribution.
a) Obtain a joint PDF of X and Y.
b) Identify and interpret the marginal distribution of X.
c) Identify and interpret the marginal distribution of Y.
d) Identify and interpret the correlation coefficient of X and Y.
e) Obtain, identify, and interpret the conditional distribution of Y given $X = x$ (for $x > 0$).
f) Determine and graph the regression equation for predicting arsenic concentration from nitrate concentration.
g) Predict the arsenic concentration for a nitrate concentration of 60 micromoles per liter.

10.106 Refer to Examples 10.21 and 10.22 on pages 611 and 615, respectively.
a) Predict the height of an eldest son whose mother is $5'\,8''$ tall.
b) Predict the height of a mother whose eldest son is $5'\,8''$ tall.
c) What is the probability that the eldest son of a $5'\,2''$-tall mother will be taller than $6'$?
d) Determine a 95% *prediction interval* for the height of the eldest son of a $5'\,2''$-tall mother. In other words, find an interval which is symmetric about the expected height of the eldest son of a $5'\,2''$-tall mother and is such that the probability is 0.95 that the son's height will lie in that interval.

10.107 Let X and Y be bivariate normal random variables.
a) Show that any nonzero linear combination of X and Y is also normally distributed. *Hint:* Refer to Equations (10.55) on page 608.
b) Use part (a) and properties of means, variances, and covariances to determine the probability distributions of the random variables $X + Y$ and $Y - X$.

10.108 Refer to Example 10.21 on page 611, where we considered the joint distribution of the heights of mothers and eldest sons.
a) Identify the probability distribution of the average height of a mother and her eldest son. *Hint:* Refer to Exercise 10.107.
b) Identify the probability distribution of the difference between the height of an eldest son and his mother.
c) On average, how much taller is an eldest son than his mother?
d) Find the probability that an eldest son is taller than his mother.
e) Find the probability that an eldest son's height exceeds 110% of his mother's height.

10.109 Let X and Y be standard normal random variables that are bivariate normal with correlation coefficient ρ. Obtain $\mathcal{E}\big(\max\{X, Y\}\big)$. *Hint:* Rotate the coordinate axes by $45°$.

10.110 Simulation: Explain how to simulate bivariate normal random variables by using a basic random number generator. *Hint:* Refer to the Box–Müller transformation presented in Equations (9.59) on page 550.

Theory Exercises

10.111 Regression model: Let X and ϵ be independent random variables, both with finite nonzero variances, and assume that ϵ has mean 0 and variance σ^2. Let β_0 and β_1 be real numbers, and set $Y = \beta_0 + \beta_1 X + \epsilon$. This equation is often referred to as the *regression model*. Verify the equations presented in parts (a)–(h), where $\rho = \rho(X, Y)$.

a) $\mu_Y = \beta_0 + \beta_1 \mu_X$ b) $\sigma_Y^2 = \beta_1^2 \sigma_X^2 + \sigma^2$

c) $\rho = \beta_1 \sigma_X / \sigma_Y$ d) $\sigma_Y^2 (1 - \rho^2) = \sigma^2$

e) $\beta_0 = \mu_Y - (\rho \sigma_Y / \sigma_X) \mu_X$ f) $\beta_1 = \rho \sigma_Y / \sigma_X$

g) $\mathcal{E}(Y \mid X = x) = \beta_0 + \beta_1 x$ h) $\mathrm{Var}(Y \mid X = x) = \sigma^2$

10.112 Refer to Exercise 10.111. Assume now that X and ϵ are normally distributed.
a) Explain why the random variable Y is normally distributed.
b) Without doing any computations, explain why the random variables X and Y have a bivariate normal distribution. *Hint:* Express X and Y in the form of Equations (10.55) on page 608, where Z_1 and Z_2 are independent standard normal random variables.
c) Determine and identify the conditional distribution of Y given $X = x$, expressing any parameters in terms of β_0, β_1, and σ.

Advanced Exercises

10.113 Let $A = \{(x, y) : x < -c \text{ and } y > 0\}$, $B = \{(x, y) : -c < x < c \text{ and } y < 0\}$, and $C = \{(x, y) : x > c \text{ and } y > 0\}$, where $c = \Phi^{-1}(3/4)$. Define $f(x, y) = \pi^{-1} e^{-(x^2 + y^2)/2}$ if $(x, y) \in A \cup B \cup C$, and $f(x, y) = 0$ otherwise.
a) Verify that f is a joint PDF of some pair of random variables—say, X and Y—by showing that it satisfies properties (a) and (b) of Proposition 9.5 on page 502.
b) Show that X and Y are each standard normal random variables.
c) Show that X and Y are uncorrelated.
d) Show that X and Y aren't independent.
e) Deduce that X and Y aren't bivariate normal.

10.114 Let $(X, Y) \sim \mathcal{BVN}(0, 1, 0, 1, \rho)$.
a) Determine a conditional PDF of X given $Y > 0$ in terms of $f_{X,Y}$. *Hint:* First obtain the conditional CDF.
b) Calculate $\mathcal{E}(X \mid Y > 0)$.

10.115 Matrix representation of a bivariate normal PDF: Suppose that X and Y are bivariate normal random variables. Let $\mathbf{x}$ be the column vector whose entries are x and y, let μ be the column vector whose entries are μ_X and μ_Y, and let

$$\Sigma = \begin{bmatrix} \mathrm{Var}(X) & \mathrm{Cov}(X, Y) \\ \mathrm{Cov}(Y, X) & \mathrm{Var}(Y) \end{bmatrix}.$$

Also, let $\mathbf{A}'$ denote the transpose of a matrix $\mathbf{A}$.
a) Explain why Σ is called the *covariance matrix* of X and Y.

b) Show that Σ is symmetric (i.e., $\Sigma' = \Sigma$) and positive definite (i.e., $\mathbf{x}'\Sigma\mathbf{x} > 0$ for all $\mathbf{x} \neq \mathbf{0}$).

c) Show that the joint PDF of X and Y can be expressed in matrix form as

$$f_{X,Y}(x, y) = \frac{1}{2\pi(\det \Sigma)^{1/2}}\, e^{-\frac{1}{2}(\mathbf{x}-\mu)'\Sigma^{-1}(\mathbf{x}-\mu)}, \tag{$*$}$$

where $\det \mathbf{A}$ denotes the determinant of a square matrix $\mathbf{A}$.

d) Show that the PDF of a univariate normal random variable can be expressed in a form similar to that of Equation $(*)$.

e) By referring to parts (c) and (d), make an educated guess at the matrix form of the joint PDF of an m-variate normal distribution.

CHAPTER REVIEW

Summary

In this chapter, we extended the idea of expected value to include continuous random variables. Conceptually, as we noted in Chapter 7, the expected value of a random variable is its long-run-average value in repeated independent observations; this meaning is the long-run-average interpretation of expected value. Mathematically, the expected value of a random variable is a weighted average of its possible values, weighted by probability. In the continuous case, this weighted average is obtained by integrating the possible values of the random variable against its probability density function. Other commonly used terms for expected value are expectation, mean, and first moment.

One of the most important results associated with the concept of expected value is the fundamental expected-value formula (FEF). In the continuous case, this formula provides a method for obtaining the expected value of a function of one or more continuous random variables by using a (joint) PDF of those random variables.

We noted that, by applying the continuous version of the FEF, we can show, as in the discrete case, that expected value has the linearity property—the expected value of a linear combination of continuous random variables equals the linear combination of the expected values. We can also show that the expected value of a product of independent continuous random variables equals the product of the expected values. In fact, by using advanced mathematics, we can show that these and other basic properties of expected value hold for all types or combinations of types of random variables.

Variance, covariance, and correlation for continuous random variables are defined in terms of expected value in exactly the same way as for discrete random variables. Consequently, properties of variance, covariance, and correlation that hold in the discrete case also hold in the continuous case—and, in fact, hold for all types or combinations of types of random variables. In particular, then, for all types or combinations of types of random variables, the variance of a sum of independent random variables equals the sum of the variances.

We also extended the concept of conditional expectation to continuous random variables. The conditional expectation of a continuous random variable is the expected value of that random variable relative to its conditional PDF, given the value of another continuous random variable.

One use of conditional expectation is as a tool for obtaining the expected value of a random variable by conditioning on the value of another random variable. This process is implemented by using the law of total expectation. Likewise, we showed how to obtain the variance of a random variable by applying the law of total variance, a formula that expresses the variance of the random variable in terms of its conditional variances and conditional expectations relative to another random variable.

As an application of conditional expectation, we discussed prediction. In the simplest case, the task is to find the best predictor of the value of one random variable, based on knowing the value of another random variable. The prediction theorem shows that, with minimum mean square error as the criterion for "best," the best possible predictor of a random variable Y based on observing the random variable X is $\mathcal{E}(Y \mid X)$.

Finally, we presented one of the most important bivariate probability distributions, the bivariate normal distribution. A bivariate normal distribution is determined by five parameters: the means and variances of the two random variables and their correlation coefficient. Bivariate normal random variables have several interesting and useful properties. For instance, their marginal and conditional distributions are all normal, and they are independent if and only if they are uncorrelated.

You Should Be Able To ...

1. state and understand the long-run-average interpretation of expected value.
2. define, apply, and interpret the expected value of a continuous random variable.
3. state and apply the fundamental expected-value formula (FEF) for continuous random variables.
4. state, interpret, and apply the basic properties of expected value.
5. use tail probabilities to obtain the expected value of a nonnegative continuous random variable.
6. obtain moments and central moments of a continuous random variable.
7. define, apply, and interpret the variance.
8. state, interpret, and apply the basic properties of variance.
9. define, apply, and interpret the covariance between two random variables.
10. state, interpret, and apply the basic properties of covariance.
11. obtain the variance of the sum of random variables, knowing their variances and covariances.
12. define, apply, and interpret the correlation coefficient of two random variables.
13. define, apply, and interpret conditional expectation.
14. state and apply both forms of the law of total expectation for continuous random variables.
15. define, apply, and interpret conditional variance.
16. state and apply the law of total variance.
17. state and apply the prediction theorem.
18. define bivariate normal random variables.
19. state and apply the basic properties of bivariate normal random variables.

Key Terms

Chapter Review Exercises

Basic Exercises

10.116 Suppose that the random variable X has the triangular distribution on the interval (a, b). Find the mean and variance of X. *Hint:* First solve this problem when $a = -1$ and $b = 1$ and then refer to Exercise 8.125 on page 462.

10.117 Pareto random variable: Let X have the exponential distribution with parameter α. Define $Y = \beta e^X$, where β is a positive real number. The random variable Y is said to be a *Pareto random variable* and to have the *Pareto distribution with parameters α and β.*
a) Show that Y has finite expectation if and only if $\alpha > 1$.

Assume now that $\alpha > 1$ so that Y has finite expectation. Determine the expected value of Y
b) by first obtaining its PDF and then applying the definition of expected value.
c) by using the FEF.
d) by using the tail probabilities of Y.
e) An actuary models the loss amounts of a certain policy by the Pareto distribution with parameters $\alpha = 2$ and $\beta = 3$, where loss amounts are measured in thousands of dollars. Determine the expected loss amount.

10.118 Weibull random variable: Let X have the exponential distribution with parameter α. Define $Y = X^{1/\beta}$, where β is a positive real number. The random variable Y is said to be a *Weibull random variable* and to have the *Weibull distribution with parameters α and β.*
a) Determine the mean and variance of Y.
b) Suppose that the lifetime of a component has the Weibull distribution with parameters $\alpha = 2$ and $\beta = 3$, where time is measured in years. Obtain the expected lifetime of the component and the standard deviation of its lifetime.

10.119 Lognormal random variable: Let $Y = e^X$, where $X \sim \mathcal{N}(\mu, \sigma^2)$. The random variable Y is said to be a *lognormal random variable* and to have the *lognormal distribution with parameters μ and σ^2.*
a) Determine the mean and variance of Y.
b) Claim amounts, in dollars, for a certain insurance policy are modeled by a lognormal distribution with parameters $\mu = 7$ and $\sigma^2 = 0.25$. Obtain the mean and standard deviation of claim amounts for this policy.

10.120 The three velocity components, X, Y, and Z, of a gas molecule are independent and identically distributed normal random variables with common mean 0 and variance σ^2. Its speed is $S = \sqrt{X^2 + Y^2 + Z^2}$. Determine the mean speed of a gas molecule by

a) applying the trivariate form of the FEF to obtain $\mathcal{E}(\sqrt{X^2 + Y^2 + Z^2})$.

b) first identifying the probability distribution of $(X^2 + Y^2 + Z^2)/\sigma^2$.

c) using the PDF of S, as found in Example 9.18 on page 533.

10.121 Let X be uniformly distributed on the interval $(0, 1)$, and set $Y = X(1 - X)$.

a) Determine $\mathcal{E}(Y)$ and $\mathcal{E}(XY)$ by using the FEF and the fact that a beta PDF integrates to 1.

b) Determine $\mathcal{E}(Y)$ and $\mathcal{E}(XY)$ without using the FEF, but rather by using properties of expected value and moments of X.

c) Show that $\mathcal{E}(XY) = \mathcal{E}(X)\mathcal{E}(Y)$, but that X and Y aren't independent.

d) What does the result of part (c) say about the relationship between independence and uncorrelatedness?

10.122 Let X and Y be uniformly distributed on the triangle $x \geq 0$, $y \geq 0$, and $x + y \leq 1$.

a) Determine $\mathcal{E}(X)$ and $\mathcal{E}(Y)$.

b) Determine $\mathrm{Var}(X)$ and $\mathrm{Var}(Y)$.

c) Without doing any computations, make an educated guess at the sign of the correlation coefficient of X and Y. Explain your reasoning.

d) Find the correlation coefficient of X and Y.

10.123 A manufacturer's annual losses follow a distribution with a probability density function $f(x) = 2.5(0.6)^{2.5}/x^{3.5}$ for $x > 0.6$, and $f(x) = 0$ otherwise. To cover its losses, the manufacturer purchases an insurance policy with an annual deductible of 2. Determine the expected amount of the manufacturer's annual losses not paid by the insurance policy.

10.124 At a service center, two successive tasks are required. Let X and Y denote the times at which the first and second tasks, respectively, are completed. A joint PDF of X and Y is $f_{X,Y}(x, y) = 6e^{-(x+2y)}$ for $0 < x < y$, and $f_{X,Y}(x, y) = 0$ otherwise. Determine how long, on average, it takes to complete the two successive tasks

a) by first finding a marginal PDF of Y and then applying the definition of expected value.

b) without first finding a marginal PDF of Y.

10.125 The lifetime of a printer costing $200 is exponentially distributed with mean 2 years. The manufacturer agrees to pay a full refund to a buyer if the printer fails during the first year following its purchase, and a one-half refund if it fails during the second year. No refund is made after the second year of use. If the manufacturer sells 100 printers, how much should it expect to pay in refunds?

10.126 A device that continuously measures and records seismic activity is placed in a remote region. The time to failure of this device is exponentially distributed with mean 3 years. If the device won't be monitored during its first 2 years of service, determine the expected time to discovery of its failure.

10.127 An insurance policy is written to cover a loss that is uniformly distributed between $0 and $1000. At what level must a deductible be set so that the expected payment is 25% of what it would be with no deductible?

10.128 A company has two gasoline generators for producing electricity. The time until failure for each generator follows an exponential distribution with mean 10 years. The company begins using the second generator immediately after the first one fails. What is the standard deviation of the total time that the generators produce electricity? State any assumptions that you make.

10.129 Let X and Y be independent random variables with finite expectation. Then you know that $X + Y$, $X - Y$, and XY all have finite expectation. Quotients, however, don't behave so nicely. Provide an example of two independent random variables X and Y with finite expectation such that Y/X doesn't have finite expectation.

10.130 An insurance policy is written to cover a loss with density $3x^2/8$ for $0 < x < 2$ and 0 otherwise. The time, in hours, to process a claim of size x is uniformly distributed on the interval from x to $2x$. Determine the expected processing time of a randomly chosen claim on this policy by
a) first obtaining a PDF of the processing time and then using the definition of expected value.
b) using the law of total expectation.

10.131 Suppose that the remaining lifetimes, in years, of a husband and wife are independent and uniformly distributed on the interval $(0, 40)$. An insurance company offers two products to married couples: One pays when the husband dies, and one pays when both the husband and wife have died. Calculate the covariance of the two payment times.

10.132 Let $X \sim \mathcal{N}(0, \sigma^2)$ and set $Y = X^2$.
a) Show that $\mathcal{E}(XY) = \mathcal{E}(X)\mathcal{E}(Y)$.
b) Does part (a) imply that X and Y are independent random variables? Justify your answer.
c) Are X and Y independent random variables?

10.133 An electronic device has two components working simultaneously and fails when both components fail. The lifetimes, X and Y, in thousands of hours, of the two components have a joint PDF given by $f_{X,Y}(x, y) = (1/64)ye^{-(x+y)/4}$ for positive x and y, and $f_{X,Y}(x, y) = 0$ otherwise. The operating cost for the first component (with lifetime X) is \$3 per 1000 hours and that for the second component (with lifetime Y) is \$5 per 1000 hours. Additionally, there is a startup cost of \$65 for running the device.
a) Determine the expected lifetime of the device by first obtaining a PDF of the lifetime and then applying the definition of expected value.
b) Determine the expected lifetime of the device by using the FEF.
c) Find the expected cost of operating the device until its failure.
d) Find the standard deviation of the cost of operating the device until its failure.

10.134 Let X and Y denote the values of two stocks at the end of a 5-year period. The random variable X is uniformly distributed on the interval $(0, 12)$ and, given that $X = x$, the random variable Y is uniformly distributed on the interval $(0, x)$. Determine $\text{Cov}(X, Y)$
a) by first obtaining a joint PDF of X and Y and then applying the FEF.
b) by using the law of total expectation.

10.135 Let X denote the amount of an item stocked at the beginning of a day and let Y denote the amount of the item sold during the day. A joint PDF of X and Y is $f_{X,Y}(x, y) = 8xy$ for $0 < y < x < 1$, and $f_{X,Y}(x, y) = 0$ otherwise. Determine $\mathcal{E}(Y)$ and $\text{Var}(Y)$ by using
a) the provided joint PDF of X and Y.
b) a marginal PDF of Y.
c) the laws of total expectation and variance. *Hint:* First obtain and identify the probability distribution of X.

10.136 Refer to Exercise 10.135.
a) If an amount 0.5 is stocked, what is the predicted amount sold?
b) If an amount 0.4 is sold, what is the best guess for the amount stocked?
c) If an amount 0.1 is left unsold, what is the best guess for the amount sold?

10.137 Let $X_1 \sim \mathcal{U}(0, 1)$ and, for $n \geq 2$, let $X_n \sim \mathcal{U}(0, X_{n-1})$. Find a formula for $\mathcal{E}(X_n)$.

10.138 Let X denote the age (in years) and let Y denote the price (in hundreds of dollars) of a car called the Orion. Assume that X and Y have a bivariate normal distribution with parameters $\mu_X = 5.3$, $\sigma_X = 1.4$, $\mu_Y = 88.6$, $\sigma_Y = 31.2$, and $\rho = -0.9$.
a) Obtain a joint PDF of X and Y.
b) Identify and interpret the marginal distribution of X.
c) Identify and interpret the marginal distribution of Y.
d) Identify and interpret the correlation coefficient of X and Y.
e) Obtain, identify, and interpret the conditional distribution of Y given $X = x$ (for $x > 0$).
f) Determine the regression equation for predicting the price of an Orion based on its age.
g) Predict the price of a 3-year-old Orion.

10.139 Arizona State University (ASU) keeps records of several variables relating to students in its ASU Focus Student Database. Let X and Y denote the high school GPA and cumulative GPA at the end of the sophomore year, respectively, of a randomly selected student. Assume that $(X, Y) \sim \mathcal{BVN}(3.0, 0.25, 2.7, 0.36, 0.5)$.
a) Obtain a joint PDF of X and Y.
b) Predict the cumulative GPA at the end of the sophomore year for an ASU student whose high school GPA was 3.5.
c) Determine a 90% prediction interval for the cumulative GPA at the end of the sophomore year for an ASU student whose high school GPA was 3.5.

10.140 At a health-food store, the weekly demand, in hundred of pounds, for unbleached enriched whole wheat flour has a beta distribution with parameters 2 and 2. At the beginning of the week, the store manager stocks $100a$ lb of this flour. The flour is purchased by the store for 10¢/lb and sold for 21¢/lb. Any flour unsold by the end of the week is discarded. How much of the flour should the manager stock to maximize the expected weekly profit?

10.141 Let $X_1, \ldots, X_n$ be a random sample from a $\mathcal{U}(\theta - 0.5, \theta + 0.5)$ distribution, where θ is an unknown parameter, and let M and $\bar{X}_n$ denote the midrange and sample mean, respectively, of the random sample. In solving some of the following problems, you will want to refer to Examples 10.11 and 10.14 on pages 584 and 589, respectively.
a) Show that $\bar{X}_n$ is an unbiased estimator of θ.
b) Show that M is an unbiased estimator of θ. *Hint:* Begin by expressing the X_js in terms of $\mathcal{U}(0, 1)$ random variables.
c) Show that $\text{Var}(M) \leq \text{Var}(\bar{X}_n)$ for all n and that strict inequality holds when $n \geq 3$.
d) Which estimator of θ do you think is better, M or $\bar{X}_n$? Explain your answer.

Theory Exercises

10.142 Stochastically larger: We say that the random variable X is *stochastically larger* than the random variable Y if $F_X(t) \leq F_Y(t)$ for all $t \in \mathcal{R}$. Let X and Y be two continuous random variables with PDFs and finite expectation. Show that, if X is stochastically larger than Y, then $\mathcal{E}(X) \geq \mathcal{E}(Y)$. *Note:* This result actually holds for any two random variables with finite expectations.

10.143 Schwarz's inequality: Let X and Y be random variables defined on the same sample space and having finite variance.
a) Prove *Schwarz's inequality:* $[\mathcal{E}(XY)]^2 \leq \mathcal{E}(X^2)\,\mathcal{E}(Y^2)$. *Hint:* Consider the quadratic function $h(t) = \mathcal{E}((X + tY)^2)$.
b) Use Schwarz's inequality to prove that, if X and Y are random variables defined on the same sample space and having finite nonzero variances, then $|\rho(X, Y)| \leq 1$.

10.144 Let X and Y be random variables defined on the same sample space and having finite means. The monotonicity property of expected value states that, if $X \leq Y$, then $\mathcal{E}(X) \leq \mathcal{E}(Y)$. Prove the strict version of this result, that is, if $X < Y$, then $\mathcal{E}(X) < \mathcal{E}(Y)$. *Hint:* Use tail probabilities.

10.145 Jensen's inequality: Let I be an interval and let g be a real-valued function that is twice differentiable on I. We say that g is *convex on I* if $g''(x) \geq 0$ for all $x \in I$. Now let X be a random variable having finite mean and whose range is contained in I.
a) Prove *Jensen's inequality:* If g is convex on I, then $g(\mathcal{E}(X)) \leq \mathcal{E}(g(X))$. *Hint:* Use Taylor's formula to show that $g(x) \geq g(\mu_X) + g'(\mu_X)(x - \mu_X)$ for all $x \in I$.
b) A real-valued function g is called *concave on I* if and only if $-g$ is convex on I. Prove that, if g is concave on I, then $g(\mathcal{E}(X)) \geq \mathcal{E}(g(X))$.
c) An investor has a choice of two investments. One is a certificate of deposit, which is guaranteed to yield \$$r$; the other is an equity investment, which has a mean yield of \$$r$. The investor decides to make her choice by maximizing the expected utility. What choice will she make if her utility function is $u(x) = \sqrt{x}$? if her utility function is $u(x) = x^{3/2}$? Justify your answers.

Advanced Exercises

10.146 Let X and Z be independent random variables with $X \sim \mathcal{N}(\mu, \sigma^2)$ and $Z \sim \mathcal{N}(0, 1)$. Set $Y = X + Z$.
a) Show that X and Y are bivariate normal random variables by applying the bivariate transformation theorem.
b) Show that X and Y are bivariate normal random variables by referring to Equations (10.55) on page 608.
c) Determine the best predictor of X based on observing Y.

10.147 Suppose that the heights of 25-year-old men are normally distributed with mean 69 inches and standard deviation 2.5 inches. A measuring device reports a man's height to be his actual height plus an error that is normally distributed with mean 0 inches and standard deviation 1 inch, independent of the man or his height. Determine the best predictor of the man's height if the reported height is 72 inches. *Hint:* Refer to Exercise 10.146.

10.148 Let X_1, X_2, ... be independent and identically distributed random variables with a PDF. Let N denote the first n such that $X_n > X_1$. Decide whether N has finite expectation.

10.149 Suppose that X_1, X_2, ... are independent $\mathcal{U}(0, 1)$ random variables. Let N denote the first positive integer n such that $X_1 + \cdots + X_n$ exceeds 1.
a) Determine the mean of N.
b) Determine the variance of N.

10.150 Simulation: This problem requires a computer or statistical calculator. Use the result of Exercise 10.149(a) to estimate the value of e by simulation. *Hint:* Use the long-run-average interpretation of expected value.

10.151 Loss functions and risk: Let Y be a continuous random variable with a PDF and having finite variance. When predicting the value of Y by a constant c, we can introduce a *loss function* $L(Y, c)$ with *risk* defined by $R(c) = \mathcal{E}(L(Y, c))$. Determine the value of c that minimizes the risk and the minimum value of the risk for each of the following loss functions.
a) $L(Y, c) = |Y - c|$ (proportional error)
b) $L(Y, c) = (Y - c)^2$ (squared error)

10.152 Let X and Y be independent random variables with $X \sim \mathcal{U}(0, 1)$ and $Y \sim \mathcal{B}(1, 0.5)$. Set $W = (X + Y)/2$.
a) Show that $W \sim \mathcal{U}(0, 1)$. *Hint:* Obtain the CDF of W by conditioning on the value of Y.
b) Find $\mathcal{E}(W \mid Y = 0)$ and $\mathcal{E}(W \mid Y = 1)$.
c) Use the result of part (b) to find the probability distribution of $\mathcal{E}(W \mid Y)$.
d) Use the result of part (c) to find $\mathrm{Var}\big(\mathcal{E}(W \mid Y)\big)$.
e) Show that $\mathrm{Var}(W \mid Y = 0) = \frac{1}{4} \mathrm{Var}(X) = \mathrm{Var}(W \mid Y = 1)$.
f) Use part (a) and the law of total variance to show that $\mathrm{Var}(X) = \frac{1}{4} \mathrm{Var}(X) + \frac{1}{16}$ and thence derive the value of $\mathrm{Var}(X)$.

10.153 Branching process: Consider a population that originates with one individual. This individual and all of its descendents produce j direct offspring with probability p_j, for $j = 0, 1, 2, \ldots$. The direct offspring of the original individual make up the first generation, those of all the individuals of the first generation make up the second generation, and so on. The individuals of each generation act independently of one another. Use X_n to denote the size of the nth generation ($X_0 = 1$). Assume that the number of direct offspring of any given individual has finite mean μ and finite variance σ^2.
a) Express μ and σ^2 in terms of the p_j.
b) Show that $\mathcal{E}(X_n) = \mu^n$.
c) Show that

$$\mathrm{Var}(X_n) = \begin{cases} n\sigma^2, & \text{if } \mu = 1; \\ \sigma^2 \mu^{n-1} \left(\dfrac{1 - \mu^n}{1 - \mu} \right), & \text{if } \mu \neq 1. \end{cases}$$

d) The *extinction probability*, denoted π_0, is defined to be the probability that there are no direct offspring for some generation. Show that $\pi_0 = \sum_{j=0}^{\infty} p_j \pi_0^j$; in other words, π_0 is a fixed point of the probability generating function P (as defined in Exercise 7.52 on page 352) of the number-of-offspring distribution. *Note:* It can be shown that if $\mu \leq 1$, then $\pi_0 = 1$, whereas, if $\mu > 1$, then π_0 is the unique fixed point of P in the interval $(0, 1)$.

Martingales: Let $\{X_n\}_{n=1}^{\infty}$ and $\{Y_n\}_{n=1}^{\infty}$ be sequences of random variables defined on the same sample space, where X_n has finite expectation for all n. We say that $\{X_n\}_{n=1}^{\infty}$ is a *martingale with respect to* $\{Y_n\}_{n=1}^{\infty}$ if $\mathcal{E}(X_{n+1} \mid Y_1, \ldots, Y_n) = X_n$ for all $n \in \mathcal{N}$. If $Y_n = X_n$ for all $n \in \mathcal{N}$, we say simply that $\{X_n\}_{n=1}^{\infty}$ is a *martingale*. In Exercises 10.154–10.158, you are to examine martingales.

10.154 Show that, if $\{X_n\}_{n=1}^{\infty}$ is a martingale with respect to $\{Y_n\}_{n=1}^{\infty}$, then $\mathcal{E}(X_n) = \mathcal{E}(X_1)$ for all $n \in \mathcal{N}$.

10.155 Let $\xi_1, \xi_2, \ldots$ be a sequence of independent random variables each having mean 0, and set $X_n = \sum_{j=1}^{n} \xi_j$ for each $n \in \mathcal{N}$. Show that $\{X_n\}_{n=1}^{\infty}$ is a martingale.

10.156 Refer to the discussion of branching processes in Exercise 10.153.
a) Show that, if $\mu = 1$, then $\{X_n\}_{n=1}^{\infty}$ is a martingale.
b) More generally than in part (a), show that $\{X_n/\mu^n\}_{n=1}^{\infty}$ is a martingale.

10.157 Let X be a random variable with finite expectation and let $\{Y_n\}_{n=1}^{\infty}$ be a sequence of random variables. Define $X_n = \mathcal{E}(X \mid Y_1, \ldots, Y_n)$. Prove that $\{X_n\}_{n=1}^{\infty}$ is a martingale with respect to $\{Y_n\}_{n=1}^{\infty}$.

10.158 Refer to Pólya's urn scheme on page 141 and assume that $c = 1$. Let Y_n be the indicator of the event that the nth ball drawn is red and let X_n denote the proportion of red balls in the urn after n draws. Show that $\{X_n\}_{n=1}^{\infty}$ is a martingale with respect to $\{Y_n\}_{n=1}^{\infty}$.

PART FOUR

Limit Theorems and Applications

William Feller *1906–1970*

William Feller was born in Zagreb, Croatia, on July 7, 1906. He was educated by private tutors and had no formal secondary schooling. Feller was awarded his first degree in 1925 from the University of Zagreb. One year later he received his Ph.D. from the University of Göttingen. Feller remained at Göttingen for two more years, at which time he accepted a position at the University of Kiel.

Hitler's policies toward Jews forced Feller to leave Germany in 1933. He stayed in Copenhagen until 1934 when he took a position at the University of Stockholm and joined the probability group. In 1939, Feller left Europe for the United States and became Professor of Mathematics at Brown University. Six years later, in 1945, he accepted a professorship at Cornell University. He worked there for 5 years until he was appointed Eugene Professor of Mathematics at Princeton University in 1950.

Feller worked on mathematical probability, using Kolmogorov's axiomatic approach. Although his main thrust was purely mathematical, he also studied and made important contributions to applications of probability. Notable accomplishments by Feller include research in Markov processes and the mathematical theory of Brownian motion and diffusion processes. One of Feller's most well-known works is his two-volume set *Introduction to Probability Theory and Its Applications*.

The famous probabilist J. L. Doob wrote the following tribute to Feller: "Those who knew him personally remember Feller best for his gusto, the pleasure with which he met life, and the excitement with which he drew on his endless fund of anecdotes about life and its absurdities, particularly the absurdities involving mathematics and mathematicians. To listen to him lecture was a unique experience, for no one else could lecture with such intense excitement."

Feller was president of the Institute of Mathematical Statistics and was a member of the Royal Statistical Society in the United Kingdom. He was awarded the 1969 National Medal for Science. Feller died on January 14, 1970, in New York.

CHAPTER ELEVEN

Generating Functions and Limit Theorems

INTRODUCTION

In this chapter, we introduce generating functions and discuss the two most important results in probability theory—namely, the *law of large numbers* and the *central limit theorem*.

Generating functions are a type of transform. Transforms are commonly used objects in mathematics and its applications. Two well-known and often applied transforms are the Laplace transform and the Fourier transform.

In Section 11.1, we introduce two of the most widely used transforms in probability theory, the *moment generating function* and the *characteristic function*. We concentrate on the moment generating function, a transform similar to the Laplace transform. In Section 11.2, we examine the multivariate version of moment generating functions, called *joint moment generating functions*.

In Section 11.3, we discuss laws of large numbers, or mathematically precise versions of what is commonly known as the law of averages. These theorems provide us with mathematical confirmation of our intuitive notions of probability and expected value.

In the final section of the chapter, Section 11.4, we present the central limit theorem. This remarkable result shows that sums (or averages) of independent and identically distributed random variables with finite nonzero variance are always asymptotically normally distributed, regardless of the common probability distribution of the summands.

11.1 Moment Generating Functions

As we mentioned in the introduction to this chapter, the moment generating function is a type of transform. Generally, transforms are used to simplify the process of obtaining certain kinds of results. We use the moment generating function to

- efficiently obtain moments of a random variable,
- identify probability distributions, and
- establish limit theorems.

Here now is the definition of the **moment generating function.**

DEFINITION 11.1   Moment Generating Function

The **moment generating function** of a random variable X, denoted M_X, is defined by

$$M_X(t) = \mathcal{E}\left(e^{tX}\right), \tag{11.1}$$

for all values of $t \in \mathcal{R}$ for which the random variable e^{tX} has finite expectation. We use **MGF** as an abbreviation for "moment generating function."

Note: Because the random variable $e^{0 \cdot X} = 1$ has finite expectation, the MGF is defined at $t = 0$ for any random variable X. Moreover, $M_X(0) = \mathcal{E}(1) = 1$.

Applying Definition 11.1 and the FEFs for discrete and continuous random variables, respectively, with the function $g(x) = e^{tx}$, we obtain the following formulas.

- If X is a discrete random variable, the MGF of X is given by

$$M_X(t) = \sum_x e^{tx} p_X(x). \tag{11.2}$$

- If X is a continuous random variable with a PDF, the MGF of X is given by

$$M_X(t) = \int_{-\infty}^{\infty} e^{tx} f_X(x)\, dx. \tag{11.3}$$

We emphasize, though, that the concept of the moment generating function applies to any type of random variable—discrete, continuous, or otherwise.

EXAMPLE 11.1 *Moment Generating Function*

Number of Siblings Refer to Example 5.1 on page 176, where we considered the number of siblings for each of the 40 students in one of Professor Weiss's classes. Let X denote the number of siblings of a randomly selected student. Then X is a discrete random variable. Its PMF is given in Table 5.5 on page 186. Determine the MGF of X.

Solution   Applying Equation (11.2) to the PMF of X, we get

$$M_X(t) = \sum_x e^{tx} p_X(x)$$

$$= e^{t \cdot 0} p_X(0) + e^{t \cdot 1} p_X(1) + e^{t \cdot 2} p_X(2) + e^{t \cdot 3} p_X(3) + e^{t \cdot 4} p_X(4)$$

$$= e^{t \cdot 0} \cdot 0.200 + e^{t \cdot 1} \cdot 0.425 + e^{t \cdot 2} \cdot 0.275 + e^{t \cdot 3} \cdot 0.075 + e^{t \cdot 4} \cdot 0.025,$$

or

$$M_X(t) = 0.200 + 0.425 \, e^t + 0.275 \, e^{2t} + 0.075 \, e^{3t} + 0.025 \, e^{4t}.$$

As e^{tX} has finite expectation for all $t \in \mathcal{R}$, the MGF of X is defined for all $t \in \mathcal{R}$.   ∎

EXAMPLE 11.2   *Moment Generating Function*

Bacteria on a Petri Dish   A petri dish of unit radius, containing nutrients upon which bacteria can multiply, is smeared with a uniform suspension of bacteria. Subsequently, spots indicating colonies of bacteria will appear. Suppose that we observe the location of the center of the first spot to appear. Let Z denote the distance of the center of the first spot from the center of the petri dish. Determine the MGF of the random variable Z.

Solution   We obtained the PDF of Z in Example 8.8 on page 421, where we found that $f_Z(z) = 2z$ for $0 \le z < 1$, and $f_Z(z) = 0$ otherwise. Applying Equation (11.3) to the PDF of Z and using integration by parts, we get, for $t \ne 0$,

$$M_Z(t) = \int_{-\infty}^{\infty} e^{tz} f_Z(z) \, dz = \int_0^1 e^{tz} 2z \, dz = \frac{2}{t^2} \left(1 - e^t + t \, e^t\right).$$

Hence

$$M_Z(t) = \begin{cases} \dfrac{2}{t^2} \left(1 - e^t + t \, e^t\right), & \text{if } t \ne 0; \\ 1, & \text{if } t = 0, \end{cases}$$

as required.   ∎

Moment Generating Function for Some Families of Random Variables

In Examples 11.3–11.5, we obtain the MGFs for some important families of random variables, namely, the binomial, gamma, and normal families.

EXAMPLE 11.3   *Moment Generating Function of a Binomial Random Variable*

Determine the MGF of a binomial random variable with parameters n and p.

Solution   Let $X \sim \mathcal{B}(n, p)$. Recall that the PMF of X is

$$p_X(x) = \binom{n}{x} p^x (1 - p)^{n-x}, \qquad x = 0, 1, \ldots, n,$$

and $p_X(x) = 0$ otherwise. Substituting this PMF into Equation (11.2), we obtain

$$M_X(t) = \sum_x e^{tx} p_X(x) = \sum_{x=0}^n e^{tx} \binom{n}{x} p^x (1-p)^{n-x}$$

$$= \sum_{x=0}^n \binom{n}{x} (pe^t)^x (1-p)^{n-x} = (pe^t + 1 - p)^n,$$

where the last equation follows from the binomial theorem. Note that this MGF is defined for all $t \in \mathcal{R}$. ∎

EXAMPLE 11.4   *Moment Generating Function of a Gamma Random Variable*

Determine the MGF of a gamma random variable with parameters α and λ.

Solution   Let $X \sim \Gamma(\alpha, \lambda)$. Recall that the PDF of X is

$$f_X(x) = \frac{\lambda^\alpha}{\Gamma(\alpha)} x^{\alpha-1} e^{-\lambda x}, \qquad x > 0,$$

and $f_X(x) = 0$ otherwise. Substituting this PDF into Equation (11.3), we obtain

$$M_X(t) = \int_0^\infty e^{tx} \frac{\lambda^\alpha}{\Gamma(\alpha)} x^{\alpha-1} e^{-\lambda x}\, dx$$

$$= \frac{\lambda^\alpha}{(\lambda - t)^\alpha} \int_0^\infty \frac{(\lambda - t)^\alpha}{\Gamma(\alpha)} x^{\alpha-1} e^{-(\lambda-t)x}\, dx = \left(\frac{\lambda}{\lambda - t}\right)^\alpha,$$

where the last equation follows from the fact that the integral of a gamma PDF with parameters α and $\lambda - t$ equals 1. Note that this MGF is defined only for $t < \lambda$. ∎

As we demonstrated in Section 8.6 (see page 456), several important probability distributions are special cases of the gamma distribution. For such distributions, we can apply the result of Example 11.4 to obtain their MGFs.

- *Exponential distribution.* The exponential distribution with parameter λ is also the gamma distribution with parameters 1 and λ. Thus the MGF of an exponential random variable with parameter λ is $\lambda/(\lambda - t)$.

- *Erlang distribution.* The Erlang distribution with parameters r and λ, where r is a positive integer, is also the gamma distribution with those two parameters. So the MGF of an Erlang random variable with parameters r and λ is $\left(\lambda/(\lambda - t)\right)^r$.

- *Chi-square distribution.* The chi-square distribution with ν degrees of freedom is also the gamma distribution with parameters $\nu/2$ and $1/2$. Thus the MGF of a chi-square random variable with ν degrees of freedom is

$$\left(\frac{1/2}{1/2 - t}\right)^{\nu/2} = (1 - 2t)^{-\nu/2}.$$

Next we obtain the MGF of a normal random variable. Before doing so, however, it's useful (although not necessary) to establish one of the several important properties of MGFs. As presented in Proposition 11.1, this property relates the MGF of a linear function of a random variable to the MGF of the random variable itself.

◆◆◆   **Proposition 11.1**   **MGF of a Linear Function of a Random Variable**

Let X be a random variable and let a and b be real numbers. Then the MGF of the random variable $a + bX$ is given by $M_{a+bX}(t) = e^{at} M_X(bt)$.

Proof   Applying basic properties of expected value, we get

$$M_{a+bX}(t) = \mathcal{E}\left(e^{t(a+bX)}\right) = \mathcal{E}\left(e^{at} e^{btX}\right) = e^{at}\mathcal{E}\left(e^{(bt)X}\right) = e^{at} M_X(bt),$$

as required.   ◆

EXAMPLE 11.5   *Moment Generating Function of a Normal Random Variable*

Determine the MGF of a normal random variable.

Solution   Let $X \sim \mathcal{N}(\mu, \sigma^2)$. From Proposition 8.10 on page 441, $Z = (X - \mu)/\sigma \sim \mathcal{N}(0, 1)$, which has PDF $\phi(z) = \left(1/\sqrt{2\pi}\right)e^{-z^2/2}$ for $-\infty < z < \infty$. Substituting this PDF into Equation (11.3) and using the technique of completing the square, we get

$$M_Z(t) = \int_{-\infty}^{\infty} e^{tz} \frac{1}{\sqrt{2\pi}} e^{-z^2/2}\, dz = e^{t^2/2} \int_{-\infty}^{\infty} \frac{1}{\sqrt{2\pi}} e^{-(z-t)^2/2}\, dz = e^{t^2/2},$$

where the last equation follows from the fact that the integral of a normal PDF with parameters t and 1 equals 1. As $X = \mu + \sigma Z$, Proposition 11.1 implies that

$$M_X(t) = M_{\mu+\sigma Z}(t) = e^{\mu t} M_Z(\sigma t) = e^{\mu t} e^{(\sigma t)^2/2} = e^{\mu t + \sigma^2 t^2/2}.$$

Note that this MGF is defined for all $t \in \mathcal{R}$.   ■

Using arguments similar to those in Examples 11.3–11.5, we can obtain the moment generating functions for several other families of discrete and continuous random variables, as shown in Table 11.1 at the top of the next page.

Moment Generation Property of MGFs

We can use the MGF of a random variable to obtain its moments. To illustrate, let's consider again the number of siblings, X, of a randomly selected student from one of Professor Weiss's classes. As we discovered in Example 11.1 on page 630,

$$M_X(t) = \sum_x p_X(x)e^{tx} = 0.200 + 0.425\, e^t + 0.275\, e^{2t} + 0.075\, e^{3t} + 0.025\, e^{4t}.$$

Taking the derivative of M_X, we get

$$M_X'(t) = \sum_x x p_X(x)e^{tx}$$

$$= 0 \cdot 0.200 + 1 \cdot 0.425\, e^t + 2 \cdot 0.275\, e^{2t} + 3 \cdot 0.075\, e^{3t} + 4 \cdot 0.025\, e^{4t}.$$

Table 11.1 Moment generating functions for selected families of random variables

Family	Parameter(s)	MGF	Family	Parameter(s)	MGF
Binomial	n and p	$(pe^t + 1 - p)^n$	Exponential	λ	$\dfrac{\lambda}{\lambda - t}$
Poisson	λ	$e^{\lambda(e^t - 1)}$	Normal	μ and σ^2	$e^{\mu t + \sigma^2 t^2 / 2}$
Geometric	p	$\dfrac{pe^t}{1 - (1-p)e^t}$	Gamma	α and λ	$\left(\dfrac{\lambda}{\lambda - t}\right)^\alpha$
Negative binomial	r and p	$\left(\dfrac{pe^t}{1 - (1-p)e^t}\right)^r$	Erlang	r and λ	$\left(\dfrac{\lambda}{\lambda - t}\right)^r$
Indicator	E	$P(E)e^t + 1 - P(E)$	Chi-square	ν	$(1 - 2t)^{-\nu/2}$
Uniform	a and b	$\dfrac{e^{bt} - e^{at}}{(b-a)t}$			

And setting $t = 0$ yields

$$M_X'(0) = \sum_x x p_X(x) = 0 \cdot 0.200 + 1 \cdot 0.425 + 2 \cdot 0.275 + 3 \cdot 0.075 + 4 \cdot 0.025.$$

The previous equation shows that $M_X'(0) = \mathcal{E}(X)$. Taking the second derivative of M_X and setting $t = 0$ shows that $M_X''(0) = \mathcal{E}(X^2)$. Similarly, $M_X'''(0) = \mathcal{E}(X^3)$, and so on. These results hold true in general, as we now demonstrate.

Let X be a random variable. As you know, its MGF is defined by

$$M_X(t) = \mathcal{E}\left(e^{tX}\right), \tag{11.4}$$

for all values of t for which e^{tX} has finite expectation. We restrict our attention to random variables X whose MGF is defined in some open interval containing 0; that is, there is a positive number t_0 such that $M_X(t)$ is defined for all t with $-t_0 < t < t_0$. This assumption implies that X has moments of all orders. It also implies that the formal derivations about M_X that follow are indeed mathematically valid.

Differentiating both sides of Equation (11.4) and interchanging the order of differentiation and expectation, we find that, for $-t_0 < t < t_0$,

$$M_X'(t) = \frac{d}{dt} M_X(t) = \frac{d}{dt} \mathcal{E}\left(e^{tX}\right) = \mathcal{E}\left(\frac{\partial}{\partial t} e^{tX}\right) = \mathcal{E}\left(X e^{tX}\right).$$

Proceeding inductively, we obtain

$$M_X^{(n)}(t) = \mathcal{E}\left(X^n e^{tX}\right), \qquad -t_0 < t < t_0, \tag{11.5}$$

for all $n \in \mathcal{N}$. And setting $t = 0$ in Equation (11.5), we get the result presented in Proposition 11.2.

◆◆◆ **Proposition 11.2   Moment Generation Property of MGFs**

Suppose that the MGF of a random variable X is defined in some open interval containing 0. Then X has moments of all orders, and we have

$$\mathcal{E}(X^n) = M_X^{(n)}(0), \qquad n \in \mathcal{N}. \tag{11.6}$$

In words, the nth moment of a random variable equals the nth derivative of its moment generating function evaluated at 0.

Proposition 11.2 provides an alternative method for finding moments of a random variable that is often more efficient or straightforward than the direct method of summation or integration used for discrete and continuous random variables, respectively. In Examples 11.6 and 11.7, we use the MGF approach of Proposition 11.2 to find the mean and variance of the binomial and normal families.

EXAMPLE 11.6   *Moment Generation Property of MGFs*

Mean and Variance of a Binomial Random Variable   Obtain the mean and variance of a binomial random variable from its MGF.

Solution   Let $X \sim \mathcal{B}(n, p)$. From Example 11.3, $M_X(t) = (pe^t + 1 - p)^n$. To use Proposition 11.2 to obtain the mean and variance of X, we need the first two derivatives of M_X. Applying differentiation rules from calculus and doing some simple algebra, we get

$$M_X'(t) = npe^t(pe^t + 1 - p)^{n-1}$$

and

$$M_X''(t) = npe^t(pe^t + 1 - p)^{n-1} + n(n-1)p^2e^{2t}(pe^t + 1 - p)^{n-2}.$$

Setting $t = 0$ in these two equations and referring to Proposition 11.2, we have

$$\mathcal{E}(X) = M_X'(0) = np \qquad \text{and} \qquad \mathcal{E}(X^2) = M_X''(0) = np + n(n-1)p^2.$$

Thus $\mathcal{E}(X) = np$ and

$$\text{Var}(X) = \mathcal{E}(X^2) - (\mathcal{E}(X))^2 = np + n(n-1)p^2 - (np)^2 = np(1 - p).$$

Compare the work done here to obtain the mean and variance of a binomial random variable to that in Examples 7.2 and 7.13 on pages 331 and 357, respectively. ∎

EXAMPLE 11.7   *Moment Generation Property of MGFs*

Mean and Variance of a Normal Random Variable   Obtain the mean and variance of a normal random variable from its MGF.

Solution   Let $X \sim \mathcal{N}(\mu, \sigma^2)$. From Example 11.5, $M_X(t) = e^{\mu t + \sigma^2 t^2/2}$. To use Proposition 11.2 to obtain the mean and variance of X, we need the first two derivatives of M_X. Applying differentiation rules from calculus and doing some simple algebra, we get

$$M_X'(t) = (\mu + \sigma^2 t)e^{\mu t + \sigma^2 t^2/2}$$

and

$$M_X''(t) = \left(\sigma^2 + (\mu + \sigma^2 t)^2\right)e^{\mu t + \sigma^2 t^2/2}.$$

Setting $t = 0$ in these two equations and referring to Proposition 11.2, we have

$$\mathcal{E}(X) = M_X'(0) = \mu \qquad \text{and} \qquad \mathcal{E}(X^2) = M_X''(0) = \sigma^2 + \mu^2.$$

Thus $\mathcal{E}(X) = \mu$ and

$$\text{Var}(X) = \mathcal{E}(X^2) - (\mathcal{E}(X))^2 = \sigma^2 + \mu^2 - (\mu)^2 = \sigma^2.$$

Compare the work done here to obtain the mean and variance of a normal random variable to that in Examples 10.3 and 10.12 on pages 567 and 586, respectively.   ∎

Other Important Properties of MGFs

Another essential property of moment generating functions is that they determine the probability distribution. Specifically, we have Proposition 11.3 whose proof we omit, as it's beyond the scope of this book.

◆◆◆ **Proposition 11.3   Uniqueness Property of MGFs**

> *If the MGFs of random variables X and Y are equal in some open interval containing 0, then X and Y have the same probability distribution.*

Because of Proposition 11.3, if we can recognize the MGF of a random variable X as that of a particular probability distribution, then X must have that probability distribution. For instance, suppose that $M_X(t) = \sqrt{3/(3 - t)}$. From Table 11.1, M_X is the MGF of the gamma distribution with parameters $1/2$ and 3. Hence, $X \sim \Gamma(1/2, 3)$.

Before presenting further applications of the uniqueness property of MGFs, we examine the relationship between sums of independent random variables and MGFs. This relationship is stated in Proposition 11.4.

◆◆◆ **Proposition 11.4   Multiplication Property of MGFs**

> *If X and Y are independent random variables, then*
>
> $$M_{X+Y} = M_X M_Y. \tag{11.7}$$
>
> *More generally, if $X_1, \ldots, X_m$ are independent random variables, then*
>
> $$M_{X_1 + \cdots + X_m} = M_{X_1} \cdots M_{X_m}. \tag{11.8}$$
>
> *In words, the moment generating function of the sum of independent random variables is the product of their moment generating functions.*

Proof   We prove Equation (11.7) and leave the proof of Equation (11.8) to you as Exercise 11.13. Because X and Y are independent random variables, Proposition 6.9 on page 291 implies that e^{tX} and e^{tY} are independent random variables for each $t \in \mathcal{R}$. For independent random variables, the expected value of the product equals the product of the expected values. Hence,

$$M_{X+Y}(t) = \mathcal{E}\left(e^{t(X+Y)}\right) = \mathcal{E}\left(e^{tX} e^{tY}\right) = \mathcal{E}\left(e^{tX}\right) \mathcal{E}\left(e^{tY}\right) = M_X(t) M_Y(t),$$

as required.   ◆

By applying, in turn, the multiplication and uniqueness properties of MGFs, we can more easily derive results for sums of independent random variables that we obtained directly in Chapters 6 and 9. To illustrate, we use those two properties of MGFs to identify the sum of independent binomial random variables with the same success-probability parameter and the sum of independent normal random variables.

EXAMPLE 11.8   *Multiplication and Uniqueness Properties of MGFs*

Sum of Independent Binomial Random Variables Let $X_1, \ldots, X_m$ be independent binomial random variables, all having the same success-probability parameter—say, $X_j \sim \mathcal{B}(n_j, p)$ for $1 \le j \le m$. Identify the probability distribution of the random variable $X_1 + \cdots + X_m$.

Solution   From Table 11.1, $M_{X_j}(t) = (pe^t + 1 - p)^{n_j}$ for $1 \le j \le m$. Because $X_1, \ldots, X_m$ are independent, we apply Proposition 11.4 and conclude that

$$M_{X_1 + \cdots + X_m}(t) = M_{X_1}(t) \cdots M_{X_m}(t) = (pe^t + 1 - p)^{n_1} \cdots (pe^t + 1 - p)^{n_m}$$

$$= (pe^t + 1 - p)^{n_1 + \cdots + n_m}.$$

Again, from Table 11.1, the MGF of $X_1 + \cdots + X_m$ is that of a binomial random variable with parameters $n_1 + \cdots + n_m$ and p. Therefore, by the uniqueness property of MGFs, $X_1 + \cdots + X_m \sim \mathcal{B}(n_1 + \cdots + n_m, p)$.   ∎

EXAMPLE 11.9   *Multiplication and Uniqueness Properties of MGFs*

Sum of Independent Normal Random Variables Let $X_1, \ldots, X_m$ be independent normal random variables—say, $X_j \sim \mathcal{N}(\mu_j, \sigma_j^2)$ for $1 \le j \le m$. Identify the probability distribution of the random variable $X_1 + \cdots + X_m$.

Solution   From Table 11.1, $M_{X_j}(t) = e^{\mu_j t + \sigma_j^2 t^2 / 2}$ for $1 \le j \le m$. As $X_1, \ldots, X_m$ are independent, we apply Proposition 11.4 and conclude that

$$M_{X_1 + \cdots + X_m}(t) = M_{X_1}(t) \cdots M_{X_m}(t) = e^{\mu_1 t + \sigma_1^2 t^2 / 2} \cdots e^{\mu_m t + \sigma_m^2 t^2 / 2}$$

$$= e^{(\mu_1 + \cdots + \mu_m)t + (\sigma_1^2 + \cdots + \sigma_m^2)t^2 / 2}.$$

Again, from Table 11.1, the MGF of $X_1 + \cdots + X_m$ is that of a normal random variable with parameters $\mu_1 + \cdots + \mu_m$ and $\sigma_1^2 + \cdots + \sigma_m^2$. So, by the uniqueness property of MGFs, $X_1 + \cdots + X_m \sim \mathcal{N}(\mu_1 + \cdots + \mu_m, \sigma_1^2 + \cdots + \sigma_m^2)$.   ∎

Some of the most important applications of moment generating functions are in proving *limit theorems,* such as the central limit theorem. For such applications, we need the result presented in Proposition 11.5, often referred to as the **continuity theorem** because it shows that distribution functions depend continuously on their MGFs. The proof of the continuity theorem is beyond the scope of this book.

◆◆◆ **Proposition 11.5   Continuity Theorem**

Let $X, X_1, X_2, \ldots$ be random variables whose MGFs are all defined in some open interval containing 0, say, $-t_0 < t < t_0$. Suppose that

$$\lim_{n \to \infty} M_{X_n}(t) = M_X(t), \qquad -t_0 < t < t_0. \tag{11.9}$$

Then

$$\lim_{n \to \infty} F_{X_n}(x) = F_X(x), \qquad x \in \mathcal{C}_{F_X}, \tag{11.10}$$

where $\mathcal{C}_{F_X}$ denotes the set of all $x \in \mathcal{R}$ at which F_X is continuous.

A sequence of random variables $\{X_n\}_{n=1}^{\infty}$ is said to **converge in distribution** to the random variable X if Equation (11.10) holds. With this terminology, the continuity theorem can be stated as follows: *If in some open interval containing 0 the MGFs of $X_1, X_2, \ldots$ converge pointwise to the MGF of X, then $\{X_n\}_{n=1}^{\infty}$ converges in distribution to X.*

We note that, in view of Proposition 8.2 on page 412, if $\{X_n\}_{n=1}^{\infty}$ converges in distribution to X, then $\lim_{n \to \infty} P(a < X_n \le b) = P(a < X \le b)$ for all $a, b \in \mathcal{C}_{F_X}$. We also note that, although the continuity theorem is often used to establish convergence in distribution, it's sometimes easier to just verify Equation (11.10) directly.

You are asked to apply the continuity theorem and the concept of convergence in distribution in the exercises for this section. In Section 11.4, we apply the continuity theorem and other properties of MGFs to prove the all-important central limit theorem.

Characteristic Functions

As we demonstrated, moment generating functions are useful for many aspects of probabilistic analysis. One major drawback of MGFs, however, is that they aren't always defined for all values of t. For instance, the MGF of a Cauchy random variable is defined only at $t = 0$, and the MGF of a gamma random variable is defined only when $t < \lambda$.

An alternative to the moment generating function is the **characteristic function,** which, for a random variable X, is defined by

$$\phi_X(t) = \mathcal{E}\left(e^{itX}\right), \qquad -\infty < t < \infty, \tag{11.11}$$

where $i = \sqrt{-1}$. The characteristic function is defined for all values of t for every random variable. Other than requiring the use of complex numbers, characteristic functions are simpler and theoretically superior to moment generating functions. In advanced probability courses, the characteristic function is almost always the transform of choice. You are asked to consider characteristic functions in Exercises 11.17–11.20.

EXERCISES 11.1   Basic Exercises

11.1 Let $X_1, X_2,$ and X_3 be a random sample from a discrete distribution with probability mass function given by $p(0) = 1/3$, $p(1) = 2/3$, and $p(x) = 0$ otherwise. Determine the moment generating function of the random variable $Y = X_1 X_2 X_3$.

11.2 Let X have the Poisson distribution with parameter λ.

a) Determine the MGF of X. *Hint:* Use the exponential series, Equation (5.26) on page 222.

b) Use the result of part (a) to obtain the mean and variance of X.

c) Suppose that X and Y are independent random variables with $X \sim \mathcal{P}(\lambda)$ and $Y \sim \mathcal{P}(\mu)$. Use moment generating functions to show that $X + Y \sim \mathcal{P}(\lambda + \mu)$.

d) Extend the result of part (c) for m independent Poisson random variables.

11.3 Let X have the geometric distribution with parameter p.

a) Determine the MGF of X, including where it's defined.

b) Use the result of part (a) to obtain the mean and variance of X.

11.4 Let X have the uniform distribution on the interval (a, b).

a) Determine the MGF of X.

b) Use the result of part (a) to obtain the mean and variance of X.

11.5 A company insures homes in three cities. Because sufficient distance separates the cities, it's reasonable to assume that the losses occurring among the three cities are independent random variables. The moment generating functions for the three loss distributions are $(1 - 2t)^{-3}$, $(1 - 2t)^{-2.5}$, and $(1 - 2t)^{-4.5}$. Determine the third moment of the combined losses from the three cities.

11.6 Let X be a Cauchy random variable. Does there exist an open interval containing 0 for which M_X is defined? Explain your answer.

11.7 In Example 9.20 on page 536, we showed that, if X and Y are independent random variables with $X \sim \Gamma(\alpha, \lambda)$ and $Y \sim \Gamma(\beta, \lambda)$, then $X + Y \sim \Gamma(\alpha + \beta, \lambda)$.

a) Use MGFs to establish the result referred to and compare your work with that required in Example 9.20.

b) Generalize the result of part (a) to m independent gamma random variables with the same second parameter, thus providing a simple proof of Proposition 9.13 on page 537.

11.8 Suppose that Y has the lognormal distribution with parameters μ and σ^2, which means that $\ln Y \sim \mathcal{N}(\mu, \sigma^2)$. Determine a formula for the nth moment of Y by

a) using a PDF of Y.

b) using the FEF and a PDF of a normal random variable.

c) recalling the formula for the MGF of a normal random variable.

d) For what values of t is the MGF of Y defined? Justify your answer.

e) As we mentioned on page 634, if the MGF of a random variable is defined (exists) in some open interval containing 0, the random variable has moments of all orders. Is the converse of this statement true? Explain your answer.

11.9 For each $n \in \mathcal{N}$, let X_n have the discrete uniform distribution on $\{-1 + 1/n, 1 - 1/n\}$.

a) Heuristically, what is the probability distribution of a random variable X to which $\{X_n\}_{n=1}^\infty$ converges in distribution?

b) Mathematically verify your heuristics in part (a) by showing that $M_{X_n}(t) \to M_X(t)$ as $n \to \infty$ for all $t \in \mathcal{R}$.

c) Mathematically verify your heuristics in part (a) by showing that $F_{X_n}(x) \to F_X(x)$ as $n \to \infty$ for all $x \in \mathcal{R}$ at which F_X is continuous.

11.10 Let $X_1, X_2, \ldots$ be independent random variables, each having a $\mathcal{U}(0, 1)$ distribution. For each $n \in \mathcal{N}$, let $U_n = \min\{X_1, \ldots, X_n\}$ and let $V_n = \max\{X_1, \ldots, X_n\}$.

a) Heuristically, what happens to the probability distribution of U_n as $n \to \infty$? Explain your reasoning.

b) Show that $\{U_n\}_{n=1}^\infty$ converges in distribution and identify its limiting distribution.

c) Repeat parts (a) and (b) for $\{V_n\}_{n=1}^\infty$.

11.11 Let $X_1, X_2, \ldots$ be independent and identically distributed random variables, each having mean μ, variance σ^2, and moment generating function M defined for $-t_0 < t < t_0$.

a) Show that $M_{\bar{X}_n}(t) = \left(M(t/n)\right)^n$ for each $n \in \mathcal{N}$.

b) Use part (a) to obtain the mean and variance of $\bar{X}_n$ in terms of μ, σ^2, and n.

c) Use part (a) to prove that $\lim_{n\to\infty} M_{\bar{X}_n}(t) = e^{\mu t}$ for all $t \in (-t_0, t_0)$. *Hint:* Take natural logarithms and apply L'Hôpital's rule.

d) Deduce from part (c) that $\{\bar{X}_n\}_{n=1}^{\infty}$ converges in distribution to μ.

e) Relate the result of part (d) to the long-run-average interpretation of expected value.

Theory Exercises

11.12 Bounded random variable: A random variable X is said to be *bounded* if there is a positive real number M such that $P(|X| \le M) = 1$.

a) Show that the MGF of a bounded discrete random variable is defined for all $t \in \mathcal{R}$.

b) Show that the MGF of a bounded continuous random variable with a PDF is defined for all $t \in \mathcal{R}$. *Hint:* Refer to Exercise 10.19 on page 570.

c) The results of parts (a) and (b) hold for any type of random variable. Deduce that a bounded random variable has finite moments of all orders.

11.13 Complete the proof of Proposition 11.4 on page 636 by showing that, if $X_1, \ldots, X_m$ are independent random variables, then $M_{X_1+\cdots+X_m} = M_{X_1} \cdots M_{X_m}$.

11.14 Poisson approximation to the binomial distribution: Proposition 5.7 on page 220 provides an informal formulation of the Poisson approximation to the binomial distribution. Use the continuity theorem to prove the following formal version of that result. Let λ be a positive constant and suppose that $X_n \sim \mathcal{B}(n, p_n)$ for each $n \in \mathcal{N}$, where $p_n = \lambda/n$.

a) Show that $\lim_{n\to\infty} M_{X_n}(t) = e^{\lambda(e^t - 1)}$ for all $t \in \mathcal{R}$. *Hint:* $\lim_{u\to 0} u^{-1} \ln(1 + u) = 1$.

b) Deduce from part (a) that $\{X_n\}_{n=1}^{\infty}$ converges in distribution to a Poisson random variable with parameter λ.

c) Use part (b) to conclude that $\lim_{n\to\infty} p_{X_n}(x) = e^{-\lambda} \lambda^x / x!$ for $x = 0, 1, 2, \ldots$.

d) Compare the result of part (c) with Proposition 5.7.

11.15 Normal approximation to the Poisson distribution: Suppose that $\{\lambda_n\}_{n=1}^{\infty}$ is a sequence of positive real numbers such that $\lim_{n\to\infty} \lambda_n = \infty$. For each $n \in \mathcal{N}$, let $X_n \sim \mathcal{P}(\lambda_n)$ and let Y_n be the standardized version of X_n.

a) Show that $\{Y_n\}_{n=1}^{\infty}$ converges in distribution to a standard normal random variable. *Hint:* Use the calculus result $\lim_{u\to 0}(e^u - 1 - u)/u^2 = 1/2$.

b) Let $X \sim \mathcal{P}(\lambda)$, where λ is large. Deduce from part (a) that

$$p_X(x) \approx \Phi\left(\frac{x + \frac{1}{2} - \lambda}{\sqrt{\lambda}}\right) - \Phi\left(\frac{x - \frac{1}{2} - \lambda}{\sqrt{\lambda}}\right), \qquad x = 0, 1, 2, \ldots.$$

Advanced Exercises

11.16 Power series expansion of MGFs: Let X be a random variable whose MGF is defined in an open interval containing 0 and let a_n be the coefficient of t^n in the power series expansion of $M_X(t)$ about $t = 0$.

a) Show that $\mathcal{E}(X^n) = n! \, a_n$ for each $n \in \mathcal{N}$.

b) Let X be an exponential random variable with parameter λ. Determine the power series expansion of M_X about $t = 0$ and use that expansion to obtain all moments of X.

c) Determine the power series expansion about $t = 0$ of the MGF of a normal random variable with parameters 0 and σ^2. Use that expansion to obtain all central moments of a normal random variable with parameters μ and σ^2.

Characteristic functions: The *characteristic function* (CHF) of a random variable X, denoted ϕ_X, is defined by $\phi_X(t) = \mathcal{E}(e^{itX})$, where $i = \sqrt{-1}$. In Exercises 11.17–11.20 you are to examine the concept of the characteristic function.

11.17 Prove that the CHF of any random variable X is defined for all $t \in \mathcal{R}$ by showing that the (complex-valued) random variable e^{itX} has finite expectation for all $t \in \mathcal{R}$.

11.18 State and prove the analogues of Propositions 11.1, 11.2, and 11.4 on pages 633, 635, and 636, respectively, for characteristic functions.

11.19 Let X be a random variable whose MGF is defined in some open interval containing 0.
a) Formally, what is the relationship between ϕ_X and M_X? *Note:* The formal relationship can be verified mathematically by using the concept of analytic continuation from the theory of complex variables.
b) Use your result from part (a) to obtain the CHF of a random variable with each of the following probability distributions: $\mathcal{B}(n, p)$, $\mathcal{P}(\lambda)$, $\mathcal{G}(p)$, $\mathcal{U}(a, b)$, $\mathcal{E}(\lambda)$, and $\mathcal{N}(\mu, \sigma^2)$.
c) Use your results from part (b) and the moment generation property of CHFs, as obtained in Exercise 11.18, to find the mean and variance of a random variable with each of the probability distributions specified in part (b).

11.20 Let X have the standard Cauchy distribution.
a) Show that $\phi_X(t) = e^{-|t|}$. *Note:* This result requires use of the residue theorem from complex analysis. If you aren't familiar with that result, just continue on to the remaining parts of this exercise.
b) Let $X_1, X_2, \dots$ be independent random variables all having the standard Cauchy distribution. Use CHFs to show that $\bar{X}_n$ also has the standard Cauchy distribution.
c) Discuss the result obtained in part (b) in relation to the long-run-average interpretation of expected value.
d) Obtain the CHF of a Cauchy random variable with parameters η and θ.

11.2   Joint Moment Generating Functions

When analyzing more than one random variable simultaneously, we find it useful to consider **joint moment generating functions.** Here we introduce this concept in the bivariate case. Later in this section, we discuss the general multivariate case.

DEFINITION 11.2   Joint MGF: Bivariate Case

Let X and Y be random variables defined on the same sample space. Then the **joint moment generating function** of X and Y, denoted $M_{X,Y}$, is defined by

$$M_{X,Y}(s, t) = \mathcal{E}\left(e^{sX+tY}\right). \tag{11.12}$$

for all values $s, t \in \mathcal{R}$ for which the random variable e^{sX+tY} has finite expectation. We use the phrase **joint MGF** as an abbreviation for "joint moment generating function."

Note: Because $e^{0 \cdot X + 0 \cdot Y} = 1$ has finite expectation, the joint MGF is defined at $s = t = 0$ for any two random variables X and Y. Moreover, $M_{X,Y}(0, 0) = \mathcal{E}(1) = 1$.

Applying Definition 11.2 and the bivariate forms of the FEFs for discrete and continuous random variables with the function $g(x, y) = e^{sx+ty}$, we get the following formulas.

- If X and Y are discrete random variables, the joint MGF of X and Y is given by

$$M_{X,Y}(s, t) = \sum_{(x,y)} \sum e^{sx+ty} p_{X,Y}(x, y). \qquad \textbf{(11.13)}$$

- If X and Y are continuous random variables with a joint PDF, the joint MGF of X and Y is given by

$$M_{X,Y}(s, t) = \int_{-\infty}^{\infty} \int_{-\infty}^{\infty} e^{sx+ty} f_{X,Y}(x, y) \, dx \, dy. \qquad \textbf{(11.14)}$$

We emphasize, though, that the concept of the joint moment generating function applies to all types or combinations of types of random variables.

One of the most important applications of joint MGFs is to normal distributions. In Example 11.10, we obtain the joint MGF of bivariate normal random variables.

EXAMPLE 11.10   *Joint MGF of Bivariate Normal Random Variables*

Let $(X, Y) \sim \mathcal{BVN}(\mu_X, \sigma_X^2, \mu_Y, \sigma_Y^2, \rho)$. Show that the joint MGF of X and Y is given by

$$M_{X,Y}(s, t) = e^{\mu_X s + \mu_Y t + \frac{1}{2}(\sigma_X^2 s^2 + 2\rho\sigma_X\sigma_Y st + \sigma_Y^2 t^2)}, \qquad s, t \in \mathcal{R}. \qquad \textbf{(11.15)}$$

Solution   An efficient way to obtain the joint MGF of X and Y is to first condition on X and then apply the law of total expectation in the form of Equation (10.50) on page 602:

$$M_{X,Y}(s, t) = \mathcal{E}\left(e^{sX+tY}\right) = \mathcal{E}\left(\mathcal{E}\left(e^{sX+tY} \mid X\right)\right). \qquad \textbf{(11.16)}$$

From Proposition 10.10(d) on page 616, the conditional distribution of Y given $X = x$ is a normal distribution with mean $\mu_Y + \rho\frac{\sigma_Y}{\sigma_X}(x - \mu_X)$ and variance $\sigma_Y^2(1 - \rho^2)$. Therefore, by Example 11.5 on page 633,

$$\mathcal{E}\left(e^{tY} \mid X\right) = e^{\left(\mu_Y + \rho\frac{\sigma_Y}{\sigma_X}(X-\mu_X)\right)t + \sigma_Y^2(1-\rho^2)t^2/2}$$

$$= e^{\left(\mu_Y - \rho\frac{\sigma_Y}{\sigma_X}\mu_X\right)t + \sigma_Y^2(1-\rho^2)t^2/2} e^{\left(\rho\frac{\sigma_Y}{\sigma_X}t\right)X}.$$

Next we apply Equation (10.49) on page 602 and the previous equation to conclude that

$$\mathcal{E}\left(e^{sX+tY} \mid X\right) = \mathcal{E}\left(e^{sX}e^{tY} \mid X\right) = e^{sX}\mathcal{E}\left(e^{tY} \mid X\right)$$

$$= e^{sX} e^{\left(\mu_Y - \rho\frac{\sigma_Y}{\sigma_X}\mu_X\right)t + \sigma_Y^2(1-\rho^2)t^2/2} e^{\left(\rho\frac{\sigma_Y}{\sigma_X}t\right)X}$$

$$= e^{\left(\mu_Y - \rho\frac{\sigma_Y}{\sigma_X}\mu_X\right)t + \sigma_Y^2(1-\rho^2)t^2/2} e^{\left(s+\rho\frac{\sigma_Y}{\sigma_X}t\right)X}.$$

From this last equation, Equation (11.16), and Example 11.5, we get

$$M_{X,Y}(s,t) = \mathcal{E}\left(e^{\left(\mu_Y - \rho\frac{\sigma_Y}{\sigma_X}\mu_X\right)t + \sigma_Y^2(1-\rho^2)t^2/2}e^{\left(s + \rho\frac{\sigma_Y}{\sigma_X}t\right)X}\right)$$

$$= e^{\left(\mu_Y - \rho\frac{\sigma_Y}{\sigma_X}\mu_X\right)t + \sigma_Y^2(1-\rho^2)t^2/2}\mathcal{E}\left(e^{\left(s + \rho\frac{\sigma_Y}{\sigma_X}t\right)X}\right)$$

$$= e^{\left(\mu_Y - \rho\frac{\sigma_Y}{\sigma_X}\mu_X\right)t + \sigma_Y^2(1-\rho^2)t^2/2}M_X\left(s + \rho\frac{\sigma_Y}{\sigma_X}t\right)$$

$$= e^{\left(\mu_Y - \rho\frac{\sigma_Y}{\sigma_X}\mu_X\right)t + \sigma_Y^2(1-\rho^2)t^2/2}e^{\mu_X\left(s + \rho\frac{\sigma_Y}{\sigma_X}t\right) + \sigma_X^2\left(s + \rho\frac{\sigma_Y}{\sigma_X}t\right)^2/2}$$

$$= e^{\left(\mu_Y - \rho\frac{\sigma_Y}{\sigma_X}\mu_X\right)t + \sigma_Y^2(1-\rho^2)t^2/2 + \mu_X\left(s + \rho\frac{\sigma_Y}{\sigma_X}t\right) + \sigma_X^2\left(s + \rho\frac{\sigma_Y}{\sigma_X}t\right)^2/2}.$$

Performing some algebra on the previous expression, we obtain Equation (11.15). ∎

Moment Generation Property of Joint MGFs

From Definition 11.2, the joint MGF of random variables X and Y is defined for all values of $s, t \in \mathcal{R}$ for which e^{sX+tY} has finite expectation. We restrict our attention to random variables X and Y whose joint MGF is defined in some open rectangle containing $(0,0)$; that is, there are positive numbers s_0 and t_0 such that $M_{X,Y}(s,t)$ is defined for all s and t with $-s_0 < s < s_0$ and $-t_0 < t < t_0$. This assumption implies that X and Y have individual and joint moments of all orders. It also implies that the following formal derivations about $M_{X,Y}$ are indeed mathematically valid.

By taking appropriate partial derivatives of the joint MGF, we can obtain individual and joint moments of X and Y. Specifically, differentiating both sides of Equation (11.12) and interchanging the order of differentiation and expectation yields

$$\frac{\partial^{j+k}M_{X,Y}}{\partial s^j \partial t^k}(s,t) = \frac{\partial^{j+k}}{\partial s^j \partial t^k}\mathcal{E}\left(e^{sX+tY}\right) = \mathcal{E}\left(\frac{\partial^{j+k}}{\partial s^j \partial t^k}e^{sX+tY}\right) = \mathcal{E}\left(X^j Y^k e^{sX+tY}\right),$$

for $-s_0 < s < s_0$ and $-t_0 < t < t_0$. And, setting $s = t = 0$, we deduce Proposition 11.6.

◆◆◆ **Proposition 11.6   Moment Generation Property of Joint MGFs**

Suppose that the joint MGF of X and Y is defined in some open rectangle containing $(0,0)$. Then X and Y have individual and joint moments of all orders and

$$\mathcal{E}\left(X^j Y^k\right) = \frac{\partial^{j+k}M_{X,Y}}{\partial s^j \partial t^k}(0,0), \qquad j, k = 0, 1, 2, \ldots. \qquad \textbf{(11.17)}$$

Proposition 11.6 provides an alternative method for finding individual and joint moments of two random variables that is often more efficient or straightforward than the direct method of summation or integration used for discrete and continuous random variables, respectively. In Example 11.11, we use the MGF approach of Proposition 11.6 to find the means, variances, and covariance of bivariate normal random variables.

EXAMPLE 11.11   *Moments of Bivariate Normal Random Variables*

Use the joint MGF to obtain the means, variances, and covariance of bivariate normal random variables.

Solution   From Example 11.10, $M_{X,Y}(s, t) = e^{\mu_X s + \mu_Y t + \frac{1}{2}(\sigma_X^2 s^2 + 2\rho\sigma_X\sigma_Y st + \sigma_Y^2 t^2)}$. Applying elementary calculus, we get

$$\frac{\partial M_{X,Y}}{\partial s}(s, t) = \left(\mu_X + \frac{1}{2}\left(2\sigma_X^2 s + 2\rho\sigma_X\sigma_Y t\right)\right) M_{X,Y}(s, t);$$

$$\frac{\partial M_{X,Y}}{\partial t}(s, t) = \left(\mu_Y + \frac{1}{2}\left(2\rho\sigma_X\sigma_Y s + 2\sigma_Y^2 t\right)\right) M_{X,Y}(s, t);$$

$$\frac{\partial^2 M_{X,Y}}{\partial s^2}(s, t) = \left[\left(\mu_X + \frac{1}{2}\left(2\sigma_X^2 s + 2\rho\sigma_X\sigma_Y t\right)\right)^2 + \sigma_X^2\right] M_{X,Y}(s, t);$$

$$\frac{\partial^2 M_{X,Y}}{\partial t^2}(s, t) = \left[\left(\mu_Y + \frac{1}{2}\left(2\rho\sigma_X\sigma_Y s + 2\sigma_Y^2 t\right)\right)^2 + \sigma_Y^2\right] M_{X,Y}(s, t);$$

$$\frac{\partial^2 M_{X,Y}}{\partial s \partial t}(s, t) = \left[\left(\mu_Y + \frac{1}{2}\left(2\rho\sigma_X\sigma_Y s + 2\sigma_Y^2 t\right)\right)\left(\mu_X + \frac{1}{2}\left(2\sigma_X^2 s + 2\rho\sigma_X\sigma_Y t\right)\right)\right.$$
$$\left. + \rho\sigma_X\sigma_Y\right] M_{X,Y}(s, t).$$

Referring to Proposition 11.6 and recalling that $M_{X,Y}(0, 0) = 1$, we conclude that

$$\mathcal{E}(X) = \mathcal{E}\left(X^1 Y^0\right) = \frac{\partial M_{X,Y}}{\partial s}(0, 0) = \mu_X;$$

$$\mathcal{E}(Y) = \mathcal{E}\left(X^0 Y^1\right) = \frac{\partial M_{X,Y}}{\partial t}(0, 0) = \mu_Y;$$

$$\mathcal{E}(X^2) = \mathcal{E}\left(X^2 Y^0\right) = \frac{\partial^2 M_{X,Y}}{\partial s^2}(0, 0) = \mu_X^2 + \sigma_X^2;$$

$$\mathcal{E}(Y^2) = \mathcal{E}\left(X^0 Y^2\right) = \frac{\partial^2 M_{X,Y}}{\partial t^2}(0, 0) = \mu_Y^2 + \sigma_Y^2;$$

$$\mathcal{E}(XY) = \mathcal{E}\left(X^1 Y^1\right) = \frac{\partial^2 M_{X,Y}}{\partial s \partial t}(0, 0) = \mu_X\mu_Y + \rho\sigma_X\sigma_Y.$$

From these equations, $\mathcal{E}(X) = \mu_X$, $\mathcal{E}(Y) = \mu_Y$,

$$\text{Var}(X) = \mathcal{E}(X^2) - (\mathcal{E}(X))^2 = \mu_X^2 + \sigma_X^2 - (\mu_X)^2 = \sigma_X^2,$$
$$\text{Var}(Y) = \mathcal{E}(Y^2) - (\mathcal{E}(Y))^2 = \mu_Y^2 + \sigma_Y^2 - (\mu_Y)^2 = \sigma_Y^2,$$

and

$$\text{Cov}(X, Y) = \mathcal{E}(XY) - \mathcal{E}(X)\mathcal{E}(Y) = \mu_X\mu_Y + \rho\sigma_X\sigma_Y - \mu_X\mu_Y = \rho\sigma_X\sigma_Y.$$

These results are consistent with those in Proposition 10.10(a)–(c) on page 616. ∎

Marginal MGFs

From the joint MGF of two random variables X and Y, we can obtain the individual MGFs of X and Y, each called, in the joint context, a **marginal moment generating function** or, simply, marginal MGF. We have

$$M_X(s) = \mathcal{E}\left(e^{sX}\right) = \mathcal{E}\left(e^{sX+0 \cdot Y}\right) = M_{X,Y}(s, 0),$$

and, likewise, $M_Y(t) = M_{X,Y}(0, t)$. In summary, we present Proposition 11.7.

◆◆◆ **Proposition 11.7   Obtaining Marginal MGFs from the Joint MGF**

Let X and Y be random variables defined on the same sample space. Then

$$M_X(s) = M_{X,Y}(s, 0) \qquad and \qquad M_Y(t) = M_{X,Y}(0, t). \tag{11.18}$$

Thus the marginal MGF of X is obtained by setting $t = 0$ in the joint MGF, and the marginal MGF of Y is obtained by setting $s = 0$ in the joint MGF.

EXAMPLE 11.12   *Marginal MGFs of Bivariate Normal Random Variables*

Obtain the marginal MGFs of bivariate normal random variables and use them to identify the marginal distributions.

Solution   Let $(X, Y) \sim \mathcal{BVN}(\mu_X, \sigma_X^2, \mu_Y, \sigma_Y^2, \rho)$. Equation (11.15) on page 642 provides the joint MGF of X and Y. Applying Proposition 11.7 to that joint MGF, we get

$$M_X(s) = M_{X,Y}(s, 0) = e^{\mu_X s + \mu_Y \cdot 0 + \frac{1}{2}\left(\sigma_X^2 s^2 + 2\rho\sigma_X\sigma_Y s \cdot 0 + \sigma_Y^2 \cdot 0^2\right)} = e^{\mu_X s + \sigma_X^2 s^2/2}.$$

From the uniqueness property of MGFs and Table 11.1 on page 634, we conclude that $X \sim \mathcal{N}(\mu_X, \sigma_X^2)$. Similarly, we find that $M_Y(t) = M_{X,Y}(0, t) = e^{\mu_Y t + \sigma_Y^2 t^2/2}$ and, hence, $Y \sim \mathcal{N}(\mu_Y, \sigma_Y^2)$. These results for the marginal distributions of X and Y are consistent with those presented in Proposition 10.10(a) and (b) on page 616.   ■

Other Important Properties of Joint MGFs

Another essential property of joint moment generating functions is that they determine the joint probability distribution. Specifically, we have Proposition 11.8 whose proof we omit, as it is beyond the scope of this book.

◆◆◆ **Proposition 11.8   Uniqueness Property of Joint MGFs**

If the joint MGF of X and Y equals the joint MGF of U and V in some open rectangle containing $(0, 0)$, then X and Y have the same joint probability distribution as U and V.

Because of Proposition 11.8, if we can recognize the joint MGF of random variables X and Y as that of a particular joint probability distribution, then X and Y must have that joint probability distribution. For instance, suppose that $M_{X,Y}(s, t) = e^{\frac{1}{2}\left(s^2 - \frac{6}{5}st + t^2\right)}$. From Equation (11.15) on page 642, $M_{X,Y}$ is the joint MGF of bivariate normal random variables—specifically, $\mathcal{BVN}(0, 1, 0, 1, -0.6)$. Therefore, in view of Proposition 11.8, we conclude that $(X, Y) \sim \mathcal{BVN}(0, 1, 0, 1, -0.6)$.

We have shown many times that independence of random variables reflects itself in one or another type of multiplication property. For instance, from Proposition 6.10, two discrete random variables are independent if and only if their joint PMF equals the product of their marginal PMFs. This phenomenon also holds for MGFs.

◆◆◆ **Proposition 11.9   Independence and Joint MGFs**

Let X and Y be random variables defined on the same sample space. Then X and Y are independent if and only if

$$M_{X,Y}(s, t) = M_X(s)M_Y(t). \tag{11.19}$$

In words, two random variables are independent if and only if their joint MGF equals the product of their marginal MGFs.

Proof   Suppose that X and Y are independent random variables. Proposition 6.9 on page 291 implies that e^{sX} and e^{tY} are also independent random variables for each $s, t \in \mathcal{R}$. Now using the fact that, for independent random variables, the expected value of the product equals the product of the expected values, we conclude that

$$M_{X,Y}(s, t) = \mathcal{E}\left(e^{sX+tY}\right) = \mathcal{E}\left(e^{sX}e^{tY}\right) = \mathcal{E}\left(e^{sX}\right)\mathcal{E}\left(e^{tY}\right) = M_X(s)M_Y(t).$$

Hence Equation (11.19) holds.

   Conversely, suppose that Equation (11.19) holds. Let U and V be independent random variables with the same (marginal) probability distributions as X and Y, respectively. Then, by Equation (11.19) and what we just proved,

$$M_{X,Y}(s, t) = M_X(s)M_Y(t) = M_U(s)M_V(t) = M_{U,V}(s, t).$$

Hence the joint MGF of X and Y is the same as that of U and V. Therefore, by the uniqueness property of joint MGFs, the joint probability distribution of X and Y is the same as that of U and V. Because U and V are independent random variables, it now follows (see Exercise 11.33) that X and Y also are independent random variables.   ◆

EXAMPLE 11.13   *Independence of Uncorrelated Bivariate Normal Random Variables*

According to Proposition 10.9 on page 613, bivariate normal random variables are independent if and only if they are uncorrelated. Prove that result by using joint MGFs.

Solution   Let $(X, Y) \sim \mathcal{BVN}(\mu_X, \sigma_X^2, \mu_Y, \sigma_Y^2, \rho)$. Then, as we know, ρ is the correlation coefficient of X and Y. Hence, in view of Proposition 11.9, we want to show that $M_{X,Y}(s, t) = M_X(s)M_Y(t)$ if and only if $\rho = 0$. From Equation (11.15) on page 642,

$$M_{X,Y}(s, t) = e^{\mu_X s + \mu_Y t + \frac{1}{2}(\sigma_X^2 s^2 + 2\rho\sigma_X\sigma_Y st + \sigma_Y^2 t^2)}, \tag{11.20}$$

and, from Example 11.12 on page 645,

$$M_X(s)M_Y(t) = e^{\mu_X s + \sigma_X^2 s^2/2} e^{\mu_Y t + \sigma_Y^2 t^2/2} = e^{\mu_X s + \mu_Y t + \frac{1}{2}(\sigma_X^2 s^2 + \sigma_Y^2 t^2)}.$$

Hence $M_{X,Y}(s, t) = M_X(s)M_Y(t)$ if and only if $\rho\sigma_X\sigma_Y st = 0$ for all s and t, which is true if and only if $\rho = 0$.   ■

The MGF of the sum of two random variables, X and Y, can be obtained from their joint MGF. Indeed, we have $M_{X+Y}(t) = \mathcal{E}\left(e^{t(X+Y)}\right) = \mathcal{E}\left(e^{tX+tY}\right) = M_{X,Y}(t, t)$. This useful result is summarized in Proposition 11.10.

◆◆◆ **Proposition 11.10**   **MGF of the Sum of Two Random Variables**

Let X and Y be random variables defined on the same sample space. Then

$$M_{X+Y}(t) = M_{X,Y}(t, t). \tag{11.21}$$

EXAMPLE 11.14   *Distribution of the Sum of Bivariate Normal Random Variables*

Determine the probability distribution of the sum of bivariate normal random variables.

Solution   Let $(X, Y) \sim \mathcal{BVN}(\mu_X, \sigma_X^2, \mu_Y, \sigma_Y^2, \rho)$. From Equation (11.15) on page 642,

$$M_{X,Y}(s, t) = e^{\mu_X s + \mu_Y t + \frac{1}{2}\left(\sigma_X^2 s^2 + 2\rho\sigma_X\sigma_Y st + \sigma_Y^2 t^2\right)}.$$

Applying Proposition 11.10 to this joint MGF, we obtain

$$M_{X+Y}(t) = M_{X,Y}(t, t) = e^{\mu_X t + \mu_Y t + \frac{1}{2}\left(\sigma_X^2 t^2 + 2\rho\sigma_X\sigma_Y t^2 + \sigma_Y^2 t^2\right)}$$

$$= e^{(\mu_X + \mu_Y)t + \left(\sigma_X^2 + 2\rho\sigma_X\sigma_Y + \sigma_Y^2\right)t^2/2}.$$

From Table 11.1 on page 634, the MGF of $X + Y$ is that of a normal distribution with parameters $\mu_X + \mu_Y$ and $\sigma_X^2 + 2\rho\sigma_X\sigma_Y + \sigma_Y^2$. Hence, by the uniqueness property of MGFs, $X + Y \sim \mathcal{N}(\mu_X + \mu_Y, \sigma_X^2 + 2\rho\sigma_X\sigma_Y + \sigma_Y^2)$. ∎

Joint MGF: Multivariate Case

So far, we have concentrated on joint MGFs for two random variables—that is, in the bivariate case. The definition of the joint MGF in the multivariate case of m random variables is a straightforward generalization of that in the bivariate case.

DEFINITION 11.3   **Joint MGF: Multivariate Case**

Let $X_1, \ldots, X_m$ be random variables defined on the same sample space. Then the **joint moment generating function** of $X_1, \ldots, X_m$, denoted $M_{X_1,\ldots,X_m}$, is defined by

$$M_{X_1,\ldots,X_m}(t_1, \ldots, t_m) = \mathcal{E}\left(e^{t_1 X_1 + \cdots + t_m X_m}\right), \tag{11.22}$$

for all values $t_1, \ldots, t_m \in \mathcal{R}$ for which the random variable $e^{t_1 X_1 + \cdots + t_m X_m}$ has finite expectation. We use the phrase **joint MGF** as an abbreviation for "joint moment generating function."

The basic properties of multivariate joint MGFs are similar to those for bivariate joint MGFs. We leave the statements of these properties to you as Exercise 11.29.

EXERCISES 11.2   **Basic Exercises**

11.21 Let X and Y denote the x and y coordinates of a point selected at random from the vertices of the unit square.
a) Determine the joint MGF of X and Y.
b) Use your result from part (a) to determine the means and variances of X and Y and their covariance.
c) Use your result from part (a) to obtain the marginal MGFs of X and Y.
d) Use your results from parts (a) and (c) to decide whether X and Y are independent random variables.
e) Use your result from part (c) to identify the marginal probability distributions of X and Y.

11.22 Let X and Y denote the x and y coordinates of a point selected at random from the unit square.
a) Determine the joint MGF of X and Y.
b) Use your result from part (a) to determine the means and variances of X and Y and their covariance.
c) Use your result from part (a) to obtain the marginal MGFs of X and Y.
d) Use your results from parts (a) and (c) to decide whether X and Y are independent.
e) Use your result from part (c) to identify the marginal probability distributions of X and Y.

11.23 In Example 9.19 on page 535, we showed that the sum of two independent uniform random variables on the interval $(0, 1)$ has the triangular distribution on the interval $(0, 2)$. Use that result and results from Exercise 11.22 to show that the MGF of a triangular distribution on the interval $(0, 2)$ is given by $2e^t t^{-2}(\cosh t - 1)$, where $\cosh t = (e^t + e^{-t})/2$, the *hyperbolic cosine* of t.

11.24 Suppose that X and Y are random variables defined on the same sample space and set $\psi_{X,Y}(s, t) = \ln M_{X,Y}(s, t)$.
a) Show that $\text{Cov}(X, Y) = \frac{\partial^2 \psi_{X,Y}}{\partial s \partial t}(0, 0)$.
b) Show that, if X and Y are bivariate normal, then $\text{Cov}(X, Y) = \frac{\partial^2 \psi_{X,Y}}{\partial s \partial t}(s, t)$ for all s and t.

11.25 Suppose that the joint MGF of X and Y is $M_{X,Y}(s, t) = (pe^s + qe^t)^n$, where p and q are positive numbers whose sum is 1 and n is a positive integer.
a) Obtain the marginal MGFs of X and Y and use them to identify the (marginal) probability distributions of X and Y.
b) Use your results from part (a) to decide whether X and Y are independent.
c) Use MGFs to identify the probability distribution of $X + Y$.

11.26 Let X and Y be bivariate normal random variables.
a) Use MGFs to show that any nonzero linear combination of X and Y is normally distributed.
b) Use the MGF of a nonzero linear combination of X and Y, as obtained in part (a), to identify its mean and variance.

11.27 Let X and Y have joint PDF given by $f_{X,Y}(x, y) = |x - y|e^{-(x+y)}$ for positive x and y, and $f_{X,Y}(x, y) = 0$ otherwise.
a) Obtain the joint MGF of X and Y.
b) Use your result from part (a) to determine the means and variances of X and Y and their covariance.
c) Use your result from part (a) to obtain the marginal MGFs of X and Y.
d) Use your results from parts (a) and (c) to decide whether X and Y are independent.
e) Use your result from part (a) to find and identify the probability distribution of $X + Y$.

Theory Exercises

11.28 Let X and Y be random variables defined on the same sample space and let a, b, c, and d be real constants. Establish the following results.
a) $M_{aX+bY}(t) = M_{X,Y}(at, bt)$
b) $M_{a+bX,c+dY}(s, t) = e^{as+ct} M_{X,Y}(bs, dt)$
c) $M_{aX+bY,cX+dY}(s, t) = M_{X,Y}(as + ct, bs + dt)$

11.29 State the multivariate analogues of Propositions 11.6–11.10 for m random variables, $X_1, \ldots, X_m$, defined on the same sample space.

11.30 Proposition 11.7 on page 645 shows how to obtain the (univariate) marginal MGFs of two random variables from their joint MGF. Suppose now that $X_1, \ldots, X_m$ are random variables defined on the same sample space. Let $1 \le j \le m$ and let $k_1 < \cdots < k_j$ be integers between 1 and m, inclusive. Explain how to obtain the joint MGF of $X_{k_1}, \ldots, X_{k_j}$ from the joint MGF of $X_1, \ldots, X_m$.

Advanced Exercises

11.31 Let X and Y be bivariate normal random variables.
a) Use MGFs to show that $X + Y$ and $X - Y$ are bivariate normal random variables.
b) When are $X + Y$ and $X - Y$ independent? Justify your answer.

11.32 Multinomial distribution: Let $X_1, \ldots, X_m$ have the multinomial distribution with parameters n and $p_1, \ldots, p_m$.
a) Obtain the joint MGF of $X_1, \ldots, X_m$.
b) Use your result from part (a) and the multivariate analogue of Proposition 11.7 on page 645 to determine and identify the univariate marginal distributions of $X_1, \ldots, X_m$.
c) Use your result from part (a) and the multivariate analogue of Proposition 11.6 on page 643 to obtain $\text{Cov}(X_k, X_\ell)$ for $k \ne \ell$. Compare your answer with Equation (∗) on page 375.
d) Obtain $\rho(X_k, X_\ell)$ for all integers k and ℓ between 1 and m.

11.33 Suppose that U and V are independent random variables and that (X, Y) and (U, V) have the same joint probability distribution; that is, $P(X \in A, Y \in B) = P(U \in A, V \in B)$ for all subsets A and B of $\mathcal{R}$. Show that X and Y are independent random variables.

11.34 Matrix representation of a bivariate normal MGF: Suppose that X and Y are bivariate normal random variables. Let $\mathbf{t}$ be the column vector whose entries are s and t, let $\boldsymbol{\mu}$ be the column vector whose entries are μ_X and μ_Y, and let

$$\Sigma = \begin{bmatrix} \text{Var}(X) & \text{Cov}(X, Y) \\ \text{Cov}(Y, X) & \text{Var}(Y) \end{bmatrix}.$$

a) Show that the joint MGF of X and Y can be expressed in matrix form as

$$M_{X,Y}(s, t) = e^{\boldsymbol{\mu}'\mathbf{t}+\frac{1}{2}\mathbf{t}'\Sigma\mathbf{t}}, \tag{∗}$$

where $\mathbf{A}'$ denotes the transpose of a matrix $\mathbf{A}$.
b) Show that the MGF of a univariate normal random variable can be expressed in the form of Equation (∗).
c) By referring to parts (a) and (b), make an educated guess at the matrix form of the joint MGF of an m-variate normal distribution.

Joint characteristic functions: Let X and Y be random variables defined on the same sample space. Then the *joint characteristic function* (joint CHF) of X and Y, denoted $\phi_{X,Y}$, is defined by $\phi_{X,Y}(s, t) = \mathcal{E}\left(e^{i(sX+tY)}\right)$, where $i = \sqrt{-1}$. In Exercises 11.35–11.38, you are to examine the concept of the joint characteristic function. You may want to refer to the exercises about univariate characteristic functions on page 641.

11.35 Prove that the joint CHF of any two random variables, X and Y, is defined for all $s, t \in \mathcal{R}$ by showing that the (complex-valued) random variable $e^{i(sX+tY)}$ has finite expectation for all $s, t \in \mathcal{R}$.

11.36 State and prove the analogues of Propositions 11.6, 11.7, 11.9, and 11.10 for bivariate characteristic functions. [The analogue of Proposition 11.8 is also true.]

11.37 Let X and Y be random variables whose joint MGF is defined in some open rectangle containing $(0, 0)$.
a) Formally, what is the relationship between $\phi_{X,Y}$ and $M_{X,Y}$?
b) Use your result from part (a) to obtain the joint CHF of bivariate normal random variables.
c) Use your results from part (b) and the moment generation property of joint CHFs, as obtained in Exercise 11.36, to find the means, variances, and covariance of bivariate normal random variables.

11.38 Suppose that $X_1, \ldots, X_m$ are random variables defined on the same sample space.
a) Define the joint CHF of $X_1, \ldots, X_m$.
b) State the analogues of Propositions 11.6–11.10 for multivariate characteristic functions.

11.3 Laws of Large Numbers

At the beginning of Section 7.1, we introduced the *long-run-average interpretation of expected value* for the purpose of developing the formal definition of the expected value of a random variable. That interpretation construes the expected value of a random variable to be the long-run-average value of the random variable in repeated independent observations.

More formally, for n independent repetitions of a random experiment, let $X_1, \ldots, X_n$ represent the n values of a random variable X with mean (expected value) μ. The long-run-average interpretation of expected value is that, for large n, the average value of $X_1, \ldots, X_n$ will approximately equal μ:

$$\frac{X_1 + \cdots + X_n}{n} \approx \mu, \qquad \text{for large } n. \tag{11.23}$$

As we noted in Section 7.1, although all attempts to use the long-run-average interpretation as a definition of expected value have failed, that interpretation is invaluable for developmental purposes.

In this section, we come full circle and establish a mathematically precise version of Relation (11.23) as a theorem. Actually, several related theorems are relevant, all of which are known as **laws of large numbers.** Roughly speaking, these theorems are phrased in the following way. Let $X_1, X_2, \ldots$ be independent and identically distributed random variables with common mean μ. Then

$$\frac{X_1 + \cdots + X_n}{n} \to \mu, \qquad \text{as } n \to \infty. \qquad \textbf{(11.24)}$$

The question now is, in what sense do we take the convergence in Relation (11.24)? Although we might want the relation to be true for all possible outcomes, that's too much to expect, as Example 11.15 shows.

EXAMPLE 11.15 *Laws of Large Numbers*

Coin Tossing Suppose that we independently and indefinitely toss a balanced coin (i.e., independently and indefinitely repeat the random experiment of tossing a balanced coin once). Let $X_j = 1$ or 0, depending on whether the jth toss of the coin comes up a head or a tail, respectively. Then $X_1, X_2, \ldots$ are independent and identically distributed random variables with common mean

$$\mu = \mathcal{E}(X_j) = 0 \cdot \frac{1}{2} + 1 \cdot \frac{1}{2} = \frac{1}{2}.$$

Show that Relation (11.24) doesn't always hold.

Solution In this context, Relation (11.24) becomes

$$\frac{X_1 + \cdots + X_n}{n} \to \frac{1}{2}, \qquad \text{as } n \to \infty, \qquad \textbf{(11.25)}$$

which says simply that, in the long run, the coin comes up a head half the time. However, Relation (11.25) clearly doesn't always hold—that is, it isn't true for every possible infinite sequence of heads and tails. For instance, if every toss comes up a head, the limit is 1; if every toss comes up a tail, the limit is 0; if every third toss comes up a head and all other tosses come up tails, the limit is 1/3. ∎

Weak Law of Large Numbers

Example 11.15 shows that expecting the convergence in Relation (11.24) always to hold is unreasonable. Nonetheless, several valid alternative formulations do exist. Historically, the first such formulation, known as the **weak law of large numbers**, is as presented in Theorem 11.1.

◆◆◆ **Theorem 11.1** **Weak Law of Large Numbers**

Let $X_1, X_2, \ldots$ be independent and identically distributed random variables with common finite mean μ. Then

$$\lim_{n \to \infty} P\left(\left| \frac{X_1 + \cdots + X_n}{n} - \mu \right| < \epsilon \right) = 1, \qquad \textbf{(11.26)}$$

for each $\epsilon > 0$.

Proof   For simplicity, we prove the theorem with the additional assumption that the random variables have common finite variance σ^2. A proof without that assumption can be accomplished by using characteristic functions, as discussed in Exercise 11.51.

For convenience, set $S_n = X_1 + \cdots + X_n$. From basic properties of expected value and variance, we find that

$$\mathcal{E}\left(\frac{S_n}{n}\right) = \mu \qquad \text{and} \qquad \text{Var}\left(\frac{S_n}{n}\right) = \frac{\sigma^2}{n}.$$

Applying the nonnegativity property of a probability measure and Chebyshev's inequality (Proposition 7.11 on page 360), we conclude that, for each $\epsilon > 0$ and $n \in \mathcal{N}$,

$$0 \le P\left(\left|\frac{S_n}{n} - \mu\right| \ge \epsilon\right) \le \frac{\text{Var}(S_n/n)}{\epsilon^2} = \frac{\sigma^2/n}{\epsilon^2} = \frac{\sigma^2}{n\epsilon^2}.$$

Hence, for each $\epsilon > 0$,

$$\lim_{n \to \infty} P\left(\left|\frac{S_n}{n} - \mu\right| \ge \epsilon\right) = 0.$$

Applying the complementation rule shows that Equation (11.26) holds.   ◆

We interpret the weak law of large numbers as follows: *When n is large, the average value of $X_1, \ldots, X_n$ is likely to be near μ.*

EXAMPLE 11.16   *Weak Law of Large Numbers*

Roulette   A (U.S.) roulette wheel contains 38 numbers of which 18 are red, 18 are black, and 2 are green. When the roulette wheel is spun, the ball is equally likely to land on any of the 38 numbers. Suppose that a gambler repeatedly bets $1 on red; that is, she repeatedly bets $1 that the ball will land on a red number. If the ball lands on a red number, she wins $1; otherwise, she loses $1.
a) In the long run, how much does the gambler lose per play?
b) Estimate the gambler's losses after 10,000 plays.

Solution   Let X_j denote the loss of the gambler on the jth play. Then $X_1, X_2, \ldots$ are independent and identically distributed random variables with common mean

$$\mu = \mathcal{E}(X_j) = 1 \cdot \frac{20}{38} - 1 \cdot \frac{18}{38} = \frac{2}{38} = \frac{1}{19}. \qquad \textbf{(11.27)}$$

In other words, the expected loss per play is $\$\frac{1}{19}$, or roughly 5.26¢.
a) After n plays, the average loss per play is $(X_1 + \cdots + X_n)/n$. According to the weak law of large numbers, for large n, that average is likely to be near μ. From Equation (11.27), we conclude that, in the long run, the gambler loses $\$\frac{1}{19}$ per play (roughly 5.26¢ per play).
b) From part (a), we know that the average loss per play is $\$\frac{1}{19}$. Hence, in 10,000 plays, we estimate that the gambler will lose a total of about $10,000 \cdot \$\frac{1}{19}$, or roughly $526. We make this result more precise in Section 11.4.   ■

Before presenting our next application of the weak law of large numbers, we need to introduce the concept of a *consistent estimator*. Let X be a random variable, and let $X_1, \ldots, X_n$ be a random sample of size n from the distribution of X. Recall that a *statistic* is any function of $X_1, \ldots, X_n$ whose numerical value can be determined from knowing only the values of the random sample.

If θ is an unknown parameter of the distribution of X, a statistic $T_n = g(X_1, \ldots, X_n)$ is said to be a **consistent estimator** of θ if, for each $\epsilon > 0$, $P\left(|T_n - \theta| < \epsilon\right) \to 1$ as $n \to \infty$. Thus, for large random samples, a consistent estimator is likely to be close to the parameter it's estimating.

EXAMPLE 11.17   *Weak Law of Large Numbers*

Consistency of the Sample Mean   Consider a random variable X with finite mean. Suppose that we don't know the mean of X but want to estimate it. To do so, we take a random sample, $X_1, \ldots, X_n$, from the distribution of X and form the *sample mean*, $\bar{X}_n = (X_1 + \cdots + X_n)/n$. Show that $\bar{X}_n$ is a consistent estimator of μ_X.

Solution   Because, for each $n \in \mathcal{N}$, $X_1, \ldots, X_n$ is a random sample from the distribution of X, we know that $X_1, X_2, \ldots$ are independent and identically distributed random variables with common mean $\mu = \mu_X$. Thus, by the weak law of large numbers,

$$\lim_{n \to \infty} P\left(|\bar{X}_n - \mu_X| < \epsilon\right) = \lim_{n \to \infty} P\left(\left|\frac{X_1 + \cdots + X_n}{n} - \mu\right| < \epsilon\right) = 1,$$

for each $\epsilon > 0$. Hence $\bar{X}_n$ is a consistent estimator of μ_X.   ∎

The weak law of large numbers provides mathematical confirmation (as a consequence of the axioms of probability) of our intuitive notion of expected value as the long-run-average value of a random variable in repeated independent observations. It also confirms mathematically our intuitive notion of the meaning of probability—the *frequentist interpretation of probability*—which construes the probability of an event to be the long-run proportion of times that the event occurs in independent repetitions of the random experiment. Specifically, we have Corollary 11.1, historically the first version of the weak law of large numbers, proved by Jacob Bernoulli and published posthumously in his *Ars Conjectandi* (The Art of Conjecturing) in 1713.

◆◆◆   **Corollary 11.1**   **Bernoulli's Weak Law of Large Numbers**

Let E be an event associated with a random experiment and let $P(E)$ denote its probability. For n independent repetitions of the random experiment, let $n(E)$ denote the number of times that event E occurs. Then

$$\lim_{n \to \infty} P\left(\left|\frac{n(E)}{n} - P(E)\right| < \epsilon\right) = 1, \tag{11.28}$$

for each $\epsilon > 0$.

Proof   For each $j \in \mathcal{N}$, let $X_j = 1$ or 0, depending on whether event E occurs or doesn't occur, respectively, on the jth repetition of the experiment. Because the repetitions of the experiment are independent, the random variables $X_1, X_2, \ldots$ are independent and identically distributed. Their common mean is $\mu = 0 \cdot (1 - P(E)) + 1 \cdot P(E) = P(E)$. As $n(E) = X_1 + \cdots + X_n$, Equation (11.28) follows immediately from the weak law of large numbers.   ◆

Note: Compare the statement of Bernoulli's weak law of large numbers to that of the frequentist interpretation of probability, as presented in Section 1.1 on page 5, specifically, to Relation (1.1).

Strong Law of Large Numbers

Let $X_1, X_2, \ldots$ be independent and identically distributed random variables with common finite mean μ. The weak law of large numbers shows that, for any particular large positive integer n, the average value of $X_1, \ldots, X_n$, that is, $(X_1 + \cdots + X_n)/n$, is likely to be near μ. But, even if the average value of $X_1, \ldots, X_n$ is near μ, the weak law of large numbers doesn't say that it will remain near μ for all subsequent n.

Thus the weak law of large numbers doesn't rule out the possibility of deviations of the average value of $X_1, \ldots, X_n$ from μ in excess of any specified amount for infinitely many n. Our intuition (in the form of the long-run-average interpretation of expected value), however, suggests that such deviations should occur for only finitely many n or, equivalently, that the average value of $X_1, \ldots, X_n$ should converge to μ with probability 1. This phenomenon is the thrust of the **strong law of large numbers.**

◆◆◆   **Theorem 11.2   Strong Law of Large Numbers**

Let $X_1, X_2, \ldots$ be independent and identically distributed random variables with common finite mean μ. Then

$$\lim_{n \to \infty} \frac{X_1 + \cdots + X_n}{n} = \mu, \tag{11.29}$$

with probability 1.

Note: The proof of the strong law of large numbers is an intricate application of analysis, which would take us from our main path. We therefore omit its proof and instead refer the interested reader to the book *A Course in Real Analysis* by John N. McDonald and Neil A. Weiss (San Diego: Academic Press, 1999, pp. 301–310).

The strong law of large numbers implies (i.e., is stronger than) the weak law of large numbers. In fact, an equivalent formulation of the strong law of large numbers is

$$\lim_{n \to \infty} P\left(\bigcap_{k=n}^{\infty} \left\{ \left| \frac{X_1 + \cdots + X_k}{k} - \mu \right| < \epsilon \right\} \right) = 1, \tag{11.30}$$

for each $\epsilon > 0$. (See Exercise 11.52.)

Equation (11.30) clearly implies Equation (11.26) on page 651, the weak law of large numbers. It also shows that we can interpret the strong law of large numbers as follows: *When n is large, the average value of $X_1, \ldots, X_k$ is likely to be near μ for all $k \geq n$.*

The strong law of large numbers provides the mathematical confirmation (as a consequence of the axioms of probability) that we really want of our intuitive notion of expected value as the long-run-average value of a random variable in repeated independent observations. Indeed, the strong law of large numbers shows that, with probability 1, the average value of a random variable in repeated independent observations converges to the expected value of the random variable.

As a special case, the strong law of large numbers gives mathematical confirmation of our intuitive notion of the meaning of probability—the *frequentist interpretation of probability*—which construes the probability of an event to be the long-run proportion of times that the event occurs in independent repetitions of the random experiment. Specifically, we have Corollary 11.2, historically the first version of the strong law of large numbers, proved by Émile Borel in 1909.

◆◆◆   **Corollary 11.2   Borel's Strong Law of Large Numbers**

Let E be an event associated with a random experiment and let $P(E)$ denote its probability. For n independent repetitions of the random experiment, let $n(E)$ denote the number of times that event E occurs. Then

$$\lim_{n \to \infty} \frac{n(E)}{n} = P(E), \tag{11.31}$$

with probability 1.

Proof   For each $j \in \mathcal{N}$, let $X_j = 1$ or 0, depending on whether event E occurs or doesn't occur, respectively, on the jth repetition of the experiment. Because the repetitions of the experiment are independent, the random variables $X_1, X_2, \ldots$ are independent and identically distributed. Their common mean is $\mu = 0 \cdot (1 - P(E)) + 1 \cdot P(E) = P(E)$. As $n(E) = X_1 + \cdots + X_n$, Equation (11.31) follows immediately from the strong law of large numbers.   ◆

Note: Compare the statement of Borel's strong law of large numbers to that of the frequentist interpretation of probability, as presented in Section 1.1 on page 5, specifically, to Relation (1.1).

Other Laws of Large Numbers

In this section, we discussed the classic versions of the weak and strong laws of large numbers. Other versions of these laws exist, and we examine some of them in the exercises for this section.

EXERCISES 11.3   Basic Exercises

11.39 Refer to Example 11.15 on page 651, where $X_1, X_2, \ldots$ are independent and identically distributed random variables with common mean $\mu = 1/2$. We noted there that the relation $(X_1 + \cdots + X_n)/n \to 1/2$ as $n \to \infty$ doesn't always hold. Explain why this property isn't in conflict with the strong law of large numbers.

11.40 Consistency of sample moments: Let X be a random variable with finite rth moment. Suppose that you don't know the rth moment of X but want to estimate it. To do so, you take a random sample, $X_1, \ldots, X_n$, from the distribution of X and form the *rth sample moment*, $(X_1^r + \cdots + X_n^r)/n$. Show that the rth sample moment is a consistent estimator of the rth moment.

11.41 A particle of initial size s is subjected to repeated impacts. After impact j, a proportion X_j of the particle remains; that is, if Y_j denotes the size of the particle after impact j, then $Y_j = X_j Y_{j-1}$. Assume that $X_1, X_2, \ldots$ are independent random variables all with the same probability distribution as a random variable X whose natural logarithm has finite mean.
a) Show that $\lim_{n\to\infty} (Y_n/s)^{1/n} = e^{\mathcal{E}(\ln X)}$ with probability 1.
b) Deduce from part (a) that, roughly, $Y_n \approx s e^{n\mathcal{E}(\ln X)}$ for large n.
c) Specialize the results of parts (a) and (b) to the cases where $X \sim \mathcal{U}(0, 1)$, $X \sim \text{Beta}(2, 1)$, $X \sim \text{Beta}(1, 2)$, and $X \sim \text{Beta}(2, 2)$.

11.42 Empirical distributions: Let $X_1, X_2, \ldots$ be independent random variables, all having the same probability distribution as a random variable X. For each $n \in \mathcal{N}$, we define the *empirical distribution function* based on the sample of size n to be

$$\hat{F}_n(x) = \frac{N(\{1 \le j \le n : X_j \le x\})}{n}, \qquad x \in \mathcal{R},$$

where $N(S)$ denotes the number of elements of a finite set S. Show that, for each $x \in \mathcal{R}$, $\lim_{n\to\infty} \hat{F}_n(x) = F_X(x)$ with probability 1. Interpret this result.

11.43 Monte Carlo integration: Suppose that you want to obtain the value of $\int_0^1 g(x)\,dx$, where g is a Riemann integrable function on the interval $[0, 1]$. Further assume that no simple formula exists for the antiderivative of g.
a) Let $X \sim \mathcal{U}(0, 1)$. Explain why the random variable $g(X)$ has finite expectation.
b) Let $X_1, X_2, \ldots$ be independent and identically distributed $\mathcal{U}(0, 1)$ random variables. Show that $\lim_{n\to\infty} (g(X_1) + \cdots + g(X_n))/n = \int_0^1 g(x)\,dx$ with probability 1.
c) Explain how to use a basic random number generator—that is, a random number generator that simulates a $\mathcal{U}(0, 1)$ random variable—to approximate the value of $\int_0^1 g(x)\,dx$.
d) **Simulation:** This part requires access to a computer or graphing calculator. Apply your procedure from part (c) to estimate $\int_0^1 x^3\,dx$ based on 1000 uniform random numbers. Compare your answer to the exact value of the integral.
e) **Simulation:** This part requires access to a computer or graphing calculator. Let Z have the standard normal distribution. Apply your procedure from part (c) to estimate $P(0 \le Z \le 1)$ based on 1000 uniform random numbers. Compare your answer to that obtained by consulting Table I in the Appendix.
f) **Simulation:** Repeat parts (d) and (e) based on 10,000 uniform random numbers.

11.44 Monte Carlo integration (continued): Suppose that you want to determine the value of $\int_a^b g(x)\,dx$, where g is a Riemann integrable function on the interval $[a, b]$. Further assume that no simple formula exists for the antiderivative of g.
a) Explain how to use a basic random number generator—that is, a random number generator that simulates a $\mathcal{U}(0, 1)$ random variable—to approximate the value of $\int_a^b g(x)\,dx$. *Hint:* Refer to Exercise 11.43.
b) **Simulation:** This part requires access to a computer or graphing calculator. Apply your procedure from part (a) to estimate $\int_1^3 x^3\,dx$ based on 1000 uniform random numbers. Compare your answer to the exact value of the integral.

c) **Simulation:** This part requires access to a computer or graphing calculator. Let Z have the standard normal distribution. Apply your procedure from part (a) to estimate $P(-1 \le Z \le 2)$ based on 1000 uniform random numbers. Compare your answer to that obtained by consulting Table I in the Appendix.

d) **Simulation:** Repeat parts (b) and (c) based on 10,000 uniform random numbers.

11.45 Let $X_1, X_2, \ldots$ be independent random variables, each having the standard Cauchy distribution. Exercise 11.20 on page 641 shows that, for each $n \in \mathcal{N}$, the random variable $(X_1 + \cdots + X_n)/n$ also has the standard Cauchy distribution. Let c be any real number and let $\epsilon > 0$.

a) Evaluate

$$\lim_{n \to \infty} P\left(\left| \frac{X_1 + \cdots + X_n}{n} - c \right| < \epsilon \right). \tag{*}$$

b) From part (a), the limit in Expression (*) isn't 1 for any real number c. Discuss this phenomenon in view of the weak law of large numbers.

11.46 Simulation: This exercise requires access to a computer or graphing calculator.

a) Use a basic random number generator—that is, a random number generator that simulates a $\mathcal{U}(0, 1)$ random variable—to obtain 10,000 observations of a uniform random variable on the interval $(0, 1)$.

b) Without doing further computations, roughly what should be the average value of the 10,000 numbers obtained in part (a)? Explain your reasoning.

c) Determine the average value of the 10,000 numbers obtained in part (a) and compare your result to your answer in part (b).

11.47 Simulation: This exercise requires access to statistical software.

a) Use a normal random number generator—that is, a random number generator that simulates observations of a normal random variable—to obtain 10,000 observations of a standard normal random variable.

b) Without doing further computations, roughly what should be the average value of the 10,000 numbers obtained in part (a)? Explain your reasoning.

c) Determine the average value of the 10,000 numbers obtained in part (a) and compare your result to your answer in part (b).

11.48 Simulation: This exercise requires access to a computer or graphing calculator.

a) Use a basic random number generator to obtain 10,000 observations of a standard Cauchy random variable. *Hint:* Refer to Proposition 8.16 on page 471.

b) Can you guess the average value of the 10,000 numbers obtained in part (a)? Explain your reasoning. *Hint:* Refer to Exercise 11.20(b) on page 641.

c) Repeat the simulation in part (a) 20 times and find the average value of the 10,000 numbers obtained for each simulation. Discuss your results in view of the strong law of large numbers.

Theory Exercises

11.49 Markov's weak law of large numbers: Suppose that $X_1, X_2, \ldots$ are random variables defined on the same sample space and having finite variances. Further suppose that $\lim_{n \to \infty} \mathrm{Var}(X_1 + \cdots + X_n)/n^2 = 0$. Prove that

$$\lim_{n \to \infty} P\left(\left| \frac{X_1 + \cdots + X_n}{n} - \frac{\mathcal{E}(X_1) + \cdots + \mathcal{E}(X_n)}{n} \right| < \epsilon \right) = 1,$$

for each $\epsilon > 0$.

11.50 Chebyshev's weak law of large numbers: Suppose that $X_1, X_2, \ldots$ are uncorrelated random variables having uniformly bounded variances. Prove that

$$\lim_{n \to \infty} P\left(\left| \frac{X_1 + \cdots + X_n}{n} - \frac{\mathcal{E}(X_1) + \cdots + \mathcal{E}(X_n)}{n} \right| < \epsilon\right) = 1,$$

for each $\epsilon > 0$.

Advanced Exercises

11.51 The weak law of large numbers, as stated in Theorem 11.1 on page 651, requires only that the independent and identically distributed random variables, $X_1, X_2, \ldots$, have finite mean. However, our proof of that theorem assumed that they have finite variance. In this exercise, you are to use characteristic functions (see page 641) to obtain a proof of the weak law of large numbers without the assumption of finite variance. Let ϕ be the common CHF of $X_1, X_2, \ldots$ and let μ be their common mean. Also, denote by ϕ_n the CHF of $(X_1 + \cdots + X_n)/n$.

a) Show that $\phi_n(t) = \left(\phi(t/n)\right)^n$ for all $t \in \mathcal{R}$ and $n \in \mathcal{N}$.

b) Show that $\lim_{n \to \infty} \phi_n(t) = e^{i\mu t}$ for all $t \in \mathcal{R}$. *Hint:* Take natural logarithms and apply L'Hôpital's rule.

c) Deduce from part (b) that $(X_1 + \cdots + X_n)/n$ converges in distribution to μ.

d) Use part (c) to prove the weak law of large numbers.

11.52 Let $X_1, X_2, \ldots$ be independent and identically distributed random variables with common finite mean μ.

a) Show that

$$\left\{ \lim_{n \to \infty} \frac{X_1 + \cdots + X_n}{n} = \mu \right\} = \bigcap_{m=1}^{\infty}\left(\bigcup_{N=1}^{\infty}\left(\bigcap_{n=N}^{\infty} \left\{ \left| \frac{X_1 + \cdots + X_n}{n} - \mu \right| < \frac{1}{m} \right\} \right) \right).$$

Hint: Recall the definition of $\lim_{n \to \infty} a_n = a$.

b) Use part (a) to show that an equivalent formulation of the strong law of large numbers is

$$\lim_{n \to \infty} P\left(\bigcap_{k=n}^{\infty}\left\{ \left| \frac{X_1 + \cdots + X_k}{k} - \mu \right| < \epsilon \right\} \right) = 1,$$

for each $\epsilon > 0$.

c) Deduce from part (b) that the strong law of large numbers is stronger than (i.e., implies) the weak law of large numbers.

d) Show that, with probability 1, deviations of the average value of $X_1, \ldots, X_n$ from μ in excess of any specified amount occur for only finitely many n.

11.53 Normal numbers: Each real number between 0 and 1 has a decimal expansion and, except for numbers of the form $m/10^n$, the expansion is unique. For definiteness, take the unique terminating expansion for numbers of the form $m/10^n$.

a) Let $x \in [0, 1]$. For each $n \in \mathcal{N}$ and $k \in \{0, 1, \ldots, 9\}$, use $n_k(x)$ to denote the number of the first n decimal digits of x that equal k. Then x is said to be a *normal number* if $n_k(x)/n \to 1/10$ as $n \to \infty$ for each k. Interpret *normal number*.

b) Let x denote a randomly selected number from the interval $[0, 1]$ and define $X_n(x) = x_n$, where x_n is the nth decimal digit of x. Show that the random variables $X_1, X_2, \ldots$ are independent and identically distributed, all having the discrete uniform distribution on the set $\{0, 1, \ldots, 9\}$. *Hint:* Note that

$$\{X_n = k\} = \bigcup_{k_1=0}^{9} \cdots \bigcup_{k_{n-1}=0}^{9} \{X_1 = k_1, \ldots, X_{n-1} = k_{n-1}, X_n = k\}$$

and that each set in the union is an interval.
c) Prove that, with probability 1, a randomly selected number from the interval $[0, 1]$ is a normal number. *Hint:* For each decimal digit k, consider the sequence of random variables $I_{\{X_1=k\}}, I_{\{X_2=k\}}, \ldots$.

11.54 Refer to Exercise 11.53, but now consider binary expansions.
a) Explain how binary expansions of real numbers between 0 and 1 provide a model for the random experiment of repeatedly tossing a balanced coin and observing at each toss whether the result is a head or a tail.
b) Prove the corresponding result obtained in Exercise 11.53 for binary expansions. Interpret this result in terms of coin tossing.

11.4   The Central Limit Theorem

In Section 11.3, we discussed laws of large numbers. Those laws indicate that, for independent and identically distributed random variables, $X_1, X_2, \ldots$, with common finite mean μ, the average value of $X_1, \ldots, X_n$ is likely to be near μ when n is large. Those laws, however, don't indicate *how likely* it is for the average value of $X_1, \ldots, X_n$ to be near μ. To obtain that information, we must first identify the probability distribution of $X_1 + \cdots + X_n$.

If the common probability distribution of $X_1, X_2, \ldots$ is a normal distribution, then Proposition 9.14 on page 540 implies that $X_1 + \cdots + X_n$ is also normally distributed for each $n \in \mathcal{N}$. Remarkable as it may seem, this result holds approximately, regardless of the common probability distribution of $X_1, X_2, \ldots$, provided only that n is large and that the common variance is finite. In other words, if $X_1, X_2, \ldots$ are independent and identically distributed random variables with finite variance, then $X_1 + \cdots + X_n$ is approximately normally distributed for large n. This result is the essence of the celebrated *central limit theorem*, which, along with the law of large numbers, constitute the two most important results in probability theory.

In 1733, Abraham De Moivre paved the way for the central limit theorem by stating it for the special case of Bernoulli trials with $p = 1/2$. Subsequently, in 1812, Pierre Laplace proved and improved on De Moivre's conjecture by establishing the central limit theorem for Bernoulli trials with arbitrary values of p. For purposes of motivating the central limit theorem, we first examine Laplace's result, now known as the *integral De Moivre–Laplace theorem*.

The Integral De Moivre–Laplace Theorem

In Section 8.5, we discussed a method for approximating a binomial distribution by a normal distribution. Specifically, we presented Proposition 8.12 on page 445—also known as the *local De Moivre–Laplace theorem*—which shows how to approximate individual binomial probabilities by using an appropriate normal PDF. The local De Moivre–Laplace theorem applies when n is large and p is moderate and is useful when we want to estimate only a few individual binomial probabilities.

If we want to estimate the probability that a binomial random variable takes a value in a range of possible integers, the integral De Moivre–Laplace theorem is more appropriate. We can use the local De Moivre–Laplace theorem to prove the integral De Moivre–Laplace theorem. Here, as before, we give a heuristic proof.

Suppose that X has the binomial distribution with parameters n and p, where we assume that n is large and p is moderate. As we know from Proposition 8.12, the approximating normal distribution has parameters np and $np(1 - p)$; that is, the approximating normal distribution is the one having the same mean and variance as the binomial distribution being approximated. Let Y be a random variable with that normal distribution. Then, according to Proposition 8.12, $p_X(x) \approx f_Y(x)$ for $x = 0, 1, \ldots, n$.

In Figure 11.1, the height of the dashed line is $f_Y(x)$, which also equals the area of the crosshatched rectangle. However, the area of the crosshatched rectangle approximately equals the shaded area under the normal curve, or $P\left(x - \frac{1}{2} < Y < x + \frac{1}{2}\right)$. Referring now to Proposition 8.11 on page 443, we conclude that

$$
p_X(x) \approx f_Y(x) \approx P\left(x - \tfrac{1}{2} < Y < x + \tfrac{1}{2}\right)
$$

$$
= \Phi\left(\frac{x + \frac{1}{2} - np}{\sqrt{np(1 - p)}}\right) - \Phi\left(\frac{x - \frac{1}{2} - np}{\sqrt{np(1 - p)}}\right). \tag{11.32}
$$

for $x = 0, 1, \ldots, n$.

Figure 11.1

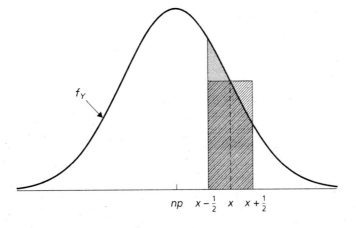

$$np \quad x - \tfrac{1}{2} \quad x \quad x + \tfrac{1}{2}$$

Now let a and b be integers with $0 \le a \le b \le n$. From the FPF and Relation (11.32),

$$P(a \le X \le b) = \sum_{x=a}^{b} p_X(x) \approx \sum_{x=a}^{b} \left[\Phi\left(\frac{x + \frac{1}{2} - np}{\sqrt{np(1-p)}} \right) - \Phi\left(\frac{x - \frac{1}{2} - np}{\sqrt{np(1-p)}} \right) \right]$$

$$= \Phi\left(\frac{b + \frac{1}{2} - np}{\sqrt{np(1-p)}} \right) - \Phi\left(\frac{a - \frac{1}{2} - np}{\sqrt{np(1-p)}} \right).$$

Therefore we have the result presented in Proposition 11.11, known as the **integral De Moivre–Laplace theorem.**

◆◆◆   **Proposition 11.11**   **Normal Approximation to the Binomial: Integral Form**

Suppose that X has the binomial distribution with parameters n and p. Then, for integers a and b, with $0 \le a \le b \le n$,

$$P(a \le X \le b) \approx \Phi\left(\frac{b + \frac{1}{2} - np}{\sqrt{np(1-p)}} \right) - \Phi\left(\frac{a - \frac{1}{2} - np}{\sqrt{np(1-p)}} \right). \qquad \textbf{(11.33)}$$

The approximation works well if n is large and p is moderate (i.e., not too far from 0.5).

The $\frac{1}{2}$s that appear on the right in Relation (11.33) arise from approximating a discrete distribution (the binomial) by a continuous distribution (the normal). Their inclusion is referred to as the **continuity correction** and generally yields a better approximation than that provided by Relation (11.33) without the $\frac{1}{2}$s. However, as n becomes increasingly large, there is very little difference in the results obtained with and without the continuity correction. (Can you explain why?) Hence, instead of Relation (11.33), we often use an alternative relation that omits the continuity correction:

$$P(a \le X \le b) \approx \Phi\left(\frac{b - np}{\sqrt{np(1-p)}} \right) - \Phi\left(\frac{a - np}{\sqrt{np(1-p)}} \right). \qquad \textbf{(11.34)}$$

EXAMPLE 11.18   *Normal Approximation to the Binomial*

Teen Pregnancy   According to a recent issue of the periodical *Zero Population Growth*, the United States leads the industrialized world in teen pregnancy rates, with 40% of U.S. females getting pregnant at least once before reaching the age of 20. Use the integral De Moivre–Laplace theorem (in the form of Proposition 11.11) to find the approximate probability that, from among 50 U.S. 20-year-old females, the number who have been pregnant at least once before the age of 20 will be

a) exactly 21.

b) between 19 and 21, inclusive.

c) Use the appropriate binomial PMF to obtain the actual probabilities in parts (a) and (b).

d) In Example 8.15 on page 445, we used the local De Moivre–Laplace theorem (in the form of Proposition 8.12 on page 445) to approximate the probabilities required in parts (a) and (b). Compare the results obtained there to those obtained here by using the integral theorem and to the actual binomial probabilities found in part (c).

Solution   Let X denote the number of the 50 females who have been pregnant at least once before the age of 20. Then the random variable X has the binomial distribution with parameters $n = 50$ and $p = 0.4$. Thus

$$p_X(x) = \binom{50}{x}(0.4)^x(0.6)^{50-x}, \qquad x = 0, 1, \ldots, 50. \tag{11.35}$$

Note, however, that $n = 50$ is (relatively) large and that $p = 0.4$ is moderate. Thus, instead of applying Equation (11.35), we can use a normal approximation to obtain the required probabilities more easily. Specifically, noting that $np = 20$ and $np(1 - p) = 12$, we can, in view of Proposition 11.11, use the approximation

$$P(a \leq X \leq b) \approx \Phi\left(\frac{b + \frac{1}{2} - 20}{\sqrt{12}}\right) - \Phi\left(\frac{a - \frac{1}{2} - 20}{\sqrt{12}}\right). \tag{11.36}$$

a) We want $P(X = 21)$. Applying Relation (11.36) with $a = b = 21$, we get

$$P(X = 21) = P(21 \leq X \leq 21)$$

$$\approx \Phi\left(\frac{21 + \frac{1}{2} - 20}{\sqrt{12}}\right) - \Phi\left(\frac{21 - \frac{1}{2} - 20}{\sqrt{12}}\right) = 0.1101.^\dagger$$

There is about an 11.0% chance that exactly 21 of the 50 women have been pregnant at least once before the age of 20.

b) We want $P(19 \leq X \leq 21)$. Using Relation (11.36) with $a = 19$ and $b = 21$, we get

$$P(19 \leq X \leq 21) \approx \Phi\left(\frac{21 + \frac{1}{2} - 20}{\sqrt{12}}\right) - \Phi\left(\frac{19 - \frac{1}{2} - 20}{\sqrt{12}}\right) = 0.3350.^\dagger$$

Chances are roughly 33.5% that between 19 and 21, inclusive, of the 50 women have been pregnant at least once before the age of 20.

c) Applying the FPF and the binomial PMF given in Equation (11.35), we find that

$$P(X = 21) = p_X(21) = \binom{50}{21}(0.4)^{21}(0.6)^{50-21} = 0.1091$$

and

$$P(19 \leq X \leq 21) = \sum_{19 \leq x \leq 21} p_X(x) = p_X(19) + p_X(20) + p_X(21) = 0.3345.$$

d) For comparison purposes, we constructed Table 11.2. It reveals that both the local and integral normal approximations provide excellent results. ∎

---

†Although Table I in the Appendix can be used to evaluate the standard normal CDF at the required values, we used statistical software to get more accurate results.

Table 11.2 Comparison of binomial probabilities with local and integral normal approximations (each parenthetical value gives the absolute percentage error)

	Binomial probability	Local normal approximation	Integral normal approximation
$P(X = 21)$	0.1091	0.1105 (1.3%)	0.1101 (0.9%)
$P(19 \leq X \leq 21)$	0.3345	0.3361 (0.5%)	0.3350 (0.1%)

Proposition 11.11 shows that we can approximate a binomial distribution by a normal distribution. The approximation works well when n is large and p is moderate. Qualitatively, these conditions on n and p ensure that the binomial distribution under consideration is reasonably close to being bell-shaped, thus indicating that a normal approximation is appropriate.

Figure 11.2, on the next page, displays probability histograms for nine different binomial distributions. As portrayed in Figures 11.2(a) and 11.2(c), a binomial distribution with $p \neq 0.5$ is skewed. For small n, the skewness is enough to preclude using a normal approximation. However, as n increases, the skewness decreases and the binomial distribution becomes sufficiently bell-shaped to permit a normal approximation. In contrast, as illustrated in Figure 11.2(b), a binomial distribution with $p = 0.5$ is symmetric, regardless of the number of trials. Nonetheless, such a distribution won't be sufficiently bell-shaped to permit a normal approximation if n is too small.

The customary rule of thumb for using the normal approximation is that *both np and $n(1 - p)$ are 5 or greater.* As suggested in Figure 11.2, this restriction indicates that the farther the success probability is from 0.5 (in either direction), the larger the number of trials must be to justify using the normal approximation.

The Central Limit Theorem

Proposition 11.11 provides a normal approximation to the binomial distribution. Essentially, it states that, for large n, the distribution of a binomial random variable can be approximated by a normal distribution with the same mean and variance. For purposes of getting to the central limit theorem, it's useful to rephrase that result.

We recall that a binomial random variable X with parameters n and p gives the number of successes in n Bernoulli trials with success probability p. For each $1 \leq j \leq n$, let X_j equal 1 or 0, depending on whether the jth trial results in success or failure, respectively. Each X_j is a Bernoulli random variable with parameter p, and $X_1, \ldots, X_n$ are independent and identically distributed with common mean $\mu = p$ and common variance $\sigma^2 = p(1 - p)$. Furthermore, $X = X_1 + \cdots + X_n$, which has mean $n\mu = np$ and variance $n\sigma^2 = np(1 - p)$.

So, we can rephrase Proposition 11.11 as follows: If $X_1, X_2, \ldots$ are independent and identically distributed Bernoulli random variables, then, for large n, the probability distribution of $X_1 + \cdots + X_n$ is approximately a normal distribution with the same mean and variance; symbolically, $X_1 + \cdots + X_n \sim \mathcal{N}(n\mu, n\sigma^2)$, approximately. Recalling

Figure 11.2 Nine different binomial distributions

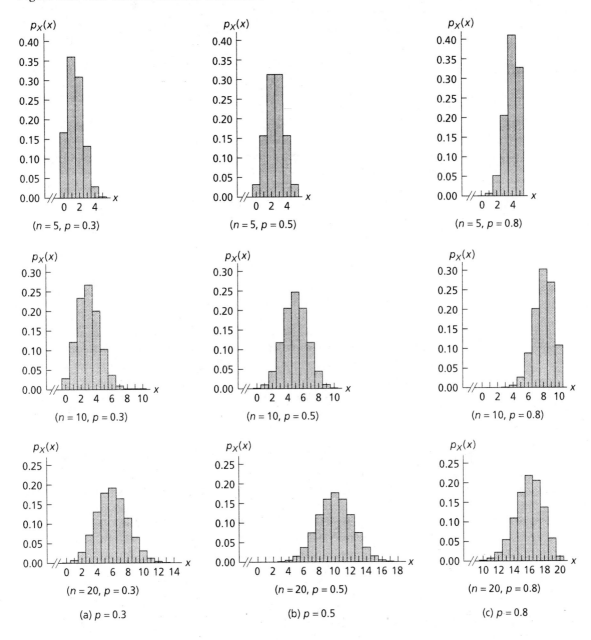

(a) $p = 0.3$ (b) $p = 0.5$ (c) $p = 0.8$

that the standardized version of a normally distributed random variable has the standard normal distribution, we conclude that $(X_1 + \cdots + X_n - n\mu)/\sigma\sqrt{n}$ has approximately the standard normal distribution for large n. A precise mathematical justification of this result is established by verifying that

$$\lim_{n \to \infty} P\left(\frac{X_1 + \cdots + X_n - n\mu}{\sigma\sqrt{n}} \le x\right) = \Phi(x), \qquad \textbf{(11.37)}$$

for all $x \in \mathcal{R}$—that is, by proving that the standardized versions of sums of independent and identically distributed Bernoulli random variables converge in distribution to the standard normal.

Remarkably, Equation (11.37) holds not just for independent and identically distributed Bernoulli random variables, but for any independent and identically distributed random variables with finite variance. This fact is the content of the **central limit theorem.** To prove it, we first establish Lemma 11.1.

◆◆◆ **Lemma 11.1**

Let X be a random variable whose MGF is defined in some open interval containing 0. Then

$$\lim_{t \to 0} \frac{\ln M_X(t) - \mu_X t}{t^2} = \frac{\sigma_X^2}{2}. \qquad \textbf{(11.38)}$$

Proof   We recall that $M_X(0) = 1$ and, from Proposition 11.2 on page 635,

$$M_X'(0) = \mathcal{E}(X) = \mu_X \quad \text{and} \quad M_X''(0) = \mathcal{E}(X^2) = \mu_X^2 + \sigma_X^2. \qquad \textbf{(11.39)}$$

We now apply L'Hôpital's rule twice and then use Equations (11.39) to obtain

$$\lim_{t \to 0} \frac{\ln M_X(t) - \mu_X t}{t^2} = \lim_{t \to 0} \frac{M_X'(t)/M_X(t) - \mu_X}{2t} = \lim_{t \to 0} \frac{M_X'(t) - \mu_X M_X(t)}{2t M_X(t)}$$

$$= \lim_{t \to 0} \frac{M_X'(t) - \mu_X M_X(t)}{2t} = \lim_{t \to 0} \frac{M_X''(t) - \mu_X M_X'(t)}{2}$$

$$= \frac{M_X''(0) - \mu_X M_X'(0)}{2} = \frac{\mu_X^2 + \sigma_X^2 - \mu_X^2}{2} = \frac{\sigma_X^2}{2},$$

as required.   ◆

◆◆◆ **Theorem 11.3   Central Limit Theorem**

Let $X_1, X_2, \ldots$ be independent and identically distributed random variables with common finite mean μ and common finite nonzero variance σ^2. Then

$$\lim_{n \to \infty} P\left(\frac{X_1 + \cdots + X_n - n\mu}{\sigma\sqrt{n}} \le x\right) = \Phi(x), \qquad \textbf{(11.40)}$$

for all $x \in \mathcal{R}$. In words, the standardized versions of sums of independent and identically distributed random variables with finite nonzero variance converge in distribution to the standard normal.

Proof   We assume that the common MGF of $X_1, X_2, \ldots$ is defined in some open interval containing 0. By using characteristic functions instead of moment generating functions, we can provide an essentially identical proof without that assumption. (See Exercise 11.75.)

For convenience, set $S_n = X_1 + \cdots + X_n$ and let Y_n denote the standardized version of $X_1 + \cdots + X_n$. Then

$$Y_n = \frac{X_1 + \cdots + X_n - n\mu}{\sigma\sqrt{n}} = \frac{S_n - n\mu}{\sigma\sqrt{n}} = -n\frac{\mu}{\sigma\sqrt{n}} + \frac{1}{\sigma\sqrt{n}}S_n.$$

Applying Propositions 11.1 and 11.4 on pages 633 and 636, respectively, we get

$$M_{Y_n}(t) = e^{-n\frac{\mu}{\sigma\sqrt{n}}t} M_{S_n}\left(t/\sigma\sqrt{n}\right)$$

$$= e^{-n\frac{\mu}{\sigma\sqrt{n}}t}\left(M\left(t/\sigma\sqrt{n}\right)\right)^n = \left(e^{-\frac{\mu}{\sigma\sqrt{n}}t}M\left(t/\sigma\sqrt{n}\right)\right)^n,$$

where M is the common MGF of $X_1, X_2, \ldots$. Taking natural logarithms yields

$$\ln M_{Y_n}(t) = n\left(\ln M\left(t/\sigma\sqrt{n}\right) - \frac{\mu}{\sigma\sqrt{n}}t\right) = \frac{t^2}{\sigma^2}\cdot\frac{\ln M\left(t/\sigma\sqrt{n}\right) - \mu\cdot\left(t/\sigma\sqrt{n}\right)}{\left(t/\sigma\sqrt{n}\right)^2}.$$

It follows from Lemma 11.1 that

$$\lim_{n\to\infty}\ln M_{Y_n}(t) = \frac{t^2}{\sigma^2}\cdot\frac{\sigma^2}{2} = \frac{t^2}{2},$$

or, equivalently,

$$\lim_{n\to\infty}M_{Y_n}(t) = e^{t^2/2}. \tag{11.41}$$

Table 11.1 on page 634 shows that the function on the right of Equation (11.41) is the MGF of a standard normal random variable. Equation (11.40) now follows from the continuity theorem, Proposition 11.5 on page 638.   ◆

Some remarks are in order concerning the central limit theorem.

- It can be shown that the convergence in Equation (11.40) is uniform for all real numbers x. This property justifies the use of the standard normal distribution as an approximation to the distribution of $(X_1 + \cdots + X_n - n\mu)/\sigma\sqrt{n}$ when n is sufficiently large.

- Approximately, for large n, we have $(X_1 + \cdots + X_n - n\mu)/\sigma\sqrt{n} \sim \mathcal{N}(0, 1)$ or, equivalently, $X_1 + \cdots + X_n \sim \mathcal{N}(n\mu, n\sigma^2)$. Thus, for each pair of real numbers a and b, with $-\infty < a \le b < \infty$,

$$P(a \le X_1 + \cdots + X_n \le b) \approx \Phi\left(\frac{b - n\mu}{\sigma\sqrt{n}}\right) - \Phi\left(\frac{a - n\mu}{\sigma\sqrt{n}}\right). \tag{11.42}$$

It is Relation (11.42) that we actually use for approximating the probability distribution of $X_1 + \cdots + X_n$.

- The central limit theorem is valid and, hence, Relation (11.42) holds, regardless of the common distribution of $X_1, X_2, \ldots$ (provided that common distribution has finite nonzero variance). In particular, we don't need to know the common distribution of $X_1, X_2, \ldots$ to approximate probabilities for $X_1 + \cdots + X_n$, an essential attribute in many applications.

- The rate of convergence in Equation (11.40) and hence the accuracy of the approximation in Relation (11.42) depends on the common distribution of the random variables $X_1, X_2, \ldots$. A rule of thumb is that the approximation in Relation (11.42) is adequate, provided that $n \geq 30$. Generally, the more skewed the common distribution, the larger n must be to obtain a good approximation.

- Compare Relation (11.34) on page 661 to Relation (11.42) when $X_1, X_2, \ldots$ are independent and identically distributed Bernoulli random variables with common success probability p.

- The central limit theorem can be rephrased in terms of sample means. Suppose that $X_1, \ldots, X_n$ is a random sample from a distribution with finite mean μ and finite nonzero variance σ^2. Recalling that $\bar{X}_n = (X_1 + \cdots + X_n)/n$, we can rewrite Equation (11.40) as

$$\lim_{n \to \infty} P\left(\frac{\bar{X}_n - \mu}{\sigma/\sqrt{n}} \leq x\right) = \Phi(x),$$

for all $x \in \mathcal{R}$. This result shows that the sample mean is asymptotically normally distributed, which is of particular importance in statistics.

We have already illustrated the central limit theorem when the common probability distribution is Bernoulli. In Example 11.19, we illustrate the central limit theorem for three additional probability distributions—the normal, exponential, and uniform.

EXAMPLE 11.19   *The Central Limit Theorem*

Normal, Exponential, and Uniform Distributions   Let $X_1, X_2, \ldots$ be independent and identically distributed random variables. Discuss the central limit theorem if the common distribution of $X_1, X_2, \ldots$ is
a) normal.   b) exponential.   c) uniform.
d) Construct graphs that illustrate the central limit theorem for each of these three families of distributions.

Solution   For all three cases, the central limit theorem says that $X_1 + \cdots + X_n$ is approximately normally distributed for large n. However, we can be more explicit here.

a) *Normal case:* As we know from Proposition 9.14 on page 540, if $X_1, X_2, \ldots$ are independent and identically distributed normal random variables with common parameters μ and σ^2, then $X_1 + \cdots + X_n \sim \mathcal{N}(n\mu, n\sigma^2)$, exactly. Hence, in this case, $(X_1 + \cdots + X_n - n\mu)/\sigma\sqrt{n} \sim \mathcal{N}(0, 1)$, exactly, for each $n \in \mathcal{N}$. In particular, then, each probability in the sequence on the left of Equation (11.40) equals $\Phi(x)$ and, for each n, the approximation in Relation (11.42) is exact.

b) *Exponential case:* We recall that an exponential random variable with parameter λ is also a gamma random variable with parameters 1 and λ. Therefore we can deduce from Proposition 9.13 on page 537 that, if $X_1, X_2, \ldots$ are independent and identically distributed exponential random variables with common parameter λ, then $X_1 + \cdots + X_n \sim \Gamma(n, \lambda)$. It therefore follows from the central limit theorem that, for large n, the distribution of a $\Gamma(n, \lambda)$-random variable is approximately normal.

c) *Uniform case:* Let $X_1, X_2, \ldots$ be independent and identically distributed uniform random variables. From Example 9.19 on page 535, $X_1 + X_2$ has a triangular distribution. The central limit theorem guarantees, however, that $X_1 + \cdots + X_n$ is approximately normally distributed for large n. In this (uniform) case, the normal approximation actually works well for n as small as 10.

d) Figure 11.3 provides graphs that illustrate the results obtained in parts (a)–(c). For each graph, the dashed curve is the standard normal PDF. The first graph in each column shows the standardized version of the distribution under consideration or, equivalently, the distribution of $(X_1 + \cdots + X_n - n\mu)/\sigma \sqrt{n}$ when $n = 1$. The next three graphs in each column show the distribution of $(X_1 + \cdots + X_n - n\mu)/\sigma \sqrt{n}$ for $n = 2$, 10, and 30. Each column provides a concrete illustration of the central limit theorem in action. ∎

Let $X_1, X_2, \ldots$ be independent and identically distributed random variables with common mean μ. The laws of large numbers indicate that the average value of $X_1, \ldots, X_n$ is likely to be near μ when n is large—but they don't indicate *how near* and *how likely*. We can use the central limit theorem to obtain the latter information.

Suppose then that $X_1, X_2, \ldots$ are independent and identically distributed random variables with common finite mean μ and common finite nonzero variance σ^2. From Relation (11.42) on page 666, for large n,

$$P\left(\left|\frac{X_1 + \cdots + X_n}{n} - \mu\right| \leq \epsilon\right) = P\left(|X_1 + \cdots + X_n - n\mu| \leq n\epsilon\right)$$

$$= P\left(n\mu - n\epsilon \leq X_1 + \cdots + X_n \leq n\mu + n\epsilon\right)$$

$$\approx \Phi\left(\frac{n\mu + n\epsilon - n\mu}{\sigma \sqrt{n}}\right) - \Phi\left(\frac{n\mu - n\epsilon - n\mu}{\sigma \sqrt{n}}\right)$$

$$= \Phi\left(\frac{\sqrt{n}\,\epsilon}{\sigma}\right) - \Phi\left(-\frac{\sqrt{n}\,\epsilon}{\sigma}\right),$$

or, in view of Equation (8.37) on page 444,

$$P\left(\left|\frac{X_1 + \cdots + X_n}{n} - \mu\right| \leq \epsilon\right) \approx 2\Phi\left(\frac{\sqrt{n}\,\epsilon}{\sigma}\right) - 1. \tag{11.43}$$

Heuristically, when n is large, Relation (11.43) provides us with a good approximation of how likely it is that the average value of $X_1, \ldots, X_n$ will be within ϵ of μ, for any prescribed positive real number ϵ.

Figure 11.3 The central limit theorem in action for (a) normal, (b) exponential, and (c) uniform distributions

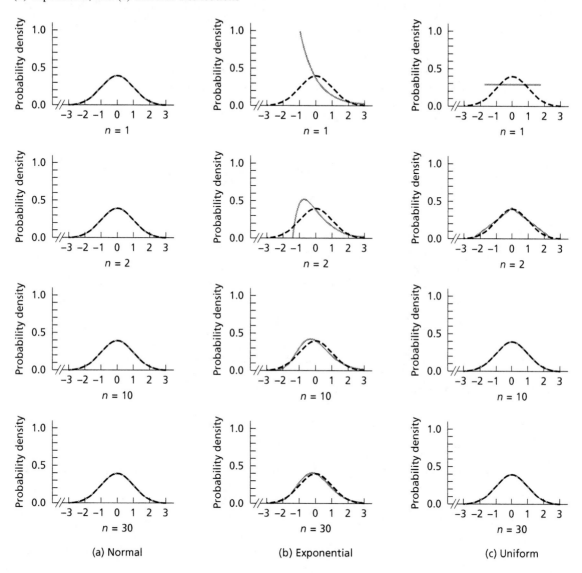

(a) Normal   (b) Exponential   (c) Uniform

In the next three examples, we apply the central limit theorem in various forms to three different situations. We begin with Example 11.20 by returning to the roulette problem first encountered in Example 11.16. Then, in the following two examples, we examine applications of the central limit theorem to roundoff error and to the estimation of the mean of a distribution by using a confidence interval.

EXAMPLE 11.20   *The Central Limit Theorem*

Roulette   Refer to Example 11.16 on page 652, where a gambler repeatedly bets $1 on red. There we applied the weak law of large numbers to conclude that, in the long run, the gambler loses $\$\frac{1}{19}$ per play and, consequently, that, in 10,000 plays, the gambler will lose a total of about $10,000 \cdot \$\frac{1}{19}$, or roughly $526. Using the central limit theorem—specifically, Relation (11.43)—we can obtain more precise results.

a) In 10,000 plays, determine the probability that the average loss per play will be within 1¢ of the expected loss per play of $\$\frac{1}{19}$.

b) Use part (a) to estimate the gambler's losses after 10,000 plays.

Solution   Let X_j denote the loss (in dollars) of the gambler on the jth play. Then $X_1, X_2, \ldots$ are independent and identically distributed random variables. Simple computations show that the common mean and variance of $X_1, X_2, \ldots$ are, respectively,

$$\mu = \frac{1}{19} \quad \text{and} \quad \sigma^2 = \frac{360}{361}. \tag{11.44}$$

For convenience, we set $n_0 = 10,000$, $\epsilon_0 = 0.01$ (1¢), and $S_n = X_1 + \cdots + X_n$—the total loss in n plays.

a) In n_0 plays, the average loss per play is S_{n_0}/n_0. Applying Relation (11.43) with $n = n_0$ and $\epsilon = \epsilon_0$, and noting that $\sqrt{n_0}\,\epsilon_0/\sigma = \sqrt{361/360}$, we find that

$$P\left(\left|\frac{S_{n_0}}{n_0} - \mu\right| \le \epsilon_0\right) \approx 2\Phi\left(\frac{\sqrt{n_0}\,\epsilon_0}{\sigma}\right) - 1 = 2\Phi\left(\sqrt{\frac{361}{360}}\right) - 1 = 0.683.$$

Thus, in 10,000 plays, chances are about 68.3% that the average loss per play will be within 1¢ of the expected loss per play of $\$\frac{1}{19}$.

b) Applying the results of part (a), we have

$$P\left(\left|S_{n_0} - n_0\mu\right| \le n_0\epsilon_0\right) = P\left(\left|\frac{S_{n_0}}{n_0} - \mu\right| \le \epsilon_0\right) \approx 0.683.$$

We note that $n_0\epsilon_0 = 100$ and that $n_0\mu = 526$, approximately. Thus, in 10,000 plays, chances are about 68.3% that the gambler will lose between $426 and $626.   ∎

EXAMPLE 11.21   *The Central Limit Theorem*

Roundoff Error   Ten numbers are rounded to the nearest integer and then summed. Determine the probability that the sum of the rounded numbers will equal the rounded sum of the unrounded numbers.

Solution   Let X_j denote the roundoff error for the jth number—that is, the difference between the jth number and its rounded value. The sum of the rounded numbers equals the rounded sum of the unrounded numbers if and only if $|X_1 + \cdots + X_{10}| \le 0.5$. We want to determine the probability of that event.

We can reasonably assume that $X_1, \ldots, X_{10}$ are independent and identically distributed random variables with common distribution $\mathcal{U}(-0.5, 0.5)$. The mean and variance of that distribution are $\mu = 0$ and $\sigma^2 = 1/12$, respectively. As we noted in the solution to Example 11.19(c), for uniform distributions, the normal approximation, as guaranteed by the central limit theorem, works well for n as small as 10. Consequently, from Relation (11.42) on page 666,

$$P\left(|X_1 + \cdots + X_{10}| \leq 0.5\right) = P\left(-0.5 \leq X_1 + \cdots + X_{10} \leq 0.5\right)$$

$$\approx \Phi\left(\frac{0.5 - 10 \cdot 0}{\sqrt{1/12}\sqrt{10}}\right) - \Phi\left(\frac{-0.5 - 10 \cdot 0}{\sqrt{1/12}\sqrt{10}}\right)$$

$$= 2\Phi\left(\frac{0.5}{\sqrt{10/12}}\right) - 1 = 0.416.$$

Thus the probability is about 0.416 that the sum of 10 rounded numbers will equal the rounded sum of the 10 unrounded numbers, less than a 50% chance. ∎

EXAMPLE 11.22   *The Central Limit Theorem*

Confidence Intervals for a Mean   Consider a random variable X having finite mean μ and finite nonzero variance σ^2. Suppose that we don't know μ but want to estimate it. To do so, we take a random sample, $X_1, \ldots, X_n$, from the distribution of X and set $\bar{X}_n = (X_1 + \cdots + X_n)/n$. Use the central limit theorem to obtain a formula for a large-sample confidence interval for μ when σ^2 is known.

Solution   Let $0 < \gamma < 1$. We want to obtain a formula for a large-sample $100\gamma\%$ **confidence interval** for μ; that is, we want to find E such that, for large n,

$$P\left(\bar{X}_n - E \leq \mu \leq \bar{X}_n + E\right) = \gamma, \tag{11.45}$$

at least approximately. The number γ is called the **confidence coefficient.**

Because, for each $n \in \mathcal{N}$, $X_1, \ldots, X_n$ is a random sample from the distribution of X, we know that $X_1, X_2, \ldots$ are independent and identically distributed random variables with common finite mean μ (which is unknown) and common finite nonzero variance σ^2 (which we are assuming is known). As n is large, we can apply the central limit theorem in the form of Relation (11.43) on page 668 and conclude that

$$P\left(|\bar{X}_n - \mu| \leq E\right) \approx 2\Phi\left(\frac{\sqrt{n}\,E}{\sigma}\right) - 1,$$

or, equivalently, that

$$P\left(\bar{X}_n - E \leq \mu \leq \bar{X}_n + E\right) \approx 2\Phi\left(\frac{\sqrt{n}\,E}{\sigma}\right) - 1. \tag{11.46}$$

Referring to Equation (11.45) and Relation (11.46), we see that E must be chosen so that $2\Phi(\sqrt{n}\,E/\sigma) - 1 = \gamma$, or $E = \Phi^{-1}\big((1 + \gamma)/2\big) \cdot \sigma/\sqrt{n}$. Consequently, if we let $\bar{x}_n$ denote the observed value of $\bar{X}_n$, then a $100\gamma\%$ confidence interval for μ is from

$$\bar{x}_n - \Phi^{-1}\left(\tfrac{1}{2}(1 + \gamma)\right) \cdot \frac{\sigma}{\sqrt{n}} \qquad \text{to} \qquad \bar{x}_n + \Phi^{-1}\left(\tfrac{1}{2}(1 + \gamma)\right) \cdot \frac{\sigma}{\sqrt{n}}. \tag{11.47}$$

We can be roughly $100\gamma\%$ confident that μ lies somewhere between the two numbers in Expression (11.47). ∎

Other Central Limit Theorems

In this section, we examined the central limit theorem in its classical form. Many other versions of the central limit theorem have now been established and can be found in the extensive literature on the subject. These versions weaken, in one way or another, the independence and/or identically-distributed assumptions of the classical central limit theorem, thus providing normal approximations in contexts where the random variables under consideration may not be independent or may not be identically distributed.

EXERCISES 11.4   **Basic Exercises**

11.55 As reported by a spokesperson for Southwest Airlines, the no-show rate for reservations is 16%—that is, the probability is 0.16 that a person making a reservation will not take the flight. For a certain flight, 42 people have reservations. For each part, determine and compare the exact probability by using the appropriate binomial PMF and an approximate probability by using the integral De Moivre–Laplace theorem (in the form of Proposition 11.11 on page 661). The probability that the number of people who don't take the flight is
a) exactly 5.   **b)** between 9 and 12, inclusive.   **c)** at least 1.   **d)** at most 2.

11.56 In Exercise 8.108 on page 449, you were asked to determine and compare the exact probability of each event in Exercise 11.55 by using the appropriate binomial PMF and an approximate probability by using the local De Moivre–Laplace theorem (in the form of Proposition 8.12 on page 445).
a) If you didn't previously do Exercise 8.108, do it now.
b) Compare the results for all three methods of obtaining the probabilities by constructing a table similar to Table 11.2 on page 663.
c) Discuss your results from part (b) in light of the rule of thumb for using the normal approximation to the binomial distribution.

11.57 Refer to the roulette illustration of Example 11.20 on page 670. Use the integral De Moivre–Laplace theorem to estimate the probability that the gambler will be ahead after
a) 100 bets.   **b)** 1000 bets.   **c)** 5000 bets.

11.58 Some boys want to play football at a park that doesn't have a formal football field. In an attempt to approximate the length of a football field, one of the boys tries to step-off 100 yards. In actuality, his steps are independent and identically distributed random variables with mean 0.95 yard and standard deviation 0.08 yard. Determine the approximate probability that the distance stepped-off by the boy is within
a) 4 yards of the length of a football field.
b) 6 yards of the length of a football field.

11.59 A brand of flashlight battery has normally distributed lifetimes with a mean of 30 hours and a standard deviation of 5 hours. A supermarket purchases 500 of these batteries from the manufacturer. What is the probability that at least 80% of them will last longer than 25 hours? Use the integral De Moivre–Laplace theorem.

11.60 The checkout times at the local food market are independent random variables having mean 3.5 minutes and standard deviation 1.5 minutes.
a) Determine the probability that it will take at least 6 hours to check out 100 customers.
b) Find the probability that the mean checkout time for 100 customers is less than 3.4 minutes.

11.61 In a large city, annual household incomes have a mean of \$35,216 and a standard deviation of \$3,134.

a) What is the probability that the average income of 160 randomly chosen households is below \$34,600?

b) Strictly speaking, in part (a), are the assumptions for the central limit theorem satisfied? Explain your answer.

c) Why is it permissible to use the central limit theorem to solve part (a)?

11.62 The claim amount for a health insurance policy follows a distribution with density function $f(x) = (1/1000)e^{-x/1000}$ for $x > 0$, and $f(x) = 0$ otherwise. The premium for the policy is set at 100 over the expected claim amount. Suppose that 100 policies are sold and that claim amounts are independent of one another.

a) Identify the exact probability distribution of the total claim amount for the 100 policies.

b) Use your answer from part (a) to obtain an expression that gives the exact probability that the insurance company will have claims exceeding the premiums collected.

c) Use the central limit theorem to obtain the approximate probability that the insurance company will have claims exceeding the premiums collected.

d) If you have access to statistical software, use it and your answer from part (a) to obtain the probability that the insurance company will have claims exceeding the premiums collected. Compare this probability to that found in part (c).

11.63 Let X and Y denote the number of hours that a randomly selected person watches movies and sporting events, respectively, during a 3-month period. Assume that $\mathcal{E}(X) = 50$, $\mathcal{E}(Y) = 20$, $\mathrm{Var}(X) = 50$, $\mathrm{Var}(Y) = 30$, and $\mathrm{Cov}(X, Y) = 10$. If 100 people are randomly selected and observed for 3 months, what is the probability that the total number of hours that they watch movies or sporting events is at most 7100?

11.64 A hardware manufacturer knows from experience that 95% of the screws produced by his company are within tolerance specifications. Each shipment of 10,000 screws comes with a warranty that promises a complete refund if more than r screws aren't within tolerance specifications. How small can r be chosen so that no more than 1% of shipments will require complete refunds? Solve this problem by

a) using Chebyshev's inequality.   **b)** using the central limit theorem.

c) If you have access to statistical software, use it to obtain the exact value of r and compare your answer to those found in parts (a) and (b).

11.65 An air-conditioning contractor plans to offer service contracts on the brand of compressor used in all of the units her company installs. First she must estimate how long those compressors last, on average. To that end, the contractor consults records on the lifetimes of 250 previously used compressors. She plans to use the sample mean lifetime of those compressors as her estimate for the mean lifetime of all such compressors. If the lifetimes of this brand of compressor have a standard deviation of 40 months, what is the probability that the contractor's estimate will be within 5 months of the true mean?

11.66 A particle of initial size s is subjected to repeated impacts. After impact j, a proportion X_j of the particle remains; that is, if Y_j denotes the size of the particle after impact j, then $Y_j = X_j Y_{j-1}$. The random variables $X_1, X_2, \ldots$ are independent, all having the same probability distribution as a random variable X.

a) Use the central limit theorem to identify the approximate probability distribution of Y_n for large n. State explicitly any assumptions that you make.

b) Specialize the results obtained in part (a) to the cases where $X \sim \mathcal{U}(0, 1)$, $X \sim \mathrm{Beta}(2, 1)$, $X \sim \mathrm{Beta}(1, 2)$, and $X \sim \mathrm{Beta}(2, 2)$.

11.67 The subterranean coruro (*Spalacopus cyanus*) is a social rodent that lives in large colonies in underground burrows that can reach lengths of up to 600 meters. Zoologists Sabine Begall and Milton H. Gallardo studied the characteristics of the burrow systems of the subterranean coruro in central Chile and published their findings in the *Journal of Zoology, London,* (2000, Vol. 251, pp. 53–60). A random sample of 51 burrows had the following depths, in centimeters (cm).

15.1	16.0	18.3	18.8	13.9	15.8	14.2	12.3	11.8
12.1	17.9	16.6	16.5	16.0	12.8	14.7	15.9	13.9
17.2	12.2	18.2	16.9	13.3	14.4	15.0	12.1	11.0
16.7	17.4	8.2	19.3	17.4	15.3	15.6	19.7	14.5
12.5	12.8	13.3	16.8	17.5	14.0	14.9	16.7	12.0
15.0	16.2	9.7	15.4	18.9	14.9			

a) Find an approximate 90% confidence interval for the mean depth of all subterranean coruro burrows. Assume that $\sigma = 2.4$ cm.

b) Interpret your answer from part (a).

c) Find an approximate 99% confidence interval for the mean depth of all subterranean coruro burrows.

d) Why is the confidence interval that you found in part (c) longer than the one in part (a)?

11.68 Simulation: This exercise requires access to a computer. Desert Samaritan Hospital in Mesa, Arizona, keeps records of its emergency-room traffic. Those records indicate that, beginning at 6:00 P.M. on any given day, the elapsed times between arriving patients are independent random variables, all having exponential distributions with parameter $\lambda = 6.9$, where time is measured in hours. Let W_n denote the elapsed time from 6:00 P.M. until the arrival of the nth patient.

a) Identify the exact probability distribution of the random variable W_n.

b) Use a basic random number generator—that is, a random number generator that simulates a $\mathcal{U}(0, 1)$ random variable—to obtain 5000 observations of W_4. *Hint:* Express W_4 as a sum of independent exponential random variables and refer to Proposition 8.16 on page 471.

c) Would you expect a histogram of the 5000 observations from part (b) to be roughly bell-shaped? If not, what shape would you expect? Explain your answer.

d) Obtain a histogram of the 5000 observations from part (b).

e) Repeat parts (b)–(d) for W_{60}.

11.69 George and Julia both work for the campus coffee shop. Sales are slow and management is considering letting one of them go. To decide which employee to keep, each is timed on 40 independent trials of making a double decaf skim milk latte. If the sample mean times differ by more than 3 seconds, the person with the larger sample mean will be let go; otherwise, both will be kept. The standard deviation of the time it takes each person to make a double decaf skim milk latte is 4 seconds. If the mean times for both George and Julia are actually the same, what is the probability that George will be let go?

11.70 A light is continually kept on. Each time a bulb burns out, it's immediately replaced by another bulb. There is a supply of 20 bulbs and the lifetime of each bulb is exponentially distributed with mean 300 hours, independent of the lifetimes of the other bulbs.

a) Use the central limit theorem to approximate the probability that all bulbs have burned out by 7000 hours.

b) If you have access to statistical software, use it to obtain the exact probability that all bulbs have burned out by 7000 hours. Compare your answer with that in part (a) and explain any discrepancy.

Theory Exercises

11.71 Margin of error: Refer to Example 11.22 on page 671, where we discussed confidence intervals for the mean, μ, of a random variable (or distribution).

a) The quantity $\Phi^{-1}\left(\frac{1}{2}(1+\gamma)\right) \cdot \sigma/\sqrt{n}$ appearing in Expression (11.47) is called the *margin of error* (or *maximum error of the estimate*) for the estimate of μ. Explain why that terminology is used.

b) Derive a formula for the sample size required for a $100\gamma\%$ confidence interval for μ with a specified margin of error, E.

Advanced Exercises

11.72 Refer to Exercises 11.71 and 11.67.

a) Obtain the margin of error for the estimate of the mean depth of all subterranean coruro burrows for a 90% confidence interval; a 99% confidence interval.

b) Find the sample size required to have a margin of error of 0.5 cm and a confidence coefficient of 0.90; of 0.99.

11.73 From past experience, it's known that purchase amounts at a certain store have a mean of \$68 with a standard deviation of \$11. In an attempt to increase average purchase amounts, management decided to advertise more aggressively and to develop a Web site that is capable of handling online sales. After implementation of this new strategy, 44 sales averaged \$73. Is there reason to believe that the new strategy is effective? Explain your reasoning and state explicitly any assumptions that you make.

11.74 Use the central limit theorem to show that $\lim_{n\to\infty} e^{-n} \sum_{k=0}^{n} n^k/k! = 1/2$.

11.75 Our proof of the central limit theorem, as presented on page 666, is based on the assumption that the common MGF of $X_1, X_2, \ldots$ is defined in some open interval containing 0.

a) What does that assumption imply about the moments of the common distribution of the random variables $X_1, X_2, \ldots$?

b) Provide a proof of the central limit theorem by using characteristic functions. *Note:* Refer to page 641.

c) Why is a proof of the central limit theorem by using characteristic functions preferable to that by using moment generating functions?

CHAPTER REVIEW

Summary

The moment generating function (MGF) of a random variable X, denoted M_X, is the expected value of the random variable e^{tX}—that is, $M_X(t) = \mathcal{E}\left(e^{tX}\right)$. More generally, the joint moment generating function (joint MGF) of m random variables $X_1, \ldots, X_m$, denoted $M_{X_1,\ldots,X_m}$, is the expected value of the random variable $e^{t_1 X_1 + \cdots + t_m X_m}$—that is, $M_{X_1,\ldots,X_m}(t_1, \ldots, t_m) = \mathcal{E}\left(e^{t_1 X_1 + \cdots + t_m X_m}\right)$.

We examined several important and useful properties of MGFs, including methods for obtaining individual and joint moments of random variables by taking appropriate derivatives of the (joint) moment generating function. One of the most essential properties of MGFs is that they determine the probability distribution. Thus, if two random variables have the same MGF, they have the same probability distribution; and, if two collections of m random variables have the same joint MGF, they have the same joint probability distribution. Two other useful properties of MGFs hold for independent random variables: (1) the MGF of their sum equals the product of their MGFs, and (2) their joint MGF equals the product of their marginal MGFs.

We discussed and applied laws of large numbers for independent and identically distributed random variables, $X_1, X_2, \ldots,$ with finite mean μ. These mathematical laws confirm our intuitive notions of expected value and probability by verifying that the average value of $X_1, \ldots, X_n$ approaches μ as $n \to \infty$. Specifically, the weak law of large numbers states that, when n is large, the average value of $X_1, \ldots, X_n$ is likely to be near μ; the strong law of large numbers states that, when n is large, the average value of $X_1, \ldots, X_n$ is likely to be near μ and to stay near μ for all subsequent n. Clearly, the strong law of large numbers is stronger than (implies) the weak law of large numbers.

Finally we presented and applied the central limit theorem for independent and identically distributed random variables, $X_1, X_2, \ldots,$ with finite mean μ and finite nonzero variance σ^2. This theorem states that, when n is large, the sum of $X_1, \ldots, X_n$ (or the average value of $X_1, \ldots, X_n$) is approximately normally distributed, regardless of the common distribution of $X_1, X_2, \ldots$.

You Should Be Able To ...

1. define, state the basic properties of, and obtain the moment generating function of a random variable.

2. use the moment generating function of a random variable to obtain the moments of the random variable.

3. identify the probability distribution of a random variable from its moment generating function.

4. obtain the moment generating function of the sum of independent random variables from their moment generating functions.

5. identify the probability distribution of the sum of independent random variables by using moment generating functions.

6. state and apply the continuity theorem for moment generating functions.

7. define, state the basic properties of, and obtain the joint moment generating function of two or more random variables.

8. use the joint moment generating function of two or more random variables to obtain moments and joint moments.

9. obtain the marginal moment generating functions from the joint moment generating function.

10. identify the joint probability distribution of m random variables from their joint moment generating function.

11. obtain the joint moment generating function of independent random variables from their marginal moment generating functions.

12. obtain the moment generating function of the sum of random variables from their joint moment generating function.

13. state, interpret, and apply the weak law of large numbers.

14. state, interpret, and apply the strong law of large numbers.

15. explain the difference between the weak and strong laws of large numbers.

16. state, interpret, and apply the integral De Moivre–Laplace theorem.

17. state, interpret, and apply the central limit theorem.

Key Terms

central limit theorem, *665*

characteristic function, *638*

confidence coefficient, *671*

confidence interval, *671*

consistent estimator, *653*

continuity correction, *661*

continuity theorem, *638*

convergence in distribution, *638*

integral De Moivre–Laplace

  theorem, *661*

joint moment generating

  function, *641, 647*

laws of large numbers, *650*

marginal moment generating

  function, *645*

moment generating function, *630*

strong law of large numbers, *654*

weak law of large numbers, *651*

Chapter Review Exercises

Basic Exercises

11.76 In Example 11.4 on page 632, we obtained the MGF of a gamma random variable with parameters α and λ. Use that MGF to find the mean and variance of such a random variable.

11.77 An actuary determines that the claim size for a certain class of accidents is a random variable X with moment generating function $M_X(t) = 1/(1 - 2500t)^4$. Find the standard deviation of the claim size for this class of accidents

a) by applying the moment generating property of MGFs.

b) without doing any computations, but rather by referring to the result of Exercise 11.76.

11.78 Proceed as follows to show that the MGF of a triangular random variable on the interval (a, b) is given by

$$M(t) = \frac{8e^{(a+b)t/2}}{(b-a)^2 t^2}\left(\cosh\frac{(b-a)t}{2} - 1\right),$$

where $\cosh t = (e^t + e^{-t})/2$, the *hyperbolic cosine* of t.

a) Obtain the MGF of a random variable with a triangular distribution on the interval $(-1, 1)$.

b) Use part (a) to determine the MGF of a $T(a, b)$ random variable. *Hint:* Refer to Exercise 8.125 on page 462.

c) In Example 9.19 on page 535, we showed that the sum of two independent $U(0, 1)$ random variables has the triangular distribution on the interval $(0, 2)$. Use moment generating functions to obtain that result more easily.

11.79 Let $T \sim T(-1, 1)$ and let $X \sim T(a, b)$.

a) Use your result from Exercise 11.78(a) to find the mean and variance of T.

b) Determine the mean and variance of X.

11.80 Let X have the negative binomial distribution with parameters r and p.

a) Obtain the MGF of X. *Hint:* Use the alternative form of a negative-binomial PMF given in Equation (5.44) and the binomial series given in Equation (5.45), both on page 241.

b) Use part (a) to find the mean and variance of X.

11.81 Refer to Exercise 11.80.

a) Use properties of MGFs to show that the sum of independent negative binomial random variables with the same success-probability parameter is also a negative binomial random variable. Specifically, if $X_1, \ldots, X_m$ are independent random variables with $X_j \sim \mathcal{NB}(r_j, p)$ for $1 \le j \le m$, then $X_1 + \cdots + X_m \sim \mathcal{NB}(r_1 + \cdots + r_m, p)$.

b) Deduce from part (a) that the sum of independent and identically distributed geometric random variables has a negative binomial distribution. Specifically, if $X_j \sim \mathcal{G}(p)$ for $1 \le j \le m$, then $X_1 + \cdots + X_m \sim \mathcal{NB}(m, p)$.

11.82 Let X and Y be independent normal random variables and let a and b be constants, at least one of which isn't 0. Use MGFs to show that $aX + bY$ is normally distributed.

11.83 Let $X_n \sim \mathcal{U}(-1/n, 1/n)$ for each $n \in \mathcal{N}$.

a) Heuristically, what is the probability distribution of the random variable X to which the sequence of random variables $\{X_n\}_{n=1}^{\infty}$ converges in distribution?

b) Mathematically verify your heuristics in part (a) by showing that $M_{X_n}(t) \to M_X(t)$ as $n \to \infty$ for all $t \in \mathcal{R}$.

c) Mathematically verify your heuristics in part (a) by showing that $F_{X_n}(x) \to F_X(x)$ as $n \to \infty$ for all $x \in \mathcal{R}$ at which F_X is continuous.

11.84 For each positive integer n, let X_n be the constant random variable equal to $1/n$; that is, $P(X_n = 1/n) = 1$. Also, let X be the constant random variable equal to 0.

a) Show that $\{X_n\}_{n=1}^{\infty}$ converges in distribution to X but that $P(X_n = 0)$ doesn't converge to $P(X = 0)$ as $n \to \infty$.

b) Your naive classmate finds the results of part (a) paradoxical. Resolve your classmate's confusion.

11.85 At a service center, two successive tasks are required. Let X and Y denote the times at which the first and second tasks are completed, respectively. A joint PDF of X and Y is $f_{X,Y}(x, y) = 6e^{-(x+2y)}$ for $0 < x < y$, and $f_{X,Y}(x, y) = 0$ otherwise.

a) Obtain the joint MGF of X and Y.

b) Use your result from part (a) to identify the marginal distribution of X.

c) Use MGFs to identify the probability distribution of the time required to complete the second task.

d) Use MGFs to show that the times required to complete the two tasks are independent random variables.

e) Use MGFs to find time required, on average, to complete the two successive tasks.

11.86 Let X and Y be discrete random variables with joint probability mass function given by $p_{X,Y}(x, y) = (4/9)2^{-(x+y)}$ for $(x, y) \in \{0, 1\}$, and $p_{X,Y}(x, y) = 0$ otherwise.

a) Obtain the joint MGF of X and Y.

b) Use your result from part (a) to determine the means and variances of X and Y and their covariance.

c) Use your result from part (a) to obtain and identify the marginal distributions of X and Y.

d) Use your result from part (a) to obtain and identify the probability distribution of $X + Y$.

e) Use MGFs to decide whether X and Y are independent random variables.

11.87 Let X and Y be discrete random variables with joint probability mass function given by $p_{X,Y}(x, y) = (4/7)2^{-(x+y)}$ for $(x, y) \in \{0, 1\}$ and $x \ge y$, and $p_{X,Y}(x, y) = 0$ otherwise.

a) Obtain the joint MGF of X and Y.

b) Use your result from part (a) to determine the means and variances of X and Y and their covariance.

c) Use your result from part (a) to obtain and identify the marginal distributions of X and Y.

d) Use your result from part (a) to obtain the probability distribution of $X + Y$.

e) Does $X + Y$ have a binomial distribution? Justify your answer.

f) Use MGFs to decide whether X and Y are independent random variables.

11.88 Simulation: This exercise requires access to a computer or graphing calculator with statistical software.

a) Use a Poisson random number generator—that is, a random number generator that simulates observations of a Poisson random variable—to obtain 10,000 observations of a $\mathcal{P}(3)$ random variable.

b) Without doing further computations, roughly what should be the average value of the 10,000 numbers obtained in part (a)? Explain your reasoning.

c) Determine the average value of the 10,000 numbers obtained in part (a) and compare your result to your answer in part (b).

11.89 Monte Carlo integration: In this exercise, you are to extend the technique of Monte Carlo integration considered in Exercises 11.43 and 11.44 on page 656. Suppose that you want to determine the value of $\int_a^b g(x)\, dx$, where g is a Riemann integrable function on the interval $[a, b]$. Further assume that no simple formula exists for the antiderivative of g. Let f be a PDF such that $f(x) > 0$ for $x \in [a, b]$, and $f(x) = 0$ for $x \notin [a, b]$.

a) Let X be a random variable with probability density function f. Show that the random variable $g(X)/f(X)$ has finite expectation.

b) Let $X_1, X_2, \ldots$ be independent and identically distributed random variables with common PDF f. Set

$$\hat{I}_n(f, g) = \frac{1}{n} \sum_{k=1}^{n} \frac{g(X_k)}{f(X_k)}.$$

Show that $\lim_{n \to \infty} \hat{I}_n(f, g) = \int_a^b g(x)\, dx$ with probability 1.

c) Determine when $\hat{I}_n(f, g)$ has finite variance and, in such a case, obtain the variance.

d) Show that, if $a = 0$, $b = 1$, and f is the PDF of a $\mathcal{U}(0, 1)$ random variable, the result of part (b) reduces to the Monte Carlo technique examined in Exercise 11.43. Obtain the variance of $\hat{I}_n(f, g)$ in this case.

11.90 A machine produces a large number of items per day. The quality control engineer samples 100 items from the daily output and declares the machine "out of control" and in need of service if 12 or more of the sampled items are defective. If the machine is actually producing 8% defective items, what is the probability that it will be declared "out of control"?

11.91 The second leading genetic cause of mental retardation is Fragile X Syndrome, named for the fragile appearance of the tip of the X chromosome in affected individuals. Worldwide, 1 in every 1500 males is affected, with no ethnic bias. For a sample of 10,000 males, use the integral De Moivre–Laplace theorem (in the form of Proposition 11.11 on page 661) to determine the probability that the number who have Fragile X Syndrome

a) exceeds 7.   **b)** is at most 10.

11.92 The probabilities in parts (a) and (b) of Exercise 11.91 were obtained in Exercise 8.111 on page 450 by using the local De Moivre–Laplace theorem (in the form of Proposition 8.12 on page 445) and in Exercise 5.80 on page 226 by using the Poisson approximation to the binomial distribution.

a) If you didn't previously do Exercises 5.80 and 8.111, do them now.

b) Compare your probability estimates for the three methods (local and integral De Moivre–Laplace approximations and the Poisson approximation).

11.93 Acute rotavirus diarrhea is the leading cause of death among children under age 5, killing an estimated 4.5 million annually in developing countries. Scientists from Finland and

Belgium claim that a new oral vaccine is 80% effective against rotavirus diarrhea. Assuming that the claim is correct, use the integral De Moivre–Laplace theorem to find the probability that, of 1500 cases, the vaccine will be effective in
a) exactly 1225 cases.   **b)** at least 1175 cases.
c) between 1150 and 1250 cases, inclusive.

11.94 As we stated on page 667, a rule of thumb for using the normal approximation is that $n \geq 30$. However, sometimes the approximation works well for quite small n, as we noted in Example 11.19(c) on page 667. Let X_1, X_2, and X_3 be independent and identically distributed random variables, all having the discrete uniform distribution on the set $\{1, 2, 3\}$.
a) By using either a direct listing or counting, determine the PMF of the random variable $X_1 + X_2 + X_3$.
b) Use the result of part (a) to obtain the exact value of $P(X_1 + X_2 + X_3 \leq 7)$.
c) Construct a probability histogram for the random variable $X_1 + X_2 + X_3$ and comment on its degree of resemblance to a normal curve.
d) Obtain an approximate value of the probability required in part (b) by using the normal approximation with a continuity correction. Compare this value to that obtained in part (b).

11.95 Experience has shown that scores on a certain aptitude test have a mean of 73 points and a standard deviation of 5 points. If the aptitude test is given to 140 people, what is the probability that their mean score will exceed 74?

11.96 In Exercise 11.70 on page 674, you were asked to consider the following problem: "A light is continually kept on. Each time a bulb burns out, it's immediately replaced by another bulb. There is a supply of 20 bulbs and the lifetime of each bulb is exponentially distributed with mean 300 hours, independent of the lifetimes of the other bulbs. What is the probability that all bulbs have burned out by 7000 hours?" Answer that same question if replacement time isn't immediate but takes a mean of 0.5 hour with a standard deviation of 0.1 hour, and successive replacement times are independent of one another and of the lifetimes of the bulbs.

11.97 In an analysis of healthcare data, based on a random sample of 48 people, ages have been rounded to the nearest multiple of 5 years. What is the approximate probability that the mean of the rounded ages is within 0.25 year of the mean of the true ages? State explicitly any assumptions that you make.

11.98 An insurance company issues 1250 vision care insurance policies. The number of claims filed by a policyholder under a vision care insurance policy during a 1-year period is a Poisson random variable with mean 2. Assume that the numbers of claims filed by distinct policyholders are independent of one another.
a) Provide an expression for the exact probability that the total number of claims filed during a 1-year period is between 2450 and 2600, inclusive.
b) Use the central limit theorem to approximate the probability that the total number of claims filed during a 1-year period is between 2450 and 2600, inclusive.

11.99 Suppose that $X_1, X_2, \ldots$ are independent and identically distributed positive random variables. Further suppose that $\ln X_j$ has finite nonzero variance. Show that, for large n, the random variable $\prod_{j=1}^{n} X_j$ has approximately a lognormal distribution. *Note:* Refer to Exercise 8.154 on page 474.

11.100 According to Scarborough Research, more than 85% of working adults commute by car. Of all U.S. cities, Washington, D.C., and New York City have the longest mean commute times. Suppose that 30 randomly obtained commute times in Washington, D.C., have a mean of 27.97 minutes. Assuming that the standard deviation of all such commute times is 10.04 minutes, determine a 90% confidence interval for the mean commute time in Washington, D.C. Interpret your result in words.

Theory Exercises

11.101 Bernstein polynomials: The *Weierstrass approximation theorem* states that any continuous function on a bounded closed interval can be uniformly approximated by polynomials. It suffices to prove this result for the interval $[0, 1]$. Here you are to consider a proof due to Serge Bernstein. Let f be a continuous function on $[0, 1]$ and let $X_1, X_2, \ldots$ be independent and identically distributed Bernoulli random variables with success probability p.

a) Prove that $\lim_{n\to\infty} f((X_1 + \cdots + X_n)/n) = f(p)$ with probability 1.

b) Show that

$$\mathcal{E}\left(f\left(\frac{X_1 + \cdots + X_n}{n} \right) \right) = \sum_{k=0}^{n} f\left(\frac{k}{n} \right) \binom{n}{k} p^k (1 - p)^{n-k}.$$

The right side of this equation is a polynomial in p, called a *Bernstein polynomial,* and is denoted $B_n(p)$.

c) Prove that $\lim_{n\to\infty} B_n(x) = f(x)$ uniformly for $x \in [0, 1]$. *Hint:* A continuous function on a bounded closed interval is uniformly continuous and bounded. Use Chebyshev's inequality.

11.102 Convergence in probability: Suppose that X and $X_1, X_2, \ldots$ are random variables all defined on the same sample space. We say that $\{X_n\}_{n=1}^{\infty}$ *converges in probability* to X if $\lim_{n\to\infty} P(|X - X_n| \geq \epsilon) = 0$ for each $\epsilon > 0$.

a) Restate the weak law of large numbers in terms of convergence in probability.

b) Restate *consistent estimator* (see page 653) in terms of convergence in probability.

c) Let X, $X_1, X_2, \ldots$, and Y, $Y_1, Y_2, \ldots$ be random variables all defined on the same sample space, and let $c_1, c_2, \ldots$ be real constants. If $\{X_n\}_{n=1}^{\infty}$ converges in probability to X, $\{Y_n\}_{n=1}^{\infty}$ converges in probability to Y, and $c_n \to c$, then $\{X_n + Y_n\}_{n=1}^{\infty}$ converges in probability to $X + Y$, $\{X_n Y_n\}_{n=1}^{\infty}$ converges in probability to XY, and $\{c_n X_n\}_{n=1}^{\infty}$ converges in probability to cX. Prove the first of these three results. *Hint:* Use the triangle inequality and Boole's inequality (Exercise 2.29 on page 47).

11.103 Consistency of the sample variance: Let X be a random variable with finite variance. Suppose that you don't know the variance of X and want to estimate it. To do so, you take a random sample, $X_1, \ldots, X_n$, from the distribution of X and form the *sample variance,* $S_n^2 = (n - 1)^{-1} \sum_{j=1}^{n} (X_j - \bar{X}_n)^2$. Show that S_n^2 is a consistent estimator of σ_X^2. *Hint:* Refer to Example 11.17 on page 653, Exercise 11.40 on page 656, and Exercise 11.102.

Advanced Exercises

11.104 Let $X \sim \Gamma(\alpha, \lambda)$. Show that $\mathcal{E}(X^n) = \alpha(\alpha + 1) \cdots (\alpha + n - 1)/\lambda^n$ for $n = 1, 2, \ldots$ by expanding the MGF of X in a power series about $t = 0$.

11.105 Let X have a PDF given by $f_X(x) = (1/2)e^{-|x|}$ for all $x \in \mathcal{R}$.

a) Obtain the MGF of X.

b) Use the result of part (a) to obtain a formula for the moments of X.

11.106 Let Z have the standard normal distribution and let $X \sim \mathcal{N}(\mu, \sigma^2)$.

a) From Example 11.5 on page 633, $M_Z(t) = e^{t^2/2}$. Expand this MGF in a power series about $t = 0$ to obtain a formula for the moments of Z.

b) Use the result of part (a) to obtain a formula for the central moments of X.

c) Use the result of part (a) to obtain a formula for the moments of X.

11.107 Cumulant generating function: Let X be a random variable whose MGF is defined in some open interval containing 0. The *cumulant generating function* of X, denoted K_X, is defined by $K_X(t) = \ln M_X(t)$. The *rth cumulant* of X, denoted κ_r, is defined to be the rth derivative of K_X evaluated at $t = 0$; that is, $\kappa_r = K_X^{(r)}(0)$.

a) Show that $\kappa_1 = \mathcal{E}(X)$ and $\kappa_2 = \text{Var}(X)$.

b) Let a_n be the coefficient of t^n in the power series expansion of $K_X(t)$ about $t = 0$. Show that $\kappa_n = n! a_n$ for each $n \in \mathcal{N}$.

c) Find all cumulants of X in case $X \sim \mathcal{N}(\mu, \sigma^2)$; $X \sim \mathcal{P}(\lambda)$; $X \sim \mathcal{E}(\lambda)$.

11.108 Suppose that $X_1, X_2, \ldots$ are independent $\mathcal{U}(0, 1)$ random variables. Let N denote the first positive integer n such that $X_1 + \cdots + X_n$ exceeds 1.

a) Obtain the MGF of N.

b) Use your result from part (a) to obtain the mean and variance of N.

11.109 Let $X_n \sim \mathcal{U}(-n, n)$ for each $n \in \mathcal{N}$.

a) Heuristically, does $\{X_n\}_{n=1}^{\infty}$ converge in distribution to some random variable X? Explain your answer.

b) Determine $\lim_{n \to \infty} F_{X_n}(x)$ for each $x \in \mathcal{R}$.

c) Relate your answers in parts (a) and (b).

11.110 Let N be a nonnegative-integer valued random variable.

a) Find the relationship between the MGF and PGF (probability generating function) of N. *Note:* Probability generating functions were introduced in Exercise 7.52 on page 352.

b) Obtain formulas for the mean and variance of N in terms of its PGF.

11.111 Refer to Exercise 11.110. Let $X_1, X_2, \ldots$ be independent random variables having the same probability distribution as a random variable X with mean μ and variance σ^2. And let N be a nonnegative-integer valued random variable, independent of the X_js. Set $S_0 = 0$ and $S_n = X_1 + \cdots + X_n$ for $n \geq 1$. Then S_N is a random variable obtained as the sum of a random number of random variables.

a) Obtain the MGF of S_N in terms of the PGF of N and the MGF of X. *Hint:* Condition on N and apply the law of total expectation.

b) Use the result of part (a) to obtain the mean and variance of S_N in terms of the means and variances of N and X.

11.112 Let $X_1, \ldots, X_n$ be a random sample from a $\mathcal{N}(\mu, \sigma^2)$ distribution. In Section 12.4, we show that $\bar{X}_n$ is independent of the random variables $X_1 - \bar{X}_n, \ldots, X_n - \bar{X}_n$. Here, use MGFs to show that result is true when $n = 2$, $\mu = 0$, and $\sigma^2 = 1$.

11.113 Suppose that $X_1, X_2, \ldots$ are random variables defined on the same sample space, that they have uniformly bounded variances, and that

$$\lim_{n \to \infty} \frac{1}{n} \sum_{k=1}^{n} \text{Cov}(X_k, X_n) = 0. \tag{$*$}$$

a) Prove that $X_1, X_2, \ldots$ satisfy the weak law of large numbers—that is, for each $\epsilon > 0$,

$$\lim_{n \to \infty} P\left(\left| \frac{X_1 + \cdots + X_n}{n} - \frac{\mathcal{E}(X_1) + \cdots + \mathcal{E}(X_n)}{n} \right| < \epsilon \right) = 1.$$

Hint: Assume the following result, known as *Toeplitz's lemma:* If $\sum_{n=1}^{\infty} a_n$ is a divergent series of positive real numbers and $\{s_n\}_{n=1}^{\infty}$ is a convergent sequence of real numbers, converging to s, then $\sum_{k=1}^{n} a_k s_k / \sum_{k=1}^{n} a_k \to s$ as $n \to \infty$.

b) By definition, random variables $X_1, X_2, \ldots$, defined on the same sample space, are *asymptotically uncorrelated* if $\text{Cov}(X_i, X_j) \to 0$ as $|i - j| \to \infty$. Prove that asymptotically uncorrelated random variables with uniformly bounded variances satisfy Equation $(*)$ and hence the weak law of large numbers. *Hint:* Use the fact that the correlation coefficient of two random variables is at most 1 in absolute value.

Characteristic functions and inversion formulas: Refer to the discussion of characteristic functions (CHFs) on page 641. An important application of characteristic functions is in obtaining *inversion formulas*—that is, formulas that express a PMF or PDF in terms of its CHF.

11.114 Let X be an integer-valued random variable.
a) Let $k \in \mathcal{Z}$. Show that $\int_{-\pi}^{\pi} e^{ikt}\, dt$ equals 2π if $k = 0$ and equals 0 otherwise.
b) Use part (a) to prove that

$$p_X(x) = \frac{1}{2\pi} \int_{-\pi}^{\pi} e^{-ixt} \phi_X(t)\, dt, \qquad x \in \mathcal{Z}.$$

Note: You may assume that it's permissible to interchange the order of integration and summation.
c) Let $X_1, \ldots, X_m$ be independent and identically distributed integer-valued random variables with common CHF ϕ. Show that

$$p_{X_1 + \cdots + X_m}(x) = \frac{1}{2\pi} \int_{-\pi}^{\pi} e^{-ixt} \big(\phi(t)\big)^m\, dt, \qquad x \in \mathcal{Z}.$$

11.115 Let X be a random variable whose CHF is integrable, that is, $\int_{-\infty}^{\infty} |\phi_X(t)|\, dt < \infty$. It can be proved that, under those circumstances, X is a continuous random variable with PDF given by the inversion formula

$$f_X(x) = \frac{1}{2\pi} \int_{-\infty}^{\infty} e^{-ixt} \phi_X(t)\, dt, \qquad x \in \mathcal{R}. \tag{$**$}$$

a) Show directly that Equation $(**)$ holds when $X \sim \mathcal{N}(0, \sigma^2)$.
b) Let X be any random variable, let c be a positive constant, and let Z be independent of X and have the standard normal distribution. Show that $X + cZ$ is a continuous random variable with PDF given by

$$f_{X+cZ}(y) = \frac{1}{2\pi} \int_{-\infty}^{\infty} e^{-ity} \phi_X(t) e^{-c^2 t^2 / 2}\, dt, \qquad y \in \mathcal{R}.$$

11.116 Let Y have PDF given by $f_Y(y) = (1/2)e^{-|y|}$ for all $y \in \mathcal{R}$.
a) Show that $\phi_Y(t) = 1/(1 + t^2)$ for all $t \in \mathcal{R}$.
b) Let X have the standard Cauchy distribution. Use part (a) and Equation $(**)$ to show that $\phi_X(t) = e^{-|t|}$ for all $t \in \mathcal{R}$.

Sir Ronald Fisher *1890—1962*

Ronald Fisher was born on February 17, 1890, in London, England; he was a surviving twin in a family of eight children; his father was a prominent auctioneer. Fisher graduated from Cambridge in 1912 with degrees in mathematics and physics.

From 1912 to 1919, Fisher worked at an investment house, did farm chores in Canada, and taught high school. In 1919, he took a position as a statistician at Rothamsted Experimental Station in Harpenden, West Hertford, England. His charge: sort and reassess 66 years of data on manurial field trials and weather records.

Fisher's work at Rothamsted during the next 15 years earned him the reputation as the leading statistician of his day and as a top-ranking geneticist. It was there, in 1925, that he published *Statistics for Research Workers*, a book that remained in print for 50 years. Fisher made important contributions to analysis of variance (ANOVA), exact tests of significance for small samples, and maximum-likelihood solutions. He developed experimental designs to address issues in biological research, such as small samples, variable materials, and fluctuating environments.

Fisher has been described as "slight, bearded, eloquent, reactionary, and quirkish; genial to his disciples and hostile to his dissenters." He was also a prolific writer—over a span of 50 years, he wrote an average of one paper every 2 months!

Due to his many significant contributions, Fisher is considered one of the founders of modern statistics. He was elected a Fellow of the Royal Society (1929) and, subsequently, received the following awards: Royal Medal of the Society (1938); Darwin Medal of the Society (1948); and the Copley Medal of the Royal Society (1955).

In 1933, Fisher became Galton professor of Eugenics at University College in London and, in 1943, Balfour professor of genetics at Cambridge. In 1952, he was knighted. Fisher "retired" in 1959, moved to Australia, and spent the last 3 years of his life working at the Division of Mathematical Statistics of the Commonwealth Scientific and Industrial Research Organization. He died in 1962 in Adelaide, Australia.

CHAPTER TWELVE

Applications of Probability Theory

INTRODUCTION

In this chapter, we present some of the myriad applications of probability theory. Several choices are available for the selection of application areas; we have chosen stochastic processes, operations research, and statistics. And, from among the many possible topics in each of these three application areas, we have selected the Poisson process, basic queueing theory, and the multivariate normal distribution and sampling distributions, respectively.

In Section 12.1, we introduce the *Poisson process*, one of the most basic and most important stochastic processes. The Poisson process is used extensively as a model for the number of occurrences of a specified event in continuous time.

As a representative topic of the field of operations research, we present, in Section 12.2, some basic models in the theory of queues. The models we discuss all employ the Poisson process for the arrival process and the exponential distribution for the service process.

The *multivariate normal distribution*, which we examine in Section 12.3, extends the bivariate normal distribution of Section 10.5 to the general multivariate setting. In Section 12.4, we discuss some of the most important *sampling distributions* occurring in statistics, including the *sampling distribution of the sample mean* and the *sampling distribution of the ratio of two independent sample variances*.

12.1 The Poisson Process

For a specified event that occurs randomly in continuous time, an important application of probability theory is in modeling the number of times such an event occurs. The following are several examples of such random phenomena.

- The number of patients that arrive at a hospital emergency room. Here the specified event is the arrival of a patient at the hospital emergency room.
- The number of customers that enter at a bank. Here the specified event is a customer entering the bank.
- The number of accidents that occur on a particular stretch of highway. Here the specified event is the occurrence of an accident on the stretch of highway.
- The number of alpha particles emitted by a radioactive substance. Here the specified event is the emission of an alpha particle by the radioactive substance.

For each real number $t \geq 0$, let $N(t)$ denote the number of times a specified event occurs by time t. Note that $N(t)$ is a nonnegative-integer valued random variable for each $t \geq 0$. The collection of random variables $\{N(t) : t \geq 0\}$ is called a **counting process** because it counts the number of times the specified event occurs in time. Here we consider the *Poisson process,* one of the most widely used counting processes.

Introducing the Poisson Process

Consider a specified event that occurs randomly and homogeneously in continuous time at an average rate of, say, λ per unit time. As previously, it's convenient to refer to the occurrence of the specified event as a **success.**

We begin counting successes at time 0 and, for each $t \geq 0$, let

$$N(t) = \text{number of successes by time } t. \qquad (12.1)$$

We note that $\{N(t) : t \geq 0\}$ is a counting process. Because we begin counting successes at time 0,

$$N(0) = 0. \qquad (12.2)$$

We also note that

$$N(t) - N(s) = \text{number of successes in the time interval } (s, t], \qquad (12.3)$$

for $0 \leq s < t < \infty$.

To model the counting process $\{N(t) : t \geq 0\}$, we consider Bernoulli trials. We recall that, in Bernoulli trials, a specified event (success) occurs randomly and homogeneously in discrete time. Thus, if we consider Bernoulli trials in which successes occur at the same average rate as in the continuous-time process—λ per unit of continuous time—and if discrete time is quantized finely enough, then the number of successes in the Bernoulli-trials process should behave essentially like $\{N(t) : t \geq 0\}$. More precisely, if Bernoulli trials with success probability λ/n (see Exercise 12.1) are performed at times $\frac{1}{n}, \frac{2}{n}, \ldots$ (i.e., time is quantized into units of duration $\frac{1}{n}$) then, for large n, the Bernoulli-trials process and the continuous-time process should be roughly the same.

Because Bernoulli trials are independent, the numbers of successes in disjoint time intervals are independent random variables. Thus, as a first consequence of our Bernoulli-trials approach, we can reasonably impose the condition of **independent increments** on our continuous-time counting process: For each $r \in \mathcal{N}$ and $0 \leq t_1 < t_2 < \cdots < t_r$,

$$N(t_j) - N(t_{j-1}), \quad 2 \leq j \leq r, \quad \text{are independent random variables.} \quad \textbf{(12.4)}$$

Our next goal is to identify, for each $0 \leq s < t < \infty$, the probability distribution of the random variable $N(t) - N(s)$, the number of successes in the time interval $(s, t]$. We first observe that, if time is quantized into units of duration $\frac{1}{n}$, then n Bernoulli trials are performed per unit of continuous time. Consequently, if we let p denote the success probability, the number of successes per unit of continuous time has the binomial distribution with parameters n and p.

We then conclude that the expected number of successes per unit of continuous time is np, which must also be the average rate of occurrence, λ. Hence, for each n, we must have $np = \lambda$, which means that the success probability is directly proportional to the duration of quantized time with proportionality constant λ, or

$$p = \frac{\lambda}{n}, \quad n \in \mathcal{N}. \quad \textbf{(12.5)}$$

We can now provide a heuristic argument to identify the probability distribution of the random variable $N(t) - N(s)$. To begin, we set $k = \lfloor ns \rfloor + 1$ and $m = \lfloor nt \rfloor - \lfloor ns \rfloor$, where $\lfloor x \rfloor$ denotes the *floor function*—the greatest integer smaller than or equal to x. Figure 12.1 shows the relationship between s, t, k, and m. See Exercise 12.2 for details.

Figure 12.1 Graph showing the relationship between $s, t, k = \lfloor ns \rfloor + 1$, and $m = \lfloor nt \rfloor - \lfloor ns \rfloor$

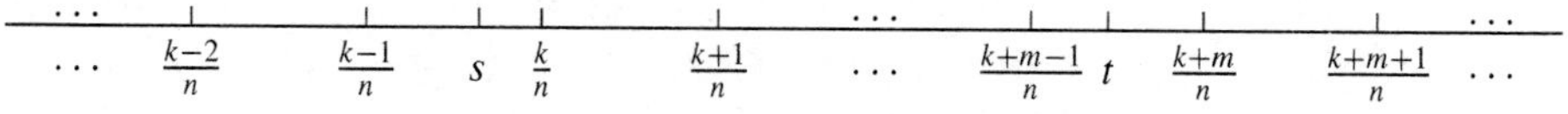

We note that m Bernoulli trials are performed between times s and t. Now we assume that n is large. From Equation (12.5), we see that $np^2 = \lambda^2/n$ and, hence, np^2 is small. Noting also that $m \approx n(t - s)$, we conclude from Proposition 5.7 on page 220 (Poisson approximation to the binomial) and Equation (12.5) that

$$P\big(N(t) - N(s) = j\big) \approx \binom{m}{j} p^j (1-p)^{m-j} \approx e^{-mp} \frac{(mp)^j}{j!}$$

$$\approx e^{-n(t-s)p} \frac{(n(t-s)p)^j}{j!} = e^{-\lambda(t-s)} \frac{(\lambda(t-s))^j}{j!}.$$

Hence we have shown heuristically that

$$N(t) - N(s) \sim \mathcal{P}\big(\lambda(t-s)\big), \quad 0 \leq s < t < \infty. \quad \textbf{(12.6)}$$

That is, for $0 \leq s < t < \infty$, the random variable $N(t) - N(s)$ has the Poisson distribution with parameter $\lambda(t - s)$.

In view of Relations (12.2), (12.4), and (12.6), we now define a **Poisson process.**

> **DEFINITION 12.1   Poisson Process**
>
> A counting process $\{N(t) : t \geq 0\}$ is said to be a **Poisson process with rate** λ if the following three conditions hold:
> a) $N(0) = 0$.
> b) $\{N(t) : t \geq 0\}$ has independent increments.
> c) $N(t) - N(s) \sim \mathcal{P}(\lambda(t - s))$ for $0 \leq s < t < \infty$.

Condition (c) indicates that the probability distribution of $N(t) - N(s)$ depends only on the difference between s and t. Any process with that property is said to have **stationary increments.**

Note: The definition of a Poisson process can be based on conditions different from, but equivalent to, those given in Definition 12.1. You are asked to investigate one set of equivalent conditions in Exercises 12.111 and 12.112.

EXAMPLE 12.1   *The Poisson Process*

Emergency Room Traffic   Desert Samaritan Hospital in Mesa, Arizona, keeps records of its emergency room traffic. Those records show that, on average, 6.9 patients arrive per hour and, moreover, that the number of arriving patients constitutes a Poisson process. For a given day, starting at midnight,
a) find the probability that exactly 12 patients arrive between 6:00 P.M. and 8:00 P.M.
b) given that exactly five patients arrive between 4:00 P.M. and 5:00 P.M., determine the probability that exactly 12 patients arrive between 6:00 P.M. and 8:00 P.M.
c) obtain the probability that at least one patient arrives in each of the time intervals 4:00 P.M.–5:00 P.M. and 6:00 P.M.–8:00 P.M.
d) determine the probability distribution of the number of patients that arrive either between 4:00 P.M. and 5:00 P.M. or between 6:00 P.M. and 8:00 P.M.

Solution   Let $N(t)$ denote the number of patients that arrive by time t. By assumption, we know that $\{N(t) : t \geq 0\}$ is a Poisson process with rate $\lambda = 6.9$ patients per hour. Note that time is being measured in hours starting at midnight; so, for instance, 6:00 P.M. is time 18.
a) Referring to Definition 12.1(c), we see that the number of patients that arrive between 6:00 P.M. and 8:00 P.M., which is $N(20) - N(18)$, has the Poisson distribution with parameter $6.9 \cdot (20 - 18) = 13.8$. Therefore, the probability that exactly 12 patients arrive between 6:00 P.M. and 8:00 P.M. is

$$P\big(N(20) - N(18) = 12\big) = e^{-13.8} \frac{(13.8)^{12}}{12!} = 0.101,$$

or roughly a 10% chance.
b) Referring to Definition 12.1(b), we know that the numbers of patients that arrive in disjoint time intervals are independent random variables. Hence the conditional probability that exactly 12 patients arrive between 6:00 P.M. and 8:00 P.M., given that

exactly five patients arrive between 4:00 P.M. and 5:00 P.M., equals the (unconditional) probability that exactly 12 patients arrive between 6:00 P.M. and 8:00 P.M. By part (a), that probability is 0.101.

c) Referring to Definition 12.1(c), the number of patients that arrive between 4:00 P.M. and 5:00 P.M., which is $N(17) - N(16)$, has the Poisson distribution with parameter $6.9 \cdot (17 - 16) = 6.9$; and the number of patients that arrive between 6:00 P.M. and 8:00 P.M., which is $N(20) - N(18)$, has the Poisson distribution with parameter $6.9 \cdot (20 - 18) = 13.8$. Furthermore, because the two time intervals are disjoint, Definition 12.1(b) implies that $N(17) - N(16)$ and $N(20) - N(18)$ are independent random variables. Consequently, the probability that at least one patient arrives in each of the time intervals 4:00 P.M.–5:00 P.M. and 6:00 P.M.–8:00 P.M. is

$$P\big(N(17) - N(16) \geq 1, \; N(20) - N(18) \geq 1\big)$$
$$= P\big(N(17) - N(16) \geq 1\big) P\big(N(20) - N(18) \geq 1\big)$$
$$= \Big(1 - P\big(N(17) - N(16) = 0\big)\Big)\Big(1 - P\big(N(20) - N(18) = 0\big)\Big)$$
$$= \left(1 - e^{-6.9} \frac{(6.9)^0}{0!}\right)\left(1 - e^{-13.8} \frac{(13.8)^0}{0!}\right) = 0.999,$$

or virtually a certainty.

d) The number of patients that arrive either between 4:00 P.M. and 5:00 P.M. or between 6:00 P.M. and 8:00 P.M. is the sum of $N(17) - N(16)$ and $N(20) - N(18)$. From part (c), we know that $N(17) - N(16)$ and $N(20) - N(18)$ are independent Poisson random variables with parameters 6.9 and 13.8, respectively. In view of Proposition 6.20(b) on page 311, the sum of those two random variables has the Poisson distribution with parameter $6.9 + 13.8 = 20.7$.   ■

Waiting and Interarrival Times

Two other important quantities in the study of counting processes are *waiting times* and *interarrival times*. For each $n \in \mathcal{N}$, we let W_n denote the time of the occurrence of the nth event—that is, the time at which the nth success occurs. The random variable W_n is called the nth **waiting time.** The elapsed time between the occurrence of the $(n - 1)$st and nth events is denoted I_n and is called the nth **interarrival time.** Clearly,

$$W_n = I_1 + \cdots + I_n \qquad \text{and} \qquad I_n = W_n - W_{n-1}, \tag{12.7}$$

for each $n \in \mathcal{N}$, where we define $W_0 = 0$. Proposition 12.1 provides the probability distributions of the waiting times for a Poisson process.

◆◆◆   **Proposition 12.1   Waiting-Time Distributions for a Poisson Process**

Let $\{N(t) : t \geq 0\}$ be a Poisson process with rate λ. Then the nth waiting time, W_n, has the Erlang distribution with parameters n and λ.

Proof   Let $t > 0$. For the nth waiting time to exceed t means that the nth success occurs after time t, which happens if and only if the number of successes by time t is at most $n - 1$;

symbolically, $\{W_n > t\} = \{N(t) \leq n - 1\}$. Knowing that $\{N(t) : t \geq 0\}$ is a Poisson process with rate λ and applying the FPF, we get

$$P(W_n > t) = P\big(N(t) \leq n - 1\big) = \sum_{k=0}^{n-1} e^{-\lambda t} \frac{(\lambda t)^k}{k!},$$

or, equivalently,

$$F_{W_n}(t) = 1 - e^{-\lambda t} \sum_{k=0}^{n-1} \frac{(\lambda t)^k}{k!}. \tag{12.8}$$

Referring now to the note in the middle of page 454, we conclude that $W_n \sim \Gamma(n, \lambda)$ or, equivalently, that W_n has the Erlang distribution with parameters n and λ.   ◆

From Proposition 12.1, W_1 has the Erlang distribution with parameters 1 and λ, which is also the exponential distribution with parameter λ. As $I_1 = W_1$, we see that the first interarrival time has the exponential distribution with parameter λ. Our next task is to show, in fact, that all of the interarrival times have the exponential distribution with parameter λ and that they are independent random variables. To that end, we first obtain a joint PDF of the first n waiting times. Although a rigorous derivation is beyond the scope of this book, the following heuristic development provides the required result.

Let $0 < t_1 < \cdots < t_n$, and let $\Delta t_1, \ldots, \Delta t_n$ represent small positive numbers. The event $\{t_1 \leq W_1 \leq t_1 + \Delta t_1, \ldots, t_n \leq W_n \leq t_n + \Delta t_n\}$ occurs if and only if no successes occur in the time interval from 0 to t_1, one success occurs in the time interval from t_1 to $t_1 + \Delta t_1$, no successes occur in the time interval from $t_1 + \Delta t_1$ to t_2, one success occurs in the time interval from t_2 to $t_2 + \Delta t_2$, $\ldots$, no successes occur in the time interval from $t_{n-1} + \Delta t_{n-1}$ to t_n, and at least one success occurs in the time interval from t_n to $t_n + \Delta t_n$. See Figure 12.2.

Figure 12.2 Graphical portrayal of the event $\{t_1 \leq W_1 \leq t_1 + \Delta t_1, \ldots, t_n \leq W_n \leq t_n + \Delta t_n\}$

Successes:	0	1	0	1	$\cdots$	≥ 1

Time:	0	t_1 $t_1 + \Delta t_1$	t_2 $t_2 + \Delta t_2$	$\cdots$	t_n $t_n + \Delta t_n$

Referring to Figure 12.2 and applying the properties of a Poisson process as given in Definition 12.1, we find that

$$P(t_1 \leq W_1 \leq t_1 + \Delta t_1, \ldots, t_n \leq W_n \leq t_n + \Delta t_n)$$

$$= P\big(N(t_1) = 0, \, N(t_1 + \Delta t_1) - N(t_1) = 1, \, N(t_2) - N(t_1 + \Delta t_1) = 0,$$

$$\ldots, \, N(t_n) - N(t_{n-1} + \Delta t_{n-1}) = 0, \, N(t_n + \Delta t_n) - N(t_n) \geq 1\big)$$

$$= P\big(N(t_1) = 0\big) P\big(N(t_1 + \Delta t_1) - N(t_1) = 1\big) P\big(N(t_2) - N(t_1 + \Delta t_1) = 0\big)$$

$$\cdots P\big(N(t_n) - N(t_{n-1} + \Delta t_{n-1}) = 0\big) P\big(N(t_n + \Delta t_n) - N(t_n) \geq 1\big)$$

$$= e^{-\lambda t_1} \cdot e^{-\lambda \Delta t_1} \lambda \Delta t_1 \cdot e^{-\lambda(t_2 - t_1 - \Delta t_1)} \cdots e^{-\lambda(t_n - t_{n-1} - \Delta t_{n-1})} \cdot (1 - e^{-\lambda \Delta t_n}).$$

Simplifying the previous expression and using Relation (5.18) on page 219, we get

$$P(t_1 \leq W_1 \leq t_1 + \Delta t_1, \ldots, t_n \leq W_n \leq t_n + \Delta t_n) \approx \lambda^n e^{-\lambda t_n} \Delta t_1 \cdots \Delta t_n.$$

Therefore a joint PDF of the random variables $W_1, \ldots, W_n$ is given by

$$f_{W_1, \ldots, W_n}(t_1, \ldots, t_n) = \lambda^n e^{-\lambda t_n}, \qquad 0 < t_1 < \cdots < t_n, \qquad \textbf{(12.9)}$$

and $f_{W_1, \ldots, W_n}(t_1, \ldots, t_n) = 0$ otherwise.

With Equation (12.9) in mind, we now obtain in Proposition 12.2 a fundamental result concerning the interarrival times of a Poisson process.

◆◆◆   Proposition 12.2   **Interarrival-Time Distributions for a Poisson Process**

Let $\{N(t) : t \geq 0\}$ be a Poisson process with rate λ. Then the interarrival times, $I_1, I_2, \ldots,$ are independent and identically distributed exponential random variables with common parameter λ.

Proof   We recall that $I_1 = W_1$ and $I_n = W_n - W_{n-1}$ for $n \geq 2$. As we ask you to show in Exercise 12.15, the Jacobian determinant of the transformation $t_1 = w_1, t_2 = w_2 - w_1,$ $\ldots, t_n = w_n - w_{n-1}$ is identically 1—that is, $J(w_1, \ldots, w_n) = 1$. Solving the equations $t_1 = w_1, t_2 = w_2 - w_1, \ldots, t_n = w_n - w_{n-1}$ for $w_1, w_2, \ldots, w_n$, yields the inverse transformation $w_1 = t_1, w_2 = t_1 + t_2, \ldots, w_n = t_1 + \cdots + t_n$. Applying the multivariate transformation theorem and Equation (12.9), we obtain, for positive $t_1, t_2, \ldots, t_n$,

$$
\begin{aligned}
f_{I_1, I_2, \ldots, I_n}(t_1, t_2, \ldots, t_n) &= \frac{1}{|J(w_1, w_2, \ldots, w_n)|} f_{W_1, W_2, \ldots, W_n}(w_1, w_2, \ldots, w_n) \\
&= f_{W_1, W_2, \ldots, W_n}(t_1, t_1 + t_2, \ldots, t_1 + t_2 + \cdots + t_n) \\
&= \lambda^n e^{-\lambda(t_1 + t_2 + \cdots + t_n)}.
\end{aligned}
$$

Thus a joint PDF of $I_1, \ldots, I_n$ is

$$f_{I_1, \ldots, I_n}(t_1, \ldots, t_n) = \lambda e^{-\lambda t_1} \cdots \lambda e^{-\lambda t_n}, \qquad t_1, \ldots, t_n > 0, \qquad \textbf{(12.10)}$$

and $f_{I_1, \ldots, I_n}(t_1, \ldots, t_n) = 0$ otherwise. As you are asked to show in Exercise 12.15, it follows easily from Equation (12.10) that $I_1, \ldots, I_n$ are independent and identically distributed exponential random variables with common parameter λ. And from that result, we conclude that the proposition holds.   ◆

EXAMPLE 12.2   *Waiting and Interarrival Times*

Emergency Room Traffic   Refer to Example 12.1 on page 688.

a) Identify the probability distribution of the arrival time of the third patient and obtain its PDF and CDF.

b) Find the probability that the third patient arrives between 12:30 A.M. and 1:00 A.M.

c) Identify the probability distribution of the elapsed time between the arrivals of the third and fourth patients. Obtain its PDF and CDF.

d) Find the probability that the elapsed time between the arrivals of the third and fourth patients exceeds 10 minutes.

e) Find the probability that the elapsed time between the arrivals of the third and fourth patients exceeds 10 minutes, given that the third patient arrives between 12:30 A.M. and 1:00 A.M.

f) What is the expected elapsed time between successive arriving patients?

Solution   Let $N(t)$ denote the number of patients that arrive by time t. By assumption, we know that $\{N(t) : t \geq 0\}$ is a Poisson process with rate $\lambda = 6.9$ patients per hour. Note that time is being measured in hours starting at midnight; so, for instance, 6:00 P.M. is time 18.

a) Proposition 12.1 indicates that the arrival time of the third patient, W_3, has an Erlang distribution with parameters 3 and 6.9 or, equivalently, a $\Gamma(3, 6.9)$ distribution. A PDF therefore is

$$f_{W_3}(t) = \frac{(6.9)^3}{\Gamma(3)} t^{3-1} e^{-6.9t} = 164.2545 t^2 e^{-6.9t}, \qquad t > 0.$$

From Equation (12.8) on page 690, the CDF is

$$F_{W_3}(t) = 1 - e^{-6.9t} \sum_{k=0}^{2} \frac{(6.9t)^k}{k!} \qquad t \geq 0.$$

b) The probability that the third patient arrives between 12:30 A.M. and 1:00 A.M. is given by $P(0.5 \leq W_3 \leq 1)$. Referring to part (a), we conclude that

$$P(0.5 \leq W_3 \leq 1) = F_{W_3}(1) - F_{W_3}(0.5)$$

$$= e^{-6.9 \cdot 0.5} \sum_{k=0}^{2} \frac{(6.9 \cdot 0.5)^k}{k!} - e^{-6.9 \cdot 1} \sum_{k=0}^{2} \frac{(6.9 \cdot 1)^k}{k!} = 0.298.$$

c) From Proposition 12.2, the elapsed time between the arrivals of the third and fourth patients, I_4, has an exponential distribution with parameter 6.9. Therefore a PDF of I_4 is $f_{I_4}(t) = 6.9 e^{-6.9t}$ for $t > 0$, and $f_{I_4}(t) = 0$ otherwise; and the CDF of I_4 is $F_{I_4}(t) = 1 - e^{-6.9t}$ for $t \geq 0$, and $F_{I_4}(t) = 0$ otherwise.

d) The probability that the elapsed time between the arrivals of the third and fourth patients exceeds 10 minutes is $P(I_4 > 1/6)$. From part (c),

$$P(I_4 > 1/6) = 1 - P(I_4 \leq 1/6) = 1 - F_{I_4}(1/6) = e^{-6.9/6} = 0.317.$$

e) Given that the third patient arrives between 12:30 A.M. and 1:00 A.M., the probability that the elapsed time between the arrivals of the third and fourth patients exceeds 10 minutes is $P(I_4 > 1/6 \mid 0.5 \leq W_3 \leq 1)$. From Equation (12.7) on page 689, $W_3 = I_1 + I_2 + I_3$ and, from Proposition 12.2, $I_1, I_2, \ldots$ are independent random variables. Hence, I_4 and W_3 are also independent. Therefore, in view of part (d),

$$P(I_4 > 1/6 \mid 0.5 \leq W_3 \leq 1) = P(I_4 > 1/6) = 0.317.$$

f) The elapsed time between successive patient arrivals has the exponential distribution with parameter 6.9. Because the mean of an exponential random variable is the reciprocal of its parameter, the expected elapsed time between successive arriving patients is $\frac{1}{6.9}$ hour, or roughly 8.7 minutes. ∎

Nonhomogeneous Poisson Process

For a Poisson process with rate λ, we have $N(0) = 0$ and $N(t) \sim \mathcal{P}(\lambda t)$ for $t > 0$. The expected number of events (successes) that occur by time t is called the **mean function,** denoted μ; hence, $\mu(t) = \mathcal{E}(N(t))$ for $t \geq 0$. Because the expected value of a Poisson random variable equals its parameter, $\mu(t) = \lambda t$; thus the mean function is linear. The **intensity function,** denoted ν, is defined to be the derivative of the mean function. We have $\nu(t) = \mu'(t) = \lambda$; thus the intensity function is a constant equal to the rate.

For a Poisson process, the constant λ gives the rate per unit time at which the specified event occurs. In many applications, however, the rate at which a specified event occurs depends on time. To account for such phenomena, we can often use a **nonhomogeneous Poisson process**, where the intensity function ν depends on time and, hence, the mean function μ is nonlinear. Specifically, we obtain a nonhomogeneous Poisson process by retaining conditions (a) and (b) of Definition 12.1 and replacing condition (c) by $N(t) - N(s) \sim \mathcal{P}(\mu(t) - \mu(s))$, for $0 \leq s < t < \infty$, where $\mu(t) = \int_0^t \nu(s)\, ds$.

To distinguish a Poisson process from a nonhomogeneous Poisson process, the former is often referred to as a **homogeneous Poisson process.** Although a homogeneous Poisson process has stationary increments, a nonhomogeneous Poisson process doesn't, as you are asked to verify in Exercise 12.18.

Compound Poisson Process

Another generalization of a (homogeneous) Poisson process occurs when a random variable is associated with the specified event. For instance, the specified event might be a customer entering a bank for the purpose of making a deposit and the associated random variable could be the amount of money deposited by the customer. If the customer arrival-process is a Poisson process and deposit amounts among customers are independent and identically distributed random variables, also independent of the arrival process, then the total amount deposited by time t $(t \geq 0)$ is an example of a *compound Poisson process.*

In general, let $\{N(t) : t \geq 0\}$ be a homogeneous Poisson process and let $X_1, X_2, \ldots$ be independent and identically distributed random variables, which are also independent of $\{N(t) : t \geq 0\}$. Let $Y(t) = 0$ if $N(t) = 0$ and let $Y(t) = \sum_{k=1}^{N(t)} X_k$ if $N(t) \geq 1$. The process $\{Y(t) : t \geq 0\}$ is called a **compound Poisson process.**

EXERCISES 12.1 **Basic Exercises**

12.1 In our development of the Poisson process from a Bernoulli-trials process in which discrete time is quantized into units of duration $\frac{1}{n}$, we chose the success probability p so that successes occur at the same average rate as in the continuous-time process—λ per unit of continuous time. Use the strong law of large numbers (Theorem 11.2 on page 654) to show that this condition implies that $p \approx \lambda/n$ for large n, thus providing another verification of Equation (12.5) on page 687.

12.2 Refer to Figure 12.1 on page 687.
a) For $t > 0$, what does $\lfloor nt \rfloor$ represent in terms of the number of Bernoulli trials performed?

b) Verify the relationship between $s, t, k = \lfloor ns \rfloor + 1$, and $m = \lfloor nt \rfloor - \lfloor ns \rfloor$ shown in Figure 12.1.

c) Show that $m/n \to t - s$ as $n \to \infty$ and, thereby, justify the assertion on page 687 that $m \approx n(t - s)$ for large n.

12.3 Refer to the emergency-room illustration in Example 12.1 on page 688. Determine the probability that

a) exactly 7 patients arrive between 4:00 A.M. and 5:00 A.M.

b) exactly 14 patients arrive between 6:00 A.M. and 8:00 A.M.

c) exactly 7 patients arrive between 4:00 A.M. and 5:00 A.M. and exactly 14 patients arrive between 6:00 A.M. and 8:00 A.M.

d) either exactly 7 patients arrive between 4:00 A.M. and 5:00 A.M. or exactly 14 patients arrive between 6:00 A.M. and 8:00 A.M. (or both).

e) exactly 20 patients arrive between 4:00 A.M. and 7:00 A.M.

f) Obtain an expression for the probability that exactly 20 patients arrive between 4:00 A.M. and 7:00 A.M. and that exactly 14 patients arrive between 6:00 A.M. and 8:00 A.M. *Hint:* Use the law of partitions (Proposition 2.8 on page 68).

g) If you have access to a scientific calculator, evaluate the expression in part (f).

h) Find the probability that either exactly 20 patients arrive between 4:00 A.M. and 7:00 A.M. or exactly 14 patients arrive between 6:00 A.M. and 8:00 A.M. (or both).

12.4 Refer to the emergency-room illustration in Example 12.1 on page 688. Determine the probability distribution of the number of patients that arrive

a) either between 4:00 A.M. and 5:00 A.M. or between 6:00 A.M. and 8:00 A.M.

b) either between 4:00 A.M. and 7:00 A.M. or between 6:00 A.M. and 8:00 A.M.

12.5 Queries to a computer database constitute a Poisson process and occur at an average rate of four per minute.

a) For $0 \le s < t$, identify the probability distribution of the number of queries that occur between times s and t.

b) On average, how many queries occur during a 2-minute interval?

c) Determine the probability that between three and five queries, inclusive, occur during a 2-minute interval.

d) What is the probability distribution of the elapsed time between successive queries?

e) On average, how long does it take between successive queries?

f) What is the probability distribution of the time elapsed until the nth query?

g) On average, how long does it take for the nth query to occur?

h) On average, how long does it take between the third and fifth queries?

i) Determine the probability that exactly five queries occur in the first minute and exactly seven queries occur in the next 2 minutes.

12.6 Customers arrive at a service facility according to a Poisson process, $\{N(t) : t \ge 0\}$, with rate λ. Each customer is, independent of all other customers, of Type 1 with probability p and of Type 2 with probability $1 - p$. Let $\{N_1(t) : t \ge 0\}$ and $\{N_2(t) : t \ge 0\}$ be the arrival processes of Type 1 and Type 2 customers, respectively.

a) For fixed $t > 0$, show that $N_1(t)$ and $N_2(t)$ are independent Poisson random variables with parameters $p\lambda t$ and $(1 - p)\lambda t$, respectively.

b) Show that $\{N_1(t) : t \ge 0\}$ and $\{N_2(t) : t \ge 0\}$ are independent Poisson processes with rates $p\lambda$ and $(1 - p)\lambda$, respectively.

12.7 Splitting a Poisson process: Generalize the results of Exercise 12.6 as follows. Suppose that $\{N(t) : t \ge 0\}$ is a Poisson process with rate λ. Further suppose that each time the specified event occurs, it's classified as having one of m mutually exclusive attributes—say, $a_1, \ldots, a_m$ with probabilities $p_1, \ldots, p_m$, respectively—independent of other occurrences of the specified event. For each $j = 1, \ldots, m$, let $N_j(t)$ denote the number of times that a specified event with attribute a_j occurs by time t.

a) For fixed $t > 0$, show that $N_1(t), \ldots, N_m(t)$ are independent Poisson random variables with parameters $p_1\lambda t, \ldots, p_m\lambda t$, respectively.

b) Show that $\{N_1(t) : t \geq 0\}, \ldots, \{N_m(t) : t \geq 0\}$ are independent Poisson processes with rates $p_1\lambda, \ldots, p_m\lambda$, respectively.

12.8 Verify in detail the argument leading to Equation (12.9) on page 691.

12.9 In this exercise, you are to consider the simulation of a Poisson process.

a) Explain how a basic random number generator—that is, a random number generator that simulates a $\mathcal{U}(0, 1)$ random variable—can be used to simulate a Poisson process with rate λ up to and including the occurrence of the nth event, where n is a positive integer. *Hint:* Refer to Example 8.22 on page 472.

b) Simulation: This part requires access to statistical software. Apply your procedure from part (a) to obtain 10 observations of a Poisson process with rate 6.9 up to the fifth success. For each observation, provide a graph of $N(t)$ against t.

12.10 Let $\{N(t) : t \geq 0\}$ be a Poisson process with rate λ and let $0 < s < t$. Determine and identify the conditional probability distribution of the number of events that occur by time s, given that exactly n events occur by time t.

12.11 A restaurant is open daily from 10:00 A.M. to 10:00 P.M. Let time be measured relative to the number of hours after midnight (i.e., military time). Customers arrive at a rate of $4t - 40$ for $10 \leq t < 12$, $-4t + 56$ for $12 \leq t < 14$, $5t - 70$ for $14 \leq t < 18$, and $-5t + 110$ for $18 \leq t < 22$. Furthermore, the arrival process constitutes a nonhomogeneous Poisson process.

a) Determine and graph the intensity function for a 24-hour period starting at midnight.

b) Determine and graph the mean function for a 24-hour period starting at midnight.

c) On average, how many customers arrive at the restaurant per day?

d) What is the mean and standard deviation of the number of customers that arrive between 11:00 A.M. and 7:00 P.M.?

e) Find the probability that between 13 and 15 customers, inclusive, arrive at the restaurant between 11:00 A.M. and 1:00 P.M.

12.12 Refer to the discussion of a compound Poisson process on page 693. By appropriate choice of the random variables $X_1, X_2, \ldots$, show that a compound Poisson process does indeed generalize the concept of a (homogeneous) Poisson process.

12.13 Let $\{Y(t) : t \geq 0\}$ be a compound Poisson process. Suppose that the random variables $X_1, X_2, \ldots$ have finite mean μ and variance σ^2.

a) Obtain the mean function, $\mu(t) = \mathcal{E}(Y(t))$, and the variance function, $\sigma^2(t) = \text{Var}(Y(t))$.

b) Find the correlation coefficient of $N(t)$ and $Y(t)$.

12.14 Customers entering a bank for the purpose of making a deposit arrive according to a Poisson process at an average rate of 25.8 per hour. The amounts deposited by customers are independent and identically distributed random variables with mean \$574 and standard deviation \$3167.

a) What is the expected amount of total deposits by time t? *Hint:* Refer to Exercise 12.13.

b) What is the standard deviation of total deposit amounts by time t?

Theory Exercises

12.15 Complete the proof of Proposition 12.2 on page 691 by establishing each of the following results.

a) Show that the transformation, $t_1 = w_1, t_2 = w_2 - w_1, \ldots, t_n = w_n - w_{n-1}$, has Jacobian determinant identically equal to 1.

b) Show that Equation (12.10) on page 691 implies that $I_1, \ldots, I_n$ are independent and identically distributed exponential random variables with common parameter λ.

c) Show that part (b) implies that $I_1, I_2, \ldots$ are independent and identically distributed exponential random variables with common parameter λ.

Advanced Exercises

12.16 Let $\{N(t) : t \geq 0\}$ be a Poisson process with rate λ.

a) Determine and identify the conditional distribution of W_1 given $N(t) = 1$ by obtaining its conditional CDF.

b) Determine and identify the conditional distribution of W_1 given $N(t) = 1$ by using a heuristic argument similar to that used in obtaining Equation (12.9) on page 691.

c) Determine and identify the joint conditional distribution of W_1 and W_2 given $N(t) = 2$ by first obtaining its conditional joint CDF.

d) Determine and identify the joint conditional distribution of W_1 and W_2 given $N(t) = 2$ by using a heuristic argument similar to that used in obtaining Equation (12.9).

e) Determine a conditional joint PDF of $W_1, \ldots, W_n$ given $N(t) = n$ by using a heuristic argument similar to that used in obtaining Equation (12.9).

f) Explain why part (e) shows that, given $N(t) = n$, the times at which the n events occur, considered as unordered random variables, are independent and uniformly distributed on the interval $(0, t)$. *Hint:* Refer to Exercise 9.34 on page 501.

12.17 This exercise provides an alternative verification of Proposition 12.2 on page 691.

a) Prove that $P(I_2 > t \mid I_1 = s) = P(N(I_1 + t) - N(I_1) = 0 \mid I_1 = s)$ for all $s, t > 0$.

b) Explain why the value of I_1 depends only on the values of $\{N(u) : u \geq 0\}$ in the interval $0 \leq u \leq I_1$.

c) Deduce from parts (a) and (b) that I_1 and I_2 are independent exponential random variables with common parameter λ.

d) Prove that, for each positive integer n, $P(I_{n+1} > t \mid I_1 = s_1, \ldots, I_n = s_n) = e^{-\lambda t}$ for all $s_1, \ldots, s_n, t > 0$.

e) Conclude from part (d) that the random variables $I_1, I_2, \ldots$ are independent and identically distributed exponential random variables with common parameter λ.

12.18 Show that a nonhomogeneous Poisson process doesn't have stationary increments.

12.19 A small bank has one teller. Customers requiring service from the teller arrive at the bank according to a Poisson process with rate λ. The first customer who arrives goes immediately to the teller. The time it takes the teller to serve a customer is a random variable X, independent of customer arrivals. Let Y denote the number of customers waiting when the first customer completes service.

a) Use the law of total expectation to obtain $\mathcal{E}(Y)$.

b) Determine a formula for the PMF of Y in terms of f_X and λ. *Hint:* Refer to the discussion of mixed bivariate distributions on pages 509 and 522.

c) Use your result from part (b) to obtain $\mathcal{E}(Y)$. Compare your answer with that obtained in part (a).

d) Specialize your results for parts (a) and (b) when $X \sim \mathcal{E}(\mu)$.

12.2   Basic Queueing Theory

One of many important applications of probability theory is to **queueing theory,** which provides mathematical models for the analysis of **queueing systems.** Customers enter a queueing system to obtain service at a service facility. The basic queueing process is as follows.

1. Each arriving customer begins service immediately if a server is available; otherwise, the customer joins a **queue**—that is, a waiting line.

2. When a server becomes available, service begins for the customer. In this book, we assume that service is provided on a first-come-first-served (FCFS) basis.

3. Upon completion of service, the customer leaves the queueing system (and the next customer in the queue commences service).

Figure 12.3 illustrates the basic queueing process.

Figure 12.3 Basic queueing process

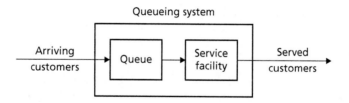

Note that the customers need not be people, nor need the servers be people. Here are some typical queueing systems.

- Patients (the customers) arrive at a hospital emergency room (the service facility) for the purpose of being treated (the service) by a doctor (the server).

- Airplanes (the customers) arrive at an airport (the service facility) for the purpose of landing (the service) on one of several runways (the servers).

- Cases (the customers) are logged into the judicial system (the service facility) for the purpose of being heard (the service) by one of many judges (the servers).

- Requests (the customers) arrive at a web site (the service facility) for the purpose of being processed (the service) by the web server (the server).

As you can imagine, there are countless types of queueing systems. Nonetheless, most of them can be modeled and analyzed by using the same fundamental principles. In this section, we examine some of the simplest types of queueing systems as applications of many of the concepts that you have studied in this book.

The M/M/1 Queue

Consider a queueing system in which there is one server—a **single-server queueing system.** Suppose that customers arrive at the queueing system according to a Poisson process with rate λ and that the successive times it takes the server to serve customers are independent exponential random variables with parameter μ, independent of arrivals as well.[†] Such a queueing system is called an ***M/M/1* queue.** From Proposition 12.2 on page 691, the interarrival times for a Poisson process with rate λ are independent exponential random variables with parameter λ. Thus we have the following definition.

DEFINITION 12.2   *M/M/1* Queueing System

For an ***M/M/1* queue:**
a) There is a single server.
b) Successive interarrival times are independent exponential random variables with parameter λ.
c) Successive service times are independent exponential random variables with parameter μ.
d) Arrival and service times are independent.

Thus, an $M/M/1$ queue is a single-server queueing system with **Poisson input** and **exponential service times.** An $M/M/1$ queue is a *single-server, infinite queueing system.*

Note: The "1" in "$M/M/1$" indicates that the service facility has one server; the "M"s stand for *memoryless* or *Markovian.* (Recall from Proposition 8.9 on page 434 that exponential random variables have the lack-of-memory property and are the only positive continuous random variables with that property.)

Because the mean of an exponential random variable is the reciprocal of its parameter, the mean time between arriving customers is $\frac{1}{\lambda}$ time units and the mean service time for each customer is $\frac{1}{\mu}$ time units. Consequently, the parameters λ and μ also represent rates: Arrivals occur at an average rate of λ customers per unit time and the server, when occupied, serves at an average rate of μ customers per unit time.

EXAMPLE 12.3   *The M/M/1 Queue*

Emergency Room Traffic   At Desert Samaritan Hospital, arrivals at the emergency room constitute a Poisson process with rate 6.9 patients per hour. Suppose that one doctor is on duty and independently spends an exponential amount of time with each patient, seeing 7.4 patients per hour, on average. Describe this emergency room as a queueing system.

[†] Although we have previously used μ to represent the mean of a random variable, standard terminology in queueing theory is to use μ to denote the service rate.

Solution   There is a single server, the one doctor. As patients (customers) arrive according to a Poisson process at a rate of 6.9 patients per hour, the successive interarrival times are independent exponential random variables with parameter 6.9. Successive service times (the times spent by the doctor with the patients) are independent exponential random variables with parameter 7.4, independent of arrivals as well. Hence, this emergency-room queueing system is an $M/M/1$ queue.

Because patients arrive at an average rate of 6.9 per hour, the mean time between arriving patients is $\frac{1}{6.9}$ hour, or roughly 8.7 minutes. And, because the doctor sees an average of 7.4 patients per hour, she spends a mean of $\frac{1}{7.4}$ hour with each patient, or roughly 8.1 minutes.   ■

Birth-and-Death Queueing Systems

We now examine more general queueing systems, called **birth-and-death queueing systems,** that have as a special case the $M/M/1$ queueing system. In doing so, we use the term **state** to refer to any specific number of customers in the queueing system. We can then describe birth-and-death queueing systems as those with Poisson input and exponential service times, but whose arrival and service rates may be state dependent.

DEFINITION 12.3   Birth-and-Death Queueing System

For a **birth-and-death queueing system:**
a) Successive interarrival times—the elapsed times from one arriving customer to the next—are independent exponential random variables; when the system is in state n, the exponential random variable has parameter λ_n.
b) Successive service times—the elapsed times from one service completion to the next—are independent exponential random variables; when the system is in state n, the exponential random variable has parameter μ_n.
c) Arrival and service times are independent.

An $M/M/1$ queue is a birth-and-death queueing system with $\lambda_n = \lambda$ for $n \geq 0$, and $\mu_n = \mu$ for $n \geq 1$; its arrival and service rates are state independent. Other birth-and-death queueing systems include the following.

- **M/M/s queue:** This queueing system is like the $M/M/1$ queue, except s servers are working independently and in parallel, each serving at the same exponential rate μ. An $M/M/s$ queue is a *multiple-server, infinite queueing system.*
- **M/M/1/N queue:** This queueing system is like the $M/M/1$ queue, except there is a maximum capacity of N customers. An $M/M/1/N$ queue is a *single-server, finite queueing system.*
- **M/M/s/N queue:** This queueing system is like the $M/M/s$ queue, except there is a maximum capacity of N customers. An $M/M/s/N$ queue is a *multiple-server, finite queueing system.*

To analyze birth-and-death queueing systems, we need the result stated in Proposition 9.11(a) on page 532: The minimum of independent exponential random variables is also an exponential random variable with parameter equal to the sum of the parameters of the component exponential random variables.

EXAMPLE 12.4   *Birth-and-Death Queueing Systems*

Obtain the arrival and service rates for an
a) $M/M/s$ queue.   b) $M/M/1/N$ queue.   c) $M/M/s/N$ queue.

Solution   Let n denote the state of the system—that is, the number of customers in the queueing system at any particular time.

a) For $n \leq s - 1$, the elapsed time until the next service completion is the minimum of the service times of the n busy servers—that is, the minimum of n independent exponential random variables with common parameter μ, which is an exponential random variable with parameter $n\mu$. For $n \geq s$, the elapsed time until the next service completion is the minimum of the service times of the s busy servers—that is, the minimum of s independent exponential random variables with common parameter μ, which is an exponential random variable with parameter $s\mu$. Consequently, the arrival and service rates for the $M/M/s$ queue are

$$\lambda_n = \lambda, \quad \text{for } n \geq 0. \qquad \mu_n = \begin{cases} n\mu, & \text{for } 1 \leq n \leq s - 1; \\ s\mu, & \text{for } n \geq s. \end{cases} \tag{12.11}$$

b) For this queueing system, no arrivals can occur unless there are fewer than N customers in the system. Thus the arrival and service rates for the $M/M/1/N$ queue are

$$\lambda_n = \begin{cases} \lambda, & \text{for } 0 \leq n \leq N - 1; \\ 0, & \text{for } n \geq N. \end{cases} \qquad \mu_n = \mu, \quad \text{for } n \geq 1. \tag{12.12}$$

c) As in parts (a) and (b), the arrival and service rates for the $M/M/s/N$ queue are

$$\lambda_n = \begin{cases} \lambda, & \text{for } 0 \leq n \leq N - 1; \\ 0, & \text{for } n \geq N. \end{cases} \qquad \mu_n = \begin{cases} n\mu, & \text{for } 1 \leq n \leq s - 1; \\ s\mu, & \text{for } n \geq s. \end{cases} \tag{12.13}$$

We return to these queueing systems later in this section.   ■

EXAMPLE 12.5   *The M/M/2 Queue*

Emergency Room Traffic   At Desert Samaritan Hospital, arrivals at the emergency room constitute a Poisson process with rate 6.9 patients per hour. In Example 12.3, we considered the case where one doctor is on duty and independently spends an exponential amount of time with each patient, seeing 7.4 patients per hour, on average. As we discovered there, that queueing system is an $M/M/1$ queue. Now suppose that an additional doctor, who works at the same rate, is assigned to the emergency room to see patients in parallel with, and independently of, the first doctor. Describe this emergency room as a queueing system.

Solution   With the addition of a second doctor, the emergency-room queueing system is an $M/M/2$ queue. In view of Equation (12.11),

$$\lambda_n = 6.9, \quad \text{for } n \geq 0. \qquad \mu_n = \begin{cases} 7.4, & \text{for } n = 1; \\ 14.8, & \text{for } n \geq 2. \end{cases}$$

When one patient is in the emergency room, patients are seen at a rate of 7.4 patients per hour; when two or more patients are in the emergency room, patients are seen at a rate of 14.8 patients per hour.   ∎

Steady-State Probabilities

Associated with any queueing system are several quantities that indicate system performance. Of particular importance are

- the expected number of customers in the system, and
- an arriving customer's expected waiting time until service completion.

These and other related quantities are usually studied when the queueing system is in **steady state,** meaning, in the long run, when the probability distribution of the number of customers in the system has become essentially independent of elapsed time. We now examine steady-state properties of birth-and-death queueing systems.

In steady state, we let X denote the number of customers in the queueing system and set $P_n = P(X = n)$, the probability that exactly n customers are in the queueing system. To obtain the steady-state probabilities $\{P_n : n \geq 0\}$—the PMF of the random variable X—we first derive an important general principle in queueing theory.

Let n be a nonnegative integer. To reenter state n, the queueing system must first leave that state. So, in each time interval $[0, t]$, the number of times the queueing system enters state n must equal or exceed by 1 the number of times the queueing system leaves state n. Hence, in the long run (as $t \to \infty$), the average number of times the queueing system enters state n must equal the average number of times it leaves state n. Thus we have the following principle, aptly called the **rate-in = rate-out principle.**

Rate-in = Rate-out Principle

The average rate at which the queueing system enters a state equals the average rate at which the queueing system leaves that state.

To apply the rate-in = rate-out principle to birth-and-death queueing systems, we first construct a **rate transition diagram,** which shows the states of the queueing system and the rates at which transitions occur from one state to another. Figure 12.4, at the top of the next page, provides a rate transition diagram for birth-and-death queueing systems. For ease in applying it, we set $\lambda_{-1} = 0$ and $\mu_0 = 0$.

Now we assume that the queueing system is in steady state and consider state n. The system is in state n a proportion P_n of the time. Figure 12.4 shows that, from state n, the system can transition only to state $n + 1$ (which it does at rate λ_n) or to state $n - 1$ (which it does at rate μ_n). Thus the average rate at which the queueing system leaves state n is $\lambda_n P_n + \mu_n P_n$, or $(\lambda_n + \mu_n)P_n$.

Figure 12.4 Rate transition diagram for a birth-and-death queueing system

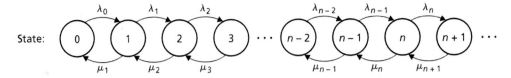

Also, the system can transition into state n only from state $n - 1$ (which it does at rate λ_{n-1}) or from state $n + 1$ (which it does at rate μ_{n+1}). As the system is in state $n - 1$ a proportion P_{n-1} of the time and is in state $n + 1$ a proportion P_{n+1} of the time, the average rate at which the queueing system enters state n is $\lambda_{n-1}P_{n-1} + \mu_{n+1}P_{n+1}$.

Applying the rate-in = rate-out principle gives

$$\lambda_{n-1}P_{n-1} + \mu_{n+1}P_{n+1} = (\lambda_n + \mu_n)P_n, \qquad n \geq 0, \tag{12.14}$$

where, again, we use the conventions $\lambda_{-1} = 0$ and $\mu_0 = 0$. Equations (12.14) are called the **balance equations** because they show the balance between the average rates of entering and leaving each state. We note in passing that there are methods for obtaining the balance equations that don't rely on the rate-in = rate-out principle.

We now solve the balance equations to obtain the steady-state probabilities of being in the various states. Using simple algebra, Equations (12.14) can be rewritten as

$$\mu_1 P_1 - \lambda_0 P_0 = 0$$

$$\mu_{n+1}P_{n+1} - \lambda_n P_n = \mu_n P_n - \lambda_{n-1}P_{n-1}, \qquad n \geq 1.$$

From these equations and mathematical induction, we find that $\mu_n P_n - \lambda_{n-1}P_{n-1} = 0$ for $n \geq 1$. Equivalently,

$$\frac{P_n}{P_{n-1}} = \frac{\lambda_{n-1}}{\mu_n}, \qquad n \geq 1.$$

From these equations, we get

$$\frac{P_n}{P_0} = \prod_{k=1}^{n} \frac{P_k}{P_{k-1}} = \prod_{k=1}^{n} \frac{\lambda_{k-1}}{\mu_k},$$

or

$$P_n = \left(\prod_{k=1}^{n} \frac{\lambda_{k-1}}{\mu_k} \right) P_0, \qquad n \geq 1. \tag{12.15}$$

Because a PMF must sum to 1, we have, in view of Equations (12.15), that

$$1 = \sum_{n=0}^{\infty} P_n = P_0 + \sum_{n=1}^{\infty} P_n = P_0 + \sum_{n=1}^{\infty} \left(\prod_{k=1}^{n} \frac{\lambda_{k-1}}{\mu_k} \right) P_0 = \left(1 + \sum_{n=1}^{\infty} \left(\prod_{k=1}^{n} \frac{\lambda_{k-1}}{\mu_k} \right) \right) P_0,$$

or

$$P_0 = \left(1 + \sum_{n=1}^{\infty} \left(\prod_{k=1}^{n} \frac{\lambda_{k-1}}{\mu_k} \right) \right)^{-1}. \tag{12.16}$$

Equations (12.15) and (12.16) yield a probability distribution if and only if the infinite series in Equation (12.16) converges. Consequently, we have established Proposition 12.3.

◆◆◆   **Proposition 12.3   Steady-State Probabilities for a Birth-and-Death Queue**

A birth-and-death queueing system has a steady-state distribution if and only if

$$\sum_{n=1}^{\infty}\left(\prod_{k=1}^{n}\frac{\lambda_{k-1}}{\mu_k}\right) < \infty, \tag{12.17}$$

in which case the steady-state distribution of the number of customers in the system is

$$P_0 = \left(1 + \sum_{n=1}^{\infty}\left(\prod_{k=1}^{n}\frac{\lambda_{k-1}}{\mu_k}\right)\right)^{-1} \quad and \quad P_n = \left(\prod_{k=1}^{n}\frac{\lambda_{k-1}}{\mu_k}\right)P_0, \quad n \geq 1. \tag{12.18}$$

We now apply Proposition 12.3 to obtain the steady-state distributions for the $M/M/1$ and $M/M/s$ queues. In Exercises 12.26 and 12.34, you are asked to do likewise for the $M/M/1/N$ and $M/M/s/N$ queues, respectively.

EXAMPLE 12.6   *Steady-State Probabilities*

M/M/1 and M/M/s **Queues**   For each of the $M/M/1$ and $M/M/s$ queues, obtain a condition on the arrival and service rates equivalent to the existence of a steady-state distribution. When such a condition is satisfied, find the steady-state distribution of the number of customers in the system.

Solution   We apply Proposition 12.3.

$M/M/1$ *Queue:* The arrival and service rates for the $M/M/1$ queue are $\lambda_n = \lambda$ for $n \geq 0$, and $\mu_n = \mu$ for $n \geq 1$. Therefore

$$\prod_{k=1}^{n}\frac{\lambda_{k-1}}{\mu_k} = \prod_{k=1}^{n}\frac{\lambda}{\mu} = (\lambda/\mu)^n, \qquad n \geq 1. \tag{12.19}$$

Because a geometric series converges if and only if its ratio is less than 1 in absolute value, we conclude that Equation (12.17) holds if and only if $\lambda < \mu$. Thus an $M/M/1$ queue has a steady-state distribution if and only if the service rate exceeds the arrival rate, which makes sense.

Assume now that $\lambda < \mu$. From Equations (12.18) and (12.19),

$$P_0 = \left(1 + \sum_{n=1}^{\infty}\left(\prod_{k=1}^{n}\frac{\lambda_{k-1}}{\mu_k}\right)\right)^{-1} = \left(1 + \sum_{n=1}^{\infty}(\lambda/\mu)^n\right)^{-1}$$

$$= \left(\sum_{n=0}^{\infty}(\lambda/\mu)^n\right)^{-1} = \left(\frac{1}{1 - \lambda/\mu}\right)^{-1} = 1 - \lambda/\mu.$$

Applying Equations (12.18) and (12.19) again, we get

$$P_n = \left(\prod_{k=1}^{n} \frac{\lambda_{k-1}}{\mu_k} \right) P_0 = (\lambda/\mu)^n \, (1 - \lambda/\mu) , \qquad n \geq 1.$$

From this equation and the previous one, we conclude that

$$P_n = (1 - \lambda/\mu) \, (\lambda/\mu)^n , \qquad n \geq 0. \tag{12.20}$$

Recalling that, in steady state, X denotes the number of customers in the system, we see that the PMF of the random variable X is given by Equation (12.20).

M/M/s Queue: The arrival and service rates for the $M/M/s$ queue are given in Equation (12.11) on page 700, from which we obtain

$$\prod_{k=1}^{n} \frac{\lambda_{k-1}}{\mu_k} =
\begin{cases}
\dfrac{\lambda^n}{n!\mu^n}, & \text{for } 1 \leq n \leq s - 1; \\[2ex]
\dfrac{\lambda^s}{s!\mu^s} \cdot \left(\dfrac{\lambda}{s\mu} \right)^{n-s}, & \text{for } n \geq s.
\end{cases}$$

$$=
\begin{cases}
\dfrac{(\lambda/\mu)^n}{n!}, & \text{for } 1 \leq n \leq s - 1; \\[2ex]
\dfrac{(\lambda/\mu)^n}{s! s^{n-s}}, & \text{for } n \geq s.
\end{cases} \tag{12.21}$$

From Equation (12.21),

$$\sum_{n=1}^{\infty} \left(\prod_{k=1}^{n} \frac{\lambda_{k-1}}{\mu_k} \right) = \sum_{n=1}^{s-1} \frac{(\lambda/\mu)^n}{n!} + \sum_{n=s}^{\infty} \frac{(\lambda/\mu)^n}{s! s^{n-s}}$$

$$= \sum_{n=1}^{s-1} \frac{(\lambda/\mu)^n}{n!} + \frac{(\lambda/\mu)^s}{s!} \sum_{n=s}^{\infty} (\lambda/s\mu)^{n-s}.$$

Therefore Equation (12.17) holds if and only if $\lambda < s\mu$; an $M/M/s$ queue has a steady-state distribution if and only if the combined service rate of all s servers exceeds the arrival rate, which makes sense.

Assume now that $\lambda < s\mu$. From Equation (12.18) and the preceding equation,

$$P_0 = \left(1 + \sum_{n=1}^{\infty} \left(\prod_{k=1}^{n} \frac{\lambda_{k-1}}{\mu_k} \right) \right)^{-1} = \left(1 + \sum_{n=1}^{s-1} \frac{(\lambda/\mu)^n}{n!} + \frac{(\lambda/\mu)^s}{s!} \sum_{n=s}^{\infty} (\lambda/s\mu)^{n-s} \right)^{-1}.$$

Using the formula for a geometric series, we find that

$$P_0 = \left(\sum_{n=0}^{s-1} \frac{(\lambda/\mu)^n}{n!} + \frac{(\lambda/\mu)^s}{s!(1 - \lambda/s\mu)} \right)^{-1}. \tag{12.22}$$

Applying Equations (12.18) and (12.21), we conclude that

$$P_n = \begin{cases} \dfrac{(\lambda/\mu)^n}{n!} P_0, & \text{for } 1 \le n \le s - 1; \\[3mm] \dfrac{(\lambda/\mu)^n}{s!\, s^{n-s}} P_0, & \text{for } n \ge s; \end{cases} \tag{12.23}$$

where P_0 is given by Equation (12.22). Observe that the two expressions on the right of the previous display agree when $n = s - 1$, so, you can use either one in that case. You should verify that, in the single-server case ($s = 1$), the results given by Equations (12.22) and (12.23) are identical to those given by Equation (12.20). ■

In Example 12.7, we apply the results of Example 12.6 to the emergency-room illustrations from Examples 12.3 and 12.5.

EXAMPLE 12.7 *Steady-State Probabilities*

Emergency Room Traffic   Refer to Examples 12.3 and 12.5 on pages 698 and 700, respectively. Obtain the steady-state distribution for the number of patients in the emergency room when
a) one doctor is on duty.
b) two doctors are on duty.
c) Compare the steady-state distributions obtained in parts (a) and (b).

Solution   a) From Example 12.3, when one doctor is on duty, we have an $M/M/1$ queue with $\lambda = 6.9$ and $\mu = 7.4$. Applying Equation (12.20), we find that the steady-state probabilities for the number of patients in the emergency room are

$$P_n = (1 - 6.9/7.4)(6.9/7.4)^n \approx 0.068 \cdot (0.932)^n, \qquad n \ge 0.$$

b) From Example 12.5, when two doctors are on duty, we have an $M/M/2$ queue with $\lambda = 6.9$ and $\mu = 7.4$. Applying Equation (12.22) with $s = 2$, we get

$$P_0 = \left(\sum_{n=0}^{1} \frac{(6.9/7.4)^n}{n!} + \frac{(6.9/7.4)^2}{2!\left(1 - 6.9/(2 \cdot 7.4)\right)} \right)^{-1} \approx 0.364.$$

Using this value of P_0 in Equation (12.23) yields

$$P_n = \frac{(6.9/7.4)^n}{2!\, 2^{n-2}} P_0 \approx 0.728 \cdot (0.466)^n, \qquad n \ge 1.$$

c) To compare the steady-state distributions obtained in parts (a) and (b), we construct Table 12.1 and Figure 12.5. In both we provide, to three decimal places, the steady-state probabilities for as many as 10 patients. Both Table 12.1 and Figure 12.5 indicate that more patients tend to be in the emergency room when one doctor is present than when two doctors are present, which is just what we would expect. ■

Table 12.1 Steady-state probabilities for the number of patients in the emergency room from $n = 0$ to $n = 10$

Number of patients, n	One doctor Probability, P_n	Two doctors Probability, P_n
0	0.068	0.364
1	0.063	0.339
2	0.059	0.158
3	0.055	0.074
4	0.051	0.034
5	0.048	0.016
6	0.044	0.007
7	0.041	0.003
8	0.039	0.002
9	0.036	0.001
10	0.034	0.000

Figure 12.5 Probability histogram of steady-state probabilities for the number of patients in the emergency room from $n = 0$ to $n = 10$: (a) one doctor; (b) two doctors

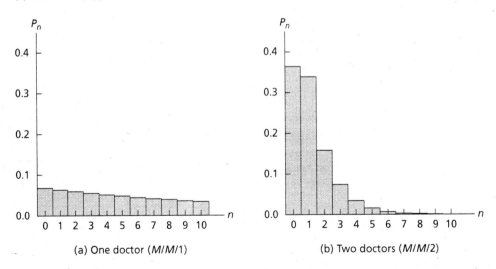

(a) One doctor ($M/M/1$)   (b) Two doctors ($M/M/2$)

Expected Number of Customers

We mentioned earlier that, associated with any queueing system, are several quantities that indicate system performance and that such quantities are usually studied in steady state. Following are some of the most important indicators of system performance and the notations that are used to represent them in steady state.

- L: Expected number of customers in the queueing system, $\mathcal{E}(X)$.
- L_q: Expected queue length.
- W: Expected waiting time in the queueing system for each individual customer.
- W_q: Expected waiting time in the queue for each individual customer.
- p_b: Probability that a specified server is busy.

In Example 12.8, we determine L for the $M/M/1$ and $M/M/s$ queues. And, in Example 12.9, we apply those results to the emergency-room illustration.

EXAMPLE 12.8   *Expected Number of Customers*

M/M/1 **and** *M/M/s* **Queues**   For each of the $M/M/1$ and $M/M/s$ queues, find the steady-state expected number of customers in the queueing system.

Solution   Although the required result for the $M/M/1$ queue can be obtained as a special case of that for the $M/M/s$ queue, treating the two queues separately is instructive. As previously, let X denote the steady-state number of customers in the queueing system. The problem is to find $L = \mathcal{E}(X)$.

M/M/1 Queue: The steady-state distribution of X is given in Equation (12.20) on page 704. We note that $X + 1$ has the geometric distribution with parameter $1 - \lambda/\mu$. As the mean of a geometric random variable is the reciprocal of its parameter, we have

$$L = \mathcal{E}(X) = \mathcal{E}\big((X+1) - 1\big) = \mathcal{E}(X+1) - 1 = \frac{1}{1 - \lambda/\mu} - 1,$$

or

$$L = \frac{\lambda}{\mu - \lambda}. \tag{12.24}$$

M/M/s Queue: For convenience, we set $r = \lambda/\mu$ and $\rho = \lambda/s\mu = r/s$. The steady-state distribution of X is given in Equations (12.22) and (12.23) on pages 704 and 705, respectively. From those equations, we get

$$L = \mathcal{E}(X) = \sum_{n=0}^{\infty} n P_n = \sum_{n=1}^{s-1} n \frac{r^n}{n!} P_0 + \sum_{n=s}^{\infty} n \frac{r^n}{s! \, s^{n-s}} P_0$$

$$= \left(\sum_{n=1}^{s} n \frac{r^n}{n!} + \sum_{n=s+1}^{\infty} n \frac{r^n}{s! \, s^{n-s}} \right) P_0. \tag{12.25}$$

We have

$$\sum_{n=1}^{s} n \frac{r^n}{n!} = r \sum_{n=1}^{s} \frac{r^{n-1}}{(n-1)!} = r \sum_{n=0}^{s-1} \frac{r^n}{n!}. \tag{12.26}$$

Using the identity $\sum_{n=1}^{\infty} na^n = a/(1-a)^2$ and the formula for the sum of a geometric series, we obtain, after some algebra, that

$$\sum_{n=s+1}^{\infty} n \frac{r^n}{s!\, s^{n-s}} = \frac{\rho r^s}{s!(1-\rho)^2} + r \frac{r^s}{s!(1-\rho)}. \tag{12.27}$$

From Equations (12.25)–(12.27) and Equation (12.22), we conclude that

$$L = \left(r \sum_{n=0}^{s-1} \frac{r^n}{n!} + r \frac{r^s}{s!(1-\rho)} + \frac{\rho r^s}{s!(1-\rho)^2} \right) P_0$$

$$= \left(r P_0^{-1} + \frac{\rho r^s}{s!(1-\rho)^2} \right) P_0 = r + \frac{\rho r^s}{s!(1-\rho)^2} P_0,$$

or

$$L = \frac{\lambda}{\mu} + \frac{(\lambda/s\mu)(\lambda/\mu)^s}{s!(1-\lambda/s\mu)^2} P_0. \tag{12.28}$$

Note that Equation (12.28) reduces to Equation (12.24) when $s = 1$.   ■

EXAMPLE 12.9   *Expected Number of Customers*

Emergency Room Traffic   Consider once more the illustration of emergency room traffic. Obtain the steady-state expected number of patients in the emergency room when
a) one doctor is on duty.
b) two doctors are on duty.
c) Compare the results found in parts (a) and (b).

Solution   a) From Example 12.3, when one doctor is on duty, we have an $M/M/1$ queue with $\lambda = 6.9$ and $\mu = 7.4$. Applying Equation (12.24) yields

$$L = \frac{6.9}{7.4 - 6.9} = 13.8.$$

Thus the steady-state expected number of patients in the emergency room is 13.8; that is, on average, the emergency room contains 13.8 patients.

b) From Example 12.5, when two doctors are on duty, we have an $M/M/2$ queue with $\lambda = 6.9$ and $\mu = 7.4$. Recalling from Example 12.7(b) that $P_0 \approx 0.364$ and applying Equation (12.28) with $s = 2$, we conclude that

$$L = \frac{6.9}{7.4} + \frac{(6.9/(2 \cdot 7.4))(6.9/7.4)^2}{2!(1 - 6.9/(2 \cdot 7.4))^2} P_0 \approx 1.2.$$

Thus the steady-state expected number of patients in the emergency room is about 1.2; that is, on average, the emergency room contains roughly 1.2 patients.

c) The results in parts (a) and (b) show that the addition of one more doctor in the emergency room drastically reduces the average number of patients in the system—from 13.8 patients with one doctor to 1.2 patients with two doctors.   ■

Two powerful, but simply stated, formulas—called **Little's formulas**—allow us to relate the first four quantities in the bulleted list on page 707 under very general conditions on the queueing system. When those conditions are satisfied, once one of the four quantities is known, the other three can be easily obtained. You are asked to examine and apply Little's formulas in Exercises 12.35–12.40.

In this section, we discussed the basic ideas of queueing theory. However, we've just scratched the surface of this important topic. Entire books and courses are devoted to queueing theory and its applications. See, for instance, the book *Fundamentals of Queueing Theory* by Donald Gross and Carl M. Harris (New York: Wiley, 1998).

EXERCISES 12.2   Basic Exercises

12.20 Regarding the rate-in = rate-out principle on page 701:
a) Provide a detailed argument explaining why, in each time interval $[0, t]$, the number of times the queueing system enters state n must equal or exceed by 1 the number of times the queueing system leaves state n.
b) Use part (a) to deduce the rate-in = rate-out principle.

12.21 Customers arrive at the drive-up station of a bank according to a Poisson process at an average rate of 10 per hour. One teller is on duty and independently spends an exponential amount of time with each customer, serving 11 customers per hour, on average.
a) Describe this drive-up station as a queueing system.
b) Determine whether a steady-state distribution exists for the number of customers in this queueing system and, if it does, obtain it.
c) Find the steady-state expected number of customers in this queueing system.
d) Find the steady-state expected number of customers in the queue.

12.22 Refer to Exercise 12.21, but now assume that there are two tellers working independently and in parallel, each serving customers at an exponential rate of 11 per hour.
a) Describe this drive-up station as a queueing system.
b) Determine whether a steady-state distribution exists for the number of customers in this queueing system and, if it does, obtain it.
c) Compare your results in part (b) to those in Exercise 12.21(b) by constructing a table and graph similar to those in Table 12.1 and Figure 12.5 on page 706.
d) Find the steady-state expected number of customers in this queueing system and compare your result to that in Exercise 12.21(c).
e) Find the steady-state expected number of customers in the queue and compare your result to that in Exercise 12.21(d).

12.23 Refer to Exercise 12.21, but now assume that there are three tellers working independently and in parallel, each serving customers at the rate of 11 per hour.
a) Describe this drive-up station as a queueing system.
b) Determine whether a steady-state distribution exists for the number of customers in this queueing system and, if it does, obtain it.
c) Compare your results in part (b) to those in Exercises 12.21(b) and 12.22(b) by constructing a table and graph similar to those in Table 12.1 and Figure 12.5 on page 706.
d) Find the steady-state expected number of customers in this queueing system and compare your result to those in Exercises 12.21(c) and 12.22(d).
e) Find the steady-state expected number of customers in the queue and compare your result to those in Exercises 12.21(d) and 12.22(e).

12.24 Explain why it's permissible to assume that $s \leq N$ for an $M/M/s/N$ queueing system.

12.25 The sole barber in a one-man barbershop gives haircuts at a rate of two per hour, and the elapsed times for successive haircuts are independent exponential random variables. Potential customers arrive according to a Poisson process at a rate of three per hour, but the waiting room accommodates at most three customers.
a) Describe this barbershop as a queueing system.
b) Obtain the steady-state distribution for the number of customers in the barbershop.
c) Obtain the steady-state expected number of customers in the barbershop.

12.26 Consider an $M/M/1/N$ queueing system.
a) Show that a steady-state distribution always exists.
b) Obtain the steady-state distribution for the number of customers in the queueing system. Consider the cases $\lambda = \mu$ and $\lambda \neq \mu$ separately.
c) Obtain the steady-state expected number of customers in the queueing system.
d) Apply your results from parts (b) and (c) to solve parts (b) and (c) of Exercise 12.25.

12.27 A convenience store has one checkout counter, where customers arrive according to a Poisson process with rate λ and checkout times are independent exponential random variables. The checker works at rate μ and, when three or more customers are at the counter, the checker is assisted by another person, making the service rate ν, where $\nu > \lambda$.
a) Obtain the steady-state distribution of the number of customers at the checkout counter.
b) Find the steady-state mean number of customers in the queue.
c) Apply your result in part (b) when $\lambda = 20$ per hour, $\mu = 16$ per hour, and $\nu = 24$ per hour.

12.28 Suppose that customers arrive at a service facility according to a Poisson process with rate λ. Which is more efficient: an $M/M/2$ queue with each server serving at rate μ or an $M/M/1$ queue with the one server serving at rate 2μ? Assume that $\lambda < 2\mu$.

Theory Exercises

12.29 For an $M/M/1$ queue, find the steady-state
a) distribution of the number of customers in the queue.
b) expected queue length, L_q.

12.30 For an $M/M/1$ queue, find the steady-state
a) distribution of the waiting time in the queueing system (i.e., the elapsed time from arrival to service completion) for each individual customer. *Hint:* Condition on X, the steady-state number of customers in the queueing system.
b) expected waiting time in the queueing system for each individual customer, W.

12.31 For an $M/M/1$ queue, find the steady-state
a) distribution of the waiting time in the queue (i.e., the elapsed time from arrival to service commencement) for each individual customer.
b) expected waiting time in the queue for each individual customer, W_q.

12.32 Consider an $M/M/1$ queue in steady-state. Given that a customer must join the queue (i.e., must wait for service commencement), determine the conditional distribution and conditional expectation of the waiting time in the queue. *Hint:* Refer to Exercise 12.31.

12.33 Consider a queueing system with a steady state distribution.
a) Show that, for a single-server system, $L_q = L - (1 - P_0)$.
b) Apply part (a) and Equation (12.24) on page 707 to obtain L_q for an $M/M/1$ queue. Compare your answer to that obtained in Exercise 12.29(b).
c) Generalize part (a) for an s-server system.

Advanced Exercises

12.34 Consider an $M/M/s/N$ queueing system.
a) Show that a steady-state distribution always exists.
b) Obtain the steady-state distribution for the number of customers in the queueing system.
Hint: Consider the cases $\lambda = s\mu$ and $\lambda \neq s\mu$ separately.
c) Assuming that $\lambda < s\mu$, show that, as $N \to \infty$, the appropriate steady-state distribution obtained in part (b) approaches that found for an $M/M/s$ queue in Example 12.6 on page 703.
d) Obtain the steady-state expected queue length.

Little's formulas: In the early 1960s, John D. C. Little developed some powerful formulas for queueing systems in steady state that hold under very general conditions—in particular, they apply to all queueing systems considered in this book. *Little's formulas* are $L = \bar{\lambda}W$ and $L_q = \bar{\lambda}W_q$, where $\bar{\lambda}$ denotes the long-run-average arrival rate. In Exercises 12.35–12.40, you are to investigate and apply Little's formulas.

12.35 Let λ_n denote the arrival rate when the queueing system is in state n.
a) Explain why $\bar{\lambda} = \sum_{n=0}^{\infty} \lambda_n P_n$.
b) Show that $\bar{\lambda} = \lambda$ for the $M/M/1$ and $M/M/s$ queues.
c) Show that $\bar{\lambda} = (1 - P_N)\lambda$ for the $M/M/1/N$ and $M/M/s/N$ queues.

12.36 Let μ denote the service rate of each individual server.
a) Use an expected-value argument to show that $W = W_q + 1/\mu$.
b) Deduce from part (a) and Little's formulas that $L = L_q + \bar{\lambda}/\mu$.
c) Explain how to obtain the four quantities L, L_q, W, and W_q when one of them is known.
d) From Equation (12.24) on page 707, $L = \lambda/(\mu - \lambda)$ for an $M/M/1$ queue. Use that result and your answer for part (c) to obtain L_q, W, and W_q. Compare your answers to those found in Exercises 12.29(b), 12.30(b), and 12.31(b), respectively.

12.37 Refer to Exercise 12.36 and the results of Example 12.9 on page 708. When one doctor is on duty, determine the steady-state
a) expected number of patients waiting to be treated.
b) expected waiting time until treatment commences.
c) expected elapsed time from arrival at the emergency room until treatment is completed.

12.38 Refer to Exercise 12.36 and the results of Example 12.9 on page 708. When two doctors are on duty, determine the steady-state
a) expected number of patients waiting to be treated.
b) expected waiting time until treatment commences.
c) expected elapsed time from arrival at the emergency room until treatment is completed.
d) Compare your answers here to those in Exercise 12.37.

12.39 For each of Exercises 12.21–12.23, determine the steady-state
a) expected time a customer spends in the queueing system.
b) expected time a customer waits before being served.
c) Compare your results for the three queueing systems.

12.40 Refer to Exercise 12.25. Determine the steady-state
a) expected number of customers in the waiting room.
b) expected time a customer waits from the time he enters the barbershop until the time his haircut is completed.
c) expected time a customer waits in the waiting room.

12.3   The Multivariate Normal Distribution

In Section 8.5, we discussed the univariate normal distribution and, in Section 10.5, we examined the bivariate normal distribution. Now we generalize those concepts to the *multivariate normal distribution.* In doing so, we use vectors and matrices. Note that a vector can be considered a matrix. Column vectors are represented by lowercase boldface letters and matrices are represented by uppercase boldface letters.

The **transpose** of a matrix $\mathbf{A}$ is denoted $\mathbf{A}'$. Now suppose that $\mathbf{A}$ is a square matrix, say, $m \times m$. The **determinant** of $\mathbf{A}$ is denoted $\det \mathbf{A}$. If $\mathbf{A}$ is a nonsingular matrix, its **inverse** is denoted $\mathbf{A}^{-1}$. If $\mathbf{A}' = \mathbf{A}$, we say that $\mathbf{A}$ is **symmetric**. A symmetric matrix $\mathbf{A}$ is called **nonnegative definite** if $\mathbf{x}'\mathbf{A}\mathbf{x} \geq 0$ for all $\mathbf{x} \in \mathcal{R}^m$ and is called **positive definite** if $\mathbf{x}'\mathbf{A}\mathbf{x} > 0$ for all $\mathbf{x} \neq \mathbf{0}$. A positive definite matrix is nonsingular.

Suppose that $X_1, \ldots, X_m$ are random variables defined on the same sample space and having finite variances. Then we let

$$\mathbf{X} = \begin{bmatrix} X_1 \\ \vdots \\ X_m \end{bmatrix}, \qquad \boldsymbol{\mu}_{\mathbf{X}} = \begin{bmatrix} \mu_{X_1} \\ \vdots \\ \mu_{X_m} \end{bmatrix},$$

and

$$\boldsymbol{\Sigma}_{\mathbf{X}} = \begin{bmatrix} \operatorname{Cov}(X_1, X_1) & \cdots & \operatorname{Cov}(X_1, X_m) \\ \vdots & \ddots & \vdots \\ \operatorname{Cov}(X_m, X_1) & \cdots & \operatorname{Cov}(X_m, X_m) \end{bmatrix}.$$

We refer to $\boldsymbol{\mu}_{\mathbf{X}}$ and $\boldsymbol{\Sigma}_{\mathbf{X}}$ as the **mean vector** and **covariance matrix,** respectively, of the random variables $X_1, \ldots, X_m$ or, equivalently, of the random vector $\mathbf{X}$. Note that, when $m = 1$ (i.e., in one dimension), $\boldsymbol{\mu}_{\mathbf{X}}$ and $\boldsymbol{\Sigma}_{\mathbf{X}}$ are simply μ_X and σ_X^2, respectively. In Exercise 12.41, you are asked to show that a covariance matrix is nonnegative definite.

Independent Multivariate Normal Random Variables

As in the presentation of the bivariate normal distribution, we begin our examination of the multivariate normal distribution with the simplest case—namely, that of independent normal random variables—say, $X_1, \ldots, X_m$. Because $X_1, \ldots, X_m$ are independent, their joint PDF equals the product of their marginal PDFs:

$$f_{X_1, \ldots, X_m}(x_1, \ldots, x_m)$$

$$= \frac{1}{\sqrt{2\pi}\,\sigma_{X_1}} e^{-\frac{1}{2}\left(\frac{x_1 - \mu_{X_1}}{\sigma_{X_1}}\right)^2} \cdots \frac{1}{\sqrt{2\pi}\,\sigma_{X_m}} e^{-\frac{1}{2}\left(\frac{x_m - \mu_{X_m}}{\sigma_{X_m}}\right)^2} \tag{12.29}$$

$$= \frac{1}{(2\pi)^{m/2}\sigma_{X_1} \cdots \sigma_{X_m}} e^{-\frac{1}{2}\left\{\left(\frac{x_1 - \mu_{X_1}}{\sigma_{X_1}}\right)^2 + \cdots + \left(\frac{x_m - \mu_{X_m}}{\sigma_{X_m}}\right)^2\right\}},$$

where $-\infty < x_j < \infty$ for $1 \leq j \leq m$. The joint distribution of $X_1, \ldots, X_m$, whose PDF is given in Equation (12.29), is an example of a multivariate normal distribution.

Because $X_1, \ldots, X_m$ are independent, they are uncorrelated, so $\text{Cov}(X_i, X_j) = 0$ for $i \neq j$. Consequently, the covariance matrix of $X_1, \ldots, X_m$ is

$$\Sigma_{\mathbf{X}} = \begin{bmatrix} \sigma_{X_1}^2 & \cdots & 0 \\ \vdots & \ddots & \vdots \\ 0 & \cdots & \sigma_{X_m}^2 \end{bmatrix}. \tag{12.30}$$

Clearly, $\det \Sigma_{\mathbf{X}} = \sigma_{X_1}^2 \cdots \sigma_{X_m}^2$. Moreover, simple matrix calculations yield

$$(\mathbf{x} - \mu_{\mathbf{X}})' \Sigma_{\mathbf{X}}^{-1} (\mathbf{x} - \mu_{\mathbf{X}}) = \left(\frac{x_1 - \mu_{X_1}}{\sigma_{X_1}} \right)^2 + \cdots + \left(\frac{x_m - \mu_{X_m}}{\sigma_{X_m}} \right)^2.$$

Therefore, in view of Equation (12.29), the joint PDF of independent normal random variables can be expressed compactly in the form

$$f_{\mathbf{X}}(\mathbf{x}) = \frac{1}{(2\pi)^{m/2} (\det \Sigma_{\mathbf{X}})^{1/2}} \, e^{-\frac{1}{2}(\mathbf{x} - \mu_{\mathbf{X}})' \Sigma_{\mathbf{X}}^{-1} (\mathbf{x} - \mu_{\mathbf{X}})}, \qquad \mathbf{x} \in \mathcal{R}^m. \tag{12.31}$$

The Nonsingular Multivariate Normal Case

We also want to consider correlated normal random variables. The question then is: What is the general form of the joint PDF of a multivariate normal distribution? To answer this question, we refer to our introduction of the bivariate normal distribution on pages 608 and 609. There we discovered that the general bivariate normal distribution can be obtained from two independent standard normal random variables, Z_1 and Z_2, by using an appropriate transformation of the form

$$X = \mu_1 + a Z_1 + b Z_2$$
$$Y = \mu_2 + c Z_1 + d Z_2.$$

We can express these two equations in matrix form as

$$\mathbf{X} = \mathbf{a} + \mathbf{B}\mathbf{Z},$$

where

$$\mathbf{X} = \begin{bmatrix} X \\ Y \end{bmatrix}, \qquad \mathbf{a} = \begin{bmatrix} \mu_1 \\ \mu_2 \end{bmatrix}, \qquad \mathbf{B} = \begin{bmatrix} a & b \\ c & d \end{bmatrix}, \qquad \text{and} \qquad \mathbf{Z} = \begin{bmatrix} Z_1 \\ Z_2 \end{bmatrix}.$$

Taking these relations as our cue, we let $Z_1, \ldots, Z_m$ be independent standard normal random variables and let μ_i ($1 \leq i \leq m$) and b_{ij} ($1 \leq i \leq m$, $1 \leq j \leq m$) be real numbers. Then we define the random variables $X_1, \ldots, X_m$ by

$$X_1 = \mu_1 + b_{11} Z_1 + \cdots + b_{1m} Z_m$$

$$\vdots \tag{12.32}$$

$$X_m = \mu_m + b_{m1} Z_1 + \cdots + b_{mm} Z_m,$$

which we can write in matrix form as

$$X = a + BZ, \tag{12.33}$$

where

$$a = \begin{bmatrix} \mu_1 \\ \vdots \\ \mu_m \end{bmatrix}, \qquad B = \begin{bmatrix} b_{11} & \cdots & b_{1m} \\ \vdots & \ddots & \vdots \\ b_{m1} & \cdots & b_{mm} \end{bmatrix}, \qquad \text{and} \qquad Z = \begin{bmatrix} Z_1 \\ \vdots \\ Z_m \end{bmatrix}. \tag{12.34}$$

We leave it for you as Exercise 12.42(b) to show that

$$\mu_X = a \qquad \text{and} \qquad \Sigma_X = BB'. \tag{12.35}$$

Because $Z_1, \ldots, Z_m$ are independent normal random variables, it follows from Proposition 9.14 on page 540 that $X_1, \ldots, X_m$ are marginally normally distributed. To obtain a joint PDF of $X_1, \ldots, X_m$, we apply the multivariate transformation theorem (Exercise 9.160 on page 553). In doing so, we assume that the matrix B is nonsingular.[†]

As $Z_1, \ldots, Z_m$ are independent standard normal random variables, their mean vector is the zero vector and their covariance matrix is the identity matrix. Hence, in view of Equation (12.31), the joint PDF of $Z_1, \ldots, Z_m$ is given by

$$f_Z(z) = \frac{1}{(2\pi)^{m/2}} e^{-\frac{1}{2}z'z}, \qquad z \in \mathcal{R}^m. \tag{12.36}$$

Clearly, the Jacobian determinant of the transformation $x = a + Bz$ equals $\det B$—that is, $J(z) = \det B$. From Equations (12.35) and the fact that a matrix and its transpose have the same determinant, we get

$$\det \Sigma_X = \det \left(BB' \right) = \det B \cdot \det B' = (\det B)^2 = \left(J(z) \right)^2.$$

Consequently,

$$|J(z)| = (\det \Sigma_X)^{1/2}. \tag{12.37}$$

Solving $x = a + Bz$ for z, we get the inverse transformation $z = B^{-1}(x - a)$. Applying the multivariate transformation theorem and keeping Equations (12.36) and (12.37) in mind, we find that

$$f_X(x) = \frac{1}{|J(z)|} f_Z(z) = \frac{1}{(\det \Sigma_X)^{1/2}} \frac{1}{(2\pi)^{m/2}} e^{-\frac{1}{2}z'z}.$$

However, because of Equations (12.35),

$$z'z = \left(B^{-1}(x - a) \right)' \left(B^{-1}(x - a) \right) = (x - a)' \left(B^{-1} \right)' B^{-1}(x - a)$$

$$= (x - a)' \left(BB' \right)^{-1} (x - a) = (x - \mu_X)' \Sigma_X^{-1} (x - \mu_X).$$

[†] If B is singular, $X_1, \ldots, X_m$ don't have a joint PDF. We address the general case, where B may or may not be nonsingular, later in this section.

Therefore,

$$f_{\mathbf{X}}(\mathbf{x}) = \frac{1}{(2\pi)^{m/2}(\det \Sigma_{\mathbf{X}})^{1/2}}\, e^{-\frac{1}{2}(\mathbf{x}-\mu_{\mathbf{X}})'\Sigma_{\mathbf{X}}^{-1}(\mathbf{x}-\mu_{\mathbf{X}})}, \qquad \mathbf{x} \in \mathcal{R}^m. \tag{12.38}$$

Based on Equation (12.38), we define **nonsingular multivariate normal random variables** and **nonsingular multivariate normal distribution** as follows.

DEFINITION 12.4   Nonsingular Multivariate Normal Distribution

Continuous random variables $X_1, \ldots, X_m$ are called **nonsingular multivariate normal random variables** if they have a joint probability density function of the form

$$f_{\mathbf{X}}(\mathbf{x}) = \frac{1}{(2\pi)^{m/2}(\det \Sigma)^{1/2}}\, e^{-\frac{1}{2}(\mathbf{x}-\mu)'\Sigma^{-1}(\mathbf{x}-\mu)}, \qquad \mathbf{x} \in \mathcal{R}^m,$$

where μ is an $m \times 1$ vector and Σ is an $m \times m$ positive definite matrix. We say that $X_1, \ldots, X_m$ have the **nonsingular multivariate normal distribution with parameters μ and Σ.** We often write $(X_1, \ldots, X_m) \sim \mathcal{N}_m(\mu, \Sigma)$ or $\mathbf{X} \sim \mathcal{N}_m(\mu, \Sigma)$ to indicate that $X_1, \ldots, X_m$ have that joint distribution.

We can obtain each univariate marginal distribution for nonsingular multivariate normal random variables by integrating out the joint PDF in Definition 12.4 on the remaining $(m - 1)$ variables. However, there's an easier way. Because a joint distribution determines the marginal distributions, the discussion leading to Definition 12.4 shows the following about nonsingular multivariate normal random variables $X_1, \ldots, X_m$ with parameters μ and Σ.

- $X_i \sim \mathcal{N}\big([\mu]_i, [\Sigma]_{ii}\big)$ for each $1 \le i \le m$, and
- $\mathrm{Cov}\big(X_i, X_j\big) = [\Sigma]_{ij}$ for each $1 \le i \le m$ and $1 \le j \le m$,

where $[\mu]_i$ denotes the ith entry of the column vector μ and $[\Sigma]_{ij}$ denotes the entry in the ith row and jth column of the matrix Σ. Together the two bulleted items state that the univariate marginal distributions of nonsingular multivariate normal random variables $X_1, \ldots, X_m$ are all normal and that the parameters μ and Σ are their mean vector and covariance matrix, respectively—that is, $\mu_{\mathbf{X}} = \mu$ and $\Sigma_{\mathbf{X}} = \Sigma$. See Exercise 12.45 for details.

The General Multivariate Normal Case

From Definition 12.4, a necessary condition for random variables $X_1, \ldots, X_m$ to have a nonsingular multivariate normal distribution is that their covariance matrix be nonsingular. As we now demonstrate, multivariate normal random variables can still be defined when the covariance matrix is singular, although such random variables don't have a joint PDF. To that end, we first obtain the joint MGF of nonsingular multivariate normal random variables, which requires Lemma 12.1.

◆◆◆   **Lemma 12.1   Joint MGF of Independent Standard Normals**

The joint MGF of independent standard normal random variables $Z_1, \ldots, Z_m$ is given by

$$M_{\mathbf{Z}}(\mathbf{t}) = e^{\frac{1}{2}\mathbf{t}'\mathbf{t}}, \qquad \mathbf{t} \in \mathcal{R}^m. \tag{12.39}$$

Proof   Applying the multivariate version of Proposition 11.9 on page 646 and using the fact that the MGF of a standard normal random variable is $e^{t^2/2}$, we get

$$M_{\mathbf{Z}}(\mathbf{t}) = M_{Z_1, \ldots, Z_m}(t_1, \ldots, t_m) = M_{Z_1}(t_1) \cdots M_{Z_m}(t_m)$$

$$= e^{t_1^2/2} \cdots e^{t_m^2/2} = e^{\frac{1}{2}\sum_{i=1}^m t_i^2} = e^{\frac{1}{2}\mathbf{t}'\mathbf{t}},$$

as required.   ◆

Before obtaining the joint MGF of nonsingular multivariate normal random variables, we recall some further results from linear algebra. If $\mathbf{A}$ is a nonnegative definite matrix, there is a unique nonnegative definite matrix $\mathbf{R}$ such that $\mathbf{R}^2 = \mathbf{A}$. We write $\mathbf{A}^{1/2}$ for $\mathbf{R}$. Furthermore, if $\mathbf{A}$ is positive definite, so is $\mathbf{A}^{1/2}$, and we write its inverse as $\mathbf{A}^{-1/2}$.

◆◆◆   **Proposition 12.4   Joint MGF of Nonsingular Multivariate Normals**

The joint MGF of nonsingular multivariate normal random variables $X_1, \ldots, X_m$ with parameters $\boldsymbol{\mu}$ and $\boldsymbol{\Sigma}$ is given by

$$M_{\mathbf{X}}(\mathbf{t}) = e^{\boldsymbol{\mu}'\mathbf{t} + \frac{1}{2}\mathbf{t}'\boldsymbol{\Sigma}\mathbf{t}}, \qquad \mathbf{t} \in \mathcal{R}^m. \tag{12.40}$$

Proof   We know that $\boldsymbol{\Sigma}$ is positive definite. Let

$$\mathbf{Z} = \boldsymbol{\Sigma}^{-1/2}(\mathbf{X} - \boldsymbol{\mu}). \tag{12.41}$$

As you are asked to verify in Exercise 12.48, the random variables $Z_1, \ldots, Z_m$ are independent standard normal. For convenience, set $\mathbf{R} = \boldsymbol{\Sigma}^{1/2}$. Applying Lemma 12.1 and basic properties of joint MGFs (see Exercise 12.47) and matrices, we get

$$M_{\mathbf{X}}(\mathbf{t}) = M_{\boldsymbol{\mu}+\mathbf{R}\mathbf{Z}}(\mathbf{t}) = e^{\boldsymbol{\mu}'\mathbf{t}} M_{\mathbf{Z}}(\mathbf{R}'\mathbf{t}) = e^{\boldsymbol{\mu}'\mathbf{t}} e^{\frac{1}{2}(\mathbf{R}'\mathbf{t})'(\mathbf{R}'\mathbf{t})} = e^{\boldsymbol{\mu}'\mathbf{t}} e^{\frac{1}{2}\mathbf{t}'\mathbf{R}\mathbf{R}'\mathbf{t}}.$$

Because $\mathbf{R}$ is a symmetric matrix whose square is $\boldsymbol{\Sigma}$, we have $\mathbf{R}\mathbf{R}' = \mathbf{R}^2 = \boldsymbol{\Sigma}$. Thus Equation (12.40) holds.   ◆

The expression in Equation (12.40) makes sense regardless of whether $\boldsymbol{\Sigma}$ is a nonsingular matrix. With that fact in mind, we now define **multivariate normal random variables** and **multivariate normal distribution** in a way that doesn't require a nonsingular covariance matrix.

> **DEFINITION 12.5 Multivariate Normal Distribution**
>
> Continuous random variables $X_1, \ldots, X_m$ are called **multivariate normal random variables** if their joint moment generating function is of the form
>
> $$M_X(t) = e^{\mu' t + \frac{1}{2} t' \Sigma t}, \qquad t \in \mathcal{R}^m,$$
>
> where μ is an $m \times 1$ vector and Σ is an $m \times m$ nonnegative definite matrix. We say that $X_1, \ldots, X_m$ have the **multivariate normal distribution with parameters μ and Σ.** We often write $(X_1, \ldots, X_m) \sim \mathcal{N}_m(\mu, \Sigma)$ or $X \sim \mathcal{N}_m(\mu, \Sigma)$ to indicate that $X_1, \ldots, X_m$ have that joint distribution.

For multivariate normal random variables $X_1, \ldots, X_m$ with parameters μ and Σ, as defined in Definition 12.5, the univariate marginal distributions are all normal, and μ and Σ are their mean vector and covariance matrix, respectively. See Exercise 12.54.

Proposition 12.4 reveals that nonsingular multivariate normal random variables, as defined in Definition 12.4 on page 715, are also multivariate normal random variables in the sense of Definition 12.5. However, the converse isn't true because multivariate normal random variables (as defined in Definition 12.5) with a singular covariance matrix aren't nonsingular multivariate normal random variables.

If multivariate normal random variables aren't nonsingular, we say that they are **singular multivariate normal random variables** and that they have a **singular multivariate normal distribution.** Proposition 12.5 provides useful equivalent conditions for multivariate normal random variables to be nonsingular.

◆◆◆ Proposition 12.5

For multivariate normal random variables $X_1, \ldots, X_m$ with parameters μ and Σ, the following conditions are equivalent.

a) $X_1, \ldots, X_m$ are nonsingular multivariate normal random variables.

b) $X_1, \ldots, X_m$ have a joint PDF.

c) Σ is a nonsingular matrix.

Proof $(a) \Rightarrow (b)$: Suppose that $X_1, \ldots, X_m$ are nonsingular multivariate normal random variables. Then, by Definition 12.4, the random variables $X_1, \ldots, X_m$ have a joint PDF.

$(b) \Rightarrow (c)$: Suppose that $X_1, \ldots, X_m$ have a joint PDF. Let $x \in \mathcal{R}^m$ be nonzero. It's easy to see that $x' \Sigma x = \text{Var}(x'X)$. Because $X_1, \ldots, X_m$ have a joint PDF, any nonzero linear combination of them is a continuous random variable. Thus, $P(x'X = c) = 0$ for all $c \in \mathcal{R}$. This fact and Proposition 7.7 on page 354 imply that $\text{Var}(x'X) > 0$, or $x' \Sigma x > 0$. Consequently, Σ is positive definite and hence nonsingular.

$(c) \Rightarrow (a)$: Suppose that Σ is nonsingular. Let $Y_1, \ldots, Y_m$ be nonsingular multivariate normal random variables with parameters μ and Σ. From Definition 12.5 and Proposition 12.4, we see that $X_1, \ldots, X_m$ and $Y_1, \ldots, Y_m$ have the same joint MGF and therefore the same joint distribution. Hence, $X_1, \ldots, X_m$ are nonsingular multivariate normal random variables. ◆

Proposition 12.6 gives a useful equivalent condition for random variables to be multivariate normal. In stating it, we make the convention that the zero random variable is normally distributed (with mean and variance 0).

◆◆◆ **Proposition 12.6**

$X_1, \ldots, X_m$ are multivariate normal random variables if and only if each linear combination of them is a normally distributed random variable.

Proof Suppose that each linear combination of $X_1, \ldots, X_m$ is a normally distributed random variable. Let $t_1, \ldots, t_m$ be real numbers. Then, by assumption, the random variable $Y = \sum_{i=1}^{m} t_i X_i$ is normally distributed. By applying basic properties of expected value and covariance, we find that

$$\mu_Y = \mu_{\mathbf{X}}' \mathbf{t} \quad \text{and} \quad \sigma_Y^2 = \mathbf{t}' \Sigma_{\mathbf{X}} \mathbf{t}. \tag{12.42}$$

Because Y is normal, $M_Y(t) = e^{\mu_Y t + \frac{1}{2}\sigma_Y^2 t^2}$. From this fact and Equations (12.42), we get

$$M_{\mathbf{X}}(\mathbf{t}) = M_{X_1,\ldots,X_m}(t_1, \ldots, t_m) = \mathcal{E}\left(e^{\sum_{i=1}^{m} t_i X_i}\right)$$

$$= \mathcal{E}\left(e^Y\right) = M_Y(1) = e^{\mu_Y + \frac{1}{2}\sigma_Y^2} = e^{\mu_{\mathbf{X}}' \mathbf{t} + \frac{1}{2}\mathbf{t}' \Sigma_{\mathbf{X}} \mathbf{t}}.$$

Hence, from Definition 12.5, $X_1, \ldots, X_m$ are multivariate normal random variables.

Conversely, suppose that $X_1, \ldots, X_m$ are multivariate normal random variables. Let $t_1, \ldots, t_m$ be real numbers and set $Y = \sum_{i=1}^{m} t_i X_i$. The mean and variance of Y are given in Equations (12.42), and we have, in view of Definition 12.5, that

$$M_Y(t) = \mathcal{E}\left(e^{tY}\right) = \mathcal{E}\left(e^{t\sum_{i=1}^{m} t_i X_i}\right) = \mathcal{E}\left(e^{\sum_{i=1}^{m}(t t_i)X_i}\right) = M_{\mathbf{X}}(t\mathbf{t})$$

$$= e^{\mu_{\mathbf{X}}'(t\mathbf{t}) + \frac{1}{2}(t\mathbf{t})'\Sigma_{\mathbf{X}}(t\mathbf{t})} = e^{(\mu_{\mathbf{X}}'\mathbf{t})t + \frac{1}{2}(\mathbf{t}'\Sigma_{\mathbf{X}}\mathbf{t})t^2} = e^{\mu_Y t + \frac{1}{2}\sigma_Y^2 t^2}.$$

Because this last expression is the MGF of a normally distributed random variable, Y is normally distributed. ◆

EXAMPLE 12.10 *A Singular Multivariate Normal Distribution*

Let Z_1 and Z_2 be independent standard normal random variables and set $X_1 = Z_1$, $X_2 = Z_2$, and $X_3 = Z_1 + Z_2$. Show that X_1, X_2, X_3 are singular multivariate normal random variables.

Solution Let Y be a linear combination of $X_1, X_2,$ and X_3. As $X_1, X_2,$ and X_3 are each a linear combination of Z_1 and Z_2, it follows that Y is a linear combination of Z_1 and Z_2. Because a linear combination of independent normal random variables is also normal, Y is normally distributed. Therefore, each linear combination of $X_1, X_2,$ and X_3 is normally distributed and hence Proposition 12.6 implies that X_1, X_2, X_3 are multivariate normal random variables.

To show that X_1, X_2, X_3 are singular, we first obtain their covariance matrix by applying standard properties of covariance:

$$\Sigma_{\mathbf{X}} = \begin{bmatrix} 1 & 0 & 1 \\ 0 & 1 & 1 \\ 1 & 1 & 2 \end{bmatrix}.$$

Simple calculations show that $\det \Sigma_{\mathbf{X}} = 0$ and, therefore, $\Sigma_{\mathbf{X}}$ is a singular matrix. Hence we conclude from Proposition 12.5 on page 717 that X_1, X_2, X_3 are singular multivariate normal random variables. ∎

Proposition 8.15 on page 468 shows that a linear function of a normal random variable is also a normal random variable—specifically, if $X \sim \mathcal{N}(\mu, \sigma^2)$ and a and $b \neq 0$ are real numbers, then $a + bX \sim \mathcal{N}(a + b\mu, b^2\sigma^2)$. This result generalizes to multivariate normal random variables, as Proposition 12.7 shows.

◆◆◆ **Proposition 12.7**   **Linear Function of Multivariate Normal Random Variables**

Suppose that $\mathbf{X} \sim \mathcal{N}_m(\boldsymbol{\mu}, \Sigma)$ and that $k \in \mathcal{N}$. Let $\mathbf{a}$ be a $k \times 1$ vector and let $\mathbf{B}$ be a $k \times m$ matrix. Then $\mathbf{a} + \mathbf{BX} \sim \mathcal{N}_k(\mathbf{a} + \mathbf{B}\boldsymbol{\mu}, \mathbf{B}\Sigma\mathbf{B}')$. In other words, if $\mathbf{X}$ has the multivariate normal distribution with parameters $\boldsymbol{\mu}$ and Σ, then $\mathbf{a} + \mathbf{BX}$ has the multivariate normal distribution with parameters $\mathbf{a} + \mathbf{B}\boldsymbol{\mu}$ and $\mathbf{B}\Sigma\mathbf{B}'$.

Proof   First we show that $\mathbf{B}\Sigma\mathbf{B}'$ is nonnegative definite. As Σ is nonnegative definite, it's symmetric. Hence, $(\mathbf{B}\Sigma\mathbf{B}')' = (\mathbf{B}')'\Sigma'\mathbf{B}' = \mathbf{B}\Sigma\mathbf{B}'$; so, $\mathbf{B}\Sigma\mathbf{B}'$ is symmetric. For $\mathbf{x} \in \mathcal{R}^k$,

$$\mathbf{x}'(\mathbf{B}\Sigma\mathbf{B}')\mathbf{x} = (\mathbf{x}'\mathbf{B})\Sigma(\mathbf{B}'\mathbf{x}) = (\mathbf{B}'\mathbf{x})'\Sigma(\mathbf{B}'\mathbf{x}) \geq 0,$$

where the inequality follows from the fact that Σ is nonnegative definite. Applying now basic properties of joint MGFs (see Exercise 12.47) and referring to Definition 12.5 on page 717, we get

$$M_{\mathbf{a}+\mathbf{BX}}(\mathbf{t}) = e^{\mathbf{a}'\mathbf{t}}M_{\mathbf{X}}(\mathbf{B}'\mathbf{t}) = e^{\mathbf{a}'\mathbf{t}}e^{\boldsymbol{\mu}'(\mathbf{B}'\mathbf{t})+\frac{1}{2}(\mathbf{B}'\mathbf{t})'\Sigma(\mathbf{B}'\mathbf{t})} = e^{(\mathbf{a}+\mathbf{B}\boldsymbol{\mu})'\mathbf{t}+\frac{1}{2}\mathbf{t}'(\mathbf{B}\Sigma\mathbf{B}')\mathbf{t}}.$$

Therefore, $\mathbf{a} + \mathbf{BX} \sim \mathcal{N}_k(\mathbf{a} + \mathbf{B}\boldsymbol{\mu}, \mathbf{B}\Sigma\mathbf{B}')$. ◆

Marginal Distributions of a Multivariate Normal

We already noted that all univariate marginal distributions of multivariate normal random variables are normal. The same is true for the multivariate marginals—all multivariate marginal distributions of multivariate normal random variables are normal, as Proposition 12.8 shows.

◆◆◆ **Proposition 12.8**   **Marginal Distributions of a Multivariate Normal**

Suppose that $X_1, \ldots, X_m$ are multivariate normal random variables with parameters $\boldsymbol{\mu}$ and Σ. Let k be a positive integer between 1 and m, inclusive, and let $i_1 < \cdots < i_k$ be distinct positive integers between 1 and m, inclusive. Then $X_{i_1}, \ldots, X_{i_k}$ have the multivariate normal distribution with parameters $\boldsymbol{\mu}^*$ and Σ^*, where $\boldsymbol{\mu}^*$ is the $k \times 1$ vector obtained by deleting all rows of $\boldsymbol{\mu}$ other than rows $i_1, \ldots, i_k$ and Σ^* is the $k \times k$ matrix obtained by deleting all rows and columns of Σ other than rows and columns $i_1, \ldots, i_k$.

Proof Let **B** be the $k \times m$ matrix obtained from the $m \times m$ identity matrix by deleting all rows other than rows $i_1, \ldots, i_k$. Set $\mathbf{Y} = \mathbf{BX}$. We observe that **Y** is the $k \times 1$ random vector whose entries are $X_{i_1}, \ldots, X_{i_k}$. From Proposition 12.7, we know that **Y** has the multivariate normal distribution with parameters $\mathbf{B\mu}$ and $\mathbf{B\Sigma B'}$. However, it's easy to see that $\mathbf{B\mu} = \mu^*$ and $\mathbf{B\Sigma B'} = \Sigma^*$.   ◆

It's also useful to note the following facts about multivariate normal random variables $X_1, \ldots, X_m$, which you are asked to verify in Exercises 12.56 and 12.58.

- If the random variables $X_1, \ldots, X_m$ are reordered, then the resulting random variables are also multivariate normal.
- The conditional distribution of any subcollection of $X_1, \ldots, X_m$, given any other subcollection of $X_1, \ldots, X_m$, is multivariate normal.

EXERCISES 12.3   Basic Exercises

Note: Because this section is theoretically oriented. we have included the basic theory exercises in the basic exercises.

12.41 Let $X_1, \ldots, X_m$ be random variables defined on the same sample space and having finite variances.
a) Show that the covariance matrix of $X_1, \ldots, X_m$ is nonnegative definite.
b) Provide an example where the covariance matrix is positive definite.
c) Provide an example where the covariance matrix isn't positive definite.

12.42 Let **a** be a $k \times 1$ vector and let **B** be a $k \times m$ matrix.
a) Suppose that $X_1, \ldots, X_m$ are random variables defined on the same sample space and having finite variances. Set $\mathbf{Y} = \mathbf{a} + \mathbf{BX}$. Show that $\mu_\mathbf{Y} = \mathbf{a} + \mathbf{B}\mu_\mathbf{X}$ and $\Sigma_\mathbf{Y} = \mathbf{B}\Sigma_\mathbf{X}\mathbf{B'}$.
b) Use part (a) to verify Equations (12.35) on page 714.

12.43 Let X_1, X_2, and X_3 be random variables with $\mathcal{E}(X_1) = -1$, $\mathcal{E}(X_2) = 0$, $\mathcal{E}(X_3) = 5$, $\mathrm{Var}(X_1) = 10$, $\mathrm{Var}(X_2) = 16$, $\mathrm{Var}(X_3) = 4$, $\mathrm{Cov}(X_1, X_2) = -5$, $\mathrm{Cov}(X_1, X_3) = 2$, and $\mathrm{Cov}(X_2, X_3) = -3$. Set $Y_1 = 3 + 2X_1 - 4X_2 + X_3$ and $Y_2 = -1 + X_2 - 4X_3$.
a) Use basic properties of means, variances, and covariances (e.g., bilinearity of covariance) to obtain the mean vector and covariance matrix of the random variables Y_1 and Y_2.
b) Obtain the mean vector and covariance matrix of X_1, X_2, and X_3.
c) Use your result from part (b) and Exercise 12.42(a) to obtain the mean vector and covariance matrix of Y_1 and Y_2.
d) Compare and reconcile your calculations in parts (a) and (c), noting especially the correspondence between the two methods.

12.44 Decide in each case whether multivariate normal random variables with the specified covariance matrix are nonsingular.

a) $\begin{bmatrix} 1 & -1 \\ -1 & 2 \end{bmatrix}$ **b)** $\begin{bmatrix} 1 & -1 \\ -1 & 1 \end{bmatrix}$ **c)** $\begin{bmatrix} 6 & 0 & -1 \\ 0 & 5 & 1 \\ -1 & 1 & 2 \end{bmatrix}$ **d)** $\begin{bmatrix} 1 & -2 & -1 \\ -2 & 4 & 2 \\ -1 & 2 & 1 \end{bmatrix}$

12.45 Suppose that $X_1, \ldots, X_m$ are nonsingular multivariate normal random variables with parameters μ and Σ. Referring to the discussion leading to Definition 12.4 on page 715, show that the univariate marginal distributions of $X_1, \ldots, X_m$ are all normal and that $\mu_\mathbf{X} = \mu$ and $\Sigma_\mathbf{X} = \Sigma$. *Hint:* From linear algebra, because Σ is positive definite, there exists a unique $m \times m$ positive definite matrix **R** such that $\mathbf{R}^2 = \Sigma$.

12.46 Verify that the concept of nonsingular multivariate normal random variables general-izes the concepts of univariate and bivariate normal random variables. Proceed as follows.

a) Show that a random variable X has a nonsingular univariate normal distribution in the sense of Definition 12.4 on page 715 if and only if it has a univariate normal distribution in the sense of Definition 8.6 on page 440.

b) Show that random variables X and Y have a nonsingular bivariate normal distribution in the sense of Definition 12.4 on page 715 if and only if they are bivariate normal random variables in the sense of Definition 10.4 on page 610.

12.47 Recall from Definition 11.3 that the joint MGF of random variables $X_1, \ldots, X_m$ defined on the same sample space is given by $M_{X_1,\ldots,X_m}(t_1, \ldots, t_m) = \mathcal{E}\left(e^{t_1 X_1 + \cdots + t_m X_m}\right)$.

a) Show that the joint MGF can be expressed in matrix form as $M_{\mathbf{X}}(\mathbf{t}) = \mathcal{E}\left(e^{\mathbf{X}'\mathbf{t}}\right)$.

b) Let $\mathbf{a}$ be a $k \times 1$ vector and let $\mathbf{B}$ be a $k \times m$ matrix. Show that $M_{\mathbf{a}+\mathbf{BX}}(\mathbf{t}) = e^{\mathbf{a}'\mathbf{t}} M_{\mathbf{X}}(\mathbf{B}'\mathbf{t})$.

12.48 Let $X_1, \ldots, X_m$ be nonsingular multivariate normal random variables with parameters $\boldsymbol{\mu}$ and $\boldsymbol{\Sigma}$, and set $\mathbf{Z} = \boldsymbol{\Sigma}^{-1/2}(\mathbf{X} - \boldsymbol{\mu})$. Show that $Z_1, \ldots, Z_m$ are independent standard normal random variables by using

a) the multivariate transformation theorem.

b) moment generating functions.

c) Proposition 12.7 on page 719.

12.49 Of what univariate concept is Equation (12.41) on page 716 the multivariate analogue?

12.50 Suppose that $X_1, \ldots, X_m$ have the nonsingular multivariate normal distribution with parameters $\boldsymbol{\mu}$ and $\boldsymbol{\Sigma}$. Show that the random variable $(\mathbf{X} - \boldsymbol{\mu})'\boldsymbol{\Sigma}^{-1}(\mathbf{X} - \boldsymbol{\mu})$ has the chi-square distribution with m degrees of freedom. *Hint:* Refer to Exercise 12.48.

12.51 Suppose that $X_1, \ldots, X_m$ have a nonsingular multivariate normal distribution. Show that there is an orthogonal matrix $\mathbf{B}$ such that $\mathbf{Y} = \mathbf{BX}$ has a multivariate normal distribution and $Y_1, \ldots, Y_m$ are independent random variables. *Hint:* From linear algebra, if $\mathbf{A}$ is a real symmetric matrix, there is an orthogonal matrix $\mathbf{P}$ such that $\mathbf{PAP}'$ is a diagonal matrix.

12.52 Let $\mathbf{B}$ be any matrix (or vector). Show that $\mathbf{BB}'$ is a covariance matrix—that is, $\mathbf{BB}'$ is nonnegative definite.

12.53 Let m and n be positive integers and let $Z_1, \ldots, Z_n$ be independent standard normal random variables. Also, let μ_i ($1 \le i \le m$) and b_{ij} ($1 \le i \le m$, $1 \le j \le n$) be real numbers. Set $X_i = \mu_i + \sum_{j=1}^{n} b_{ij} Z_j$ for $1 \le i \le m$. Without using Proposition 12.7 on page 719,

a) show that $X_1, \ldots, X_m$ are multivariate normal random variables.

b) determine the mean vector and covariance matrix of $X_1, \ldots, X_m$.

c) obtain a necessary and sufficient condition in terms of the matrix $[b_{ij}]$ for $X_1, \ldots, X_m$ to be singular.

12.54 Suppose that $X_1, \ldots, X_m$ are multivariate normal random variables with parameters $\boldsymbol{\mu}$ and $\boldsymbol{\Sigma}$, as defined in Definition 12.5 on page 717. Show that the univariate marginal distributions of $X_1, \ldots, X_m$ are all normal and that $\boldsymbol{\mu}_{\mathbf{X}} = \boldsymbol{\mu}$ and $\boldsymbol{\Sigma}_{\mathbf{X}} = \boldsymbol{\Sigma}$. *Hint:* Refer to Exercise 11.29 on page 649.

12.55 Let X_1, X_2, X_3, and X_4 have the multivariate normal distribution with parameters

$$\boldsymbol{\mu} = \begin{bmatrix} 2 \\ 1 \\ 0 \\ -1 \end{bmatrix} \quad \text{and} \quad \boldsymbol{\Sigma} = \begin{bmatrix} 3 & 0 & -1 & 0 \\ 0 & 6 & 3 & 2 \\ -1 & 3 & 6 & 1 \\ 0 & 2 & 1 & 2 \end{bmatrix}.$$

Obtain and identify all marginal distributions of X_1, X_2, X_3, and X_4.

Advanced Exercises

12.56 Let $X_1, \ldots, X_m$ be multivariate normal random variables with parameters μ and Σ and let $i_1, \ldots, i_m$ be a permutation of $1, \ldots, m$, denoted $\mathcal{P}$.

a) Prove that $X_{i_1}, \ldots, X_{i_m}$ are multivariate normal with parameters μ^* and Σ^*, where μ^* is the vector obtained by permuting the rows of μ according to $\mathcal{P}$ and Σ^* is the matrix obtained by permuting both the rows and columns of Σ according to $\mathcal{P}$.

b) Prove that, if $X_1, \ldots, X_m$ are nonsingular, then so are $X_{i_1}, \ldots, X_{i_m}$.

12.57 Marginals of multivariate normals: Let $\mathbf{X} \sim \mathcal{N}_m(\mu, \Sigma)$ and let j and k be positive integers such that $j + k = m$. Suppose that $\mathbf{X}$, μ, and Σ are partitioned as

$$\mathbf{X} = \begin{bmatrix} \mathbf{X}_1 \\ \mathbf{X}_2 \end{bmatrix}, \qquad \mu = \begin{bmatrix} \mu_1 \\ \mu_2 \end{bmatrix}, \qquad \text{and} \qquad \Sigma = \begin{bmatrix} \Sigma_{11} & \Sigma_{12} \\ \Sigma_{21} & \Sigma_{22} \end{bmatrix},$$

where $\mathbf{X}_1$ and μ_1 are $j \times 1$, $\mathbf{X}_2$ and μ_2 are $k \times 1$, Σ_{11} is $j \times j$, Σ_{12} is $j \times k$, Σ_{21} is $k \times j$, and Σ_{22} is $k \times k$.

a) Use MGFs to prove that $\mathbf{X}_1 \sim \mathcal{N}_j(\mu_1, \Sigma_{11})$.

b) Explain how to use the result of part (a) to obtain the marginal distribution of any subcollection of $X_1, \ldots, X_m$. *Hint:* Consider permutation.

c) Use MGFs to prove that $\mathbf{X}_1$ and $\mathbf{X}_2$ are independent if and only if $\Sigma_{12} = \mathbf{0}$.

12.58 Conditionals of multivariate normals: Let $\mathbf{X} \sim \mathcal{N}_m(\mu, \Sigma)$ and consider the partitioning in Exercise 12.57.

a) Prove that, if Σ_{22} is nonsingular, then the conditional distribution of $\mathbf{X}_1$ given $\mathbf{X}_2 = \mathbf{x}_2$ is multivariate normal with parameters $\mu_1 + \Sigma_{12}\Sigma_{22}^{-1}(\mathbf{x}_2 - \mu_2)$ and $\Sigma_{11} - \Sigma_{12}\Sigma_{22}^{-1}\Sigma_{21}$. *Note:* It can be shown that the conditional distribution of $\mathbf{X}_1$ given $\mathbf{X}_2 = \mathbf{x}_2$ is multivariate normal regardless of whether Σ_{22} is nonsingular.

b) Explain how to use the result of part (a) to obtain the conditional distribution of any subcollection of $X_1, \ldots, X_m$, given any other subcollection of $X_1, \ldots, X_m$. *Hint:* Consider permutation.

c) Show that the result of part (a) is consistent with that obtained for bivariate normal random variables in Proposition 10.10 on page 616.

12.59 Refer to Exercises 12.55 and 12.58. Obtain and identify the conditional distribution of

a) X_1 given $X_2 = x_2$, $X_3 = x_3$, and $X_4 = x_4$.

b) X_1 and X_2 given $X_3 = x_3$ and $X_4 = x_4$.

c) X_1, X_2, and X_3 given $X_4 = x_4$.

12.4 Sampling Distributions

Many aspects of probability theory are applied to mathematical statistics. In this section, we discuss some of the most important statistics and their probability distributions.

To begin, we recall the definition of a *statistic*. Let X be a random variable, and let $X_1, \ldots, X_n$ be a random sample of size n from the distribution of X—that is, $X_1, \ldots, X_n$ are independent random variables, all with the same probability distribution as X. A **statistic** is any function of $X_1, \ldots, X_n$ whose numerical value can be determined by knowing only the values of the random sample. Thus a statistic is a random variable of the form $g(X_1, \ldots, X_n)$, where $g: \mathcal{R}^n \to \mathcal{R}$ doesn't depend on any unknown parameters.

Generally, we devise a statistic for the purpose of making an inference (e.g., construct-ing a *confidence interval* or performing a *hypothesis test*) about an unknown parameter of the distribution of X. In doing so, we need to know the probability distribution of the statistic, which is called its **sampling distribution.** Here we examine sampling distri-butions associated with statistics that are used to make inferences about one mean, one variance, and the ratio of two variances.

Sampling Distributions for Mean and Variance

Consider a random variable X having a normal distribution with unknown mean μ and variance σ^2 (which may or may not be known). An important problem in statistics is to make inferences about μ based on a random sample from the distribution of X.

As we discussed earlier, the statistic used to obtain a numerical estimate—called a *point estimate*—of an unknown mean, μ, is the **sample mean,** $\bar{X}_n$, defined by

$$\bar{X}_n = \frac{X_1 + \cdots + X_n}{n} = \frac{1}{n} \sum_{j=1}^{n} X_j. \tag{12.43}$$

The sample mean—actually, a function of the sample mean—is also used to make inferences about μ. The exact form of the function depends on whether σ^2 is known or unknown. When σ^2 is known, inferences about μ are based on the random variable $Z = (\bar{X}_n - \mu)/(\sigma/\sqrt{n})$, whose probability distribution is obtained in Proposition 12.9.

◆◆◆   **Proposition 12.9   Standardized Version of the Sample Mean**

Let $X_1, \ldots, X_n$ be a random sample from a normal distribution with mean μ and variance σ^2. Then the random variable

$$Z = \frac{\bar{X}_n - \mu}{\sigma/\sqrt{n}} \tag{12.44}$$

has the standard normal distribution.

Proof   Because $X_1, \ldots, X_n$ are a random sample from a normal distribution, Proposition 9.14 on page 540 implies that $\bar{X}_n$ is also normally distributed. From Example 7.16 on page 369,

$$\mathcal{E}(\bar{X}_n) = \mu \quad \text{and} \quad \text{Var}(\bar{X}_n) = \sigma^2/n. \tag{12.45}$$

Referring now to Proposition 8.10 on page 441, we conclude that the standardized version of $\bar{X}_n$—that is, the random variable in Equation (12.44)—has the standard normal distribution.   ◆

When σ^2 is unknown, the appearance of its square root in the standardized version of $\bar{X}_n$, given in Equation (12.44), renders that random variable unsuitable for use in mak-ing inferences about μ. Instead, we replace the unknown parameter σ—the distribution standard deviation—by the random variable S_n—the **sample standard deviation**—which is defined by

$$S_n = \sqrt{\frac{1}{n-1} \sum_{k=1}^{n} (X_k - \bar{X}_n)^2}. \tag{12.46}$$

Thus, when σ^2 is unknown, inferences about μ are based on the random variable

$$T_n = \frac{\bar{X}_n - \mu}{S_n/\sqrt{n}}. \tag{12.47}$$

To determine the probability distribution of T_n, we first establish Proposition 12.10, which shows that, for normal distributions, the sample mean and sample variance (square of the sample standard deviation) are independent random variables.

◆◆◆ **Proposition 12.10**

Let $X_1, \ldots, X_n$ be a random sample from a normal distribution. Then the random variables $\bar{X}_n$ and S_n^2 are independent.

Proof   We begin by showing that the sample mean and each deviation from the sample mean are uncorrelated random variables—that is,

$$\text{Cov}(\bar{X}_n, X_k - \bar{X}_n) = 0, \qquad 1 \leq k \leq n. \tag{12.48}$$

Because $X_1, \ldots, X_n$ are independent random variables, each having variance σ^2,

$$\text{Cov}(X_j, X_k) = \begin{cases} \sigma^2, & \text{if } j = k; \\ 0, & \text{if } j \neq k. \end{cases} \tag{12.49}$$

Referring to Equations (12.49) and (12.45) and applying properties of covariance, we get

$$\text{Cov}(\bar{X}_n, X_k - \bar{X}_n) = \text{Cov}(\bar{X}_n, X_k) - \text{Cov}(\bar{X}_n, \bar{X}_n)$$

$$= \text{Cov}\left(\frac{1}{n}\sum_{j=1}^{n} X_j, X_k\right) - \text{Var}(\bar{X}_n)$$

$$= \frac{1}{n}\sum_{j=1}^{n} \text{Cov}(X_j, X_k) - \text{Var}(\bar{X}_n) = \frac{1}{n}\sigma^2 - \frac{\sigma^2}{n} = 0,$$

for each k. Hence Equation (12.48) holds.

Next we show that $\bar{X}_n, X_1 - \bar{X}_n, \ldots, X_n - \bar{X}_n$ are multivariate normal random variables. Let W be a linear combination of $\bar{X}_n, X_1 - \bar{X}_n, \ldots, X_n - \bar{X}_n$. As each of those random variables is a linear combination of the independent normal random variables $X_1, \ldots, X_n$, so is W. Hence, from Proposition 9.14 on page 540, W is normally distributed. Applying Proposition 12.6 on page 718, we conclude that $\bar{X}_n, X_1 - \bar{X}_n, \ldots, X_n - \bar{X}_n$ are multivariate normal random variables.

We now let Y be a normal random variable, independent of $X_1, \ldots, X_n$, and having the same mean and variance as $\bar{X}_n$—namely, μ and σ^2/n, respectively. Arguing as in the preceding paragraph, we find that $Y, X_1 - \bar{X}_n, \ldots, X_n - \bar{X}_n$ are multivariate normal random variables. Moreover, because Y is independent of $X_1, \ldots, X_n$, it's also independent of $X_1 - \bar{X}_n, \ldots, X_n - \bar{X}_n$. Thus $\text{Cov}(Y, X_k - \bar{X}_n) = 0$ for $1 \leq k \leq n$.

Equations (12.45) and (12.48) reveal that the multivariate normal random variables $\bar{X}_n, X_1 - \bar{X}_n, \ldots, X_n - \bar{X}_n$ and the multivariate normal random variables $Y, X_1 - \bar{X}_n, \ldots, X_n - \bar{X}_n$ have the same mean vector and covariance matrix. However, as we see from Definition 12.5 on page 717, the joint MGF—and hence the joint probability distribution—of multivariate normal random variables is completely determined by the mean vector and covariance matrix. Thus, $\bar{X}_n, X_1 - \bar{X}_n, \ldots, X_n - \bar{X}_n$ have the same joint distribution as $Y, X_1 - \bar{X}_n, \ldots, X_n - \bar{X}_n$. In particular, then, $\bar{X}_n$ is independent of $X_1 - \bar{X}_n, \ldots, X_n - \bar{X}_n$. From this fact and Proposition 6.13 on page 297, we conclude that $\bar{X}_n$ and S_n^2 are independent random variables. ◆

Note: For an alternate proof of Proposition 12.10, see Exercise 12.71.

The next step in obtaining the probability distribution of T_n is to identify the probability distribution of the sample variance—actually, of the random variable $(n-1)S_n^2/\sigma^2$—which we do in Proposition 12.11.[†]

◆◆◆ **Proposition 12.11**

Let $X_1, \ldots, X_n$ be a random sample from a normal distribution with variance σ^2. Then $(n-1)S_n^2/\sigma^2$ has the chi-square distribution with $n-1$ degrees of freedom.

Proof Let μ denote the mean of the normal distribution. Applying simple algebra and using the definitions of S_n^2 and $\bar{X}_n$ gives

$$(n-1)S_n^2 = \sum_{k=1}^{n}(X_k - \bar{X}_n)^2 = \sum_{k=1}^{n}\left((X_k - \mu) - (\bar{X}_n - \mu)\right)^2$$

$$= \sum_{k=1}^{n}(X_k - \mu)^2 - 2(\bar{X}_n - \mu)\sum_{k=1}^{n}(X_k - \mu) + n(\bar{X}_n - \mu)^2$$

$$= \sum_{k=1}^{n}(X_k - \mu)^2 - n(\bar{X}_n - \mu)^2.$$

or

$$(n-1)S_n^2 + n(\bar{X}_n - \mu)^2 = \sum_{k=1}^{n}(X_k - \mu)^2.$$

Dividing both sides of this equation by σ^2, we get

$$\frac{(n-1)S_n^2}{\sigma^2} + \left(\frac{\bar{X}_n - \mu}{\sigma/\sqrt{n}}\right)^2 = \sum_{k=1}^{n}\left(\frac{X_k - \mu}{\sigma}\right)^2. \tag{12.50}$$

To proceed, we need the following previously established results.

- The square of a standard normal random variable is a chi-square random variable with 1 degree of freedom, as stated in Proposition 8.13 on page 466.
- The sum of m independent chi-square random variables, each with 1 degree of freedom, is a chi-square random variable with m degrees of freedom, which follows from the second bulleted item on page 537.
- The MGF of a chi-square random variable with ν degrees of freedom is given by $(1 - 2t)^{-\nu/2}$, as stated in the third bulleted item on page 632.

[†]Inferences about an unknown variance, σ^2, of a normal distribution are also based on the random variable $(n-1)S_n^2/\sigma^2$. See Exercise 12.103.

Applying in turn the three bulleted items, we find that the random variable on the right in Equation (12.50) has the chi-square distribution with n degrees of freedom and, hence, has MGF $(1 - 2t)^{-n/2}$. And, referring to Proposition 12.9 on page 723 and the first and third bulleted items, we conclude that the second random variable in Equation (12.50) has the chi-square distribution with 1 degree of freedom and, hence, has MGF $(1 - 2t)^{-1/2}$.

Because of Proposition 12.10 and Proposition 6.9 on page 291, the two random variables on the left in Equation (12.50) are independent. Therefore, by Proposition 11.4 on page 636, the MGF of their sum equals the product of their MGFs. Setting $Y_n = (n - 1)S_n^2/\sigma^2$, we now conclude that

$$M_{Y_n}(t) \cdot (1 - 2t)^{-1/2} = (1 - 2t)^{-n/2},$$

or $M_{Y_n}(t) = (1 - 2t)^{-(n-1)/2}$. However, as the function on the right in this equation is the MGF of a chi-square random variable with $n - 1$ degrees of freedom, Y_n has that probability distribution. ◆

Let Z and Y be independent random variables such that Z has the standard normal distribution and Y has the chi-square distribution with ν degrees of freedom. Then, from Exercise 9.164 on page 553, the random variable $T = Z/\sqrt{Y/\nu}$ has a PDF given by

$$f_T(t) = \frac{\Gamma\big((\nu + 1)/2\big)}{\sqrt{\nu\pi}\,\Gamma(\nu/2)} (1 + t^2/\nu)^{-(\nu+1)/2}, \qquad \infty < t < \infty. \tag{12.51}$$

A random variable with this PDF is said to have the **Student's t-distribution with ν degrees of freedom.**

We are now in a position to identify the probability distribution of the random variable T_n in Equation (12.47) on page 724, which we do in Proposition 12.12.

◆◆◆ **Proposition 12.12** **Studentized Version of the Sample Mean**

Let $X_1, \ldots, X_n$ be a random sample from a normal distribution with mean μ. Then the random variable

$$T_n = \frac{\bar{X}_n - \mu}{S_n/\sqrt{n}} \tag{12.52}$$

has the Student's t-distribution with $n - 1$ degrees of freedom.

Proof Let σ^2 denote the variance of the normal distribution and, for convenience, set

$$Z = \frac{\bar{X}_n - \mu}{\sigma/\sqrt{n}} \qquad \text{and} \qquad Y_n = \frac{(n - 1)S_n^2}{\sigma^2}.$$

From Propositions 12.9–12.11, we know that Z has the standard normal distribution, that Z and Y_n are independent random variables, and that Y_n has the chi-square distribution with $n - 1$ degrees of freedom. Therefore, $Z/\sqrt{Y_n/(n - 1)}$ has the Student's t-distribution with $n - 1$ degrees of freedom. However,

$$\frac{Z}{\sqrt{Y_n/(n - 1)}} = \frac{\bar{X}_n - \mu}{\sigma/\sqrt{n}} \Bigg/ \sqrt{\frac{(n - 1)S_n^2}{\sigma^2} \Bigg/ (n - 1)} = \frac{\bar{X}_n - \mu}{S_n/\sqrt{n}}.$$

Thus the proposition is established. ◆

Sampling Distribution for Two Variances

Another important inference in statistics involves comparing the variances of two normal distributions, based on independent random samples. Specifically, we consider two normal distributions with unknown variances σ_1^2 and σ_2^2, and let $S_{n_1}^2$ and $S_{n_2}^2$ denote the sample variances obtained from independent random samples of sizes n_1 and n_2, respectively, from the two normal distributions. Inferences for comparing σ_1^2 and σ_2^2 are based on the random variable

$$F_{n_1, n_2} = \frac{S_{n_1}^2/\sigma_1^2}{S_{n_2}^2/\sigma_2^2}. \tag{12.53}$$

To obtain the probability distribution of F_{n_1, n_2}, we first establish Lemma 12.2.

◆◆◆ **Lemma 12.2   Quotient of Two Independent Chi-Square Random Variables**

Let U_1 and U_2 be independent chi-square random variables with ν_1 and ν_2 degrees of freedom, respectively. Then a PDF of the random variable $W = (U_1/\nu_1)/(U_2/\nu_2)$ is

$$f_W(w) = \frac{(\nu_1/\nu_2)^{\nu_1/2}\Gamma\big((\nu_1 + \nu_2)/2\big)}{\Gamma(\nu_1/2)\Gamma(\nu_2/2)} \frac{w^{\nu_1/2 - 1}}{\big(1 + (\nu_1/\nu_2)w\big)^{(\nu_1 + \nu_2)/2}}, \quad w > 0, \tag{12.54}$$

and $f_W(w) = 0$ otherwise.

Proof   We have $U_1 \sim \Gamma(\nu_1/2, 1/2)$ and $U_2 \sim \Gamma(\nu_2/2, 1/2)$. Applying Exercise 9.133(a) on page 544 and Equation (8.54) on page 456, we get, for $z > 0$,

$$f_{U_1/U_2}(z) = \frac{1}{B\left(\frac{\nu_2}{2}, \frac{\nu_1}{2}\right)} \frac{z^{\frac{\nu_1}{2} - 1}}{(z + 1)^{\frac{\nu_2}{2} + \frac{\nu_1}{2}}} = \frac{\Gamma\big((\nu_1 + \nu_2)/2\big)}{\Gamma(\nu_1/2)\Gamma(\nu_2/2)} \frac{z^{\nu_1/2 - 1}}{(1 + z)^{(\nu_1 + \nu_2)/2}}. \tag{12.55}$$

Now let $V = U_1/U_2$. Then $W = (\nu_2/\nu_1)V$. Applying first the univariate transformation theorem (Proposition 8.14 on page 466) and then Equation (12.55), we get, for $w > 0$,

$$f_W(w) = (\nu_1/\nu_2)f_V\big((\nu_1/\nu_2)w\big)$$

$$= (\nu_1/\nu_2)\frac{\Gamma\big((\nu_1 + \nu_2)/2\big)}{\Gamma(\nu_1/2)\Gamma(\nu_2/2)} \frac{\big((\nu_1/\nu_2)w\big)^{\nu_1/2 - 1}}{\big(1 + (\nu_1/\nu_2)w\big)^{(\nu_1 + \nu_2)/2}}$$

$$= \frac{(\nu_1/\nu_2)^{\nu_1/2}\Gamma\big((\nu_1 + \nu_2)/2\big)}{\Gamma(\nu_1/2)\Gamma(\nu_2/2)} \frac{w^{\nu_1/2 - 1}}{\big(1 + (\nu_1/\nu_2)w\big)^{(\nu_1 + \nu_2)/2}},$$

as required.   ◆

A random variable that has a PDF given by Equation (12.54) is said to have the **F-distribution with degrees of freedom ν_1 and ν_2.** Such random variables are of major importance in statistics. As we now show in Proposition 12.13, the random variable F_{n_1, n_2} in Equation (12.53) has an F-distribution.

◆◆◆   **Proposition 12.13   Quotient of Two Independent Sample Variances**

Let $S_{n_1}^2$ and $S_{n_2}^2$ be the sample variances of independent random samples of sizes n_1 and n_2, respectively, from normal distributions with variances σ_1^2 and σ_2^2, respectively. Then the random variable

$$F_{n_1, n_2} = \frac{S_{n_1}^2 / \sigma_1^2}{S_{n_2}^2 / \sigma_2^2} \tag{12.56}$$

has the F-distribution with degrees of freedom $n_1 - 1$ and $n_2 - 1$.

Proof   We have

$$F_{n_1, n_2} = \frac{S_{n_1}^2 / \sigma_1^2}{S_{n_2}^2 / \sigma_2^2} = \frac{\left[(n_1 - 1) S_{n_1}^2 / \sigma_1^2 \right] / (n_1 - 1)}{\left[(n_2 - 1) S_{n_2}^2 / \sigma_2^2 \right] / (n_2 - 1)}. \tag{12.57}$$

Applying Proposition 12.11 (page 725) and Proposition 6.9 (page 291), we conclude that $(n_1 - 1) S_{n_1}^2 / \sigma_1^2$ and $(n_2 - 1) S_{n_2}^2 / \sigma_2^2$ are independent chi-square random variables with degrees of freedom $n_1 - 1$ and $n_2 - 1$, respectively. The required result now follows from Equation (12.57) and Lemma 12.2.   ◆

EXERCISES 12.4   Basic Exercises

12.60 Let T have the Student's t-distribution with ν degrees of freedom. Determine and identify the probability distribution of the random variable T^2.

12.61 Let $X_1, \ldots, X_n$ be a random sample from a normal distribution with variance σ^2. Referring to Proposition 12.11 on page 725, show that S_n^2 is
a) an unbiased estimator of σ^2.
b) a consistent estimator of σ^2. *Hint:* Apply Chebyshev's inequality.

12.62 Let $X_1, \ldots, X_n$ be a random sample from a normal distribution with variance σ^2.
a) Obtain a formula for $\mathcal{E}(S_n)$, the expected value of the sample standard deviation.
b) Is S_n an unbiased estimator of σ? Explain your answer.

12.63 Confidence intervals for a mean: Suppose that a random sample of size n is taken from a normal distribution with mean μ.
a) For $0 < p < 1$, let $t_{p, \nu} = F_T^{-1}(p)$, where T is a random variable having the Student's t-distribution with ν degrees of freedom. Show that, for $0 < \gamma < 1$,

$$P\left(\bar{X}_n - t_{(1+\gamma)/2, n-1} \cdot S_n / \sqrt{n} \leq \mu \leq \bar{X}_n + t_{(1+\gamma)/2, n-1} \cdot S_n / \sqrt{n} \right) = \gamma.$$

b) Conclude from part (a) that a $100\gamma\%$ confidence interval for μ is from

$$\bar{x}_n - t_{(1+\gamma)/2, n-1} \cdot s_n / \sqrt{n} \qquad \text{to} \qquad \bar{x}_n + t_{(1+\gamma)/2, n-1} \cdot s_n / \sqrt{n},$$

where $\bar{x}_n$ and s_n are the observed values of $\bar{X}_n$ and S_n, respectively. Interpret this result.

12.64 A paper by Cho et al. in the May 2000 issue of *The Journal of Pediatrics* (Vol. 136(5), pp. 587–592) presented the results of research on various characteristics in children of diabetic mothers. Past studies have shown that maternal diabetes results in obesity, blood pressure, and glucose tolerance complications in the offspring. Following are the arterial blood pressures, in millimeters of mercury (mm Hg), for a random sample of 16 children of diabetic mothers.

81.6	84.1	87.6	82.8	82.0	88.9	86.7	96.4
84.6	104.9	90.8	94.0	69.4	78.9	75.2	91.0

a) Use these data and Exercise 12.63(b) to obtain a 95% confidence interval for the mean arterial blood pressure for all children of diabetic mothers. *Note:* If T has the Student's t-distribution with 15 degrees of freedom, then $F_T(2.13) = 0.975$. Also, for the sample data, $\bar{x}_n = 86.2$ mm Hg and $s_n = 8.5$ mm Hg.

b) Interpret your answer from part (a).

c) What assumptions did you make in applying Exercise 12.63(b) to obtain the confidence interval required here?

12.65 The mean retail price for bananas in 1998 was 51.0 cents per pound, as reported by the U.S. Department of Agriculture in *Food Cost Review*. Recently, a random sample of 15 markets gave the following prices for bananas, in cents per pound.

56	53	55	53	50	57	58	54
48	47	50	57	57	51	55	

a) The random variable T_n in Equation (12.52) on page 726 is frequently referred to as the *one-sample-t statistic*. Determine an expression in terms of the CDF of an appropriate Student's t-distribution that gives the probability of observing a value of the one-sample-t statistic at least as large in magnitude as that actually observed if, in fact, the mean retail price of bananas hasn't changed from that in 1998. State explicitly any assumptions that you make. *Note:* For the sample data, $\bar{x}_n = 53.4$ and $s_n = 3.5$.

b) Using statistical software, we find that the probability expression required in part (a) equals 0.019. Can you reasonably conclude that the current mean retail price of bananas is different from the 1998 mean of 51.0 cents per pound? Explain your reasoning.

12.66 Confidence intervals for the ratio of two variances: Suppose that independent random samples of sizes n_1 and n_2 are taken from normal distributions with variances σ_1^2 and σ_2^2, respectively, and let $S_{n_1}^2$ and $S_{n_2}^2$ denote the corresponding sample variances.

a) For $0 < p < 1$, let $f_{p,v_1,v_2} = F_W^{-1}(p)$, where W has the F-distribution with degrees of freedom v_1 and v_2. Show that, for $0 < \gamma < 1$,

$$P\left(\frac{S_{n_1}^2/S_{n_2}^2}{f_{(1+\gamma)/2, n_1-1, n_2-1}} \le \frac{\sigma_1^2}{\sigma_2^2} \le \frac{S_{n_1}^2/S_{n_2}^2}{f_{(1-\gamma)/2, n_1-1, n_2-1}} \right) = \gamma.$$

b) Conclude from part (a) that a $100\gamma\%$ confidence interval for σ_1^2/σ_2^2 is from

$$\frac{s_{n_1}^2/s_{n_2}^2}{f_{(1+\gamma)/2, n_1-1, n_2-1}} \quad \text{to} \quad \frac{s_{n_1}^2/s_{n_2}^2}{f_{(1-\gamma)/2, n_1-1, n_2-1}},$$

where $s_{n_1}^2$ and $s_{n_2}^2$ are the observed values of $S_{n_1}^2$ and $S_{n_2}^2$, respectively.

12.67 Variation within a method used for testing a product is an essential factor in deciding whether the method should be employed. Indeed, when the variation of such a test is high, ascertaining the true quality of a product is difficult. In the article "Using Repeatability and Reproducibility Studies to Evaluate a Destructive Test Method" (*Quality Engineering*, 1997, Vol. 10(2), pp. 283–290), A. Phillips et al. studied the variability of the Elmendorf tear test. That test is used to evaluate material strength for fiberglass shingles, paper quality, and other manufactured products. In one aspect of the study, the researchers independently and randomly obtained data on Elmendorf tear strength for three different vinyl floor coverings. The following table provides the data, in grams, for two of the three vinyl floor coverings.

Brand A					Brand B				
2288	2384	2368	2304	2528	2592	2384	2512	2432	2576
2240	2144	2208	2160	2112	2112	2176	2288	2304	2752

a) Referring to Exercise 12.66(b), find a 95% confidence interval for the ratio, σ_1^2/σ_2^2, of the variances of tear strength for Brand A and Brand B vinyl floor coverings. *Note:* If W has the F-distribution with degrees of freedom 9 and 9, then $F_W(0.248) = 0.025$ and $F_W(4.026) = 0.975$. Also, for the sample data, $s_{n_1} = 128.3$ g and $s_{n_2} = 199.7$ g.
b) Interpret your result in part (a).
c) What assumptions did you make in obtaining the confidence interval in part (a)?

12.68 Let T have the Student's t-distribution with ν degrees of freedom.
a) Determine when T has finite expectation.
b) Obtain $\mathcal{E}(T)$ when T has finite expectation.
c) Determine when T has finite variance.
d) Obtain $\text{Var}(T)$ when T has finite variance. *Hint:* Use integration by parts and then make the substitution $u = t\sqrt{(\nu - 2)/\nu}$.

12.69 Let $X_1, \ldots, X_n$ be a random sample from a normal distribution with mean μ and variance σ^2. Determine the mean and variance of
a) $(\bar{X}_n - \mu)/(\sigma/\sqrt{n})$.   b) $(\bar{X}_n - \mu)/(S_n/\sqrt{n})$. *Hint:* Refer to Exercise 12.68.
c) Compare your results in parts (a) and (b).

12.70 Simulation: In this exercise, you are to perform simulations to compare the standardized and studentized versions of the sample mean, as presented in Propositions 12.9 and 12.12 on pages 723 and 726, respectively.
a) Simulate 5000 samples of size 4 each from a normal distribution with mean 15 and standard deviation 0.8.
b) For each of the 5000 samples obtained in part (a), determine the observed values of both the standardized and studentized versions of the sample mean.
c) Obtain histograms of both the 5000 observed values of the standardized version of the sample mean and the 5000 observed values of the studentized version of the sample mean.
d) From your histograms, discuss the common features of the standardized and studentized versions of the sample mean.
e) From your histograms, identify an important difference between the standardized and studentized versions of the sample mean. Explain the reason for this difference intuitively and, by referring to Exercise 12.69, explain the reason for this difference mathematically.
f) Why would the histograms have the same look if the simulations were carried out with a different normal distribution?

Theory Exercises

12.71 Provide an alternate proof of Proposition 12.10 on page 724, by proceeding as follows.
a) Let $Y_1 = \bar{X}_n$, $Y_2 = X_1 - \bar{X}_n, \ldots, Y_{n+1} = X_n - \bar{X}_n$. Express $\mathbf{Y}$ in the form $\mathbf{Y} = \mathbf{BX}$.
b) Use Exercise 12.57(c) on page 722 and your result from part (a) to show that $\bar{X}_n$ is independent of $X_1 - \bar{X}_n, \ldots, X_n - \bar{X}_n$.
c) Deduce from part (b) that $\bar{X}_n$ and S_n^2 are independent random variables.

Advanced Exercises

Pooled-t procedures: Suppose that independent random samples of sizes n_1 and n_2 are taken from normal distributions with unknown means μ_1 and μ_2, respectively, and the same unknown variance, σ^2. Inferences about the two distribution means can be made by using *pooled-t procedures*, which you are to examine and apply in Exercises 12.72–12.76.

12.72 Pooled sample variance: To estimate σ^2, you could use either $S_{n_1}^2$ or $S_{n_2}^2$ (the sample variances for the two samples), both of which are unbiased estimators of σ^2, as shown in Exercise 12.61(a). Alternatively, the information from both samples could be "pooled" to obtain an estimator of σ^2 of the form $c S_{n_1}^2 + (1-c) S_{n_2}^2$, where $0 < c < 1$.
a) Show that $c S_{n_1}^2 + (1-c) S_{n_2}^2$ is an unbiased estimator of σ^2 for each $0 \le c \le 1$.
b) Determine the value of c that yields the estimator with the minimum variance.
c) Use your result from part (b) to conclude that, among the estimators of σ^2 considered in part (a),

$$ S_p^2 = \frac{(n_1 - 1)S_{n_1}^2 + (n_2 - 1)S_{n_2}^2}{n_1 + n_2 - 2} $$

has minimum variance. The random variable S_p^2 is called the *pooled sample variance.*

12.73 Pooled-t statistic: Suppose that independent random samples of sizes n_1 and n_2 are taken from $\mathcal{N}(\mu_1, \sigma^2)$ and $\mathcal{N}(\mu_2, \sigma^2)$ distributions, respectively, and let S_p^2 be the pooled variance, as defined in Exercise 12.72.
a) Show that $(n_1 + n_2 - 2)S_p^2/\sigma^2$ has the chi-square distribution with $n_1 + n_2 - 2$ degrees of freedom.
b) Let $\bar{X}_{n_1}$ and $\bar{X}_{n_2}$ denote the corresponding sample means. Prove that

$$ T = \frac{(\bar{X}_{n_1} - \bar{X}_{n_2}) - (\mu_1 - \mu_2)}{S_p \sqrt{(1/n_1) + (1/n_2)}} $$

has the Student's t-distribution with $n_1 + n_2 - 2$ degrees of freedom. The random variable T is called the *pooled-t statistic.*

12.74 Confidence intervals for the difference between two means: Suppose that independent random samples of sizes n_1 and n_2 are taken from $\mathcal{N}(\mu_1, \sigma^2)$ and $\mathcal{N}(\mu_2, \sigma^2)$ distributions, respectively. Apply Exercise 12.73 to obtain a formula for a $100\gamma\%$ confidence interval for the difference between μ_1 and μ_2.

12.75 Philosophical and health issues are prompting an increasing number of Taiwanese to switch to a vegetarian lifestyle. A study by Lu et al., published in the *Journal of Nutrition* (2000, Vol. 130, pp. 1591–1596), compared the daily intake of nutrients by vegetarians and omnivores living in Taiwan. A random sample of 51 female vegetarians had a mean daily protein intake of 39.04 g with a standard deviation of 18.82 g; and an independent random sample of 53 female omnivores had a mean daily protein intake of 49.92 g with a standard deviation of 18.97 g.
a) Referring to Exercise 12.74, obtain a 99% confidence interval for the difference between the mean daily protein intakes of female vegetarians and female omnivores. State explicitly any assumptions that you make. *Note:* For a random variable T having the Student's t-distribution with 102 degrees of freedom, $F_T(2.625) = 0.995$.
b) Based on your result in part (a), what conclusions would you draw regarding the two distribution means? Explain your reasoning.

12.76 The American Association of University Professors (AAUP) conducts salary studies of college professors and publishes its findings in *AAUP Annual Report on the Economic Status of the Profession.* In the year 2000, a random sample of 30 faculty members in public institutions had a mean annual salary of $57.48 thousand with a standard deviation of $23.95 thousand; and an independent random sample of 35 faculty members in private institutions had a mean annual salary of $66.39 thousand with a standard deviation of $22.26 thousand.
a) Referring to Exercise 12.73, find an expression in terms of the CDF of an appropriate Student's t-distribution that gives the probability of observing a value of the pooled-t statistic at least as large in magnitude as that actually observed if, in fact, the two salary distributions have the same mean. State explicitly any assumptions that you make.
b) Using statistical software, we find that the probability expression in part (a) equals 0.1252. Does this result provide reasonably strong evidence that a difference exists in mean annual salaries for faculty in public and private institutions? Explain your answer.

CHAPTER REVIEW

Summary

In this chapter, we examined a few of the myriad applications of probability theory. As a representative topic of the field of stochastic processes, we discussed, in Section 12.1, the Poisson process, probably the most important counting process. For a specified event (success) that occurs randomly and homogeneously in continuous time, the Poisson process provides a probability model for the frequency of the event's occurrence.

A Poisson process with rate λ can be characterized as a counting process in which counting begins at time 0, the number of successes in disjoint time intervals are independent random variables (independent increments), and the number of successes that occur in the time interval $(s, t]$ has the Poisson distribution with parameter $\lambda(t - s)$. An essential property of a Poisson process is that the interarrival times (elapsed times between successes) are independent exponential random variables with parameter λ.

Next, as a representative of the field of operations research, we presented, in Section 12.2, the fundamentals of queueing theory. We concentrated on birth-and-death queueing systems, characterized by Poisson input and exponential service times. Specifically, the elapsed times between one arriving customer and the next are independent exponential random variables; the elapsed times from one service completion to the next are independent exponential random variables; and arrival and service times are independent random variables. In a birth-and-death queueing system, the parameters (rates) of the exponential random variables are permitted to be state dependent—that is, to depend on the number of customers currently in the queueing system.

To analyze the long-run (steady-state) behavior of birth-and-death queueing systems, we applied the *rate-in = rate-out principle* to obtain the balance equations, which relate the steady-state probabilities to the arrival and service rates of the queueing system. As an application of the steady-state probabilities, we obtained the steady-state expected number of customers in the queueing system for the $M/M/1$ and $M/M/s$ queues.

As representatives of the field of statistics, we investigated, in Sections 12.3 and 12.4, the multivariate normal distribution and sampling distributions, respectively. The multivariate normal distribution provides a generalization of the univariate normal distribution (Section 8.5) and the bivariate normal distribution (Section 10.5), and is the most important multivariate distribution in statistics. To examine the multivariate normal distribution effectively, we introduced the concepts of the mean vector and covariance matrix of multivariate random variables. An essential fact about multivariate normal random variables is that their joint distribution is completely determined by their mean vector and covariance matrix.

In Section 12.4, we examined some of the most important sampling distributions—that is, probability distributions of statistics. We discovered that, for random samples from normal distributions, the standardized version of the sample mean has the standard normal distribution and the studentized version of the sample mean has a Student's t-distribution. Also, when suitably normalized, the sample variance has a chi-square distribution and the quotient of two independent sample variances has an F-distribution.

You Should Be Able To ...

1. define, state the basic properties of, and apply the Poisson process.
2. identify and apply the distributions of the waiting times and interarrival times for a Poisson process.
3. state the basic components of a queueing system.
4. define and apply the $M/M/1$ queue.
5. define a birth-and-death queueing system.
6. state and apply the *rate-in = rate-out principle*.
7. construct a rate transition diagram for a birth-and-death queueing system.
8. use the *rate-in = rate-out principle* to derive the balance equations for a birth-and-death queueing system.
9. determine the steady-state probabilities for a birth-and-death queueing system.
10. determine the steady-state expected number of customers in the system for an $M/M/1$ or $M/M/s$ queue.
11. define and apply the mean vector and covariance matrix.
12. define and state the basic properties of nonsingular multivariate normal random variables.
13. identify the MGF of nonsingular multivariate normal random variables.
14. define and state the basic properties of multivariate normal random variables.
15. define the concept of a sampling distribution.
16. identify the probability distributions of the standardized and studentized versions of the sample mean for random samples from a normal distribution.
17. identify the probability distribution of a suitably-normalized version of the sample variance for random samples from a normal distribution.
18. identify the probability distribution of a suitably-normalized version of the quotient of two sample variances for independent random samples from two normal distributions.

Key Terms

balance equations, *702*

birth-and-death queueing
    system, *699*

compound Poisson process, *693*

counting process, *686*

covariance matrix, *712*

determinant of a matrix, *712*

exponential service times, *698*

F-distribution, *727*

homogeneous Poisson
    process, *693*

independent increments, *687*

intensity function, *693*

interarrival time, *689*

inverse of a matrix, *712*

Little's formulas, *709*

$M/M/1$ queue, *698*

$M/M/1/N$ queue, *699*

$M/M/s$ queue, *699*

$M/M/s/N$ queue, *699*

mean function, *693*

mean vector, *712*

multivariate normal
    distribution, *717*

multivariate normal random
    variables, *717*

nonhomogeneous Poisson
    process, *693*

nonnegative definite matrix, *712*

nonsingular multivariate normal
    distribution, *715*

nonsingular multivariate normal
    random variables, *715*

Poisson input, *698*

Poisson process, *688*

positive definite matrix, *712*

queue, *697*

queueing systems, *697*

queueing theory, *697*

rate-in = rate-out principle, *701*

rate transition diagram, *701*

sample mean, *723*

sample standard deviation, *723*

sampling distribution, *723*

single-server queueing
    system, *698*

singular multivariate normal
    distribution, *717*

singular multivariate normal
    random variables, *717*

state, *699*

stationary increments, *688*

statistic, *722*

steady state, *701*

Student's *t*-distribution, *726*

success, *686*

symmetric matrix, *712*

transpose of a matrix, *712*

waiting time, *689*

Chapter Review Exercises

Basic Exercises

12.77 The following table shows the analogy between Bernoulli trials with success probability p and a Poisson process with rate λ. Supply the four missing entries.

Bernoulli trials	Poisson process
Expected number of successes per trial	Expected number of successes per unit time λ
Number of trials between successes Independent $\mathcal{G}(p)$	Elapsed time between successes
Number of trials until the rth success	Elapsed time until the rth success $\Gamma(r, \lambda)$
Number of successes in n trials $\mathcal{B}(n, p)$	Number of successes in $(s, t]$

12.78 Refer to Examples 12.1 and 12.2 on pages 688 and 691, respectively, where patients arrive at a hospital emergency room according to a Poisson process with rate 6.9 per hour. Assume that no patients are in the emergency room at midnight.

a) Determine the probability that there is no arrival-free span of 8 minutes or more up to the time when the third patient arrives.

b) Suppose that it takes 3 minutes for a patient to check in, following which the patient sees a doctor immediately. What is the probability that no patients have been seen by a doctor by 12:20 A.M.?

c) Answer part (b) if the check-in time has a triangular distribution on the interval $(2, 4)$ minutes, independent of the arrival times.

d) Suppose that check-in time has a triangular distribution on the interval $(2, 4)$ minutes and that the time a patient spends with a doctor has a uniform distribution on the interval $(5, 10)$ minutes. Further suppose that arrival times, check-in times, and service times are all independent random variables. What is the probability that the first patient to arrive after midnight is still in the emergency room after 12:20 A.M.?

12.79 Consider a Poisson process with unknown rate λ. Suppose that you want to estimate λ and that you have as data the value of $N(t)$, where t is some specified time.

a) Show that $N(t)/t$ is an unbiased estimator of λ: $\mathcal{E}(N(t)/t) = \lambda$ for each $t > 0$.

b) Show that $N(t)/t$ is a consistent estimator of λ:

$$\lim_{t \to \infty} P\left(\left| \frac{N(t)}{t} - \lambda \right| < \epsilon \right) = 1,$$

for each $\epsilon > 0$.

12.80 Let $\{N(t) : t \geq 0\}$ be a Poisson process with rate λ.

a) Show that $\lim_{n \to \infty} n/W_n = \lambda$ with probability 1.

b) Suppose that λ is unknown and that you want to estimate it. If you know only that the fourth success occurred at time 15, what would be your estimate?

12.81 Customers arrive at a service facility according to a Poisson process with rate λ. Two servers are available to serve arriving customers. The servers work independently of one another and in parallel, each with service times that are exponentially distributed with rate μ. However, if only one customer is in the system, both servers work independently on the required task, until one of them completes it.

a) Determine the steady-state distribution for the number of customers in the system.

b) Obtain the steady-state expected number of customers in the system.

12.82 Let $\{N(t) : t \geq 0\}$ be a Poisson process with rate λ and let $s, t \geq 0$.

a) Determine $\text{Cov}(N(s), N(t))$.

b) Determine the correlation coefficient of $N(s)$ and $N(s + t)$.

12.83 Beginning at 7:45 A.M., passengers arrive at a bus stop according to a Poisson process at a rate of one passenger per minute. The bus is scheduled to arrive at 8:00 A.M., but the actual arrival time, in hours after midnight, is a random variable T such that $4T - 32$ has the beta distribution with parameters 2 and 2. Determine the mean and standard deviation of the number of passengers who are waiting when the bus arrives.

12.84 Securities orders received at a large brokerage house can be classified as taxable or nontaxable. Taxable orders arrive according to a Poisson process $\{N_1(t) : t \geq 0\}$ with rate λ_1. Nontaxable orders arrive according to a Poisson process $\{N_2(t) : t \geq 0\}$ with rate λ_2, independent of the arrival of taxable orders. For each $t \geq 0$, let $N(t) = N_1(t) + N_2(t)$.

a) Interpret $\{N(t) : t \geq 0\}$.

b) Show that $\{N(t) : t \geq 0\}$ is a Poisson process with rate $\lambda_1 + \lambda_2$.

c) What is the probability that a securities order is taxable? Justify your answer. *Hint:* Refer to Proposition 9.11(b) on page 532.

d) Find the probability that two taxable orders are received before the first nontaxable order.

e) Determine the probability that n_1 taxable orders are received before n_2 nontaxable orders are received, where n_1 and n_2 are positive integers.

12.85 Combining Poisson processes: Generalize the result of Exercise 12.84 as follows. Suppose that $\{N_1(t) : t \geq 0\}, \ldots, \{N_m(t) : t \geq 0\}$ are independent Poisson processes with rates $\lambda_1, \ldots, \lambda_m$, respectively. For each each $t \geq 0$, let $N(t) = N_1(t) + \cdots + N_m(t)$. Also, for $1 \leq j \leq m$, refer to an event that occurs in the $\{N_j(t) : t \geq 0\}$ process as a Type j event.

a) Interpret $\{N(t) : t \geq 0\}$.

b) Show that $\{N(t) : t \geq 0\}$ is a Poisson process with rate $\lambda_1 + \cdots + \lambda_m$.

c) What is the probability that an event occurring in the $\{N(t) : t \geq 0\}$ process is a Type j event $(1 \leq j \leq m)$? Justify your answer. *Hint:* Refer to Proposition 9.11(b) on page 532.

12.86 Consider a compound Poisson process $\{Y(t) : t \geq 0\}$, where $X_1, X_2, \ldots$ are independent and identically distributed Bernoulli random variables with common parameter p. Show that $\{Y(t) : t \geq 0\}$ is a Poisson process with parameter $p\lambda$, where λ is the rate of the Poisson process $\{N(t) : t \geq 0\}$.

12.87 In Example 12.3 on page 698, we described the emergency room at Desert Samaritan Hospital as an $M/M/1$ queue.

a) In addition to the assumptions given in the statement of the example, what assumption is being made tacitly regarding the capacity of the emergency room?

b) In view of your answer to part (a), which birth-and-death queueing system would provide a better model for the emergency room?

c) Suppose that, in reality, the emergency room has a capacity of 40. Compare the steady-state distributions of the number of patients in the emergency room for the $M/M/1$ queue and the $M/M/1/N$ queue. *Note:* You may want to refer to Example 12.7(a) and Exercise 12.26 on pages 705 and 710, respectively.

d) Solve part (c) if the emergency room has a capacity of 100.

12.88 A factory has a very large number of identical machines, which arrive for repair at an automated service facility according to a Poisson process at a rate of five per day. The service facility can attend to only one machine at a time and repair time is exponentially distributed with a mean of 4 hours. It costs $247 per day to operate the repair facility, and, in addition, the company loses $50 per day for each day a machine isn't working.

a) What is the average cost per day to the company due to down machines, including the cost of operating the service facility?

b) A new service facility—whose mean repair time is 3 hours—can be used as a replacement. What is the maximum daily allowable operating cost for the new service facility before it's no longer economically feasible?

12.89 A mathematics department has two photocopying machines. One is in the faculty coffee room and is used by 12 people per hour, on average. The other is in the graduate student lounge and is used by 7 people per hour, on average. For both machines, the average job takes 4 minutes. Would the photocopying facilities be more effectively used if they were in the same room? Assume Poisson input and exponential service times.

12.90 A full-service gasoline station has three pumps and two attendants. Potential customers arrive according to a Poisson process at a rate of 25 cars per hour. If all the pumps are occupied, a potential customer won't enter. Service times are exponential and average 6 minutes.

a) Determine the expected number of customers in the gas station.

b) What percentage of potential customers is lost?

c) What percentage of the time are both attendants idle?

12.91 Infinite server queue ($M/M/\infty$): An $M/M/\infty$ queue is like an $M/M/s$ queue except that there are infinitely many servers.

a) Identify a queueing system in which it's reasonable to use an $M/M/\infty$ queue.

b) Obtain the arrival and service rates for an $M/M/\infty$ queue.

c) Determine and identify the steady-state distribution of the number of customers in the queueing system.

d) Obtain the steady-state expected number of customers in the queueing system.

e) Without doing any computations, determine L_q, W, and W_q.

12.92 A finite source queue: A factory has M machines working independently and subject to breakdown. There is one repairman who works on one machine at a time. The repair times are exponentially distributed with rate μ. Machines that break down when the repairman is busy wait their turn. Breakdown time for each machine is exponentially distributed with parameter λ. Define the state of the queueing system to be the number of down machines.

a) Obtain the arrival and service rates for this queueing system.

b) Determine the steady-state distribution for the number of down machines.

c) Show that the expected number of working machines is $\mu(1 - P_0)/\lambda$. *Hint:* Note that we can write $n = M - (M - n)$.

d) Apply your results in parts (a)–(c) when there are 10 machines, the mean time to breakdown is 6 hours, and the mean repair time is 1.5 hours.

12.93 Consider an $M/M/2$ queue with $\lambda < 2\mu$.

a) Show that $P_0 = (2\mu - \lambda)/(2\mu + \lambda)$.

b) Show that the steady-state expected number of customers in the queueing system is given by $L = 4\lambda\mu/(2\mu - \lambda)(2\mu + \lambda)$.

c) Refer to Example 12.5 on page 700. Use parts (a) and (b) to obtain P_0 and the steady-state expected number of patients in the emergency room when two doctors are on duty. Compare your answers to those found in Examples 12.7(b) and 12.9(b) on pages 705 and 708, respectively.

12.94 Two servers, each serving at exponential rate μ, are available to serve customers that arrive according to a Poisson process with rate λ, where $\lambda < 2\mu$. Determine the steady-state expected number of customers in the service facility if

a) the two servers are used to form an $M/M/2$ queueing system.

b) a single $M/M/1$ queueing system is formed where each server works independently on the required task and the customer is considered served as soon as one of the two servers completes the task.

c) two parallel $M/M/1$ queueing systems are formed with the two servers and each arriving customer is randomly (with equal probability 0.5) assigned to one of the two queueing systems. *Hint:* Refer to Exercise 12.6 on page 694.

d) Compare the results that you obtained in parts (a)–(c) to decide which configuration is most efficient; least efficient.

12.95 Let $X_1, \ldots, X_m$ be random variables with mean vector μ and covariance matrix $\mathbf{0}$ (the zero $m \times m$ matrix).

a) Determine the joint MGF of $X_1, \ldots, X_m$.

b) Obtain and identify the univariate marginal distributions of $X_1, \ldots, X_m$ and their joint distribution.

12.96 Let $\mathbf{X} \sim \mathcal{N}_m(\mu, \Sigma)$ be nonsingular and set $\mathbf{Y} = \mathbf{a} + \mathbf{BX}$, where $\mathbf{a}$ is an $m \times 1$ vector and $\mathbf{B}$ is an $m \times m$ nonsingular matrix. Show that $Y_1, \ldots, Y_m$ are nonsingular multivariate normal random variables with parameters $\mathbf{a} + \mathbf{B}\mu$ and $\mathbf{B}\Sigma\mathbf{B}'$.

12.97 Let X_1, X_2, X_3, and X_4 be random variables mean vector and covariance matrix given by

$$\mu = \begin{bmatrix} 1 \\ 0 \\ -2 \\ 0 \end{bmatrix} \quad \text{and} \quad \Sigma = \begin{bmatrix} 1.0 & 0.1 & 0.4 & 0.6 \\ 0.1 & 1.0 & 1.0 & 1.0 \\ 0.4 & 1.0 & 4.0 & 2.4 \\ 0.6 & 1.0 & 2.4 & 4.0 \end{bmatrix},$$

respectively. Find the correlation coefficient of $Y_1 = X_1 - 2X_2$ and $Y_2 = X_3 - 3X_4$.

12.98 Let X_1, X_2, and X_3 have a multivariate normal distribution with mean vector and covariance matrix given by

$$\mu = \begin{bmatrix} -1 \\ 0 \\ 2 \end{bmatrix} \quad \text{and} \quad \Sigma = \begin{bmatrix} 4 & 1 & 2 \\ 1 & 1 & 0 \\ 2 & 0 & 3 \end{bmatrix}.$$

a) Find c so that X_1 and $X_2 - cX_1$ are independent random variables.
b) Identify the joint distribution of X_1 and X_3.
c) Identify the distribution of $X_1 - 2X_2 + 3X_3$.
d) Are the random variables $X_1 + X_2 - X_3$ and $X_1 - 2X_2$ nonsingular bivariate normal random variables? Explain your answer.

12.99 Let X_1, X_2, and X_3 denote cholesterol level, systolic blood pressure, and diastolic blood pressure, respectively. In a certain population, X_1, X_2, and X_3 can be modeled as having a trivariate normal distribution with mean vector and covariance matrix given by

$$\mu = \begin{bmatrix} 150 \\ 120 \\ 80 \end{bmatrix} \quad \text{and} \quad \Sigma = \begin{bmatrix} 400 & 60 & 72 \\ 60 & 25 & 27 \\ 72 & 27 & 36 \end{bmatrix}.$$

a) Determine and interpret $\rho(X_1, X_2)$, $\rho(X_1, X_3)$, and $\rho(X_2, X_3)$.
Identify and interpret the
b) three univariate marginal distributions.
c) the three bivariate marginal distributions.
d) the conditional distribution of X_1 given $X_2 = 130$ and $X_3 = 90$. *Hint:* Refer to Exercise 12.58(a) on page 722.
e) the conditional distribution of X_2 and X_3 given $X_1 = 200$.

12.100 Compare the standard normal distribution and Student's t-distribution as follows.
a) On the same graph, plot the standard normal PDF and the PDFs of Student's t-distributions with degrees of freedom 1, 6, and 15.
b) What does your graph in part (a) indicate about the relationship among the standard normal distribution and Student's t-distributions?
c) Referring to Propositions 12.9 and 12.12 on pages 723 and 726, respectively, provide a heuristic argument to explain why a Student's t-distribution approaches the standard normal distribution as the number of degrees of freedom increases without bound. *Hint:* The sample standard deviation is a consistent estimator of the distribution standard deviation.

12.101 Let F have the F-distribution with degrees of freedom v_1 and v_2.
a) Without doing any calculations, explain why $1/F$ has the F-distribution with degrees of freedom v_2 and v_1.
b) Use the univariate transformation theorem (Proposition 8.14 on page 466) to show that $1/F$ has the F-distribution with degrees of freedom v_2 and v_1.

12.102 Let F have the F-distribution with degrees of freedom v_1 and v_2. Determine and identify the probability distribution of the random variable $1/\big(1 + (v_1/v_2)F\big)$.

12.103 Confidence intervals for a variance: Suppose that a random sample of size n is taken from a normal distribution with variance σ^2.

a) For $0 < p < 1$, let $\chi^2_{p,\nu} = F_X^{-1}(p)$, where X is a random variable having the chi-square distribution with ν degrees of freedom. Show that, for $0 < \gamma < 1$,

$$P\left(\frac{(n-1)S_n^2}{\chi^2_{(1+\gamma)/2,n-1}} \leq \sigma^2 \leq \frac{(n-1)S_n^2}{\chi^2_{(1-\gamma)/2,n-1}} \right) = \gamma.$$

b) Conclude from part (a) that a $100\gamma\%$ confidence interval for σ^2 is from

$$\frac{(n-1)s_n^2}{\chi^2_{(1+\gamma)/2,n-1}} \qquad \text{to} \qquad \frac{(n-1)s_n^2}{\chi^2_{(1-\gamma)/2,n-1}},$$

where s_n^2 is the observed value of S_n^2. Interpret this result.

12.104 Gas mileage estimates for cars and light-duty trucks are determined and published by the Environmental Protection Agency (EPA). The mileage estimate given for one 2003 model is 23 mpg on the highway. A random sample of 12 cars of this model yields the following highway mileages.

24.1	23.3	22.5	23.2	22.3	21.1
21.4	23.4	23.5	22.8	24.5	24.3

a) Referring to Exercise 12.103(b), find a 95% confidence interval for the standard deviation of highway gas mileages for all cars of the model and year in question. *Note:* If X has the chi-square distribution with 11 degrees of freedom, then $F_X(3.816) = 0.025$ and $F_X(21.920) = 0.975$.

b) Interpret your result in part (a).

c) What assumptions are you making in obtaining the confidence interval in part (a)?

12.105 Paired-t confidence intervals for the difference between two means: Suppose that a random sample of size n is taken from the distribution of (X, Y), where X and Y are random variables defined on the same sample space and having means μ_X and μ_Y, respectively. Further suppose that the random variable $D = X - Y$ is normally distributed.

a) For $0 < p < 1$, let $t_{p,\nu} = F_T^{-1}(p)$, where T is a random variable having the Student's t-distribution with ν degrees of freedom. Show that, for $0 < \gamma < 1$,

$$P\left(\bar{D}_n - t_{(1+\gamma)/2,n-1} \cdot S_n/\sqrt{n} \leq \mu_X - \mu_Y \leq \bar{D}_n + t_{(1+\gamma)/2,n-1} \cdot S_n/\sqrt{n} \right) = \gamma,$$

where $S_n^2 = (n-1)^{-1} \sum_{k=1}^{n} (D_k - \bar{D}_n)^2$.

b) Conclude from part (a) that a $100\gamma\%$ confidence interval for $\mu_X - \mu_Y$ is from

$$\bar{d}_n - t_{(1+\gamma)/2,n-1} \cdot s_n/\sqrt{n} \qquad \text{to} \qquad \bar{d}_n + t_{(1+\gamma)/2,n-1} \cdot s_n/\sqrt{n},$$

where $\bar{d}_n$ and s_n are the observed values of $\bar{D}_n$ and S_n, respectively.

c) Do bivariate normal random variables, X and Y, satisfy the conditions required for use of the confidence-interval procedure in part (b)? Explain your answer.

12.106 Published by the U.S. Bureau of the Census, *Current Population Reports* presents data on the ages of married people. Ten married couples are randomly selected and have the ages, in years, shown in the following table.

Husband	54	21	32	78	70	33	68	32	54	52
Wife	53	22	33	74	64	35	67	28	41	44

a) Referring to Exercise 12.105(b), find a 95% confidence interval for the difference between the mean age of married men and the mean age of married women. *Note:* If T has the Student's t-distribution with nine degrees of freedom, then $F_T(2.262) = 0.975$.

b) Obtain a 95% confidence interval for the mean age difference of married couples.

c) Regarding the distributional condition for the procedure that you used in parts (a) and (b), is it necessary that ages of married men and ages of married women are both normally distributed? Explain your answer.

d) Regarding the distributional condition for the procedure that you used in parts (a) and (b), is it sufficient that ages of married men and ages of married women are both normally distributed? Explain your answer.

Theory Exercises

12.107 Let $\{N(t) : t \geq 0\}$ be a Poisson process with rate λ and let $0 < t_1 < \cdots < t_m$. Find the joint PMF of the random variables $N(t_1), \ldots, N(t_m)$.

12.108 Let $\mathbf{X} \sim \mathcal{N}_m(\mu, \Sigma)$. Prove that the following statements are equivalent.

a) $X_1, \ldots, X_m$ are (mutually) independent.

b) Σ is a diagonal matrix.

c) X_i and X_j are uncorrelated for $i \neq j$.

d) $X_1, \ldots, X_m$ are pairwise independent.

Advanced Exercises

12.109 Consider a counting process $\{N(t) : t \geq 0\}$ with independent increments such that the expected number of successes is λt for any time interval of length t, where λ is a positive constant. Prove that, for $0 \leq s < t < \infty$, the random variable $N(t) - N(s)$ is unbounded; that is, there is no real number M such that $P\big(N(t) - N(s) \leq M\big) = 1$.

12.110 In Exercise 12.16 on page 696, you were asked to show that, in a Poisson process, events are distributed uniformly and independently in time. Here you are asked to provide another verification of that result.

a) Suppose that $X_1, \ldots, X_n$ are independent $\mathcal{U}(0, t)$ random variables. Let $m \in \mathcal{N}$ and let $0 = t_0 < t_1 < \cdots < t_m = t$ be a partition of the interval $[0, t]$. For $1 \leq j \leq m$, set $Y_j = N(\{k \in \{1, \ldots, n\} : X_k \in (t_{j-1}, t_j]\})$, where $N(E)$ denotes the number of elements of a finite set E. Find the joint PMF of $Y_1, \ldots, Y_m$.

b) Suppose that $\{N(t) : t \geq 0\}$ is a Poisson process with rate λ. For $t > 0$ and $m \in \mathcal{N}$, let $0 = t_0 < t_1 < \cdots < t_m = t$ be a partition of the interval $[0, t]$. For $1 \leq j \leq m$, set $Z_j = N(t_j) - N(t_{j-1})$. Find the conditional joint PMF of $Z_1, \ldots, Z_m$ given $N(t) = n$.

c) Compare and interpret your results from parts (a) and (b).

Equivalent conditions for a Poisson process: In Exercises 12.111 and 12.112, you are asked to show that a counting process $\{N(t) : t \geq 0\}$ is a Poisson process with rate λ if and only if it satisfies these five conditions.

    (1) $N(0) = 0$.

    (2) $\{N(t) : t \geq 0\}$ has independent increments.

    (3) $\{N(t) : t \geq 0\}$ has stationary increments.

    (4) $P(N(h) = 1) = \lambda h + o(h)$.

    (5) $P(N(h) \geq 2) = o(h)$.

Note: A function g is said to be $o(h)$ (read "little o of h") as $h \to 0$ if $\lim_{h \to 0} g(h)/h = 0$. Observe that a finite linear combination of $o(h)$ functions is $o(h)$.

12.111 Prove that a Poisson process with rate λ satisfies conditions (1)–(5). *Hint:* First use L'Hôpital's rule to show that $e^{-\lambda h} = 1 - \lambda h + o(h)$.

12.112 Prove that a counting process that satisfies conditions (1)–(5) is a Poisson process with rate λ. Use the following steps.
a) Explain why it suffices to prove that $N(t) \sim \mathcal{P}(\lambda t)$ for each $t > 0$.
b) Show that $P_n(t + h) = \lambda h P_{n-1}(t) + (1 - \lambda h) P_n(t) + o(h)$ for each nonnegative integer n and each $h > 0$, where $P_n(t) = P\big(N(t) = n\big)$.
c) Conclude from part (b) that $P'_n(t) = -\lambda P_n(t) + \lambda P_{n-1}(t)$.
d) Use part (c) and mathematical induction to prove that $P_n(t) = e^{-\lambda t} (\lambda t)^n / n!$, thereby establishing that $N(t) \sim \mathcal{P}(\lambda t)$.

12.113 Let $\{N(t) : t \geq 0\}$ be a Poisson process with rate λ. Fix $t > 0$ and let Z denote the length of the interarrival interval that contains t.
a) Make an educated guess for $\mathcal{E}(Z)$.
b) Let X denote the elapsed time from t to the next event. Determine the probability distribution of X and use it to obtain $\mathcal{E}(X)$.
c) Let Y denote the elapsed time from the last event that occurs before time t to t. Determine the probability distribution of Y and use it to obtain $\mathcal{E}(Y)$. *Hint:* Refer to Proposition 10.5 on page 577, which holds for any type of random variable.
d) Use your results from parts (b) and (c) to obtain $\mathcal{E}(Z)$. Compare your answer to your educated guess from part (a).
e) Roughly, for large t, what is the expected length of the interarrival interval that contains t? How does this expected length compare with the common expected length of the interarrival intervals?

12.114 Strong law for Poisson processes: Let $\{N(t) : t \geq 0\}$ be a Poisson process with rate λ. Prove that
a) $W_{N(t)} \leq t < W_{N(t)+1}$ for all $t \geq 0$.
b) $\lim_{n \to \infty} W_n / n = 1/\lambda$ with probability 1.
c) $\lim_{t \to \infty} N(t) = \infty$ with probability 1.
d) Use parts (a)–(c) to prove that $\lim_{t \to \infty} N(t)/t = \lambda$ with probability 1.
e) Interpret the result in part (d).

12.115 In this exercise, you are to derive the balance equations for a birth-and-death queueing system by first obtaining a collection of associated differential equations. Let $X(t)$ denote the number of customers in the queueing system at time t, and set $P_n(t) = P\big(X(t) = n\big)$.
a) Prove that, for each $h > 0$ and nonnegative integer n,

$$P_n(t + h) = \lambda_{n-1} h P_{n-1}(t) + \big(1 - (\lambda_n + \mu_n)h\big) P_n(t) + \mu_{n+1} h P_{n+1}(t) + o(h),$$

where $o(h)$ is defined just before Exercise 12.111 and, by convention, $\lambda_{-1} = 0$ and $\mu_0 = 0$.
b) Use part (a) to conclude that

$$P'_n(t) = \lambda_{n-1} P_{n-1}(t) - (\lambda_n + \mu_n) P_n(t) + \mu_{n+1} P_{n+1}(t),$$

for each nonnegative integer n.
c) Use a heuristic argument to obtain the balance equations from the differential equations in part (b).

12.116 Let $X_1, \ldots, X_m$ have a nonsingular multivariate normal distribution, let k be a positive integer at most m, let $\mathbf{a}$ be a $k \times 1$ vector, and let $\mathbf{B}$ be a $k \times m$ matrix having rank k. Show that $\mathbf{Y} = \mathbf{a} + \mathbf{BX}$ has a nonsingular multivariate normal distribution.

Little's formulas: For Exercises 12.117–12.119, refer to the discussion of Little's formulas on page 711 and the exercises that follow that discussion.

12.117 From Exercise 12.91(d), for an $M/M/\infty$ queue, the steady-state expected number of customers in the system is λ/μ. Use that result and Little's formulas to obtain and interpret L_q, W, and W_q.

12.118 Refer to the discussion of the finite source queue in Exercise 12.92.

a) Without doing any calculations, explain why $\bar{\lambda} = (M - L)\lambda$.

b) Use the definition of $\bar{\lambda}$ to obtain the result in part (a).

c) From Exercise 12.92(d), when $M = 10$, $\lambda = 1/6$, and $\mu = 2/3$, the expected number of down machines is roughly 6.02. Use that result and Little's formulas to obtain and interpret L_q, W, and W_q.

12.119 Consider a queueing system with s parallel servers and in steady-state.

a) Determine the expected number of customers in service in terms of $\bar{\lambda}$ and μ.

b) Use your result from part (a) to obtain the probability, p_b, that a specified server is busy in terms of $\bar{\lambda}$, μ, and s.

c) Deduce from part (b) that $P_0 = 1 - \bar{\lambda}/\mu$ for a single-server queue.

d) Use part (c) to obtain P_0 for the $M/M/1$ queue.

Answers to Selected Exercises

Note: Many of the probability answers are expressed as decimals rounded to three significant digits.

Chapter 1

EXERCISES 1.1

1.1 a) The five specified U.S. senators **b)** 0.6

1.3
a) Using the first letter of the last name as an abbreviation, the 20 possibilities are as follows: GB, GC, GM, GK, BG, BC, BM, BK, CG, CB, CM, CK, MG, MB, MC, MK, KG, KB, KC, KM.
b) It is equally probable. In both cases, the probability is 0.2 that Senator Graham will be chosen as chair.
c) 0.6

1.5 a) 0.105 **b)** 0.747 **c)** 0.156 **d)** 0.474
e) All murder cases during the year in question in which the murder victims were between 20 and 59 years old

1.7 a) 0.504 **b)** 0.0496

1.9 205 (Note that 2004 was a leap year.)

1.11 a) 1256 **b)** 333 **c)** 156

EXERCISES 1.2

1.15 Answers will vary.

1.17 Answers will vary.

1.19 a) $(1, 2)$ **b)** $[1, 2)$ **c)** $[1, 2]$ **d)** $\{3\}$
e) $\emptyset$ **f)** $\emptyset$ **g)** $[5, 6)$

1.21
a) $(0, 0, 0)$, $(0, 0, 1)$, $(0, 1, 0)$, $(0, 1, 1)$, $(1, 0, 0)$, $(1, 0, 1)$, $(1, 1, 0)$, $(1, 1, 1)$
b) $(0, 0, 1)$, $(0, 0, 2)$, $(0, 1, 1)$, $(0, 1, 2)$, $(1, 0, 1)$, $(1, 0, 2)$, $(1, 1, 1)$, $(1, 1, 2)$
c) (a, f), (a, g), (a, h), (b, f), (b, g), (b, h), (c, f), (c, g), (c, h), (d, f), (d, g), (d, h), (e, f), (e, g), (e, h)

d) (a, f), (a, g), (a, h), (b, f), (b, g), (b, h), (c, f), (c, g), (c, h), (d, f), (d, g), (d, h), (e, f), (e, g), (e, h)

1.23 One possibility is to take $I_n = (0, 1/n)$, $n \in \mathcal{N}$.

1.25 One possibility is $\mathcal{R} = \bigcup_{n=-\infty}^{\infty} (n - 1, n]$.

REVIEW EXERCISES

1.46 a) 0.6875 **b)** The 16 states in the South

1.47 a) 0.00893 **b)** 0.192 **c)** 0.576

1.48
a) In a large number of tosses of the die, the result will be 3 (i.e., three dots facing up) about 1/6 of the time.
b) In a large number of tosses of the die, the result will be 3 or more about 2/3 of the time.
c) 1667 **d)** 6667

1.49 $n(Y) \approx 0.6n$

1.50 a) 0, 1, and 4 **b)** $[0, 4)$

1.51 Associate 0 with a tail and 1 with a head. Then, for instance, the element $(0, 1, 1)$ of $\{0, 1\}^3$ corresponds to a tail on the first toss of the coin and heads on the second and third tosses. The probability of two heads and one tail is $3/8 = 0.375$.

1.52 A

1.53 a) $(0, 1]$ **b)** $(0, 1]$
c) The sets in the union in part (a) are pairwise disjoint; those in the union in part (b) are not pairwise disjoint, but their intersection is empty.

1.54 a) Finite **b)** Uncountable **c)** Uncountable
d) Countably infinite **e)** Uncountable
f) Countably infinite **g)** Finite

1.55 a) $\{3, 4, 5\}$ **b)** $\{3, 4, 5\}$

Chapter 2

EXERCISES 2.1

2.3 a) $[0, 1)$ **b)** $[1/4, 1/2]$
c) The spot is at most 1/3 unit from the center of the petri dish.

2.5
a) The collection of 39 trees
b) B^c is the event that the tree obtained has less than 20% seed damage; it consists of 21 trees.
c) $C \cap D$ is the event that the tree obtained has at least 50% seed damage but less than 60% seed damage; it consists of two trees.
d) $A \cup D$ is the event that the tree obtained has either less than 40% seed damage or at least 50% seed damage; it consists of 33 trees.
e) C^c is the event that the tree obtained has either less than 30% seed damage or at least 60% seed damage; it consists of 28 trees.
f) $A \cap D$ is the event that the tree obtained has less than 40% seed damage and at least 50% seed damage, which is impossible; it consists of zero trees.
g) Only A and D

2.7 $A_2 = \{(1, 1)\}$, $A_3 = \{(1, 2), (2, 1)\}$,
$A_4 = \{(1, 3), (2, 2), (3, 1)\}$,
$A_5 = \{(1, 4), (2, 3), (3, 2), (4, 1)\}$,
$A_6 = \{(1, 5), (2, 4), (3, 3), (4, 2), (5, 1)\}$,
$A_7 = \{(1, 6), (2, 5), (3, 4), (4, 3), (5, 2), (6, 1)\}$,
$A_8 = \{(2, 6), (3, 5), (4, 4), (5, 3), (6, 2)\}$,
$A_9 = \{(3, 6), (4, 5), (5, 4), (6, 3)\}$,
$A_{10} = \{(4, 6), (5, 5), (6, 4)\}$,
$A_{11} = \{(5, 6), (6, 5)\}$, $A_{12} = \{(6, 6)\}$

2.9
a) $\{T, HT, HHT, HHHT, \dots\}$
b) $\{HT, HHHT, HHHHHT, \dots\}$

2.11
a) $\Omega = \{4, 5, 6, 7, 8, 9, 10\}$
b) $A \cup B = \{4, 5, 6, 7, 8, 9, 10\}$, $A \cap B = \{6, 7, 8\}$, $A \cap B^c = \{9, 10\}$
c) A and B are not mutually exclusive; A and B^c are not mutually exclusive; A^c and B^c are mutually exclusive.

2.13 a) $A \cap B^c$ **b)** $(A \cap B^c) \cup (A^c \cap B)$
c) $(A \cap B^c \cap C^c) \cup (A^c \cap B \cap C^c) \cup (A^c \cap B^c \cap C)$
d) $A^c \cup B^c \cup C^c$

2.15 a) $(0, \infty)$ **b)** $\{0, 1, 2, \dots\}$ **c)** $A \subset B$

d) $\Omega = \{(t_1, t_2, t_3, \dots) : 0 < t_1 < t_2 < t_3 < \cdots\}$, where t_j denotes the elapsed time in hours from 6:00 P.M. until the arrival of the jth patient who arrives after 6:00 P.M.
e) False   **f)** False   **g)** False   **h)** False   **i)** False
j) True   **k)** True   **l)** False   **m)** False

EXERCISES 2.2

2.21
a) The required result follows from Proposition 2.3 with $\omega_1 = H$, $\omega_2 = T$, $p_1 = p$, and $p_2 = 1 - p$.
b) $P(\emptyset) = 0$, $P(\{H\}) = p$, $P(\{T\}) = 1 - p$, $P(\Omega) = 1$

2.23
a) All but assignment #3
b) #1: 3/4; 15/16; 63/64. #2: 2/3; 1; 1. #4: 0; 1; 1. #5: $1 - (1 - p)^2$; $1 - (1 - p)^4$; $1 - (1 - p)^6$.
c) No, because the probabilities of the individual outcomes don't sum to 1. In terms of the random experiment, $p = 0$ means that a success is impossible. Add an outcome $f, f, f, \dots$ and assign it probability 1.

2.25
a) The probability assigned to each of the 36 possible outcomes is a nonnegative number (1/36), and the sum of all the probabilities is 1 (36 · 1/36). Thus, in view of Proposition 2.3, the probability assignment is legitimate.
b) $P(A_i) = \begin{cases} (i - 1)/36, & 2 \le i \le 7; \\ (13 - i)/36, & 8 \le i \le 12. \end{cases}$
c) Answers will vary.   **d)** No. Yes.

2.27
a) $\Omega = \{(x, y) : x, y \in \{1, 2, 3, 4\}\}$
b) 1/16   **c)** 0.75

EXERCISES 2.3

2.35 a) 0.474   **b)** 0.474   **c)** 0.0526   **d)** 0.526
e) 0.526   **f)** 0.947

2.37
a) The six genotypes are equally likely.
b) 0.333
c) No, the six genotypes are not equally likely.
d) 0.123

2.39 a) 12   **b)** 94   **c)** 42   **d)** 54   **e)** 22
f) The player obtained has between 6 and 10 years of experience, inclusive; the player obtained weighs between 200 and 300 lb, inclusive; the player obtained weighs under 200 lb and has between 1 and 5 years of experience, inclusive.

g) 0.0851; 0.574; 0.0957

2.41 a) 0.167   **b)** 0.667   **c)** 0.167

2.43 a) 0.25   **b)** 0.75   **c)** 0.363   **d)** 0.315

2.45 a) 0.5   **b)** 0.75   **c)** 0.25   **d)** 0.25
e) 0.75   **f)** 0.375   **g)** 0.125

2.47 a) 0.984   **b)** 0.239   **c)** 0

2.49

a) $\begin{cases} 0, & \text{if } x < \ell\sqrt{3}/2; \\ \sqrt{4x^2 - 3\ell^2}/\ell, & \text{if } \ell\sqrt{3}/2 \le x < \ell; \\ 1, & \text{if } x \ge \ell. \end{cases}$

b) $F(y) - F(x)$, where F is the function in part (a).

2.51 Subjective

EXERCISES 2.4

2.59 Answers will vary.

2.61 0.833.

2.63 a) 0.553   **b)** 0.798   **c)** 0.532

2.65 a) 0.540   **b)** 0.490   **c)** 0.469   **d)** 0.030

2.67 a) 47/120   **b)** 7/30   **c)** 17/24   **d)** 9/10

2.69 a) 0.566   **b)** 0.546   **c)** 0.176
d) 0.071   **e)** 0.230

REVIEW EXERCISES

2.79 $\{(2, y) : y \in \{1, 2, 3, 4, 5, 6\}\}$

2.80

a) $\{\{a, b\}, \{a, c\}, \{a, d\}, \{a, e\}, \{a, f\}, \{b, c\},$
$\{b, d\}, \{b, e\}, \{b, f\}, \{c, d\}, \{c, e\}, \{c, f\},$
$\{d, e\}, \{d, f\}, \{e, f\}\}$

b) $\{\{a, b\}, \{a, c\}, \{a, d\}, \{a, e\}, \{a, f\}, \{b, f\},$
$\{c, f\}, \{d, f\}, \{e, f\}\}$

c) $\{\{a, d\}, \{a, e\}, \{a, f\}, \{b, d\}, \{b, e\}, \{b, f\},$
$\{c, d\}, \{c, e\}, \{c, f\}\};$
$\{\{a, d\}, \{a, e\}, \{a, f\}, \{b, f\}, \{c, f\}\}$

d) $\{(a, b), (a, c), (a, d), (a, e), (a, f),$
$(b, a), (b, c), (b, d), (b, e), (b, f),$
$(c, a), (c, b), (c, d), (c, e), (c, f),$
$(d, a), (d, b), (d, c), (d, e), (d, f),$
$(e, a), (e, b), (e, c), (e, d), (e, f),$
$(f, a), (f, b), (f, c), (f, d), (f, e)\};$
$\{(a, b), (a, c), (a, d), (a, e), (a, f), (b, a),$
$(b, f), (c, a), (c, f), (d, a), (d, f), (e, a),$
$(e, f), (f, a)(f, b), (f, c), (f, d), (f, e)\}$

2.81 There are six categories. For $1 \le i \le 6$, let x_i denote the number of the 100 cars observed that are in category i. A sample space for the random experiment is $\{(x_1, x_2, x_4, x_5, x_6) : x_i \in \{0, 1, 2, \ldots\}$ for $1 \le i \le 6$, and $\sum_{i=1}^{6} x_i = 100\}$.

2.82

a) Not mutually exclusive   **b)** Mutually exclusive
c) Mutually exclusive   **d)** Not mutually exclusive

2.83

a) $\{(A, B, C), (A, C, B), (B, A, C), (B, C, A),$
$(C, A, B), (C, B, A)\}$

b) $\{(A, B, C), (A, C, B)\}$

c) A is not the first alternate, B is not the second, and C is not the third; $\{(B, C, A), (C, A, B)\}$

d) A is the first alternate, B is the second, and C is the third; $\{(A, B, C)\}$

e) A is the first alternate and C is the third; $\{(A, B, C)\}$

f) Assign probability 1/6 to each of the six possible outcomes.

g) 1/3; 1/6; 1/6

2.84 Answers will vary.

2.85 The missing events in the first column of the table are: $A^c \cap B^c$; $A \cap B \cap C$; $A \cup (B \cap C)$; $A^c \cup B^c \cup C^c$; and $\bigcup_n A_n$. The missing descriptions in the second column of the table are: either A occurs or B doesn't; exactly one of A and B occurs; A occurs and at least one of B and C occurs; at least one of A, B, and C occurs; all of $A_1, A_2, \ldots$ occur.

2.86 $(A \cap B^c) \cup (A^c \cap B)$

2.87 a) True   **b)** False

2.88

a) Yes, because the numbers assigned to the individual outcomes are nonnegative and sum to 1.

b)

Event	Probability	Event	Probability
$\emptyset$	0	{HT, TH}	$2p(1 - p)$
{HH}	p^2	{HT, TT}	$1 - p$
{HT}	$p(1 - p)$	{TH, TT}	$1 - p$
{TH}	$p(1 - p)$	{HH, HT, TH}	$p(2 - p)$
{TT}	$(1 - p)^2$	{HH, HT, TT}	$1 - p + p^2$
{HH, HT}	p	{HH, TH, TT}	$1 - p + p^2$
{HH, TH}	p	{HT, TH, TT}	$1 - p^2$
{HH, TT}	$1 - 2p + 2p^2$	Ω	1

2.89

a) Take the sample space to be

$$\Omega = \{(x_1, x_2, x_3) : x_i \in \{w, b\}, i = 1, 2, 3\}$$
$$= \{(w, w, w), (w, w, b), (w, b, w), (w, b, b),$$
$$(b, w, w), (b, w, b), (b, b, w), (b, b, b)\}$$

and assign probability 1/8 to each of the eight individual outcomes.

b) 0.125   c) 0.5

2.90 a) 0.75   **b)** 0.841   **c)** 0.682

2.91

a) The numbers $1/2^n, n \in \mathcal{N}$, are nonnegative and they sum to 1.

b) 2/3

2.92

a) Because $B_6 \subset B_5$, the domination principle implies that $P(B_5) \geq P(B_6)$.

b) $\lim_{n\to\infty} P(B_n)$

c) Let B denote the event that you will never reach your goal. Observe that, if either the first three tosses are heads (which has probability 1/8) or the first two tosses are heads, the third is a tail, and the fourth, fifth, and sixth are heads (which has positive probability), then you will reach your goal. Hence, $P(B^c) > 1/8$, which implies that $P(B) < 7/8$.

2.93 5/9   **2.94** Answers will vary.

2.95 a) Empirical   **b)** Answers will vary.
c) Empirical   **d)** Answers will vary.

2.96 a) 0.55   **b)** 0.1

2.97 All frequencies are in thousands.

a) 91,621   **b)** 20,391   **c)** 4,461
d) 320   **e)** 24,532

f) C_1 is the event that the vehicle selected is in the United States; V_3 is the event that the vehicle selected is a truck; and $C_1 \cap V_3$ is the event that the vehicle selected is a truck in the United States.

g) 0.862; 0.359; 0.312   **h)** 0.908   **i)** 0.909

2.98 a) $P(A_i) = 1/3$ for $i = 1, 2, 3$ **b)** No.

2.99

a) 1/4 (*Note:* The outcome HHHH vacuously belongs to the event in question.)

b) 3/16

2.100 a) 1/6   **b)** 1/3   **c)** 2/3

2.101 a) 0.39   **b)** 0.27   **c)** 0.34

2.102 0.2

Chapter 3

EXERCISES 3.1

3.1 b) 12   **c)** 12

3.3 360

3.5 a) 100,000,000   **b)** 1,814,400
c) 47,829,690   **d)** 23,592,960

3.7 a) 80,000   **b)** 9000   **c)** 720,000,000

3.9 a) 36   **b)** 4   **c)** 6   **d)** 18

3.11 a) 1296   **b)** 360   **c)** 432   **d)** 671

3.13 a) 60   **b)** 96

3.15 a) 384   **b)** 96

3.17 $n(n-1)/2$

3.19 a) $n!$ **b)** $\binom{n}{2}n!$ **c)** $n^n - n!$

EXERCISES 3.2

3.25 $(18)_3 = 4896$

3.27 $7! = 5040$

3.29 $\binom{18}{3} = 816$

3.31 850,668

3.33 a) $\binom{80}{10} \approx 1.65 \times 10^{12}$ **b)** 184,756

3.35 351,520,000

3.37 a) 4800   **b)** 74,880   **c)** 449,280
d) 612,960   **e)** 1,224,000   **f)** 6,589,440
g) 14,826,240   **h)** 131,788,800

3.39 a) 75,287,520   **b)** 67,800,320

3.41

a) The number of ways that you can choose k objects from a collection of n objects to constitute a group (which is given by the expression on the left) equals the number of ways that you can choose $n - k$ objects from the n objects that won't be in the group (which is given by the expression on the right).

b) The number of possible ordered partitions of n objects into two distinct groups of sizes k and $n - k$ (which is given by the expression on the right) equals the number of ways that you can choose k objects from the n objects to constitute the first group (which is given by the expression on the left).

3.43 7560

3.45

a) $\binom{20}{8}(8)_3 = 42,325,920$

b) $\binom{20}{5}(15)_3 = 42,325,920$

c) $(20)_3\binom{17}{5} = 42,325,920$

d) From a list of n candidates, you must choose a committee of k and, from among those k, you must choose j officers. The three quantities presented in the equations in part (d) provide expressions for the number of ways of doing that by using the steps indicated in parts (a)–(c), respectively.

e) All three quantities presented in the equations in part (d) reduce to $n!/\big((n-k)!(k-j)!\big)$.

EXERCISES 3.3

3.57 a) 0.0000480   **b)** 0.676

3.59 a) 1/13   **b)** 0.0524

3.61 a) 0.864   **b)** $1-(365)_N/365^N$ **d)** 23

3.63 0.000216

3.65

a) $(N)_n/N^n$

b) Note that

$$\frac{(N)_n}{N^n} = \frac{N(N-1)(N-2)\cdots(N-n+1)}{N^n}$$

$$= \left(1-\frac{1}{N}\right)\left(1-\frac{2}{N}\right)\cdots\left(1-\frac{n-1}{N}\right).$$

This last expression approaches 1 as $N \to \infty$. The interpretation is that, when the population size is large relative to the sample size, then it is unlikely that a member of the population will be selected more than once for the sample.

3.67

a) $N(E) = n(N-1)_{n-1}$;
   $P(E) = n(N-1)_{n-1}/(N)_n = n/N$

b) As the kth member selected is equally likely to be any one of the N members of the population, conclude that $P(A_k) = 1/N$, for $k = 1, 2, \ldots, n$. The required result now follows because $E = \bigcup_{k=1}^{n} A_k$ and $A_1, A_2, \ldots, A_n$ are mutually exclusive.

3.69 a) 3,628,800   **b)** 0.000000276

3.71 a) $\frac{1}{N}\left(1-\frac{1}{N}\right)^{n-1}$ **b)** $1-\left(1-\frac{1}{N}\right)^{n}$

3.73

a) 0.0000154, 0.000240, 0.00144, 0.00197, 0.00392, 0.0211, 0.0475, 0.423

b) No, because for random samples without replacement, you can compute probabilities based on either ordered samples or unordered samples provided that you are consistent throughout the solution.

3.75 $\binom{n}{2}n!/n^n$

REVIEW EXERCISES

3.83 a) 151,200   **b)** 1,000,000   **c)** 0.1512

3.84 1,021,440

3.85 $(15)_3 = 2730$

3.86 8,145,060

3.87 a) 17,280   **b)** 2880

3.88 $\binom{50}{5}\binom{45}{10}\binom{35}{10} \approx 1.24 \times 10^{24}$

3.89 $\binom{100}{10}\binom{90}{7}\binom{83}{15}\binom{68}{8} \approx 1.16 \times 10^{49}$

3.90 a) 66   **b)** 1320

3.91 1680

3.92 a) 4200   **b)** 25,200   **c)** 151,200

3.93 a) 34,650   **b)** 64,864,800

3.94 0.341

3.95 a) 0.213   **b)** 0.000452   **c)** 0.0317
   d) 0.0128   **e)** 0.0511

3.96 a) 0.105   **b)** 0.250

3.97 0.369

3.98 a) 5/12   **b)** 4/9

3.99 a) 0.0889   **b)** 0.09

3.100 a) 0.154   **b)** 0.139

3.101 $\binom{k-1}{M-1}/\binom{N}{M}$

3.102 a) 0.000000191   **b)** 0.00180   **c)** 0.971
   d) 0.000000191N

3.103 0.375

3.104 0.0769

3.105 $\binom{N-k}{n-k}/\binom{N}{n}$

3.106

a) $\binom{n}{k}\binom{N-n}{m-k}/\binom{N}{m}$

b) $\binom{m}{k}\binom{N-m}{n-k}/\binom{N}{n}$

c) The expressions in parts (a) and (b) both represent the probability that the two samples have exactly k elements in common and hence the two expressions must be equal.

3.107 a) 6^n **b)** 5^{n-1} **c)** $5^{n-1}/6^n$

Chapter 4

EXERCISES 4.1

4.1 Answers will vary.

4.3 a) 1/13   **b)** 1/3   **c)** 1/13   **d)** 0   **e)** 3/13
f) 1   **g)** 3/13   **h)** 1/6

4.5 a) 0.275   **b)** 0.142
c) 27.5% of the faculty are assistant professors; 14.2% of the faculty in their 50s are assistant professors.

4.7
a)

Weight	Probability
Under 200	0.214
200–300	0.524
Over 300	0.262

b)

Experience	Rookie	1–5	6–10	10+
Probability		0.579	0.368	0.053	0.000

4.9 a) 1/4   **b)** 3/13

4.11 Assume that a child of the king is equally likely to be a boy or a girl, independent of previous births.
a) 1/2   **b)** 1/3

4.13 38.8%

4.15 a) 0.5   **b)** 0.5   **c)** 0.75

EXERCISES 4.2

4.23 a) 0.090
b) 9.0% of Americans with Internet access are regular Internet users who feel that the Web has reduced their social contact.

4.25 a) 0.0478   **b)** 0.0980   **d)** 0.245   **e)** 0.127

4.27 a) 0.216   **b)** 0.0883   **c)** 90.4%

4.29 0.108; 10.8% of Americans, 7 years old or older, play golf.

4.31 0.258

4.33 0.762

4.35 0.590

EXERCISES 4.3

4.43
a) By assumption, $P(B \mid A) = P(B)$. Applying the general multiplication rule gives

$$P(A \cap B) = P(A)P(B \mid A) = P(A)P(B),$$

so that A and B are independent in the sense of Definition 4.4.
b) By assumption, $P(A \cap B) = P(A)P(B)$ and also $P(A) > 0$. Applying the conditional probability rule gives

$$P(B \mid A) = \frac{P(A \cap B)}{P(A)} = \frac{P(A)P(B)}{P(A)} = P(B),$$

so that event B is independent of event A in the sense of Definition 4.3.

4.45 No, because

$$0.038 = P(A_2 \cap L_1) \neq P(A_2)P(L_1) = 0.051.$$

For a randomly selected U.S. citizen, 15 years of age or older, the event that the person obtained is between 25 and 44 years old and the event that the person lives alone are not independent.

4.47
a) Suppose first that $P(A) = 0$. Then, for each event B of the sample space, $P(A \cap B) = 0$. Consequently, $P(A \cap B) = P(A)P(B)$, which means that A and B are independent. Now assume that $P(A) = 1$. Then $P(A^c) = 0$, so, by what was just proved, A^c and B are independent for each event B. Applying Proposition 4.4 on page 149, conclude that A and B are independent for each event B.
b) If A and A are independent events, then

$$\left(P(A)\right)^2 = P(A)P(A) = P(A \cap A) = P(A).$$

Hence either $P(A) = 0$ or $P(A) = 1$.

4.49 False.

4.51 $2^n - n - 1$

4.55
a) 0
b) If A and B are independent events with $P(A) > 0$ and $P(B) > 0$, then $P(A \cap B) = P(A)P(B) > 0$. Hence, from part (a), A and B can't be mutually exclusive.
c) Answers will vary.

4.57

a) The event that exactly two of the four children are boys occurs if and only if the first two are boys and the second two are girls (which is represented by the first intersection of four events) or the first is a boy, the second is a girl, the third is a boy, and the fourth is a girl (which is represented by the second intersection of four events), and so on.

b) $P(S_2) = 6p^2(1-p)^2$

c) $P(S_k) = \binom{4}{k}p^k(1-p)^{4-k}$

EXERCISES 4.4

4.67 25.6%   **4.69** 0.0989

4.71

a) 1/6; 1/3; 1/2

b) Same as in part (a). The posterior probabilities of the urns, given that a red marble is obtained, are unaffected by the addition of the extra green marbles.

c) 9/24; 8/24; 7/24. 0.333; 0.333, 0.333. Thus, the posterior probabilities of urns 1 and 3, given that a green marble is obtained, are affected by the addition of the extra green marbles, whereas the posterior probability of urn 2, given that a green marble is obtained, is unaffected by the addition of the extra green marbles.

4.73 0.803   **4.75** 0.595

REVIEW EXERCISES

4.79 a) 3/8 **b)** 3/10 **c)** 9/14; 10/14 **d)** 3/7
e) 2/3 **f)** 9/14 **g)** 5/14

4.80 0.456

4.81

a) 0.352

b) No, as $P(L_3 \mid T_2) = 0.352 \neq 0.218 = P(L_3)$.

4.82 a) 0.00872 **b)** 0.0262 **c)** 0.167 **d)** 0.217

4.83 special multiplication rule

4.84 80%   **4.85** 0; 1/3

4.86 Answers will vary.

4.87 $\binom{k-1}{M-1} \big/ \binom{N}{M}$

4.88 a) 0.4 **b)** 0.4 **c)** 0.4

4.89 a) 0.518 **b)** 0.518

c) It can reasonably be assumed that the tosses of the die are independent of one another. This contextual independence permits you to obtain the probabilities required in parts (a) and (b).

4.90 a) 0.900 **b)** 0.633 **c)** 0.703 **d)** 0.658

4.91 a) 0.686 **b)** 0.0678 **c)** 0.271 **d)** 0.0402

4.92 No, because the percentages of males and females with activity limitations are not equal.

4.93 a) Independent **b)** Independent

4.94 a) Not independent **b)** Independent

4.95 0.0721. There is a 7.21% chance that at least one "criticality 1" item would fail; or, on average, at least one "criticality 1" item would fail in 7.21 of 100 such missions.

4.96 a) 0.244 **b)** 0.471 **c)** 0.402

d) Among U.S. engineers and scientists: 24.4% have a masters as their highest degree; 47.1% are engineers; 40.2% of those having a masters as their highest degree are engineers.

4.97 a) 2/3

b) The red face showing is equally likely to be any one of the three red sides, and the other side can be red in two of those three possibilities.

4.98 a) 2.3% **b)** 26.1%   **4.99** 2/3

4.100 a) $\dfrac{500p}{1+4p}\%$

Chapter 5

EXERCISES 5.1

5.1 Answers will vary.

5.3

a) Yes to both questions because X is a real-valued function on the sample space with a finite range—namely, {0, 1, 2, 3, 4, 5}.

b) At least four of the components are working:
$\{(s,s,s,s,f), (s,s,s,f,s), (s,s,f,s,s),$
$(s,f,s,s,s), (f,s,s,s,s), (s,s,s,s,s)\}$.

c) None of the components are working:
$\{(f,f,f,f,f)\}$.

d) All of the components are working:
$\{(s,s,s,s,s)\}$.

e) At least one of the components is working: This event consists of the 31 members of the sample space other than (f,f,f,f,f).

f) $\{X = 2\}$ g) $\{X \geq 2\}$

h) $\{X \leq 2\}$ i) $\{2 \leq X \leq 4\}$

5.5 Take $\Omega = \{ (x, y) : |x| \leq 3, |y| \leq 3 \}$. For brevity, set $d(x, y) = \sqrt{x^2 + y^2}$.

a) $S(x, y) = 10$ if $d(x, y) \leq 1$, $S(x, y) = 5$ if $1 < d(x, y) \leq 2$, and $S(x, y) = 0$ otherwise.

b) Yes, because S is a real-valued function on the sample space with a finite range—namely, $\{0, 5, 10\}$.

c) The archer's score is 5 points—that is, she hits the ring with inner radius 1 foot and outer radius 2 feet centered at the origin: $\{ (x, y) : 1 < d(x, y) \leq 2 \}$.

d) The archer's score is positive—that is, she hits the disk of radius 2 feet centered at the origin: $\{ (x, y) : d(x, y) \leq 2 \}$.

e) The archer's score is at most 7 points—that is, she hits outside the disk of radius 1 foot centered at the origin: $\{ (x, y) : d(x, y) > 1 \}$.

f) The archer's score exceeds 5 points but is at most 15 points—that is, she hits the disk of radius 1 foot centered at the origin: $\{ (x, y) : d(x, y) \leq 1 \}$.

g) The archer's score is less than 15 points, which is certain: Ω.

h) The archer's score is negative, which is impossible: $\emptyset$.

5.7

a) $\{(I, r, r), (I, r, w), (I, w, r),$
$(II, r, w), (II, w, r), (II, w, w)\}$.

b)

ω	$X(\omega)$
(I, r, r)	2
(I, r, w)	1
(I, w, r)	1
(II, r, w)	1
(II, w, r)	1
(II, w, w)	0

5.9

a)

ω	HHH	HHT	HTH	HTT	THH	THT	TTH	TTT
$Y(\omega)$	3	1	1	−1	1	−1	−1	−3

b) The number of heads equals the number of tails, which is impossible: $\{Y = 0\} = \emptyset$.

5.11

a) Yes, because W is a real-valued function on the sample space.

b) Yes, because the range of W is countable.

c) You don't win a prize the first week.

d) You win a prize by the tenth week.

e) It takes at least 15 weeks but less than 20 weeks before you win a prize.

EXERCISES 5.2

5.25

a)

x	$p_X(x)$	x	$p_X(x)$
0	0.00032	3	0.20480
1	0.00640	4	0.40960
2	0.05120	5	0.32768

b) 0.2624; $A = \{ x \in \mathcal{R} : 1 \leq x \leq 3 \} = [1, 3]$
c) $\{X = 5\}$; 0.32768 **d)** $\{X \geq 1\}$; 0.99968
e)

x	$p_X(x)$
0	$(1 - p)^5$
1	$5p(1 - p)^4$
2	$10p^2(1 - p)^3$
3	$10p^3(1 - p)^2$
4	$5p^4(1 - p)$
5	p^5

$5p(1 - p)^2(1 + p^2); \ p^5; \ 1 - (1 - p)^5$

5.27

a)

s	$p_S(s)$
0	0.6509
5	0.2618
10	0.0873

b) 0.2618 **c)** 0.3491 **d)** 0.9127 **e)** 0.0873
f) 1   **g)** 0

5.29

a)

x	0	1	2	3
$p_X(x)$	0.1212	0.4545	0.3636	0.0606

b) 0.8181

5.31

a) $p_W(w) = p(1 - p)^{w-1}$ for $w \in \mathcal{N}$; $p_W(w) = 0$ otherwise.

b) $(1 - p)^n$

c) You don't win a prize during any of the first n weeks.

d) $(1 - p)^n$ **e)** $1 - p$ **f)** $1 - (1 - p)^{10}$
g) $(1 - p)^{15} - (1 - p)^{20}$ **h)** $(1 - p)^{14} - (1 - p)^{19}$

5.35

a) The number of ways, $N(y)$, of getting each of the 16 possible sums are given in the following table. To obtain the PMF of Y, simply divide each $N(y)$ by 216.

y	N(y)	y	N(y)	y	N(y)	y	N(y)
3	1	7	15	11	27	15	10
4	3	8	21	12	25	16	6
5	6	9	25	13	21	17	3
6	10	10	27	14	15	18	1

EXERCISES 5.3

5.39

a) Here the random experiment consists of administering pyrantel pamoate to a child with pinworm infestation; each administration (to a different child) constitutes one trial. The specified event is that a child is cured; thus a success is a cure and a failure is a noncure. The success probability, p, is the probability that, for any particular treated child, the child is cured; the failure probability, $q = 1 - p$, is the probability that, for any particular treated child, the child is not cured. By assumption, $p = 0.9$, so $q = 0.1$. Finally, it can be reasonably assumed that the results of administering pyrantel pamoate are independent from one child to another.

b)

Outcome	Probability	Outcome	Probability
(s, s, s)	0.729	(f, s, s)	0.081
(s, s, f)	0.081	(f, s, f)	0.009
(s, f, s)	0.081	(f, f, s)	0.009
(s, f, f)	0.009	(f, f, f)	0.001

c) (s, s, f), (s, f, s), and (f, s, s)

d) 0.081. Because each probability is obtained by multiplying two success probabilities of 0.9 and one failure probability of 0.1.

e) 0.243

f)

x	0	1	2	3
$p_X(x)$	0.001	0.027	0.243	0.729

5.41 For random sampling with replacement, a member selected is returned to the population for possible reselection; hence the result of one trial has no effect on the results of other trials (i.e., the trials are independent). However, for random sampling without replacement, a member selected isn't returned to the population for possible reselection; hence the result of one trial affects the results of the other trials (i.e., the trials aren't independent).

5.43 a) 0.25 **b)** 0.763

c) Let Y denote the number of children of the five who have sickle cell anemia.

y	$p_Y(y)$	y	$p_Y(y)$
0	0.237305	3	0.087891
1	0.395508	4	0.014648
2	0.263672	5	0.000977

5.45

a) For this baseball player, successive trials at bat are independent, and the probability of a hit on any given at bat coincides with the player's batting average.

b) 0.279

5.47 a) 0.698; 7 and 8   **b)** 0.563; 6

5.49 $npq(p^{n-2} + q^{n-2})$, where $q = 1 - p$

5.51

a) Consider the experiment of randomly selecting a number from the interval $(0, 1)$, and let the specified event be that the number obtained exceeds 0.7. In three independent repetitions of this random experiment, the median of the three numbers obtained exceeds 0.7 if and only if two or more of them exceed 0.7—that is, if and only if two or more of the three trials result in success.

b) 0.3

EXERCISES 5.4

5.59 a) 35   **b)** 0   **c)** -84 **d)** 0   **e)** 0
f) 0   **g)** -36 **h)** 0   **i)** 7/128
j) 0   **k)** $-15/128$ **l)** 0   **m)** No.

5.61 a) 0.0660

b) Hypergeometric with parameters 52, 5, and 3/13; $\binom{12}{x}\binom{40}{5-x}/\binom{52}{5}$, $x = 0, 1, \ldots, 5$

5.63 a) 0.897   **b)** 0.0833

5.65 0.476

5.67

a)

x	$p_X(x)$	x	$p_X(x)$
0	0.72908522	3	0.00116115
1	0.24302841	4	0.00001873
2	0.02670642	5	0.00000008

b)

Defectives x	Hypergeometric probability	Binomial probability
0	0.72908522	0.73390402
1	0.24302841	0.23422469
2	0.02670642	0.02990102
3	0.00116115	0.00190858
4	0.00001873	0.00006091
5	0.00000008	0.00000078

EXERCISES 5.5

5.77 a) 0.224   **b)** 0.423   **c)** 0.353
d) 0.647   **e)** 0.577

5.79 a) 0.311   **b)** 0.757   **c)** 0.507
d)

x	$p_X(x)$	x	$p_X(x)$
0	0.183	4	0.064
1	0.311	5	0.022
2	0.264	6	0.006
3	0.150	7	0.001

5.81 a) 0.000000118   **b)** 0.000000123

5.83 0.905

5.85 Let Y denote the number of eggs observed in a nest. Then

$$p_Y(y) = \frac{e^{-4}4^y/y!}{1 - e^{-4}}, \qquad y \in \mathcal{N},$$

and $p_Y(y) = 0$, otherwise.

EXERCISES 5.6

5.93
a) $p_X(x) = 0.67(0.33)^{x-1}$ for $x \in \mathcal{N}$, and $p_X(x) = 0$ otherwise. $X \sim \mathcal{G}(0.67)$—that is, X has the geometric distribution with parameter 0.67.
b) 0.0730; 0.109; 0.964   **c)** 5

5.95 a) $(1 - p)/(2 - p)$ **b)** $1/(2 - p)$

c) $1 - (1 - p)^6$ **d)** $p(1 - p)^{k-1}/\left(1 - (1 - p)^n\right)$
e) $p(1 - p)^{n-k-1}/\left(1 - (1 - p)^{n-1}\right)$
5.97 a) $p(1 - p)^{k-n-1}$ **b)** $p(1 - p)^{k-n-1}$
5.99 Let $q = 1 - p$.
a) $p_X(x) = pq^{x-1}$ for $x \in \mathcal{N}$, and $p_X(x) = 0$ otherwise. Note that $X \sim \mathcal{G}(p)$.
b) $p_X(x) = Np(Nq)_{x-1}/(N)_x$ for $x = 1, 2, \ldots, N$, and $p_X(x) = 0$ otherwise.
c) $1/N$

5.101 Let $q_i = 1 - p_i$ for $i = 1, 2$.
a) $\frac{1}{2} + \frac{1}{2}(p_1 - p_2)/(1 - q_1 q_2)$ **b)** $p_1 = p_2$

EXERCISES 5.7

5.111
a) Both $I_{E \cap F}$ and $I_E \cdot I_F$ are 1 if $E \cap F$ occurs and both are 0 otherwise.
b) Events E and F are mutually exclusive.
c) $I_{E \cup F} = I_E + I_F - I_E \cdot I_F$

5.113
a) For $j = 1, 2, \ldots, n$, let E_j denote the event of a success on trial j. Then $X = \sum_{j=1}^{n} I_{E_j}$.
b) For $j = 1, 2, \ldots, n$, let X_j equal 1 or 0 if the jth trial is a success or failure, respectively. Then X_1, X_2, $\ldots$, X_n are Bernoulli random variables (each with parameter p) and $X = \sum_{j=1}^{n} X_j$.

5.115
a) $p_Y(y) = 1/10$ if $y = 0, 1, \ldots, 9$, and $p_Y(y) = 0$ otherwise. Y has the discrete uniform distribution on $\{0, 1, \ldots, 9\}$.
b) Same answer as in part (a).

5.117 a) 0.110   **b)** 0.612   **c)** 0.710

5.119 0.975

EXERCISES 5.8

5.131
a)

y	-6	-3	0	3	6	9
$p_Y(y)$	0.10	0.20	0.20	0.15	0.15	0.20

b) The transformation results in a change of scale.
c) Yes; a change of scale is a one-to-one function.

5.133

a)

y	-8	-5	-2	1	4	7
$p_Y(y)$	0.10	0.20	0.20	0.15	0.15	0.20

b)

y	$a-2b$	$a-b$	a	$a+b$	$a+2b$	$a+3b$
$p_Y(y)$	0.10	0.20	0.20	0.15	0.15	0.20

c) Yes; a simultaneous location and scale change is a one-to-one function.

5.135

a) $\{n/(n+1) : n \in \mathcal{N}\}$

b) Let R_Y denote the range of Y, obtained in part (a). $p_Y(y) = p_X\big(y/(1-y)\big)$ if $y \in R_Y$, and $p_Y(y) = 0$ otherwise.

c) $R_Z = \{(n+1)/n : n \in \mathcal{N}\}$, where R_Z denotes the range of Z; $p_Z(z) = p_X\big(1/(z-1)\big)$ if $z \in R_Z$, and $p_Z(z) = 0$ otherwise.

5.137 $p_Y(y) = 27e^{-3}\big(3^y/(3+y)! + 3^{-y}/(3-y)!\big)$ for $y = 1, 2, 3$; $p_Y(y) = 27e^{-3}3^y/(3+y)!$ for $y = 0, 4, 5, 6, \ldots$; and $p_Y(y) = 0$ otherwise.

5.139

a) $Y = 1 - 3/X$

b) $R_Y = \{1 - 3/x : x = 3, 4, 5, \ldots\}$, where R_Y is the range of Y. The PMF of Y is

$$p_Y(y) = \binom{(2+y)/(1-y)}{2} p^3(1-p)^{3y/(1-y)},$$

for $y \in R_Y$, and $p_Y(y) = 0$ otherwise.

REVIEW EXERCISES

5.142 a) 1, 2, 3, and 4 **b)** $\{X = 3\}$

c) 0.252; 25.2% of ASU undergraduates who attended the main campus (during the semester under consideration) were juniors.

d)

x	1	2	3	4
$p_X(x)$	0.191	0.210	0.252	0.347

5.143 a) $\{Y = 4\}$ **b)** $\{Y \geq 4\}$ **c)** $\{2 \leq Y \leq 4\}$
d) 0.174; 0.322; 0.646

5.144

b) Because the range of X is (countably) infinite.

c) 1/16 **d)** 1/16 **e)** 2^{-x}; 2^{-x}

5.145

a) 0.4

b)

Outcome	Probability	Outcome	Probability
(s,s,s)	0.064	(f,s,s)	0.096
(s,s,f)	0.096	(f,s,f)	0.144
(s,f,s)	0.096	(f,f,s)	0.144
(s,f,f)	0.144	(f,f,f)	0.216

c) (s,s,f), (s,f,s), and (f,s,s)

d) 0.096. Because each probability is obtained by multiplying two success probabilities of 0.4 and one failure probability of 0.6.

e) 0.288

f) Binomial with parameters $n = 3$ and $p = 0.4$

g)

x	0	1	2	3
$p_X(x)$	0.216	0.432	0.288	0.064

5.146 a) 0.0864 **b)** 0.870 **c)** 0.216 **d)** 3

e) $p_Y(y) = 0.4(0.6)^{y-1}$ if $y \in \mathcal{N}$, and $p_Y(y) = 0$ otherwise. Y has the geometric distribution with parameter $p = 0.4$.

5.147 0.962

5.148 5

5.149 a) 0.253 **b)** 0.346 **c)** 0.654 **d)** 0.746

e)

x	$p_X(x)$	x	$p_X(x)$
0	0.000796	3	0.252850
1	0.012607	4	0.400346
2	0.079847	5	0.253553

f) The PMF obtained in part (e) assumes sampling with replacement, whereas the sampling is actually without replacement. Thus the binomial approximation to the hypergeometric distribution (Proposition 5.6 on page 215) has been used.

g) $p_X(x) = \binom{0.76N}{x}\binom{0.24N}{5-x}/\binom{N}{5}$ for $x = 0, 1, 2, \ldots, 5$, and $p_X(x) = 0$ otherwise.

5.150 a) k/n **b)** k/n

5.151 a) $n(E)/n$ **b)** $n(E)/n$

5.152 a) 126 **b)** 0 **c)** 495 **d)** 0 **e)** 0
f) 0 **g)** -220 **h)** 0 **i)** $-5/128$ **j)** 0
k) 35/128 **l)** 0

5.153 Let Y denote the number of tests required under the alternative scheme. Then $p_Y(1) = (1-p)^n$, $p_Y(n+1) = 1 - (1-p)^n$, and $p_Y(y) = 0$ otherwise.

5.154 a) 0.00101 **b)** 0.569 **c)** 0.818
d) 0.136 **e)** 0.798

5.155
a) Under the assumptions, X has the binomial distribution with parameters $n = 2500$ and $p = 0.001$. The PMF of X is $p_X(x) = \binom{2500}{x}(0.001)^x(0.999)^{2500-x}$ for $x = 0, 1, \ldots, 2500$, and $p_X(x) = 0$ otherwise.
b) The Poisson distribution with parameter $\lambda = 2.5$, which has PMF $e^{-2.5}(2.5)^x/x!$ for $x = 0, 1, \ldots$.
c) 0.1087; 0.1088

5.156 a) 0.266 **b)** 0.0986 **c)** 0.826
d)

x	$p_X(x)$	x	$p_X(x)$
0	0.174	4	0.068
1	0.304	5	0.024
2	0.266	6	0.007
3	0.155	7	0.002

f) The probability distribution is right skewed. Yes, all Poisson distributions are right skewed.

5.157 a) 0.00024 **b)** 0.261; 0.692

5.158 By interchanging the roles of success and failure, Proposition 5.7 can be applied to approximate the binomial distribution with parameters n and $1 - p$ by the Poisson distribution with parameter $\lambda = n(1-p)$. By again interchanging the roles of success and failure, the required Poisson approximation to the original binomial distribution is obtained.

5.159 Let Y denote the weekly gas sales, to the nearest gallon.
a) The PMF of Y is given by $p_Y(y) = e^{-\lambda}\lambda^y/y!$ for $y = 0, 1, \ldots, m-1$, $p_Y(m) = e^{-\lambda}\sum_{k=m}^{\infty}\lambda^k/k!$, and $p_Y(y) = 0$ otherwise.
b) $e^{-\lambda}\sum_{k=m}^{\infty}\lambda^k/k!$

5.160 a) 0.0918 **b)** 0.380 **c)** 0.380

5.161
a) Both probabilities equal $(1-p)^6$.
b) Consider Bernoulli trials with success probability p. Let X denote the number of trials until the first success and let Y denote the number of successes in the first six trials. Then $X \sim \mathcal{G}(p)$ and $Y \sim \mathcal{B}(6; p)$. Event $\{X > 6\}$ occurs if and only if there are no suc-

cesses in the first six trials, which happens if and only if event $\{Y = 0\}$ occurs. Thus, $\{X > 6\} = \{Y = 0\}$.

5.162
a) X has the negative binomial distribution with parameters 3 and p. Hence, $p_X(x) = \binom{x-1}{2}p^3(1-p)^{x-3}$ for $x = 3, 4, \ldots$, and $p_X(x) = 0$ otherwise.
b) 38 terms. $P(X \le 40) = \sum_{x=3}^{40}\binom{x-1}{2}p^3(1-p)^{x-3}$.
c) Let Y denote the number of successes in the first 40 trials, and observe that $Y \sim \mathcal{B}(40, p)$. Noting that $\{X > 40\} = \{Y \le 2\}$, the complementation rule gives $P(X \le 40) = 1 - \sum_{y=0}^{2}\binom{40}{y}p^y(1-p)^{40-y}$.

5.163 0.683

Chapter 6

EXERCISES 6.1

6.1
a) $p/(2-p)$. The probability that the lifetimes of the two components are the same is $p/(2-p)$.
b) $(1-p)/(2-p)$. The probability that the first component outlasts the second is $(1-p)/(2-p)$.
c) Because the joint PMF of X and Y is symmetric in x and y.
d) $(1-p)/(2-p)$

6.3
a) $p_{X,Y}(x, y) = 1/90$ for $x, y \in S$ and $x \ne y$, and $p_{X,Y}(x, y) = 0$ otherwise.
b) 0 **c)** 0.5
d) $p_Y(y) = 1/10$ for $y \in S$, and $p_Y(y) = 0$ otherwise.
e) $p_Y(y) = 1/10$ for $y \in S$, and $p_Y(y) = 0$ otherwise.

6.5 Applying the FPF to the set $A = \{x_0\} \times \mathcal{R}$ gives

$$p_X(x_0) = P(X = x_0) = P(X = x_0, Y \in \mathcal{R})$$
$$= \sum_{x=x_0, y \in \mathcal{R}} p_{X,Y}(x, y)$$
$$= \sum_{y \in \mathcal{R}}\left(\sum_{x=x_0} p_{X,Y}(x, y)\right)$$
$$= \sum_{y} p_{X,Y}(x_0, y),$$

which is Equation (6.3) on page 263. Equation (6.4) on page 263 is obtained similarly.

6.7
a) $p_{X,Y}(x, y) = \binom{3}{x, y, 3-x-y}\left(\frac{1}{4}\right)^x\left(\frac{1}{4}\right)^y\left(\frac{1}{2}\right)^{3-x-y}$ for x and y nonnegative integers whose sum is at most 3, and $p_{X,Y}(x, y) = 0$ otherwise.

b)

	Democrats, y				
Greens, x		0	1	2	3	$p_X(x)$

Greens, x	0	1	2	3	$p_X(x)$
0	$\frac{1}{8}$	$\frac{3}{16}$	$\frac{3}{32}$	$\frac{1}{64}$	$\frac{27}{64}$
1	$\frac{3}{16}$	$\frac{3}{16}$	$\frac{3}{64}$	0	$\frac{27}{64}$
2	$\frac{3}{32}$	$\frac{3}{64}$	0	0	$\frac{9}{64}$
3	$\frac{1}{64}$	0	0	0	$\frac{1}{64}$
$p_Y(y)$	$\frac{27}{64}$	$\frac{27}{64}$	$\frac{9}{64}$	$\frac{1}{64}$	1

c) 11/32. The probability is 11/32 that more Greens than Democrats will be selected.

d) $\{X + Y = 3\}$ **e)** 1/8 **f)** $\mathcal{B}(3, 1/2)$; 1/8

g) $X \sim \mathcal{B}(3, 1/4)$. $p_X(x) = \binom{3}{x}\left(\frac{1}{4}\right)^x\left(\frac{3}{4}\right)^{3-x}$ for $x = 0, 1, 2, 3$, and $p_X(x) = 0$ otherwise.

h) $X + Y \sim \mathcal{B}(3, 1/2)$. $p_{X+Y}(z) = \binom{3}{z}/8$ for $z = 0, 1, 2, 3$, and $p_{X+Y}(z) = 0$ otherwise.

i) 3/4 **j)** 3/4

6.9 a) $X \sim \mathcal{G}(p)$ **b)** $Y \sim \mathcal{NB}(2, p)$

c) $p_{X,Y}(x, y) = p^2(1 - p)^{y-2}$ for $x, y \in \mathcal{N}$, $y > x$, and $p_{X,Y}(x, y) = 0$ otherwise.

d) $p_X(x) = p(1 - p)^{x-1}$ for $x \in \mathcal{N}$, and $p_X(x) = 0$ otherwise; and $p_Y(y) = (y - 1)p^2(1 - p)^{y-2}$ for $y = 2, 3, \ldots$, and $p_Y(y) = 0$ otherwise.

e) $Y - X$ is the number of trials from the first success to the second success.

f) $Y - X \sim \mathcal{G}(p)$

g) $P(Y - X = z) = p(1 - p)^{z-1}$ for $z \in \mathcal{N}$.

6.11 Let $q = 1 - p$.

a) {HHH, HHT, HTH, HTT, THH, THT, TTH, TTT}

b)

		y		
	−3	−1	1	3
−2	q^3	pq^2	0	0
x 0	0	$2pq^2$	$2p^2q$	0
2	0	0	p^2q	p^3

c) q

d) q. Event $\{X > Y\}$ means that the third toss is a tail, which has probability q.

e) $p_X(-2) = q^2$, $p_X(0) = 2pq$, $p_X(2) = p^2$, and $p_X(x) = 0$ otherwise.

f) Write $X = 2V - 2$, where V is the number of heads in the first two tosses. As $V \sim \mathcal{B}(2, p)$, it follows that $p_X(x) = \binom{2}{1+x/2}p^{1+x/2}q^{1-x/2}$ if $x = -2, 0, 2$, and $p_X(x) = 0$ otherwise.

g) $p_Y(-3) = q^3$, $p_Y(-1) = 3pq^2$, $p_Y(1) = 3p^2q$, $p_Y(3) = p^3$, and $p_Y(y) = 0$ otherwise.

h) Write $Y = 2W - 3$, where W is the number of heads in the first three tosses. As $W \sim \mathcal{B}(3, p)$, it follows that $p_Y(y) = \binom{3}{(3+y)/2}p^{(3+y)/2}q^{(3-y)/2}$ if $y = -3, -1, 1, 3$, and $p_Y(y) = 0$ otherwise.

i) $Y - X$ equals −1 or 1, respectively, if the third toss is a tail or a head; so, $p_{Y-X}(-1) = q$, $p_{Y-X}(1) = p$, and $p_{Y-X}(z) = 0$ otherwise.

6.13

a) $p_X(x) = e^{-\lambda}\lambda^x/x!$ if $x = 0, 1, 2, \ldots$; $p_X(x) = 0$ otherwise. $p_Y(y) = e^{-\mu}\mu^y/y!$ if $y = 0, 1, 2, \ldots$; $p_Y(y) = 0$ otherwise. $X \sim \mathcal{P}(\lambda)$ and $Y \sim \mathcal{P}(\mu)$.

b) The joint PMF of X and Y equals the product of the marginal PMFs.

6.15

a) $p_X(x) = \left(\sum_y r(y)\right)q(x)$ for all $x \in \mathcal{R}$; $p_Y(y) = \left(\sum_x q(x)\right)r(y)$ for all $y \in \mathcal{R}$.

b) From Proposition 6.1(c), $\left(\sum_x q(x)\right)\left(\sum_y r(y)\right) = 1$. This equation and the results of part (a) show that $p_{X,Y}(x, y) = p_X(x)p_Y(y)$ for all $x, y \in \mathcal{R}$. Thus the joint PMF of X and Y equals the product of the marginal PMFs.

c) $\sum_x q(x) = 1$ or $\sum_y r(y) = 1$. (If one sum equals 1, so does the other sum.)

EXERCISES 6.2

6.21 $2^m - 2$

6.23

a) $p_{X,Y,Z}(x, y, z) = 1/720$ if $x, y, z \in S$, $x \neq y \neq z$, and $p_{X,Y,Z}(x, y, z) = 0$ otherwise.

b) 1/6 **c)** 1/6

d) Univariate marginals: $p_X(x) = 1/10$ if $x \in S$, and $p_X(x) = 0$ otherwise; each of Y and Z has the same univariate marginal distribution as X.

Bivariate marginals: $p_{X,Y}(x, y) = 1/90$ if $x, y \in S$ with $x \neq y$, and $p_{X,Y}(x, y) = 0$ otherwise; each of (X, Z) and (Y, Z) has the same bivariate marginal distribution as (X, Y).

6.25

a) The random variables X, Y, Z have the multinomial distribution with parameters 3 and 1/4, 1/4, 1/2. So, $p_{X,Y,Z}(x, y, z) = \binom{3}{x,y,z}\left(\frac{1}{4}\right)^x\left(\frac{1}{4}\right)^y\left(\frac{1}{2}\right)^z$ if x, y, and z are nonnegative integers whose sum is 3, and $p_{X,Y,Z}(x, y, z) = 0$ otherwise.

b) 3/64, 5/16, 19/64

c) $X \sim \mathcal{B}(3, 1/4)$, $Y \sim \mathcal{B}(3, 1/4)$, $Z \sim \mathcal{B}(3, 1/2)$

d) $X + Y \sim \mathcal{B}(3, 1/2)$, $X + Z \sim \mathcal{B}(3, 3/4)$, $Y + Z \sim \mathcal{B}(3, 3/4)$

e) $X + Y$ and Z have the multinomial distribution with parameters 3 and 1/2, 1/2; $p_{X+Y,Z}(u, v) = \binom{3}{u}/8$ if u and v are nonnegative integers whose sum is 3, and $p_{X+Y,Z}(u, v) = 0$ otherwise.

f) 9/16 g) 9/16

h) $p_{X+Y+Z}(3) = 1$, and $p_{X+Y+Z}(w) = 0$ if $w \neq 3$.

6.27 a) 0.00135 b) 0.00135 c) 0.00135

6.29

a) $X_2 \sim \mathcal{B}(n, p_2)$ b) $X_3 + X_5 \sim \mathcal{B}(n, p_3 + p_5)$

c) $X_1 + X_4 + X_5 \sim \mathcal{B}(n, p_1 + p_4 + p_5)$

d) $X_1 + X_2$ and $X_3 + X_4 + X_5$ are multinomial with parameters n and $p_1 + p_2$, $p_3 + p_4 + p_5$.

e) X_1, $X_2 + X_3$, and $X_4 + X_5$ are multinomial with parameters n and p_1, $p_2 + p_3$, $p_4 + p_5$.

f) X_1, $X_2 + X_3 + X_4$, and X_5 are multinomial with parameters n and p_1, $p_2 + p_3 + p_4$, p_5.

6.33

a) Multiple hypergeometric with parameters 40, 5, and 0.15, 0.375, 0.3, 0.175.

b) 0.0581 c) 0.264 d) 0.178

EXERCISES 6.3

6.45

a) See the rows of the table in the answer to part (b).

b)

Sisters, y

	0	1	2	3	Total
0	1.000	0.000	0.000	0.000	1.000
1	0.471	0.529	0.000	0.000	1.000
2	0.091	0.727	0.182	0.000	1.000
3	0.000	0.333	0.667	0.000	1.000
4	0.000	0.000	0.000	1.000	1.000
$p_Y(y)$	0.425	0.450	0.100	0.025	1.000

(left axis: Siblings, x)

c) 90.9%

d) See the columns of the table in the answer to part (e).

e)

Sisters, y

	0	1	2	3	$p_X(x)$
0	0.471	0.000	0.000	0.000	0.200
1	0.471	0.500	0.000	0.000	0.425
2	0.059	0.444	0.500	0.000	0.275
3	0.000	0.056	0.500	0.000	0.075
4	0.000	0.000	0.000	1.000	0.025
Total	1.000	1.000	1.000	1.000	1.000

(left axis: Siblings, x)

f) 50.0%

6.47

a)

y

		0	1	Total
x	0	$P(F^c \mid E^c)$	$P(F \mid E^c)$	1
	1	$P(F^c \mid E)$	$P(F \mid E)$	1
$p_{1_F}(y)$		$P(F^c)$	$P(F)$	1

b) Proposition 4.1 on page 134.

6.49

a) For each $x \in \mathcal{N}$, $p_{Y \mid X}(y \mid x) = p(1 - p)^{y-x-1}$ if $y = x + 1$, $x + 2$, ..., and $p_{Y \mid X}(y \mid x) = 0$ otherwise.

b) Same answer as for part (a).

c) For each $y = 2, 3, \ldots$, $p_{X \mid Y}(x \mid y) = 1/(y - 1)$ if $x = 1, 2, \ldots, y - 1$, and $p_{X \mid Y}(x \mid y) = 0$ otherwise. Thus, $X_{\mid Y=y}$ has the discrete uniform distribution on $\{1, 2, \ldots, y - 1\}$ for each $y = 2, 3, \ldots$.

d) Same answer as for part (c).

e) For each $x \in \mathcal{N}$, $p_{Y-X \mid X}(z \mid x) = p(1 - p)^{z-1}$ if $z \in \mathcal{N}$, and $p_{Y-X \mid X}(z \mid x) = 0$ otherwise. Consequently, $(Y - X)_{\mid X=x} \sim \mathcal{G}(p)$ for each $x \in \mathcal{N}$.

f) Same answer as for part (e).

6.51

a) $p_{Y \mid X}(y \mid x) = p_Y(y)$. The conditional PMF of Y given $X = x$ is identical to the marginal PMF of Y.

b) $p_{X \mid Y}(x \mid y) = p_X(x)$. The conditional PMF of X given $Y = y$ is identical to the marginal PMF of X.

6.53

a) For $x = 0, 1, \ldots, n$,

$$p_{Y,Z|X}(y, z \mid x) = \binom{n-x}{y, z}\left(\frac{q}{1-p}\right)^y\left(\frac{r}{1-p}\right)^z$$

if y and z are nonnegative integers with sum $n - x$, and $p_{Y,Z|X}(y, z \mid x) = 0$ otherwise. Consequently, $(Y, Z)_{|X=x}$ has the multinomial distribution with parameters $n - x$ and $q/(1-p)$, $r/(1-p)$.

b) Let x and y be nonnegative integers whose sum is at most n. Given that $X = x$ and $Y = y$, it must be that $Z = n - x - y$. So, $p_{Z|X,Y}(z \mid x, y) = 1$ if $z = n - x - y$, and $p_{Z|X,Y}(z \mid x, y) = 0$ otherwise.

6.55

a) For $x_m = 0, 1, \ldots, Np_m$,

$$p_{X_1,\ldots,X_{m-1}|X_m}(x_1, \ldots, x_{m-1} \mid x_m)$$

$$= \binom{Np_1}{x_1}\cdots\binom{Np_{m-1}}{x_{m-1}} \bigg/ \binom{N(1-p_m)}{n-x_m},$$

if $x_1, \ldots, x_{m-1}$ are nonnegative integers whose sum is $n - x_m$; $p_{X_1,\ldots,X_{m-1}|X_m}(x_1, \ldots, x_{m-1} \mid x_m) = 0$ otherwise. Consequently, $(X_1, \ldots, X_{m-1})_{|X_m=x_m}$ has the multiple hypergeometric distribution with parameters $N(1 - p_m)$, $n - x_m$, and $p_1/(1 - p_m)$, $\ldots, p_{m-1}/(1 - p_m)$

b) $p_{X_1,X_2,X_3|X_4}(x_1, x_2, x_3 \mid 1) = \binom{6}{x_1}\binom{15}{x_2}\binom{12}{x_3}/\binom{33}{4}$ if x_1, x_2, and x_3 are nonnegative integers whose sum is 4, and $p_{X_1,X_2,X_3|X_4}(x_1, x_2, x_3 \mid 1) = 0$ otherwise; multiple hypergeometric with parameters 33, 4, and 2/11, 5/11, 4/11.

EXERCISES 6.4

6.61 No. $p_{X,Y}(x, y) = p_X(x)p_Y(y)$ doesn't hold for all $x, y \in \mathcal{R}$. For instance, that equation doesn't hold when $x = 2$ and $y = 1$.

6.63 No. That X and Y aren't independent can be seen in several ways. For instance, $p_{Y|X}(4 \mid 5) = 0$, whereas, $p_Y(4) > 0$. Therefore, from Proposition 6.11 on page 295, the random variables X and Y aren't independent.

6.65 No, because if $X_1, \ldots, X_m$ are independent random variables, so are the random variables obtained from a subcollection of those random variables.

6.67

a) The random variables X_1 and X_2 aren't independent because, for instance, although, unconditionally, X_2 can equal 1, given that $X_1 = 5$, X_2 must equal 0.

b) If X_1, X_2, X_3, and X_4 were independent, then so would be X_1 and X_2. Thus, in view of part (a), X_1, X_2, X_3, and X_4 aren't independent.

6.69 Yes. Let S denote the set of decimal digits. The joint PMF of X and Y is $p_{X,Y}(x, y) = 1/100$ if $x, y \in S$, and $p_{X,Y}(x, y) = 0$ otherwise. The marginal PMFs of X and Y are both uniform on S. From these results, it follows that $p_{X,Y}(x, y) = p_X(x)p_Y(y)$ for all $x, y \in \mathcal{R}$; so X and Y are independent random variables.

6.71 a) 0.551 **b)** 0.623

c) Let Z denote the total time that a student takes to complete both exams. Then the PMF of the random variable Z is $p_Z(z) = 0.1\left((0.98)^{z-1} - (0.975)^{z-1}\right)$ for $z = 2, 3, 4, \ldots$, and $p_Z(z) = 0$ otherwise.

d) 74.9%

6.73

a) Let Z denote the total number of patients that arrive each hour at both emergency rooms combined. Then $p_Z(z) = e^{-9.5}(9.5)^z/z!$ for $z = 0, 1, 2, \ldots$, and $p_Z(z) = 0$ otherwise; $Z \sim \mathcal{P}(9.5)$.

b) 9.5 **c)** 0.392

EXERCISES 6.5

6.85

a)

z	4	5	6	7	8	9
$p_{X+Y}(z)$	0.06	0.28	0.28	0.26	0.10	0.02

b)

z	−1	0	1	2
$p_{X-Y}(z)$	0.06	0.40	0.50	0.04

c)

z	2	3	4	5
$p_{\max X, Y}(z)$	0.06	0.52	0.40	0.02

d)

z	2	3	4
$p_{\min X, Y}(z)$	0.38	0.50	0.12

6.87

a)

z	0	1	2	3	4	5
$p_{Y-X}(z)$	$\frac{1}{6}$	$\frac{5}{18}$	$\frac{2}{9}$	$\frac{1}{6}$	$\frac{1}{9}$	$\frac{1}{18}$

b) The magnitude of the difference between the two faces showing.

6.89 Let $Z = |Y - X|$, the magnitude of the difference between X and Y.

a) The PMF of Z is given by $p_Z(0) = 1/(N+1)$, $p_Z(z) = 2(N+1-z)/(N+1)^2$ if $z = 1, 2, \ldots, N$, and $p_Z(z) = 0$ otherwise.

b) Same answer as in part (a).

6.91 Let $Z = \min\{X, Y\}$.

a) $p_Z(z) = (p + q - pq)\big(1 - (p + q - pq)\big)^{z-1}$ for $z \in \mathcal{N}$, and $p_Z(z) = 0$ otherwise.

b) Same as part (a). *Note:* $Z \sim \mathcal{G}(p + q - pq)$.

c) If $p = q$, then $p + q - pq = 2p - p^2$.

EXERCISES 6.6

6.101

a) The mean of the two faces showing.

b) $p_M(m) = (2m - 1)/36$ if $m = 1, 1.5, 2, 2.5, 3, 3.5$; $p_M(m) = (13 - 2m)/36$ if $m = 4, 4.5, 5, 5.5, 6$; and $p_M(m) = 0$ otherwise.

6.103 No. The sum of two independent geometric random variables with the same success probability has the negative binomial distribution with parameters 2 and the common success probability.

6.105

a) Geometric with parameter 0.5

b) Negative binomial with parameters 2 and 0.5

c) $p_{X+Y}(z) = (z + 1)2^{-(z+2)}$ for $z = 0, 1, 2, \ldots$, and $p_{X+Y}(z) = 0$ otherwise.

6.107 Negative binomial with parameters r and p

6.109

a) For $z = r + s, r + s + 1, \ldots$,
$$p_{X \mid X+Y}(x \mid z) = \binom{x-1}{r-1}\binom{z-x-1}{s-1}\Big/\binom{z-1}{r+s-1}$$
if $x = r, r + 1, \ldots, z - s$, and $p_{X \mid X+Y}(x \mid z) = 0$ otherwise.

b) This result follows immediately from part (a) and the fact that a conditional PMF is a PMF (Proposition 6.8 on page 285) and therefore that its values sum to 1.

c) In view of part (b) and Proposition 6.19 on page 309, for $z = r + s, r + s + 1, \ldots$,
$$p_{X+Y}(z) = \sum_x p_X(x) p_Y(z - x)$$
$$= p^{r+s}(1 - p)^{z-(r+s)} \sum_{x=r}^{z-s} \binom{x-1}{r-1}\binom{z-x-1}{s-1}$$
$$= \binom{z-1}{r+s-1} p^{r+s}(1 - p)^{z-(r+s)}.$$
Hence, $X + Y \sim \mathcal{NB}(r + s, p)$.

6.111

a) For $z = 0, \ldots, m + n$,
$$p_{X+Y}(z) = (1 - p)^m q^z (1 - q)^{n-z}$$
$$\times \sum_{x=0}^{z} \binom{m}{x}\binom{n}{z-x}\left(\frac{p(1-q)}{q(1-p)}\right)^x,$$
and $p_{X+Y}(z) = 0$ otherwise.

b) From Example 6.24, $X + Y \sim \mathcal{B}(m + n, p)$.

c) Substituting p for q in the result of part (a) and applying Vandermonde's identity yields
$$p_{X+Y}(z) = \binom{m+n}{z} p^z (1 - p)^{m+n-z},$$
for $z = 0, \ldots, m + n$, and $p_{X+Y}(z) = 0$ otherwise. Thus, $X + Y \sim \mathcal{B}(m + n, p)$.

REVIEW EXERCISES

6.122

a)

	College, y		
	1	2	$p_X(x)$
Gender, x 1	0.171	0.257	0.429
Gender, x 2	0.229	0.343	0.571
$p_Y(y)$	0.400	0.600	1.000

b) 15

c) From part (b), there are 15 different joint PMFs with the same marginal PMFs as those shown in the answer to part (a).

6.123

a) 0.14. The probability is 0.14 that the family selected owns three televisions and six radios.

b) See Table A.1.   **c)** $\{Y = 2X\}$; 0.172

d) $\{Y \geq X + 5\}$; 0.156

Table A.1 Joint and marginal PMFs for Exercise 6.123(b)

Televisions, x \ Radios, y	3	4	5	6	7	8	$p_X(x)$
0	0.000	0.004	0.004	0.004	0.004	0.000	0.016
1	0.000	0.024	0.060	0.072	0.008	0.000	0.164
2	0.004	0.032	0.128	0.176	0.056	0.004	0.400
3	0.000	0.024	0.148	0.140	0.036	0.004	0.352
4	0.000	0.004	0.028	0.024	0.012	0.000	0.068
$p_Y(y)$	0.004	0.088	0.368	0.416	0.116	0.008	1.000

6.124

a) See the rows of Table A.2.
b) See Table A.2. **c)** 41%
d) See the columns of Table A.3.
e) See Table A.3. **f)** 47.8%

6.125

a) $Y_{|X=0} \equiv 0$ and $Y_{|X=x} \sim \mathcal{B}(x, \alpha)$ for $x = 1, 2, \ldots, n$.
b) $p_{X,Y}(x, y) = \binom{n}{x}\binom{x}{y} p^x \alpha^y (1-p)^{n-x}(1-\alpha)^{x-y}$ if $y = 0, \ldots, x$ and $x = 0, \ldots, n$; and $p_{X,Y}(x, y) = 0$ otherwise.
c) $Y \sim \mathcal{B}(n, \alpha p)$
d) Let $r = (1-\alpha)p/(1-\alpha p)$. For $y = 0, 1, \ldots, n$, $p_{X\,|\,Y}(x \mid y) = \binom{n-y}{x-y} r^{x-y}(1-r)^{n-x}$ if $x = y, \ldots, n$, and $p_{X\,|\,Y}(x \mid y) = 0$ otherwise.
e) $(X - Y)_{|Y=y} \sim \mathcal{B}\big(n - y, (1-\alpha)p/(1-\alpha p)\big)$.

6.126 Let X, Y, and Z denote the number of holes that Jan wins, that Jean wins, and that are tied, respectively. $p_{X,Y,Z}(x, y, z) = \binom{18}{x,y,z} p^x q^y r^z$ if x, y, z are nonnegative integers whose sum is 18, and $p_{X,Y,Z}(x, y, z) = 0$ otherwise. The random variables X, Y, Z have the multinomial distribution with parameters 18 and p, q, r.

6.127 $2(1-p)^3/(2-p)$

Table A.2 Conditional PMFs for Exercise 6.124(b)

Televisions, x \ Radios, y	3	4	5	6	7	8	Total
0	0.000	0.250	0.250	0.250	0.250	0.000	1.000
1	0.000	0.146	0.366	0.439	0.049	0.000	1.000
2	0.010	0.080	0.320	0.440	0.140	0.010	1.000
3	0.000	0.068	0.420	0.398	0.102	0.011	1.000
4	0.000	0.059	0.412	0.353	0.176	0.000	1.000
$p_Y(y)$	0.004	0.088	0.368	0.416	0.116	0.008	1.000

Table A.3 Conditional PMFs for Exercise 6.124(e)

		Radios, y					
		3	4	5	6	7	8	$p_X(x)$

Televisions, x	3	4	5	6	7	8	$p_X(x)$
0	0.000	0.045	0.011	0.010	0.034	0.000	0.016
1	0.000	0.273	0.163	0.173	0.069	0.000	0.164
2	1.000	0.364	0.348	0.423	0.483	0.500	0.400
3	0.000	0.273	0.402	0.337	0.310	0.500	0.352
4	0.000	0.045	0.076	0.058	0.103	0.000	0.068
Total	1.000	1.000	1.000	1.000	1.000	1.000	1.000

6.128 Let $Z = |Y - X|$, the magnitude of the difference between the lifetimes of the two components.
a) $p_Z(0) = p/(2 - p)$, $p_Z(z) = 2p(1 - p)^z/(2 - p)$ for $z \in \mathcal{N}$, and $p_Z(z) = 0$ otherwise.
b) $2(1 - p)^3/(2 - p)$

6.129 Let $p = \pi/36$, $q = \pi/12$, and $r = 1 - p - q$.
a) $p_{X,Y,Z}(x, y, z) = \binom{4}{x,y,z} p^x q^y r^z$ if x, y, z are non-negative integers with sum 4; $p_{X,Y,Z}(x, y, z) = 0$ otherwise. X, Y, Z have the multinomial distribution with parameters 4 and p, q, r.
b) $X \sim \mathcal{B}(4, p)$, $Y \sim \mathcal{B}(4, q)$, and $Z \sim \mathcal{B}(4, r)$.
c) $X + Y \sim \mathcal{B}(4, p + q)$ and represents the number of times the archer scores in four attempts.
d) $T = 10X + 5Y + 0Z$. The PMF of T is:

t	$p_T(t)$	t	$p_T(t)$
0	0.1795346	25	0.0218368
5	0.2888283	30	0.0048621
10	0.2705220	35	0.0006959
15	0.1628840	40	0.0000580
20	0.0707783		

6.131 For convenience, set $U = U_{0.3}$ and $V = U_{0.6}$. The joint and marginal PMFs of U and V are provided in the following table.

			v		
		0	1	2	$p_U(u)$
u	0	0.16	0.24	0.09	0.49
	1	0.00	0.24	0.18	0.42
	2	0.00	0.00	0.09	0.09
	$p_V(v)$	0.16	0.48	0.36	1.00

a) See the last row of the preceding table.
b) See the last column of the preceding table.
c) See the cells of the preceding table.
d) The conditional PMFs of U for each possible value of V are provided in the cell columns of the following table.

			v		
		0	1	2	$p_U(u)$
u	0	1.00	0.50	0.25	0.49
	1	0.00	0.50	0.50	0.42
	2	0.00	0.00	0.25	0.09
	Total	1.00	1.00	1.00	1.00

e) The conditional PMFs of V for each possible value of U are provided in the cell rows of the following table.

		v			
		0	1	2	Total
u	0	0.327	0.490	0.184	1.000
	1	0.000	0.571	0.429	1.000
	2	0.000	0.000	1.000	1.000
	$p_V(v)$	0.160	0.480	0.360	1.000

6.132 a) 0.0944   **b)** 0.503

6.133 a) 0.00673   **b)** 0.00000624   **c)** 0.00260

6.134 Poisson with parameter 8.24.

6.135

a) $p_X(x) = 1/M$ if $x = 1, 2, \ldots, M$, and $p_X(x) = 0$ otherwise. The random variable X has the discrete uniform distribution on $\{1, 2, \ldots, M\}$.

b) $p_Y(y) = (y!M)^{-1} \sum_{x=1}^{M} x^y e^{-x}$ if $y = 0, 1, \ldots$, and $p_Y(y) = 0$ otherwise.

c) For $x = 1, 2, \ldots, M$, $p_{Y \mid X}(y \mid x) = e^{-x} x^y / y!$ if $y = 0, 1, \ldots$, and $p_{Y \mid X}(y \mid x) = 0$ otherwise. Thus, $Y_{\mid X=x} \sim \mathcal{P}(x)$ for $x = 1, 2, \ldots, M$.

d) No, because the conditional PMF of Y given $X = x$ is not identical to the PMF of Y.

6.136 Let X_1, X_2, X_3, and X_4 denote, respectively, the numbers of the n items sampled that are in the four specified categories. Then the random variables X_1, X_2, X_3, X_4 have the multinomial distribution with parameters n and $(1 - p)(1 - q)$, $(1 - p)q$, $p(1 - q)$, pq.

6.137 Let the notation be as in the answer to Exercise 6.136.

a) The random variables X_1, X_2, X_3, X_4 have the multiple hypergeometric distribution with parameters N, n and $(1 - p)(1 - q)$, $(1 - p)q$, $p(1 - q)$, pq.

b) When the lot size, N, is large relative to the sample size, n. Refer to Exercise 6.41 on page 282.

6.138 The conditional distribution of the number of undetected defectives (i.e., defective-uninspected), given that the number of detected defectives (i.e., defective-inspected) equals k, is binomial with parameters $n - k$ and $p(1 - q)/(1 - pq)$.

6.139

a) For $x = 0, 1, \ldots$,

$$p_{N \mid X}(n \mid x) = e^{-\lambda(1-p)} \big(\lambda(1 - p)\big)^{n-x} / (n - x)!$$

if $n = x, x + 1, \ldots$; $p_{N \mid X}(n \mid x) = 0$ otherwise.

b) $(N - X)_{\mid X=x} \sim \mathcal{P}\big((1 - p)\lambda\big)$

c) Yes, because the conditional PMF of $N - X$ given $X = x$ doesn't depend on x.

d) Note that $N - X$ is the number of people who don't make a deposit. Hence the numbers of people who do and don't make a deposit are independent random variables.

6.140 Given a total of n customers during the hour, the numbers of customers requesting the three specified options during the hour have the multinomial distribution with parameters n and $\lambda/(\lambda + \mu + \nu)$, $\mu/(\lambda + \mu + \nu)$, $\nu/(\lambda + \mu + \nu)$.

6.141 Let X denote the number of successes in the n Bernoulli trials and let Y denote the trial on which the first success occurs. Then, for $k = 1, 2, \ldots, n$,

$$p_{Y \mid X}(y \mid k) = \binom{n - y}{k - 1} \Big/ \binom{n}{k}$$

if $y = 1, 2, \ldots, n - k + 1$; $p_{Y \mid X}(y \mid k) = 0$ otherwise.

6.142

a) For $y \in S$,

$$p_{X \mid Y}(x \mid y) = \frac{1}{10 - x} \Big/ \sum_{k=0}^{y} \frac{1}{10 - k},$$

if $x = 0, 1, \ldots, y$, and $p_{X \mid Y}(x \mid y) = 0$ otherwise.

b)

x	$p_{X \mid Y}(x \mid 6)$	x	$p_{X \mid Y}(x \mid 6)$
0	0.0913	4	0.1521
1	0.1014	5	0.1825
2	0.1141	6	0.2282
3	0.1304		

6.143

b) $T \sim \mathcal{G}(1 - m/N)$

c) Consider the random experiment of selecting a number at random from the first N positive integers. Let the specified event (i.e., a success) be that the number selected exceeds m. Independent repetitions of this experiment constitute Bernoulli trials with success probability $1 - m/N$. The random variable T is the time of the first success, which has the geometric distribution with parameter $1 - m/N$.

6.144 For $u = 2, 3, 4, \ldots$,

$$p_U(u) = \frac{1}{N}\left(\delta_{u2} + \sum_{k=2}^{N}\left(1 - \frac{k-1}{N}\right)\left(\frac{k-1}{N}\right)^{u-2}\right),$$

and $p_U(u) = 0$ otherwise. Here δ_{ij} is Kronecker's delta.

6.145 Multinomial with parameters z and $p_1, \ldots, p_m$, where $p_j = \lambda_j/(\lambda_1 + \cdots + \lambda_m)$ for $j = 1, 2, \ldots, m$.

6.146
a) $X_m \sim \mathcal{B}(n, p_m)$
b) Multinomial distribution with parameters $n - x_m$ and $p_1/(1 - p_m), \ldots, p_{m-1}/(1 - p_m)$.

6.147
a) Binomial distribution with parameters $m + n$ and p.
b) Hypergeometric distribution with parameters $m + n$, z, and $m/(m + n)$, where z is the combined total of defective components.

6.148 Let ℓ denote the daily output for Factory C.
a) Binomial with parameters $m + n + \ell$ and p.
b) Hypergeometric with parameters $m + n + \ell$, w, and $m/(m + n + \ell)$, where w is the combined total of defective components.

6.149 Let $N = n_1 + \cdots + n_m$.
a) Multiple hypergeometric distribution with parameters N, z, and $n_1/N, \ldots, n_m/N$.
b) Use the fact that the jth univariate marginal of a multiple hypergeometric distribution with parameters N, n, and $p_1, \ldots, p_m$ has the hypergeometric distribution with parameters N, n, and p_j.

6.150
a) Bernoulli distribution with parameter p.
b) Bernoulli distribution with parameter x/n.
c) Bernoulli distribution with parameter x/n.

Chapter 7

EXERCISES 7.1

7.1 7

7.3 a) 1 **b)** 0.88

7.5 Note that $\sum_{k=1}^{m}(x_k - \bar{x})p_k = 0$. But,

$$\sum_{k=1}^{m}(x_k - \bar{x})p_k = \sum_{k=1}^{m}x_k p_k - \bar{x}\sum_{k=1}^{m}p_k = \mathcal{E}(X) - \bar{x}.$$

Therefore, $\bar{x} = \mathcal{E}(X)$.

7.7 a) 82.4 **b)** 82.4n

7.9 18 **7.11** 1.25

7.13 0.3. On average, 0.3 TVs of the 5 TVs sampled will be defective.

7.15
a) $\lambda = 15$, because the mean of a Poisson distribution equals its parameter.
b) 63.5%

7.17 Assuming that the customers act independently, the expected number per hour who buy a café mocha with no whipped cream is 9.3.

7.19 If E is an event, then $I_E \sim \mathcal{B}(1, P(E))$. Therefore, $\mathcal{E}(I_E) = 1 \cdot P(E) = P(E)$.

7.21
a) Answers will vary. **b)** 4.5

EXERCISES 7.2

7.29
a) $\mathcal{E}(Y) = \mathcal{E}(Z) = 3.5$
b) $X = Y + Z; \mathcal{E}(X) = \mathcal{E}(Y) + \mathcal{E}(Z) = 7$

7.31 a) 91/36 **b)** 91/36

7.33
a) $\frac{1}{2}\left(1 - (1 - 2p)^n\right)$
b) Same answer as in part (a).

7.35 1 **7.37** 14.7

7.39 a) 16/17 **b)** 169/17 **c)** 104/17
d) The sum of the three expected values in parts (a), (b), and (c) must equal the total number of items, which is 17.

7.41
a) 19.836 minutes
b) 40.164 minutes
c) 20.328 minutes

7.43 Applying the linearity property of expected value yields the equation

$$\mathcal{E}\big((X - \mathcal{E}(X))(Y - \mathcal{E}(Y))\big) = \mathcal{E}(XY) - \mathcal{E}(X)\mathcal{E}(Y).$$

a) This result follows immediately from the preceding equation.
b) Yes, by the preceding equation and Proposition 7.5 on page 346.
c) No, not necessarily. The random variables with joint PDF shown in Table 7.8 on page 347 are linearly uncorrelated but not independent.

EXERCISES 7.3

7.55 10.512   **7.57** $72/121 \approx 0.595$

7.59 a) 600   **b)** 3000

7.61 $\sqrt{15}/4 \approx 0.968$

7.63 0.271   **7.65** 15

7.67 $4(1 - 5e^{-4})/(1 - e^{-4})^2 \approx 3.771$

7.69 Recall that $\mathcal{E}(I_E) = P(E)$.

$$\mathrm{Var}(I_E) = \mathcal{E}(I_E^2) - \big(\mathcal{E}(I_E)\big)^2 = \mathcal{E}(I_E) - \big(\mathcal{E}(I_E)\big)^2$$
$$= P(E) - \big(P(E)\big)^2 = P(E)\big(1 - P(E)\big).$$

7.73

a) The probability that the observed value of a random variable is within k standard deviations of its mean is at least $1 - 1/k^2$.

b) $k = 2$: Chances are at least 75% that X will take a value within two standard deviations of its mean.

$k = 3$: Chances are at least 88.8% that X will take a value within three standard deviations of its mean.

c) Applying the complementation rule and Proposition 7.11 on page 360 yields

$$P(|X - \mu_X| < k\sigma_X) = 1 - P(|X - \mu_X| \geq k\sigma_X)$$
$$\geq 1 - \mathrm{Var}(X)/(k^2 \sigma_X^2)$$
$$= 1 - 1/k^2.$$

7.75 a) $\mathcal{E}(Y)$ **b)** $\mathrm{Var}(Y)$

EXERCISES 7.4

7.87 Let p, q, and r denote the probabilities that the archer hits regions A, B, and C, respectively. Then $p = \pi/36$, $q = \pi/12$, and $r = 1 - p - q$.
a) Multinomial with parameters 4 and p, q, r.
b) $T = 10X + 5Y + 0Z$; $\mathcal{E}(T^2) = 118.202$
c) 42.048   **d)** 42.048
e) First find the PMF of T by using the FPF and the joint PMF of X, Y, and Z, and then obtain $\mathrm{Var}(T)$ by using the definition of the variance of a random variable.

7.89 1   **7.91** 38.99

7.95 $n\big(p_1(1 - p_1) + 9p_2(1 - p_2) - 6p_1 p_2\big)$

7.97 $35/73 \approx 0.479$

7.101 Let X denote the value of the variable for a randomly selected member from the population and let $X_1, \ldots, X_n$ denote, respectively, the values of the variable for the n members of the sample. As the sampling is with replacement, the random variables $X_1, \ldots, X_n$ are independent. In fact, they constitute a random sample of size n from the distribution of X. Therefore, by Example 7.16(b) on page 369, $\mathrm{Var}(\bar{X}_n) = \mathrm{Var}(X)/n$. The result now follows from Proposition 7.9 on page 356.

7.103
a) All odd moments of X are zero, so

$$\mathrm{Cov}(X, Y) = \mathcal{E}(XY) - \mathcal{E}(X)\mathcal{E}(Y)$$
$$= \mathcal{E}(X^3) - \mathcal{E}(X)\mathcal{E}(Y) = 0.$$

Hence, $\rho(X, Y) = 0$.
b) The correlation coefficient is a measure of *linear* association.

EXERCISES 7.5

7.113 $1/p$

7.115
a)

Siblings x	Conditional expectation $\mathcal{E}(Y \mid X = x)$
0	0.000
1	0.529
2	1.091
3	1.667
4	3.000

b)

Siblings x	Conditional variance $\mathrm{Var}(Y \mid X = x)$
0	0.000
1	0.249
2	0.265
3	0.222
4	0.000

7.117 For $x \in S$,
a) $\mathcal{E}(Y \mid X = x) = (45 - x)/9$.
b) $\mathrm{Var}(Y \mid X = x) = (540 + 90x - 10x^2)/81$.

7.119 $1939.046   **7.121** $6.28

REVIEW EXERCISES

7.134 31.4, because of the long-run-average interpretation of expected value.

7.135 a) 0.29 ($290)   **b)** 2.9 ($2900)

7.136

a) $420; $76.68

b) Sales attempts by the salesman constitute Bernoulli trials.

7.137 a) $573,089.75   **b)** $698,899.60

7.138

a) No, because the mean returns are equal.

b) The second investment, because the standard deviation of its return is smaller than that of the first investment.

7.139 a) λ^n **b)** $\lambda; \lambda$

7.140 $(1 - (1 - p)^m)/p$

7.141 a) 50 minutes   **b)** 35.122 minutes

7.142

a)

Televisions x	Conditional expectation $\mathcal{E}(Y \mid X = x)$
0	5.500
1	5.391
2	5.650
3	5.562
4	5.646

b)

Televisions x	Conditional variance $\text{Var}(Y \mid X = x)$
0	1.250
1	0.628
2	0.788
3	0.651
4	0.699

7.143

a) 5.574

b) 5.576. The difference between the results in parts (a) and (b) is due to rounding error.

c) 0.723

d) 0.724. The difference between the results in parts (c) and (d) is due to rounding error.

7.144 a) 91/36   **b)** 91/36

7.145 a) -30 **b)** -119 **c)** 109   **d)** 229
e) 1809   **f)** 2520

7.146 a) $7500   **b)** $2121.32

7.147 $178.67   **7.148** 0.4995; 0.2887

7.149 $1414.41

7.150 Let X and Y denote the numbers of red balls obtained when the balls are drawn with and without replacement, respectively.

a) $X \sim \mathcal{B}(2, 0.6)$ **b)** $Y \sim \mathcal{H}(10, 2, 0.6)$

c) $\mathcal{E}(X) = 1.2, \mathcal{E}(Y) = 1.2$

d) $\mathcal{E}(X) = 1.2, \mathcal{E}(Y) = 1.2$

e) $\text{Var}(X) = 0.48, \text{Var}(Y) = 0.427$

f) $\text{Var}(X) = 0.48, \text{Var}(Y) = 0.427$

7.151

a) $1 + n(1 - (1 - p)^n)$

b) $p < 1 - (1/n)^{1/n}$

7.152

a) $m + n(1 - (1 - p)^{n/m})$

b) Separate: 100. All pooled: 64.4. Grouped: 24.9.

7.153 $n(1 - 1/n)^n$

7.154

a) $1/(p + q - pq)$

b) $1/p + 1/q - 1/(p + q - pq)$

c) $1/(2p - p^2); (3 - 2p)/(2p - p^2)$

7.156 mean $= (n - k)p(1 - q)/(1 - pq)$;
variance $= (n - k)p(1 - p)(1 - q)/(1 - pq)^2$

7.157

a) $\mu_X = 77.7''; \sigma_X = 2.759''$

b) $\mu_Y = -2.642$ cm; $\sigma_Y = 7.007$ cm

c) 1

7.158 $\mathcal{E}(1/X) = -p \ln p/(1 - p); 1/\mathcal{E}(X) = p$. Observe that $\mathcal{E}(1/X) \neq 1/\mathcal{E}(X)$.

7.159 a) 1   **b)** 1   **c)** $2(1 - p)$

7.160 Because $\lim_{n \to \infty} np_n(1 - p_n) = \lambda$, conjecture that the mean of a Poisson random variable with parameter λ equals λ.

7.161 a) $np(1 - p)$ **b)** $np(1 - p)$

7.162

a) $zm/(m + n)$

b) $zmn(m + n - z)/((m + n - 1)(m + n)^2)$

Chapter 8

EXERCISES 8.1

8.3 $P(X = x) = |\{\arctan x\}|/\pi = 0/\pi = 0$, for each $x \in \mathcal{R}$, where $|\ |$ denotes length.

8.5 For $y \in \mathcal{R}$, let L_y denote the intersection of the interior of the triangle with the line y units from the base of the triangle. Then $P(Y = y) = 2|L_y|/(bh) = 0$, where $|\,|$ denotes area.

EXERCISES 8.2

8.15
a) $F_Z(z) = 0$ if $z < 0$, $F_Z(z) = z^3/r^3$ if $0 \le z < r$, and $F_Z(z) = 1$ if $z \ge r$.
c) Z is a continuous random variable because F_Z is an everywhere continuous function.

8.17
a) $F_Y(y) = 0$ if $y < 0$, $F_Y(y) = (2hy - y^2)/h^2$ if $0 \le y < h$, and $F_Y(y) = 1$ if $y \ge h$.
c) Y is a continuous random variable because F_Y is an everywhere continuous function.

8.19
a) $F_Y(y) = 0$ if $y < 0$, $F_Y(y) = 1 - P(E)$ if $0 \le y < 1$, and $F_Y(y) = 1$ if $y \ge 1$.
c) Y is not a continuous random variable because F_Y is not an everywhere continuous function.

8.21
a) $F_X(x) = 0$ if $x < 0$, $F_X(x) = 4/33$ if $0 \le x < 1$, $F_X(x) = 19/33$ if $1 \le x < 2$, $F_X(x) = 31/33$ if $2 \le x < 3$, and $F_X(x) = 1$ if $x \ge 3$.
c) X is not a continuous random variable because F_X is not an everywhere continuous function.

8.23
a) $F_X(x) = 0$ if $x < 1$, $F_X(x) = \lfloor x \rfloor/N$ if $1 \le x < N$, and $F_X(x) = 1$ if $x \ge N$.
b) $p_X(x) = 1/N$ for $x = 1, 2, \ldots, N$, and $p_X(x) = 0$ otherwise.

8.25
a) $0.375, 0, 0.375, 0.75$ **b)** $0.75, 0.375, 0.375, 0.75$
c) $0.375, 0.375, 0.75, 0.75$ **d)** $0.75, 0.75, 0.75, 0.75$

8.27 a) All 0.6 **b)** All 0.64 **c)** All 0.96
d) All 1 **e)** All 0 **f)** All 0
g) Because Z is a continuous random variable.

8.29
a) $F_X(x) = \int_{-\infty}^{x} f(t)\, dt$ for all $x \in \mathcal{R}$.
b) $F_X'(x) = f(x)$ for all x at which f is continuous.
c) All equal $\int_a^b f(t)\, dt$.
d) Because X is a continuous random variable or, equivalently, because F_X is an everywhere continuous function.

8.31 a) Yes; discrete   **b)** Yes; discrete
c) No; property (d) is violated.

d) Yes; discrete   **e)** Yes; continuous
f) No; property (d) is violated.   **g)** Yes; mixed

8.33
a) $F_Y(y) = \left(F_X(y)\right)^m$
b) $F_Z(z) = 1 - \left(1 - F_X(z)\right)^m$

EXERCISES 8.3

8.41
a) $f_Z(z) = 3z^2/r^3$ if $0 \le z < r$, and $f_Z(z) = 0$ otherwise.
b) Roughly, for $0 \le z < r$, the larger the value of z, the greater is its likelihood. More precisely, the distance from the center of the sphere to the point chosen is more likely to be large than small.
c) $1/8$; $63/64$; $7/64$ **d)** $1/8$; $63/64$; $7/64$

8.43
a) $f_X(x) = \left(\pi(1 + x^2)\right)^{-1}$ for all $x \in \mathcal{R}$.
b) Roughly, the smaller the magnitude of x, the greater is its likelihood. More precisely, the tangent of the angle chosen is more likely to be small in magnitude than large in magnitude.
c) 0.75 **d)** 0.75

8.45
a) $f_X(x) = (2/\pi)\sqrt{1 - x^2}$ if $-1 < x < 1$; $f_X(x) = 0$ otherwise.
b) Roughly, for $-1 < x < 1$, the smaller the magnitude of x, the greater is its likelihood. More precisely, the x-coordinate of the center of the first spot to appear is more likely to be small in magnitude than large in magnitude.
c) By symmetry, X and Y have the same distribution. Thus, $f_Y(y) = (2/\pi)\sqrt{1 - y^2}$ if $-1 < y < 1$, and $f_Y(y) = 0$ otherwise.

8.47 f is a PDF of X

8.49
a) Yes. $F(x) = 0$ if $x < 0$; $F(x) = 1 - e^{-\lambda x}$ if $x \ge 0$.
b) Yes. $F(x) = 0$ if $x < a$, $F(x) = (x - a)/(b - a)$ if $a \le x < b$, and $F(x) = 1$ if $x \ge b$.
c) No. Property (a) is violated.
d) No. Property (b) is violated.
e) Yes.

$$F(x) = \begin{cases} 0, & \text{if } x < -b; \\ \frac{1}{2}(1 + x/b)^2, & \text{if } -b \le x < 0; \\ 1 - \frac{1}{2}(1 - x/b)^2, & \text{if } 0 \le x < b; \\ 1, & \text{if } x \ge b. \end{cases}$$

8.51 Answers will vary.

8.53

a) $F_Y(y) = 0$ if $y < 2$, $F_Y(y) = (y - 2)/3$
if $2 \le y < 5$, and $F_Y(y) = 1$ if $y \ge 5$.
$f_Y(y) = 1/3$ if $2 < y < 5$, and $f_Y(y) = 0$
otherwise.

b) $F_Z(z) = 0$ if $z < a$, $F_Z(z) = (z - a)/(b - a)$
if $a \le z < b$, and $F_Z(z) = 1$ if $z \ge b$.
$f_Z(z) = 1/(b - a)$ if $a < z < b$, and $f_Z(z) = 0$
otherwise.

EXERCISES 8.4

8.67

a) $Y \sim \mathcal{U}(c, c + d)$.

b) $f_Y(y) = 1/d$ if $c < y < c + d$, and $f_Y(y) = 0$ otherwise. $Y \sim \mathcal{U}(c, c + d)$.

c) $c = a$ and $d = b - a$.

8.71 In view of Exercise 8.70, $F_X(x) = 0$ if $x < 0$, and
$F_X(x) = 1 - e^{-6.9x}$ if $x \ge 0$. Thus,

$$P(1/4 < X < 1/2) = F_X(1/2) - F_X(1/4) = 0.146$$

and

$$P(X < 1) = F_X(1) = 0.999.$$

8.75 a) 0.250   **b)** 0.247   **c)** 0.0150   **d)** 0.247

8.77 0.134

8.79

a) $P(X > t) = 1 - t$.

b) Applying the conditional probability rule yields

$$P(X > s + t \mid X > s) = (1 - s - t)/(1 - s)$$
$$< 1 - t = P(X > t).$$

8.81 $f_R(r) = 5/(2r^2)$ if $5/6 < r < 5/4$; $f_R(r) = 0$
otherwise.

8.83

c) Let $X = I_E$ and $p = P(E)$. For $p < 0.5$, $p > 0.5$,
and $p = 0.5$, respectively, the median(s) are 0, 1,
and [0, 1].

e) $(a + b)/2$ **f)** $(a + b)/2$ **g)** $\lambda^{-1} \ln 2$

EXERCISES 8.5

8.93 a) True   **b)** False

8.97 They are the same because two normal random
variables with the same μ and σ^2 parameters have the
same PDF and hence the same probability distribution.
In other words, μ and σ^2 together completely determine
a normal distribution.

8.99 Using Table I, we obtain

a) 0.0594.   **b)** 0.2699.

8.101 a) 98.32%   **b)** 96.64%

c) The probability is about 0.004, or less than a one-half
of one percent chance, that the defendant could have
fathered the child. Although possible, it's highly
unlikely that the defendant fathered the child.

8.103 0.09 mm

8.105 $f_{1/X^2}(y) = \left(\alpha/\sqrt{2\pi y^3}\right) e^{-\alpha^2/2y}$ if $y > 0$,
and $f_{1/X^2}(y) = 0$ otherwise.

8.107 87.0%

8.109

a) $\mathcal{H}(N, 500, 0.56)$, where N is the number of Anchorage residents.

b) $\mathcal{B}(500, 0.56)$

c) $\mathcal{N}(280, 123.2)$

d) 0.0359

e) 0.107

8.111 a) 0.375   **b)** 0.930

c) The Poisson approximation is preferable to the normal approximation because the assumptions for the
former (large n and small p) are better met than those
for the latter (large n and moderate p).

EXERCISES 8.6

8.117

a) For $x > 0$,

$$f_X(x) = \frac{\lambda^r}{(r - 1)!} x^{r-1} e^{-\lambda x},$$

and $f_X(x) = 0$ otherwise.

b) For $x > 0$,

$$f_X(x) = \frac{1}{2^{v/2}\Gamma(v/2)} x^{v/2-1} e^{-x/2},$$

and $f_X(x) = 0$ otherwise.

8.119 The required result follows immediately from
Equation (8.55) on page 457.

8.123 a) 0.3483   **b)** 0.1285

8.125 Let ψ denote the PDF of a $T(-1, 1)$ random
variable and let $X \sim T(a, b)$. Then

$$f_X(x) = \frac{2}{b - a} \psi\left(\frac{2}{b - a}(x - (a + b)/2)\right).$$

8.127

a) For $x > 0$,

$$f_X(x) = \frac{1}{24 \times 10^5} x^4 e^{-x/10},$$

and $f_X(x) = 0$ otherwise. Also, for $x \geq 0$,

$$F_X(x) = 1 - e^{-x/10} \sum_{j=0}^{4} \frac{(x/10)^j}{j!},$$

and $F_X(x) = 0$ otherwise.

b) 0.715 **c)** 0.188

8.129 Let Y denote the loss amount.

a) $f_Y(y) = 18/y^3$ if $y > 3$, and $f_Y(y) = 0$ otherwise. Also, $F_Y(y) = 1 - 9/y^2$ if $y \geq 3$, and $F_Y(y) = 0$ otherwise.

b) 0.141 **c)** 0.2025

8.131 Let Y denote the lifetime of the component.

a) $f_Y(y) = 6y^2 e^{-2y^3}$ if $y > 0$; $f_Y(y) = 0$ otherwise. $F_Y(y) = 1 - e^{-2y^3}$ if $y \geq 0$; $F_Y(y) = 0$ otherwise.

b) 0.221; 0.643

EXERCISES 8.7

8.141 Let $Y = c + dX$. The transformation method is appropriate. $f_Y(y) = 1/d$ if $c < y < c + d$; $f_Y(y) = 0$ otherwise. So, $c + dX \sim \mathcal{U}(c, c + d)$.

8.143 Let $Y = -\ln X$. The transformation method is appropriate. $f_Y(y) = e^{-y}$ if $y > 0$; $f_Y(y) = 0$ otherwise. So, $-\ln X \sim \mathcal{E}(1)$.

8.145 The transformation method isn't appropriate because the function $g(x) = x^2$ isn't monotone on the range of X, which is $(-\infty, \infty)$.

8.147 Let $Y = X^{1/\alpha}$. The transformation method is appropriate. $f_Y(y) = \alpha y^{\alpha-1}$ if $0 < y < 1$; $f_Y(y) = 0$ otherwise. So, $X^{1/\alpha}$ has the beta distribution with parameters α and 1.

8.149 Let $Y = X^{1/\beta}$. The transformation method is appropriate. $f_Y(y) = \alpha\beta y^{\beta-1} e^{-\alpha y^\beta}$ if $y > 0$; $f_Y(y) = 0$ otherwise. So, $X^{1/\beta}$ has the Weibull distribution with parameters α and β.

8.151 Let $g(x) = a + bx$. Note that $g'(x) \equiv b$ and $g^{-1}(y) = (y - a)/b$. From the transformation method, $f_{a+bX}(y) = |g'(x)|^{-1} f_X(x) = |b|^{-1} f_X\big((y - a)/b\big)$.

8.153

a) Because $\psi(y) = \big(\pi(1 + y^2)\big)^{-1}$,

$$f_X(x) = \frac{\theta}{\pi\left(\theta^2 + (x - \eta)^2\right)}$$

$$= \frac{1}{\theta} \frac{1}{\pi\left(1 + \left(\frac{x-\eta}{\theta}\right)^2\right)} = \frac{1}{\theta} \psi\left(\frac{x - \eta}{\theta}\right).$$

b) First apply Equation (8.64) (or Exercise 8.151) with $a = -\eta/\theta$ and $b = 1/\theta$, and then use part (a).

8.155

a) $f_Y(y) = \left(\pi\sqrt{1 - y^2}\right)^{-1}$ if $-1 < y < 1$; $f_Y(y) = 0$ otherwise.

b) Same answer as in part (a).

8.157

a) The range of Y is the interval (a, b).

b) For $a < y < b$,

$$f_Y(y) = \frac{(y - a)^{\alpha-1}(b - y)^{\beta-1}}{(b - a)^{\alpha+\beta-1} B(\alpha, \beta)},$$

and $f_Y(y) = 0$ otherwise.

8.159 $R^2 \sim \mathcal{E}(1/2\sigma^2)$

8.161 Let $Y = 1/(1 + X^2)$. $f_Y(y) = \left(\pi\sqrt{y - y^2}\right)^{-1}$ if $0 < y < 1$, and $f_Y(y) = 0$ otherwise. Y has the beta distribution with parameters 1/2 and 1/2 or, equivalently, the arcsine distribution.

8.163

a) $F_X(x) = (x - a)/(b - a)$ for $a < x < b$. So, by Proposition 8.16(a), $(X - a)/(b - a) \sim \mathcal{U}(0, 1)$.

b) $F_X^{-1}(u) = a + (b - a)u$ for $0 < u < 1$. Therefore, by Proposition 8.16(b), $a + (b - a)U$ has the same probability distribution as X, which is $\mathcal{U}(a, b)$.

8.165 $\lfloor m + (n - m + 1)X \rfloor$ has the discrete uniform distribution on $\{m, m + 1, \ldots, n\}$.

REVIEW EXERCISES

8.175 Your colleague should have written

$$F_X(x) = \begin{cases} 0, & \text{if } x < 1; \\ 1/3, & \text{if } 1 \leq x < 2; \\ 2/3, & \text{if } 2 \leq x < 3; \\ 1, & \text{if } x \geq 3. \end{cases}$$

8.176

a) $F_W(w) = 0$ if $w < 0$, $F_W(w) = 2w - w^2$ if $0 \leq w < 1$, and $F_W(w) = 1$ if $w \geq 1$.

b) $f_W(w) = 2(1 - w)$ if $0 < w < 1$, and $f_W(w) = 0$ otherwise.

c) W has the beta distribution with parameters 1 and 2.

8.177

a) $F_Y(y) = 0$ if $y < 0.2$, $F_Y(y) = y$ if $0.2 \le y < 0.9$, and $F_Y(y) = 1$ if $y \ge 0.9$.

b) Y is neither a discrete random variable nor a continuous random variable.

c) Set $V = Y_{|\{Y \in S\}}$. Then $F_V(v) = 0$ if $v < 0.2$, $F_V(v) = 2/3$ if $0.2 \le v < 0.9$, and $F_V(v) = 1$ if $v \ge 0.9$. V is a discrete random variable (and not a continuous random variable).

d) Set $W = Y_{|\{Y \notin S\}}$. Then $F_W(w) = 0$ if $w < 0.2$, $F_W(w) = (w - 0.2)/0.7$ if $0.2 \le w < 0.9$, and $F_W(w) = 1$ if $w \ge 0.9$. W is a continuous random variable (and not a discrete random variable).

e) $a = 0.3$ and $b = 0.7$

f) X: 0.7, 0.7, 0.7, 0.7. Y: 0.9, 0.7, 0.8, 1. V: $\frac{2}{3}, 0, \frac{1}{3}, 1$. W: 1, 1, 1, 1.

8.178 Yes. The CDF can be obtained from the tail probabilities, which, according to the fourth bulleted item on page 407, determines the probability distribution.

8.179 Let Y denote weekly sales, in gallons.

a) $F_Y(y) = 0$ if $y < 0$, $F_Y(y) = 1 - e^{-\lambda y}$ if $0 \le y < m$, and $F_Y(y) = 1$ if $y \ge m$.

b) Y is neither discrete nor continuous.

c) $e^{-\lambda m}$

8.180

a) $\{0, 1/2, 1\}$　**b)** none　**c)** 2/3

d)
$$F_X(x) = \begin{cases} 0, & \text{if } x < 0; \\ 2x/3, & \text{if } 0 \le x < 1/2; \\ (4x - 1)/3, & \text{if } 1/2 \le x < 1; \\ 1, & \text{if } x \ge 1. \end{cases}$$

8.181 0.048

8.182 If cg were a CDF, then
$$1 = \int_{-\infty}^{\infty} cg(x)\, dx = c \int_0^1 (1/x)\, dx,$$
which is impossible because $\int_0^1 (1/x)\, dx = \infty$.

8.183

a) X has the beta distribution with parameters 3 and 2.

b) 12　**c)** 12　**d)** 0.3125　**e)** 0.1625

8.184 a) $x > 0$　**b)** 0

c) $F_X\left(\sqrt{x}\right) - F_X\left(-\sqrt{x}\right) = 2F_X\left(\sqrt{x}\right) - 1$

8.185

a) Z has the beta distribution with parameters 2 and 1.

b) $F_\Theta(\theta) = 0$ if $\theta < 0$, $F_\Theta(\theta) = \theta/2\pi$ if $0 \le \theta < 2\pi$, and $F_\Theta(\theta) = 1$ if $\theta \ge 2\pi$.

c) $\Theta \sim \mathcal{U}(0, 2\pi)$

8.186 a) $3.8t$　**b)** $N(t) = 0$. $N(t) \ge 1$.

c) $e^{-3.8t}$　**d)** $W_1 \sim \mathcal{E}(3.8)$

e) $W_3 \sim \Gamma(3, 3.8)$　**f)** $W_n \sim \Gamma(n, 3.8)$

8.187 Let X denote the computed area of the circle.

a) A PDF of X is:
$$f_X(x) = \frac{1}{2\sqrt{\pi \epsilon}} x^{-1/2},$$
if $\pi(d - \epsilon)^2/4 < x < \pi(d + \epsilon)^2/4$, and $f_X(x) = 0$ otherwise.

b) A PDF of X is:
$$f_X(x) = \frac{1}{\pi \epsilon \sqrt{2x}} \left(e^{-(2\sqrt{x} - \sqrt{\pi}d)^2/2\pi \epsilon^2} + e^{-(2\sqrt{x} + \sqrt{\pi}d)^2/2\pi \epsilon^2} \right),$$
if $x > 0$, and $f_X(x) = 0$ otherwise.

8.188 a) $e^{-6.9t}$　**b)** 0.0317　**c)** 0.0317

8.189 a) $X \sim \mathcal{E}(\ln b)$　**b)** $X \sim \mathcal{E}(\ln b)$

8.190 0.122

8.191 The corresponding PDFs are

a) $nf F^{n-1}$.　**b)** $nf(1 - F)^{n-1}$.

8.192 Let $Y = |X(t)|$.

a) $f_Y(y) = \sqrt{2/\pi \sigma^2 t}\, e^{-y^2/2\sigma^2 t}$ if $y > 0$; $f_Y(y) = 0$ otherwise.

b) $Y \sim \mathcal{U}(0, t)$

8.193 Let $Y = X^+(t) = \max\{X(t), 0\}$.

a) $F_Y(y) = 0$ if $y < 0$; $F_Y(y) = \Phi\left(y/\sigma\sqrt{t}\right)$ if $y \ge 0$.

b) $F_Y(y) = 0$ if $y < 0$, $F_Y(y) = (y + t)/2t$ if $0 \le y < t$, and $F_Y(y) = 1$ if $y \ge t$.

c) No in both cases.

8.194

a) $f_K(x) = \left(a\sqrt{2x}/m^{3/2}\right)e^{-2bx}$ if $x > 0$; $f_K(x) = 0$ otherwise.

b) $K \sim \Gamma(3/2, 2b)$　**c)** $a = 4(bm)^{3/2}/\sqrt{\pi}$

8.195

a) Because a normal random variable can be negative, whereas a gestation period can't.

b) The probability that a normal random variable with parameters $\mu = 266$ and $\sigma^2 = 256$ will be negative is essentially zero (2.30×10^{-62}).

8.196 a) 76.5%   **b)** 0.03%

8.197

a) 0.601

b) $c = 3.2$. Eighty percent of 20-year-old men have heights that are within 3.2 inches of the mean height of 69 inches.

8.198 a) 0.0799   **b)** 0.237

8.199 a) 0.0525   **b)** 0.355

8.200 Note that $Y = \lfloor 100X \rfloor$. Y has the discrete uniform distribution on $\{00, 01, \ldots, 99\}$.

8.201 α/λ and $(\alpha + 1)\alpha/\lambda^2$

8.202

a) $X \sim \Gamma(2, 1)$

b) $F_X(x) = 0$ if $x < 0$, and $F_X(x) = 1 - e^{-x} - xe^{-x}$ if $x \geq 0$.

c) 73.6

Chapter 9

EXERCISES 9.1

9.1 $F_{U,V}(u, v) = F_{X,Y}\big((u - a)/b, (v - c)/d\big)$

9.3

a) $F_X(x) = 0$ if $x < 0$, $F_X(x) = x$ if $0 \leq x < 1$, and $F_X(x) = 1$ if $x \geq 1$.

b) $X \sim \mathcal{U}(0, 1)$

c) $F_Y(y) = 0$ if $y < 0$, $F_Y(y) = y$ if $0 \leq y < 1$, and $F_Y(y) = 1$ if $y \geq 1$. $Y \sim \mathcal{U}(0, 1)$.

9.5

a)
$$
F_{X,Y}(x, y) = \begin{cases} 0, & \text{if } x < 0 \text{ or } y < 0; \\ 1/4, & \text{if } 0 \leq x < 1 \ \& \ 0 \leq y < 1; \\ 1/2, & \text{if } 0 \leq x < 1 \ \& \ y \geq 1, \\ & \text{or } x \geq 1 \ \& \ 0 \leq y < 1; \\ 1, & \text{if } x \geq 1 \ \& \ y \geq 1. \end{cases}
$$

b) $F_X(x) = 0$ if $x < 0$, $F_X(x) = 1/2$ if $0 \leq x < 1$, and $F_X(x) = 1$ if $x \geq 1$.

c) Bernoulli with parameter 1/2.

d) $F_Y(y) = 0$ if $y < 0$, $F_Y(y) = 1/2$ if $0 \leq y < 1$, and $F_Y(y) = 1$ if $y \geq 1$. Bernoulli with parameter 1/2.

e) $F_{X,Y}(x, y) = F_X(x)F_Y(y)$ for all $x, y \in \mathcal{R}$.

9.7

a) $F_X(x) = 0$ if $x < 0$, and $F_X(x) = 1 - e^{-\lambda x}$ if $x \geq 0$. $F_Y(y) = 0$ if $y < 0$, and $F_Y(y) = 1 - e^{-\mu y}$ if $y \geq 0$.

b) $F_{X,Y}(x, y) = 0$ if $x < 0$ or $y < 0$, and $F_{X,Y}(x, y) = 1 - e^{-\lambda x} - e^{-\mu y} + e^{-\lambda x - \mu y}$ if $x \geq 0$ and $y \geq 0$.

9.9

a) $F_V(v) = F_{X,Y}(v, v)$

b) $F_U(u) = F_X(u) + F_Y(u) - F_{X,Y}(u, u)$

c) $F_U(u) = F_{X,Y}(u, \infty) + F_{X,Y}(\infty, u) - F_{X,Y}(u, u)$

EXERCISES 9.2

9.19 The domination principle and the definition of a continuous random variable give
$$
0 \leq P(X = x, \ Y = y) \leq P(X = x) = 0,
$$
which implies that $P(X = x, \ Y = y) = 0$.

9.21 For brevity, write $f = f_{X,Y}$. Then
$$
f_{U,V}(u, v) = \frac{1}{4\sqrt{uv}} \Big(f\big(\sqrt{u}, \sqrt{v}\big) + f\big(-\sqrt{u}, \sqrt{v}\big)
$$
$$
+ f\big(\sqrt{u}, -\sqrt{v}\big) + f\big(-\sqrt{u}, -\sqrt{v}\big) \Big),
$$
if $u > 0$ and $v > 0$, and $f_{U,V}(u, v) = 0$ otherwise.

9.23 Let H denote the upper half of the unit disk.

a) $f_{X,Y}(x, y) = 2/\pi$ if $(x, y) \in H$, and $f_{X,Y}(x, y) = 0$ otherwise.

b) Because the appropriate probability model is a geometric probability model—a point is being selected at random from H.

9.25 A joint PDF of X and Y is:
$$
f_{X,Y}(x, y) = n(n - 1)\lambda^2 e^{-\lambda(x+y)} \big(e^{-\lambda x} - e^{-\lambda y}\big)^{n-2},
$$
if $0 < x < y$, and $f_{X,Y}(x, y) = 0$ otherwise.

EXERCISES 9.3

9.39 Choice (b)

9.41

a) $f_{X,Y}(x, y) = 1$ if $0 < x < 1$ and $0 < y < 1$, and $f_{X,Y}(x, y) = 0$ otherwise.

c) 7/16; 7/16

9.43

a) $f_{X_1,\ldots,X_m}(x_1, \ldots, x_m) = 1/|S|$ if $(x_1, \ldots, x_m) \in S$, and $f_{X_1,\ldots,X_m}(x_1, \ldots, x_m) = 0$ otherwise.

b) $P\big((X_1, \ldots, X_m) \in A\big) = |A \cap S|/|S|$.

9.45 a) 0.5   **b)** 0.488

9.47 Applying the FPF for two continuous random variables gives

$$P(Y = X) = \iint_{y=x} f_{X,Y}(x, y)\, dx\, dy$$

$$= \int_{-\infty}^{\infty} \left(\int_{x}^{x} f_{X,Y}(x, y)\, dy \right) dx = 0.$$

9.49

a) A joint PDF of X and Y is:

$$f_{X,Y}(x, y) = n(n-1)\lambda^2 e^{-\lambda(x+y)} \left(e^{-\lambda x} - e^{-\lambda y} \right)^{n-2}$$

if $0 < x < y$, and $f_{X,Y}(x, y) = 0$ otherwise.

b) $1 - \left(1 - e^{-\lambda} \right)^{n-1}$

c) $f_R(r) = (n - 1)\lambda e^{-\lambda r} \left(1 - e^{-\lambda r} \right)^{n-2}$ if $r > 0$, and $f_R(r) = 0$ otherwise.

d) $1 - \left(1 - e^{-\lambda} \right)^{n-1}$

EXERCISES 9.4

9.63

a) For each $x \in (0, 1)$, $Y_{|X=x} \sim \mathcal{U}(0, 1)$; and for each $y \in (0, 1)$, $X_{|Y=y} \sim \mathcal{U}(0, 1)$.

b) From part (a), conclude that the conditional PDFs of Y given $X = x$ are identical to each other and to the marginal PDF of Y. Thus, in this case, knowing the value of the random variable X imparts no information about the distribution of Y; in other words, knowing the value of X doesn't affect the probability distribution of Y.

9.65

a) $f_X(x) = (2/\pi)\sqrt{1 - x^2}$ if $-1 < x < 1$; $f_X(x) = 0$ otherwise.

b) $f_Y(y) = (4/\pi)\sqrt{1 - y^2}$ if $0 < y < 1$; $f_Y(y) = 0$ otherwise.

c) For $-1 < x < 1$, $Y_{|X=x} \sim \mathcal{U}\left(0, \sqrt{1 - x^2}\right)$.

d) For $0 < y < 1$, $X_{|Y=y} \sim \mathcal{U}\left(-\sqrt{1 - y^2}, \sqrt{1 - y^2}\right)$.

9.67 The required result follows from the observation that $f_X(x) = nx^{1-1}(1 - x)^{n-1}$ if $0 < x < 1$.

9.69

a) The methods are probabilistically different. With Method II, $X \sim \mathcal{U}(0, 1)$. However, with Method I, values of X near 1 are more likely than values of X near 0—so that $X \not\sim \mathcal{U}(0, 1)$.

b) Method I: $f_{X,Y}(x, y) = 2$ if $(x, y) \in T$, and $f_{X,Y}(x, y) = 0$ otherwise.

Method II: $f_{X,Y}(x, y) = 1/x$ if $(x, y) \in T$, and $f_{X,Y}(x, y) = 0$ otherwise.

c) Method I: $X \sim \text{Beta}(2, 1)$. $Y \sim \text{Beta}(1, 2)$.

Method II: $X \sim \mathcal{U}(0, 1)$. $f_Y(y) = -\ln y$ if $0 < y < 1$, and $f_Y(y) = 0$ otherwise.

d) Method I: For $0 < x < 1$, $Y_{|X=x} \sim \mathcal{U}(0, x)$. For $0 < y < 1$, $X_{|Y=y} \sim \mathcal{U}(y, 1)$.

Method II: For $0 < x < 1$, $Y_{|X=x} \sim \mathcal{U}(0, x)$. For $0 < y < 1$, $f_{X|Y}(x \mid y) = -1/(x \ln y)$ if $y < x < 1$, and $f_{X|Y}(x \mid y) = 0$ otherwise.

9.71

a) The lifetimes of the components (i.e, the random variables $X_1, \ldots, X_n$) can be considered a random sample of size n from an exponential distribution with parameter λ.

b) $X \sim \mathcal{E}(n\lambda)$.

c) $f_Y(y) = n\lambda e^{-\lambda y} \left(1 - e^{-\lambda y} \right)^{n-1}$ if $y > 0$; $f_Y(y) = 0$ otherwise.

d) For $x > 0$,

$$f_{Y|X}(y \mid x) = (n - 1)\lambda e^{-\lambda(y-x)} \left(1 - e^{-\lambda(y-x)} \right)^{n-2},$$

if $y > x$, and $f_{Y|X}(y \mid x) = 0$ otherwise.

e) Same answer as in part (d).

f) For $y > 0$,

$$f_{X|Y}(x \mid y) = \frac{(n - 1)\lambda e^{-\lambda x} \left(e^{-\lambda x} - e^{-\lambda y} \right)^{n-2}}{\left(1 - e^{-\lambda y} \right)^{n-1}}.$$

if $0 < x < y$, and $f_{X|Y}(x \mid y) = 0$ otherwise.

9.73 8/9

9.75

a) $f_Y(y) = \alpha\beta^\alpha / (\beta + y)^{\alpha+1}$ if $y > 0$, and $f_Y(y) = 0$ otherwise.

b) $\Lambda_{|Y=y} \sim \Gamma(\alpha + 1, \beta + y)$.

EXERCISES 9.5

9.89 From Example 9.11, each conditional PDF of Y given $X = x$ is a PDF of Y. Independence of X and Y now follows from Proposition 9.10(a) on page 527.

9.91 Yes, X and Y are independent.

9.93 X and Y are independent if and only if $\rho = 0$.

9.95

a) $f_X(x) = \left(\int_{-\infty}^{\infty} h(y)\, dy \right) g(x)$ for all $x \in \mathcal{R}$.

b) $f_Y(y) = \left(\int_{-\infty}^{\infty} g(x)\, dx \right) h(y)$ for all $y \in \mathcal{R}$.

c) The required result follows from Proposition 9.5(b) on page 502.

d) The required result follows from parts (a)–(c) and Proposition 9.9 on page 523.

9.97

b) No. Exercise 9.95 requires that a joint PDF of X and Y can be expressed as a product of a function of x alone with a function of y alone, for *all* $x, y \in \mathcal{R}$.

c) No.

9.99

a) Because of the general multiplication rule, Equation (9.34) on page 517.

b) Independence of X and Y, as indicated by Equation (9.38) on page 523.

9.101 0.19

9.103

a) By symmetry, each of the 3! possible orderings of the random variables X, Y, and Z are equally likely, so each has probability 1/6.

c) $1/n!$

9.105 a) $1/27$ **b)** $d \le 1/2$ **c)** $(1 - 2d)^3$

9.107 a) $1/n$ **b)** $1/n$

EXERCISES 9.6

9.117 Let Y denote system lifetime. Then $F_Y(y) = 0$ if $y < 0$, and $F_Y(y) = \prod_{k=1}^{n}\left(1 - e^{-\lambda_k y}\right)$ if $y \ge 0$. So,

$$f_Y(y) = \left(\prod_{k=1}^{n}\left(1 - e^{-\lambda_k y}\right)\right)\left(\sum_{k=1}^{n} \frac{\lambda_k e^{-\lambda_k y}}{1 - e^{-\lambda_k y}}\right),$$

if $y > 0$, and $f_Y(y) = 0$ otherwise.

9.119 $f_Z(z) = \frac{1}{2}\lambda e^{-\lambda|z|}$ for all $z \in \mathcal{R}$.

9.121 a) 0.4 **b)** 0.4

9.123 $X + Y \sim \Gamma(\alpha + \beta + 2, \lambda)$

9.125

a) $(X_1 + \cdots + X_n - n\mu)/\sqrt{n}\sigma$; the standard normal distribution.

b) $\bar{X}_n \sim \mathcal{N}\left(\mu, \sigma^2/n\right)$

9.127 16   **9.129** $n\lambda/(n\lambda + m\mu)$

9.133

b) Z has the beta distribution with parameters α and β.

c) From part (b), the proportion of time until system failure that the original component is working has the beta distribution with parameters 1 and 1, which is the uniform distribution on the interval (0, 1).

EXERCISES 9.7

9.147

a) $f_{S,T}(s, t) = \lambda \mu t e^{-(\mu + (\lambda - \mu)s)t}$ if $0 < s < 1, t > 0$; $f_{S,T}(s, t) = 0$ otherwise.

b) $f_S(s) = \lambda \mu/\left(\mu + (\lambda - \mu)s\right)^2$ if $0 < s < 1$; $f_S(s) = 0$ otherwise.

c) $f_T(t) = \lambda \mu\left(e^{-\mu t} - e^{-\lambda t}\right)/(\lambda - \mu)$ if $t > 0$; $f_T(t) = 0$ otherwise.

d) No; S and T are not independent random variables.

9.149

a) $f_{R,\Theta}(r, \theta) = r/\pi$ if $0 < r < 1, 0 < \theta < 2\pi$; $f_{R,\Theta}(r, \theta) = 0$ otherwise.

b) $R \sim \text{Beta}(2, 1)$ and $\Theta \sim \mathcal{U}(0, 2\pi)$.

c) Yes; R and Θ are independent random variables.

9.151

a) A joint PDF of R and M is:

$$f_{R,M}(r, m) = n(n - 1)f\left(m - \tfrac{r}{2}\right)f\left(m + \tfrac{r}{2}\right)$$
$$\times \left(F\left(m + \tfrac{r}{2}\right) - F\left(m - \tfrac{r}{2}\right)\right)^{n-2},$$

if $r > 0$ and $-\infty < m < \infty$, and $f_{R,M}(r, m) = 0$ otherwise.

b) A marginal PDF of R is:

$$f_R(r) = n(n - 1)\int_{-\infty}^{\infty} f(u)f(r + u)$$
$$\times \left(F(r + u) - F(u)\right)^{n-2} du,$$

if $r > 0$, and $f_R(r) = 0$ otherwise.

c) A marginal PDF of M is:

$$f_M(m) = 2n(n - 1)\int_{-\infty}^{m} f(u)f(2m - u)$$
$$\times \left(F(2m - u) - F(u)\right)^{n-2} du,$$

for all $m \in \mathcal{R}$.

d) No; R and M are not independent random variables.

9.153 Applying the bivariate transformation theorem yields

$$f_{X+Y, Y/X}(u, v) = \frac{\lambda^{\alpha+\beta}}{\Gamma(\alpha)\Gamma(\beta)} \frac{u^{\alpha+\beta-1}v^{\beta-1}}{(1 + v)^{\alpha+\beta}} e^{-\lambda u},$$

if $u > 0$ and $v > 0$, and $f_{X+Y, Y/X}(u, v) = 0$ otherwise. From this result, it can now be shown in various ways that $X + Y$ and Y/X are independent random variables.

9.155

a) $f_{R(t),\Theta(t)}(r, \theta) = \left(r/2\pi\sigma^2 t\right)e^{-r^2/2\sigma^2 t}$ if $r > 0$ and $0 < \theta < 2\pi$, and $f_{R(t),\Theta(t)}(r, \theta) = 0$ otherwise.

b) $R(t)$ has the Rayleigh distribution (as defined in Exercise 8.159 on page 475) with parameter $\sigma^2 t$; also, $\Theta(t) \sim \mathcal{U}(0, 2\pi)$.

c) $R(t)$ and $\Theta(t)$ are independent random variables.

9.157 Get two (independent) numbers—say, u and v—from a basic random number generator. In view of the Box–Müller transformation and Proposition 8.15 on page 468, the number $x = \mu + \sigma\sqrt{-2\ln u}\cos(2\pi v)$ is a simulated observation of a normal random variable with parameters μ and σ^2.

9.159

a) A joint PDF of X and Y is:

$$f_{X,Y}(x, y) = \frac{1}{2\pi\sqrt{1 - \rho^2}} e^{-(x^2 - 2\rho xy + y^2)/2(1-\rho^2)},$$

for all $x, y \in \mathcal{R}$.

b) Obtain numbers, u and v, from a basic random number generator. The required pair (x, y) is

$$x = \sqrt{-2\ln u}\cos(2\pi v)$$

$$y = \rho\sqrt{-2\ln u}\cos(2\pi v)$$

$$+ \sqrt{1 - \rho^2}\sqrt{-2\ln u}\sin(2\pi v).$$

REVIEW EXERCISES

9.166

a) Apply successively the complementation rule, the general addition rule, and the definitions of univariate and bivariate CDFs.

b) Use part (a) and Proposition 9.2 on page 490.

9.167 a) 0.42 **b)** 0.42 **c)** 0.42 **d)** 0.42

9.168

a) $F_{X,Y}(x, y)$ is

$$\begin{cases} 0, & \text{if } x < a \text{ or } y < c; \\ \frac{(x-a)(y-c)}{(b-a)(d-c)}, & \text{if } a \le x < b \text{ and } c \le y < d; \\ \frac{x-a}{b-a}, & \text{if } a \le x < b \text{ and } y \ge d; \\ \frac{y-c}{d-c}, & \text{if } x \ge b \text{ and } c \le y < d; \\ 1, & \text{if } x \ge b \text{ and } y \ge d. \end{cases}$$

b) $F_X(x) = 0$ if $x < a$, $F_X(x) = (x - a)/(b - a)$ if $a \le x < b$, and $F_X(x) = 1$ if $x \ge b$. $F_Y(y) = 0$ if $y < c$, $F_Y(y) = (y - c)/(d - c)$ if $c \le y < d$, and $F_Y(y) = 1$ if $y \ge d$.

c) $X \sim \mathcal{U}(a, b)$ and $Y \sim \mathcal{U}(c, d)$.

d) The joint CDF of X and Y equals the product of the marginal CDFs of X and Y.

e) Yes, X and Y are independent random variables.

9.169 $f_{X,Y}(x, y) = \left((b - a)(d - c)\right)^{-1}$ if $a < x < b$ and $c < y < d$, and $f_{X,Y}(x, y) = 0$ otherwise.

9.170 Suppose to the contrary that F is a joint CDF—say, of the random variables X and Y. Then, from Proposition 9.1, $P(-1 < X \le 1, -1 < Y \le 1) = -1$, which is impossible.

9.171 Yes; $f(x, y) = \left(1/\sqrt{3}\pi\right)g(x, y)$.

9.172 0.708

9.173

a) $X \sim \mathcal{E}(1)$ and $Y \sim \Gamma(2, 1)$.

b) X and Y are not independent random variables.

c) $f_{X+Y}(z) = e^{-z/2} - e^{-z}$ if $z > 0$, and $f_{X+Y}(z) = 0$ otherwise.

d) For $y > 0$, $X_{|Y=y} \sim \mathcal{U}(0, y)$.

e) 0.0308 **f)** 0.4

9.174 a) $\int_{-\infty}^{\infty}\left(\int_{y}^{y+5} f_X(x)\,dx\right)f_Y(y)\,dy$ **b)** 1/6

9.175

a) $\left(F(b) - F(a)\right)^n$

b) $\binom{n}{k}\left(F(b) - F(a)\right)^k\left(1 - F(b) + F(a)\right)^{n-k}$

c)

k	$P(A_k)$
0	0.0081
1	0.0756
2	0.2646
3	0.4116
4	0.2401

d)

k	$P(A_k)$
0	0.1194
1	0.3349
2	0.3522
3	0.1646
4	0.0289

9.176

a) $3|E \cap S|/4\pi$, where $|\,\,|$ denotes volume.

b) $f_{X,Y,Z}(x, y, z) = \frac{3}{4\pi}$ if $(x, y, z) \in S$, and $f_{X,Y,Z}(x, y, z) = 0$ otherwise.

d) $f_X(x) = \frac{3}{4}(1 - x^2)$ if $-1 < x < 1$, and $f_X(x) = 0$ otherwise. By symmetry, conclude that each of Y and Z has the same PDF as X.

e) $f_{X,Y}(x, y) = \frac{3}{2\pi}\sqrt{1 - x^2 - y^2}$ if $x^2 + y^2 < 1$, and $f_{X,Y}(x, y) = 0$ otherwise. By symmetry, conclude that each of (X, Z) and (Y, Z) has the same joint PDF as (X, Y).

f) For $x^2 + y^2 < 1$, $Z_{|X=x, Y=y} \sim \mathcal{U}(I_{x,y})$, where $I_{x,y} = \left(-\sqrt{1 - x^2 - y^2}, \sqrt{1 - x^2 - y^2}\right)$.

g) For $-1 < x < 1$, $(Y, Z)_{|X=x} \sim \mathcal{U}(D_x)$, where $D_x = \left\{(y, z) : y^2 + z^2 < 1 - x^2\right\}$.

h) No; X, Y, and Z are not independent.

9.177 The distance of the point obtained from the origin has the beta distribution with parameters 3 and 1.

9.178 (e)   **9.179** 0.417

9.180

a)
$$F_{X,Y}(x, y) = \begin{cases} 0, & \text{if } x < 0 \text{ or } y < 0; \\ y, & \text{if } 0 \leq y < \min\{1, x\}; \\ x, & \text{if } 0 \leq x < \min\{1, y\}; \\ 1, & \text{if } x \geq 1 \text{ and } y \geq 1. \end{cases}$$

b) $F_X(x) = 0$ if $x < 0$, $F_X(x) = x$ if $0 \leq x < 1$, and $F_X(x) = 1$ if $x \geq 1$. By symmetry, Y has the same CDF as X.

c) $X \sim \mathcal{U}(0, 1)$ and $Y \sim \mathcal{U}(0, 1)$.

9.181 0.414

9.182 a) 0.125   **b)** 0.384   **c)** 0.343

9.183

a) $f_{X,Y}(x, y) = 1$ if $(x, y) \in C$, and $f_{X,Y}(x, y) = 0$ otherwise.

b) $f_X(x) = f(x)$ for all $x \in \mathcal{R}$.

c) Let $| \ |$ denote length and let $A(y) = \{x : f(x) > y\}$. $f_Y(y) = |A(y)|$ if $y > 0$; $f_Y(y) = 0$ otherwise.

d) For $f(x) > 0$, $Y_{|X=x} \sim \mathcal{U}(0, f(x))$.

e) For $f_Y(y) > 0$, $X_{|Y=y} \sim \mathcal{U}(A(y))$.

9.184 0.602   **9.185** 0.781

9.187 With time measured in minutes,

$$f_T(t) = \begin{cases} \frac{1}{2}(1 - e^{-(t-1)/600}), & 1 < t < 3; \\ \frac{1}{2}(e^{-(t-3)/600} - e^{-(t-1)/600}), & t > 3; \\ 0, & \text{otherwise.} \end{cases}$$

9.188 $\Theta(t) \sim \mathcal{U}(-\pi/2, \pi/2)$ for all $t > 0$.

9.189 Let Z denote the fraction of harmful impurities in the product, and note that $Z = XY$.

a) $f_Z(z) = \int_0^1 (1/x) f_X(x) f_Y(z/x) \, dx$ if $0 < z < 1$; $f_Z(z) = 0$ otherwise.

b) $f_Z(z) = -20 \ln(20z)$ if $0 < z < 1/20$; $f_Z(z) = 0$ otherwise.

9.190

a) $f_X(x) = 4x^3$ if $0 < x < 1$, and $f_X(x) = 0$ otherwise. $X \sim \text{Beta}(4, 1)$.

b) $f_Y(y) = 4y(1 - y^2)$ if $0 < y < 1$, and $f_Y(y) = 0$ otherwise.

c) Let Z denote the amount of the item remaining at the end of the day. Then $f_Z(z) = 4z^3/3 - 4z + 8/3$ if $0 < z < 1$, and $f_Z(z) = 0$ otherwise.

d) 5/9

9.191 0.274   **9.192** 0.223

9.193 For both parts (a) and (b), $f_W(w) = 6(1 - \sqrt{w})^2$ if $0 < w < 1$, and $f_W(w) = 0$ otherwise.

9.195 Let $U = \sum_{k=1}^n X_k$ and $V = \sum_{k=1}^m X_k$. Then, for all $u, v \in \mathcal{R}$,

$$f_{U,V}(u, v) = \frac{1}{2\pi \sqrt{n}\sigma \sqrt{m}\sigma \sqrt{1 - m/n}} e^{-\frac{1}{2}Q(u,v)},$$

where

$$Q(u, v) = \frac{1}{1 - m/n} \left\{ \left(\frac{u - n\mu}{\sqrt{n}\sigma} \right)^2 - 2\sqrt{\frac{m}{n}} \left(\frac{u - n\mu}{\sqrt{n}\sigma} \right) \right.$$
$$\left. \times \left(\frac{v - m\mu}{\sqrt{m}\sigma} \right) + \left(\frac{v - m\mu}{\sqrt{m}\sigma} \right)^2 \right\}.$$

Chapter 10

EXERCISES 10.1

10.1 $3r/4$

10.3 $\mathcal{E}(Y) = h/3$. On average, the point chosen will be a distance of $h/3$ from the base of the triangle.

10.5 $\mathcal{E}(X) = 15$. On average, John will wait 15 minutes for the train.

10.7 $\frac{5}{7}$ sec

10.11 Some possible beta distributions are:

a) $\alpha = \beta = 1$; $\mu_X = 0.5$; $P(X \geq \mu_X) = 0.5$

b) $\alpha = 3, \beta = 1$; $\mu_X = 0.75$; $P(X \geq \mu_X) = 0.578$

c) $\alpha = 1, \beta = 3$; $\mu_X = 0.25$; $P(X \geq \mu_X) = 0.422$

10.13 1/2   **10.17** 5.73

EXERCISES 10.2

10.23

a) The required result is a consequence of Propositions 8.10, 8.13, 6.13, and 9.13 on pages 441, 466, 297, and 537, respectively.

b) $2\sqrt{2/\pi}\,\sigma$

10.25 $\mathcal{E}(X) = 8/5$, $\mathcal{E}(Y) = 8/15$

10.27 1.9   **10.29** \$5644.23

10.31 a) $1/(n + 1)$ **b)** $n/(n + 1)$
c) $(n - 1)/(n + 1)$ **d)** 1/2

10.33

a) From Example 9.25(b), $S \sim \mathcal{U}(0, 1)$; consequently, $\mathcal{E}(S) = 1/2 = 50\%$.

10.37
b) No. Independence of random variables is not a necessary condition for the expected value of a product to equal the product of the expected values.
c) No.

10.41
a) $\mathcal{E}(1/\ell)$, due to the lack-of-memory property of the exponential distribution.
b) ℓ **c)** ℓ

10.43
a) $F_X(x) = 0$ if $x < 0$, and $F_X(x) = p + (1 - p)F(x)$ if $x \geq 0$.
b) $(1 - p) \int_0^\infty \left(1 - F(x)\right) dx$ **c)** $(1 - p)/\lambda$

EXERCISES 10.3

10.51 a) 20%   **b)** 44%   **c)** 20%

10.53 75   **10.55** $\frac{5}{7}$ sec

10.59 a) 2.23   **b)** 2.23

10.63 a) 0   **b)** 0

10.65 (b) is correct because of Equation (10.32) on page 589 and the fact that $\text{Cov}(X, Y) < 0$.

10.67 a) $1/n$ **c)** 0   **d)** No.   **e)** No.

EXERCISES 10.4

10.77
a) For $0 < x < 1$, $\mathcal{E}(Y \mid X = x) = x/2$ and $\text{Var}(Y \mid X = x) = x^2/12$.
b) $\mathcal{E}(Y) = 1/4$ and $\text{Var}(Y) = 7/144$.

10.79 8/5

10.81 a) $\mathcal{E}(Y)$ **b)** $\mathcal{E}(Y)$ **c)** $\text{Var}(Y)$ **d)** $\text{Var}(Y)$

10.83 $2\lambda/3$ and $\lambda(\lambda + 12)/18$

10.85 a) $g(X)$ **b)** $g(X)$

10.87 a) $1/2 + c - c^2$ **b)** 1/2

10.89 mean $= p\mu_1 + (1 - p)\mu_2$,
variance $= p\sigma_1^2 + (1 - p)\sigma_2^2 + p(1 - p)(\mu_2 - \mu_1)^2$.

10.91 $328.20

EXERCISES 10.5

10.101 $\mu_W = 4$ and $\sigma_W = 8.93$.

10.103
a) We find that the discriminant of a level curve is proportional to $\rho^2 - 1$, which is less than 0. Hence each level curve is an ellipse (or a circle).

b) The axes of the level curves are parallel to the coordinate axes if and only if $\rho = 0$—that is, if and only if X and Y are independent random variables.

10.105
a) A joint PDF of X and Y is obtained by substituting the values $\mu_X = 49.8$, $\sigma_X = 30.7$, $\mu_Y = 2.26$, $\sigma_Y = 1.18$, and $\rho = 0.732$ in the equations in Definition 10.4 on page 610.
b) $X \sim \mathcal{N}(49.8, 30.7^2)$. Nitrate concentration is normally distributed with a mean of 49.8 micromoles per liter and a standard deviation of 30.7 micromoles per liter.
c) $Y \sim \mathcal{N}(2.26, 1.18^2)$. Arsenic concentration is normally distributed with a mean of 2.26 nanomoles per liter and a standard deviation of 1.18 nanomoles per liter.
d) $\rho(X, Y) = 0.732$. A relatively high positive correlation exists between nitrate concentration and arsenic concentration.
e) $Y_{|X=x} \sim \mathcal{N}(0.859 + 0.028x, 0.646)$. Given a nitrate concentration of x micromoles per liter, arsenic concentration is normally distributed with a mean of $0.859 + 0.028x$ nanomoles per liter and a standard deviation of 0.804 nanomoles per liter.
f) $y = 0.859 + 0.028x$ **g)** 2.54 nanomoles per liter

10.107
b) $X + Y \sim \mathcal{N}\left(\mu_X + \mu_Y, \sigma_X^2 + 2\rho\sigma_X\sigma_Y + \sigma_Y^2\right)$ and $Y - X \sim \mathcal{N}\left(\mu_Y - \mu_X, \sigma_X^2 - 2\rho\sigma_X\sigma_Y + \sigma_Y^2\right)$.

10.109 $\sqrt{(1 - \rho)/\pi}$

REVIEW EXERCISES

10.116 $\mathcal{E}(X) = (a + b)/2$, $\text{Var}(X) = (b - a)^2/24$

10.117 b) $\alpha\beta/(\alpha - 1)$ **c)** $\alpha\beta/(\alpha - 1)$
d) $\alpha\beta/(\alpha - 1)$ **e)** $6000

10.118
a) $\mathcal{E}(Y) = \alpha^{-1/\beta}\Gamma(1 + 1/\beta)$,
  $\text{Var}(Y) = \alpha^{-2/\beta}\left(\Gamma(1 + 2/\beta) - \left(\Gamma(1 + 1/\beta)\right)^2\right)$
b) 0.71 yr; 0.26 yr

10.119
a) $\mathcal{E}(Y) = e^{\mu + \sigma^2/2}$, $\text{Var}(Y) = e^{2\mu + \sigma^2}\left(e^{\sigma^2} - 1\right)$
b) $1242.65; $662.26

10.120 a) $2\sqrt{2/\pi}\,\sigma$ **b)** $2\sqrt{2/\pi}\,\sigma$ **c)** $2\sqrt{2/\pi}\,\sigma$

10.121
a) $\mathcal{E}(Y) = 1/6$, $\mathcal{E}(XY) = 1/12$

b) $\mathcal{E}(Y) = 1/6$, $\mathcal{E}(XY) = 1/12$

d) Two random variables can be uncorrelated without being independent. Consequently, by referring to Proposition 7.16(b) on page 372, conclude that independence is a sufficient, but not necessary, condition for uncorrelatedness.

10.122

a) $\mathcal{E}(X) = \mathcal{E}(Y) = 1/3$

b) $\text{Var}(X) = \text{Var}(Y) = 1/18$

c) Negative   **d)** $-1/2$

10.123 0.934   **10.124 a)** 5/6   **b)** 5/6

10.125 \$10,255.90   **10.126** 3.54 yr

10.127 \$500   **10.128** 14.1 yr

10.129 Answers will vary.

10.130 a) 2.25   **b)** 2.25   **10.131** 200/3

10.132 b) No.   **c)** No.

10.133 a) 9000 hr   **b)** 9000 hr
c) \$117   **d)** \$30.72

10.134 a) 6   **b)** 6

10.135 $\mathcal{E}(Y) = 8/15$, $\text{Var}(Y) = 11/225$

10.136 a) 1/3   **b)** 26/35   **c)** 93/140

10.137 $\mathcal{E}(X_n) = 2^{-n}$

10.138

a) A joint PDF of X and Y is obtained by substituting $\mu_X = 5.3$, $\sigma_X = 1.4$, $\mu_Y = 88.6$, $\sigma_Y = 31.2$, and $\rho = -0.9$ into the equations in Definition 10.4 on page 610.

b) $X \sim \mathcal{N}(5.3, 1.4^2)$. The ages of Orions are normally distributed with a mean of 5.3 years and a standard deviation of 1.4 years.

c) $Y \sim \mathcal{N}(88.6, 31.2^2)$. The prices of Orions are normally distributed with a mean of \$8860 and a standard deviation of \$3120.

d) $\rho(X, Y) = -0.9$. There is a strong negative correlation between age and price of Orions.

e) $Y_{|X=x} \sim \mathcal{N}(194.90 - 20.06x, 184.95)$. The prices of x-year-old Orions are normally distributed with a mean of \$$(19490 - 2006x)$ and a standard deviation of \$1360.

f) $y = 194.90 - 20.06x$ **g)** \$13,472

10.139

a) A joint PDF of X and Y is obtained by substituting $\mu_X = 3.0$, $\sigma_X = 0.5$, $\mu_Y = 2.7$, $\sigma_Y = 0.6$, and $\rho = 0.5$ into the equations in Definition 10.4 on page 610.

b) 3.0   **c)** 2.15 to 3.85

10.140 51.59 lb

Chapter 11

EXERCISES 11.1

11.1 $M_X(t) = \frac{19}{27} + \frac{8}{27}e^t$, $t \in \mathcal{R}$.

11.3

a) $M_X(t) = pe^t / \left(1 - (1 - p)e^t\right)$, $t < -\ln(1 - p)$.

b) $\mathcal{E}(X) = 1/p$, $\text{Var}(X) = (1 - p)/p^2$

11.5 10,560

11.9 a) Discrete uniform distribution on $\{-1, 1\}$

11.11

b) $\mathcal{E}(\bar{X}_n) = \mu$, $\text{Var}(\bar{X}_n) = \sigma^2/n$

d) If $X \equiv \mu$, then $M_X(t) = e^{\mu t}$. Referring to part (c) and the continuity theorem (Proposition 11.5), deduce that $\{\bar{X}_n\}_{n=1}^{\infty}$ converges in distribution to μ.

EXERCISES 11.2

11.21

a) $M_{X,Y}(s, t) = \frac{1}{4}\left(1 + e^s + e^t + e^{s+t}\right)$, $s, t \in \mathcal{R}$.

b) We have $\mathcal{E}(X) = 1/2$, $\text{Var}(X) = 1/4$, $\mathcal{E}(Y) = 1/2$, $\text{Var}(Y) = 1/4$, and $\text{Cov}(X, Y) = 0$.

c) $M_X(s) = \frac{1}{2} + \frac{1}{2}e^s$, $M_Y(t) = \frac{1}{2} + \frac{1}{2}e^t$

d) X and Y are independent random variables.

e) X and Y are Bernoulli random variables each with parameter 1/2 or, equivalently, discrete uniform random variables on $\{0, 1\}$.

11.25

a) $M_X(s) = \left(pe^s + (1 - p)\right)^n$; $X \sim \mathcal{B}(n, p)$.
$M_Y(t) = \left(qe^t + (1 - q)\right)^n$; $Y \sim \mathcal{B}(n, q)$.

b) X and Y are not independent random variables.

c) $p_{X+Y}(n) = 1$, and $p_{X+Y}(z) = 0$ otherwise.

11.27

a) $M_{X,Y}(s, t) = (2 - s - t)^{-1}\left((1 - s)^{-2} + (1 - t)^{-2}\right)$, for $s, t < 1$.

b) $\mathcal{E}(X) = \mathcal{E}(Y) = 3/2$, $\text{Var}(X) = \text{Var}(Y) = 9/4$, and $\text{Cov}(X, Y) = -3/4$.

c) $M_X(s) = (2 - s)^{-1}\left(1 + (1 - s)^{-2}\right)$, $s < 1$;
$M_Y(t) = (2 - t)^{-1}\left(1 + (1 - t)^{-2}\right)$, $t < 1$

d) X and Y are not independent random variables.

e) $X + Y \sim \Gamma(3, 1)$

EXERCISES 11.3

11.39 In this context, the strong law of large numbers states that $\lim_{n\to\infty}(X_1 + \cdots + X_n)/n = 1/2$ with probability 1, not that the limit always equals 1/2.

11.41
c) The values of $\mathcal{E}(\ln X)$ for the four specified distributions are -1, $-1/2$, $-3/2$, and $-5/6$, respectively.

11.43
a) Because a Riemann integrable function on $[0, 1]$ is necessarily bounded thereon, that $g(X)$ has finite expectation follows from the FEF: specifically, from Equation (10.6) on page 572.
c) For a large $n \in \mathcal{N}$, use a basic random number generator to obtain n (independent) numbers between 0 and 1—say, $x_1, \ldots, x_n$. In view of the result in part (b), $\int_0^1 g(x)\,dx \approx \big(g(x_1) + \cdots + g(x_n)\big)/n$.

11.45
a) $\frac{1}{\pi}\big(\arctan(c + \epsilon) - \arctan(c - \epsilon)\big)$
b) From part (a), the weak law of large numbers doesn't hold for the standard Cauchy distribution, the reason being that this distribution doesn't have finite mean.

11.47
b) Roughly 0, by the law of large numbers and the fact that the mean of a standard normal random variable equals 0.

EXERCISES 11.4

Note: For this section, we have used statistical software to obtain any required standard normal probabilities. If you use Table I instead, your answers may differ somewhat from those shown here.

11.55 For each part, the first and second probabilities are obtained from the binomial PMF and normal approximation, respectively.
a) 0.1408; 0.1288   b) 0.2087; 0.2194
c) 0.9993; 0.9956   d) 0.0266; 0.0367

11.57 a) 0.265   **b)** 0.0447   **c)** 0.0000917

11.59 0.995

11.61
a) 0.00646
b) No; the independence assumption isn't strictly satisfied when sampling is without replacement.
c) Because the population size is large relative to the sample size, there is little difference between sampling without and with replacement and, in the latter case, the independence assumption is satisfied.

11.63 0.841   **11.65** 0.952

11.67
a) 14.50 cm to 15.60 cm
b) We can be 90% confident that the mean depth of all subterranean coruro burrows is somewhere between 14.50 cm and 15.60 cm.
c) 14.19 cm to 15.92 cm
d) If you want to be more confident that μ lies in your confidence interval, then you must settle for a greater interval.

11.69 0.000398

REVIEW EXERCISES

Note: See the note at the beginning of the answers for Section 11.4.

11.76 mean $= \alpha/\lambda$, variance $= \alpha/\lambda^2$

11.77 a) 5000   **b)** 5000

11.78 a) $2t^{-2}(\cosh t - 1)$

11.79
a) $\mathcal{E}(T) = 0$, $\mathrm{Var}(T) = 1/6$
b) $\mathcal{E}(X) = (a + b)/2$, $\mathrm{Var}(X) = (b - a)^2/24$

11.80
a) For $t < -\ln(1 - p)$,
$$M_X(t) = \left(\frac{pe^t}{1 - (1 - p)e^t}\right)^r.$$
b) $\mathcal{E}(X) = r/p$, $\mathrm{Var}(X) = r(1 - p)/p^2$

11.83
a) $p_X(0) = 1$, and $p_X(x) = 0$ otherwise.

11.85
a) $M_{X,Y}(s, t) = 6\big((3 - s - t)(2 - t)\big)^{-1}$ if $t < 2$ and $s + t < 3$.
b) $X \sim \mathcal{E}(3)$ c) $Y - X \sim \mathcal{E}(2)$ e) 5/6

11.86
a) $M_{X,Y}(s, t) = \frac{4}{9} + \frac{2}{9}e^s + \frac{2}{9}e^t + \frac{1}{9}e^{s+t}$.
b) $\mathcal{E}(X) = \mathcal{E}(Y) = 1/3$, $\mathrm{Var}(X) = \mathrm{Var}(Y) = 2/9$, and $\mathrm{Cov}(X, Y) = 0$.
c) Both X and Y are Bernoulli with parameter 1/3.
d) $X + Y \sim \mathcal{B}(2, 1/3)$
e) X and Y are independent random variables.

11.87
a) $M_{X,Y}(s, t) = \frac{4}{7} + \frac{2}{7}e^s + \frac{1}{7}e^{s+t}$.
b) $\mathcal{E}(X) = 3/7$, $\mathcal{E}(Y) = 1/7$, $\mathrm{Var}(X) = 12/49$, $\mathrm{Var}(Y) = 6/49$, and $\mathrm{Cov}(X, Y) = 4/49$.

c) X and Y are Bernoulli with parameters 3/7 and 1/7, respectively.

d) $p_{X+Y}(0) = 4/7$, $p_{X+Y}(1) = 2/7$, $p_{X+Y}(2) = 1/7$, and $p_{X+Y}(z) = 0$ otherwise.

e) No; $\sqrt{p_{X+Y}(0)} + \sqrt{p_{X+Y}(2)} \neq 1$.

f) X and Y are not independent random variables.

11.89

a) Because a Riemann integrable function on $[a, b]$ is necessarily bounded thereon, that $g(X)/f(X)$ has finite expectation follows from the FEF: specifically, from Equation (10.6) on page 572.

c) The random variable $\hat{I}_n(f, g)$ has finite variance if and only if $\int_a^b ((g(x))^2/f(x)) \, dx < \infty$, in which case, the variance equals

$$\frac{1}{n} \left(\int_a^b \frac{(g(x))^2}{f(x)} \, dx - \left(\int_a^b g(x) \, dx \right)^2 \right).$$

11.90 Approximately 0.0985

11.91 a) 0.373 **b)** 0.928

11.92

b)

Probability	Local	Integral	Poisson
$P(X > 7)$	0.375	0.373	0.352
$P(X \leq 10)$	0.930	0.928	0.923

11.93 a) 0.00701 **b)** 0.950 **c)** 0.999

11.94

a) For convenience, set $S = X_1 + X_2 + X_3$.

s	3	4	5	6	7	8	9
$p_S(s)$	$\frac{1}{27}$	$\frac{3}{27}$	$\frac{6}{27}$	$\frac{7}{27}$	$\frac{6}{27}$	$\frac{3}{27}$	$\frac{1}{27}$

b) $\frac{23}{27} \approx 0.852$ **d)** 0.856

11.95 0.00898

11.96 0.770

11.97 0.770

11.98

a) $e^{-2500} \sum_{k=2450}^{2600} (2500)^k/k!$ **b)** 0.819

11.100 24.955 to 30.985. We can be 90% confident that the mean commute time in Washington, D.C., is somewhere between 24.955 and 30.985 minutes.

Chapter 12

EXERCISES 12.1

12.3 a) 0.149 **b)** 0.106 **c)** 0.0158 **d)** 0.239
e) 0.0878

f) $e^{-27.6}(6.9)^{14} \sum_{k=0}^{14} \frac{(13.8)^{20-k}}{(20-k)! \, k! \, (14-k)!}$

g) 0.0101 **h)** 0.183

12.5 a) $\mathcal{P}(4(t-s))$ **b)** 8 **c)** 0.177
d) $\mathcal{E}(4)$ **e)** 15 sec **f)** $\Gamma(n, 4)$
g) $n/4$ min **h)** 30 sec **i)** 0.0218

12.9

a) Use a basic random number generator to simulate n independent observations of an exponential random variable with parameter λ, say, $I_1, \ldots, I_n$, as explained in the solution to part (c) of Example 8.22 on page 472. Set $W_k = I_1 + \cdots + I_k$ for $1 \leq k \leq n$. Then the W_ks represent the times of occurrences of the first n successes (events).

12.11

a) The intensity function, $\nu(t)$, equals 0 if $0 \leq t < 10$ or $22 \leq t < 24$, and equals the rates given in the statement of the problem otherwise.

b)
$$\mu(t) = \begin{cases} 0, & 0 \leq t < 10; \\ 2t^2 - 40t + 200, & 10 \leq t < 12; \\ -2t^2 + 56t - 376, & 12 \leq t < 14; \\ \frac{5}{2}t^2 - 70t + 506, & 14 \leq t < 18; \\ -\frac{5}{2}t^2 + 110t - 1114, & 18 \leq t < 22; \\ 96, & 22 \leq t < 24. \end{cases}$$

c) 96 **d)** 71.5; 8.46 **e)** 0.268

12.13 Let λ denote the rate of the homogenous Poisson process.

a) $\mu(t) = \lambda t \mu$; $\sigma^2(t) = \lambda t (\mu^2 + \sigma^2)$

b) $\mu/\sqrt{\mu^2 + \sigma^2}$

EXERCISES 12.2

12.21

a) $M/M/1$ queue with $\lambda = 10$ and $\mu = 11$.

b) The steady-state distribution exists because $\lambda < \mu$. $P_n = \frac{1}{11} \left(\frac{10}{11} \right)^n$ for $n \geq 0$.

c) 10 **d)** 100/11

12.23

a) $M/M/3$ queue with $\lambda = 10$ and $\mu = 11$.

b) The steady-state distribution exists because $\lambda < 3\mu$. $P_0 \approx 0.400$, $P_1 \approx 0.363$, $P_2 \approx 0.165$, and $P_n \approx 1.799(0.303)^n$ for $n \geq 3$.

c) The following table compares the steady-state distributions for the number of customers in the queueing systems for one, two, and three tellers, from $n = 0$ to $n = 12$.

n	One teller	Two tellers	Three tellers
0	0.090909	0.375000	0.399684
1	0.082645	0.340909	0.363349
2	0.075131	0.154959	0.165159
3	0.068301	0.070436	0.050048
4	0.062092	0.032016	0.015166
5	0.056447	0.014553	0.004596
6	0.051316	0.006615	0.001393
7	0.046651	0.003007	0.000422
8	0.042410	0.001367	0.000128
9	0.038554	0.000621	0.000039
10	0.035049	0.000282	0.000012
11	0.031863	0.000128	0.000004
12	0.028966	0.000058	0.000001

d) The steady-state expected number of customers in the three-teller system is 0.940 compared with 10 and 1.146 in the one-teller and two-teller systems, respectively.

e) The steady-state expected number of customers in the queue in the three-teller system is 0.031 compared with 9.091 and 0.237 in the one-teller and two-teller systems, respectively.

12.25
a) $M/M/1/4$ queue with $\lambda = 3$ and $\mu = 2$.
b) $P_n = \frac{16}{211}\left(\frac{3}{2}\right)^n$ if $0 \leq n \leq 4$, and $P_n = 0$ otherwise.
c) 2.758

12.27
a) $P_0 = \left(1 + \frac{\lambda}{\mu} + \left(\frac{\lambda}{\mu}\right)^2 \frac{\nu}{\nu - \lambda}\right)^{-1}$, $P_1 = \frac{\lambda}{\mu} P_0$, and
$P_n = \left(\frac{\lambda}{\mu}\right)^2 \left(\frac{\lambda}{\nu}\right)^{n-2} P_0$ if $n \geq 2$.
b) $L_q = \left(\frac{\lambda \nu}{\mu(\nu - \lambda)}\right)^2 P_0$
c) $P_0 = \frac{8}{93}$, $P_1 = \frac{10}{93}$, and $P_n = \frac{25}{186}\left(\frac{5}{6}\right)^{n-2}$ for $n \geq 2$. Also, $L_q = 150/31 \approx 4.8$.

EXERCISES 12.3

12.43
a) $\mu_Y = \begin{bmatrix} 6 \\ -21 \end{bmatrix}$ and $\Sigma_Y = \begin{bmatrix} 412 & -157 \\ -157 & 104 \end{bmatrix}$

b) $\mu_X = \begin{bmatrix} -1 \\ 0 \\ 5 \end{bmatrix}$ and $\Sigma_X = \begin{bmatrix} 10 & -5 & 2 \\ -5 & 16 & -3 \\ 2 & -3 & 4 \end{bmatrix}$

c) $\mu_Y = \begin{bmatrix} 6 \\ -21 \end{bmatrix}$ and $\Sigma_Y = \begin{bmatrix} 412 & -157 \\ -157 & 104 \end{bmatrix}$

12.49 Standardizing

12.53 Let μ be the $m \times 1$ vector whose ith entry is μ_i and let $\mathbf{B}$ be the $m \times n$ matrix whose i-j entry is b_{ij}.
a) Apply Proposition 12.6 on page 718.
b) $\mu_X = \mu$ and $\Sigma_X = \mathbf{BB}'$
c) $\det(\mathbf{BB}') = 0$—that is, $\mathbf{BB}'$ is singular.

12.55 See Proposition 12.8 on page 719.

EXERCISES 12.4

12.65
a) $2(1 - F(2.656))$, where F is the CDF of the Student's t-distribution with 14 degrees of freedom.
b) Yes.

12.67
a) 0.10 to 1.66
b) We can be 95% confident that the ratio of the variances of tear strength for Brand A and Brand B vinyl floor coverings is somewhere between 0.10 and 1.66.
c) Tear strengths for both Brands A and B are normally distributed.

12.69
a) mean = 0; variance = 1
b) mean = 0 if $n \geq 3$;
variance = $(n - 1)/(n - 3)$ if $n \geq 4$

REVIEW EXERCISES

12.77 The missing entries in the first column are "p" and "$\mathcal{N}\mathcal{B}(r, p)$," and the missing entries in the second column are "Independent $\mathcal{E}(\lambda)$" and "$\mathcal{P}(\lambda(t - s))$."

12.78 a) 0.218 b) 0.142 c) 0.142 d) 0.341

12.80 b) $4/15 \approx 0.267$

12.81 Assuming that $\lambda < 2\mu$,
a) $P_n = (1 - \lambda/2\mu)(\lambda/2\mu)^n$ for $n \geq 0$.
b) $L = \lambda/(2\mu - \lambda)$.

12.82 a) $\lambda \min\{s, t\}$ b) $\sqrt{s/(s + t)}$

12.83 mean = 22.5 passengers;
standard deviation = 5.81 passengers

12.84
a) For each $t \geq 0$, $N(t)$ is the total number of securities orders received at the brokerage house by time t.
c) $\lambda_1/(\lambda_1 + \lambda_2)$ d) $\left(\lambda_1/(\lambda_1 + \lambda_2)\right)^2$
e) $\sum_{k=n_1}^{n_1+n_2-1} \binom{n_1+n_2-1}{k} \left(\frac{\lambda_1}{\lambda_1+\lambda_2}\right)^k \left(\frac{\lambda_2}{\lambda_1+\lambda_2}\right)^{n_1+n_2-1-k}$

12.85

a) For each $t \geq 0$, $N(t)$ is the total number of events of all m types that occur by time t.

c) $\lambda_j/(\lambda_1 + \cdots + \lambda_m)$

12.87 a) It's infinite.   **b)** $M/M/1/N$ queue

c) For the $M/M/1$ queue,

$$P_n = \left(1 - \frac{6.9}{7.4}\right)\left(\frac{6.9}{7.4}\right)^n \approx 0.06757 \cdot (0.93243)^n,$$

for $n \geq 0$. For the $M/M/1/40$ queue,

$$P_n = \frac{\left(1 - \frac{6.9}{7.4}\right)\left(\frac{6.9}{7.4}\right)^n}{1 - \left(\frac{6.9}{7.4}\right)^{41}} \approx 0.07164 \cdot (0.93243)^n,$$

for $0 \leq n \leq 40$.

d) For the $M/M/1/100$ queue,

$$P_n = \frac{\left(1 - \frac{6.9}{7.4}\right)\left(\frac{6.9}{7.4}\right)^n}{1 - \left(\frac{6.9}{7.4}\right)^{101}} \approx 0.06763 \cdot (0.93243)^n,$$

for $0 \leq n \leq 100$.

12.88 a) $497   **b)** $413.67

12.89 Yes, $L = 4.875$ if the photocopying machines are in different rooms and $L = 2.115$ if the photocopying machines are in the same room.

12.90 a) 1.944   **b)** 37.1%   **c)** 9.5%

12.91

a) Answers will vary.

b) $\lambda_n = \lambda$ and $\mu_n = n\mu$.

c) $\mathcal{P}(\lambda/\mu)$ **d)** λ/μ **e)** $0, 1/\mu, 0$

12.92

a) $\lambda_n = (M - n)\lambda$ if $0 \leq n < M$, and $\lambda_n = 0$ otherwise; $\mu_n = \mu$.

b) $P_0 = \left(\sum_{n=0}^{M}(M)_n(\lambda/\mu)^n\right)^{-1}$, $P_n = (M)_n(\lambda/\mu)^n P_0$ for $1 \leq n \leq M$, and $P_n = 0$ otherwise.

d) $\lambda_n = (10 - n)/6$ if $0 \leq n < 10$, and $\lambda_n = 0$ otherwise; $\mu_n = 2/3$. The steady-state distribution for the number of down machines is as follows.

n	P_n	n	P_n	n	P_n
0	0.00531	4	0.10449	8	0.14694
1	0.01327	5	0.15674	9	0.07347
2	0.02985	6	0.19592	10	0.01837
3	0.05971	7	0.19592		

The expected number of working machines is 3.979.

12.93 c) $P_0 = 0.364$; $L = 1.2$

12.94

a) $4\lambda\mu/(4\mu^2 - \lambda^2)$ **b)** $\lambda/(2\mu - \lambda)$

c) $2\lambda/(2\mu - \lambda)$

d) The second configuration is always most efficient, and the third configuration is always least efficient.

12.95 Let μ_i denote the ith entry of $\boldsymbol{\mu}$.

a) $M_{X_1,\ldots,X_m}(t_1, \ldots, t_m) = e^{\sum_{i=1}^{m}\mu_i t_i}$

b) For $1 \leq i \leq m$, $p_{X_i}(\mu_i) = 1$, and $p_{X_i}(x) = 0$ otherwise. Also, $p_{X_1,\ldots,X_m}(\mu_1, \ldots, \mu_m) = 1$, and $p_{X_1,\ldots,X_m}(x_1, \ldots, x_m) = 0$ otherwise.

12.97 0.24

12.98

a) $c = 1/4$

b) $(X_1, X_3) \sim \mathcal{N}_2\left(\begin{bmatrix} -1 \\ 2 \end{bmatrix}, \begin{bmatrix} 4 & 2 \\ 2 & 3 \end{bmatrix}\right)$

c) $X_1 - 2X_2 + 3X_3 \sim \mathcal{N}(5, 43)$

d) Yes, by Propositions 12.7 and 12.5 on pages 719 and 717, respectively.

12.99

a) 0.6, 0.6, 0.9

b) $X_1 \sim \mathcal{N}(150, 400)$, $X_2 \sim \mathcal{N}(120, 25)$, $X_3 \sim \mathcal{N}(80, 36)$

c) $(X_1, X_2) \sim \mathcal{N}_2\left(\begin{bmatrix} 150 \\ 120 \end{bmatrix}, \begin{bmatrix} 400 & 60 \\ 60 & 25 \end{bmatrix}\right)$

$(X_1, X_3) \sim \mathcal{N}_2\left(\begin{bmatrix} 150 \\ 80 \end{bmatrix}, \begin{bmatrix} 400 & 72 \\ 72 & 36 \end{bmatrix}\right)$

$(X_2, X_3) \sim \mathcal{N}_2\left(\begin{bmatrix} 120 \\ 80 \end{bmatrix}, \begin{bmatrix} 25 & 27 \\ 27 & 36 \end{bmatrix}\right)$

d) $X_{1|X_2=130, X_3=90} \sim \mathcal{N}(173.2, 248.4)$

e) $(X_2, X_3)_{|X_1=200} \sim \mathcal{N}_2\left(\begin{bmatrix} 127.5 \\ 89 \end{bmatrix}, \begin{bmatrix} 16 & 16.2 \\ 16.2 & 23.0 \end{bmatrix}\right)$

12.102 Beta with parameters $v_2/2$ and $v_1/2$.

12.104

a) 0.76 to 1.82

b) We can be 95% confident that the standard deviation of highway gas mileages for all cars of the model and year in question is somewhere between 0.76 mpg and 1.82 mpg.

c) That highway gas mileages for all cars of the model and year in question are normally distributed.

12.105

c) Yes, because a linear combination of multivariate normal random variables is normally distributed.

12.106 a) -0.07 to 6.67 yr   **b)** -0.07 to 6.67 yr

c) No.   **d)** No.

Table I Values of Φ, the standard normal CDF

z	0.00	0.01	0.02	0.03	0.04	0.05	0.06	0.07	0.08	0.09
					Second decimal place in z					
0.0	0.5000	0.5040	0.5080	0.5120	0.5160	0.5199	0.5239	0.5279	0.5319	0.5359
0.1	0.5398	0.5438	0.5478	0.5517	0.5557	0.5596	0.5636	0.5675	0.5714	0.5753
0.2	0.5793	0.5832	0.5871	0.5910	0.5948	0.5987	0.6026	0.6064	0.6103	0.6141
0.3	0.6179	0.6217	0.6255	0.6293	0.6331	0.6368	0.6406	0.6443	0.6480	0.6517
0.4	0.6554	0.6591	0.6628	0.6664	0.6700	0.6736	0.6772	0.6808	0.6844	0.6879
0.5	0.6915	0.6950	0.6985	0.7019	0.7054	0.7088	0.7123	0.7157	0.7190	0.7224
0.6	0.7257	0.7291	0.7324	0.7357	0.7389	0.7422	0.7454	0.7486	0.7517	0.7549
0.7	0.7580	0.7611	0.7642	0.7673	0.7704	0.7734	0.7764	0.7794	0.7823	0.7852
0.8	0.7881	0.7910	0.7939	0.7967	0.7995	0.8023	0.8051	0.8078	0.8106	0.8133
0.9	0.8159	0.8186	0.8212	0.8238	0.8264	0.8289	0.8315	0.8340	0.8365	0.8389
1.0	0.8413	0.8438	0.8461	0.8485	0.8508	0.8531	0.8554	0.8577	0.8599	0.8621
1.1	0.8643	0.8665	0.8686	0.8708	0.8729	0.8749	0.8770	0.8790	0.8810	0.8830
1.2	0.8849	0.8869	0.8888	0.8907	0.8925	0.8944	0.8962	0.8980	0.8997	0.9015
1.3	0.9032	0.9049	0.9066	0.9082	0.9099	0.9115	0.9131	0.9147	0.9162	0.9177
1.4	0.9192	0.9207	0.9222	0.9236	0.9251	0.9265	0.9279	0.9292	0.9306	0.9319
1.5	0.9332	0.9345	0.9357	0.9370	0.9382	0.9394	0.9406	0.9418	0.9429	0.9441
1.6	0.9452	0.9463	0.9474	0.9484	0.9495	0.9505	0.9515	0.9525	0.9535	0.9545
1.7	0.9554	0.9564	0.9573	0.9582	0.9591	0.9599	0.9608	0.9616	0.9625	0.9633
1.8	0.9641	0.9649	0.9656	0.9664	0.9671	0.9678	0.9686	0.9693	0.9699	0.9706
1.9	0.9713	0.9719	0.9726	0.9732	0.9738	0.9744	0.9750	0.9756	0.9761	0.9767
2.0	0.9772	0.9778	0.9783	0.9788	0.9793	0.9798	0.9803	0.9808	0.9812	0.9817
2.1	0.9821	0.9826	0.9830	0.9834	0.9838	0.9842	0.9846	0.9850	0.9854	0.9857
2.2	0.9861	0.9864	0.9868	0.9871	0.9875	0.9878	0.9881	0.9884	0.9887	0.9890
2.3	0.9893	0.9896	0.9898	0.9901	0.9904	0.9906	0.9909	0.9911	0.9913	0.9916
2.4	0.9918	0.9920	0.9922	0.9925	0.9927	0.9929	0.9931	0.9932	0.9934	0.9936
2.5	0.9938	0.9940	0.9941	0.9943	0.9945	0.9946	0.9948	0.9949	0.9951	0.9952
2.6	0.9953	0.9955	0.9956	0.9957	0.9959	0.9960	0.9961	0.9962	0.9963	0.9964
2.7	0.9965	0.9966	0.9967	0.9968	0.9969	0.9970	0.9971	0.9972	0.9973	0.9974
2.8	0.9974	0.9975	0.9976	0.9977	0.9977	0.9978	0.9979	0.9979	0.9980	0.9981
2.9	0.9981	0.9982	0.9982	0.9983	0.9984	0.9984	0.9985	0.9985	0.9986	0.9986
3.0	0.9987	0.9987	0.9987	0.9988	0.9988	0.9989	0.9989	0.9989	0.9990	0.9990
3.1	0.9990	0.9991	0.9991	0.9991	0.9992	0.9992	0.9992	0.9992	0.9993	0.9993
3.2	0.9993	0.9993	0.9994	0.9994	0.9994	0.9994	0.9994	0.9995	0.9995	0.9995
3.3	0.9995	0.9995	0.9995	0.9996	0.9996	0.9996	0.9996	0.9996	0.9996	0.9997
3.4	0.9997	0.9997	0.9997	0.9997	0.9997	0.9997	0.9997	0.9997	0.9997	0.9998
3.5	0.9998	0.9998	0.9998	0.9998	0.9998	0.9998	0.9998	0.9998	0.9998	0.9998
3.6	0.9998	0.9998	0.9999[†]							

† To four decimal places, $\Phi(z) = 0.9999$ for $3.62 \leq z \leq 3.89$, and $\Phi(z) = 1.0000$ for $z \geq 3.90$.

Index

Selected Discrete Distributions

BINOMIAL	Parameters:	n and p $\quad (n \in \mathcal{N}; 0 < p < 1)$
$\mathcal{B}(n, p)$	PMF:	$\binom{n}{x} p^x (1-p)^{n-x}, \qquad x = 0, 1, \ldots, n$
	Mean:	np
	Variance:	$np(1-p)$
	MGF:	$(pe^t + 1 - p)^n$
HYPERGEOMETRIC	Parameters:	$N, n,$ and p $\quad (N, n, Np \in \mathcal{N}; n \leq N; 0 < p < 1)$
$\mathcal{H}(N, n, p)$	PMF:	$\binom{Np}{x}\binom{N(1-p)}{n-x} \Big/ \binom{N}{n}, \qquad x = 0, 1, \ldots, n$
	Mean:	np
	Variance:	$\left(\dfrac{N-n}{N-1}\right) np(1-p)$
	MGF:	——
POISSON	Parameters:	$\lambda \quad (\lambda > 0)$
$\mathcal{P}(\lambda)$	PMF:	$e^{-\lambda} \dfrac{\lambda^x}{x!}, \qquad x = 0, 1, 2, \ldots$
	Mean:	λ
	Variance:	λ
	MGF:	$e^{\lambda(e^t - 1)}$

(continued on the next page and inside the back cover)

Selected Discrete Distributions (cont.)

GEOMETRIC	Parameters:	$p \quad (0 < p < 1)$
$\mathcal{G}(p)$	PMF:	$p(1 - p)^{x-1}, \qquad x = 1, 2, 3, \ldots$
	Mean:	$\dfrac{1}{p}$
	Variance:	$\dfrac{1 - p}{p^2}$
	MGF:	$\dfrac{pe^t}{1 - (1 - p)e^t}, \qquad t < -\ln(1 - p)$

NEGATIVE BINOMIAL	Parameters:	r and $p \quad (r \in \mathcal{N}; 0 < p < 1)$
$\mathcal{NB}(r, p)$	PMF:	$\binom{x - 1}{r - 1} p^r (1 - p)^{x-r}, \qquad x = r, r + 1, \ldots$
	Mean:	$\dfrac{r}{p}$
	Variance:	$\dfrac{r(1 - p)}{p^2}$
	MGF:	$\left(\dfrac{pe^t}{1 - (1 - p)e^t} \right)^r, \qquad t < -\ln(1 - p)$

INDICATOR	Parameters:	$E \quad (E \text{ an event})$
	PMF:	$\big(P(E)\big)^x \big(1 - P(E)\big)^{1-x}, \qquad x = 0, 1$
	Mean:	$P(E)$
	Variance:	$P(E)\big(1 - P(E)\big)$
	MGF:	$P(E)e^t + 1 - P(E)$

(continued on the preceding page and inside the back cover)

Selected Continuous Distributions

UNIFORM	Parameters:	a and b $(a, b \in \mathcal{R}; a < b)$
$\mathcal{U}(a, b)$	PDF:	$\dfrac{1}{b-a}, \quad a < x < b$
	Mean:	$\dfrac{a+b}{2}$
	Variance:	$\dfrac{(b-a)^2}{12}$
	MGF:	$\dfrac{e^{bt} - e^{at}}{(b-a)t}, \quad t \neq 0 \quad (1, \text{if } t = 0)$

EXPONENTIAL	Parameters:	$\lambda \quad (\lambda > 0)$
$\mathcal{E}(\lambda)$	PDF:	$\lambda e^{-\lambda x}, \quad x > 0$
	Mean:	$\dfrac{1}{\lambda}$
	Variance:	$\dfrac{1}{\lambda^2}$
	MGF:	$\dfrac{\lambda}{\lambda - t}, \quad t < \lambda$

NORMAL	Parameters:	μ and σ^2 $(\mu \in \mathcal{R}; \sigma > 0)$
$\mathcal{N}(\mu, \sigma^2)$	PDF:	$\dfrac{1}{\sqrt{2\pi}\,\sigma} e^{-(x-\mu)^2/2\sigma^2}, \quad -\infty < x < \infty$
	Mean:	μ
	Variance:	σ^2
	MGF:	$e^{\mu t + \sigma^2 t^2 / 2}$

(continued on the next page and inside the front cover)

Selected Continuous Distributions (cont.)

GAMMA	Parameters:	α and λ $\quad (\alpha > 0$ and $\lambda > 0)$
$\Gamma(\alpha, \lambda)$	PDF:	$\dfrac{\lambda^{\alpha}}{\Gamma(\alpha)} x^{\alpha-1} e^{-\lambda x}, \qquad x > 0$
	Mean:	$\dfrac{\alpha}{\lambda}$
	Variance:	$\dfrac{\alpha}{\lambda^2}$
	MGF:	$\left(\dfrac{\lambda}{\lambda - t}\right)^{\alpha}, \qquad t < \lambda$
BETA	Parameters:	α and β $\quad (\alpha > 0$ and $\beta > 0)$
$\text{Beta}(\alpha, \beta)$	PDF:	$\dfrac{1}{B(\alpha, \beta)} x^{\alpha-1}(1 - x)^{\beta-1}, \qquad 0 < x < 1$
	Mean:	$\dfrac{\alpha}{\alpha + \beta}$
	Variance:	$\dfrac{\alpha\beta}{(\alpha + \beta)^2(\alpha + \beta + 1)}$
	MGF:	——
TRIANGULAR	Parameters:	a and b $\quad (a, b \in \mathcal{R}; a < b)$
$\mathcal{T}(a, b)$	PDF:	$\dfrac{2}{b - a}\left(1 - \dfrac{|a + b - 2x|}{b - a}\right), \qquad a < x < b$
	Mean:	$\dfrac{a + b}{2}$
	Variance:	$\dfrac{(b - a)^2}{24}$
	MGF:	$\dfrac{8e^{(a+b)t/2}}{(b - a)^2 t^2}\left(\cosh\dfrac{(b - a)t}{2} - 1\right), \qquad t \neq 0 \quad (1, \text{ if } t = 0)$

(continued on the preceding page and inside the front cover)